M. Eslami

Theory of Sensitivity in Dynamic Systems

An Introduction

With 35 Figures

Springer-Verlag
Berlin Heidelberg New York
London Paris Tokyo
Hong Kong Barcelona Budapest

Prof. Mansour Eslami
University of Illinois at Chicago
Dept. of Electrical Engineering
and Computer Science (M/C 154)
College of Engineering, Room 1120
851 South Morgan Street
Chicago, IL 60607-7053
USA

ISBN 3-540-54761-4 Springer-Verlag Berlin Heidelberg New York
ISBN 0-387-54761-4 Springer-Verlag New York Berlin Heidelberg

Library of Congress Cataloging-in-Publication Data
Eslami, Mansour
Theory of sensitivity in dynamic systems: an introduction / Mansour Eslami
Includes bibliographical references and indexes.
ISBN 0-387-54761-4
1. Sensitivity theory (Mathematics) 2. Differentiable dynamical systems.
3. State-space methods
I. Title.
QA402.3.E85   1994   94-19386

This work is subject to copyright. All rights are reserved, whether the whole or part of the material is concerned, specifically the rights of translation, reprinting, reuse of illustrations, recitation, broadcasting, reproduction on microfilm or in other ways, and storage in data banks. Duplication of this publication or parts thereof is permitted only under the provisions of the German Copyright Law of September 9, 1965, in its current version, and permission for use must always be obtained from Springer-Verlag. Violations are liable for prosecution act under German Copyright Law.

© Springer-Verlag Berlin Heidelberg 1994
Printed in Germany

The use of general descriptive names, registered names, trademarks, etc. in this publication does not imply, even in the absence of a specific statement, that such names are exempt from the relevant protective laws and regulations and therefore free for general use.

Typesetting: Camera-ready by author
SPIN:10029991   61/3020-5 4 3 2 1 0 - Printed on acid -free paper

Sensitivity [or lack of it] may either be a virtue or a curse.

*In
the memory
of my parents*

and

my brother

PREFACE

The primary purpose of this book is to provide a guided entry to the emerging field of sensitivity and robustness theory in dynamical systems modeled with state-space representations. It is intended that both research students and practicing engineers in general areas of control system theory use the book as a textbook or as a reference book. In preparing this book some basic background in linear systems and linear optimal control theory as well as intermediate-level mathematics is assumed.

On the overall objectives of this book, it is intended to provide a broad scope of sensitivity theory as developed in the last several decades pertaining to electrical engineering. In this regard, preserving the past is one of the main features of this project. The given list of references, and Appendix B which in effect is a sensitivity tree, shows where most topics are, and this feature is particularly useful for new researchers who may not have followed the overall development of this theory and may need a broad view of these issues before they launch their research activities in this area. Indeed this service provides an opportunity for new researchers to review thoroughly the earlier work. This author believes that the statute of limitations on scientific work is eternity. Along the same line there are many interesting issues which are thought by different groups of researchers in different disciplines who are independently pursuing their own research and may not be aware of the similar work done by others, on exactly the same topic; this fact became evident to the author when reading the volume of papers reported in this book. Thus I hope this book serves as a reference to bring together different research communities whose interests are in sensitivity theory. Yet again, the main purpose of this book is to structure the areas of research as best as this author is capable of in order to provide a coherent media for future researchers.

On the technical aspects of the book, it is intended to provide a trajectory which covers most essential features of sensitivity theory. It is the opinion of this author that sensitivity in its most generic sense is synonymous with the notion of distance. For instance, the "distance" between the desired behavior and an actual situation represents sensitivity. The more sophisticated and the more abstract the desired and actual situations become, the more sophisticated and the more abstract mathematics are needed to *measure* and/or to *compensate* this distance. The book is prepared with the intention to direct the readers towards the advanced aspects of this theory. Having faced a dilemma of including as much work from the literature as I can, and the allotted space, I was forced to skip many features very important in their own rights. That decision perhaps has resulted in some inadequacies in certain areas, but in most cases I have made efforts to give a picture for the underlying topic that seems sufficient for its future pursuit. I take full responsibility for shortcoming of any kind in this book and with my inherent optimism, I am convinced that any shortcoming will be considered as an opportunity for other researchers whose interests brush these areas and wish to work on that. It is my utmost desire to spur interests among critics to point out constructive changes to improve the quality of the book and to guide the readers to work on those issues. I believe the book provides ample opportunities for others to contribute towards completing the drawing of the sensitivity tree, and gives them a chance to prune many branches.

I am grateful to many of my friends who have been very supportive, encouraging, and instrumental in completion of this project. In particular, Professor Chi-Tsong Chen whom I have known since Sept. 1979, when I started to teach at the State University of New York at Stony Brook and who has been very influential in my professional career. Also my old friend Dr. H. Sadreddinamin who has been always very supportive. So has been Professor M. Jamshidi.

I am also very grateful to my colleagues for their support and encouragement, in particular Professor Wai-Kai Chen, Professor C.K. Sanathanan, Professor T.G. Moher, Professor R.V. Kenyon, and Professor D. Ucci.

I have been extremely fortunate to be associated with a number of world-class scientists and educators who have been very encouraging and supportive throughout my career and I owe them heartfelt gratitude for helping me to complete this project. It is indeed my greatest honor and distinct pleasure to acknowledge their support. Among them are Professor K.S. Narendra, Professor F.M. Reza, Professor S.S.L. Chang, Professor T.J. Tarn, Professor J.B. Cruz, Jr., Professor W.R. Perkins, Professor M.G. Safonov, Professor S. Rappaport, and especially Professor A.H. Zemanian. Often this project seemed unmanageable, and finally when I finished it I realized that without the constant support of these individuals, I just could not have made it. Thank you all.

The initial influence and words of my professors at the University of Wisconsin-Madison are very present in this book. Many thanks are extended to all of them, and in particular, to Professor R.S. Marleau, Professor D.L. Russell, Professor T.J. Higgins, and Professor W. Dickey. I was very fortunate in being able to attend Professor W. Rudin's class during 1976-1977.

Through the years of teaching various graduate-level control courses, I have presented portions of this book to my former students. I want to thank them all for making many constructive suggestions to improve my presentation. Primarily, they are the reason why this book is written. In particular, I would like to acknowledge several suggestions to improve this book provided by Dr. D.S. Bayard.

Special thanks are due to my very good friends: Ralph Orlick who painstakingly has kept the computer system and all its peripheral networks in the most efficient working conditions and without his knowledge and assistance I would have had many more sleepless nights; to Tyrone Tolliver, who skillfully typed, retyped, and retyped again the book; to Leland Luecke, who often assisted me with my computer; and to Anne Lange who has been always most encouraging and helpful with many aspects of this project.

Mansour Eslami
Chicago, Illinois
September 9, 1994

DEDICATION

I have been blessed with the most supportive and the kindest family that anyone has ever had. This project is by and large a family enterprise and the author is indebted to his family for their many sacrifices and collective support throughout the long and often tedious course of its completion. Were it not for their inspiration I would have never been able to complete this task. This work is dedicated to all of them and in particular my dear father Mr. Ali-Asghar Eslami who lived with the highest standards of morality, integrity, dignity, and independence. He had an uncompromising dedication to excellence and perfection in every facet of life and he was a brilliant jurist and legal scholar. My dear loving mother, Mrs. Shamimeh Eslami, was an inspiring and courageous woman whose infinite optimism was the guiding light of my life. My parents taught me tenacity, patience and perseverance and my life was much richer when I was with them. My uncle, Mr. Mohammad-Sadegh Eslami, was a learned man and together with my father left their families in their teens to go to major cities for their education in law and this unconventional act opened the door for the youth in my family to do the same. Also my uncle, Mr. Aghil Eslami who was the kindest man I have ever met, always accepted me with all my flaws.

Finally, I am indebted to my eldest brother, Mr. Ali-Mohammad Eslami, whose regal manner, style and supreme kindness were my inspiration. Although they all are gone, they have never left my heart and they are the true authors of this book.

Mansour Eslami

TABLE OF CONTENTS

LIST OF ACRONYMS

CHAPTER ONE

INTRODUCTION

1.1. OVERVIEW AND SCOPE OF THE BOOK, 1
1.2. HISTORICAL REMARKS ON THE THEORY OF SENSITIVITY, 4
1.3. SUMMARY OF THE BOOK, 7
1.4. A SUMMARY OF ISSUES NOT INCLUDED IN THE BOOK, 9
1.5. REFERENCES, 13

CHAPTER TWO

THE PRINCIPAL ASPECTS OF SENSITIVITY THEORY

2.1. INTRODUCTION, 17
2.2. ESSENTIAL DEFINITIONS FOR PARAMETRIC SENSITIVITY, 17
2.3. THE FUNDAMENTAL ASPECTS OF SENSITIVITY ANALYSIS, 23
2.4. SENSITIVITY FUNCTIONS GENERATION, 25
2.5. THE PRELIMINARY MATHEMATICS, 27
2.5.1. Introduction, 27
2.5.2. Mappings, 27
2.5.3. Semigroups, Groups, Rings, and Fields, 28
2.5.4. Complete Mathematical Spaces, 30
2.5.5. Basics of Set Theory and Topology, 34
2.5.6. Mappings Continued, 37
2.5.7. Classes of L^∞ – and H^p – Functions, 40
2.5.8. Matrix and Determinant, 46
2.5.9. Vector and Matrix Norms, 47

2.6.	ASPECTS OF SYSTEM THEORY, 48	
2.6.1.	Introduction, 48	
2.6.2.	State – Space Realization of Transfer Matrices, 49	
2.6.3.	Classes of Signals, 57	
2.6.3.1.	Introduction, 57	
2.6.3.2.	Time – Domain Signals, 57	
2.6.3.3.	Frequency – Domain Signals, 58	
2.6.4.	Computation of Transfer Matrix Norms, 60	
2.6.4.1.	The H_2–Norm, 60	
2.6.4.2.	The H_∞–Norm, 61	
2.7.	OVERVIEW OF GENERAL SENSITIVITY MINIMIZATION, 62	
2.7.1.	Introduction, 62	
2.7.2.	The Tracking and Stabilization Problems, 63	
2.7.3.	The Model Matching Problem, 65	
2.8.	CONCLUDING REMARKS, 65	
2.9.	REFERENCES, 67	

CHAPTER THREE

SENSITIVITY FUNCTIONS GENERATION

3.1.	INTRODUCTION, 74
3.2.	STRUCTURAL – PARAMETER SENSITIVITY FUNCTIONS GENERATION, 76
3.2.1.	Introduction, 76
3.2.2.	Open-Loop Reconstructible System, 76
3.2.3.	Closed-Loop Reconstructible System, 87
3.3.	PHYSICAL – PARAMETER SENSITIVITY FUNCTIONS GENERATION, 90
3.3.1.	Introduction, 90
3.3.2.	Formal Analysis for Open-Loop Reconstructible System, 90
3.3.2.1.	Higher-Order Sensitivity Functions, 91
3.3.2.2.	Direct Computation of Sensitivity Functions, 91
3.3.2.3.	Alternative Method To Compute Sensitivity Functions, 96
3.3.3.	Formal Analysis for Closed-Loop Reconstructible System, 97
3.4.	SENSITIVITY FUNCTION GENERATION: FREQUENCY DOMAIN, 97
3.5.	AUGMENTED STATE AND STATE – SENSITIVITY DIFFERENTIAL EQUATIONS: SYSTEM MODEL, 98
3.5.1.	Introduction, 98
3.5.2.	Open-Loop System Model, 98
3.6.	SENSITIVITY INVARIANCE, 100
3.6.1.	Introduction, 100
3.6.2.	Sensitivity Measure Invariants for the Linear System, 100
3.7.	REFERENCES, 108

CHAPTER FOUR

ALGEBRAIC PROPERTIES OF SYSTEM MODELS TERMINAL CONDITIONS INSENSITIVITY

4.1.	INTRODUCTION, 114
4.2.	STABILITY OF SYSTEM MODEL, 115
4.3.	CONTROLLABILITY OF SYSTEM MODEL, 117
4.3.1.	Introduction, 117
4.3.2.	Analyses, 118
4.4.	TERMINAL CONDITIONS INSENSITIVITY, 125
4.4.1.	Introduction, 125
4.4.2.	Formulation and Analysis, 126
4.5.	GUIDANCE THEORY AS PERTAINS TO TERMINAL CONDITIONS INSENSITIVITY, 129
4.6.	REFERENCES, 130

CHAPTER FIVE

STABILITY ROBUSTNESS ANALYSIS MANEUVERABILITY, AND SENSITIVITY IN THE LARGE

5.1.	INTRODUCTION, 132
5.1.1.	Stability of Linear Time-Invariant System, 133
5.2.	ROOT SENSITIVITY AND POLE PLACEMENT, 135
5.2.1.	Introduction, 135
5.2.2.	Certain Formulae On Root Sensitivity, 136
5.3.	EIGENVALUE AND EIGENVECTOR SENSITIVITY ANALYSIS, 139
5.4.	STABILITY ROBUSTNESS ANALYSIS: ISSUES AND TECHNIQUES, 140
5.4.1.	Introduction, 141
5.4.2.	Stability Robustness Analysis Formulation, 143
5.5.	CONSTRUCTION OF ROBUST LINEAR DYNAMIC SYSTEMS, 149
5.5.1.	Lyapunov Direct Method, 149
5.5.2.	Catalog of Structured Perturbation Matrices, 151
5.5.3.	Matrix Inequalities, 153
5.6.	COMPUTATIONAL ISSUES, 154
5.6.1.	Introduction, 154
5.6.2.	Single – Parameter Variations, 155
5.6.3.	Multiple – Parameter Variations, 162
5.7.	EXTENSIONS TO HURWITZ POLYNOMIALS, 164
5.7.1.	Introduction, 164
5.7.2.	Brief Review, 165

5.7.3.	Alternative Analysis, **169**	
5.8.	NUMERICAL EXAMPLES, **174**	
5.8.1.	Introduction, **174**	
5.8.2.	Concluding Remarks on Numerical Examples, **209**	
5.9.	CONSTRUCTION OF ROBUSTLY STABLE LARGE – SCALE SYSTEMS, **211**	
5.9.1.	Introduction, **211**	
5.9.2.	Special Coupled Systems, **213**	
5.10.	CONSTRUCTION OF ROBUSTLY STABLE (CONVERGENT) DISCRETE – TIME SYSTEMS, **216**	
5.10.1.	Introduction, **216**	
5.10.2.	Discrete Lyapunov Method, **218**	
5.10.3.	Computational Issues, **220**	
5.10.3.1.	Single – Parameter Variations, **220**	
5.10.3.2.	Multiple – Parameter Variations, **223**	
5.10.4.	Extensions to Polynomials, **224**	
5.10.5.	Numerical Examples, **226**	
5.11.	CERTAIN EXTENSIONS TO NONLINEAR SYSTEMS, **241**	
5.11.1.	Continuous – Time Nonlinear Systems, **241**	
5.11.2.	Discrete – Time Nonlinear Systems, **253**	
5.11.3.	Final Remarks, **254**	
5.12.	REFERENCES, **255**	

CHAPTER SIX

SENSITIVITY REDUCTION AND ROBUSTNESS

6.1.	INTRODUCTION, **297**	
6.2.	TRAJECTORY – SENSITIVITY COMPARISON AND CERTAIN FEEDBACK PROPERTIES OF LINEAR SYSTEMS, **302**	
6.2.1.	Introduction, **302**	
6.2.2.	Analyses and Algorithms, **304**	
6.3.	PERFORMANCE – INDEX SENSITIVITY COMPARISON, **310**	
6.4.	OPTIMALITY – INDEX SENSITIVITY: ANALYSIS AND SYNTHESIS, **311**	
6.4.1.	Introduction, **311**	
6.4.2.	PI – Small – Sensitivity Analysis, **312**	
6.4.3.	PI – Large – Sensitivity Analysis, **314**	
6.4.4.	PI – Small – Sensitivity Synthesis, **318**	
6.5.	TRAJECTORY – SENSITIVITY OPTIMIZATION, **322**	
6.5.1.	Introduction and Review of the Existing Methods, **322**	
6.5.2.	Alternative Approach to Generate an Optimal Controller Gain for Trajectory – Sensitivity Minimization, **324**	

6.6.		SENSITIVITY OPTIMIZATION WITH ADAPTIVE CONTROLLER, **332**
6.6.1.		Introduction and Problem Statement, **332**
6.7.		SENSITIVITY – MEASURE OPTIMIZATION FOR SYSTEMS WITH LARGE PARAMETER VARIATIONS, **337**
6.7.1.		Introduction, **337**
6.7.2.		Problem Statement, **339**
6.7.3.		Optimal Controller Gain with Trajectory Feedback, **340**
6.7.4.		Optimal Controller Gain with Both Trajectory and Model Feedback, **347**
6.7.5.		Numerical Example, **352**
6.8.		COMMONALTIES IN OPTIMAL INPUT FOR IDENTIFICATION AND OPTIMAL AUXILIARY INPUT FOR ROBUSTIFICATION, **357**
6.8.1.		Introduction, **357**
6.8.2.		Problem Statement, **361**
6.8.3.		Solution of Problem 6.8-1, **364**
6.8.3.1.		Derivation of the Optimal Auxiliary Input v(t) by the Maximum Principle, **364**
6.8.3.2.		Derivation of the Optimal Controller Gain, **368**
6.8.4.		Numerical Algorithm, **371**
6.8.5.		Numerical Examples, **374**
6.9.		SENSITIVITY MINIMIZATION IN THE HARDY SPACES, **379**
6.9.1.		Introduction, **379**
6.9.2.		Riccati Algebraic – Matrix Equation, **382**
6.9.3.		The H^2 – and H^∞ – Optimization Schemes, **385**
6.9.3.1.		Introduction, **385**
6.9.3.2.		General Transfer Matrix, **386**
6.9.3.3.		Special Transfer Matrices, **388**
6.9.3.4.		The Preliminary Analysis, **390**
6.9.3.5.		Special Structures Revisited, **398**
6.9.3.6.		Solution of Problem 6.9-1, **404**
6.9.3.7.		Solution of Problem 6.9-2, **407**
6.9.3.8.		Lifting Constraints *via* Scaling and Loop Shifting, **417**
6.9.3.9.		General Remarks, **431**
6.10.		THE THEORY OF INVARIANCE, **434**
6.10.1.		Introduction, **434**
6.10.2.		The Preliminary Analysis, **436**
6.10.3.		The Basic Steps in the Development of Invariance Theory, **441**
6.10.3.1.		Introduction, **441**
6.10.3.2.		Partial Compensation for Disturbances and Conditions for Invariance up to ε, and Realizability Condition, **442**
6.10.3.3.		Extensions to Multidimensional Systems, **445**
6.10.3.4.		Extensions to Nonlinear Systems, **448**
6.10.4.		Variational Approach to Invariance Problem, **449**
6.10.4.1.		Introduction, **449**
6.10.4.2.		Rozonoer's Analysis for Linear System, **452**

6.10.4.3.		Extensions of Rozonoer's Analysis for Linear System, 455
6.10.4.4.		Extensions to Linear Time-Varying Systems, 457
6.11.	BOUNDED UNCERTAINTY AND DISTURBANCE REJECTION, 459	
6.11.1.		Introduction, 459
6.11.2.		Deterministic Approaches to Meet Uncertainty, 460
6.11.2.1.		Generalized Dynamical Systems, 462
6.11.2.2.		Asymptotic Stability in Generalized Dynamical Systems, 465
6.11.2.3.		Solution of Problem 6.11-1, 465
6.11.2.4.		Specializations, Applications and Extensions, 470
6.11.3.		Disturbance Rejection, 473
6.12.	REFERENCES, 477	

CHAPTER SEVEN

THE THEORY OF SENSITIVITY IN NETWORKS

7.1.	INTRODUCTION, 519
7.2.	GENERAL NETWORK SENSITIVITY: DEFINITIONS AND ANALYSES, 520
7.3.	TOLERANCES OR SENSITIVITY IN THE LARGE OF NETWORKS, 520
7.4.	SENSITIVITY MODEL OF CIRCUITS, 522
7.5.	NETWORK ELEMENTS SENSITIVITY AND ENERGY STORED IN THE NETWORK, VRATSANOS' THEOREM, 523
7.6.	NETWORK SENSITIVITY INVARIANTS AND BOUNDS ON SENSITIVITY FUNCTIONS, 523
7.7.	CONCEPT OF CONTINUOUSLY EQUIVALENT NETWORKS SYNTHESIS PROCEDURES WITH SENSITIVITY INVARIANTS, 524
7.8.	NETWORK SYNTHESIS WITH MINIMUM OR PRESCRIBED SENSITIVITY MEASURE, 525
7.9.	SENSITIVITY OF THE COST FUNCTION IN THE OPTIMAL CIRCUIT DESIGN, 526
7.10.	SENSITIVITY ANALYSIS AND SYNTHESIS IN ACTIVE NETWORKS, 526
7.10.1.	The Pole-Zero Sensitivity and Certain Other General Sensitivity Measures (Gain and Phase) of Active Network, 527
7.10.2.	The Q and ω_o As Sensitivity Measures of the Active Networks with Corresponding Comparison and Minimization of These Measures, 528
7.10.3.	Active Distributed Parameter RC Networks, 528
7.10.4.	Active Network Sensitivity Analysis, Statistically Oriented, 529
7.11.	SENSITIVITY ANALYSIS IN DIGITAL FILTERS, 529
7.12.	SENSITIVITY THEORY – STATISTICALLY ORIENTED, 530
7.13.	TELLEGEN'S THEOREM: THE GENERALIZED ADJOINT NETWORK AND NETWORK SENSITIVITY, 530
7.13.1.	The General Adjoint Network, 531
7.13.2.	General Network Sensitivity, 534

7.14. NETWORK SENSITIVITY FUNCTION GENERATION: COMPUTATIONAL PROCEDURES, **535**
7.15. SENSITIVITY ANALYSIS OF NONLINEAR NETWORKS, **536**
7.16. REFERENCES, **535**

APPENDIX A

SELECTED NUMERICAL APPLICATIONS

A.1. INTRODUCTION, **557**
A.2. POSITION SERVO SYSTEM, **557**
A.3. THE PRELIMINARY ANALYSIS, **560**
A.4. SYSTEM SENSITIVITY: ANALYSIS AND SYNTHESIS, **563**
A.5. PROJECT, **575**

APPENDIX B

LIST OF THE REFERENCES: A SENSITIVITY TREE, **576**

APPENDIX C

JOURNAL ABBREVIATIONS AND ACRONYMS, **579**

AUTHOR INDEX, **583**

SUBJECT INDEX, **597**

LIST OF ACRONYMS

CRM	Convergence robustness measure.
DCLTIS	Deterministic, continuous, linear, time-invariant system.
DCLTI – GS	DCLTI – Generating system.
GDS	Generalized dynamical systems.
LFT	Linear fractional transformation.
MSM	Matrices of system model.
ODE	Ordinary differential equation.
PI	Performance index.
PPSF	Physical-parameter sensitivity function.
PS	Parameter space.
PVS	Parameter variations space.
SPSF	Structural-parameter sensitivity function.
SRA	Stability robustness analysis.
SRM	Stability robustness measure.
TCI	Terminal conditions insensitivity.
TSF	Total sensitivity function.

CHAPTER ONE

INTRODUCTION

1.1. OVERVIEW AND SCOPE OF THE BOOK

Historically speaking, to maintain system stability has been the prime concern in design of any practical system. Presently, however, much effort is being also devoted to the evaluations of possible system parameter variations, and their effects on system performance or "output". By output we mean any functional from the system quantities which can be measured or sensed. In this book, however, we mainly emphasize the system trajectory as system output, and we investigate the corresponding variations caused by parameter changes, if any, and the effects of these changes on system stability.

It is a common knowledge that there always exists a certain discrepancy between an actual (real-operating) and the nominal trajectories of any system. This discrepancy is partly due to various inherently approximational schemes in system identification, and partly due to possible further parameter variations stimulated by environmental changes; slow aging of components; variations in raw products, or system components in process control applications; inaccuracy of measurement devices or methods, *etc.*. Here we must unequivocally state that in spite of terms such as "parameter variations," we mean static changes, i.e., we do not associate neither time-dependency nor any statistical properties to these variations whatsoever. That is *the* fundamental assumption in the theory of sensitivity. If we associate to each parameter variation an explicit time dependency, or if we associate to each parameter variation certain probability measure (specially, a time-varying probability measure), then that study is in the realm of the adaptive or the stochastic adaptive control theory. Thus we maintain the forthcoming analyses within the *static-parameter changes,* which this idea is more or less an anticipatory concept than actual physical changes. In spite of these parameter changes, the today's technology expects system design to be more reliable and self-correcting than ever before. To address partially the issues pertaining to system sensitivity and primarily with respect to the internal static-parameter changes is the main goal of this book. However, we also address certain results for sensitivity reduction of exogenous signals.

Hence systematic worst case procedures in order to anticipate the extreme variations in system output corresponding to change(s) in parameter(s) of the system must be developed. Based on this information we must obtain certain specific bounds on system parameters in order to hold its output within certain bounds, and/or in order to know when system operation does not follow its index of optimality or approaches instability. Since it is often unclear which one of the many system parameters is going to change and when this change will occur, it is important to develop a theory that enables us to anticipate change(s) that might occur in system operation due to the change in any system parameter. It is further essential to develop techniques to ensure that the designed system responds in the neighborhood of a prespecified mode of operation or stays in a *hyperbox* (an abstraction of a square or a rectangle to that in the higher-dimensional space) in the parameter variations space; i.e., the system operation will be such that its index of optimality will remain within certain prespecified bounds [hyperbox] and the system will stay stable for all operational conditions which correspond to that hyperbox.

To respond to such vast demands, sensitivity consideration (analysis and/or synthesis) of systems has been advanced and is commonly employed. Sensitivity considerations play the central role in the overall study of system and is at the *heart* of any engineering practice. It is indeed the subject of sensitivity theory to cope with such conflicts as discrepancies in system operation and to provide the system with more reliable operation than before: in the sense of having the minimum discrepancy between an actual and the nominal pattern of system operations, in spite of several possible disturbing stimuli which may exist. In other words, the goal of sensitivity is a model matching, or sensitivity is a *distance* between an actual and the desired situations.

As is evident from the large number of papers on different aspects of sensitivity theory, it is not known exactly why the parameters of a system change or how we can predict the worst possible change(s); nor is it clear when these changes will occur. The sensitivity problem that confronts us, in general, is many-fold and can be divided into the following categories.

I: Given a system such that *a* mathematical model is known (identified) and the nominal values of system parameters are numerically specified. Then:

(a) Determination of system behavior (i.e., the rate of change in the output) as parameters change will require a systematic procedure to analyze system sensitivity.

(b) Determination of which parameters are changing and anticipation of the worst situation that may occur and possibly its time of occurrence also requires a system sensitivity analysis.

(c) Furthermore, to utilize the above information in order to hold system operation in the neighborhood of its desired operational mode is the final step in system sensitivity analysis and synthesis.

1.1. Overview and Scope of The Book

II: For those systems whose parameters (or mathematical models) are not yet specified or finalized, certainly the above steps will become more complicated than before, because in this situation we need to be concerned with the accuracy of the nominal values as well.

III: In the case of a completely unidentified system, we must be additionally concerned with the accuracy of system mathematical model as well. This additional approximation of system model will further complicate the corresponding sensitivity problems.

Clearly in all the above general categories, we must be also concerned with either the *small-* or the *large-* parameter variations. The small variation means the rate of change in the parameter over a short period of measurement is infinitesimal; i.e., the rate of change approaches zero (differentiable) (or continual, in the terminology of Belanger in [6F. 1]).[1] On the other hand, the large variation means that the rate of change in the parameter is finite (or intermittent, inferring step variations in the system parameters [6F. 1]).

Traditionally, much of the sensitivity analysis has been carried out under the assumption of small changes in system parameters (cf., Section 2.3), in this sense a simple Taylor-series expansion is used frequently to complete the analysis. On the other hand, the large-parameter variations and the corresponding sensitivity considerations are of much importance in engineering practice. In this book several algorithms which are useful for stability analysis and for compensating the troublesome effects of such *large*-parameter variations on the overall system performance are presented under the name of *robustness*. We must also point out that a variety of system-sensitivity measures are in common use today, but there is no unique best sensitivity measure for all systems. We concentrate our studies mainly on the system-trajectory sensitivity function, unless stated otherwise, for the following reason.

Much of the traditional sensitivity analysis for linear time-invariant system has been carried out in the frequency domain. In spite of generality of the frequency-domain analysis, this approach does not provide adequate insight to many other control problems of equal importance and even in the same class. On that note we emphasize the time-domain sensitivity analysis for the sake of completeness and the breadth of application. The time-domain approach or the state-space representation is essentially prerequisite to on-line measurement and control of a dynamical system with any complexity. By emphasizing the time-domain approach we do not imply that the frequency-domain analysis is of insignificant value: rather we simply say that in many problems the time-domain techniques are more straightforward and generally applicable than the corresponding frequency-domain techniques. As we further elaborate on these differences subsequently in Sections 2.7 and 6.9, many advanced algorithms for the frequency-domain sensitivity synthesis are recast

---

[1] Numbers in [ ] refer to the references. Here 6F means Section *[F]* of the references in Chapter Six, and 1 is the reference number in that section. If 6 were not present, then we mean Section *[F]* of the present chapter, *per se*.

by the familiar equations from the time-domain optimization settings. Thus the *scope* of this book is to study *primarily* issues pertaining to the time-domain sensitivity of a dynamical system and we consider the frequency-domain approach only peripherally. The book is prepared at an intermediate graduate level, for those with experience in linear system theory and linear optimal control theory, and as a reference textbook.

1.2. HISTORICAL REMARKS ON THE THEORY OF SENSITIVITY

From a fascinating account of the history of electrical bridge, by Hague [A. 17, pp. 2-6],[2] it can be clearly seen that sensitivity analysis was applied in electrical engineering as early as the work of Wheatstone [A. 25] in 1843, who applied the bridge principle, first introduced by Christie [A. 14] in 1833, for the resistance measurement of resistors. Hence, it is clearly seen that the beginning of sensitivity analysis (particularly in electrical engineering) was not initiated in the 1937 to 1947 era; contrary to many papers in the theory of sensitivity which recall such a date. However, the formative years of the development of system-sensitivity theory are considered to be 1937 to 1947. In general, most of the development in this interval treated a system with one parameter, single-input and single-output. Other sample historical references on this subject are presented in Section *[A]* of the references for this chapter (i.e., § 1.5.*A*).

Since system stability was one of the most crucial concerns in the earlier stage of automatic control, and is of no less consequence now, there was much interest among researchers in the early 1940's, in developing techniques which enable them to compensate for the change in system parameters that might cause difficulties in maintaining system stability. Correspondingly, the study of sensitivity in automatic control advanced with the origin of *feedback system theory*. The primary reason for introducing feedback was in its usefulness in cancelling to some extent the effects of external disturbances (or unwanted signals) and parameter variations. H.W. Bode [A. 5], is responsible primarily for the beginning of the modern theory of *feedback amplifiers*. Bode advanced a logarithmic derivative as a measure of sensitivity. This is now used as a logarithmic derivative of a function in terms of certain parameters (cf., Remark 2.2-4). The difference between Bode's sensitivity measure and that used in dc−bridges, several decades earlier, is simply the weighting factor.

The eminent earlier theoretical work in sensitivity theory is that of Bykhovskiy [A. 10 & 11], which is advanced in [A. 12]. This is a pioneering and

---

[2] This reference was brought to the attention of this author by the Emeritus Professor T.J. Higgins of the University of Wisconsin-Madison.

1.2. Historical Remarks on the Theory of Sensitivity

novel piece of research on sensitivity theory. Tikhonov's classical paper [A. 23], concerning the application of singular perturbations in dynamical systems is another early work on this subject. The contents of the earlier works of Andronov and Pontryagin [A. 1], and of Andronov et al. [A. 2], on certain mathematical results from the theory of differential equations, which, together with the work of Miller and Murray [A. 20] and Bateman [A. 3] effected the establishment of a rigorous foundation for the study of sensitivity analysis.

Research in this field has grown rapidly since 1940's, and nowadays several thousands of papers in various theoretical and practical aspects of this theory are being documented as we have [partially] reported them in this book. Before proceeding further, we must point out that the advancement of singular-perturbation theory and its applications in automatic control system theory diverges from sensitivity theory. Singular perturbation is an independent theoretical concept with extensive research activities of its own and it is not studied herein. Going back to the historical remarks on the theory of sensitivity, we must note that Kokotovic and Rutman [C. 3] present one of the earliest survey papers (originally in Russian in 1964: an English translation appeared in 1965) on the foundation of sensitivity theory and its applications in automatic control. In this paper there are 157 references, providing a comprehensive picture of researchers' efforts in this area from 1937 to 1964.

According to Kokotovic and Rutman [C. 3], the earliest reference in the East is Bruevich [A. 8], in 1941, which details the importance of the exactness for the value of each component in an electrical circuit, and the consequent problems which arise if these values deviate from the desired ones. As a matter of fact, the first place that sensitivity analysis became an important part of the design and measurement procedures is in network theory, where the accuracy of the specified elements is of great importance. In spite of all these developments since 1937, many years later Brown and Campbell [A. 7], in 1948, and Chestnut and Mayer [A. 13], in 1951, and Truxal [A. 24], in 1955, are among the first authors who placed sensitivity analysis in academic textbooks. Later Horowitz emphasized sensitivity analysis in automatic control [A. 19], and subsequently Tomovic published his textbook [B. 24] on this theory, which has been revised by himself and Vukobratovic [B. 25].

In general, sensitivity analysis is related to a variation which does not change the order of the system, nor its initial condition, in the terminology of Miller and Murray [A. 20], α–variation. Variations of the initial condition is called β–variation, and that leading to changes in the order of system is called γ–variation [A. 20]. The latter type of variations are due to neglecting certain small parameters in a system mathematical model.

The next survey paper by Sobral [C. 6] appeared four years later in 1968, which emphasized the contributions on sensitivity aspects of certain optimal control problems namely: the performance sensitivity; the trajectory sensitivity; and the combination of these two issues.

As the number of papers on sensitivity theory grew, two more survey papers appeared relating to: *(i)* sensitivity and error analysis in optimal estimation by Sage [*C*. 5] in 1968; and *(ii)* sensitivity of automatic control systems by Ngo [*C*. 4] in 1970 which classified the pertinent papers chronologically from 1965 to 1970. Ngo [*C*. 4] emphasizes the sensitivity analysis of overshoot and the synthesis of control systems considering the information stemming from sensitivity analysis. This paper also presents many applications of sensitivity theory in automatic control problems.

Sesak's Ph.D. thesis [*B*. 22] also contains a rich bibliography of the papers on the theory of sensitivity especially as pertinent to automatic control theory from 1940 through the early 1974.

The publications on sensitivity theory are, however, growing in one form or another. There have been several international symposia held, since 1960, on the theory of sensitivity and its applications. A number of books, monographs, and the conference proceedings have been appeared, which serve in various ways to support the development of this theory (cf., Section *[B]* of the references) and the list is progressing.

To close this section, we must point out that to classify the documented research on the theory of sensitivity as applied in electrical engineering, we must study the pertinent documents in circuit theory as well as those treating sensitivity theory as applied in automatic control systems. One can not be done without the other and this distinction is important. Nowadays almost every component of a control system is built or is effected by electrical circuits which are manufactured with certain tolerances. We recall that, in general, in network synthesis we assume that we can obtain the nominal values of the components (or elements) in an electrical circuit, but this expectation is generally unrealistic, and as a matter of fact, and because of these inaccuracies in element values, practicing engineers became increasingly aware of the idea of tolerance and subsequently of the general concept of sensitivity. The differences between actual element values of a control device and analog or digital-measurement devices may resonate each other; hence result in substantial differences between the analytical and the experimental responses of a system. Therefore thought has to be given to the accuracy of measurement devices and to all electrical circuits incorporated in the interfacing control system components. On the other hand, many ideas in system sensitivity theory are direct generalizations of those used in circuit theory. Also the trends of, and the concerns for, sensitivity of a network are different enough to be studied separately. Thus we have included Chapter Seven on sensitivity of network which is the revised version of two survey papers on the sensitivity theory as applied to circuit theory [*C*. 1 & 2].

1.3. SUMMARY OF THE BOOK

Chapter Two summarizes a list of definitions pertinent to the theory of sensitivity as applied in automatic control theory, and also reviews the fundamental aspects of this theory. This chapter includes a section on the preliminary mathematics which reviews topics on mappings, semigroups, groups, rings, and fields, followed by definitions of complete mathematical spaces such as Banach and Hilbert spaces. Basics of sets theory and topology and certain topics on functions and mappings such as continuity, differentiability, convexity, contraction, and boundedness are also reviewed. In this section topics on classes of L^p- and H^p-functions and their properties are presented. Finally, in this section, a few properties of matrix and determinant and the Hadamard's inequality, as well as computation of norms for constant vector and matrix are described. The chapter continues with a review of those topics in system theory which are utilized and/or developed in conjunction with recent sensitivity literature on the $H^\infty-$optimization methodology. Included in this section are topics on state-space realization of transfer matrices, linear fractional transformations, classes of time-domain and frequency-domain signals, computation of the H_2- and $H_\infty-$norms of a transfer matrix. The chapter closes with a section on an overview of sensitivity minimization particularly in the $H^\infty-$spaces, where three standard problems (tracking, stabilization, and model matching) are reviewed. The concluding remarks on this theory and our thought on a plan of work are presented next. This and every chapter end with a detailed and categorized list of references.

Chapter Three treating deterministic, continuous, linear, time-invariant systems *(DCLTIS)*, advances analytical expressions for sensitivity functions with distinctions between analysis for system-structural parameter (Definition 2.2-10) and system-physical parameter (Definition 2.2-11). This chapter includes general remarks about sensitivity functions and sensitivity functions generation for *DCLTIS*, in both time-domain and frequency-domain analysis. Special distinction between sensitivity function of an open-loop and a closed-loop system, as well as reconstructible and unreconstructible systems are cited. The concept of low-order sensitivity functions and complete simultaneity properties (for higher-order sensitivity functions) are discussed. A sensitivity measure called system − *TSF* is introduced, in order to demonstrate various applications of this theory. The details of a system model are also included in this chapter.

Additionally in Chapter Three we concurrently explore certain basic techniques available for the analysis of so-called sensitivity in the large. Effort is also being made to establish two important sensitivity invariants for linear systems in the final part of this chapter.

In Chapter Four some algebraic properties for the corresponding matrices in a system model *(MSM)*, and the requirements on the terminal conditions insensitivity *(TCI)* in a linear system for several different system-sensitivity measures, are established. In particular, the two important theorems of Gupta and Mehra on rank computation of controllability matrix corresponding to the *MSM* are described, which these theorems advance the computational efficiency of generating all trajectory-sensitivity functions.

Chapter Five primarily advances the correlation between the theory of stability and the theory of sensitivity with respect to constructing a robustly stable system stemming from the Lyapunov stability theory. The eigenvalue and pole-zero (root) sensitivity of linear, time-invariant systems are revisited. Bounds on the large variations in system parameters which assure stability are developed. Certain additional supporting information, such as the allowable variations in the coefficients of a stable (and a convergent) polynomial for continuous- (and discrete-) time linear control systems, is also established. The chapter ends with some extensions of these results to a special class of nonlinear systems – the concept of absolute stability.

Chapter Six advances several new algorithms for designing a system with a minimum sensitivity measure. After a brief general remark about various aspects of sensitivity reduction, this chapter opens with a qualitative discussion on the properties of a linear feedback control law in reducing the system-trajectory sensitivity function. Then a concise review of performance-index sensitivity comparison precedes the analysis and the synthesis of the optimality-index sensitivity, with several algorithms. In the next section; the trajectory-sensitivity function is considered and various propositions for minimizing its quadratic norm are presented in several algorithms. An algorithm in regard to the sensitivity minimization with adaptive controller gain is given subsequently, which shows that its preceding algorithm is an especial case (within certain desired degree of accuracy) of the present algorithm. Next, several algorithms for designing a system with a minimum quadratic sensitivity measure in spite of the large-parameter variations are presented. In this regard, the concept of utilizing an auxiliary input to robustify system performance is presented in conjunction with its applications to system identification. In the following section sensitivity minimization in the Hardy spaces of H^2 and H^∞, for treating exogenous signals in a system which is described by its state-space model, is presented. Discussions on the so-called *the theory of invariance* and its current standing are included in the subsequent section. Among topics covered in that section are the classical problems in the theory of invariance, the basic steps of its development, and the variational (or Rozonoer's) approaches. This chapter closes with a section on bounded uncertainty and disturbance rejection; where an overview of deterministic approaches based on the Lyapunov stability theory to meet these uncertainties, generalized dynamical systems and their asymptotic stability; and a brief discussion on the disturbance rejection are reviewed.

Chapter Seven primarily presents a concise account of sensitivity theory as applied in network theory, by categorizing various areas of active research in network-sensitivity theory into 14 groups.

The book closes with three appendices: Appendix A – Selected Numerical Applications; Appendix B – List of the References, A Sensitivity Tree; and Appendix C – Journal Abbreviations and Acronyms; and ends with author and subject indices.

1.4. A SUMMARY OF ISSUES NOT INCLUDED IN THE BOOK

Sensitivity is one of the most versatile words in engineering as well as social sciences. There are so many variations of this word used by researchers in different disciplines, that are too numerous to count. Thus in this section we explicitly state a partial list of topics which are not presented in this book, although they constitute research disciplines of their own, and we have provided a group of references for them. But first, we must acknowledge that by providing this list we are not suggesting that the augmentation of topics in Sections 1.3 and 1.4 will be the whole picture of whatever the word sensitivity means to different scientists. The following list, of course, is in addition to many other sets of topics inside the book that we state are not covered herein.

1. The System Classifications: According to the flow-chart in Table 3.1-1, we are mainly concerned with the deterministic, continuous, linear, time-invariant systems, where the input u(t) can only be a linear function of system state x(t). Therefore topics that are pertinent to the sensitivity considerations in discrete-time, nonlinear, delay (special class of nonlinear discrete-time), and distributed-parameter systems (i.e., issues corresponding to the references in § 6.12.*L* to § 6.12.*O*) are not considered in this book. Also statistical analyses corresponding to the references in § 6.12.*L,* and all issues pertaining to probabilistic modeling of parameter variations as well as stochastic systems are not included. We are not, however, suggesting that we have exhaustively covered all aspects of the above mentioned linear systems either, although we have given sufficient information on the sensitivity matters corresponding to this class of systems, which serves as a limiting condition for more advanced classes of systems.

2. Modeling Inaccuracies: One of the fundamental issues in control theory that sets an appropriate frame of reference for studying a given system, is to have an access to a meaningful system model. By meaningful we mean a model that represents the class of system with a sufficiently accurate set of parameters. Here we are not demanding an *exact* model of the physical system, because it is well known that in the real-world situation that demand is hardly met. But it is absolutely essential to model a system with time-lag, *per se,* as such, otherwise there is no merit in error analysis, when the system model is metaphorically speaking inadequate. We note that by model we mean more than one point or a very narrow local representation of a system. However, no matter how accurate or exact these models become, any model is an approximation model. If this fact is accepted, then that opens a host of new questions which must be considered for evaluating the accuracy and/or the exactness of the results and so forth. These are very important issues and they are subjects of extensive research of their own rights. We suggest [B. 27] as a starting point for this line of research. In this reference which uses advanced analysis, a thorough discussion on the choice of an appropriate model for a dynamical system and its sensitivity as well as sensitivity analysis of the [traditional] optimal control theory are included. The objective of sensitivity analysis in this reference is an analysis for the variations in the minimal value of the performance index subject to *small*-parameter variations.

3. Computational Sensitivity: Upon finalizing the modeling choice and the related sensitivity matters, we face a number of other new and challenging issues that must be considered in the overall sensitivity considerations of a dynamical system. Namely, issues regarding to such matters as the computational sensitivity, which are multidisciplinaries and they are of great importance both in theory and in practice. From the most simplest case of linear equations to the most exotic optimization methods, sensitivity and computational stability as defined appropriately are of paramount importance. There exists a very active research community in this area and a body of knowledge has been already developed, and more to be explored in these multidisciplinary areas of research which provide ample research opportunities for years to come.

In this regard we cite references in § 6.12.*P*, in particular [6] and [10] which are also given in § 1.5.*B*. Starting with [*B*. 6], this work very well covers the subject of sensitivity analysis as related to linear equations, namely, the first basic problem in linear systems which is a set of linear equations. Here the problem of executing perturbational analysis of linear equations are presented, while making reference to some relevant applications. This well-prepared textbook has over 300 references.[3]

---

[3] "All in all, methods of mathematical optimization rely one way or the other on relative sensitivities, under a different title in each, ranging from the gradient methods to model tracking or self-learning systems. Even the simple task of fitting data to a curve usually involves sensitivity calculations. As for social scientist, economist, as well as for many other disciplines, sensitivity and perturbation techniques can provide valuable information about the amount of inaccuracy in the behaviour of a model as related to the inaccuracies in the system's data. If the data gathered by field study or experimental testing falls within certain tolerance limits, the tolerances may well be amplified and widened in the output results obtained. The question might then arise to how uncertain the results are − or how unreliable − in relation to data's uncertainties. In this instance, perturbation analysis can provide valuable information about the regions of compatibility and admissibility of solutions. An alternate use might also be to determine the allowable data tolerances in a parameter for the results to sustain a certain level of accuracy; and so forth.

Perturbation techniques and sensitivity analysis are of course no new terms, nor are they recently explored fields. As a mathematical discipline, however, a unified body of knowledge rather than an elementary application, they are rather young. Only during this last century have celestial mechanics withnessed an era of rapid progress; the three-body problem − in contrast with Newton's two body problem − becoming the new challenge. Workers in the field opted to consider this third body as a perturbation in the field. In this context also, Lagrange's ingenious method of variation of elements was introduced, and Poincare's theory of asymptotic expansions enabled the summing up of a few terms of a divergent series to yield almost exactly its sum.

Perturbation theory in linear algebra is an even more recent branch. In 1948, Turing's famous paper [Rounding-off errors in matrix processes, *Quart. J. Mechanics and Applied Mathematics*,

1.4. A Summary of Issues not Included in the Book

Certainly, as the set of references in § 6.12.*P* suggests there are many research areas pertaining to sensitivity and stability analysis in parametric programming. "It seems plausible that a significant impediment to extensive use of nonlinear programming in applications is the lack of standardized user-friendly software for calculating useful sensitivity and stability information. The remarkable developments of the past decade have shown that characterizations of optimality and algorithmic convergence properties are significantly extended and deepened through a perturbational analysis. Typically, a parametric solution or error bounds on a solution with perturbed data are of great interest, both in practical applications and in theoretical characterizations [*B*. 10]." In this reference, an array of papers, which are concisely and authoritatively covering different aspects of sensitivity analysis as related to various mathematical programmings, are surveyed, and this reference should serve as an excellent starting point for further research in this area. Clearly the main thought in this area is along the same line as our concern in engineering applications of sensitivity theory, however, there is yet a gap (though not unbridgeable) between these two disciplines.

4. *Imbedding and/or Embedding Problems:* There are two groups of problems in system engineering which fall under these categories: *First*, there are many cases that we artificially imbed a parameter into the system in order to transform the original system into a parametric family of systems, such that the original problem becomes only one member of this family. For example, consider the following system.

$$\dot{x}(t) = \begin{bmatrix} A_{11} & A_{12} \\ A_{21} & A_{22} \end{bmatrix} x(t) + \begin{bmatrix} B_1 \\ B_2 \end{bmatrix} u(t). \qquad (1.4\text{-}1)$$

Now, let us consider

$$\dot{x}(t) = \begin{bmatrix} A_{11} & \gamma A_{12} \\ \gamma A_{21} & A_{22} \end{bmatrix} x(t) + \begin{bmatrix} B_1 \\ B_2 \end{bmatrix} u(t), \qquad (1.4\text{-}2)$$

where γ is an imbedded parameter into the system (1.4-1) which is considered as the original system. Clearly, when $\gamma = 1$, the (1.4-2) becomes the same as (1.4-1) which is a set of coupled differential equations. On the other hand, when $\gamma = 0$,

---

vol. 1, pp. 287-308] triggered interest in problem of sensitivity of solutions of linear equations to round-off errors. In this paper, Turing laid down the definition of a condition number by which a small input error in the data can be drastically amplified in the solution. Numerical analysts then acknowledged this number as the major factor affecting computational accuracy, and have tried since then to control it while working out any new numerical procedure. But an ill-conditioned system can only be cured up to a certain extent, and no matter how cunning, skilled or elaborate one is, Turing's number or a variant of it will eventually hinder our illusion [*B*. 6]."

then (1.4-2) becomes a decoupled system that is perhaps computationally easier to analyze than (1.4-1). Thus the original system of (1.4-1) is only one member of the family of systems corresponding to (1.4-2). Choosing $\gamma \in [0, 1]$, we can perform a sensitivity analysis for this type of problems very analogously to what we describe in this book and thus we can infer various information from this analysis. We can use many other system "outputs" for this purpose and define similar situations. Thus much of the analyses in this book can be extended to this class of problems. However, this sort of problems is not developed herein, and an interested reader may consult [B. 4] for typical studies of this type.

Secondly, there are cases in which we have a parametric problem entailing a set of varying parameters, and we embed from that set a particular parameter in order to perform an optimization or some other design study. This type of analyses requires a special searching procedure (for example a gradient analysis) that is very much similar to our sensitivity analysis. Here again we do not explicitly study these matters.

In summary and as mentioned at the outset of this section, the above list is by no means the whole story, but for the sake of time and space we close this list. Certainly and in the future, many more new areas for applications will be explored and the boundaries of these existing areas for research will also be refined. Incidentally, and in spite of efforts to classify various research opportunities that exist in different scientific disciplines pertaining to the theory of sensitivity, these boundaries should not be considered very rigid and indeed they are quite fuzzy.

Finally, considering all these different areas of research supported by a large volume of [recently] documented works, we conclude that the theory of sensitivity as a whole is like a very healthy tree with many strong branches: some are pruned neatly but perhaps some more need to be pruned; while this tree will grow indefinitely to give us more and more branches.

1.5. REFERENCES

This research is being conducted primarily at the Kurt F. Wendt Physical Sciences and Engineering Library of the University of Wisconsin-Madison.

In preparing this book, efforts were made to classify the references that this author had *mostly* access to (almost all in English and French, but none in Russian and German) as narrowly as possible – in order to provide an entry guide for the future research. Therefore the list is by no means complete, and that has not been our intention, although we have tried to bring together the global research community. Occasionally, we have cited some of these references more than once due to their special contributions. It is also quite possible that certain corresponding authors may not agree with our classifications; we regret if there is an omission on our part. Clearly, a few of these references are only of historical values and they are probably out of print, but most of them can be found in the libraries and it is interesting to note that some of them are still very inspiring and full of unexplored ideas, and quite a few of them are not well scrutinized. It has been indeed a very difficult task to delete some of these earlier references in a given specific area (despite of many recent ones which are appearing in that area) partly because the special effects that they had on the training of this author. We hope that this list will encourage some of the future researchers to take time and to look into these earlier references before they launch their research activities in this area. To save space, we abbreviate those journals which are repeatedly cited, and for the list of these abbreviations and acronyms refer to Appendix C. However, due to a very severe budgetary constraint in the recent years, the Wendt Library has not received the current issues of quite a few foreign journals in that list and we may be missing some interesting articles in these journals.

[A] Representative Historical References

[1] A.A. Andronov and L.S. Pontryagin, "Structurally stable systems," *DOKL. AN SSSR*, vol. 14, no. 5, 1937.

[2] A.A. Andronov, A.V. Vitt and S.E. Khaikin, *Theory of Oscillations [in Russian]*. ONTI, 1937. (According to [C. 3] the name of A.V. Vitt did not appear in 1937 edition). Two English editions of this work appeared as follows. *(i)* under A.A. Andronow and C.E. Chaikin, *Theory of Oscillations*. Princeton, NJ: Princeton Univ. Press, 1949. [Translated by N. Goldowskaja, edited by S. Lefschetz.] *(ii)* under A.A. Andronov, A.A. Vitt and S.E. Khaikin, *Theory of Oscillators*. Oxford: Pergammon Press (and Addison-Wesley in U.S.A.), 1966. [Translated by F. Immirzi, edited and abridged by W. Fishwick.]

[3] H. Bateman, "The control of an elastic fluid," *Bulletin of the American Mathematical Society*, vol. 51, pp. 601-646, 1945. [This important paper detailing many references on early use of different *governors* (prior to 1943), provides a fascinating and invaluable account of control theory based upon linear equations with small perturbation. Also appeared in [4], pp. 18-64.]

[4] R. Bellman and R. Kalaba, Eds., *Selected Papers on Mathematical Trends in Control Theory*. New York: Dover, 1964. [This entry, is suggested by Professor T.J. Higgins, which provides a set of 13 pioneering reprints by such prominent contributors as: J.C. Maxwell, H. Bateman, R. Bellman, R. Kalaba, A. Hurwitz, H. Nyquist, H.W. Bode, B. van der Pol, N. Minorsky, L.A. Zadeh, J.R. Ragazzini, J.P. LaSalle, V. Boltyanskii, R. Gamkrelidze, and L. Pontryagin; that establish the very foundation of modern control theory. This author acknowledges that reading these papers was a crash course in humility for him.]

[5] H.W. Bode, *Network Analysis and Feedback Amplifier Design*. New York: Van Nostrand, 1945.

[6] E. Bradshaw, "A note on the sensitivity of the Schering bridge network," *J. of the Royal Technical College of Glasgow* (Scotland), vol. 4, pp. 144-146, 1937.

[7] G.S. Brown and D.P. Campbell, *Principles of Servomechanisms, dynamics and synthesis of closed-loop systems*. New York: Wiley, 1948. [This entry, is suggested by Professor T.J. Higgins, has additional discussions of the earlier work by Hall [18], and other students at Massachusetts Institute of Technology, with some discussions on "sensitivity factor". The textbook has also an informative bibliography of the early work in this area. "Much of our text material stems from an educational program in which we acted mainly as catalysts by stimulating students to make contributions in uncharted areas. Our viewpoint has developed largely by our work with these students, who, while primarily interested in servomechanisms, were studying in related fields. This student association has cross-fertilized various specialized branches of engineering and continues as thesis investigations and seminars are conducted on the subject."]

[8] N.G. Bruevich, *On the Accuracy of Mechanisms*. Izd-vo AN SSSR, 1941.

[9] S. Butterworth, "On the vibration Galvanometer and its application to inductance bridges," *Proc. Physical Society* (England), vol. 25, pp. 75-94, 1912.

[10] M.L. Bykhovskiy, "The accuracy of mechanisms for which the state (or position) of the components (links) are described by differential equations," *Izvestiya AN SSSR, OTN*, no. 11, 1947.

[11] _____, "Accuracy of electrical decision circuits," *Izvestiya AN SSSR, ONT*, no. 8, 1948.

[12] _____, "Sensitivity and dynamic accuracy of control systems," *ECy*,[4] vol. 2, pp. 121-134, 1964.

[13] H. Chestnut and R.W. Mayer, *Servomechanisms and Regulating System Design*. New York: Wiley, vol. I, 1951, vol. II, 1955. [This entry has a list of 117 references in chronological order, from 1934 to 1954, and has a brief discussion on sensitivity.]

[14] S.H. Christie, "Experimental determination of the laws of magneto – electric induction," *Philosophical Transactions of the Royal Society of London*, vol. 123, pp. 95-142, 1833.

[15] J. Fischer, "Eigenschaften der Wheatsoneschen Brutcke," *Elektrotechnik und Maschinenbau*, vol. 48, pp. 1060-1064, 1930.

[16] I.S. Gradshtein, "On the behavior of the solutions of a system of linear differential equations with constant coefficients, degenerating in the limit," *Izv. AN SSSR*, Ser. Matem., no. 3, 1949.

[17] B. Hague, *Alternating Current Bridge Methods*. [Revised by T.R. Foord.] London: Pitman Press, sixth edition 1971. First edition 1923.

[18] A.C. Hall, "Application of circuit theory to the design of servomechanisms," *JFI*, vol. 242, pp. 279-307, 1946. [This entry, is suggested by Professor T.J. Higgins, is adapted from a [doctoral] thesis entitled, "The analysis and synthesis of linear servomechanisms," submitted to Electrical Engineering department at Massachusetts Institute of Technology, May 1943.

---

[4] For journal abbreviations and acronyms refer to Appendix C. The page numbers for those papers which are translated into English are always relative to the translated edition.

1.5.A. Representative Historical References

This source has discussions on sensitivity, and is one of the first instances that the use of this term in "control" theory was made.]

[19] I.M. Horowitz, *Synthesis of Feedback Systems*. New York: Academic Press, 1963.

[20] K.S. Miller and F.J. Murray, "A mathematical basis for an error analysis of differential analyzers," *JMP*, vol. 32, nos. 2 and 3, pp. 136-163, July/Oct. 1953.

[21] L. Schwendler, "On the Galvanometer resistance to be employed in testing with Wheatstone's diagram," *Philosophical Magazine* (England), 4th series, vol. 31, pp. 364-368, 1866.

[22] A.C. Seletzky and L.A. Zurcher, "Sensitivity of the four-arm bridge," *Electrical Engineering* (U.S.A.), vol. 58, pp. 723-728, 1939. [Some of the conclusions were anticipated by J. Fischer in connection with d.c. bridges, a comment from [17].]

[23] A.N. Tikhonov, "On the dependency of the solutions of differential equations on a small parameter," *(Moscow) Matem. Sborn.*, vol. 22 (64), no. 2, 1948.

[24] J.G. Truxal, *Automatic Feedback Control System Synthesis*. New York: McGraw-Hill, 1955.

[25] C. Wheatstone, "An account of several new instruments and processes for determining the constants of a voltaic circuit," *Philosophical Transactions of the Royal Society of London*, vol. 133, pp. 303-327, 1843.

[B] Representative Monographs, Books, Reports and the Conference Proceedings

[1] M.J. Ashworth, *Feedback Design of Systems with Significant Uncertainty*. Chichester, England: Research Studies Press [a division of Wiley], 1982. Reviewed by I. Horowitz, *Automatica*, vol. 19, no. 5, p. 581, 1983.

[2] S.P. Bingulac, Ed., *System Sensitivity and Adaptivity*. Proc. Second IFAC Symposium (Dubrovnik, former Yugoslavia), August 26-31, 1968.

[3] S.P. Bingulac, "Sensitivity in automatic control," (Commentary) in *Proc. IFAC*, vol. 3 (of the 5th Congress), pp. C-31.1-C-31.7, 1972.

[4] J.B. Cruz, Jr., Ed., *Feedback Systems*. New York: McGraw-Hill, 1972. Reviewed by H. Kwakernaak, *IEEE-TAC*, vol. AC-17, no. 5, pp. 751-752, 1972.

[5] _____, *System Sensitivity Analysis*. Stroudsburg, PA: Dowden, Hutchinson, 1973. Reviewed by J. Sesak, *IEEE-TSMC*, vol. SMC-5, no. 2, p. 290, March 1975.

[6] A. Deif, *Sensitivity Analysis in Linear Systems*. New York: Springer – Verlag, 1986.

[7] P. Dorato, P. Kokotovic, and H. Kwakernaak, "Report on the second IFAC symposium on system sensitivity and adaptivity," *Automatica*, vol. 5, pp. 251-256, 1969.

[8] M. Eslami, "Sensitivity analysis and synthesis in automatic control systems," Ph.D. Dissertation, The University of Wisconsin-Madison, May 1978.

[9] A.V. Fiacco, *Introduction to Sensitivity and Stability Analysis in Nonlinear Programming*. New York: Academic Press, 1983. Reviewed by C.E. Lemke, *SIAM R.*, vol. 27, no. 1, pp. 114-115, 1985.

[10] A.V. Fiacco, Ed., *Sensitivity, Stability, and Parametric Analysis*. (A publication of the Mathematical Programming Society.) Amsterdam: North-Holland, 1984.

[11] P.M. Frank, *Introduction To System Sensitivity Theory*. New York: Academic Press, 1978. Reviewed by M. Eslami, *IEEE-TSMC*, vol. SMC-10, no. 6, pp. 337-338, June 1980, and by J.B. Cruz, Jr., *IEEE Circuits and Systems Magazine*, vol. 2, no. 2, p. 22, June 1980.

[12] M.S. Ghausi, P. Stavroulakis, and K.R. Laker, Eds., *Special Issue on Sensitivity*. *JFI*, vol. 312, nos. 3/4, Sept./Oct. 1981.

[13] V.I. Gorodetskii, F.M. Zakharin, E.N. Rozenvasser, and R.M. Yusupov, *Methods of Sensitivity Theory in Automatic Control [in Russian]*. Moscow: Energiya, 1971.

[14] G. Guardabassi, A. Locatelli and S. Rinaldi, Eds., *Sensitivity, Adaptivity and Optimality*. Proc. Third IFAC Symposium (Ischia, Italy), June 18-23, 1973. [Pittsburgh, Pennsylvania (Instrument Society of America)].

[15] C.A. Harvey and R.E. Pope, *Study of Synthesis Techniques for Insensitive Aircraft Control Systems (Report)*. Minneapolis, Minnesota: Honeywell Inc. Systems and Research Center, Oct. 1976.

[16] E.J. Haug, K.K. Choi, and V. Komkov, *Design Sensitivity Analysis of Structural Systems*. Orlando, FL: Academic Press, 1986. Reviewed by R.V. Kohn, *SIAM R.*, vol. 30, no. 3, pp. 518-519, 1988.

[17] B. Kouvaritakis, D.H. Owens, and M.J. Grimble, Eds., *Special Issue on Sensitivity and Robustness in Control Systems Theory and Design*. Proc. IEE, vol. 129, Pt. D, no. 6, 1982.

[18] J. Lunze, *Robust Multivariable Feedback Control*. London: Prentice-Hall, 1988. Reviewed by S. Engell, *Automatica*, vol. 27, no. 4, pp. 749-750, 1991.

[19] M. Morari and E. Zafiriou, *Robust Process Control*. Englewood Cliffs, NJ: Prentice-Hall, 1989. Reviewed by J.C. Kantor, *AIChE J.*, vol. 37, no. 12, pp. 1905-1906, 1991.

[20] L. Radanovic, Ed., *Sensitivity Methods in Control Theory*. Proc. IFAC Symposium (Dubrovnik, former Yugoslavia), Aug. 31-Sept. 5, 1964. [New York: Pergamon Press, 1966.]

[21] E.N. Rozenvasser and R.M. Yusupov, *Sensitivity of Automatic Control [in Russian]*. Moscow: Izd. Energiya, 1969.

[22] J.R. Sesak, "Sensitivity-constrained linear optimal control: Analysis and synthesis," Ph.D. Dissertation, The University of Wisconsin-Madison, Oct. 1974.

[23] D.D. Siljak, *Nonlinear Systems, the parameter analysis and design*. New York: Wiley, 1969.

[24] R. Tomovic, *Sensitivity Analysis of Dynamic Systems*. New York: McGraw-Hill, 1964.

[25] R. Tomovic and M. Vukobratovic, *General Sensitivity Theory*. New York: American Elsevier, 1972. Reviewed by W.A. Porter, *IEEE-TAC*, vol. AC-18, no. 2, pp. 199-200, April 1973, and by J. Sesak, *IEEE-TSMC*, vol. SMC-4, no. 2, p. 235, March 1974.

[26] W. Von Dinkelbach, *Sensitivitatsanalysen und parametrische Programmierung*. Berlin: Springer – Verlag, 1969. Reviewed by S.I. Gass, *SIAM R.*, vol. 12, no. 1, pp. 165-166, 1970.

[27] A. Wierzbicki, *Models and Sensitivity of Control Systems*. Amsterdam: Elsevier, 1984. [Translated by R. Miklaszewski and J. Gandelman from Polish.] Reviewed by D.H. Owens, *Automatica*, vol. 22, no. 2, pp. 260-261, March 1986.

[C] Survey Papers

[1] M. Eslami and R.S. Marleau, "Theory of sensitivity as applied to circuit theory – the state of the art," in *Proc. 21st Midwest Symp. on Circuits and Systems*, pp. 1-16, 1978.

[2] _____, "Theory of sensitivity of network: A tutorial," *IEEE-TE (Special Issue on Circuits and Systems: Research and Education)*, vol. 32, no. 3, pp. 319-334, Aug. 1989.

[3] P.V. Kokotovic and R.S. Rutman, "Sensitivity of automatic control systems," *ARC*, vol. 26, pp. 727-749, April 1965.

[4] N.T. Ngo, "Sensitivity of automatic control systems," *ARC*, vol. 32, pp. 735-762, May 1971.

[5] A.P. Sage, "A survey of sensitivity and error analysis methods in optimum estimation," in *Proc. 21st Annual Southwest IEEE Conf.*, pp. 19A1-19A8, Texas, 1968.

[6] M. Sobral, Jr., "Sensitivity in optimal control systems," *Proc. IEEE*, vol. 56, no. 10, pp. 1644-1652, Oct. 1968.

CHAPTER TWO
THE PRINCIPAL ASPECTS OF SENSITIVITY THEORY

2.1. INTRODUCTION

This chapter summarizes a list of definitions pertinent to sensitivity theory as applied in control theory. The fundamental aspects of sensitivity analysis are reviewed in Sections 2.3 and 2.4, followed by discussions on the preliminary mathematics and aspects of system theory in Sections 2.5 and 2.6. The chapter closes with a brief discussion on synthesis and the H^∞-sensitivity minimization.

The theory of sensitivity is a rational theory resulting from the behavioral studying of system performance, under parametric variations, and certain unwanted exogenous input acting on the system. In Section 1.1 a list of possible causes for system parameter variations is outlined, which comprises most conjectures concerning these causes as cited by several authors who have discussed this theory in one form or another. Generally speaking, any phenomenon which yields system behavior to be different in any way from its prespecified (or expected) pattern of operation is called noise.

The theory of sensitivity provides a systematic manner to reduce unwanted effects of *certain* noise in a system which has resulted from parameter variations and certain unwanted exogenous input acting on that system.

2.2. ESSENTIAL DEFINITIONS FOR PARAMETRIC SENSITIVITY

In order to avoid confusion in using certain terminology in the subsequent chapters, the following list of definitions is cited.

DEFINITION 2.2-1 (Generating System[1] – GS): Consider the following deterministic, continuous, linear, time-varying system described by

$$\dot{x}(t;\ \theta) = A(t;\ \theta)x(t;\ \theta) + B(t;\ \theta)u(t), \qquad (2.2\text{-}1)$$

$$z(t;\ \theta) = C(t;\ \theta)x(t;\ \theta), \qquad (2.2\text{-}2)$$

---

[1] The words generating system are chosen herein because all the subsequent definitions and theorems are *generated* from this system. This choice also confirms our conviction that this theory is based on an anticipation of changes rather than an actual physical changes. In the literature terms such as "original system", or "basic system", *etc.*, are used for similar purposes.

where $x(t; \theta) \in R^n(t; \theta)$ is the system state vector; $u(t) \in R^m(t)$ is the system input vector; $A(t; \theta)$ and $B(t; \theta)$, are $n \times n$ and $n \times m$, system matrices; $z(t; \theta) \in R^\omega(t; \theta)$ is the system output (response) or often called system observation vector; $\theta \in R^r$ is the system parameter vector (Definitions 2.2-10 to 13). These *Roman* n, m, ω, r, and θ are consistently used as above, and for the sake of time and space the former illustrations are no longer being repeated.

DEFINITION 2.2-2 (DCLTI – GS): Consider the system of Definition 2.2-1, if A, B, and C are not functions of time; i.e., these are $A \triangleq A(\theta)$, $B \triangleq B(\theta)$, and $C \triangleq C(\theta)$, then the corresponding system is called "*DCLTI – GS*", that is deterministic, continuous, linear, time-invariant – generating system.

The following three definitions are stated herein for the sake of consistency in our presentation. In the literature these are defined slightly differently.

DEFINITION 2.2-3 (System Identification): We define system identification to be relative to the process of derivation (exact or estimated) of a mathematical model – in parametrical form – which characterizes the physical behavior of system (in terms of either system input-output, or system-components interrelationships), with the parameter numerical values are not yet known.

DEFINITION 2.2-4 (System Specification): We refer to the process of evaluating either exactly *(Determination)*, or approximately *(Estimation)*, the values of system parameters, whose parametrical form has been already known (identified), system specification.

DEFINITION 2.2-5 (System Realization): We refer to the process of constructing the physical configuration of each system, whose mathematical model is already specified, system realization.

REMARK 2.2-1: Suppose that the relation between input x, and output z, of a system is "identified" as a linear equation, i.e., $z = ax + b$. Then measurements are made to evaluate a and b in order to "specify" the system. We may then take the necessary steps to realize this system from this information. However, if this system contains a complex electronic circuitry that has many elements, *per se,* we may not be able to "realize" this system by knowing only the above specified linear relationship.

DEFINITION 2.2-6 (Complete Controllability): Consider the *DCLTI – GS* of Definition 2.2-2. This system is completely controllable if

$$\text{Rank } (B, AB, \cdots, A^{n-1}B) = n. \tag{2.2-3}$$

DEFINITION 2.2-7 (Complete Observability): Consider the *DCLTI – GS* of Definition 2.2-2. This system is completely observable if

2.2. Essential Definitions

$$\text{Rank } [C^T, A^T C^T, \cdots, (A^T)^{n-1} C^T] = n. \tag{2.2-4}$$

(The superscript *Roman* T always stands for the transpose of a matrix).

DEFINITION 2.2-8 (Normal System): Consider the *DCLTI – GS* of Definition 2.2-2. This system is termed normal if it is completely controllable from any one of its inputs acting alone.

DEFINITION 2.2-9 (Nominally-Equivalent Systems): Consider two systems of equal dimension as described in Definition 2.2-2. Suppose that the first one is an open-loop design with state $x_o(t, \theta)$ (i.e., $u_o = u(t) \neq u(t; x_o(t, \theta))$), and the second one is a closed-loop design with state $x_c(t, \theta)$ (i.e., $u_c = u(t; x_c(t, \theta))$), such that for the same input function (i.e., $u_o = u_c$) they both produce the same state vector (i.e., $x_o(t, \theta) = x_c(t, \theta)$). Then these two systems are called nominally equivalent.

REMARK 2.2-2: For a meaningful perturbational or sensitivity analysis we need to distinguish among several different sets of parameters that we are facing in a comprehensive study of a dynamical system. To set the record straight we present the following case.

Consider a standard regulator system comprising two distinct sub-systems which are coupled as follows.

$$\frac{d}{dt} \begin{Bmatrix} x_1 \\ x_2 \\ x_3 \\ x_4 \end{Bmatrix} = \begin{bmatrix} 0 & 1 & c_{13} & c_{14} \\ \alpha_1 & \beta_1 & c_{23} & c_{24} \\ c_{31} & c_{32} & 0 & 1 \\ c_{41} & c_{42} & \alpha_2 & \beta_2 \end{bmatrix} \begin{Bmatrix} x_1 \\ x_2 \\ x_3 \\ x_4 \end{Bmatrix} + \begin{bmatrix} 0 & 0 \\ \gamma_1 & 0 \\ 0 & 0 \\ 0 & \gamma_2 \end{bmatrix} \begin{Bmatrix} u_1 \\ u_2 \end{Bmatrix}$$

$$\triangleq \begin{bmatrix} A_{11} & C'_{12} \\ C'_{21} & A_{22} \end{bmatrix} x + \begin{bmatrix} B_1 \\ B_2 \end{bmatrix} \begin{Bmatrix} u_1 \\ u_2 \end{Bmatrix} \triangleq Ax + Bu. \tag{2.2-5}$$

We assume that u_1 and u_2 are chosen such that the following cost is minimized.

$$J = \min \int_0^\infty [x^T W_0 x + u^T W u] d\tau. \tag{2.2-6}$$

Here the weighting matrices are $W_0 = W_0^T \geq 0$ and $W = W^T > 0$, with $x(0) = x_o$. The well-known optimal solution of this problem yields $u = Kx$, where

$$K \triangleq \begin{bmatrix} k_{11} & k_{12} & k_{13} & k_{14} \\ k_{21} & k_{22} & k_{23} & k_{24} \end{bmatrix}, \quad (2.2\text{-}7)$$

is generated *via* the corresponding Riccati matrix equation that is a function of (A, B) as well as (W_0, W). Certainly the minimum J is also a function of x_0. Upon substituting this optimal K in (2.2-5) we have:

$$\dot{x} = (A + BK)x \triangleq S(\theta)x, \quad (2.2\text{-}8)$$

where

$$S(\theta) \triangleq \begin{bmatrix} 0 & 1 & c_{13} & c_{14} \\ \alpha_1 + \gamma_1 k_{11} & \beta_1 + \gamma_1 k_{12} & c_{23} + \gamma_1 k_{13} & c_{24} + \gamma_1 k_{14} \\ c_{31} & c_{32} & 0 & 1 \\ c_{41} + \gamma_2 k_{21} & c_{42} + \gamma_2 k_{22} & \alpha_2 + \gamma_2 k_{23} & \beta_2 + \gamma_2 k_{24} \end{bmatrix}. \quad (2.2\text{-}9)$$

Here "θ" is a generic name for the vector of system *parameters* which is described by (2.2-5) to (2.2-7). This example shows how quickly we may accumulate a number of parameters in a very simple problem. We now present the following four definitions.

DEFINITION 2.2-10 (System-Structural Parameters): The minimum number of parameters that must be specified for identifying a *particular* parametrical form of *a* system mathematical model is defined as the set of system-structural parameters.

The *Roman* letter "q" is reserved for the vector whose components are elements of this set.

In (2.2-5), A and B have respectively 16 and 8 structural parameters (or entries) that are members of this set. In particular, each sub-system A_{ii} in (2.2-5b) is shown by a phase-variable canonical model and as such the 1's and 0's are fixed members of this set.

DEFINITION 2.2-11 (System-Physical Parameters): This set consists of parameters that must be specified for realizing a system.

The *Roman* letter "p" is reserved for the vector whose components are members of this set.

For each sub-system A_{ii} in (2.2-5b) the parameters $\{\alpha_i, \beta_i, \gamma_i\}$ are members of this set. In general, the above two sets of parameters are, of course, different.

DEFINITION 2.2-12 (System-Auxiliary Parameters): The set of parameters in a comprehensive study (analysis and synthesis) of a system which does not belong; neither to the set of system-structural parameters, nor to the set of system-physical parameters, but it reflects our choice for a particular design methodology is called the set of system-auxiliary parameters.

2.2. Essential Definitions

Here, W_0 and W_1 of (2.2-6) are members of this set, which their contributions have been propagated into the system, and therefore their variations must be accounted for in a comprehensive system study.

DEFINITION 2.2-13 *(System-Coupling Parameters):* Certain large-scale systems are constructed by augmenting several sub-systems and using coupling blocks that interconnect these sub-systems. The set of parameters which completes this construction is called the set of system-coupling parameters. This set is the amalgamation of the previous three sets, with perhaps a few additional *parameters*.

COMMENT 2.2-1: Whether, for example, $c_{23} + \gamma_1 k_{13}$ is considered as one coupling parameter or a function of three parameters is a separate issue. Nevertheless, their entities as members of a new parameter set are obvious.

In general we may show all or any one of the above four parameter sets by one symbol: say – θ.

DEFINITION 2.2-14 *(Nominal Value):* The nominal value of a parameter vector θ, which is shown by θ^o, is that value of θ whose corresponding experimental and analytical system responses are initially very close to each other – or, ideally, these two values are identical. (Generally speaking, by recalling θ we mean both the name of parameter vector as well as its nominal value.)

DEFINITION 2.2-15 *(Parameter Variation):* If θ changes from its *nominal* (initial) value θ^o to a new value $\theta^o + \Delta\theta$, then $\Delta\theta$ is defined as parameter variation relative to the nominal condition θ^o. It is assumed that this change takes place over a "long" interval of time.

In the case of static or time-invariant variation, this parameter variation is the same as the so-called *tolerance*.

DEFINITION 2.2-16 *(Primitive-Sensitivity Function):* Consider the *DCLTI – GS* of Definition 2.2-2. Assume that $\Delta\theta_j$ (the jth – component of $\Delta\theta$) in Definition 2.2-15 is the parameter variation for θ_j. Suppose that z_k (the kth – component of system response) changes to $z_k + \Delta z_k$ as $\theta_j \to \theta_j + \Delta\theta_j$, then the primitive-sensitivity function of z_k in terms of θ_j is defined as the ratio of $\Delta z_k/\Delta\theta_j$.

DEFINITION 2.2-17 *(Sensitivity in the Large):* For the case in which $\Delta\theta_j$ of Definition 2.2-16 is "large" and it takes place over a "short" interval of time, then Δz_k (in Definition 2.2-16) is the measure of sensitivity in the large for our system, provided that this measure is meant with respect to a certain parameter variation such as $\Delta\theta_j$.

DEFINITION 2.2-18 *(Small-Sensitivity Function):* Suppose that in Definition 2.2-17 $\Delta\theta_j$ is "small", such that $\lim_{\Delta\theta_j \to 0} (\Delta z_k/\Delta\theta_j)$ exists: then $(\partial z_k/\partial\theta_j)$ is defined as the small-sensitivity function of z_k in terms of θ_j which is evaluated at a given nominal condition θ^o.

REMARK 2.2-3: The main implication of Definitions 2.2-15 to 18 (which are evaluated at θ^o, the nominal values of system parameters) is to bring a flavor of randomness to parameter variations, because sensitivity analysis is inherently based on such consideration. Since the time-varying parameter changes require a different treatment, in the present context the actual mechanism of time dependency in parameter variations is neglected for the sake of analytical simplicity.

DEFINITION 2.2-19 *(Unnormalized-Sensitivity Function):* Suppose that the sensitivity function $(\partial z_k/\partial \theta_j)$ in Definition 2.2-18, is multiplied by θ_j: then $\theta_j(\partial z_k/\partial \theta_j)$, evaluated at θ^o, is termed the unnormalized-sensitivity function of z_k with respect to θ_j.

DEFINITION 2.2-20 *(Normalized-Sensitivity Function):* If the unnormalized-sensitivity function $\theta_j(\partial z_k/\partial \theta_j)$ is divided by z_k (or the sensitivity function $\partial z_k/\partial \theta_j$ is multiplied by a *weighting factor* θ_j/z_k), then $(\theta_j/z_k) \times (\partial z_k/\partial \theta_j)$, evaluated at θ^o, is termed the normalized-sensitivity function of z_k with respect to θ_j.

REMARK 2.2-4: Bode is the first to introduce the above *measure* (or the preceding weighting factor), in order to eliminate the physical dimension of unnormalized-sensitivity function. It is easy to show that the normalized-sensitivity function can be expressed in the form of $(\partial ln\, z_k/\partial ln\, \theta_j)$, where *ln* means the natural logarithm. Though this representation does not carry any additional information, it does provide another commonly used name: *The Bode's Logarithmic Sensitivity Function*.

DEFINITION 2.2-21 *(Vector-Sensitivity Function):* Suppose that the sensitivity function of x_i with respect to θ_j is derived for i = 1, 2, ..., n. Then

$$\frac{\partial x}{\partial \theta_j} \triangleq \begin{Bmatrix} \dfrac{\partial x_1}{\partial \theta_j} \\ \cdot \\ \cdot \\ \cdot \\ \dfrac{\partial x_n}{\partial \theta_j} \end{Bmatrix}, \qquad (2.2\text{-}10)$$

is defined as the vector-sensitivity function of x with respect to θ_j.

DEFINITION 2.2-22 *(Sensitivity Matrix):* The sensitivity matrix of $x(t; \theta) \in R^n(t; \theta)$ (with respect to $\theta \in R^r$) is an $n \times r$ matrix, whose columns are the vector sensitivity function of x with respect to θ_j, and for j = 1, ..., r.

2.3. THE FUNDAMENTAL ASPECTS OF SENSITIVITY ANALYSIS

A concise account of basic aspects of sensitivity analysis, and a statement for some major applications of sensitivity theory comprises this section.

Consider a system whose scalar output function is $z = z(t; \theta)$. Here the real variable or time is shown by t and θ is a scalar parameter. We associate to this parameter a nominal value θ^o (cf., Definition 2.2-14), and the corresponding scalar function z has also a nominal value $z^o = z(t; \theta^o)$ at each given time. Now, suppose that θ changes from θ^o to $\theta^o + \Delta\theta$. We do not know how, but we assume that parameter changes do not occur fast with respect to time. This basic assumption is at the *heart* of sensitivity analysis and its main underlying constraint. Had we known the exact mechanism of this parameter change, then we should have incorporated that information in our analysis as this is the case in adaptive control theory. For this reason, the sensitivity has sometimes been called "passive adaptive", although this phrase has certain special meaning in the stochastic adaptive control theory [C. 6]. In any event, we anticipate what may happen to z when $\theta^o \rightarrow \theta^o + \Delta\theta$, by introducing sensitivity analysis. This analysis is performed in two distinct ways depending on the magnitude of $\Delta\theta$. If $|\Delta\theta|$ is *large*, then we need to deal directly with the error $e \triangleq z(t; \theta^o + \Delta\theta) - z(t; \theta^o)$. This sort of analysis is called herein sensitivity in the large. An analysis of this sort is reported in [3A. 7]. In subsequent chapters we use this measure to develop controllers that can stand against these sorts of variations. Incidentally, we associate to those systems that can tolerate these large variations the new adjective of *robust*. On the other hand, if $|\Delta\theta| \ll 1$, then we can expand $z(t; \theta^o + \Delta\theta)$ around its nominal value by a Taylor series (that converges). Based on this fundamental assumption (on $|\Delta\theta|$) and the existence of a Taylor-series expansion we can approximate this series as follows.

$$z \approx z^o + \frac{\partial z}{\partial \theta}\bigg|_{\theta^o} \frac{(\Delta\theta)}{1!} + \frac{\partial^2 z}{\partial \theta^2}\bigg|_{\theta^o} \frac{(\Delta\theta)^2}{2!} + \cdots . \quad (2.3\text{-}1)$$

This scheme lends itself to what is known in the literature as "small signal" sensitivity analysis which is true if $|\Delta\theta| \ll 1$. If we are content with the first two-terms of the above series, then we would study the first derivative of signal z with respect to parameter θ as *a measure* of system sensitivity. Similar discussion leads to the conclusion that the coefficient of $(\Delta\theta)^n/n!$ in (2.3-1) is the nth-order sensitivity function of $z(t; \theta)$, which is simply $\partial^n z(t; \theta^o)/\partial \theta^n$, provided that this derivative exists. This argument is strongly based on the assumption that $|\Delta\theta| \ll 1$. Generally speaking, the first- and the second-order sensitivity functions are of more interest to us than any other higher-order derivatives. In other words, if $|\Delta\theta|$ is not sufficiently small requiring us to consider higher-order derivatives than two in our analysis, then we are advised to avoid (2.3-1) entirely and look for the corresponding large-sensitivity measure as described earlier in this section.

We now choose the same z with $\theta \in R^r$ as the system parameter vector. Suppose that $\theta^o \rightarrow \theta^o + \Delta\theta$, where θ^o is the nominal value of θ. Then the corresponding Taylor-series expansion for each $|\Delta\theta_j| \ll 1$, $j = 1, ..., r$, becomes

$$z(t; \theta) = z(t; \theta^o) + \sum_{j=1}^{r} \frac{\partial z}{\partial \theta_j}\bigg|_{\theta^o} \frac{\Delta\theta_j}{1!}$$

$$+ \left[\frac{\partial^2 z}{\partial \theta_1^2} \frac{(\Delta\theta_1)^2}{2!} + \frac{\partial^2 z}{\partial \theta_1 \partial \theta_2} \frac{(\Delta\theta_1)(\Delta\theta_2)}{2!} + ... + \frac{\partial^2 z}{\partial \theta_r^2} \frac{(\Delta\theta_r)^2}{2!}\right]\bigg|_{\theta^o} + \cdots, \quad (2.3\text{-}2)$$

where the second $\cdots$ refers to the higher-order derivatives of z with respect to θ_j's. If we assume that $|\Delta\theta_j| < 1$ for all j, then the first (r + 1)-terms of (2.3-2) is sufficient for approximating $z(t; \theta)$. Or equivalently for this case we can write

$$\Delta z(t; \theta) \triangleq z(t; \theta) - z(t; \theta^o) \approx \sum_{j=1}^{r} \frac{\partial z}{\partial \theta_j}\bigg|_{\theta^o} \Delta\theta_j. \quad (2.3\text{-}3)$$

Since we do not know, in general, the sign of $\Delta\theta_j$, i.e., whether the changes with respect to the parameter nominal values are in positive or in negative direction, we may select the absolute value of $\Delta\theta_j$ in (2.3-3) and we may also choose the largest estimate of each $|\frac{\partial z}{\partial \theta_j}| |\Delta\theta_j|$, say for the $j = l$, as the worst possible error. Thus we get

$$|\Delta z(t; \theta)| \leq \sum_{j=1}^{r} |\frac{\partial z}{\partial \theta_j}| |\Delta\theta_j| \leq r |\frac{\partial z}{\partial \theta_l}| |\Delta\theta_l|. \quad (2.3\text{-}4)$$

Now, consider the case in which we have several outputs — say $z \triangleq z(t; \theta) \in R^\omega(t; \theta)$. Again, we assume that $z \in C^1(t)$,[2] and $z \in C^\infty(\theta)$. Then the vector-sensitivity function of z with respect to θ is defined next. Certainly the requirement on $z \in C^1(t)$ is academic at the moment and this becomes apparent in the future chapters. Also $z \in C^\infty(\theta)$ can be replaced by $C^n(\theta)$. Therefore

$$\frac{\partial z}{\partial \theta} \triangleq [\frac{\partial z_1}{\partial \theta}, \frac{\partial z_2}{\partial \theta}, ..., \frac{\partial z_\omega}{\partial \theta}]^T. \quad (2.3\text{-}5)$$

An immediate generalization of this definition should now be apparent.

Consider a multi-output, multi-parameter system where $z \triangleq z(t; \theta) \in R^\omega(t; \theta)$ and $\theta \in R^r$. Suppose that $z_k \in C^\infty(\theta_j)$, for all $k = 1, 2, ..., \omega$, and for all $j = 1, 2, ..., r$. Then the following $\omega \times r$ matrix is called the sensitivity matrix of z in terms of θ.

---

[2] $C^n(\cdot)$ means continuous with respect to its argument and its first n derivatives exist and are also continuous.

2.3. The Fundamental Aspects of Sensitivity Analysis

$$\frac{\partial z}{\partial \theta} \triangleq \begin{bmatrix} \frac{\partial z_1}{\partial \theta_1} & \frac{\partial z_1}{\partial \theta_2} & \cdots & \frac{\partial z_1}{\partial \theta_r} \\ \frac{\partial z_2}{\partial \theta_1} & \frac{\partial z_2}{\partial \theta_2} & \cdots & \\ \vdots & & & \\ \frac{\partial z_\omega}{\partial \theta_1} & \frac{\partial z_\omega}{\partial \theta_2} & & \frac{\partial z_\omega}{\partial \theta_r} \end{bmatrix}. \qquad (2.3\text{-}6)$$

Here each entry of $\frac{\partial z}{\partial \theta}$ is the appropriate first-order sensitivity function of z_k with respect to θ_j that is evaluated at the nominal values of $\theta_1, \ldots, \theta_r$, i.e., θ^o.

We must note that this matrix is *not* the sensitivity-comparison matrix of Cruz and Perkins [6A. 7 & 8] (cf., § 6.2.2).

2.4. SENSITIVITY FUNCTIONS GENERATION

Consider a plant whose state $x(t) \in R^n(t)$ is related to plant input $u(t) \in R^m(t)$ by a nonlinear differential equation

$$\dot{x}(t) = f[x(t), u(t), t; \theta], \qquad (2.4\text{-}1)$$

where t is the real variable, and $\theta \in R^r$ is the plant parameter vector. If $x(t) \in C^1(t; \theta)$, then with proper continuity conditions we have

$$\frac{\partial x}{\partial \theta} \triangleq \begin{bmatrix} \frac{\partial x_1}{\partial \theta_1} & \frac{\partial x_1}{\partial \theta_2} & \cdots & \frac{\partial x_1}{\partial \theta_r} \\ \frac{\partial x_2}{\partial \theta_1} & \frac{\partial x_2}{\partial \theta_2} & & \\ \frac{\partial x_n}{\partial \theta_1} & & & \frac{\partial x_n}{\partial \theta_r} \end{bmatrix}, \qquad (2.4\text{-}2)$$

as the first-order sensitivity matrix of x with respect to θ, which is evaluated at the nominal values of the parameters. Here (2.4-2) is determined from the following differential equation

$$\frac{d}{dt}\frac{\partial x}{\partial \theta} = \frac{\partial f}{\partial x}\frac{\partial x}{\partial \theta} + \frac{\partial f}{\partial \theta}, \qquad (2.4\text{-}3)$$

under the assumptions that

$$\frac{\partial x}{\partial \theta}\bigg|_{t=t_0} = (0)_{n \times r}, \text{ and } \frac{\partial u}{\partial \theta} = (0)_{m \times r} \quad \text{(an open–loop system); with} \qquad (2.4\text{-}4)$$

$$\frac{\partial f}{\partial x} \triangleq \begin{bmatrix} \frac{\partial f_1}{\partial x_1} & \cdots & \frac{\partial f_1}{\partial x_n} \\ \vdots & & \vdots \\ \frac{\partial f_n}{\partial x_1} & \cdots & \frac{\partial f_n}{\partial x_n} \end{bmatrix}, \text{ and } \frac{\partial f}{\partial \theta} \triangleq \begin{bmatrix} \frac{\partial f_1}{\partial \theta_1} & \cdots & \frac{\partial f_1}{\partial \theta_r} \\ & & \\ \frac{\partial f_n}{\partial \theta_1} & \cdots & \frac{\partial f_n}{\partial \theta_r} \end{bmatrix}. \qquad (2.4\text{-}5)$$

Since (2.4-3) is linear, though time-varying, if its transition matrix is called $\Phi(t, \tau)$, then the corresponding solution, subject to (2.4-4), becomes [1A. 20]:

$$\frac{\partial x}{\partial \theta} = \int_{t_0}^{t} \Phi(t, \tau) \frac{\partial f}{\partial \theta} d\tau. \qquad (2.4\text{-}6)$$

In the case of an open-loop linear time-invariant plant, system (2.4-1) becomes

$$\dot{x}(t; \theta) = A(\theta)x(t; \theta) + B(\theta)u(t), \quad x(t_0; \theta) = x_0. \qquad (2.4\text{-}7)$$

The state sensitivity differential equation becomes

$$\frac{d}{dt}\frac{\partial x}{\partial \theta} = A(\theta)\frac{\partial x}{\partial \theta} + \frac{\partial A}{\partial \theta}x(t; \theta) + \frac{\partial B}{\partial \theta}u(t); \quad \frac{\partial x}{\partial \theta}\bigg|_{t=t_0} = 0. \qquad (2.4\text{-}8)$$

Clearly, the homogeneous part of (2.4-7) and (2.4-8) are similar, but the forcing functions are different. As we can easily see, to generate a complete set of sensitivity functions for a typical system, if needed, is a formidable task. We study this issue again in Chapter Three.

EXAMPLE 2.4-1 (Nonlinear sensitivity functions): Consider a nonlinear system described by $\dot{x} = f(x; \theta)$, where

$$\dot{x} \triangleq \begin{Bmatrix} \dot{x}_1 \\ \dot{x}_2 \end{Bmatrix} = \begin{Bmatrix} x_2 \\ -\theta_1 x_1 - \theta_2 x_2 + \theta_3 x_1 x_2 \end{Bmatrix}, \quad x(t_0) = x_0. \qquad (2.4\text{-}9)$$

Here the parameter nominal values are $\theta^o \triangleq [\theta_1^o, \theta_2^o, \theta_3^o] \equiv [1, 0, 1]$. For this system generate the corresponding first-order differential equation yielding $\dfrac{\partial x}{\partial \theta}$ evaluated at θ^o.

SOLUTION: The corresponding (2.4-3) for this problem becomes

$$\begin{bmatrix} \dfrac{\partial \dot{x}_1}{\partial \theta_1} & \dfrac{\partial \dot{x}_1}{\partial \theta_2} & \dfrac{\partial \dot{x}_1}{\partial \theta_3} \\ \dfrac{\partial \dot{x}_2}{\partial \theta_1} & \dfrac{\partial \dot{x}_2}{\partial \theta_2} & \dfrac{\partial \dot{x}_2}{\partial \theta_3} \end{bmatrix} = \begin{bmatrix} 0 & 1 \\ -1+x_2 & x_1 \end{bmatrix} \dfrac{\partial x}{\partial \theta} + \begin{bmatrix} 0 & 0 & 0 \\ -x_1 & -x_2 & x_1 x_2 \end{bmatrix}. \qquad (2.4\text{-}10)$$

▼

2.5. THE PRELIMINARY MATHEMATICS

2.5.1. Introduction

In this section we review concisely those mathematical topics which are of interest in advanced sensitivity literature. Although this information is not directly cited in most parts of the book, this level of mathematical maturity is expected in the literature. This intermediate-level presentation, due to space limitation, is of a glossary nature and is brought here only for the sake of easy access. These topics are compiled from the classical references cited in § 2.9.A guided by [A. 24] and [B. 4], which this list comprises an indispensable set of references for any engineering researcher who is mathematically inclined. The following review consists a set of well-known elements from algebra, analysis and topology. Unless defined otherwise, R means the set of all real numbers and R^+ is its positive subset; C means the set of all complex numbers; and we read $\Leftrightarrow$ as if and only if, and $\Rightarrow$ as implies.

2.5.2. Mappings

An introductory knowledge of set theory is assumed and in this regard the words *collection, family, bank* and *class* are used synonymously with set. Similarly, the words *element, member,* and *point* are used interchangeably as each becomes evident from the context.

■

Consider $y = f(x)$ where both x and y are real numbers. Each element of $f(x)$ is called the *image* of x under f. The *image set* or *range* of [a function] $f: R \to R$, and for all $x \in R$, is defined as the set of all images $f(x)$ and is shown by Im f (or *Range* (f)) $\triangleq \{f(x) : x \in R\}$. Let X be the set of all $x \in R$ and let $f: X \to R$ for all $x \in X$, then X is called the *domain* of f. The range of [a function] f is then the image set of its domain. In general, an operation between two non-empty sets X and Y is called a *mapping* and is shown by $f: X \to Y$, however, when Y is R (or C) or when $Y \subset R$ (a subset of R) (or $Y \subset C$), the mapping f is called a *function*. The set X is called the *domain* of f and the set Y is called *codomain* of f. The image set (or range) of f is $\text{Im} f \triangleq \{y \in Y : y = f(x), \text{ for some } x \in X\} \subseteq Y$. Two mappings f and g are said *equal* if they have the same domain and the same codomain and for all $x \in X$, $f(x) = g(x)$.

∎

Given $f: X \to Y$ and S as a proper subset of X such that $f(S) \subseteq T \subseteq Y$, then $g: S \to T$ such that $g(x) = f(x)$ for all $x \in S$ is called the *restriction* of f to S and T. Conversely, a mapping $f: X \to Y$ is called an *extension* of a mapping $g: S \to T$ if g is the restriction of f to S and T.

∎

A mapping $f: X \to Y$ is called an *into* mapping if $\text{Im} f \subset Y$.

∎

A mapping $f: X \to Y$ is called an *onto* or *surjective* mapping if $\text{Im} f = Y$.

∎

A mapping $f: X \to Y$ is called a *(1–1)* or *injective* mapping if given $x_1, x_2 \in X$, $x_1 \neq x_2$, then $f(x_1) \neq f(x_2)$.

∎

A mapping $f: X \to Y$ is called a *(1–1) and onto* or *bijective* mapping if f is both injective and surjective.

∎

A mapping which represents the operation of two (or more) mappings is called a *composite* mapping, and is shown by $f \circ g$. The composition of mappings follows an associative law (i.e., $(f \circ g) \circ h = f \circ (g \circ h)$) and this mapping is shown by a *commutative diagram*.

∎

An inverse mapping of f when exists is shown by f^{-1} and is defined by $f \circ f^{-1} \triangleq I$, which I is an *identity* mapping that maps a non-empty set to itself.

2.5.3. Semigroups, Groups, Rings, and Fields

Given a finite (or an infinite) set S, a binary operation "$\circ$" on two elements a and b of this set is shown by $c = a \circ b$, where c may or may not be a member of S. If for all ordered pairs (a, b), the corresponding $c \in S$, then the mapping $S \times S \to S$ is called *closed*. (Here the $S \times S$ refers to the *Cartesian product*.) This non-empty set S is called an *algebraic system* under a specific binary operation. A *semigroup* refers to a non-empty set S with a binary operation "$\circ$" such that: *(i)* S is closed under this operation; and *(ii)* this operation is associative for all a, b, c, $\in S$ (i.e., $(a \circ b) \circ c = a \circ (b \circ c)$). A semigroup *may* have an identity element under the operation "$\circ$" that is shown by I, such that $a \circ I = I \circ a = a$, which this element is

2.5. The Preliminary Mathematics

also unique. If in this semigroup there is an element b such that $a \circ b = b \circ a = I$, then b is equal to the inverse of a, and this inverse is shown by a^{-1} which is unique. A *subsemigroup* consists a subset of S with the same binary operation, but possibly a different identity element.

■

A *group* is a non-empty set G with one binary operation "$\circ$" such that under this operation G is a semigroup which has an identity element; and for each element of G there exists a unique inverse in G.

Any *subgroup* is defined relative to a subset of G.

■

An *abelian group* with an additive binary operation "+", an identity 0, and an inverse $-x$ for an element x is a group such that given $x_1, x_2, x_3 \in G$:

$x_1 + x_2 \in G$	(closed under +)
$(x_1 + x_2) + x_3 = x_1 + (x_2 + x_3)$	(associative law)
$x_1 + x_2 = x_2 + x_1$	(commutative law)
$x_1 + 0 = 0 + x_1 = x_1$	(identity 0)
$x_1 + (-x_1) = (-x_1) + x_1 = 0$	(inverse $(-x_1)$).

■

Group *isomorphic* refers to the equivalence between *two* algebraic systems that are structurally identical. Here, if G_1 and G_2 are two groups and if there exists a *bijective* mapping $f : G_1 \rightarrow G_2$, such that $f(x_1 \circ x_2) = f(x_1) \circ f(x_2)$, for all $x_1, x_2 \in G_1$, then G_1 is *isomorphic* to G_2. The two binary operations in G_1 and G_2 are generally different. The bijective mapping f is called a *(group) isomorphism* from G_1 to G_2. An isomorphism of G *onto* itself is called *automorphism*.

■

A *(group) homomorphism* from G_1 to G_2 is a mapping that satisfies $f(x_1 \circ x_2) = f(x_1) \circ f(x_2)$ for all x_1 and $x_2 \in G_1$. Here the type of mapping is not an issue, but the following distinctions are made. If the mapping is *onto*, or *(1–1)*, then we have, respectively, *epimorphism*, or *monomorphism*. A homomorphism of G into itself is called an *endomorphism*.

■

The *kernel* of a group homomorphism $f : G_1 \rightarrow G_2$ is the subset of G_1 and is given by $\operatorname{Ker} f \triangleq \{x \in G_1 : f(x) = I_2\}$, where I_2 is the identity element of G_2.

■

A *ring* is a non-empty set *R* with *two* binary operations of addition "+" and multiplication "×" which these two are related by distributive laws; and the set *R* becomes an abelian group with respect to addition and a semigroup with respect to multiplication. A *subring* of a ring *R* is a ring for a subset of *R* under these operations. Here the identity of the subring may be different from that of a ring.

■

A *commutative* ring is a ring R with a commutative multiplication. The set of polynomials is a commutative ring with identity.

∎

A *homomorphism of a ring* R_1 into a ring R_2 is a mapping $f: R_1 \to R_2$ such that it satisfies the ring operations, i.e., for all $x_1, x_2 \in R_1$, $f(x_1 + x_2) = f(x_1) + f(x_2)$, and $f(x_1 \times x_2) = f(x_1) \times f(x_2)$.

A *subring* of R is called an *ideal*, if it is closed under multiplication by the elements of R.

∎

A *field* F is a commutative ring with an identity and all of its nonzero elements have multiplicative inverses in the corresponding set. For instance, both R and C are fields.

2.5.4. Complete Mathematical Spaces

We first review several mathematical spaces, and in conjunction with discussions on convergence and boundedness, we present complete mathematical spaces which are of interest to us. Perhaps we should emphasize that by "mathematical spaces" we mean those spaces which we *synthesize* by incorporating elements from the sets of real or complex numbers and various sets of functions and/or mappings together with new and specific underlying rules for each specific space. In pure mathematical parlance this notion refers to as studying *abstract* spaces versus *concrete* spaces. Our presentation starts with the basic ideas and in the most simplest cases using only R, and advances to higher-level topics. Throughout this study we assume that the *dimension* of the underlying space is finite.

∎

A *vector (or linear) space* X over F, is shown by $V \triangleq (X, F)$ and is formed as follows. Let F be a field with 1 as its unity, and X be an additive abelian group such that for $\alpha \in F$, $x \in X$, $\alpha x \in X$, then X is called a vector space over F if multiplication by scalar is closed and the following distributive laws hold for all α, $\beta \in F$, and all $x, y \in X$: $\alpha(x + y) = \alpha x + \beta y$ and $(\alpha + \beta)x = \alpha x + \beta x$.

∎

Any vector space can be represented as linear combination of a set of vectors which are called *bases*.

A subspace of a vector space is defined relative to a subgroup of V.

A mapping $f: X \to Y$ is called a linear map if $f(x_1 + x_2) = f(x_1) + f(x_2)$ and $f(\alpha x) = \alpha f(x)$, for all $x_1, x_2 \in X$ and $\alpha \in R$. If V_1 and V_2 are two vector spaces over the same field F, and if there is a bijective mapping $f: V_1 \to V_2$ such that $f(x_1 + x_2) = f(x_1) + f(x_2)$, $x_1, x_2 \in V_1$, and $f(\alpha x) = \alpha f(x)$, $\alpha \in F$, then V_1 and V_2 are called *isomorphic* and the mapping f is an *isomorphism* of V_1 onto V_2.

∎

2.5. The Preliminary Mathematics

In a vector space, when we replace the field F with a ring R, then the corresponding algebraic system is called a *module*.

■

A generalization of the basic notion of distance (or modulus $|\cdot|$) between two points in an Euclidean space to that between the elements of a general set is now in order.

■

Given a non-empty set X with the *Cartesian product* $X \times X$, and a real-valued function $f: X \times X \to R^+$ such that for any $x_1, x_2, x_3 \in X$ we have:

(i) $f(x_1, x_2) \geq 0$, and $|x_1 - x_2| = 0 \Longleftrightarrow x_1 = x_2$,

(ii) $f(x_1, x_2) = f(x_2, x_1)$ (symmetric property),

(iii) $f(x_1, x_2) \leq f(x_1, x_3) + f(x_3, x_2)$, (triangle inequality),

then the set X and the function f form a *metric space*, and this is shown by (X, f), with f as its *metric*.

■

The length of a vector $x \in R^n$ is defined as $\sqrt{\sum_{i=1}^{n} x_i^2}$ which is shown by $\|x\|$ — called the *norm* of x. Based on this generalization we present the following space.

■

Given a vector space X over the field of R and a real-valued function $f: X \to R^+$ (is shown by $x \to \|x\|$) such that for any $x, y \in X$ and $\alpha \in R$ we have:

(i) $\|x\| \geq 0$, and $\|x\| = 0 \Longleftrightarrow x = 0$,

(ii) $\|\alpha x\| = |\alpha| \|x\|$,

(iii) $\|x + y\| \leq \|x\| + \|y\|$,

then the vector space X over R and the function f form a *normed-vector space*, and is shown by $(V, \|\cdot\|)$.

■

Clearly all norms define metrics, however, the converse is not true. We can also show that a function $f: X \times X \to R^+$ defined by $f(x, y) = \|x - y\|$ is a metric on V, and this function is called *metric induced by the given norm*.

■

An inner product between two vectors x and y is shown by $\langle x, y \rangle = x^T y$, and this leads to the next space.

■

Given a vector space X over the field of R and a linear, real-valued function $f: X \times X \to R^+$ (is shown by $(x, y) \to \langle x, y \rangle$ and is called an inner product) such that for any ordered pair of $x, y \in X$ and $\alpha \in R$ we have:

(i) $\langle x, x \rangle \geq 0$, and $\langle x, x \rangle = 0 \Longleftrightarrow x = 0$,

(ii) $\langle x, y \rangle = \langle y, x \rangle$,

(iii) $\langle \alpha x, y \rangle = \langle x, \alpha y \rangle = \alpha \langle x, y \rangle$, then the vector space X over R and the function f form an *inner-product space*, and is shown by $(V, <\cdot, \cdot>)$.

∎

Here $<x, x> = \| x \|^2$, which means the inner product induces a norm that in turn induces a metric.

∎

The above mathematical spaces which are also called the *real* spaces, can be advanced in a number of ways in order to make them more amenable to our study. *For instance*, at the moment we have no information on their completeness, which means we do not know whether a problem that is set and solved in one particular space has its final solution in the same space or not. This completeness is particularly a serious matter when the solution of a [control] problem is generated iteratively, *via* a sequence of possible solutions. This notion immediately directs us to the study of convergence in a sequence, and clearly the very first item which captures our attention is that each convergence must be defined relative to the "metric" that is induced in the given space. A generic definition for convergence is as follows. Given a sequence (finite or infinite) of real numbers $\{x_k\}$ and a small number $\varepsilon > 0$, if there exists a constant $k_o = k_o(\varepsilon) > 0$ such that for a finite l the metric $f(x_k, l)$ is less than ε whenever $k > k_o$, then this sequence is called *convergent* and l is its *limit point*. Otherwise the sequence is called *divergent*. Further advancement of the preceding spaces requires that we generalize the field of real numbers R to that of the complex numbers C, and consider the sequence of vectors and/or general functions and mappings instead of real numbers and we should also use appropriate metrics. But first we dispose a few preliminaries.

∎

Generally speaking, each *bounded* sequence [of real numbers] has some upper and lower bounds. Among all these possible bounds two are of particular interest to us, namely, the least-upper bound that is shown by "sup" (stands for *supremum*), and the greatest-lower bound that is shown by "inf" (stands for *infimum*). These two bounds may or may not be the member of the original sequence but they always exist for any bounded sequence. We note that the existence of an upper (or lower) bound, of course, depends on the definition of a real number, and here it is also implied that the sequence is ordered. An increasing (or decreasing) sequence that is bounded above (or below) converges to a real number. All convergent sequences are bounded, but not all bounded sequences are convergent. If the sequence is defined as $\{x_k(t)\}$, and for all $t \in [a, b]$, $\lim_{k \to \infty} x_k(t) = x(t)$, then the convergence is called *uniform* over $[a, b]$.

∎

Given a sequence (finite or infinite) of real numbers $\{x_k\}$ and a small number $\varepsilon > 0$, if there exists an integer $N(\varepsilon)$ such that for a metric f, $f(x_n, x_m) < \varepsilon$ whenever $n, m > N$, then this sequence is called *Cauchy*. A standard approach to show whether a sequence is Cauchy or not is to take another member of the sequence as an intermediate point and invoke the triangle property of the corresponding space,

2.5. The Preliminary Mathematics

in order to estimate the distance between two arbitrary members of the sequence. If this estimate meets our definition, then the sequence is Cauchy. A set S of the real numbers $\{x_k\}$ is called *complete* if every Cauchy sequence in S converges to a point in S. Not every Cauchy sequence is convergent in a general metric space, but every Cauchy sequence in R is convergent. A mathematical space is *complete* if every Cauchy sequence defined according to the "metric" of this space has a limit point in that space. We can certainly imbed an incomplete space to that of a complete space. ∎

Now, we proceed to replace the sequence of real numbers by a sequence of elements $\{x_k\} \in X$. Given a mathematical space with the "metric" f, a sequence of elements $\{x_k\} \in X$ is called *Cauchy* in that space, if for every small number $\varepsilon > 0$ there exists $k = k(\varepsilon) > 0$ such that $f(x_m, x_n) < \varepsilon$, whenever m, n > k. Also given a space (X, f), a sequence $\{x_k\} \in X$ is called *convergent* if $\lim_{k \to \infty} f(x_k, x) = 0$ for a "metric" f and some $x \in X$. Here x is called the limit (or accumulation) point of x_k and this is the same as saying that $\lim_{k \to \infty} x_k = x$, *if a metric is implied*. Every convergent sequence is a Cauchy sequence but the converse is not true. In the case of vectors and matrices, the convergence of elements is in the sense of componentwise.

DEFINITION 2.5-1 (Banach Space): Given a vector space X over the field of C and a real-valued function $f: X \to R^+$ (is shown by $x \to \| x \|$ and is called a norm) such that for any $x, y \in X$ and $\alpha \in C$ we have:

(i) $\| x \| \geq 0$, and $\| x \| = 0 \Longleftrightarrow x = 0$,

(ii) $\| \alpha x \| = |\alpha| \| x \|$,

(iii) $\| x + y \| \leq \| x \| + \| y \|$,

(iv) every Cauchy sequence $\{x_k\} \in X$ converges in the metric that is induced by the above norm to a limit point $x \in X$,

then this *complete, normed-vector space* X over C is called a Banach space.

DEFINITION 2.5-2 (Hilbert Space): Given a vector space X over the field of C and a *linear* function $f: X \times X \to C$ (is shown by $(x, y) \to <x, y>$ and is called an inner product) such that for any ordered pair $x, y \in X$ and $\alpha \in C$ we have:

(i) $<x, x> \geq 0$, and $<x, x> = 0 \Longleftrightarrow x = 0$,

(ii) $<x, y> = \overline{<y, x>}$, (the upper bar means complex conjugate),

(iii) $<\alpha x, y> = <x, \overline{\alpha} y> = \alpha <x, y>$,

(iv) every Cauchy sequence $\{x_k\} \in X$ converges in the norm induced by the above inner product to a limit point $x \in X$,

then this *complete, inner-product space* X over C is called a Hilbert space. ∎

Most control problems *may* be formulated in either a Banach or a Hilbert space using C or R. Indeed to solve many of our sensitivity optimization algorithms in Chapter Six, we use iterative methods to compute the final solution, realizing that in each case we are generating a Cauchy sequence for the prospective optimal controller gain which ultimately converges to its limit value and yields the corresponding optimal solution. ∎

Two vectors x and y in a Hilbert space are called *orthogonal* if $<x,y> = 0$. When this is the case, we may partition X into two subspaces S (possibly closed) and $S^\perp$ which is orthogonal complement of S with respect to X, and X is the direct sum of these two subspaces and is shown by $X = S \oplus S^\perp$.

2.5.5. Basics of Set Theory and Topology

Here we review a list of definitions pertaining to set theory and topology, but first we recall that the *distance* between two points $x, y \in R^n$ is shown by $d(x,y)$ and for the sake of discussion we may consider $d(x,y) = \| x - y \|$. We also note that the following concepts can be mostly generalized to other metrics.

DEFINITIONS 2.5-3: (i) The *distance* between a *point* $x \in R^n$ and a *set* $A \subset R^n$ is defined by $d(x, A) \triangleq \inf \{ \| x - a \| : a \in A \}$.

(ii) The *separation* of $A \subset R^n$ from $B \subset R^n$ is defined by $d^*(A, B) \triangleq \sup \{ d(a, B) : a \in A \}$.

(iii) The *distance between two sets* A and B is defined by $d(A, B) = d(B, A) \triangleq \max \{ d^*(A, B), d^*(B, A) \}$.

(iv) The *ε–neighborhood* of a set $A \subset R^n$ is $N_\varepsilon(A) \triangleq \{ x \in R^n : d(x, A) < \varepsilon \}$. This definition also refers to as an "open ball", although depending on the metric chosen the "ball" may actually be a "square". Here "open" is used because of the strict inequality $d(x, A) < \varepsilon$.

(v) A *closed- (or an open-) unit ball* refers to $d(x, A) \leq 1$ (or < 1). ∎

Now, we stack a few additional definitions and properties in the following. A set $A \subset R^n$ is called *open* if for every point $x \in A$ we can define an ε–neighborhood, i.e., these ε–neighborhoods are contained in A. Union ("∪") and finite intersection ("∩") of open sets are open in the metric space. The complement of a set $S \subset X$ is shown by S^c, which is a set such that $X = S \cup S^c$. In general, the complement of an open set is a *closed* set. As mentioned in § 2.5.4, in a closed set every sequence converges to a limit point in that set. A limit (or accumulation) point has an ε–neighborhood which intersects S, i.e., it is not, in general, a proper subset of S. The *closure* of a set S is shown by $\overline{S}$, which is a closed set that contains set S with all its limit points. A necessary and sufficient condition for a set S

2.5. The Preliminary Mathematics

to be closed is that $S = \bar{S}$. A subset $S \subset X$ is called *dense* in X if and only if $\bar{S} = X$. The definition of a bounded sequence relative to a metric f can be generalized to that of a set as follows. For instance, a non-empty set $A \subset X$ is bounded if its distance $d(A) = \sup_{x,y \in A} f(x,y)$ is finite. A set which is closed and bounded is called a *compact* set. A solution for an ordinary differential equation always exists on a compact set. The boundary set of a set S is shown by ∂S which is the set of all boundary points for S, and each member of this set has an ε-neighborhood that is extended to both S and S^c. An open set contains no members of its boundary set, but a closed set has all its boundary points. The *interior* of a set S is shown by int S which is an open set and is equal to $S - \partial S$, where the "−" refers to *exclusion*. Alternatively, a point $x \in S$ is in the interior of S if its ε-neighborhood is in S. In other words, the interior of S is the "largest" open set in S. An *exterior* point of a set is an interior point of its complement, and the set of all such points is called the *exterior* of S and is shown by ext S. When the ext S, int S, and ∂S exist, they are disjoint sets. An open set is *connected* if every pair of points can be connected by a "path" which is contained in S. A set is called *convex* if for every pair of points (x, y), their convex combination (i.e., $\lambda x + (1 - \lambda)y$ for $0 < \lambda < 1$) is contained in this set for all λ. The intersection of convex sets is convex. An optimization over a convex set yields a global minimum. Given a nonconvex set S, then the *smallest* convex set that contains S is shown by $\text{co}\, S \triangleq \{x : x = \sum_{i \in I} \lambda_i x_i, x_i \in S, \lambda_i \geq 0, \sum_{i \in I} \lambda_i = 1\}$, and is also called a convex hull (*I* is a general index set that is to be specified in each case). Most convex sets of interest to us have *corners*, or *vertexes*, or *extreme points*, and these are the only points in the set which correspond to either $\lambda = 0$ or $\lambda = 1$ of the associated convex combination machinery. Given a metric space X and $S \subset X$, and for a finite (or an infinite) collection of open sets $\{A_i : i \in I\}$ such that $S \subseteq \bigcup_{i \in I} A_i$, then the set $\{A_i : i \in I\}$ is called a *cover* (or an *open cover*) of S. When the index set *I* is a finite integer, then this cover which is a collection of finite number of open sets $\{A_i\}$ is called a *finite cover*, and this idea leads to a new definition for compact set. Namely, if every open cover of a set S consists a finite cover, then S is compact. A *null* set is a set such that for every small number $\varepsilon > 0$ there exists a cover $\{A_i : i \in I\}$ of S with $\sum_{i \in I} d(A_i) < \varepsilon$, here $d(A_i)$ is the distance of each set A_i.

∎

We present the following topological notion which generalizes some aspects of the preceding topics. An empty set is shown by $\emptyset$.

∎

A collection τ of subsets of a set X is called a *topology* in X if the following properties hold:

(i) $\emptyset \in \tau$ and $X \in \tau$,

(ii) If $T_i \in \tau$, $i \in I$, where *I* is the index set, then $\bigcup_{i \in I} T_i \in \tau$,

(iii) If $T_i \in \tau$, $i \in I'$ (*I'* is *only* a set of finite number), then $\bigcap_{i \in I'} T_i \in \tau$.

∎

If τ is a topology in X, then the pair (X, τ) is called a *topological space*, and the members of τ are called the *open sets* in X.

■

Based on this general concept we can show that, for instance, a metric space defined in § 2.5.4 is one special case of topological space. Indeed, many of the topics reviewed earlier in this and the preceding sections can be carried over, with some exception, to topological spaces by replacing the open ball with the open sets, however, these issues are not at the moment pursued herein. The elements of any topological space can be represented as the union of a special set of open sets which are called the *bases* for that space.

■

An example of the topological space is the *Hausdorff space*, which is a space (X, τ) with the following property. Given any two distinct points $x_1, x_2 \in X$, if we can find $T_1, T_2 \in \tau$ such that $x_1 \in T_1$ and $x_2 \in T_2$, then $T_1 \cap T_2 = \emptyset$. The distance between two compact sets is finite and it defines the Hausdorff metric.

■

An important definition that is used to set up a measurable set is as follows.

■

A collection m of subsets of a set X is called a σ–*algebra in* X if the following properties hold:

(i) $X \in m$,
(ii) If $S \in m$, then $S^c \in m$. Since $\emptyset = X^c, \Rightarrow \emptyset \in m$,
(iii) If $S_i \in m$ for $i \in I$, then $S = \bigcup_{i \in I} S_i \in m$.

Taking $S_{n+1} = S_{n+2} = \cdots = \emptyset$, we note that $\bigcup_{i=1}^{n} S_i \in m$ if $S_i \in m$ for $i = 1, 2, \cdots, n$. Since $\bigcap_{i=1}^{\infty} S_i = (\bigcup_{i=1}^{\infty} S_i^c)^c$, m is closed under the formation of the countable intersections. Also $S_k - S_l \in m$ if S_k and $S_l \in m$, because $S_k - S_l = S_k \cap S_l^c$.

■

Furthermore, if m is a σ–algebra in X, then X is called a *measurable space* and the members of m are called the measurable sets in X. Although measurability is an extremely important information that plays the central role in integration theory, from our control theory point of view we always assume that most variables or functions of interest to us are already members of an appropriate measurable set. A null set has measure zero.

COMMENT 2.5-1 (On terminology [A. 24]): In many occasions in mathematics, whether we are talking about a measurable space, or a topological space, *etc.*, instead of giving the full ordered set of specifications such as (X, m) or (X, τ), *etc.*,

2.5. The Preliminary Mathematics

to represent these spaces, we specify them by X alone. The rationale for this *customary* choice is that when X is specified, then a measure, *per se*, can be defined on some σ–algebra in X.

2.5.6. Mappings Continued

We now examine a few additional properties of functions and general mappings such as *continuity, differentiability, convexity, contraction,* and *boundedness*. In regard to a function or a general mapping $f: X \to Y$, we also note that many of the properties stated subsequently for a normed-vector space can be generalized to other spaces. But first we present the next two items.

(i) Two functions f and g defined on a set $A \subseteq R^n$ that differ in value only on a null set are called equal *almost everywhere*.

(ii) For any function – say $f \in C[0, 1]$ (continuous in a closed-interval $[0, 1]$), we can choose several different norms. For instance, we can choose $\|f\| = \max_{t \in [0,1]} |f(t)|$, or $\|f\| = (\int_0^1 [f(t)]^2 dt)^{1/2}$. Each norm has its own specific application and we must choose the one that is most pertinent to our case study. ∎

Continuous Functions: A function $f: t \to R$, where $t \in A \subseteq R$, is continuous at a specific point $t_0 \in A$ if given $\varepsilon > 0$, there exists $\delta = \delta(\varepsilon) > 0$ such that $|t - t_0| < \delta \Rightarrow |f(t) - f(t_0)| < \varepsilon$. This property is defined relative to the domain of f, which may be bounded or unbounded, and it may be an open or a closed set. If this property holds at every point of A, then f is continuous in its entire domain. If this property holds and δ is the same at every point, then f is *uniformly continuous* in its domain. The preceding definition of continuity can be stated as the uniform convergence of a sequence of functions to a limit point which is a continuous function. For two functions f and g that are continuous at t_0, and two scalars α and β, $\alpha f + \beta g$ is continuous at t_0. Similarly, a composition of these two functions $f \circ g$ is also continuous if g is continuous at t_0 and f is continuous at $g(t_0)$. When a function is not continuous, it is called discontinuous with either a finite (jump) or an infinite discontinuity. If a function is continuous in an interval except at a few finite points where f has finite discontinuities, then f is called *piecewise continuous*. Consider a real function $f(x)$ on a topological space. If $\{x: f(x) > \alpha\}$ is open for every real α, then f is called *lower semicontinuous*. If $\{x: f(x) < \alpha\}$ is open for every real α, then f is called *upper semicontinuous*. A real function is continuous if and only if it is both upper and lower semicontinuous. The supremum of any collection of lower-semicontinuous functions is lower semicontinuous, and the infimum of any collection of upper-semicontinuous functions is upper semicontinuous. ∎

Consider a function $f: [a, b] \to R$ and assume that the n disjoint-open subintervals (a_i, b_i) are contained in $[a, b]$. If there exists an $\varepsilon > 0$ and $\delta > 0$ such that $\sum_{i=1}^n (b_i - a_i) < \delta \Rightarrow |\sum_{i=1}^n (f(b_i) - f(a_i))| < \varepsilon$, then f is called *absolutely continuous*.

Assume that the above interval [a, b] is partitioned according to $a = t_0 < t_1 < t_2 < \cdots < t_n = b$. If there is a constant $c \geq 0$ such that $\sum_{i=1}^{n} |f(t_i) - f(t_{i-1})| < c$, then f is called a *bounded variation (BV)* on [a, b]. The corresponding *total variation function (TVF)* for f is defined as follows. $TVF \triangleq \sup_{t_i} \sum_{i=1}^{n} |f(t_i) - f(t_{i-1})|$. This concept has application in worst-case study of a system subject to discontinuous inputs and other jump variations in system.

The absolute continuity of f on [a, b] (any bounded interval) implies the uniform continuity and bounded variation of f on that interval.

■

Continuous Mappings: Consider $f : R^n \to R^m$, then we can define the continuity of f by using an appropriate norm. For instance, if there is an $\varepsilon > 0$ and $\delta = \delta(\varepsilon) > 0$ such that $\| x - y \| < \delta \Rightarrow \| f(x) - f(y) \| < \varepsilon$, then f is a continuous function. (For m = 1, we use $|f(x) - f(y)| < \varepsilon$, of course, instead of norm). For a mapping $f : X \to Y$ from one metric space to another the continuity is defined as follows. The mapping f is continuous at a point $x_0 \in X$ if given any $\varepsilon > 0$ there exists $\delta > 0$ such that $f_X(x, x_0) < \delta \Rightarrow f_Y(f(x), f(x_0)) < \varepsilon$, where f_X and f_Y are the corresponding metrics in X and Y. In a more general setting and independent from the chosen metrics the continuity of a general mapping is defined as follows. If X and Y are two topological spaces and if f is a mapping of $X \to Y$, then the mapping f is called continuous if $f^{-1}(V)$ is an open set in X for every open set V in Y.

■

Differentiable Functions and Mappings: A basic knowledge of differentiation of a function is assumed. A function is called differentiable on an open interval if it is differentiable at every point of that interval. A differentiable function in an interval is continuous on that interval, but the derivatives of a continuous function may not be continuous. We recall that given x(t), its derivative at t_0 is defined as $\dot{x}(t)\big|_{t=t_0} \triangleq \lim_{\Delta t \to 0} \{\{x(t_0 + \Delta t) - x(t_0)\}/\Delta t\} \Rightarrow \lim_{\Delta t \to 0} \{\{x(t_0 + \Delta t) - x(t_0) - l(\Delta t)\}/\Delta t\} = 0$. Here $l(\Delta t)$ [$= \dot{x}(t_0)\Delta t$] is a linear function in Δt. From this representation the derivative of a mapping is defined as follows. Given two normed-vector spaces X and Y and V as an open subset of X, then a continuous mapping $f : V \to Y$ is called differentiable at a point $x \in V$ if and only if to every $\varepsilon > 0$ there exists a $\delta > 0$ and a linear map $l : X \to Y$ such that $\| f(x + h) - f(x) - l h \| \leq \varepsilon \| h \|$ holds for all $h \in X$ with $\| h \| < \delta$. Then l is defined as the derivative of f at x.

■

Convexity and Contraction Properties of Functions and Mappings: (i) *Convexity:* A function $f : X \to R$ is called *convex* if its domain is a convex set and for all $x_1, x_2 \in X$ and $0 < \lambda < 1$, we have $f(\lambda x_1 + (1 - \lambda)x_2) \leq \lambda f(x_1) + (1 - \lambda)f(x_2)$.

If x and y are two distinct points in R^n, then the line connecting x to y is shown by $z = \lambda x + (1 - \lambda)y$, $0 < \lambda < 1$.

2.5. The Preliminary Mathematics

(ii) The mean-value theorem: Given $f: R^n \to R^m$, which is differentiable at each point of an open set $A \subseteq R^n$, and given any two points $x, y \in A$, there exists a point $z \in A$ on the line connecting x to y such that $f(x) - f(y) = \frac{\partial f}{\partial x}\Big|_{x=z} (x - y)$.

(iii) The implicit-function theorem: Given $f: X \times U \to R^m$, with $X \subseteq R^n$, $U \subseteq R^m$, such that f is continuously differentiable at each point of an open set $S \subset X \times U$. Let (x_0, u_0) be a point such that $f(x_0, u_0) = 0$ and $\partial f(x_0, u_0) / \partial u$ is nonsingular. Then there exist two open neighborhoods X_0 and U_0 for these x_0 and u_0, respectively, such that for each $x \in X_0$, $f(x, u) = 0$ has a unique solution $u = g(x)$.

(iv) Contraction and isometry: Given a metric space (X, f), then a mapping $g: X \to X$ is called a *contraction* on X if $f(g(x_1), g(x_2)) \leq \gamma f(x_1, x_2)$ for some $0 \leq \gamma < 1$ and for all $x_1, x_2 \in X$. On the other hand, if there are two metric spaces (X, f_1), (Y, f_2) with a mapping $g: X \to Y$ such that $f_2(g(x_1), g(x_2)) = f_1(x_1, x_2)$ for all $x_1, x_2 \in X$, then this mapping is called *isometric*.

(v) The contraction-mapping theorem: Let $x = f(x)$, $x \in S \subseteq X$ (S is a closed subspace of a Banach space (X, C)), and $\| f(x) - f(y) \| \leq \gamma \| x - y \|$, for $0 \leq \gamma < 1$, then there exists a unique solution (fixed point) $x^* \in S$ such that $x^* = f(x^*)$. ∎

Boundedness of Functions and Mappings: *(i)* If $f: [a, b] \to R$ is a continuous function in a closed interval $[a, b] \subset R$, then there exists a constant $k > 0$ such that $|f(t)| \leq k$ for all $t \in [a, b]$ and f is called *bounded*. This result is not generally true when the function has an open domain.

(ii) The boundedness for general mapping is defined as follows. Suppose that $f: X \to Y$, where X and Y are two normed-vector spaces with norms $\|\cdot\|_X$ and $\|\cdot\|_Y$, and f is a linear mapping. Then the mapping f is called *bounded* if there exists a $k > 0$ such that for each $x \in X$, $\| f(x) \|_Y \leq k \| x \|_X$, or $\{ \| f(x) \|_Y / \| x \|_X \} \leq k$. The *infimum* of k is the *norm* of f, which is usually shown by $\| f \| = \sup_{x \neq 0} \{ \| f(x) \| / \| x \| \}$. This representation implies that each norm is computed accordingly and relative to the corresponding domains.

(iii) When a function is continuous on a closed interval $[a, b]$, it is also uniformly continuous on that interval. The existence and uniqueness of a solution to a scalar differential equation $\dot{x} = f(x)$, *per se*, depends on the continuity of f and df/dx. The continuity of the derivative can be replaced by *Lipschitz* condition which says that if there exists an $L > 0$ such that $|f(t_1) - f(t_2)| \leq L(t_1 - t_2)$ for $t_1, t_2 \in [a, b]$, then f satisfies a Lipschitz condition. In the case of a vector differential equation $\dot{x}(t) = f(x, t)$, if $f(x, t)$ is piecewise continuous in t and satisfies the Lipschitz condition $\| f(x, t) - f(y, t) \| \leq L \| x - y \|$, with $\| f(x_0, t) \| \leq h$, for all x and y and all $t \in [t_0, t_1]$, then we have a unique solution for $x(t)$ with $x(t_0) = x_0$. Here we note that the Lipschitz property can either be *local* or *global*, however, this property is stronger than the corresponding continuity condition and is weaker than the corresponding continuous differentiability condition.

(iv) In the case of a differential equation $\dot{x}(t) = f(x,t;\theta)$ with the varying parameter $\theta \in R^r$, the additional customary assumption on the continuity and differentiability of f with respect to θ strongly guarantees the existence and uniqueness of the solution for the sensitivity differential equations with respect to these parameters as used in the book.

2.5.7. Classes of L^p − and H^p − Functions

Here we discuss certain properties of two classes of functions [or mappings] which are used extensively in the literature. Subsequently, it becomes clear that these classes of functions are [complete] mathematical spaces. Throughout this study, and unless stated otherwise, we consider an arbitrary measurable set X with a positive measure μ.

Class of L^p −Functions: Generally speaking, it is customary to show these functions by $L^p(\mu)$ or $L^p(R^n)$ if μ is the *Lebesgue measure* on R^n. Starting with $p = 1$, the $L^1(\mu)$ is defined as the collection of all complex measurable functions f on X such that $\|f\|_1 \{\triangleq \int_X |f| d\mu\} < \infty$. Here the measurability of f guarantees that of $|f|$. The set of all such functions are called *Lebesgue integrable functions* or *summable functions* (with respect to the above positive measure μ). If $0 < p < \infty$ and if f is a complex measurable function on X, then the class of $L^p(\mu)$ is defined accordingly as the set of all functions whose L_p −norm is bounded, i.e., the class of all functions f such that $\|f\|_p \{\triangleq \{\int_X |f|^p d\mu\}^{1/p}\} < \infty$. By definition for $p = \infty$, the $\|f\|_\infty$ is the *essential supremum (ess sup)* of $|f|$, and $L^\infty(\mu)$ is the class of all f whose $\|f\|_\infty \{\triangleq \underset{X}{\text{ess sup}} |f(x)|\} < \infty$.

■

Inequalities: To study various properties, particularly the completeness, of a class of L^p −functions, which we subsequently call them L^p −spaces (whose elements are *equivalence classes of functions* − not just functions [A. 24]), we need to utilize a few inequalities. We note that most inequalities stem from the *notion of convexity*. Here we review a few of them which are often used to prove the completeness of different mathematical spaces resulted from the metric induced by the L_p −norms, or certain inner product. But we should also point out that inequalities play a central role in analysis and an excellent source for many of those which are "of daily use" is [A. 10].

(i) The Jensen's inequality: Let X be such that $\mu(X) = 1$. If $f \in L^1(\mu)$ and $a < f(x) < b$ for all $x \in X$, and if ϕ is convex on (a,b) (including $a = -\infty$ and $b = \infty$), then

$$\phi(\int_X f d\mu) \leq \int_X (\phi \circ f) d\mu. \tag{2.5-1}$$

2.5. The Preliminary Mathematics

This powerful result can be used to show many other interesting inequalities, for instance, we can show that $\exp\{\int_X \log f \, d\mu\} \le \int_X f \, d\mu$ [A. 24].
∎

Consider two positive numbers p and q such that $p + q = pq$, or equivalently $\frac{1}{p} + \frac{1}{q} = 1$, then these numbers are called *conjugate exponents*. Suppose these p and q are such that $0 < p < \infty$. Let X be a measure space with a positive measure μ, and also let f and g be two measurable functions on X with range in $[0, \infty]$, then we have the following inequalities.

(ii) *The Holder's inequality*:

$$\int_X f g \, d\mu \le \left\{\int_X f^p \, d\mu\right\}^{1/p} \left\{\int_X g^q \, d\mu\right\}^{1/q}. \tag{2.5-2}$$

For $p = q = 2$, this is called the Schwarz inequality.

(iii) *The Minkowski's inequality*:

$$\left\{\int_X (f+g)^p \, d\mu\right\}^{1/p} \le \left\{\int_X f^p \, d\mu\right\}^{1/p} + \left\{\int_X g^p \, d\mu\right\}^{1/p}. \tag{2.5-3}$$

∎

L^p –*Spaces:* We now review a few pertinent properties of the class of L^p –functions which lend themselves as mathematical spaces. To prove these properties we use the preceding inequalities. Here $\|\cdot\|$ is the norm in the L^p –*sense*.

(i) For the two conjugate exponents p and q with $1 \le p \le \infty$, if $f \in L^p(\mu)$ and $g \in L^q(\mu)$, then $fg \in L^1(\mu)$ with $\|fg\|_1 \le \|f\|_p \|g\|_q$.

(ii) For the two conjugate exponents p and q with $1 \le p \le \infty$, if $f \in L^p(\mu)$ and $g \in L^p(\mu)$, then $f + g \in L^p(\mu)$.

(iii) $L^p(\mu)$ is a complete metric space for $1 \le p \le \infty$ and for every positive measure μ. Here, we show every Cauchy sequence $\{f_n\}$ in $L^p(\mu)$ converges to a limit $f \in L^p(\mu)$, i.e., $\lim_{n \to \infty} \|f_n - f\|_p = 0$, which is the same as saying $\{f_n\}$ converges to f in the mean of order p, or in the L^p –sense.

∎

Properties of Inner Product and Hilbert Spaces: (i) Let f and g be two compatible vectors belonging to an inner product space, the Schwarz (or the Cauchy – Bunyakovski – Schwarz [A. 13]) inequality is as follows.

$$|<f,g>|^2 \le <f,f><g,g> \Rightarrow <f,g> \le \|f\| \|g\|. \tag{2.5-4}$$

(ii) Any set of vectors such as $x \triangleq [x_1, \cdots, x_n] \in C^n$ over the field of C forms a Hilbert space with the inner product $<x,y> = \sum_{i=1}^{n} x_i \overline{y_i}$, when the corresponding binary operations are defined componentwise.

(iii) For any positive measure μ the set of functions $f \in L^2(\mu)$ forms a Hilbert space with the inner product $<f,g> = \int_X f \overline{g} \, d\mu$.

(iv) The vector space of all continuous functions on $[0, 1]$ is an inner product space with $<f, g> = \int_0^1 f(x)\overline{g}(x)dx$, but this is not a Hilbert space.

(v) For any $x, y \in X$ — a Hilbert space, the mappings $x \to <x, y>$, $x \to <y, x>$, and $x \to \|x\|$ are continuous functions.

(vi) In general, a Hilbert space (just as any vector space) can be represented by a set of bases and the projections of vectors in this space can be studied on that set of bases, and in particular, when the corresponding set is a maximal-orthonormal set, this study becomes more appealing than otherwise. Since, further discussion on this topic exceeds the scope of this presentation, in the following, we only study one aspect of such set.

(vii) The complete trigonometric series: The set of all complex numbers with absolute value one is shown by T. Then the $L^p(T)$, for $1 \le p < \infty$ is the class of all complex, Lebesgue measurable, 2π–periodic functions on R^1 such that $\|f\|_p \{\triangleq \{\frac{1}{2\pi}\int_{-\pi}^{\pi}|f(t)|^p dt\}^{1/p}\} < \infty$. The $L^\infty(T)$ is the class of all 2π–periodic members of $L^\infty(R^1)$ such that $\|f\|_\infty \{\triangleq \operatorname*{ess\,sup}_t |f(t)|\} < \infty$. A trigonometric polynomial is generally shown by a finite sum as follows. $f(t) = a_0 + \sum_{k=1}^{K}(a_k \cos kt + b_k \sin kt)$, which has a period of 2π. Here a_0, a_k's and b_k's, $k = 1, 2, \cdots, K$, are complex numbers and $t \in R^1$. Using the Euler formula, with $j = \sqrt{-1}$, $f(t)$ can be written as $f(t) = \sum_{k=-K}^{K} c_k e^{jkt} \triangleq \sum_{k=-K}^{K} c_k u_k$. If we show the set of all integers (positive, zero, and negative) by Z, then we can show that the set $\{u_k : k \in Z\}$ is an orthonormal set in $L^2(T)$. (This fact can be shown by computing $<f, g> = \frac{1}{2\pi}\int_{-\pi}^{\pi} f(t)\overline{g}(t)dt$, and noting that $<u_k, u_l> = 1 \Leftrightarrow k = l$, otherwise $<u_k, u_l> = 0$). Indeed as $K \to \infty$ this set becomes *a maximal-orthonormal* set which forms a complete inner-product space (or a Hilbert space). This is a *concrete* example of the *abstract* Hilbert space.

(viii) Fourier series: For any $f \in L^1(T)$ the Fourier *coefficients* of f are defined according to $\hat{f}(k) = \frac{1}{2\pi}\int_{-\pi}^{\pi} f(t)e^{-jkt}dt$, $(k \in Z)$. The Fourier *series* of f is defined as $\sum_{k=-\infty}^{\infty} \hat{f}(k)e^{jkt}$, and its *partial sums* are defined as $s_K(t) = \sum_{k=-K}^{K} \hat{f}(k)e^{jkt}$, $K = 0, 1, 2, \cdots$.

We should remind our reader that further theoretical study of Fourier series for all $f \in L^1(T)$ is beyond the scope of this presentation. However, as a consequence of the above results and the fact that $L^2(T) \subset L^1(T)$, we can define $\hat{f}(k)$ for any periodic function $f \in L^2(T)$. From these observations we have the following theorems.

(ix) The Riesz–Fischer theorem: If $\{c_k\}$ is a sequence of complex numbers and $\sum_{k=-\infty}^{\infty}|c_k|^2 < \infty$, then there exists a function $f \in L^2(T)$ such that $c_k =$

2.5. The Preliminary Mathematics

$$\frac{1}{2\pi}\int_{-\pi}^{\pi} f(t)\, e^{-jkt}\, dt.$$

  (x) The Parseval theorem: If $f \in L^2(T)$ and $g \in L^2(T)$, then $\sum_{k=-\infty}^{\infty} \hat{f}(k)\,\overline{\hat{g}(k)}$
$= \dfrac{1}{2\pi}\int_{-\pi}^{\pi} f(t)\,\overline{g(t)}\, dt$ (here the convergence is absolute), and $\lim_{K \to \infty} \| f - s_K \|_2 = 0$, where $s_K(t) = \sum_{k=-K}^{K} \hat{f}(k)\, e^{jkt}$ is the corresponding partial sum, and the preceding limit is in the L^2–sense.

COMMENT 2.5-2 [A. 24]: The Riesz–Fischer theorem and the Parseval theorem says that the mapping $f \to \hat{f}$ is a Hilbert space isomorphism of $L^2(T)$ onto $l^2(Z)$. Here we should point out that when the positive measure μ, in the $L^p(\mu)$, is the counting measure on a countable set, its corresponding L^p–space is shown by l^p–space.

∎

  Additional Properties of Banach Spaces: *(i)* Every Hilbert space is a Banach space, therefore each member of L^p–spaces for $1 \leq p \leq \infty$ forms a Banach space normed by $\|\cdot\|_p$ (considering functions which are equal almost everywhere).

  (ii) One concrete Banach space is the field of complex numbers C.

  (iii) Our considerations of bounded mappings in § 2.5.6 are valid in this space when we use appropriate norms.

  (iv) A mapping $\Lambda : X \to Y$, between two normed-vector spaces X and Y, is also called an *operator*. For any operator Λ, each of the following three conditions means the other two: (a) Λ is bounded. (b) Λ is continuous. (c) Λ is continuous on one point of X.

  (v) The open-mapping theorem: Let U and V be the unit-open balls of the Banach spaces X and Y. To every bounded linear mapping Λ of X *onto* Y, there corresponds a $\delta > 0$ such that $\Lambda(U) \supset \delta V$. Here δV is the set $\{\delta y : y \in V\}$ or the set $\{y \in Y : \|y\| < \delta\}$. In other words, the image of every open set is open, or to every set $\{y \in Y : \|y\| < \delta\}$ there corresponds a set $\{x \in X : \|x\| < 1\}$ such that $\Lambda x = y$.

  (vi) If X and Y are two Banach spaces and if Λ is a bounded-linear mapping of X *onto* Y which is also *(1–1)*, then Λ^{-1} is a bounded-linear mapping of Y onto X.

  (vii) The Hahn – Banach theorem: If M is a subspace of normed-vector space X and if f is a bounded-linear mapping on M, then f can be extended to a bounded-linear mapping F on X such that $\|F\| = \|f\|$.

∎

  Fourier Transformation: A few cautionary words on the applications of this integral transform are in order. Given f(t) its Fourier transform is defined by $\hat{f}(\omega)$
$= \dfrac{1}{\sqrt{2\pi}} \int_{-\infty}^{\infty} f(t)\, e^{-j\omega t}\, dt$, $\omega \in R^1$. There are a number of properties associated with

this transformation which are familiar to every engineer. However, to prove the inversion formula, i.e., $f(t) = \frac{1}{\sqrt{2\pi}} \int_{-\infty}^{\infty} \hat{f}(\omega) e^{j\omega t} d\omega$ requires some technical conditions on $\hat{f}$, which exceed $\hat{f} \in L^1$. For instance, if we let $f \in L^1 \cap L^2$, then we can show that $\hat{f} \in L^2$ and the above two formulae can be shown as such rather easily. Also for every $f \in L^2$, $\|\hat{f}\|_2 = \|f\|_2$ (norm in the L^p –sense). Furthermore, the mapping $f \to \hat{f}$ is a Hilbert space isomorphism of L^2 onto L^2.

∎

Class of H^p –Functions: The so-called H^p –spaces (H is for G.H. Hardy) refer to the classes of H^p –functions that have many interesting properties, among them the boundary behaviors and factorizations are of particular interest to us. We note that here we consider an analytic (or holomorphic – H) function f that is shown by $f \in H(U)$ in a *unit disc* (U), whose boundary is, of course, T. In § 2.7.3 we offer an alternative domain for these functions.

∎

(i) Let $j = \sqrt{-1}$ and $\log^+ t = \log t$ if $t \geq 1$ and $\log^+ t = 0$ if $t < 1$. If

$$M_0(f; r) = \exp\left\{\frac{1}{2\pi} \int_{-\pi}^{\pi} \log^+ |f(re^{j\theta})| d\theta\right\}, \qquad (2.5\text{-}5a)$$

$$M_p(f; r) = \left\{\frac{1}{2\pi} \int_{-\pi}^{\pi} |f(re^{j\theta})|^p d\theta\right\}^{1/p}, \qquad (2.5\text{-}5b)$$

$$M_\infty(f; r) = \sup_\theta |f(re^{j\theta})|, \qquad (2.5\text{-}5c)$$

then M_0, M_p, and M_∞ are *monotonically increasing functions* of r in [0, 1).

(ii) For any $f \in H(U)$ and for $0 \leq p \leq \infty$, we let $\|f\|_p = \lim_{r \to 1} M_p(f; r)$, where $M_p(f; r)$ is defined in (2.5-5).

DEFINITION 2.5-4 (H^p – *Hardy-Spaces*): For $0 < p \leq \infty$ the class of H^p –functions is defined to consist of all $f \in H(U)$ such that $\|f\|_p < \infty$. In particular the class of H^∞ –functions is the collection of all bounded analytic (holomorphic) functions in the unit disc, normed by $\|f\|_\infty = \sup_{z \to 1} |f(z)| < \infty$. (For $p = 0$ we refer to case *(iii)* below).

∎

(iii) Let N (N is for R. Nevanlinna) be the class of all functions f in a unit disc such that

$$\sup_{0 < r < 1} \left\{\frac{1}{2\pi} \int_{-\pi}^{\pi} \log^+ |f(re^{j\theta})| d\theta\right\} < \infty. \qquad (2.5\text{-}5d)$$

Thus the class of N–functions consists of all $f \in H(U)$ whose $\|f\|_0 < \infty$.

2.5. The Preliminary Mathematics

(iv) It is thus clear that for $0 < \alpha < \beta < \infty$, we have

$$H^\infty \subset H^\beta \subset H^\alpha \subset N. \qquad (2.5\text{-}6)$$

(v) For $1 \le p \le \infty$, $\|f\|_p$ satisfies the triangle inequality and we can show that H^p is a normed-vector space.

(vi) For $1 \le p \le \infty$, H^p is indeed a Banach space, i.e., we can show that every Cauchy sequence in H^p converges [indeed uniformly] relative to the corresponding H_p–norm, or in the H^p–sense, to a function in this class of $f \in H(U)$.

(vii) For $p < 1$, H^p remains a vector space, but the triangle inequality does not hold for the corresponding H_p–norm.

∎

Class of H^2–Functions: We now concentrate on studying a particular class of H^p–functions which has many applications in control theory, namely that of H^2–functions. This class of functions *also* forms a Hilbert space and it can be associated with a subspace of $L^2(T)$–spaces.

The norm of any function $f \in L^2(T)$ is given by $\|f\|_2 = \{\frac{1}{2\pi}\int_{-\pi}^{\pi} |f(e^{j\theta})|^2 d\theta\}^{1/2}$ and its Fourier coefficient is $\hat{f}(k) = \frac{1}{2\pi}\int_{-\pi}^{\pi} f(e^{j\theta}) e^{-jk\theta} d\theta$, $k = 0, \pm 1, \pm 2, \cdots$.

Basic Properties of H^2–Functions: (i) A function $f \in H(U)$, of the form $f(z) = \sum_{k=0}^{\infty} a_k z^k$, ($z \in U$), is in H^2 if and only if $\sum_{k=0}^{\infty} |a_k|^2 < \infty$; then $\|f\|_2 = \{\sum_{k=0}^{\infty} |a_k|^2\}^{1/2}$.

(ii) If $f \in H^2$, then f has radial limits $f^*(e^{j\theta})$ at almost every point of T and $f^* \in L^2(T)$; the kth-Fourier coefficient of f^* is a_k if $k \ge 0$ and is 0 if $k < 0$; the L^2–approximation of $\lim_{r \to 1} \frac{1}{2\pi} \int_{-\pi}^{\pi} |f^*(e^{j\theta}) - f(re^{j\theta})|^2 d\theta = 0$ holds.

(iii) The mapping $f \to f^*$ is an isometry of H^2 onto the subspace of $L^2(T)$ which consists of functions $f \in L^2(T)$ whose $\hat{f}(k) = 0$ for all $k < 0$.

∎

Boundary Behavior of H^p–Spaces: (i) *The maximum module principle:* If f is continuous on the closure of a bounded region and is analytic in this region, then its maximum occurs on the boundary of the region. This is also shown by $|f| = \|f\|_\infty$.

(ii) An *interpolation problem* refers to finding an analytic function in a set Ω such that this function has certain values at each and every point of any subset of Ω. We can also specify finitely many derivatives at these points.

∎

Factorization in H^p–Spaces: The following factorization methods also provide a pattern for the limiting behavior of the corresponding function.

(i) The Blaschke product: If $\{\alpha_n\}$ is a sequence in U such that $\alpha_n \neq 0$ and $\sum_{n=1}^{\infty} (1 - |\alpha_n|) < \infty$, if k is a nonnegative integer, then the so-called Blaschke product is as follows. $B(z) = z^k \prod_{n=1}^{\infty} \dfrac{\alpha_n - z}{1 - \overline{\alpha}_n z} \dfrac{|\alpha_n|}{\alpha_n}$, for $z \in U$, and $B(z) \in H^{\infty}$. Here α_n are zeros of B and they may repeat, and also on T the $|B(z)| = 1$. The B is also called a Blaschke product when n is finite.

(ii) Every $f \in H^p$ (except $f = 0$) can be factored into a Blaschke product and a function $g \in H^{\infty}$ which has no zeros in U.

(iii) An *inner function* refers to all $f \in H^{\infty}$ such that $|f^*| = 1$ almost everywhere on T, where f^* is the radial limit of f.

(iv) An *outer function* refers to all $g(z)$ such that

$$g(z) = c \exp\left\{\frac{1}{2\pi} \int_{-\pi}^{\pi} \frac{e^{jt}+z}{e^{jt}-z} \log f(e^{jt})\, dt\right\}, \quad z \in U, \qquad (2.5\text{-}7)$$

where c is a constant with $|c| = 1$ and f is a positive measurable function on T with $\log f \in L^1(T)$.

(v) Every Blaschke product is an inner function but there are others. A method to construct an inner function is to use

$$f(z) = cB(z) \exp\left\{-\int_{-\pi}^{\pi} \frac{e^{jt}+z}{e^{jt}-z}\, d\mu(t)\right\}, \quad z \in U, \qquad (2.5\text{-}8)$$

where $B(z)$ and c are as in *(i), (iv)*, and μ is a finite positive Borel measure [Borel sets in [a, b] are members of the smallest σ–algebra in [a, b] that contains all intervals] on T [which is singular with respect to the Lebesgue measure].

(vi) Suppose that g is an outer function related to f as in (2.5-7), then $\lim_{r \to 1} |g(re^{j\theta})| = f(e^{j\theta})$ almost everywhere on T. Also $g \in H^p$ if and only if $f \in L^p(T)$, then $\|g\|_p = \|f\|_p$.

(vii) Suppose that $0 < p \leq \infty$, $f \in H^p$, f does not vanish identically, then $\log|f^*| \in L^1(T)$, the corresponding outer function is $f_{outer} \in H^p$, and there is an inner function f_{inner} such that $f = f_{inner} f_{outer}$.

2.5.8. Matrix and Determinant

Throughout this book we extensively utilize matrix algebra. Thus we postpone specific discussions on this topic to subsequent sections as pertinent. The following properties, however, have broad applications in sensitivity analysis.

∎

2.5. The Preliminary Mathematics

(i) Eigenvalues/eigenvectors and singular values: A basic knowledge of these concepts is assumed, and we note that these and many other pertinent matrix quantities can be easily computed using already available computing algorithms. In this regard, specification of $\sigma_{max}(A)/\sigma_{min}(A)$, the maximum and minimum singular values for matrix A, indicates the desired degree of the overall computational accuracy and this ratio is called *a condition number* of A. The largest *modulus* of the eigenvalues of A is called its *spectral radius,* shown by $\rho(A)$. Trace (sum of all the eigenvalues) of matrix A is shown by $Tr(A)$.

(ii) The singular-value decomposition: Given a matrix $A \in C^{n \times n}$, we can find two orthogonal matrices U and V ($UU^* = I$, and $VV^* = I$, * means transpose conjugate) and a diagonal matrix Σ such that $A = U\Sigma V^*$, where Σ has all the singular values of A on its diagonal. This decomposition can be generalized to rectangular matrices.

(iii) The kernel and image of a matrix: Given a matrix $A \in C^{m \times n}$, the kernel or (null space) of A is defined as follows. $\text{Ker} A \triangleq \{x : Ax = 0\}$. The range or image of A is defined as follows. $\text{Im} A \triangleq \{y : y = Ax \text{ for some } x \neq 0\}$. The dimension of Im A is the rank of matrix, and that of Ker A is its nullity.

∎

Consider a 3×3 determinant whose rows (or columns) represent three independent vectors. Then the absolute value of this determinant represents an upper bound for the volume of the parallelepiped corresponding to these three vectors. An estimate of this volume is always equal to the multiplication of the length of these vectors. A generalization of this idea to the higher-order cases is given next.

(iv) The Hadamard's inequality [A. 12]: For an $n \times n$ matrix $A \triangleq (a_{ij})$, with possibly complex entries, an upper bound for its determinant is given by $(\det A)^2 \leq \prod_{i=1}^{n} \sum_{j=1}^{n} |a_{ij}|^2$.

2.5.9. Vector and Matrix Norms

In this section we review several widely used norms for vectors and matrices. Other norms may also be defined. Recall that in this study we assume all vectors and matrices are also finite dimensional.

∎

Given $x \in R^n$ we have the following norms for a vector.

(i) $\| x \|_1 = |x_1| + \cdots + |x_n|$ (Octahedral norm),

(ii) $\| x \|_p = (|x_1|^p + \cdots + |x_n|^p)^{1/p}$ (p-norm, $1 < p < \infty$),

(ii)' $\| x \|_2 = (|x_1|^2 + \cdots + |x_n|^2)^{1/2} = (x^T x)^{1/2}$,   (Euclidean norm),

(iii) $\| x \|_\infty = \max\{|x_1|, \cdots, |x_n|\}$,   (cubic norm).

All p-norms are equivalent in the sense that they are related through constants. Vector Euclidean norms are invariant under orthogonal transformations. ∎

Given $A \in R^{m \times n}$ which represents a mapping $R^n \to R^m$, we may consider the following norms:

(iv) $\| A \|_1 = \max_j \sum_{i=1}^{m} |a_{ij}|$,

(v) $\| A \|_p = \sup_{x \neq 0} \dfrac{\| Ax \|_p}{\| x \|_p}$,

(v)' $\| A \|_2 = \sup_{x \neq 0} \dfrac{\| Ax \|_2}{\| x \|_2} = (\lambda_{max}(A^T A))^{1/2} \triangleq \sigma_{max}(A)$,

(vi) $\| A \|_\infty = \max_i \sum_{j=1}^{n} |a_{ij}|$.

Here $\lambda_{max}(\cdot)$ refers to the maximum eigenvalue of $(\cdot)$ and $\sigma_{max}(\cdot)$ stands for the maximum singular value of $(\cdot)$.

(vii) $\| A \|_F = \sqrt{\sum_{i=1}^{n} \sum_{j=1}^{n} a_{ij}^2}$, the Frobenius norm of a matrix.

Based on the singular-value decomposition we have $\| A \|_F^2 = \sum_{i=1}^{n} \sigma_i^2$. ∎

As is evident from the above list, many different vector and matrix norms can be defined, however, the above norms are equivalent in the sense that the metrics induced by any two norms on a finite-dimensional vector space give rise to the same topology on that space. For the above matrix norms there are upper and lower bounds on the ratios of any two norms which remain the same for each matrix. Here $\| A \|_2$ and $\| A \|_F$ are invariant under orthogonal transformations.

2.6. ASPECTS OF SYSTEM THEORY

2.6.1. Introduction

In this section we review those special topics in system theory which are utilized and/or developed in conjunction with recent sensitivity literature on the H^∞-sensitivity minimization. Although this methodology is primarily developed for a system that is modeled in the frequency domain, our presentation focuses on those issues pertaining to a linear time-invariant system that is modeled in the state-space form. In this regard, we are considering the corresponding frequency-domain approach only on a peripheral basis. In the subsequent section an overview of a general synthesis problem and in particular the H^∞-sensitivity minimization is presented.

2.6.2. State – Space Realization of Transfer Matrices

Consider a state-space representation of a *DCLTI – GS* as follows.

$$\dot{x}(t) = Ax(t) + Bu(t), \quad x(0) \equiv 0, \qquad (2.6\text{-}1)$$

$$z(t) = Cx(t) + Du(t). \qquad (2.6\text{-}2)$$

The transfer matrix G(s) between input u(s) and output z(s) is

$$G(s) = C(sI - A)^{-1}B + D. \qquad (2.6\text{-}3)$$

This transfer matrix is depicted by the following *notation*

$$G(s) \triangleq \begin{bmatrix} A & \vdots & B \\ \cdots & \vdots & \cdots \\ C & \vdots & D \end{bmatrix}. \qquad (2.6\text{-}4)$$

The right-hand side of (2.6-4) is not partitioned in the usual way, rather this is a widely used notation that results in a quick transformation between the frequency-domain and the time-domain representation of the underlying system.

Clearly, the (2.6-4) generates a new block-diagram algebra between different blocks of an interconnected system – each block is represented by its corresponding (2.6-4). A summary of certain essential such algebra is shown in Tables 2.6-1 & 2. These values can be verified directly.

To give a point of view as how these two tables are generated based on the initial notation of (2.6-4) we compute the $G^{-1}(s)$. Let us assume that $G(s) \triangleq (A_1, B_1, C_1, D_1)$ and $G^{-1}(s) \triangleq (A_2, B_2, C_2, D_2)$. Then from Table 2.6-2 and for a series connection such as $\tilde{G}(s) = G(s)G^{-1}(s)$, we have

$$\tilde{G}(s) \triangleq \begin{bmatrix} A_1 & B_1C_2 & \vdots & B_1D_2 \\ 0 & A_2 & \vdots & B_2 \\ \cdots & \cdots & \vdots & \cdots \\ C_1 & D_1C_2 & \vdots & D_1D_2 \end{bmatrix}, \qquad (2.6\text{-}5)$$

resulting in

Table 2.6-1. Block-Diagram Algebra of A Single-Block System.

The initial notation: $G(s) = C(sI-A)^{-1}B + D$ $\triangleq \begin{bmatrix} A & B \\ \hline C & D \end{bmatrix} \triangleq (A, B, C, D)$	The initial system is under a nonsingular similarity transformation: $G(s) \triangleq \begin{bmatrix} T^{-1}AT & T^{-1}B \\ \hline CT & D \end{bmatrix}$
Transpose: $G^T(s) \triangleq \begin{bmatrix} A^T & C^T \\ \hline B^T & D^T \end{bmatrix}$	Transpose Conjugate on the $j\omega$-axis: $G^*(j\omega) \triangleq \begin{bmatrix} -A^T & -C^T \\ \hline B^T & D^T \end{bmatrix}$ $G^T(-s)$ has the same structure.

Table 2.6-2. Block-Diagram Algebra of An Interconnected System.

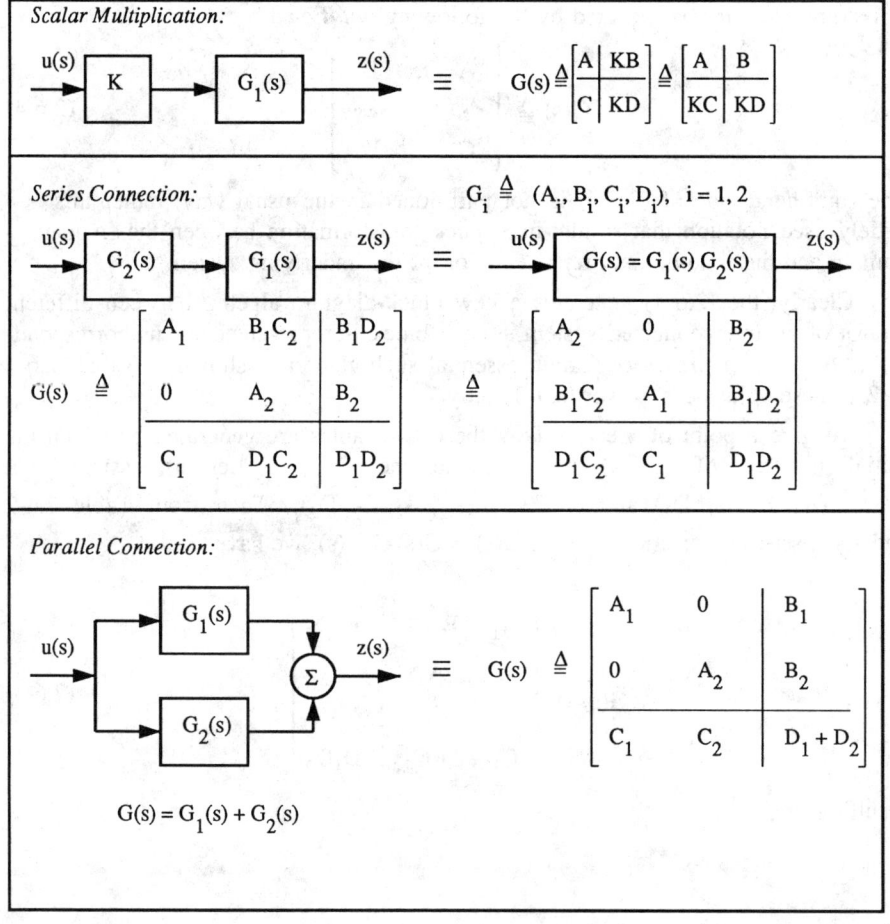

2.6. Aspects of System Theory

$$\tilde{G}(s) \triangleq (C_1, D_1C_2) \begin{bmatrix} sI - A_1 & -B_1C_2 \\ 0 & sI - A_2 \end{bmatrix}^{-1} \begin{bmatrix} B_1D_2 \\ B_2 \end{bmatrix} + D_1D_2$$

$$= (C_1, D_1C_2) \begin{bmatrix} (sI - A_1)^{-1} & (sI - A_1)^{-1}B_1C_2(sI - A_2)^{-1} \\ 0 & (sI - A_2)^{-1} \end{bmatrix} \begin{bmatrix} B_1D_2 \\ B_2 \end{bmatrix} + D_1D_2$$

$$= (C_1(sI - A_1)^{-1}, \ C_1(sI - A_1)^{-1}B_1C_2(sI - A_2)^{-1} + D_1C_2(sI - A_2)^{-1})$$

$$\times \begin{bmatrix} B_1D_2 \\ B_2 \end{bmatrix} + D_1D_2. \tag{2.6-6}$$

Suppose that D_1^{-1} exists and $D_2 = D_1^{-1}$, then $D_1D_2 = I$. In order that $\tilde{G}(s) \equiv I$, we let the remaining terms in (2.6-6) vanish, resulting in a set of choices for them. For instance, if we let $C_2 = D_1^{-1}C_1$ and $B_2 = -B_1D_1^{-1}$, then

$$\tilde{G}(s) = \left(C_1(sI - A_1)^{-1}B_1D_1^{-1} - C_1(sI - A_1)^{-1}B_1D_1^{-1}C_1(sI - A_2)^{-1}B_1D_1^{-1} \right.$$

$$\left. - C_1(sI - A_2)^{-1}B_1D_1^{-1} \right) + I$$

$$= C_1(sI - A_1)^{-1}(-A_2 - B_1D_1^{-1}C_1 + A_1)(sI - A_2)^{-1}B_1D_1^{-1} + I. \tag{2.6-7}$$

By letting $A_2 = A_1 - B_1D_1^{-1}C_1$ we have $\tilde{G}(s) \equiv I$. Thus we infer from this analysis that the inverse of a transfer matrix, when D^{-1} exists, becomes

$$G^{-1}(s) = \begin{bmatrix} A - BD^{-1}C & \vdots & -BD^{-1} \\ \cdots \cdots \cdots & \vdots & \cdots \cdots \\ D^{-1}C & \vdots & D^{-1} \end{bmatrix}. \tag{2.6-8}$$

∎

Several key-transfer matrices are presented in Table 2.6-3, which are computed based on the formula (2.6-8) for inverse of a transfer matrix. In this table, the first two cases are straightforward to verify, and we offer the following analysis to justify *Case 3*. First we recall the definitions of Δ_{12} and Δ_{21} from Table 2.6-3, followed by the facts that $D_1\Delta_{21} = \Delta_{12}D_1$, and $I + D_1\Delta_{21}D_2 = I + \Delta_{12}D_1D_2 = \Delta_{12}$. Next, using the second choice for series connection in Table 2.6-2, we have

$$(I - G_2G_1)^{-1} =$$

Table 2.6-3. Essential Transfer Matrices.

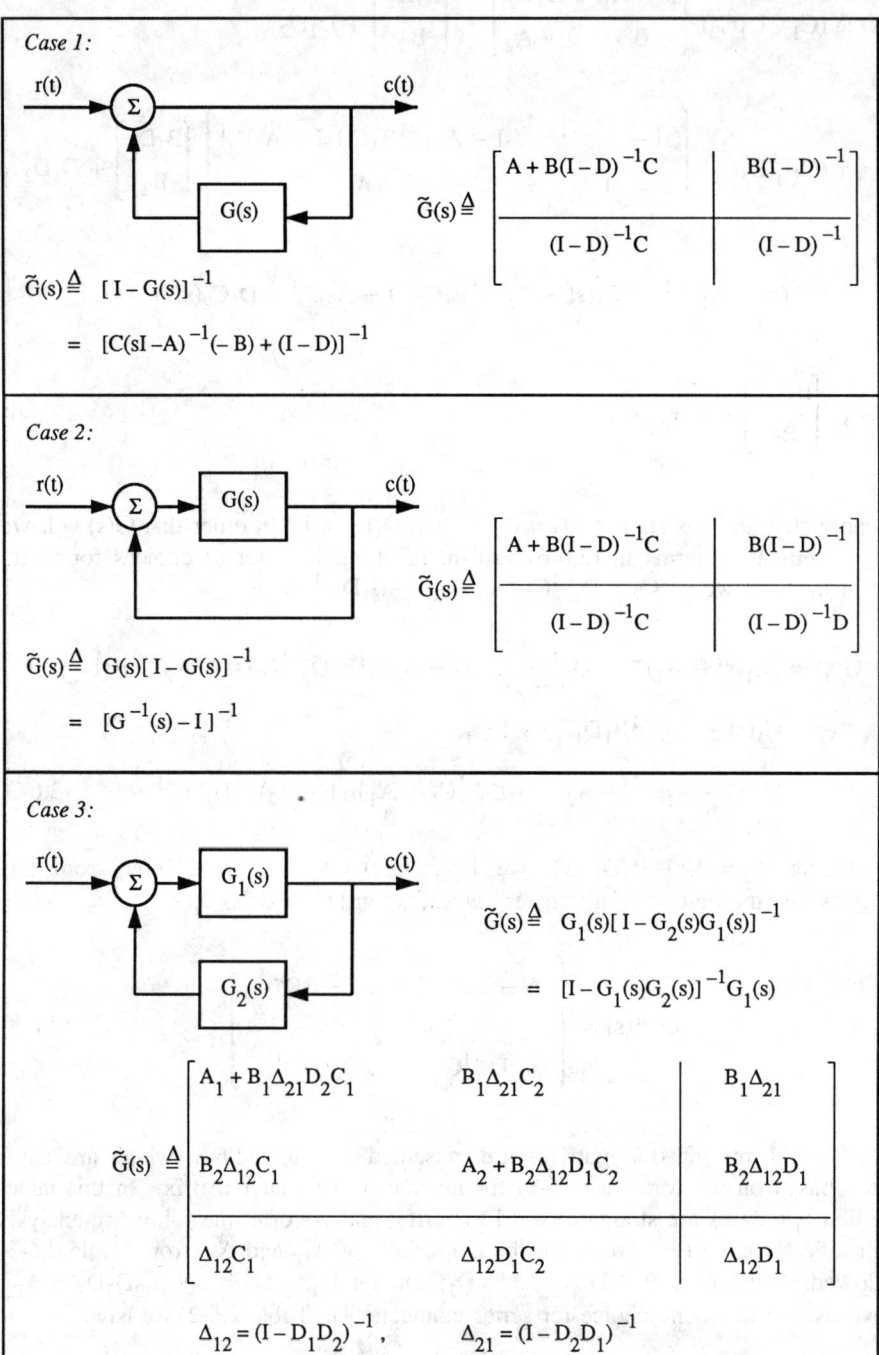

2.6. Aspects of System Theory

$$\begin{bmatrix} A_1 + B_1\Delta_{21}D_2C_1 & B_1\Delta_{21}C_2 & \vdots & B_1\Delta_{21} \\ B_2\Delta_{12}C_1 & A_2 + B_2\Delta_{12}D_1C_2 & \vdots & B_2\Delta_{12}D_1 \\ \cdots\cdots\cdots\cdots & \cdots\cdots\cdots\cdots & \vdots & \cdots\cdots \\ \Delta_{21}D_2C_1 & \Delta_{21}C_2 & \vdots & \Delta_{21} \end{bmatrix}, \quad (2.6\text{-}9)$$

$$G_1(I - G_2G_1)^{-1} =$$

$$\begin{bmatrix} A_1 + B_1\Delta_{21}D_2C_1 & B_1\Delta_{21}C_2 & 0 & \vdots & B_1\Delta_{21} \\ B_2\Delta_{12}C_1 & A_2 + B_2\Delta_{12}D_1C_2 & 0 & \vdots & B_2\Delta_{12}D_1 \\ B_1\Delta_{21}D_2C_1 & B_1\Delta_{21}C_2 & A_1 & \vdots & B_1\Delta_{21} \\ \cdots\cdots\cdots & \cdots\cdots\cdots & \cdots & \vdots & \cdots\cdots \\ \Delta_{12}C_1 - C_1 & \Delta_{12}D_1C_2 & C_1 & \vdots & \Delta_{21}D_1 \end{bmatrix}. \quad (2.6\text{-}10)$$

Since application of a similarity transformation T does not change G(s), and this method is often sought in this type analysis, we use the following similarity transformation to simplify (2.6-10).

$$T = \begin{bmatrix} I & 0 & 0 \\ 0 & I & 0 \\ I & 0 & I \end{bmatrix}, \quad T^{-1} = \begin{bmatrix} I & 0 & 0 \\ 0 & I & 0 \\ -I & 0 & I \end{bmatrix}, \quad (2.6\text{-}11)$$

then (2.6-10) becomes

$$\begin{bmatrix} A_1 + B_1\Delta_{21}D_2C_1 & B_1\Delta_{21}C_2 & 0 & \vdots & B_1\Delta_{21} \\ B_2\Delta_{12}C_1 & A_2 + B_2\Delta_{12}D_1C_2 & 0 & \vdots & B_2\Delta_{12}D_1 \\ 0 & 0 & A_1 & \vdots & 0 \\ \cdots\cdots\cdots & \cdots\cdots\cdots & \cdots & \vdots & \cdots\cdots \\ \Delta_{12}C_1 & \Delta_{12}D_1C_2 & C_1 & \vdots & \Delta_{21}D_1 \end{bmatrix}. \quad (2.6\text{-}12)$$

Clearly (2.6-12) is equivalent to $\tilde{G}(s)$ in *Case 3* of Table 2.6-3. ∎

Linear Fractional Transformations (LFT's): The recent theoretical extensions of the H^∞-optimization theory to system which is modeled by a state-space form depend on the judicious decomposition of (A, B, C, D) parameters associated with the underlying system. Furthermore, we need to parameterize the plant (initially given by a transfer matrix) in order to be able to apply different design methodologies and finally to synthesize the control problem. One such parameterization stems from the so-called linear fractional transformation and in this regard the work of Redheffer [6H1. 15 to 17] should be noted.

Consider a partitioned transfer matrix G(s) as follows.

$$z(s) \triangleq \begin{Bmatrix} z_1(s) \\ z_2(s) \end{Bmatrix} = G(s)u(s) \triangleq \begin{bmatrix} G_{11}(s) & G_{12}(s) \\ G_{21}(s) & G_{22}(s) \end{bmatrix} \begin{Bmatrix} u_1(s) \\ u_2(s) \end{Bmatrix}. \qquad (2.6\text{-}13)$$

This is the same as the following two sets of equations

$$\begin{cases} z_1(s) = G_{11}(s)u_1(s) + G_{12}(s)u_2(s) \\ z_2(s) = G_{21}(s)u_1(s) + G_{22}(s)u_2(s). \end{cases} \qquad (2.6\text{-}14)$$

We may now parameterize the system either as

$$\begin{cases} u_2(s) = K(s)z_2(s), \text{ or} \\ u_1(s) = \Delta(s)z_1(s), \end{cases} \qquad (2.6\text{-}15)$$

where $K(s)$ and $\Delta(s)$ represent controller and plant perturbations, respectively. Each parameterization has its own application, however, in either case when we substitute each one of (2.6-15) into (2.6-14) and rearrange that equation, we get a linear relationship between one input and one output which is affine in terms of K or Δ. We consider both cases and the summary of our analyses with the corresponding state-space realizations are given in Table 2.6-4.

To reach the result in *Case 1* of Table 2.6-4, we first compute $W \triangleq K(I - G_{22}K)^{-1}$ in a similar way as *Case 3* of Table 2.6-3, except that we use the other choice for the corresponding series connection in Table 2.6-2. Upon application of a similarity transformation we have

$$W \triangleq \begin{bmatrix} A + B_2\Delta_{322}D_3C_2 & B_2\Delta_{322}C_3 & \vdots & B_2\Delta_{322}D_3 \\ B_3\Delta_{223}C_2 & A_3 + B_3\Delta_{223}D_{22}C_3 & \vdots & B_3\Delta_{223} \\ \cdots & \cdots & \vdots & \cdots \\ \Delta_{322}D_3C_2 & \Delta_{322}C_3 & \vdots & \Delta_{322}D_3 \end{bmatrix}, \qquad (2.6\text{-}16)$$

where Δ_{223} and Δ_{322} are given in Table 2.6-4. Similarly,

2.6. Aspects of System Theory

Table 2.6-4. Linear Fractional Transformations.

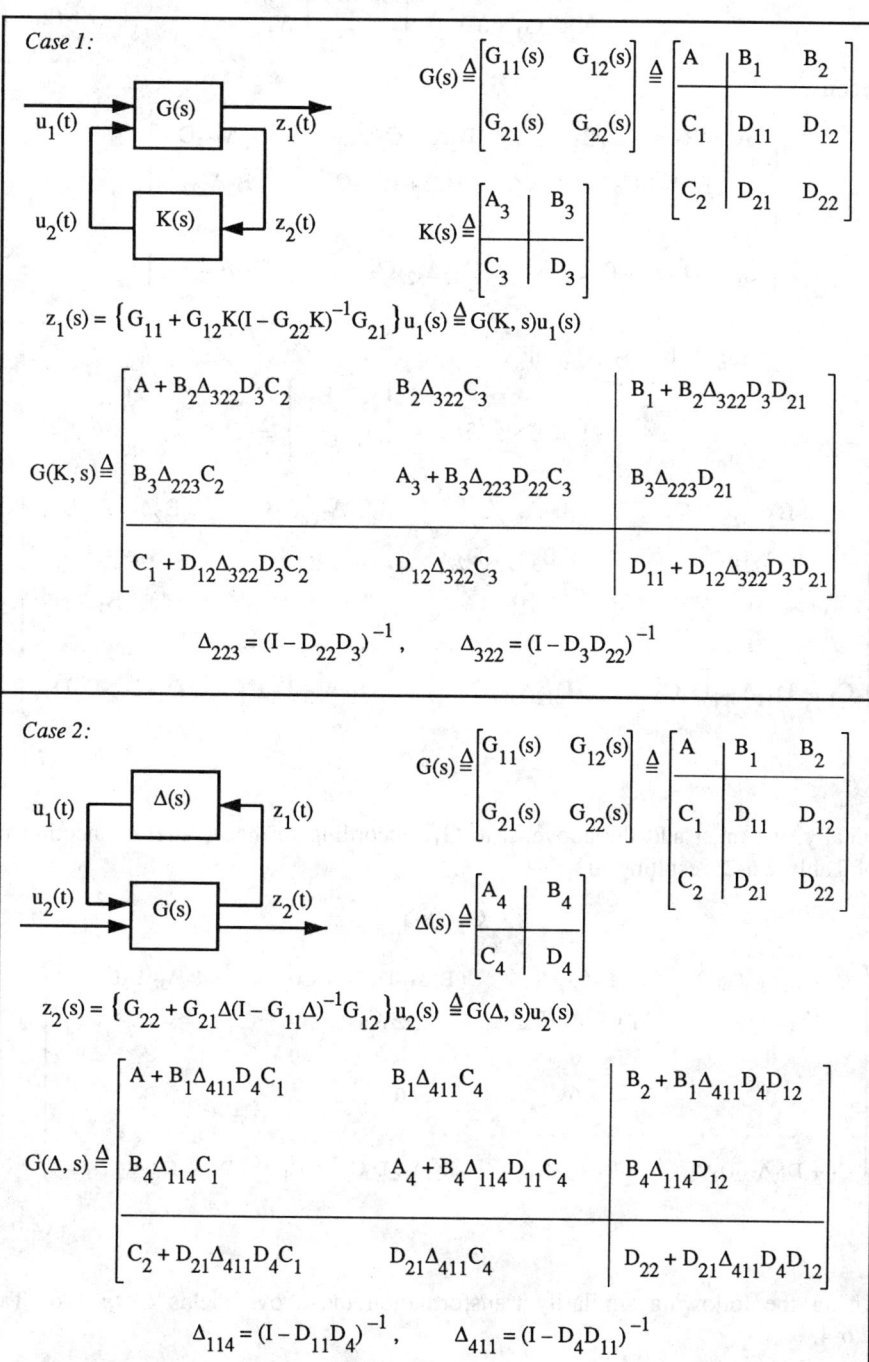

Case 1:

$$G(s) \triangleq \begin{bmatrix} G_{11}(s) & G_{12}(s) \\ G_{21}(s) & G_{22}(s) \end{bmatrix} \triangleq \left[\begin{array}{c|cc} A & B_1 & B_2 \\ \hline C_1 & D_{11} & D_{12} \\ C_2 & D_{21} & D_{22} \end{array}\right]$$

$$K(s) \triangleq \left[\begin{array}{c|c} A_3 & B_3 \\ \hline C_3 & D_3 \end{array}\right]$$

$$z_1(s) = \{G_{11} + G_{12}K(I - G_{22}K)^{-1}G_{21}\}u_1(s) \triangleq G(K,s)u_1(s)$$

$$G(K,s) \triangleq \left[\begin{array}{c|cc} A + B_2\Delta_{322}D_3C_2 & B_2\Delta_{322}C_3 & B_1 + B_2\Delta_{322}D_3D_{21} \\ B_3\Delta_{223}C_2 & A_3 + B_3\Delta_{223}D_{22}C_3 & B_3\Delta_{223}D_{21} \\ \hline C_1 + D_{12}\Delta_{322}D_3C_2 & D_{12}\Delta_{322}C_3 & D_{11} + D_{12}\Delta_{322}D_3D_{21} \end{array}\right]$$

$$\Delta_{223} = (I - D_{22}D_3)^{-1}, \quad \Delta_{322} = (I - D_3D_{22})^{-1}$$

Case 2:

$$G(s) \triangleq \begin{bmatrix} G_{11}(s) & G_{12}(s) \\ G_{21}(s) & G_{22}(s) \end{bmatrix} \triangleq \left[\begin{array}{c|cc} A & B_1 & B_2 \\ \hline C_1 & D_{11} & D_{12} \\ C_2 & D_{21} & D_{22} \end{array}\right]$$

$$\Delta(s) \triangleq \left[\begin{array}{c|c} A_4 & B_4 \\ \hline C_4 & D_4 \end{array}\right]$$

$$z_2(s) = \{G_{22} + G_{21}\Delta(I - G_{11}\Delta)^{-1}G_{12}\}u_2(s) \triangleq G(\Delta,s)u_2(s)$$

$$G(\Delta,s) \triangleq \left[\begin{array}{c|cc} A + B_1\Delta_{411}D_4C_1 & B_1\Delta_{411}C_4 & B_2 + B_1\Delta_{411}D_4D_{12} \\ B_4\Delta_{114}C_1 & A_4 + B_4\Delta_{114}D_{11}C_4 & B_4\Delta_{114}D_{12} \\ \hline C_2 + D_{21}\Delta_{411}D_4C_1 & D_{21}\Delta_{411}C_4 & D_{22} + D_{21}\Delta_{411}D_4D_{12} \end{array}\right]$$

$$\Delta_{114} = (I - D_{11}D_4)^{-1}, \quad \Delta_{411} = (I - D_4D_{11})^{-1}$$

$$X \triangleq G_{12}(s)W \triangleq \begin{bmatrix} A & B_2 \\ C_1 & D_{12} \end{bmatrix} W, \qquad (2.6\text{-}17)$$

becomes

$$X \triangleq \left[\begin{array}{ccc:c} A + B_2\Delta_{322}D_3C_2 & B_2\Delta_{322}C_3 & & B_2\Delta_{322}D_3 \\ B_3\Delta_{223}C_2 & A_3 + B_3\Delta_{223}D_{22}C_3 & & B_3\Delta_{223} \\ \hdashline \\ C_1 + D_{12}\Delta_{322}D_3C_2 & D_{12}\Delta_{322}C_3 & & D_{12}\Delta_{322}D_3 \end{array} \right]. \qquad (2.6\text{-}18)$$

Postmultiplying X by $G_{21}(s)$ yields

$$Y \triangleq XG_{21}(s) \triangleq X \begin{bmatrix} A & B_1 \\ C_2 & D_{21} \end{bmatrix} \triangleq$$

$$\left[\begin{array}{cccc:c} A + B_2\Delta_{322}D_3C_2 & B_2\Delta_{322}C_3 & B_2\Delta_{322}D_3C_2 & & B_2\Delta_{322}D_3D_{21} \\ B_3\Delta_{223}C_2 & A_3 + B_3\Delta_{223}D_{22}C_3 & B_3\Delta_{223}C_2 & & B_3\Delta_{223}D_{21} \\ 0 & 0 & A & & B_1 \\ \hdashline \\ C_1 + D_{12}\Delta_{322}D_3C_2 & D_{12}\Delta_{322}C_3 & D_{12}\Delta_{322}D_3C_2 & & D_{12}\Delta_{322}D_3D_{21} \end{array} \right].$$

$$(2.6\text{-}19)$$

Finally, we must add the above Y to G_{11} according to the parallel connection rule of Table 2.6-2, resulting in

$$G_{11} + Y \triangleq$$

$$\left[\begin{array}{ccccc:c} A + B_2\Delta_{322}D_3C_2 & B_2\Delta_{322}C_3 & B_2\Delta_{322}D_3C_2 & 0 & & B_2\Delta_{322}D_3D_{21} \\ B_3\Delta_{223}C_2 & A_3 + B_3\Delta_{223}D_{22}C_3 & B_3\Delta_{223}C_2 & 0 & & B_3\Delta_{223}D_{21} \\ 0 & 0 & A & 0 & & B_1 \\ 0 & 0 & 0 & A & & B_1 \\ \hdashline \\ C_1 + D_{12}\Delta_{322}D_3C_2 & D_{12}\Delta_{322}C_3 & D_{12}\Delta_{322}D_3C_2 & C_1 & & D_{11} + D_{12}\Delta_{322}D_3D_{21} \end{array} \right].$$

$$(2.6\text{-}20)$$

Using the following similarity transformation the above yields *Case 1* of Table 2.6-4.

2.6. Aspects of System Theory

$$T = \begin{bmatrix} I & 0 & -I & 0 \\ 0 & I & 0 & 0 \\ 0 & 0 & I & 0 \\ 0 & 0 & I & I \end{bmatrix}, \quad T^{-1} = \begin{bmatrix} I & 0 & I & 0 \\ 0 & I & 0 & 0 \\ 0 & 0 & I & 0 \\ 0 & 0 & -I & I \end{bmatrix}. \quad (2.6\text{-}21)$$

∎

The analysis for *Case 2* in Table 2.6-4 is exactly the same as above.

2.6.3. Classes of Signals

2.6.3.1. Introduction

The word signal to all intents and purposes is synonymous with the corresponding function, however, one implied distinction between these two is that signal has been physically generated. Signals may not be easily presentable or classified by mathematical forms. For instance, an unknown disturbance acting on a system is an example of a signal that we need to classify in order to structure our studies, however, any inaccurate classification of this signal puts our problem in a different category for analyses. Thus we must always justify these mathematical classifications for signals. When that classification is done, we recall from § 2.5.4 the process of constructing several complete mathematical spaces, and from § 2.5.7 the definitions for classes of L^p- and H^p-functions which *form* [complete] mathematical spaces. Correspondingly, herein we classify signals according to the preceding spaces, and without repeating various requirements which are discussed in § 2.5.4 and § 2.5.7. In this regard, we only modify Definition 2.5-4, in order to make it more amenable to the stability definition for a *DCLTIS* which is described in § 5.1.1. We should also point out that any sensitivity measure is a signal and this signal in its many facets is, in our opinion, synonymous with the notion of distance, and thus we propose to state that *sensitivity is a distance*. A distance between an actual and the *desired* situations. Therefore when incorporating such signals in design and/or optimization, we should apply appropriate metrics to measure and/or to compensate them, or otherwise no definitive conclusions can be made from the underlying sensitivity analysis. This choice of a proper metric is particularly an important issue in the H^∞-sensitivity minimization.

2.6.3.2. Time – Domain Signals

First we consider scalar-valued signals. All subsequent time-domain signals are assumed to be members of class for L^p-functions, and as indicated in § 2.5.7 they form complete mathematical spaces. The *set* of most commonly-used signals in control theory belongs to this class of functions. Starting with an absolute-integrable function, this is a member of class for L^1-functions. Certain results for systems with this class of signals are reported in § 6.12.*K2*, and this is a new area of active research. The next group of widely-used signals are the square-integrable functions which are members of class for L^2-functions. This so-called 2-norm corresponds to the square root of the energy level contained in a signal. Most results in control theory are stemming from this class of functions. Another group of signals in this set is that with a bounded magnitude which this group belongs to class for L^∞-functions. An example of the L^∞-functions that is of "the daily use" in control theory is the step function which is a bounded function on R^+, and clearly this step function does not belong to other classes for L^p-functions because their corresponding p-norms are not bounded. A closely related example of this type signal is the saturation curve. On the other hand, an exponentially-decaying sinusoidal signal becomes a member of classes for L^1, L^2, and L^∞-functions. There are other signals belonging to the class for L^p-functions which are of interest, for instance, the signals which are members of $L^1 \cap L^2$.

Second we consider matrix-valued signals. This is particularly interesting because sensitivity measures are often expressed in matrix forms. Suppose that for the sake of generality we choose f(t) ∈ $C^{m \times n}(t)$, an m×n matrix whose entries are possibly complex-valued functions of time, then the following inner-product space forms a Hilbert space.

$$<f, g> \triangleq \int_{-\infty}^{\infty} \text{Tr}[f^*(t)g(t)]\, dt. \qquad (2.6\text{-}22)$$

Here f ∈ L^2 in the above sense. In the engineering literature this space is shown by $L^2(-\infty, \infty)$. We can partition this Hilbert space into two subspaces *corresponding* to $L^2_+[0, \infty)$ that represents all signals in L^2 which are zero for t < 0 *(causal functions)*. Its orthogonal complement is shown by $L^2_-(-\infty, 0]$ that represents all signals in L^2 which are zero for t > 0 *(anticausal functions)*. Thus in time-domain representation we have $L^2 = L^2_+ \oplus L^2_-$.

Here we may again define the L^∞-functions (or signals) to consist of all essentially-bounded functions whose L_∞-norms are finite.

2.6.3.3. Frequency – Domain Signals

First of all we assume that all subsequent frequency-domain signals are members of class for H^p-functions which form [complete] mathematical spaces also called the Hardy spaces. Secondly, we replace the unit disc in Definition 2.5-4 with the corresponding s-plane and use the jω–axis as the critical boundary set. The modification for the remaining part of that definition is straightforward as elucidated below for H^2- and H^∞-spaces. However, we also recognize that the square-integrable time- and the frequency-domain signals are equivalent due to the Hilbert space isomorphism *via* the Laplace or Fourier transform and the Paley-Wiener theorem.

2.6. Aspects of System Theory

Suppose that $F(j\omega) \in C^{m \times n}(j\omega)$ is the Fourier transform of $f(t) \in L^2$, then the following inner-product space forms a Hilbert space that is shown by H_0^2, where the subscript "0" is for $\sigma = 0$ or the $j\omega$–axis.

$$<F, G> \triangleq \frac{1}{2\pi} \int_{-\infty}^{\infty} \text{Tr}[F^*(j\omega)G(j\omega)] \, d\omega. \tag{2.6-23}$$

Similarly we can define a Hilbert space H_+^2 consisting of all functions $F(s) \in C^{m \times n}(s)$, $s = \sigma + j\omega$, that are analytic on the open *right*-hand side of the s-plane (or for $\text{Re}(s) = \sigma > 0$) (no right half-plane poles) and for them

$$\sup_{\sigma > 0} \left\{ \int_{-\infty}^{\infty} \text{Tr}[F^*(\sigma + j\omega) F(\sigma + j\omega)] \, d\omega \right\} < \infty. \tag{2.6-24}$$

When $F(s)$ is a rational transfer matrix this property is the same as saying that $F(s)$ is a *stable* transfer matrix (no right half-plane poles), with $\lim_{\omega \to \infty} F(j\omega) = 0$.

Also we can define a Hilbert space H_-^2 which consists of all functions $F(s) \in C^{m \times n}(s)$, $s = \sigma + j\omega$, that are analytic on the open *left*-hand side of the s-plane (or for $\text{Re}(s) = \sigma < 0$) (no left half-plane poles) and for them

$$\sup_{\sigma < 0} \left\{ \int_{-\infty}^{\infty} \text{Tr}[F^*(\sigma + j\omega) F(\sigma + j\omega)] \, d\omega \right\} < \infty. \tag{2.6-25}$$

Here also when $F(s)$ is a rational transfer matrix this property is the same as saying that $F(s)$ represents an *antistable* transfer matrix with $\lim_{\omega \to \infty} F(j\omega) = 0$.

As stated before the square-integrable time- and frequency-domain signals are equivalent due to the Hilbert space isomorphism. Using the Fourier transform we can show that in fact L_+^2 maps *onto* H_+^2 and L_-^2 maps *onto* H_-^2, while these mappings are also *(1–1)* – bijective mappings.

In the engineering literature instead of H_+^2, H_0^2 and H_-^2; H^2 and H_l^2 are introduced to consist of square-integrable functions, respectively, on the imaginary axis with analytic continuation into the right- and the left-half plane.

The essentially-bounded matrices (bounded except on a set of measure zero) $F(j\omega)$ (or functions with bounded gain on the $j\omega$–axis) has a norm

$$\| F \|_\infty = \operatorname*{ess\,sup}_{\omega} \sigma_{\max}[F(j\omega)] < \infty. \tag{2.6-26}$$

Now, the definition for H^∞–spaces becomes the set of all complex-valued matrices $F(s)$ which are analytic and bounded when $\text{Re}(s) = \sigma > 0$, in the above sense (or in the H^∞–sense).

$$\| F \|_\infty = \sigma_{max}[F(s)] < \infty, \qquad (2.6\text{-}27)$$

for all s such that $Re(s) = \sigma > 0$. Here the $\lim_{\omega \to \infty} \| F \|_\infty$ is a nonzero finite value. This definition becomes more simpler than before when $F(s)$ is a real, rational, and stable matrix (no right-half plane poles), then by the maximum modules theorem we can replace s in (2.6-27) by $j\omega$.

2.6.4. Computation of Transfer Matrix Norms

In § 2.5.9 we discuss the computation of norms for a vector or matrix. Here we study the same computation primarily for a frequency-dependent matrix that stems from a linear time-invariant system. This computation has application in the H^∞-sensitivity minimization, since transfer matrix in such a system represents, in a generic sense, certain sensitivity measure for that system (cf., Sections 6.2 and 6.9). For the time being, however, we look at this norm computation only as a general system problem, but due to the scope of this computation we only consider the second- and the infinite-norm of the corresponding transfer matrices. The time- or frequency-domain computation always becomes evident from the context.

2.6.4.1. The H_2-Norm

Since we are studying a linear time-invariant system, its corresponding transfer matrix $G(s)$ is a rational matrix given by (2.6-3) or equivalently given by (2.6-4). The H_2-norm of this transfer matrix is

$$<G,G> = \| G \|_2^2 \triangleq \frac{1}{2\pi} \int_{-\infty}^{\infty} \text{Tr}[G^*(j\omega)G(j\omega)]\, d\omega$$

$$\triangleq \frac{1}{2\pi} \int_{-\infty}^{\infty} \| G(j\omega) \|_F^2\, d\omega, \qquad (2.6\text{-}28)$$

the Frobenius norm of G, because for any complex matrix G, $\text{Tr}(G^*G) = \| G \|_F^2$. As established before, we also have $<G,G> = <g,g>$, where $g(t)$ (the Laplace inverse of $G(s)$), is the impulse response of the system. Thus at the moment

$$<G,G> = \| G \|_2^2 = \int_{-\infty}^{\infty} \text{Tr}[g^T(t)\, g(t)]\, dt. \qquad (2.6\text{-}29)$$

However, in order to have a bounded H_2-norm we definitely must restrict our transfer matrices to cases where in (2.6-3) (or (2.6-4)) $D \equiv 0$, and A is stable (cf., § 5.1.1.), otherwise (2.6-29) does not converge. Thus under these conditions $g(t) = Ce^{At}B$ and substituting this $g(t)$ into (2.6-29) yields

2.6. Aspects of System Theory

$$<G, G> = Tr[B^T \int_{-\infty}^{\infty} e^{A^T t} C^T C e^{At} dt\, B] \triangleq Tr(B^T L_o B)$$

$$= Tr[C \int_{-\infty}^{\infty} e^{At} BB^T e^{A^T t} dt\, C^T] \triangleq Tr(C L_c C^T). \quad (2.6\text{-}30)$$

Here L_o and L_c represent the observability and controllability Gramians matrices, respectively. For more information on these matrices the reader may refer to [C. 34]. (In § 5.1.1 we point out that a matrix of the *type* L_o or L_c represents a Lyapunov equation and thus can be easily computed for any given system, provided that A is stable, cf. (6.9-18).) Furthermore the (2.6-30) is valid if and only if the system is both completely observable and completely controllable. This norm can be computed exactly.

∎

The preceding L_o and L_c depend on the state-space representation of the system, however, their product is invariant under any state-space transformations and that product results in the next definition which represents the fundamental invariants related to both gain and complexity of a linear system

DEFINITION 2.6-1 (Hankel Operator [6H4. 5]): The Hankel singular values of G(s), when A is completely stable (cf., § 5.1.1) are defined as $\sigma_i[G(s)] \triangleq \{\lambda_i(L_o L_c)\}^{1/2}$, where $\lambda_i(\cdot)$ refers to the ith-eigenvalue (or spectral radius if needed) of $(\cdot)$, and by convention $\sigma_i(\cdot) \geq \sigma_{i+1}(\cdot)$. The Hankel norm (or operator) refers to $\sigma_1(\cdot)$.

2.6.4.2. The H_∞–Norm

Given a linear time-invariant system that is represented by a transfer matrix $G(s) \triangleq (A, B, C, D)$, the H_∞–norm of G(s) is computed from

$$\| G \|_\infty = \text{ess sup}_\omega \sigma_{max}[G(j\omega)]. \quad (2.6\text{-}31)$$

One interpretation of this norm is as follows. Suppose that G(s) is a scalar-valued, analytic and bounded in $Re(s) = \sigma > 0$, and is continuous on the $j\omega$–axis, then the $\| G \|_\infty$ represents the distance from the origin in the s-plane to the farthest point on its Nyquist plot [6H2. 12]. This norm can also be interpreted as the maximum transfer matrix gain G(s) on the $j\omega$–axis and the maximum induced norm when input is an L^2–function, i.e., $\| G \|_\infty = \sup_{u \neq 0} \{ \| Gu \|_2 / \| u \|_2 \}$ (cf., § 2.5.9(v)'). Here $\| u \|_2$ corresponds to the input energy and the $\| Gu \|_2$ corresponds to the output energy. Thus the $\| G \|_\infty$ represents the maximum gain in transferring energy from input to output of a system.

∎

To compute this norm we may plot $\|G\|_\infty$ versus ω and experimentally determine its maximum value. That method, however, is not efficient and instead we review the following method which is based on the properties of a Hamiltonian matrix.

DEFINITION 2.6-2 (Hamiltonian Matrix): A Hamiltonian matrix H is a $2n \times 2n$ real matrix with the following block structure and properties

$$H \triangleq \begin{bmatrix} A & P \\ Q & -A^T \end{bmatrix}, \quad P = P^T \in R^{n \times n}, \quad Q = Q^T \in R^{n \times n}. \quad (2.6\text{-}32)$$

Here for each eigenvalue λ_i of H, there is another one with the value $-\lambda_i$. ∎

We now present the main theorem for computing the $\|G\|_\infty$.

THEOREM 2.6-1 [6H5. 4 & 5]: Given $G(s) \triangleq (A, B, C, D)$, let A be such that all its eigenvalues have negative real-parts; choose $\gamma > \sigma_{\max}(D)$; and define

$$H_\gamma \triangleq \begin{bmatrix} A - BR^{-1}D^T C & -\gamma BR^{-1}B^T \\ \gamma C^T S^{-1} C & -A^T + C^T D R^{-1} B^T \end{bmatrix}, \quad (2.6\text{-}33)$$

where $R = (D^T D - \gamma^2 I)$ and $S = (DD^T - \gamma^2 I)$. Then $\|G\|_\infty \geq \gamma \iff H_\gamma$ has at least one imaginary eigenvalue. ∎

To state Theorem 2.6-1 in an algorithmic manner, if γ is chosen small, H_γ has one or more imaginary eigenvalues. If γ is chosen large, H_γ has no eigenvalue on the jω-axis. Thus by searching on γ – using a bisection search, this procedure converges to $\|G\|_\infty$ in a reasonable number of iterations and an acceptable degree of accuracy. A version of this algorithm is also reported in [6H3. 6]. Another numerically efficient method to compute $\|G\|_\infty$ is reported in [6H5. 31]. This norm can only be computed approximately.

2.7. OVERVIEW OF GENERAL SENSITIVITY MINIMIZATION

2.7.1. Introduction

This section is intended to complement Section 2.3 by discussing a few general synthesis problems and in particular from the H^∞-sensitivity minimization point of view. Since description of the H^∞-sensitivity measure requires some preliminary analysis, we postpone specific discussions on this topic to Section 6.9, where an updated and advanced version of this methodology, based on the state-space representation of the underlying system, is presented. We also note that the

2.7. Overview of General Sensitivity Minimization

H^∞–sensitivity optimization is *primarily* developed for a finite-dimensional, linear, time-invariant system, which is modeled by its transfer matrix in the frequency domain. Since, an extended frequency-domain study of this methodology is outside the scope of this book, in the following we look only at certain conceptual issues regarding the synthesis and without getting into any specific analytical solutions, in order to point out the main thrust of the H^∞–sensitivity minimization and its relation with the methods presented in this book. In passing, however, we should also note that the H^∞–optimization approaches are being applied to a host of other control problems and as we see from the literature (cf., § 6.12.*H6*) the list of such applications is expanding.

■

At this juncture we should point out that sensitivity synthesis and in its classical sense, refers to the design of a compensator which minimizes or reduces the undesirable effects of system-parameter variations, but the aim of H^∞–sensitivity minimization is to cope primarily with undesirable effects of exogenous signals which are acting on the system (such as disturbances, various noises, and reference signals). In the H^∞–sensitivity minimization approaches *currently* the sensitivity measure is specified in terms of a given transfer matrix and no longer that specification is of a choice, while in the classical sense we have ample opportunities to choose different forms of sensitivity measures pertaining to different desirable system performances. In other words, at the moment it is not easily possible to minimize the H_∞–norm of any arbitrary sensitivity measure and use that information in design.

2.7.2. The Tracking and Stabilization Problems

Tracking and stabilization problems are cited in the literature as two standard [analysis and/or synthesis] problems [6*H2*. 13], which studying them help to initiate a review of sensitivity problems, in particular, those which are being successfully treated using the H^∞–sensitivity approach. In this regard, we first state the following analysis and/or synthesis problems from the H^∞–sensitivity point of view. In § 2.7.3 we make a few additional remarks on these issues.

The first standard problem is the *tracking* problem which is a plant whose output tracks a given reference signal, and the objective is to maintain the amplitude of this output bounded, throughout the desirable and operational frequency range. Here, if for the sake of discussion we assume that the variables are all scalar and the reference signal is sinusoidal with amplitude one, then the amplitude of output signal can be approximated by the H_∞–norm of the transfer function between output and reference signal. Thus the performance index becomes the H_∞–norm of this transfer function and that is one example where such a norm appears naturally. If we further parameterize this plant, using a host of methodologies, then we can minimize the H_∞–norm of output signal with respect to this parameter and that optimization or synthesis problem becomes one application of the H^∞-optimization methodology. We can also put a frequency-dependent weight on this transfer function – in our terminology a frequency-dependent auxiliary parameter – and find an optimal parameter, based on the optimization of frequency weighted H_∞-norm of the transfer function. ∎

The second standard problem [analysis and/or synthesis] which we cite from the literature is called the *stabilization* problem [6H2. 13]. Consider the plant shown in *Case 3* of Table 2.6-3 with $G_1(s)$ as plant and $G_2(s)$ as its feedback compensator. Suppose that the plant transfer matrix $G_1(s)$ is subject to some *additive* uncertainty, which is assumed to be a real-rational transfer matrix in s and is shown by $\Delta G_1(s)$, i.e., in reality the plant transfer matrix is $G_1(s) + \Delta G_1(s)$. Then the question which has been often raised is "how large" $\Delta G_1(s)$ can get before the system becomes unstable? (By unstable we mean all poles of the corresponding transfer matrix move to the right-half plane.) Here we *may* define the meaning of "how large" by the H_∞-norm of $\Delta G_1(s)$ and answer this question accordingly. In this case by a sequence of loop transformations the $\Delta G_1(s)$ is put in the feedforward block and the remaining terms are grouped in the feedback block of the associated configuration. The preceding question is then restated as follows. Find the class of all compensators such that the set of all "disturbed" plants remains stable for the "largest" $\Delta G_1(s)$. ∎

The preceding two examples and many more similar cases can be defined where the control problem has been parameterized and the H_∞-norm has been used to set up an optimization problem in order to find accordingly the optimal controller gain or parameter. We again emphasize that this method is used for both analysis and synthesis and when the system is subject to certain exogenous signals. Since we have not set up all the necessary background information to answer explicitly this type of optimization problems, we postpone our analysis to Section 6.9. However, in the most generic sense an H^∞-optimization (or synthesis) problem currently refers to the following situation. Given a finite-dimensional, linear, time-invariant system described by its transfer matrix G(s), suppose that a particular structure as that shown in *Case 1* (for the controller gain K(s)) or *Case 2* (for the uncertainty $\Delta(s)$) of Table 2.6-4 is specified. Now, choose, for instance, K(s) such that the corresponding transfer matrix G(K, s) is: *(i)* stable (no right half-plane poles); and *(ii)* the $\| G(K, s) \|_\infty$ is minimum. These criteria together with other considerations such as computational efficiency, the hardware realizability of the final controller, and the ultimate cost constitute our bases for acceptability of a particular design algorithm stemming from this methodology.

2.7.3. The Model Matching Problem

This is perhaps the most natural way of setting up a sensitivity minimization or synthesis problem as described next. Given a plant represented by its transfer matrix G(s), find a compensation, or a frequency-shaping, policy represented by $G_c(s)$, such that the compensated transfer matrix (an appropriate combination of G(s) and $G_c(s)$) follows or matches a desired model M(s). This problem can be formulated in a number of ways and the compensator may actually be more than one which are inserted in a number of places in the overall system. In an actual application we must also elaborate on the *mechanism* of this match and the procedure for its compensation. For instance, we may choose two compensators $G_{c1}(s)$ and $G_{c2}(s)$ and place them before and after G(s) and look for a set of conditions in order that $G_{c1}(s)G(s)G_{c2}(s)$ matches the desired model M(s) at the steady state, *per se*. For the time being we note that this type of problems concerns with the *distance* between two systems, and if, for instance, we use the H_∞-norm to *interpret* the underlying distance, then we can mathematically formulate the problem as follows. Given a real-rational transfer matrix M(s) which belongs to the H^∞-space, find another transfer matrix G(s) (herein a compensated plant) of the same class, such that $\| M(s) - G(s) \|_\infty$ is minimized, or is strictly less than a specified positive number, while the overall system remains stable (no right half-plane poles). This distance minimization between two matrices is called a Nehari problem. The procedure and solution that is proposed in § 6.9.3 for minimizing the H_∞-norm of a transfer matrix is in spirit for the same purpose. ∎

In our final thought, we recall that most synthesis results which have been developed in the H^∞-optimization methodology are intended to cope with the exogenous signals. When there are parametric uncertainties – the kind of the problems which we study in this book – and in particular, the synthesis algorithms in Sections 6.2 to 6.8, 6.10 and 6.11, as well as the explicit robust stabilization procedures of Chapter Five – it seems the corresponding H^∞-optimization methodologies are considerably more involved than those presented herein. As far as modeling the underlying system by its transfer matrix / state-space representation is concerned, we can utilize many of the results developed in this book for the same class of problems, in order to answer the stabilization, tracking, and model matching problems quite efficiently. Thus we conclude that additional studies are merited in order to merge these different approaches and that is the essence of our thought on these matters.

2.8. CONCLUDING REMARKS

One of the major issues in a comprehensive sensitivity analysis is to define a meaningful measure for system sensitivity. This task is not trivial. Certainly, we cannot claim that a particular sensitivity measure is universally acceptable for all systems and all applications. On the other hand, we shall not sit idle because of this difficulty. Therefore we select a few different sensitivity measures to study a number of problems and procedures. Perhaps it is always advisable to look into several different sensitivity measures, and to apply each and every one of them to a given problem, in order to study these measures and their behaviors, before selecting *the* sensitivity measure for a system. For example, in a 1970's (still so true) comprehensive research study that is reported by Harvey *et al.* [1*B*. 15], different [then] existing sensitivity measures are applied to the model of a C5-Galaxy airplane to find out which one of them is superior to the other ones, if any. This study shows that to select an appropriate sensitivity measure is a serious matter, and we must have some insights about the system before we can choose a proper sensitivity measure for that system. Nevertheless, the trajectory- or state-sensitivity function appears to be a reasonable choice for many dynamical systems and that is mostly our measure of system sensitivity in this study. But in general, the main aspects of sensitivity analysis and synthesis can be summarized as follows:

(a) A meaningful sensitivity measure according to the system operation and its intended performance must be selected;

(b) This measure must be constructed (on-line or off-line) for the purpose of analysis;

(c) The invariance property of this measure, if any, must be established; and

(d) Finally, this measure must be considered in the design or synthesis algorithms to develop a robust dynamical system which can stand against some degree of variations in system components and/or its operations.

As is evident from this very broad menu and because there are literally infinitely many possibilities regarding the choices for sensitivity measures and subsequently there are that many different ways to optimize them, we conclude that the utilization of any sensitivity consideration (analysis and/or synthesis) is a very expensive proposition which requires both theoretical and practical experiences. We must judiciously make decision on how to present system-sensitivity measure and how to balance a design choice versus the cost of maintaining its corresponding sensitivity measure. Despite these conflicting issues, we plan to study this subject in order to sort out some of our options as best as we can. In particular, we maintain a path for our research which by itself is on the periphery of these issues. Thus the present book is prepared to address these considerations as succinctly as possible.

2.9. REFERENCES
[A] General Mathematical References

[1] N.I. Akhiezer and I.M. Glazman, *Theory of Linear Operators in Hibert Space*, Two Volumes. [Translated from Russian by M. Nestell.] New York: Frederick Ungar Pub. Co., second printings, Volume I, 1966, Volume II, 1963.

[2] C. Caratheodory, *Algebraic Theory of Measure and Integration*. [Translated from German by F.E.J. Linton.] New York: Chelsea, 1986.

[3] E.A. Coddington and N. Levinson, *Theory of Ordinary Differential Equations*. New York: McGraw-Hill, 1955. (Reprint, R.E. Krieger, 1984).

[4] R. Courant and D. Hilbert, *Methods of Mathematical Physics*, Two Volumes. New York: Wiley (Classics Edition), 1989.

[5] N. Dunford and J.T. Schwartz, *Linear Operators*, Three Volumes. Part I: *General Theory*. Part II: *Spectral Theory*. Part III: *Spectral Operators*. New York: Wiley (Classics Edition), 1988.

[6] P.L. Duren, *Theory of H^p Spaces*. New York: Academic Press, 1970.

[7] J.N. Franklin, *Matrix Theory*. Englewood Cliffs, NJ: Prentice-Hall, 1968.

[8] F.R. Gantmacher, *The Theory of Matrices I*. New York: Chelsea, 1959.

[9] P.R. Garabedian, *Partial Differential Equations*. New York: Wiley, 1964. (Reprint, Chelsea, 1986.)

[10] G.H. Hardy, J.E. Littlewood, and G. Polya, *Inequalities*. London: Cambridge University Press, 1934.

[11] T. Hawkins, *Lebesgue's Theory of Integration, its origin and development*. New York: Chelsea, 1979.

[12] H. Hochstadt, *Integral Equations*. New York: Wiley (Classics Edition), 1989. The 1973 edition is reviewed by M. Yanowitch, *SIAM R.*, vol. 18, no. 4, p. 782, 1976.

[13] A.S. Householder, *The Theory of Matrices in Numerical Analysis*. New York: Dover, 1975.

[14] J.L. Kelley, *General Topology*. Princeton, NJ: D. Van Nostrand, 1955.

[15] S. MacLane and G. Birkhoff, *Algebra*. New York: Macmillan, 1967. (2nd edition, 1979.)

[16] W.S. Massey, *A Basic Course in Algebraic Topology*. New York: Springer – Verlag, 1991.

[17] J.R. Munkres, *Topology, a first course*. Englewood Cliffs, NJ: Prentice-Hall, 1975.

[18] R.E.A.C. Paley and N. Wiener, *Fourier Transforms in the Complex Domain*. Providence, RI: American Mathematical Society, 1934. (Fifth printing, 1964.)

[19] C.E. Pearson, Ed., *Handbook of Applied Mathematics, selected results and methods*. New York: Van Nostrand, 1974

[20] M. Reed and B. Simon, *Methods of Modern Mathematical Physics*. Volume I: *Functional Analysis*. Volume II: *Fourier Analysis, Self-Adjointness*. New York: Academic Press, Volume I, 1972, Volume II, 1975.

[21] K. Rektorys, Ed., *Survey of Applicable Mathematics*. Cambridge, MA: MIT Press, 1969.

[22] F. Riesz and B. Sz.-Nagy, *Functional Analysis*. [Translated from the 2nd French edition by L.F. Boron.] New York: Dover, 1990.

[23] W. Rudin, *Principles of Mathematical Analysis*. New York: McGraw-Hill, 1964.

[24] _____, *Real and Complex Analysis*. New York: McGraw-Hill, 1974.

[25] _____, *Fourier Analysis on Groups*. New York: Wiley (Classics Edition), 1990.

[26] _____, *Functioanl Analysis*. New York: McGraw-Hill, 2nd edition, 1991.

[27] L. Sirovich, *Techniques of Asymptotic Analysis*. New York: Springer – Verlag, 1971.

[28] I.N. Sneddon, *The Use of Integral Transforms*. New York: McGraw-Hill, 1972. Reviewed by W.L. Perry, *JFI*, vol. 300, no. 1, p. 79, 1975.

[29] E.H. Spanier, *Algebraic Topology*. New York: McGraw-Hill, 1966. [This entry is suggested by Prof. J. Wood of the University of Illinois at Chicago.]

[30] I. Stakgold, *Boundary Value Problems of Mathematical Physics*, Two Volumes. New York: Macmillan, Volume I, 6th printing, 1972, Volume II, 2nd printing, 1971. The 1968 edition of Volume II is reviewed by R.B. Guenther, *SIAM R.*, vol. 11, no. 2, p. 289, 1969.

[31] _____, *Green's Functions and Boundary Value Problems*. New York: Wiley, 1979. Reviewed by W. Kaplan, *SIAM R.*, vol. 23, no. 1, pp. 117-118, 1981.

[32] B. Van Der Pol and H. Bremmer, *Operational Calculus, based on the two-sided Laplace integral*. New York: Chelsea, 1987.

[33] A.J. Weir, *Lebesgue Integration & Measure*. Cambridge, England: C.U. Press, 1973.

[34] A.H. Zemanian, *Distribution Theory and Transform Analysis*. New York: Dover, 1987.

[35] _____, *Generalized Integral Transformations*. New York: Dover, 1987. The 1968 edition is reviewed by L. Weiss, *JFI*, vol. 288, no. 4, pp. 329-330, 1969, and by J. Korevaar, *SIAM R.*, vol. 15, no. 1, pp. 232-234, 1973.

[B] General References on System Theory

[1] B.D.O. Anderson, "A system theory criterion for positive real matrices," *SIAM J. Control*, vol. 5, no. 2, pp. 171-182, 1967.

[2] B.D.O. Anderson, M.A. Arbib, and E. Manes, "Foundations of system theory: Multidecomposable systems," *JFI*, vol. 301, no. 6, pp. 497-508, 1976.

[3] Y.N. Andreev, "Algebraic methods of state space in linear object control theory (survey of foreign literature)," *ARC*, vol. 38, no. 3, pp. 305-342, March 1977.

[4] D.J. Bell, *Mathematics of Linear and Nonlinear Systems, for engineers and applied scientists*. New York: Oxford University Press, 1990.

[5] R.W. Brockett, "Poles, zeros, and feedback: State space interpretation," *IEEE-TAC*, vol. AC-10, pp. 129-135, April 1965.

[6] _____, *Finite-Dimensional Linear Systems*. New York: Wiley, 1970. Reviewed by M. Vidyasagar, *IEEE-TSMC*, vol. SMC-1, pp. 96-97, 1971, and in *JFI*, vol. 291, no. 1, pp. 83-84, 1971, and by D. Tabak, *JFI*, vol. 292, no. 5, p. 388, 1972, and by M. Sain, *IEEE-TAC*, vol. AC-17, no. 5, pp. 753-754, 1972.

[7] C.-T. Chen, *Linear System Theory and Design*. New York: Holt, Rinehart and Winston, 1984. The 1970 edition is reviewed by A.S. Morse, *IEEE-TAC*, vol. AC-17, no. 5, pp. 748-749, 1972, and the 1984 edition is reviewed by S. Barnett, *Automatica*, vol. 22, no. 3, pp. 385-386, 1986.

[8] C.-T. Chen and C.A. Desoer, "Controllability and observability of composite systems," *IEEE-TAC*, vol. AC-12, no. 4, pp. 402-409, Aug. 1967.

[9] F.G. Csaki, "Some notes on the inversion of confluent Vandermonde matrices," *IEEE-TAC*, vol. AC-20, no. 1, pp. 154-157, Feb. 1975.

[10] E.G. Gilbert, "Controllability and observability in multivariable control systems," *SIAM J. Contr.*, vol. 2, no. 1, pp. 128-151, 1963.

[11] I.C. Goknar, "Obtaining the inverse of the generalized Vandermonde matrix of the most general type," *IEEE-TAC*, vol. AC-18, no. 5, pp. 530-532, Oct. 1973.

2.9.B. General References on System Theory

[12] D.G. Luenberger, "Observers for multivariable systems," *IEEE-TAC*, vol. AC-11, no. 2, pp. 190-197, April 1966.

[13] T. Kailath, *Linear Systems*. Englewood Cliffs, NJ: Prentice-Hall, 1980. Reviewed by S.R. Liberty, *IEEE-TAC*, vol. AC-26, no. 3, pp. 804-805, 1981, and by J.C. Willems, *Automatica*, vol. 18, no. 4, pp. 497-499, 1982, and by S.H. Zak, *SIAM R.*, vol. 26, no. 1, pp. 132-135, 1984.

[14] W.R. Perkins and J.B. Cruz, Jr., *Engineering of Dynamic Systems*. New York: Wiley, 1969. Reviewed by M. Athans, *IEEE-TAC*, vol. AC-17, no. 1, pp. 185-186, 1976.

[15] W.A. Porter, *Modern Foundations of Systems Engineering*. New York: Macmillan, 1966. Reviewed by B.D.O. Anderson, *IEEE-TAC*, vol. AC-12, no. 5, pp. 636-637, 1967.

[16] H.H. Rosenbrock, *State-Space and Multivariable Theory*. New York: Wiley, 1970. Reviewed by W. Wolovich, *IEEE-TAC*, vol. AC-17, no. 4, pp. 583-584, 1972.

[17] J.C. Willems and S.K. Mitter, "Controllability observability, pole allocation and state reconstruction," *IEEE-TAC*, vol. AC-16, no. 6, pp. 582-595, Dec. 1971.

[18] W.M. Wonham and A.S. Morse, "Decoupling and pole assignment in linear multivariable systems: A geometric approach," *SIAM J. Contr.*, vol. 8, no. 1, pp. 1-18, Feb. 1970.

[19] D.C. Youla, "On the factorization of rational matrices," *IRE-TIT*, vol. IT-7, pp. 172-189, July 1961.

[20] L.A. Zadeh and C.A. Desoer, *Linear System Theory*. New York: McGraw-Hill, 1963.

[C] Representative References on:
Advanced System Theory, Optimal Control Systems,
Calculus of Variations, and Dynamic Programming

[1] F. Albrecht, *Topics in Control Theory*. Berlin: Springer – Verlag, 1968.

[2] B.D.O. Anderson and J.B. Moore, *Linear Optimal Control*. Englewood Cliffs, NJ: Prentice-Hall, 1971.

[3] _____, *Optimal Filtering*. Englewood Cliffs, NJ: Prentice-Hall, 1979. Reviewed by J.M. Mendel, *IEEE-TAC*, vol. AC-25, no. 3, pp. 615-616, 1980, and by M. Eslami, *IEEE-TSMC*, vol. SMC-12, no. 2, pp. 235-236, 1982.

[4] M. Aoki, *Optimization of Stochastic Systems*. New York: Academic Press 1967.

[5] M. Athans and P. Falb, *Optimal Control: an introduction to the theory and its applications*. New York: McGraw-Hill, 1966. Reviewed by S.J. Kahne and E.B. Lee, *IEEE-TAC*, vol. AC-12, no. 3, pp. 345-347, 1967, and by E. Kreindler, *JFI*, vol. 283, no. 4, pp. 350-351, 1967.

[6] D.S. Bayard and M. Eslami, *Stochastic Adaptive Control for General Dynamic Systems*. New York: Marcel Dekker, 1995, to appear.

[7] G.A. Bliss, *Lectures on the Calculus of Variations*. Chicago: The University of Chicago Press, eighth impression, 1968.

[8] A.E. Bryson and Y.C. Ho, *Applied Optimal Control, optimization, estimation, and control*. Waltham, MA: Blaisdell Pub. Comp., 1969. [New York: Halsted Press Book, a division of John Wiley and Sons (2nd edition), 1975.] The 1969 edition is reviewed by Y.Z. Tsypkin, *Automatica*, vol. 6, pp. 825-826, 1970, and by J.S. Meditch, *IEEE-TAC*, vol. AC-17, no. 1, pp. 186-188, 1972, and by P.-T. Hsu, *JFI*, vol. 293, no. 1, pp. 69-70, 1972.

[9] R.S. Bucy and P.D. Joseph, *Filtering for Stochastic Processes with Application to Guidance*. New York: Interscience, 1968. (Reprint, Chelsea, 1987.)

[10] S.S.L. Chang, *Synthesis of Optimum Control Systems*. New York: McGraw-Hill, 1961.

[11] F.L. Chernousko and A.A. Lyubushin, "Method of successive approximations for solution of optimal control problems," *JOCAM*, vol. 3, pp. 101-114, 1982.

[12] S.E. Dreyfus, *Dynamic Programming and the Calculus of Variations*. New York: Academic Press, 1965.

[13] S.E. Dreyfus and A.M. Law, *The Art and Theory of Dynamic Programming*. New York: Academic Press, 1977.

[14] L.E. Elsgoc, *Calculus of Variations*. New York: Pergamon Press, Ltd., 1961.

[15] A.A. Feldbaum, *Optimal Control Systems*. [Translated from Russian by A. Kraiman.] New York: Academic Press, 1965. Reviewed by J.C. Hsu, *IEEE-TAC*, vol. AC-13, no. 2, p. 224, 1968.

[16] W.H. Fleming and R.W. Rishel, *Deterministic and Stochastic Optimal Control*. New York: Springer – Verlag, 1975. Reviewed by A.N. Michel, *IEEE-TAC*, vol. AC-22, no. 6, pp. 997-998, 1977.

[17] R.V. Gamkrelidze, *Principle of Optimal Control Theory*. [Translated from Russian by K. Makowski. Edited by L.D. Berkovitz.] New York: Plenum Press, 1978.

[18] A. Gelb, *Applied Optimal Estimation*. Cambridge, MA: MIT Press, 1974.

[19] I.M. Gelfand and S.V. Fomin, *Calculus of Variations*. Englewood Cliffs, NJ: Prentice-Hall, 1963.

[20] J.C. Gille, M.J. Pelegrin, and P. Decauline, *Theorie et Technique des Asservissements*. Paris: Dunod, 1956.

[21] B.S. Goh, *A Theory of Second Variation in Optimal Control*. Berkeley, CA: The University of Calif., Division of Applied Mechanics, 1970.

[22] D. Graupe, *Identification of Systems*. New York: Van Nostrand Reinhold, 1972. (Second rev. edition, R.E. Krieger, 1976).

[23] H. Hermes and J.P. LaSalle, *Functional Analysis and Time Optimal Control*. New York: Academic Press 1969. Reviewed by D, Bushaw, *IEEE-TAC*, vol. AC-17, no. 1, pp. 189-190, 1972.

[24] R. Howard, *Dynamic Programming and Markov Processes*. New York: Wiley, 1960.

[25] R.E. Kalman, P.L. Falb, and M. Arbib, *Topics in Mathematical System Theory*. New York: McGraw-Hill, 1969. Reviewed by W. Kaplan, *SIAM R.*, vol. 12, no. 1, pp. 157-158, 1970, and by J.C. Willems, J.L. Massey, and W.M. Wonham, *IEEE-TAC*, vol. AC-17, no. 1, pp. 181-183, 1972.

[26] D.E. Kirk, *Optimal Control Theory, an introduction*. Englewood Cliffs, NJ: Prentice-Hall, 1970. Reviewed by W.S. Levine, *IEEE-TAC*, vol. AC-17, no. 3, p. 423, 1972.

[27] B.C. Kuo, *Discrete-Data Control Systems*. Englewood Cliffs, NJ: Prentice-Hall, 1970. Reviewed by J.B. Lewis, *IEEE-TAC*, vol. AC-17, no. 3, pp. 418-419, 1972.

[28] H. Kwakernaak and R. Sivan, *Linear Optimal Control Systems*. New York: Wiley, 1972. Reviewed by A.E. Pearson, *IEEE-TAC*, vol. AC-19, no. 5, pp. 631-632, 1974.

[29] _____, *Signals and Systems*. Englewood Cliffs, NJ: Prentice-Hall, 1990.

[30] E.B. Lee and L. Markus, *Foundations of Optimal Control Theory*. New York: Wiley, 1967. (Reprint, R.E. Kreiger, 1986). Reviewed by H. Hermes, *IEEE-TAC*, vol. AC-13, no. 2, pp. 222-223, 1968, and by R. Datko, *SIAM R.*, vol. 11, no. 1, pp. 93-95, 1969.

[31] D.G. Luenberger, *Optimization by Vector Space Methods*. New York: Wiley, 1969. Reviewed by R.W. Brockett and J.C. Willems, *IEEE-TAC*, vol. AC-15, no. 1, pp. 160-161, 1970, and by J.E. Falk, *SIAM R.*, vol. 12, no. 2, pp. 315-316, 1970.

[32] L.S. Pontryagin, V.G. Boltyanskii, R.V. Gamkrelidze, and E.F. Mishchenko, *The Mathematical Theory of Optimal Process*. [Translated from Russian by K.N. Trirogoff. Edited by L.W. Neustadt.] New York: Interscience Publishers, 1962.

2.9.C. Optimal Control, Calculus of Variations, Dynamic Programming

[33] D.L. Russell, *Topics in Ordinary Differential Equations*. The University of Wisconsin-Madison, Department of Mathematics, Unpublished lecture notes, Spring 1975.

[34] _____, *Mathematics of Finite-Dimensional Control System, theory and design*. New York: Marcel Dekker, 1979. Reviewed by H. Sagan, *SIAM R.*, vol. 23, no. 1, pp. 115-116, 1981, and by M. Eslami, *IEEE-TSMC* vol. SMC-14, no. 1, pp. 166-168, 1984.

[35] A.P. Sage and C.C. White, III, *Optimum Systems Control*. Englewood Cliffs, NJ: Prentice-Hall (2nd edition), 1977. Reviewed by M. Eslami, *JFI*, vol. 305, no. 1, pp. 57-58, Jan. 1978, and by G.M. Siouris, *IEEE-TSMC*, vol. SMC-9, no. 2, pp. 102-103, 1979.

[36] N.R. Sandell and M. Athans, *Modern Control Theory (Computer Manual)*. Center for Advanced Engineering Study: MIT, 1974.

[37] F.C. Schweppe, *Uncertain Dynamic Systems*. Englewood Cliffs, NJ: Prentice-Hall, 1973.

[38] H.L. Van Trees, *Synthesis of Optimum Nonlinear Control Systems*. Cambridge, MA: MIT Press, 1962.

[39] M. Vidyasagar, *Control System Synthesis: A Factorization Approach*. Cambridge, MA: MIT Press, 1985. Reviewed by B.F. Wyman, *IEEE-TAC*, vol. AC-31, no. 11, p. 1085, 1986, and by F.M. Callier, *Automatica*, vol. 22, no. 4, pp. 500-501, 1986.

[40] J.C. Willems, *The Analysis of Feedback Systems*. Cambridge, MA: MIT Press, 1971. Reviewed by R. Saeks, *IEEE-TAC*, vol. AC-17, no. 5, pp. 745-746, 1972.

[41] W.M. Wonham, *Linear Multivariable Control: A Geometric Approach*. New York: Springer – Verlag, 1974. (Third edition, 1985.) The 1974 edition is reviewed by J.B. Pearson, *IEEE-TAC*, vol. AC-22, no. 6, pp. 1000-1001, 1977.

[42] V. Zeidan, "First and second order sufficient conditions for optimal control and the calculus of variations," *AMO*, vol. 11, pp. 209-226, 1984.

[D] Certain Papers on Optimal Deterministic Control Systems

[1] M. Athans, "The matrix minimum principle," *IC*, vol. 11, pp. 592-606, 1968.

[2] J. Cullum, "Discrete approximations to continuous optimal control problems," *SIAM J. Contr.*, vol. 7, no. 1, pp. 32-49, Feb. 1969.

[3] J.W. Daniel, "The Ritz-Galerkin method for abstract optimal control problems," *SIAM J. Contr.*, vol. 11, no. 1, pp. 53-63, Feb. 1973.

[4] R.E. Kalman, "Contributions to the theory of optimal control," *BSMM*, vol. 5, pp. 102-119, 1960.

[5] _____, "When is a linear control system optimal?" *ASME-JBE*, Series D, pp. 51-60, March 1964.

[6] D.L. Kleinman and M. Athans, "The design of suboptimal linear time-varying systems," *IEEE-TAC*, vol. AC-13, pp. 150-159, April 1968.

[7] W.S. Levine, T.L. Johnson, and M. Athans, "Optimal limited state variable feedback controllers for linear systems," *IEEE-TAC*, vol. AC-16, no. 6, pp. 785-792, Dec. 1971.

[8] D.L. Lukes, "Optimal regulation of nonlinear dynamical systems," *SIAM J. Contr.*, vol. 7, no. 1, pp. 75-100, Feb. 1969.

[9] _____, "Equilibrium feedback control in linear games with quadratic costs," *SIAM J. Contr.*, vol. 9, no. 2, pp. 234-252, May 1971.

[10] L. Markus and E.B. Lee, "On the existence of optimal controls," *ASME-JBE*, pp. 13-22, March 1962.

[11] C. Marshal, "Second-order tests in optimization theories," (Survey paper) *JOTA*, vol. 15, no. 6, pp. 633-666, 1975.

[12] E.J. Messerli and E. Polak, "On second-order necessary conditions of optimality," *SIAM J. Contr.*, vol. 7, no. 2, pp. 272-291, May 1969.

[13] L.I. Rozonoer, "L.S. Pontryagin maximum principle in the theory of optimum systems. I," *ARC*, vol. 20, no. 10, pp. 1288-1302, 1959.

[14] _____, "L.S. Pontryagin's maximum principle in optimal system theory – II," *ARC*, vol. 20, no. 11, pp. 1405-1421, 1959.

[15] _____, "The maximum principle of L.S. Pontryagin in optimal-system theory. Part III," *ARC*, vol. 20, no. 12, pp. 1517-1532, 1959.[3]

[16] K. Spingarn, "Some numerical aspects of optimal control," *JFI*, vol. 289, pp. 351-359, May 1970.

[17] _____, "A comparison of numerical methods for solving optimal control problems," *IEEE-TAES*, vol. AES-7, no. 1, pp. 73-78, Jan. 1971.

[18] L.N. Volgin, "Diophantine polynomial calculus and its application to the solution of mathematical problems in control theory (a survey)," *SJAIS*, vol. 20, no. 1, pp. 40-49, 1987.

[19] W.R. Wakeland, "A study of weighting factors of the quadratic performance index," *JFI*, vol. 287, no. 2, pp. 101-113, Feb. 1969.

[E] Representative References on the Theory of Errors Perturbation, and Approximation Theory

[1] N.I. Achieser, *Theory of Approximation.* [Translated from Russian by C.J. Hyman.] New York: Dover, 1992.

[2] W.K. Bachmann, *Theorie des Erreurs et Compensation des Triangulations Aeriennes.* Lausanne: Impr. La Concorde, 1946.

[3] Y. Beers, *Introduction to the Theory of Error.* Cambridge, MA: Addison-Wesley, 1953.

[4] A. Bjerhammar, *Theory of Errors and Generalized Matrix Inverses.* Amsterdam: Elsevier Scientific Pub. Co., 1973.

[5] E.W. Cheney, *Introduction to Approximation Theory.* New York: Chelsea, 1982. The 1969 edition is reviewed by D. Chazan, *SIAM R.*, vol. 10, no. 3, pp. 393-394, 1968.

[6] A.A. Clifford, *Multivariate Error Analysis.* New York: Halsted Press (A Division of John Wiley and Sons), 1973.

[7] G. Cullmann, *Codes Detecteurs et Correcteurs d' Erreurs.* Paris: Dunod, 1967.

[8] P.J. Davis, *Interpolation & Approximation.* New York: Dover, 1975.

[9] V.F. Dem'yanov and V.N. Malozemov, *Introduction to MINIMAX.* [Translated from Russian by D. Louvish.] New York: Dover, 1990.

[10] U. Grenander and G. Szego, *Toeplitz Forms and Their Applications.* New York: Chelsea, 1984.

[11] H. Levy and F. Lessman, *Finite Difference Equations.* New York: Dover, 1992.

---

[3] Clearly this and the preceding article are the continuation of [13]. However, we have made no attempt to editorialize these matters in this book, in other words, we report the title and the author's initials as appeared in the actual paper in spite of some apparent inconsistency in the style. (In one incident no initials were given and in several occasions these were not complete.) Furthermore, the page numbers for each paper which are translated into English are given always relative to the translated edition.

[12] G.G. Lorentz, *Approximation of Functions*. New York: Chelsea, 1986.

[13] A.H. Nayfeh, *Perturbation Methods*. New York: Wiley, 1973. Reviewed by J.D. Cole, *SIAM R.*, vol. 18, no. 1, pp. 139-140, 1976, and by V. Singh, *IEEE-TSMC*, vol. SMC-8, no. 5, pp. 417-418, 1978, and by P.D. Usher, *JFI*, vol. 298, no. 3, p. 247, 1974, and by M.L. Rasmussen, *ibid.*, vol. 307, no. 2, p. 151, 1979.

[14] J.L. Walsh, *Interpolation and Approximation, by rational function in the complex domain*. Providence, RI: American Mathematical Society, Colloquium, vol. 20, 1935. (Boston: Spaulding-Moss Comp., third edition, 1960.)

CHAPTER THREE

SENSITIVITY FUNCTIONS GENERATION

3.1. INTRODUCTION

This chapter treating a deterministic, continuous, linear, time-invariant system *(DCLTIS)*, advances analytical expressions for sensitivity functions with distinctions between analysis for system-structural parameters (Definition 2.2-10) and that of system-physical parameters (Definition 2.2-11), followed by sensitivity functions generation in the frequency domain. Special distinction between sensitivity functions of open-loop and closed-loop systems, as well as reconstructible and unreconstructible systems should be cited. The concept of low-order sensitivity functions and that of complete simultaneity properties (for higher-order sensitivity functions) are discussed. Additionally, the concept of total-sensitivity functions *(TSF)* is introduced. The details of a system model are also included, and in the last part of this chapter sensitivity invariance is presented.

Before proceeding further, it is important to distinguish between certain classes of dynamical systems as the flow-chart shown in Table 3.1-1 indicates these distinctions.

The parameter-vector θ (either q – the structural parameter – or p – the physical parameter) is always assumed to be a real quantity, unless stated otherwise. Every component of $\theta \in R^r$ must be considered as one independent parameter. In Sections 2.3 and 2.4 the importance of sensitivity functions generation is established. The sensitivity functions reflect the geometric trend of variations in the coordinates of the state-space representation in a dynamical system subject to parameter variations [A. 10]. To generate the corresponding sensitivity functions is the first step toward utilization of sensitivity theory. Although a set of qualitatively general equations (for deriving the sensitivity functions) had been previously developed by Miller and Murray [A. 13], to generate these functions *efficiently*, and in particular for a *DCLTIS,* has been of an enormous interest among researchers in this field.

Depending on how the system parameters are considered (cf., Definitions 2.2-10 & 11), the corresponding sensitivity functions, for a system with a given mathematical model, must be generated accordingly. The methods developed herein are applicable to a differential equation representation of the system and can be extended to other system representations.

3.1. Introduction

Table 3.1-1. System Classifications Flow-Chart Diagram.

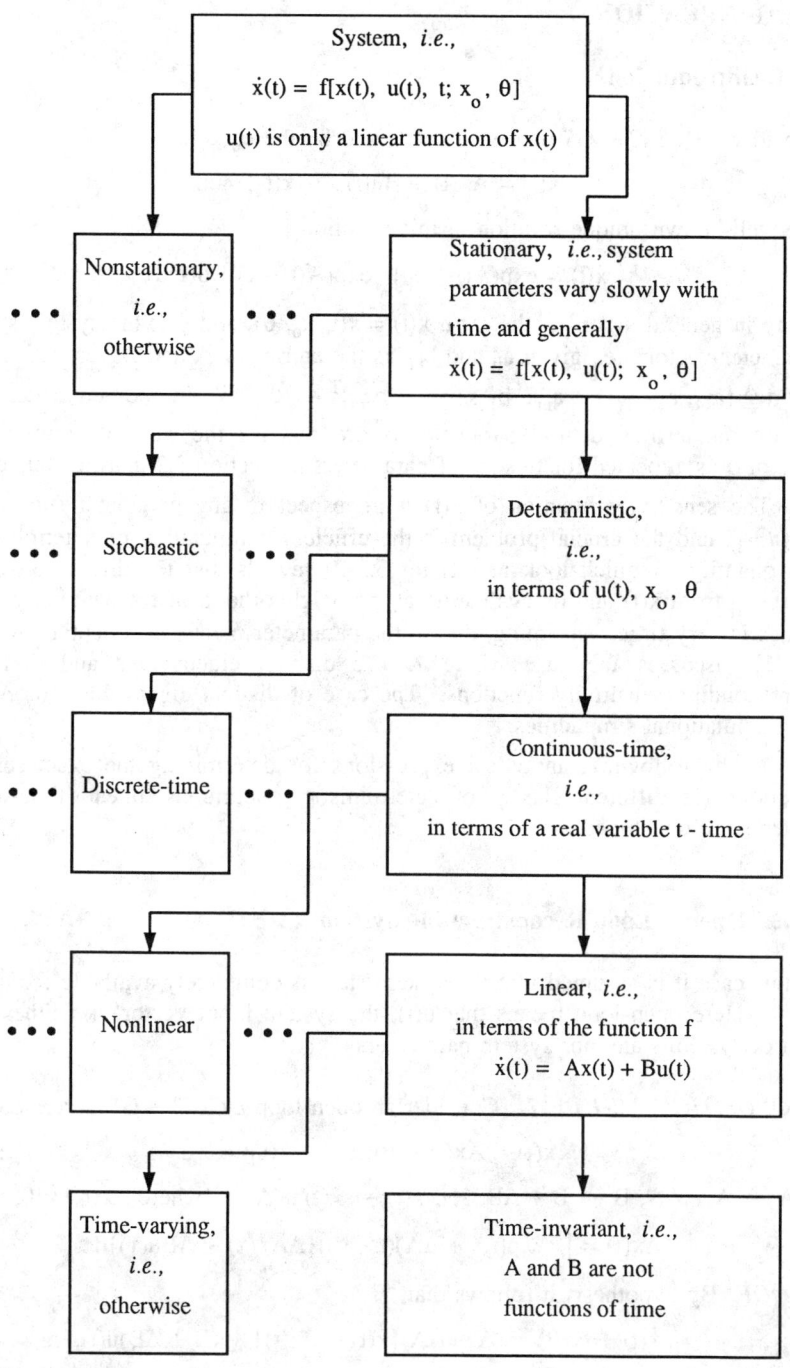

3.2. STRUCTURAL – PARAMETER SENSITIVITY FUNCTIONS GENERATION

3.2.1. Introduction

Consider a *DCLTI – GS* that is characterized by the equation

$$\dot{x}(t) = Ax(t) + Bu(t), \quad x(t_0) = x_0. \tag{3.2-1}$$

The well-known unique solution of this equation is

$$x(t) = \exp(At)x_0 + \int_{t_0}^{t} \exp[A(t - \tau)]Bu(\tau)d\tau, \tag{3.2-2}$$

where in general $x(t)$ is of the form $x(t) = x(t; x_0, \theta)$, and θ is the system-structural parameter vector, i.e., given a_{ij} and b_{kl} as the entries of A and B, respectively, then $\theta \equiv q \triangleq [a_{11}, a_{12}, \cdots, a_{nn}, b_{11}, \cdots, b_{nm}]^T \in R^{n(n+m)}$. In the remainder of this section the term structural parameter is dropped for the sake of simplicity. The symbol q as allocated for this set of parameters in Section 2.2 clarifies this choice.

The sensitivity function of $x(t)$ with respect to any parameter such as a_{ij} is $\partial x(t)/\partial a_{ij}$, and the crucial problem is the efficient computation or determination of this quantity. A quick look at $x(t)$ in (3.2-2) reveals that the direct derivation of $\partial x(t)/\partial a_{ij}$ from $x(t)$ and its evaluation at the neighborhood of the parameter nominal values is very time consuming, due to the parameter matrix exponential. Allwright [A. 1] discusses the case where A has distinct eigenvalues and derives the corresponding sensitivity functions. The case of distinct eigenvalues for A results in computational simplicities.

In the following, analytical expressions for determining the exact sensitivity functions for different classes of deterministic, continuous, linear, time-invariant systems are developed.

3.2.2. Open – Loop Reconstructible System

In this case it is assumed that the system state is completely available for measurement. Here open-loop means that $u(t)$, the system input vector, is neither a function of system state, nor system parameters.

PROPOSITION 3.2-1 [A. 7]: Consider an open-loop *DCLTI – GS* described by

$$\dot{x}(t) = Ax(t) + Bu(t), \quad x(t_0) = x_0. \tag{3.2-3}$$

If $A \to A + \Delta A$, $B \to B + \Delta B$, and $x(t) \to x(t) + \Delta x(t)$, where $\Delta x(t_0) = 0$, then

$$\Delta x(t) = \int_{t_0}^{t} \exp[(A + \Delta A)(t - \tau)]\{\Delta A x(\tau) + \Delta B u(\tau)\}d\tau. \tag{3.2-4}$$

PROOF: By hypothesis it follows that,

$$\dot{x}(t) + \Delta\dot{x}(t) = (A + \Delta A)\{x(t) + \Delta x(t)\} + (B + \Delta B)u(t). \tag{3.2-5}$$

3.2. Structural – Parameter Sensitivity Functions Generation

Subtracting (3.2-3) from (3.2-5) results in

$$\Delta\dot{x}(t) = (A + \Delta A)\Delta x(t) + \Delta A x(t) + \Delta B u(t), \qquad \Delta x(t_0) = 0. \qquad (3.2\text{-}6)$$

The solution to (3.2-6) considering $\Delta x(t_0) = 0$ is that of (3.2-4).

∎

A major objective of our study is to reduce certain norm of $\Delta x(t)$, in particular when Δx is caused by parameter variations. Several algorithms to accomplish this goal are developed in Chapter Six.

For convenience and without any loss of generality, we assume that the uncertainty or parameter variations are in effect after $t \geq 0$. (Or we let $t_0 = 0$ in (3.2-1)). The system may have been in operation for $t < 0$, but it is assumed that the system is following its nominal trajectory and up to that time. Knowing that at $t = 0$, $x(0) = x_0$, the deviation from nominal trajectory can be determined with no concern for its behavior before $t = 0$. This assumption is reasonable as an immediate consequence of the continuous dependence of the solution for a linear differential equation on its coefficients.

From the previous analysis it is clear that (3.2-4) holds for "small" as well as "large" changes in the system-parameter matrices (also refer to § 3.7.G, in particular [G. 6] for additional information on large-parameter variations). For the case of several consecutive large changes, the system sensitivity for each time interval, i.e., $[t_0, t] = [t_0, t_1) + [t_1, t_2) + \cdots + [t_n, t]$, has to be analyzed separately, but analogously to (3.2-4).

Letting $t_0 = 0$, (3.2-4) becomes

$$\Delta x(t) = \int_0^t \exp[(A + \Delta A)(t - \tau)]\{\Delta A x(\tau) + \Delta B u(\tau)\} d\tau. \qquad (3.2\text{-}7)$$

From (3.2-7) it is easy to show that if $\Delta B = 0$ and

$$\Delta A = \begin{bmatrix} 0 & 0 & 0 \\ 0 & \Delta a_{ij} & 0 \\ 0 & 0 & 0 \end{bmatrix}, \qquad (3.2\text{-}8)$$

then

$$\frac{\Delta x(t)}{\Delta a_{ij}} = \int_0^t \exp[(A + \Delta A)(t - \tau)] \begin{bmatrix} 0 & 0 & 0 \\ 0 & 1_{ij} & 0 \\ 0 & 0 & 0 \end{bmatrix} x(\tau) d\tau. \qquad (3.2\text{-}9)$$

Taking the limit in (3.2-9) as $\Delta A \to 0$, results in

$$\frac{\partial x(t)}{\partial a_{ij}} = \int_0^t \exp[A(t - \tau)] \begin{bmatrix} 0 & 0 & 0 \\ 0 & 1_{ij} & 0 \\ 0 & 0 & 0 \end{bmatrix} x(\tau) d\tau. \qquad (3.2\text{-}10)$$

Similarly if $\Delta A = 0$ and

$$\Delta B = \begin{bmatrix} 0 & 0 & 0 \\ 0 & \Delta b_{kl} & 0 \\ 0 & 0 & 0 \end{bmatrix}, \qquad (3.2\text{-}11)$$

then from (3.2-7) we obtain

$$\frac{\partial x(t)}{\partial b_{kl}} = \int_0^t \exp[A(t-\tau)] \begin{bmatrix} 0 & 0 & 0 \\ 0 & 1_{kl} & 0 \\ 0 & 0 & 0 \end{bmatrix} u(\tau) d\tau. \qquad (3.2\text{-}12)$$

Equations (3.2-10) and (3.2-12) are sensitivity functions generated from system differential equation. These results are specially effective in applications related to sensitivity analysis of a general *DCLTI – GS*. The above procedure also holds for the case wherein the system-structural parameters are linear functions of system-physical parameters, and in particular, when A and B have common parameters in several of their entries. ∎

Now, consider a special case of a *DCLTI – GS* with zero initial condition and one input. Furthermore, let the original system be represented in a phase-variable canonical form.

$$\dot{x}(t) = Ax(t) + bu(t), \qquad x(0) = 0; \qquad (3.2\text{-}13)$$

$$A \triangleq \begin{bmatrix} 0 & I_{(n-1)} \\ -a_1 & -a_2, \ldots, -a_n \end{bmatrix}_{n \times n}, \quad b \triangleq \{[0, \cdots, 0, 1]^T\}_{n \times 1}. \qquad (3.2\text{-}14)$$

Here I_n is an $n \times n$ identity matrix, and it is assumed that a_j's, $j = 1, \ldots, n$, have the same signs: Otherwise the definition of sensitivity function must be changed accordingly.

By Proposition 3.2-1, if $A \to A + \Delta A$ (here $\Delta b = 0$), the incremental change in system state vector becomes

$$\Delta x(t) = \int_0^t \exp[(A + \Delta A)(t - \tau)] \Delta A x(\tau) d\tau. \qquad (3.2\text{-}15)$$

Let

$$\Delta A = \begin{bmatrix} 0 & 0 & 0 \\ 0 & -\Delta a_j & 0 \end{bmatrix}, \qquad (3.2\text{-}16)$$

then following the same reasoning as before yields

$$\frac{\partial x(t)}{\partial a_j} = \int_0^t \exp[A(t-\tau)] \begin{bmatrix} 0 & 0 & 0 \\ 0 & -1_j & 0 \end{bmatrix} x(\tau) d\tau. \qquad (3.2\text{-}17)$$

Since $x(t)$ is the solution of system differential equation, $\dot{x}(t)$ exists. It is also assumed that $x(t)$ is a continuous function of t and the parameters, thus the order of differentiation with respect to t and parameters can be changed. Based on this fact it is easy to show that,

3.2. Structural – Parameter Sensitivity Functions Generation

$$\frac{d}{dt}\frac{\partial x(t)}{\partial a_j} = A\frac{\partial x(t)}{\partial a_j} + \begin{bmatrix} 0 & 0 & 0 \\ 0 & -1_j & 0 \end{bmatrix} x(t), \quad \frac{\partial x(t)}{\partial a_j}\bigg|_{t=0} = 0. \quad (3.2\text{-}18)$$

Let $\dfrac{\partial x(t)}{\partial a_j} = z(t)$: then (3.2-18) becomes

$$\dot{z}(t) = Az(t) + \begin{bmatrix} 0 & 0 & 0 \\ 0 & -1_j & 0 \end{bmatrix} x(t), \quad z(0) = 0. \quad (3.2\text{-}19)$$

Similarly, let $\dfrac{\partial x(t)}{\partial a_{j+1}} = w(t)$, then the same derivation as that of (3.2-19) results in

$$\dot{w}(t) = Aw(t) + \begin{bmatrix} 0 & 0 & 0 \\ 0 & -1_{j+1} & 0 \end{bmatrix} x(t), \quad w(0) = 0. \quad (3.2\text{-}20)$$

In (3.2-19), let $\dot{z}(t) = y(t)$: and as $x(0) = 0$ (by assumption), (3.2-19) yields

$$\dot{y}(t) = Ay(t) + \begin{bmatrix} 0 & 0 & 0 \\ 0 & -1_{j+1} & 0 \end{bmatrix} x(t), \quad y(0) = 0. \quad (3.2\text{-}21)$$

Notice that $\dot{x}_i(t) = x_{i+1}(t)$ for $i = 1, 2, \ldots, n-1$. Now, observation reveals that (3.2-20) and (3.2-21) are identical. Thus by the uniqueness theorem for the solution of a differential equation [2A. 3], it is clear that $y(t) \equiv w(t)$, or

$$\dot{z}(t) = w(t). \quad (3.2\text{-}22)$$

If the equivalent for $\dot{z}(t)$, from (3.2-19), as well as that for $w(t)$, from (3.2-20), is written, then

$$\frac{\partial x_{l+1}(t)}{\partial a_j} = \frac{\partial x_l(t)}{\partial a_{j+1}}, \quad \text{for } l, j = 1, \ldots, n-1. \quad (3.2\text{-}23)$$

∎

The above discussion (under the special case) has been cited here to show an application of Proposition 3.2-1, to generate the sensitivity function of a *DCLTI – GS*. Further, it furnishes a different proof to the following corollary, known as the "Total Symmetry Property" and given first by Wilkie and Perkins [B. 12], as "Low-Order Sensitivity Functions Models."

COROLLARY 3.2-1 (Total Symmetry Property [B. 12]): Consider the single-input open-loop *DCLTI – GS* described by

$$\dot{x}(t) = Ax(t) + bu(t), \quad x(0) = 0, \quad (3.2\text{-}24)$$

where A and b are given in (3.2-14) and it is assumed that all a_j, $j = 1, 2, \ldots, n$ have the same signs. Define the state sensitivity function as

$$\{\xi_j\} \triangleq \frac{\partial x(t)}{\partial a_j}, \quad j = 1, 2, \ldots, n, \quad (3.2\text{-}25)$$

and the sensitivity matrix $(\xi_{k,l})$ as

$$(\xi_{k, l}) \triangleq (\xi_1, \xi_2, \ldots, \xi_n). \qquad (3.2\text{-}26)$$

Then the sensitivity matrix $(\xi_{k, l})$ has the following total symmetry property:

$$\xi_{k, l}(t) = \xi_{k-1, l+1}(t), \quad \text{for } k = 2, \ldots, n, \text{ and } l = 1, \ldots, n-1. \qquad (3.2\text{-}27)$$

So that all elements along the antidiagonals of the $(\xi_{k, l})$ matrix are equal, as shown in (3.2-28)

$$(\xi_{k, l}) \triangleq \begin{bmatrix} \xi_{1,1} & \xi_{1,2} & \xi_{1,3}, & \ldots, & \xi_{1,n} \\ \xi_{1,2} & \xi_{1,3}, & \ldots \ldots, & \xi_{2,n} \\ \xi_{1,3} & \cdot & & & \\ \cdot & & & & \\ \cdot & & & & \\ \xi_{1,n} & \xi_{2,n}, & \ldots \ldots, & \xi_{n,n} \end{bmatrix} \qquad (3.2\text{-}28)$$

■

Now, recall (3.2-18), (3.2-19), (3.2-20) and (3.2-22). Equation (3.2-21) can be written as

$$\frac{d}{dt} \frac{\partial x(t)}{\partial a_j} = \frac{\partial x(t)}{\partial a_{j+1}}. \qquad (3.2\text{-}29)$$

Substituting (3.2-29) in the left-hand side of (3.2-18) yields

$$\frac{\partial x(t)}{\partial a_{j+1}} = A \frac{\partial x(t)}{\partial a_j} + \begin{bmatrix} 0 & 0 & 0 \\ 0 & -1_j & 0 \end{bmatrix} x(t). \qquad (3.2\text{-}30)$$

The above equation is an algebraic equation which gives the state-sensitivity function with respect to a_{j+1}, versus the state trajectory $x(t)$ and the state-sensitivity function with respect to a_j for $j = 1, \ldots, n-1$.

Using the result of Corollary 3.2-1 and (3.2-30), for $j = 1, \ldots, n-1$, all of the sensitivity functions for $x(t)$ with respect to its parameters can be recursively derived. The above derivations provides a proof to the following property given first by Wilkie and Perkins [B. 12].

COROLLARY 3.2-2 (Complete Simultaneity Property [B. 12]): For a phase-variable canonical system, all of the sensitivity functions $\partial x_i / \partial a_j$, $i, j = 1, \ldots, n$, can be obtained as linear combinations of the signals appearing in one-sensitivity model of the system together with the system state.

■

Now that the first-order sensitivity functions have been determined, consider derivation of the second–order sensitivity functions for a phase-variable canonical system.

3.2. Structural – Parameter Sensitivity Functions Generation

Considering (3.2-30), differentiation of both sides with respect to a_l ($l = 1, 2, \ldots, n$) gives

$$\frac{\partial^2 x(t)}{\partial a_l \partial a_{j+1}} = A \frac{\partial^2 x(t)}{\partial a_l \partial a_j} + \frac{\partial A}{\partial a_l} \frac{\partial x(t)}{\partial a_j} + \begin{bmatrix} 0 & 0 & 0 \\ 0 & -1_j & 0 \end{bmatrix} \frac{\partial x(t)}{\partial a_l}, \qquad (3.2\text{-}31)$$

which simplifies to

$$\frac{\partial^2 x(t)}{\partial a_l \partial a_{j+1}} = A \frac{\partial^2 x(t)}{\partial a_l \partial a_j} + \begin{bmatrix} 0 & 0 & 0 \\ 0 & -1_l & 0 \end{bmatrix} \frac{\partial x(t)}{\partial a_j} + \begin{bmatrix} 0 & 0 & 0 \\ 0 & -1_j & 0 \end{bmatrix} \frac{\partial x(t)}{\partial a_l}. \qquad (3.2\text{-}32)$$

Proceeding similarly, it is easy to show that any other higher-order sensitivity function can be specified as a linear (algebraic) combination of signals in one-system sensitivity model of the same order and one-sensitivity model of degree one lower than the order of sensitivity function.

Neuman and Sood have also studied this problem with several new results. Following property generated from (3.2-32) is also reported in [B. 6].

COROLLARY 3.2-3 (Complete Higher-Order Simultaneity Property for a Phase-Variable Canonical System [B. 6]): All higher-order sensitivity functions $\partial^h x_i / \partial^\alpha a_j \cdots \partial^\zeta a_l$, ($\alpha + \cdots + \zeta = h$, $j, \cdots, l$ are $1, \cdots, n$, and $i = 1, \cdots, n$) can be obtained as linear combinations of signals appearing in one-system sensitivity model of the same order (i.e., h) and one-system sensitivity model of order one less (i.e., h−1).
∎

The main shortcoming of this property ("low-order sensitivity model") is that not every system can be cast in the phase-variable canonical form. Wilkie and Perkins [B. 12] use a canonical transformation [2C. 28] to evaluate (in fact to construct the model) sensitivity functions for single-input and multi-input systems as follows.

Let a completely-controllable generating system be described by

$$\dot{x}(t) = Ax(t) + bu(t), \qquad (3.2\text{-}33)$$

$$z(t) = Cx(t). \qquad (3.2\text{-}34)$$

It can be shown that there exists a nonsingular transformation

$$x(t) = Tx'(t), \qquad (3.2\text{-}35)$$

such that

$$\dot{x}'(t) = \hat{A}x'(t) + \hat{b}u(t); \qquad (3.2\text{-}36)$$

where

$$\hat{A} \triangleq T^{-1}AT = \begin{bmatrix} 0 & I_{(n-1)} \\ -a_1 & -a_2, \ldots, -a_n \end{bmatrix}, \text{ and } \hat{b} \triangleq T^{-1}b = [0, \ldots, 0, 1]^T. \qquad (3.2\text{-}37)$$

Here $a_1, \ldots, a_n$ are the coefficients of characteristic equation of A. Since (3.2-36) can always be written if the pair (A, b) is completely-controllable, the result of Corollary 3.2-1 may be easily used to generate the corresponding system sensitivity functions. However, even for a single-input system, this technique is very costly due to the computational difficulties in evaluating the above T. This is yet more involved for a controllable system with multiple input, where many T_k's (corresponding to each input u_k) have to be derived such that each u_k and x takes on the form (3.2-36). We must note that after evaluating many T_k's (depending on the number of inputs – say, m), and if $x^1(t), x^2(t), ..., x^m(t)$ are the solutions of x(t) for the inputs $u_1(t), u_2(t), ..., u_m(t)$, then $x(t) = \sum_{k=1}^{m} x^k(t)$ (linear systems, the superposition rule). Each T_k and $x^k(t)$ together with $u_k(t)$ can be treated similarly as (3.2-36) (using the result of Corollary 3.2-1). Due to many parametric values of T_k, k = 1, ..., m, and the numerical sensitivity associated with these transformations, we encounter computationally an extremely complicated problem. In addition, in the process of these transformations the physical insights of the parameters will be lost, which is undesirable. In any sensitivity analysis it is always preferable to use a direct instead of a transformational approach. This issue of the minimum required sensitivity functions has been continuously studied by several researchers. The latest and perhaps the final answer to the number of maximum independent sensitivity functions is given by Gupta and Mehra in two theorems [*B.* 4] (cf., Theorems 4.3-1 & 2).

REMARK 3.2-1: To continue our sensitivity analyses, we need to define *a* measure of sensitivity that is also easily accessible. What is a *good* and globally acceptable measure of system sensitivity? We do not know the answer to this question. Certainly the choice of this measure depends on the particular situation. Since we do not know which structural parameter is changing, it seems appropriate to use some sort of an algebraic relationship among all of these varying parameters. We propose to use the sum of unnormalized, structural-sensitivity functions which we call the total-sensitivity function *(TSF)*, as our initial system-sensitivity measure. Adding a number of variables that may have different algebraic signs and perhaps different physical dimensions certainly concerns us. However, because these unnormalized-sensitivity functions have different algebraic signs, this sensitivity measure (or *TSF*) represents some sort of a *net* value for parameter variations. This aspect of the *TSF* is certainly in accordance with the practical situations in which whenever we have several parameters that are changing simultaneously and in opposite directions, then they may compensate for each other's unwanted effects on the system dynamics. The remaining issue that has to be addressed is whether this measure will ever be identically equal to zero. The answer is definitely no as we subsequently prove that fact. Therefore we believe that system – *TSF*, which is a low-order sensitivity function and is easily accessible, is a *useful academic tool* to describe a number of important issues and/or algorithms in the following.

PROPOSITION 3.2-2: Consider an open-loop *DCLTI – GS* described by (3.2-1). The total-sensitivity function *(TSF)* of x(t) with respect to all of its structural parameters is defined by

3.2. Structural – Parameter Sensitivity Functions Generation

$$TSF \triangleq \sum_{i,j} a_{ij} \frac{\partial x(t)}{\partial a_{ij}} + \sum_{k,l} b_{kl} \frac{\partial x(t)}{\partial b_{kl}}, \qquad (3.2\text{-}38)$$

where a_{ij} and b_{kl} are the entries of the coefficient matrices A and B in (3.2-1). This sensitivity function is equal to

$$TSF \triangleq \int_0^t \exp[A(t-\tau)]\{Ax(\tau) + Bu(\tau)\}d\tau. \qquad (3.2\text{-}39)$$

PROOF: It is easy to show that

$$\frac{\partial x(t)}{\partial a_{ij}} = \int_0^t \exp[A(t-\tau)] \begin{bmatrix} 0 & 0 & 0 \\ 0 & 1_{ij} & 0 \\ 0 & 0 & 0 \end{bmatrix} x(\tau)d\tau, \qquad (3.2\text{-}40)$$

and if both sides of (3.2-40) are multiplied by a_{ij} and are summed over all a_{ij}, then it follows that

$$\sum_{i,j} a_{ij} \frac{\partial x(t)}{\partial a_{ij}} = \int_0^t \exp[A(t-\tau)]Ax(\tau)d\tau. \qquad (3.2\text{-}41)$$

Similarly, we have

$$\sum_{k,l} b_{kl} \frac{\partial x(t)}{\partial b_{kl}} = \int_0^t \exp[A(t-\tau)]Bu(\tau)d\tau. \qquad (3.2\text{-}42)$$

Adding (3.2-41) and (3.2-42) results in

$$f_{o1}(t) \triangleq \sum_{i,j} a_{ij} \frac{\partial x(t)}{\partial a_{ij}} + \sum_{k,l} b_{kl} \frac{\partial x(t)}{\partial b_{kl}}$$

$$= \int_0^t \exp[A(t-\tau)]\{Ax(\tau) + Bu(\tau)\}d\tau = \int_0^t \exp[A(t-\tau)] \frac{\partial x(\tau)}{\partial \tau} d\tau. \qquad (3.2\text{-}43)$$

Here the subscript "o" refers to an open-loop system wherein the input is not a function of system parameters, and "1" means the first-order – TSF. ∎

An interpretation of this result and the fact that this measure does not identically vanish follows. Let q be the vector of structural parameters; i.e., a vector of dimension $(n^2 + nm)$, whose components are $\{a_{ij}\}$, for all i, j, and $\{b_{kl}\}$, for all k, l, respectively. Then (3.2-43) can be written in a compact form,

$$f_{o1}(t) = [\text{grad}_q x(t)]q. \qquad (3.2\text{-}44)$$

This notation implies that $f_{o1}^i(t)$ (i.e., the ith-component of $f_{o1}(t)$) is the inner product of the nominal, structural-parameter vector and the gradient of $x^i(t)$ in terms of q, and it continuously depends on t for all i. For each i, $f_{o1}^i(t)$ is written as

$$f_{o1}^i = \| \frac{\partial x^i(t)}{\partial q} \| \; \|q\| \cos\phi, \qquad (3.2\text{-}45)$$

where ϕ is the angle between nominal, structural-parameter vector q and the gradient of $x^i(t)$ in terms of q, and $\|q\|$ is the Euclidean norm of q: i.e., $\sqrt{\sum_i q_i^2}$. Since

$\|q\|$ and $\|\frac{\partial x^i(t)}{\partial q}\|$ are both non-zero (otherwise trivial), then the only way that $f_{o1}^i(t)$ can be zero is that if $\frac{\partial x^i(t)}{\partial q}$ is in a hyperplane of dimension one-less than the dimension of q and perpendicular to q; i.e., $\phi = 90°$. This requirement is in general very severe. Therefore it must be expected that a total-sensitivity function, as above, identically equal to zero seldom exists and this fact supports our motivation for looking at this function as a good candidate for system-sensitivity measure. Furthermore the main reason for nonvanishing property of this function is that given by (3.2-43), which based on the fundamental lemma from the Calculus of Variations [2C. 7], if (3.2-43) vanishes, then $\dot{x}(t) \equiv 0$ and that is a contradiction. ∎

Straightforward differentiation of (3.2-43) shows that $f_{o1}(t)$ satisfies the following equation

$$\dot{f}_{o1}(t) = Af_{o1}(t) + Ax(t) + Bu(t), \qquad f_{o1}(0) = 0. \qquad (3.2-46)$$

Because of the nonvanishing property of (3.2-43), it seems that this measure is a reasonable *qualitative* sensitivity measure of the system. Also it reveals that system − TSF as a sensitivity measure does not contradict our physical expectation from a dynamic system, which says: Every system is always sensitive to several of its parameters.

PROPOSITION 3.2-3: Consider an open-loop *DCLTI − GS* described by

$$\dot{x}(t) = Ax(t) + Bu(t), \qquad x(0) = x_o. \qquad (3.2-47)$$

Then

$$f_{o2}(t) \triangleq \sum_{i,j} \sum_{k,l} \eta_{ij}\eta_{kl} \frac{\partial^2 x(t)}{\partial \eta_{ij}\partial \eta_{kl}} = 2A \int_0^t \exp[A(t-\tau)] \sum_{e,g} \eta_{eg} \frac{\partial x(\tau)}{\partial \eta_{eg}} d\tau; \quad (3.2-48)$$

where $H \triangleq (\eta_{ij}) = (a_{ij}$ and b_{kl} for all i, j, k and l). We show this H as $A \cup B$.

PROOF: Straightforward steps as given in [A. 7]. ∎

Now that the first- and the second-order, total-sensitivity functions for an open-loop *DCLTI − GS* are available, we can show that indeed these equations satisfy the following three sets of equations.

$$\dot{x}(t) = Ax(t) + Bu(t), \qquad x(t_0) = x_o. \qquad (3.2-1)$$

$$\dot{f}_{o1}(t) = Af_{o1}(t) + Ax(t) + Bu(t), \qquad f_{o1}(0) = 0. \qquad (3.2-46)$$

$$\dot{f}_{o2}(t) = Af_{o2}(t) + 2Af_{o1}(t), \qquad f_{o2}(0) = 0. \qquad (3.2-49)$$

These equations can be easily shown by appropriate block diagram representations, wherein the number of integrators for generating $f_{o1}(t)$ is the same as that for constructing system state $x(t)$ (the same is true for $f_{o2}(t)$). Also starting from (3.2-49),

3.2. Structural – Parameter Sensitivity Functions Generation

any other higher-order, total-sensitivity function can be generated, but in any case the structure is similar to that of $f_{o1}(t)$ and/or $f_{o2}(t)$.

REMARK 3.2-2: All of the derivations so far are based on the assumption that the generating open-loop system is completely reconstructible; i.e., the complete state vector is available for measurements. On the other hand, for systems which are unreconstructible no sensitivity functions can be developed by the above means, although one certainly exists. Thus, in the following the state sensitivity function of a *DCLTI – GS* is constructed, by first constructing the system state vector using the Luenberger's theory of reduced-order observers, described next [2B. 12]. ∎

Consider an open-loop *DCLTI – GS* described by

$$\dot{x}(t) = Fx(t) + Lu(t), \quad x(0) = x_o. \quad (3.2\text{-}50)$$

Observations of state vector are available according to

$$z(t) = Hx(t), \quad (3.2\text{-}51)$$

where $H \in R^{\omega \times n}$ is assumed to be of full rank, and $\omega < n$. Thus $z(t)$ consists of ω linearly independent combinations of state vector $x(t)$. It is also assumed that the system characterized by (3.2-50) and (3.2-51) is completely observable. Thus the following standard procedure provides [2C. 18] an estimate of state vector $x(t)$ employing an $(n - \omega)$th-order observer.

Let $\zeta(t)$ be the $(n - \omega)$-dimensional vector

$$\zeta(t) = Tx(t), \quad (3.2\text{-}52)$$

such that

$$\begin{bmatrix} T \\ H \end{bmatrix}, \quad (3.2\text{-}53)$$

is a nonsingular matrix. The vector $\zeta(t)$ represents $(n - \omega)$ linear combinations of the elements in system state $x(t)$ which are independent of the measurement $z(t)$. Thus $x(t)$ may be obtained as follows.

$$x(t) = \begin{bmatrix} T \\ H \end{bmatrix}^{-1} \begin{Bmatrix} \zeta(t) \\ z(t) \end{Bmatrix}. \quad (3.2\text{-}54)$$

For convenience, we define

$$\begin{bmatrix} T \\ H \end{bmatrix}^{-1} = (A, B), \quad (3.2\text{-}55)$$

so that

$$x(t) = A\zeta(t) + Bz(t). \quad (3.2\text{-}56)$$

From (3.2-50), (3.2-52) and (3.2-56) we conclude the following.

$$\dot{\zeta}(t) = T\dot{x}(t) = T\{F(A\zeta(t) + Bz(t)) + Lu(t)\}$$

$$= TFA\zeta(t) + TFBz(t) + TLu(t). \qquad (3.2\text{-}57)$$

Now, an $(n - \omega)$th-order estimate for the transformed state vector $\zeta(t)$, which can be used to reconstruct an estimate of the original state vector $x(t)$, according to (3.2-56) is developed. To reconstruct an estimate, $\hat{\zeta}(t)$ of the vector $\zeta(t)$, it is appropriate to design an observer which models the known dynamics of $\zeta(t)$, given by (3.2-57), utilizing $z(t)$ and $u(t)$ as known inputs. Correspondingly the result is an observer of the form

$$\dot{\hat{\zeta}}(t) = TFA\hat{\zeta}(t) + TFBz(t) + TLu(t), \qquad (3.2\text{-}58)$$

$$\hat{x}(t) = A\hat{\zeta}(t) + Bz(t). \qquad (3.2\text{-}59)$$

Fig. 3.2-1 depicts the observer block diagram and its total-sensitivity function in terms of F and L

$$\dot{\hat{f}}_{o1}(t) = F\hat{f}_{o1}(t) + F\hat{x}(t) + Lu(t), \qquad \hat{f}_{o1}(0) = 0. \qquad (3.2\text{-}60)$$

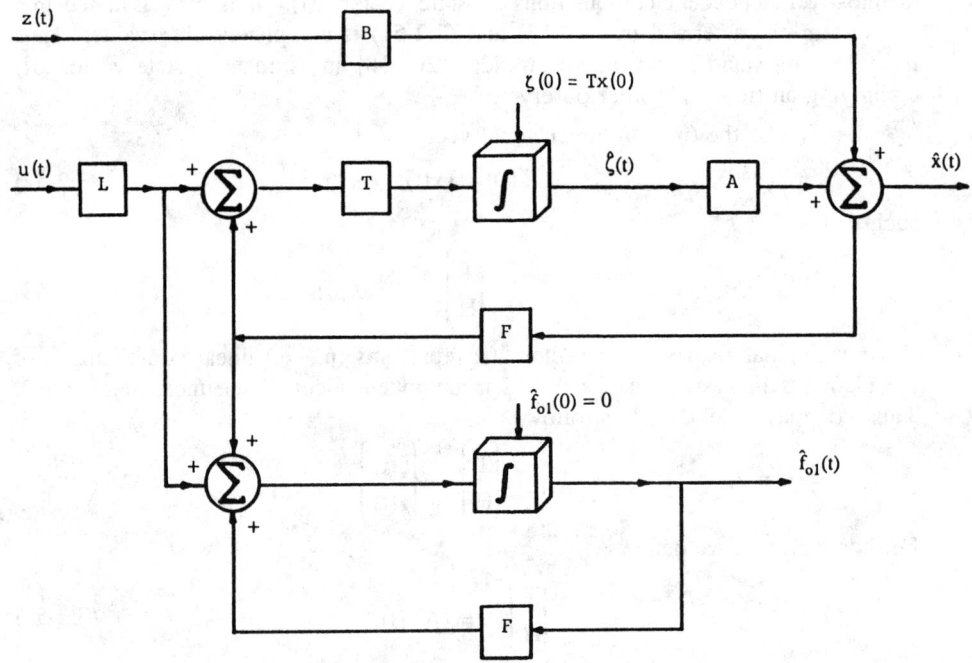

Fig. 3.2-1. The *TSF* for an open-loop observable system.

3.2.3. Closed – Loop Reconstructible System

The term closed-loop means, of course, that the input vector u(t) is a function of state vector; hence, system parameters. To distinguish between an open-loop and a closed-loop system, the input vector is shown either as u(t), for an open-loop system, or as u(t; θ), for a closed-loop system. In the case of a closed-loop reconstructible system it is necessary to consider the variation of system input with respect to system parameters and relative to determination in the rate of change in system state vector as system parameters change. The following results pertain to this case.

PROPOSITION 3.2-4: Consider a closed-loop reconstructible *DCLTI – GS* characterized by

$$\dot{x}(t) = Ax(t) + Bu(t; q), \quad x(0) = x_o. \tag{3.2-61}$$

Suppose that $A \to A + \Delta A$, $B \to B + \Delta B$, $x(t) \to x(t) + \Delta x(t)$, and $u(t; q) \to u(t; q) + \Delta u(t; q)$, where $\Delta x(0) = 0$. Then

$$\Delta x(t) = \int_0^t \exp[(A+\Delta A)(t-\tau)]\{\Delta Ax(\tau) + \Delta Bu(\tau; q) + (B+\Delta B)\Delta u(\tau; q)\}d\tau. \tag{3.2-62}$$

PROOF: By hypothesis

$$\dot{x}(t) + \Delta\dot{x}(t) = (A + \Delta A)\{x(t) + \Delta x(t)\} + (B + \Delta B)\{u(t; q) + \Delta u(t; q)\}. \tag{3.2-63}$$

Subtracting (3.2-61) from (3.2-63) results in

$$\Delta\dot{x}(t) = (A + \Delta A)\Delta x(t) + \Delta Ax(t) + \Delta Bu(t; q) + (B + \Delta B)\Delta u(t; q), \tag{3.2-64}$$

which together with $\Delta x(0) = 0$, is equivalent to (3.2-62).

∎

From (3.2-62) it is evident that in order to evaluate $\Delta x(t)$, $\Delta u(t; q)$ must be known in advance: i.e., the rate of change in u(t; q) caused by system parameter variations. For $\Delta u(t; q) \equiv 0$, (3.2-62) reduces to the result for the open-loop system. Equation (3.2-62) is used in Section 6.2 for certain design comparison and is no longer used directly to calculate any other sensitivity functions.

COROLLARY 3.2-4: For the system of Proposition 3.2-4, the total-sensitivity function of x(t) with respect to system parameters is

$$f_{c1}(t) \triangleq \sum_{i,j} \eta_{ij} \frac{\partial x(t)}{\partial \eta_{ij}} = \int_0^t \exp[A(t-\tau)]\{Ax(\tau) + Bu(\tau; q) + Bf_{ul}(\tau; q)\}d\tau; \tag{3.2-65}$$

where $H \triangleq (\eta_{ij}) = A \cup B$, and $f_{ul}(t) = \sum_{i,j} \eta_{ij} \frac{\partial u(t; q)}{\partial \eta_{ij}}$.

PROOF: Let $\Delta A = (\Delta a_{ij})$ and $\Delta B = 0$; then from (3.2-62) we have

$$\sum_{i,j} a_{ij} \frac{\partial x(t)}{\partial a_{ij}} = \int_0^t \exp[A(t-\tau)]\{Ax(\tau) + Bf_{ula_{ij}}(\tau; q)\}d\tau, \tag{3.2-66}$$

here $f_{ula_{ij}}(t; q) \triangleq \sum_{i,j} a_{ij} \dfrac{\partial u(t; q)}{\partial a_{ij}}$; i.e., the first-order, total-sensitivity function of input with respect to (a_{ij}). Similarly, let $\Delta A = 0$ and $\Delta B = (\Delta b_{ij})$. Then

$$\sum_{i,j} b_{ij}\dfrac{\partial x(t)}{\partial b_{ij}} = \int_0^t \exp[A(t-\tau)]\{Bu(\tau; q) + Bf_{ulb_{ij}}(\tau; q)\}d\tau, \qquad (3.2\text{-}67)$$

where $f_{ulb_{ij}}(t; q) \triangleq \sum_{i,j} b_{ij} \dfrac{\partial u(t; q)}{\partial b_{ij}}$; i.e., the first-order, total-sensitivity function of input with respect to (b_{ij}). Adding (3.2-66) and (3.2-67) yields (3.2-65), where $f_{ul}(t; q) = f_{ula_{ij}}(t; q) + f_{ulb_{ij}}(t; q)$, is the first-order, total-sensitivity function of the closed-loop input with respect to system-structural parameters.

COROLLARY 3.2-5: Consider the closed-loop reconstructible system of Corollary 3.2-4 with a linear feedback input as follows.

$$u(t; q) = -Kx(t; q) + Gy(t), \qquad (3.2\text{-}68)$$

where G and K are constant (gain) matrices of appropriate dimensions, and $y(t) \in R^v$ is an additional component of input vector which is not disturbed by system parameter. Then the total-sensitivity function of $x(t; q)$ with respect to the overall structural parameters in A, B, K, and G is

$$f''_{c1}(t) \triangleq \sum_{i,j} \eta''_{ij} \dfrac{\partial x(t)}{\partial \eta''_{ij}}$$

$$= \int_0^t \exp[(A - BK)(t - \tau)]\{(A - 2BK)x(\tau) + 2BGy(\tau)\}d\tau; \qquad (3.2\text{-}69)$$

where $H'' \triangleq (\eta''_{ij}) = A \cup B \cup K \cup G$.

PROOF: It is easy to show that for this system

$$f_{ul}(t; q) = -Kx(t) - Kf''_{c1}(t) + Gy(t). \qquad (3.2\text{-}70)$$

Substituting (3.2-70) for $f_{ul}(t; q)$, and (3.2-68) for $u(t; q)$, into (3.2-65), results in

$$f''_{c1}(t) = \int_0^t \exp[A(t - \tau)]\{Ax(\tau) - BKx(\tau) + BGy(\tau)$$

$$- BKx(\tau) - BKf''_{c1}(\tau) + BGy(\tau)\}d\tau, \qquad (3.2\text{-}71)$$

where the change of notation $(f_{c1} \to f''_{c1})$ is due to the change of $H \to H''$. Differentiating (3.2-71) with respect to t yields

$$\dot{f}''_{c1}(t) = (A - BK)f''_{c1}(t) + (A - 2BK)x(t) + 2BGy(t), \quad f''_{c1}(0) = 0. \qquad (3.2\text{-}72)$$

REMARK 3.2-3: For the system of Corollary 3.2-5, and a linear feedback control law

$$u(t; q) = -Kx(t; q), \qquad (3.2\text{-}73)$$

3.2. Structural – Parameter Sensitivity Functions Generation

$$\dot{f}'_{c1}(t) = (A - BK)f'_{c1}(t) + (A - 2BK)x(t), \quad f'_{c1}(0) = 0. \tag{3.2-74}$$

Here $f'_{c1} \triangleq \sum_{i,j} \eta'_{ij} \frac{\partial x(t)}{\partial \eta'_{ij}}$ and $H' \triangleq (\eta'_{ij}) = A \cup B \cup K$. However, for the case wherein $H \triangleq (\eta_{ij}) = A \cup B$, and $f_{c1} \triangleq \sum_{i,j} \eta_{ij} \frac{\partial x(t)}{\partial \eta_{ij}}$, we have

$$\dot{f}_{c1}(t) = (A - BK)f_{c1}(t) + (A - BK)x(t), \quad f_{c1}(0) = 0. \tag{3.2-75}$$

■

To synthesize the total-sensitivity function $f''_{c1}(t)$, recall system state equation, linear feedback control law, and the system – *TSF*, which are repeated here for convenience and depicted in Fig. 3.2-2.

$$\dot{x}(t) = Ax(t) + Bu(t; q), \quad x(0) = x_o, \tag{3.2-61}$$

$$u(t; q) = -Kx(t; q) + Gy(t), \tag{3.2-68}$$

$$\dot{f}''_{c1}(t) = (A - BK)f''_{c1}(t) + (A - 2BK)x(t) + 2BGy(t), \quad f''_{c1}(0) = 0. \tag{3.2-72}$$

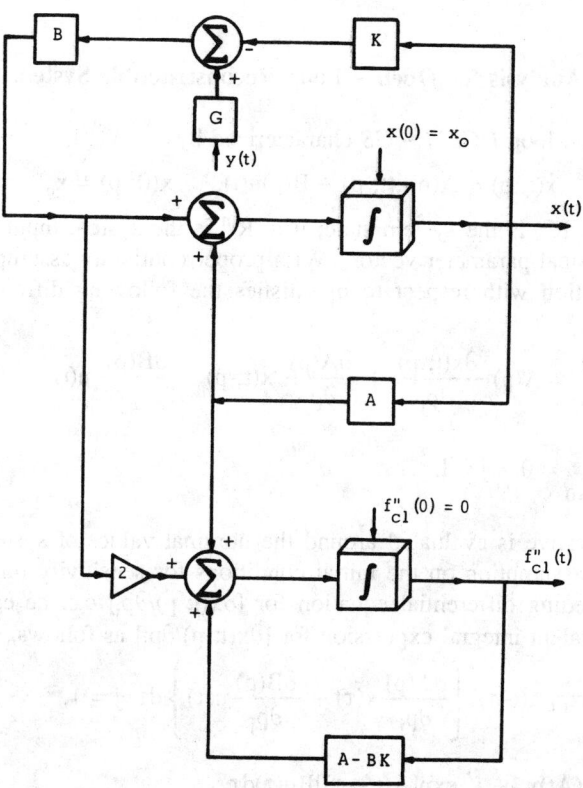

Fig. 3.2-2. The *TSF* for a closed-loop *DCLTIS*.

3.3. PHYSICAL – PARAMETER SENSITIVITY FUNCTIONS GENERATION

3.3.1. Introduction

The thoughts in physical-parameter sensitivity functions *(PPSF's)* generation are parallel to those of the structural-parameter sensitivity functions *(SPSF's)* generation. However, as stated before, distinction between the generation of the *PPSF* and that of *SPSF* is essential. Each has its own merits and the study of both cases is necessary. Perhaps the *PPSF* generation is more appealing than that of *SPSF*, if an access to an accurate system mathematical model is provided. Generally speaking, the number of physical parameters is less than that of structural parameters, and therefore a lesser number of equations must be generated in this case than that of *SPSF*. But efficiently generating these functions is still a major task. Conceptually the derivation of a sensitivity function from the parametric solution of system trajectory is straightforward. However, a novel approach is to avoid such a costly or time consuming process since this direct calculation is not always easy, specially for multi-parameter and higher-order systems.

3.3.2. Formal Analysis for Open – Loop Reconstructible System

Consider an open-loop *DCLTI – GS* characterized by

$$\dot{x}(t; p) = A(p)x(t; p) + B(p)u(t), \quad x(0; p) = x_o. \tag{3.3-1}$$

Here again $x \in R^n$, is the system state; $u \in R^m$ is the system input and $p \in R^r$ is the system physical-parameter vector. With proper continuity assumptions the state sensitivity function with respect to p_j satisfies the following differential equation [2A. 3].

$$\frac{d}{dt}\frac{\partial x(t; p)}{\partial p_j} = A(p)\frac{\partial x(t; p)}{\partial p_j} + \frac{\partial A(p)}{\partial p_j}x(t; p) + \frac{\partial B(p)}{\partial p_j}u(t),$$

$$\left.\frac{\partial x(t; p)}{\partial p_j}\right|_{t=0} = 0, \quad j = 1, ..., r. \tag{3.3-2}$$

Each derivative is evaluated around the nominal values of system parameters, and the above assumption on the initial conditions for sensitivity functions is standard. The preceding differential equation for $[\partial x(t; p)/\partial p_j]$ can be easily solved to obtain an equivalent integral expression for $[\partial x(t; p)/\partial p_j]$ as follows.

$$\frac{\partial x(t; p)}{\partial p_j} = \int_0^t \exp[A(t-\tau)]\left\{\frac{\partial A(p)}{\partial p_j}x(\tau) + \frac{\partial B(p)}{\partial p_j}u(\tau)\right\}d\tau, \, j = 1, \cdots, r, \tag{3.3-3a}$$

$$x(t) = \exp(At)x_o + \int_0^t \exp[A(t-\tau)]Bu(\tau)d\tau. \tag{3.3-3b}$$

3.3. Physical – Parameter Sensitivity Functions Generation

As is evident from (3.3-3) to compute each and every sensitivity functions in a typical system is a formidable task. Contrary to the *SPSF's* generation, it is not easy to make any general or broad statement regarding the nature of these differential equations, which would involve development of a property parallel to that of the low-order sensitivity function (cf., Corollary 3.2-1). This is difficult unless it happens that for example the two sets of parameters are identical, or some other pertinent information is available. However, it is very well possible that for a given system the physical-parameter sensitivity functions be redundant as this issue is discussed in Section 4.3. Therefore at the moment it seems that at most r more nth-order differential equations in addition to the system state vector must be solved in order to determine the first-order sensitivity functions for all system physical parameters.

3.3.2.1. Higher – Order Sensitivity Functions

To generate any higher-order sensitivity function, an access to the one-order less sensitivity function is necessary. For example, for a second-order sensitivity function of (3.3-2) with respect to p_k, and with proper continuity requirement, we have

$$\dot{x}_{k,j} \triangleq \frac{d}{dt} \frac{\partial^2 x(t; p)}{\partial p_k \partial p_j} = A x_{k,j} + A_k x_j + A_j x_k + A_{k,j} x + B_{k,j} u,$$

$$x_{k,j}(0; p) = 0, \text{ for all } k \text{ and } j. \tag{3.3-4}$$

Here $A_{k,j} \triangleq [\partial^2 A(p)/\partial p_k \partial p_j]$, similarly A_k, A_j, $B_{k,j}$, x_k, and x_j are defined. As is shown in (3.3-4) in order to solve for $x_{k,j}$; first we must determine x_k, x_j, and x, and this chain rule goes on to any other higher-order sensitivity functions.

3.3.2.2. Direct Computation of Sensitivity Functions

To compute $[\partial x(t; p)/\partial p_j]$ directly from $x(t; p)$, it is necessary to evaluate

$$\frac{\partial x(t; p)}{\partial p_j} = \frac{\partial}{\partial p_j} \{\exp(At)x_o + \int_0^t \exp[A(t - \tau)] B u(\tau) d\tau\}. \tag{3.3-5}$$

A quick look at (3.3-5) reveals that the derivation of $Y_j \triangleq [\partial \exp(At)/\partial p_j]$ is essential in order to evaluate $[\partial x(t; p)/\partial p_j]$. The following approach to generate this quantity is originated independently in [1B. 7] based on a comment made by Sesak [1B. 22]. It seems that this approach is computationally preferable (it is of lower dimension) to that proposed by Reid et al. [A. 16], and is perhaps complementing the recent result given in [A. 20] which follows subsequently.

It is easy to show that for $A_j \triangleq [\partial A(p^o)/\partial p_j]$ we have.

$$Y_j(t) = \int_0^t \exp[A(t - \tau)]A_j \exp(A\tau)d\tau. \qquad (3.3\text{-}6)$$

If A and A_j commute (which is perhaps a trivial case), then

$$Y_j(t) = \int_0^t A_j \exp(At)d\tau = tA_j \exp(At). \qquad (3.3\text{-}7)$$

But in general A and A_j do not commute and the following steps for computing Y_j are taken to that effect.

CASE I – Distinct Eigenvalues: Consider the case wherein (the nominal) A has distinct eigenvalues. Then it is well known that [2A. 27]

$$\exp(At) = \sum_{i=1}^{n} \alpha_i(t)A^{i-1}, \qquad (3.3\text{-}8)$$

where the scalar functions $\alpha_i(t)$, $i = 1, ..., n$, are solutions of the following linear equations

$$\begin{Bmatrix} \exp(\lambda_1 t) \\ \exp(\lambda_2 t) \\ \cdot \\ \cdot \\ \cdot \\ \exp(\lambda_n t) \end{Bmatrix} = \begin{bmatrix} 1 & \lambda_1 & \lambda_1^2 & \lambda_1^{n-1} \\ & & & \\ \cdot & \cdot & \cdot & \cdot \\ \cdot & \cdot & \cdot & \cdot \\ \cdot & \cdot & \cdot & \cdot \\ 1 & \cdot & \cdot & \lambda_n^{n-1} \end{bmatrix} \begin{Bmatrix} \alpha_1(t) \\ \alpha_2(t) \\ \cdot \\ \cdot \\ \cdot \\ \alpha_n(t) \end{Bmatrix} \triangleq \Lambda\alpha(t). \qquad (3.3\text{-}9)$$

Here λ_i, $i = 1, 2, ..., n$, are the eigenvalues of A. It is well known that in this case Λ^{-1} exists, and $\alpha(t)$ can be uniquely determined from [note that $\Lambda^{-1} \triangleq \Gamma \triangleq (\gamma_{ij})$].

$$\alpha(t) = \Lambda^{-1} \begin{Bmatrix} \exp(\lambda_1 t) \\ \exp(\lambda_2 t) \\ \cdot \\ \cdot \\ \cdot \\ \exp(\lambda_n t) \end{Bmatrix} \triangleq \Gamma \begin{Bmatrix} \exp(\lambda_1 t) \\ \exp(\lambda_2 t) \\ \cdot \\ \cdot \\ \cdot \\ \exp(\lambda_n t) \end{Bmatrix}. \qquad (3.3\text{-}10)$$

There are a number of ways to compute these kinds of matrices efficiently (cf., [2B. 9 and 11]).

Based on (3.3-8), $Y_j(t)$ of (3.3-7) can be expressed as

$$Y_j(t) = \int_0^t [\sum_{i=1}^{n} \alpha_i(t - \tau)A^{i-1}] A_j [\sum_{i=1}^{n} \alpha_i(\tau)A^{i-1}]d\tau. \qquad (3.3\text{-}11)$$

From (3.3-10) we have:

$$\alpha_k(t - \tau) = \sum_{i=1}^{n} \gamma_{ki} e^{\lambda_i(t - \tau)}, \quad \text{and} \quad \alpha_l(\tau) = \sum_{g=1}^{n} \gamma_{lg} e^{\lambda_g \tau}. \qquad (3.3\text{-}12)$$

Then

3.3. Physical – Parameter Sensitivity Functions Generation

$$Y_j(t) = \int_0^t [\sum_{k=1}^n \sum_{i=1}^n \gamma_{ki} e^{\lambda_i(t-\tau)} A^{k-1}] A_j [\sum_{l=1}^n \sum_{g=1}^n \gamma_{lg} e^{\lambda_g \tau} A^{l-1}] d\tau. \quad (3.3\text{-}13)$$

On changing the order of summations we have

$$Y_j(t) = \int_0^t [\sum_{i=1}^n \sum_{k=1}^n \gamma_{ki} A^{k-1} e^{\lambda_i(t-\tau)}] A_j [\sum_{g=1}^n \sum_{l=1}^n \gamma_{lg} A^{l-1} e^{\lambda_g \tau}] d\tau. \quad (3.3\text{-}14)$$

Let

$$Z_i \triangleq \sum_{k=1}^n \gamma_{ki} A^{k-1}, \quad (3.3\text{-}15)$$

then we can show that

$$Y_j(t) = \int_0^t \sum_{i=1}^n Z_i e^{\lambda_i(t-\tau)} A_j \sum_{g=1}^n Z_g e^{\lambda_g \tau} d\tau \quad (3.3\text{-}16a)$$

$$= \sum_{i=1}^n Z_i A_j Z_i t\, e^{\lambda_i t} + \sum_{i=1}^n \sum_{\substack{g=1 \\ g \neq i}}^n \frac{1}{\lambda_g - \lambda_i} Z_i A_j Z_g (e^{\lambda_g t} - e^{\lambda_i t}). \quad (3.3\text{-}16b)$$

Now, using Y_j and (3.3-8) in (3.3-5) yields

$$\frac{\partial x(t; p)}{\partial p_j} = Y_j(t)x_o + \int_0^t Y_j(t-\tau)Bu(\tau)d\tau + \int_0^t \sum_{i=1}^n \alpha_i(t-\tau)A^{i-1}\frac{\partial B(p)}{\partial p_j} u(\tau)d\tau$$

$$= Y_j(t)x_o + \int_0^t Y_j(t-\tau)Bu(\tau)d\tau + \sum_{i=1}^n A^{i-1}\frac{\partial B(p)}{\partial p_j} \int_0^t \alpha_i(t-\tau)u(\tau)d\tau.$$

$$(3.3\text{-}17)$$

Substituting (3.3-16) into (3.3-17), results in an equation that involves only scalar integrations of exponentials multiplied by the input vector. The coefficient matrices in (3.3-17) can be evaluated off-line, and they mainly involve manipulations of the eigenvalues of $A(p^o)$ which are available, and furthermore $A(p^o)$ is of a lesser dimension than that of the coefficient matrix in the augmented set of differential equations.

For generating any other sensitivity function the only change occurs in A_j term, which must be substituted by the new value – say A_k. This property facilitates the overall computations for all the sensitivity functions. The main advantage of this approach over that of Reid *et al.* [A. 16] is the choice of Λ. They have considered a generalized Λ for a higher-dimensional augmented system, whose inversion and other algebraic manipulations are more involved than that of the present approach.

CASE II – Multiple Eigenvalues: Suppose that $A(p^o)$ has one eigenvalue of multiplicity $\mu + 1$ and $(n - \mu - 1)$ other eigenvalues that are distinct. Then it is well known that [2A. 27]

$$\exp(At) = \sum_{i=1}^{n} \beta_i(t) A^{i-1}, \qquad (3.3\text{-}18)$$

here $\beta_i(t)$, $i = 1, ..., n$, are the solution of the following equations:

$$t^l \exp(\lambda_1 t) = \frac{d^l}{dz^l}[1\ z\ z^2,\ ...,\ z^{n-1}]\Big|_{z=\lambda_1} \beta(t),\ \ l = 0,1, ..., \mu, \qquad (3.3\text{-}19a)$$

$$\exp(\lambda_i t) = [1\ \lambda_i\ \lambda_i^2,\ ...,\ \lambda_i^{n-1}]\beta(t),\ \text{ and }\ i = \mu + 2, ..., n. \qquad (3.3\text{-}19b)$$

Define

$$(\Gamma')^{-1} \triangleq \begin{bmatrix} 1 & \lambda_1 & \lambda_1^2 & \cdots & \lambda_1^\mu & \lambda_1^{\mu+1} & \cdots & \lambda_1^{n-1} \\ 0 & 1 & 2\lambda_1 & \cdots & \mu\lambda_1^{\mu-1} & (\mu+1)\lambda_1^\mu & \cdots & (n-1)\lambda_1^{n-2} \\ \cdot & & & & & & & \cdot \\ \cdot & & & & & & & \cdot \\ 0 & 0 & 0 & \cdots & \mu! & \cdot & \cdots & (n-1)...(n-\mu)\lambda_1^{(n-\mu-1)} \\ 1 & \lambda_{\mu+2} & \lambda_{\mu+2}^2 & \cdots & \cdot & \cdot & \cdots & \lambda_{\mu+2}^{n-1} \\ \cdot & & & & & & & \cdot \\ \cdot & & & & & & & \cdot \\ 1 & \lambda_n & \cdot & \cdot & \cdot & \cdot & \cdot & \lambda_n^{n-1} \end{bmatrix}, \qquad (3.3\text{-}20)$$

and

$$e(t) \triangleq \left[t^0 \exp(\lambda_1 t),\ \cdots,\ t^\mu \exp(\lambda_1 t),\ \exp(\lambda_{\mu+2} t),\ \cdots,\ \exp(\lambda_n t)\right]^T. \qquad (3.3\text{-}21)$$

Then it is clear that

$$\beta(t) = \Gamma' e(t). \qquad (3.3\text{-}22)$$

From (3.3-22) we can easily deduce that

$$\beta_k(t-\tau) = \sum_{i=1}^{\mu+1} \gamma'_{ki}(t-\tau)^{i-1} e^{\lambda_1(t-\tau)} + \sum_{i=\mu+2}^{n} \gamma'_{ki} e^{\lambda_i(t-\tau)}, \qquad (3.3\text{-}23)$$

$$\beta_l(\tau) = \sum_{g=1}^{\mu+1} \gamma'_{lg} \tau^{g-1} e^{\lambda_1 \tau} + \sum_{g=\mu+2}^{n} \gamma'_{lg} e^{\lambda_g \tau}. \qquad (3.3\text{-}24)$$

Then

3.3. Physical – Parameter Sensitivity Functions Generation

$$Y'_j(t) \triangleq \frac{\partial}{\partial p_j} \left[\sum_{i=1}^{n} \beta_i(t) A^{i-1} \right]$$

$$= \int_0^t \left[\sum_{k=1}^{n} \beta_k(t-\tau) A^{k-1} \right] A_j \left[\sum_{l=1}^{n} \beta_l(\tau) A^{l-1} \right] d\tau. \tag{3.3-25}$$

The above expression, considering (3.3-24) and (3.3-25), becomes

$$Y'_j(t) = \int_0^t \left[\sum_{k=1}^{n} \left[\sum_{i=1}^{\mu+1} \gamma'_{ki}(t-\tau)^{i-1} e^{\lambda_1(t-\tau)} + \sum_{i=\mu+2}^{n} \gamma'_{ki} e^{\lambda_i(t-\tau)} \right] A^{k-1} \right] A_j$$

$$\times \left[\sum_{l=1}^{n} \left[\sum_{g=1}^{\mu+1} \gamma'_{lg} \tau^{g-1} e^{\lambda_1 \tau} + \sum_{g=\mu+2}^{n} \gamma'_{lg} e^{\lambda_g \tau} \right] A^{l-1} \right] d\tau. \tag{3.3-26}$$

Changing the order of summations in the above expression gives

$$Y'_j(t) = \int_0^t \left[\sum_{i=1}^{\mu+1} \sum_{k=1}^{n} \gamma'_{ki} A^{k-1} (t-\tau)^{i-1} e^{\lambda_1(t-\tau)} + \sum_{i=\mu+2}^{n} \sum_{k=1}^{n} \gamma'_{ki} A^{k-1} e^{\lambda_i(t-\tau)} \right] A_j$$

$$\times \left[\sum_{g=1}^{\mu+1} \sum_{l=1}^{n} \gamma'_{lg} A^{l-1} \tau^{g-1} e^{\lambda_1 \tau} + \sum_{g=\mu+2}^{n} \sum_{l=1}^{n} \gamma'_{lg} A^{l-1} e^{\lambda_g \tau} \right] d\tau. \tag{3.3-27}$$

Let

$$Z'_i \triangleq \sum_{k=1}^{n} \gamma'_{ki} A^{k-1}, \tag{3.3-28}$$

then $Y_j(t)$ can be written as

$$Y'_j(t) \triangleq I_1 + I_2 + I_3 + I_4. \tag{3.3-29}$$

Here I_k, $k = 1, \ldots, 4$, are, respectively,

$$I_1 \triangleq e^{\lambda_1 t} \sum_{i=1}^{\mu+1} Z'_i A_j \sum_{g=1}^{\mu+1} Z'_g \int_0^t (t-\tau)^{i-1} \tau^{g-1} d\tau, \tag{3.3-30}$$

$$I_2 \triangleq e^{\lambda_1 t} \sum_{i=1}^{\mu+1} Z'_i A_j \sum_{g=\mu+2}^{n} Z'_g \int_0^t (t-\tau)^{i-1} e^{(\lambda_g - \lambda_1)\tau} d\tau, \tag{3.3-31}$$

$$I_3 \triangleq \sum_{i=\mu+2}^{n} e^{\lambda_i t} Z'_i A_j \sum_{g=1}^{\mu+1} Z'_g \int_0^t \tau^{g-1} e^{(\lambda_1 - \lambda_i)\tau} d\tau, \tag{3.3-32}$$

$$I_4 \triangleq \sum_{i=\mu+2}^{n} e^{\lambda_i t} Z_i' A_j \sum_{g=\mu+2}^{n} Z_g' \int_0^t e^{(\lambda_g - \lambda_i)\tau} d\tau. \qquad (3.3-33)$$

Now, the message is clear. For A which has several eigenvalues with different multiplicities the above technique, must be used accordingly. We must point out that the integrals in I_1, I_2, I_3, I_4 are tabulated in most computer libraries and these are easily accessible. To use (3.3-30) to (3.3-33) no additional property for A is required and these equations can be generated upon knowing the eigenvalues of A which must be computed in any case. An alternative and compact representation of these sensitivity functions is derived based on the matrix calculus and its extension in [A. 5].

REMARKS 3.3-4: (i) This direct computation can also be used to compute $\partial x/\partial K$, where $u = -Kx$ is a closed-loop optimal control law for a system described by (3.3-1). (ii) For the case of sensitivity functions generation in an open-loop unreconstructible system we can follow the same procedure as that of the structural-parameter sensitivity functions generation (cf., Remark 3.2-2).

3.3.2.3. Alternative Method To Compute Sensitivity Functions

An alternative method to compute $Y_j \triangleq [\partial \exp(At)/\partial p_j]$ of (3.3-6) which is essential for generating the system trajectory sensitivity function is reported in [A. 20] for the case where A is diagonalizable. This requirement on the diagonalizability of A is not needed in the analyses of the previous section.

Suppose that there exists $V \in R^{n \times n}$ such that $A(p^o)V - V A(p^o) = 0$. Then $\exp[A(p^o)t]V = V \exp[A(p^o)t]$, and

$$\int_0^t \exp[-A(p^o)\tau] \, V \, \exp[A(p^o)\tau] d\tau = tV. \qquad (3.3-34)$$

Also for any $U \in R^{n \times n}$,

$$e^{-A(p^o)t} [A(p^o)U - U A(p^o)] e^{A(p^o)t} = -\frac{d}{dt}[e^{-A(p^o)t} U e^{A(p^o)t}]. \qquad (3.3-35)$$

PROPOSITION 3.3-1 [A. 20]: For $A(p^o)$ of (3.3-1), if there exist $V \in R^{n \times n}$ and $U \in R^{n \times n}$ such that

$$A_j \triangleq \frac{\partial A(p^o)}{\partial p_j} = V + [A(p^o)U - U A(p^o)]. \qquad (3.3-36)$$

and

$$A(p^o) V - V A(p^o) = 0. \qquad (3.3-37)$$

Then

$$Y_j(t) \triangleq \int_0^t \exp[A(t-\tau)]A_j \exp(A\tau)d\tau$$

$$= t \exp(At) V + [\exp(At) U - U \exp(At)]. \quad (3.3\text{-}38)$$

Here A means, of course, $A(p^o)$.

∎

Equation (3.3-38) which explicitly contains $\exp(At)$ is an alternative representation of (3.3-16b). Although (3.3-38) is in a compact form, we believe that to generate the corresponding V and U as well as $\exp(At)$ which are detailed in [A. 20] may become as involve as computing (3.3-16b) directly. Furthermore, this result is, of course, valid only for a diagonalizable A.

3.3.3. Formal Analysis for Closed – Loop Reconstructible System

Consider a closed-loop *DCLTI – GS* described by

$$\dot{x}(t; p) = A(p)x(t; p) + B(p)u(t; p), \quad x(0; p) = x_0. \quad (3.3\text{-}39)$$

The sensitivity function with respect to p_j is

$$\frac{d}{dt}\frac{\partial x(t; p)}{\partial p_j} = A(p)\frac{\partial x(t; p)}{\partial p_j} + \frac{\partial A(p)}{\partial p_j} x(t; p) + \frac{\partial B(p)}{\partial p_j} u(t; p) + B(p)\frac{\partial u(t; p)}{\partial p_j},$$

$$\left.\frac{\partial x(t; p)}{\partial p_j}\right|_{t=0} = 0. \quad (3.3\text{-}40)$$

It is clear that $\partial u(t; p)/\partial p_j$ must be known in advance in order to derive $\partial x(t; p)/\partial p_j$ for a closed-loop system. However, we realize that, in general, an exact value of $\partial x(t; p)/\partial p_j$ is not accessible in any closed-loop implementation of a system, unless a precise form of input function is given. Even then there is a question on the accuracy of closed-loop sensitivity function. Equation (3.3-40) is derived merely for the sake of a formal presentation. Additional comments about trajectory-sensitivity optimization involving this relationship are deferred to Section 6.5.

3.4. SENSITIVITY FUNCTION GENERATION: FREQUENCY DOMAIN

System sensitivity analysis is commonly carried out interchangeably in both time domain and frequency domain. There are, however, a number of special frequency-domain sensitivity measures that we have not yet introduced. But for the purpose of the present study Laplace transforms of the earlier time-domain sensitivity functions seem sufficient as integral transforms which map the time- (or real-) domain sensitivity functions to that of the frequency- (or complex-) domain. This method is, of course, valid when the sensitivity function in the time domain is available. Otherwise, we must start from step one to generate frequency-domain sensitivity functions more or less in the same manner as we do for their counterparts in the time domain. Finally, we must point out that this Laplace transform is easily generated because most of these results are in a convolution form whose Laplace transform is simply the product of the corresponding two Laplace transform functions.

The following list is a sample of the frequency-domain sensitivity functions for some of our earlier equations as indicated by primes.

$$\Delta x_o(s) = (sI - A - \Delta A)^{-1}\{\Delta A x_o(s) + \Delta B u_o(s)\}, \qquad (3.2\text{-}7)'$$

$$f_{o1}(s) = (sI - A)^{-1}\{A x_o(s) + B u_o(s)\}, \qquad (3.2\text{-}46)'$$

$$f_{o2}(s) = 2A(sI - A)^{-1}f_{o1}(s), \qquad (3.2\text{-}49)'$$

$$\Delta x_c(s) = (sI - A - \Delta A)^{-1}\{\Delta A x_c(s) + \Delta B u_c(s; q) + (B + \Delta B)\Delta u_c(s; q)\}, \qquad (3.2\text{-}62)'$$

$$f''_{c1}(s) = (sI - A + BK)^{-1}\{(A - 2BK)x(s) + 2BGy(s)\}, \qquad (3.2\text{-}72)'$$

$$\frac{\partial x_o(s; p)}{\partial p_j} = (sI - A)^{-1}\{\frac{\partial A(p)}{\partial p_j}x_o(s; p) + \frac{\partial B(p)}{\partial p_j}u_o(s)\}. \qquad (3.3\text{-}2a)'$$

3.5. AUGMENTED STATE AND STATE-SENSITIVITY DIFFERENTIAL EQUATIONS: SYSTEM MODEL

3.5.1. Introduction

It is now clear that in analyzing the sensitivity differential equations, system state acts as a forcing function. Therefore the system differential equation must be solved first, in order to simulate (or to solve numerically) the sensitivity differential equations. In computer-aided design entailing simulating the system state and its sensitivity functions on a computer, these sets of differential equations can be augmented (and the set of augmented differential equations is called system model), and they can be studied simultaneously. In this section these sets of equations are given by a compact notation. Further applications and the study of their properties are deferred to Chapter Four.

3.5.2. Open – Loop System Model

For an open-loop *DCLTI* reconstructible-generating system described by

$$\dot{x}(t) = Ax(t) + Bu(t), \qquad x(0) = x_o, \qquad (3.5\text{-}1)$$

the structural-parameter sensitivity function is

3.5. Augmented State and State-Sensitivity Differential Equations

$$\frac{d}{dt}\frac{\partial x(t)}{\partial \eta_{ij}} = A\frac{\partial x(t)}{\partial \eta_{ij}} + \frac{\partial A}{\partial \eta_{ij}}x(t) + \frac{\partial B}{\partial \eta_{ij}}u(t), \quad \frac{\partial x(t)}{\partial \eta_{ij}}\bigg|_{t=0} = 0, \quad (3.5\text{-}2)$$

where $H \triangleq (\eta_{ij}) = A \cup B$.

The corresponding set of augmented differential equations or system model is

$$\dot{m}_{ij}(t) = \tilde{A}_{ij}m_{ij}(t) + \tilde{B}_{ij}u(t), \quad m_{ij}(0) = m_o, \quad (3.5\text{-}3)$$

where

$$m_{ij}(t) \triangleq \left\{\begin{array}{c} x(t) \\ \dfrac{\partial x(t)}{\partial \eta_{ij}} \end{array}\right\}_{2n \times 1}, \quad \tilde{B}_{ij} \triangleq \left\{\begin{array}{c} B \\ \dfrac{\partial B}{\partial \eta_{ij}} \end{array}\right\}, \quad m_{ij}(0) = \left\{\begin{array}{c} x_o \\ 0 \end{array}\right\}, \text{ and}$$

$$\tilde{A}_{ij} \triangleq \begin{bmatrix} A & 0 \\ \dfrac{\partial A}{\partial \eta_{ij}} & A \end{bmatrix}_{2n \times 2n}. \quad (3.5\text{-}4)$$

Here if (η_{ij}) is (b_{kl}), then $\dfrac{\partial A}{\partial \eta_{ij}} = 0$, and if (η_{ij}) is (a_{kl}), then $\dfrac{\partial B}{\partial \eta_{ij}} = 0$. The $\tilde{A}_{ij}$ and $\tilde{B}_{ij}$ are matrices of system model *(MSM)*.

The system model for system *PPSF* is

$$\dot{m}_j(t) = \tilde{A}_j(p)m_j(t) + \tilde{B}_j(p)u(t), \quad m_j(0) = m_o, \quad (3.5\text{-}5a)$$

where

$$m_j(t) \triangleq \left\{\begin{array}{c} x(t) \\ \dfrac{\partial x(t)}{\partial p_j} \end{array}\right\}_{2n \times 1}, \quad \tilde{B}_j \triangleq \begin{bmatrix} B(p) \\ \dfrac{\partial B(p)}{\partial p_j} \end{bmatrix}, \quad m_j(0) = \left\{\begin{array}{c} x_o \\ 0 \end{array}\right\}, \text{ and}$$

$$\tilde{A}_j(p) \triangleq \begin{bmatrix} A(p) & 0 \\ \dfrac{\partial A(p)}{\partial p_j} & A(p) \end{bmatrix}_{2n \times 2n}. \quad (3.5\text{-}3b)$$

In general considering all *PPSF's*, the coefficient matrix of homogeneous part in the set of augmented differential equations is as follows.

$$\tilde{A}_r(p) \triangleq \begin{bmatrix} A(p) & 0 & 0 & \cdots & 0 \\ \dfrac{\partial A(p)}{\partial p_1} & A(p) & 0 & \cdots & 0 \\ \dfrac{\partial A(p)}{\partial p_2} & 0 & A(p) & \cdots & 0 \\ \cdot & \cdot & \cdot & \cdot & \cdot \\ \cdot & \cdot & \cdot & & \cdot \\ \cdot & \cdot & \cdot & & \cdot \\ \dfrac{\partial A(p)}{\partial p_r} & 0 & 0 & \cdots & A(p) \end{bmatrix} \qquad (3.5\text{-}6)$$

Similarly, the coefficient matrix of the inhomogeneous part is

$$\tilde{B}_r^T(p) \triangleq [B^T(p), \dfrac{\partial B^T(p)}{\partial p_1}, \cdots, \dfrac{\partial B^T(p)}{\partial p_r}]^T. \qquad (3.5\text{-}7)$$

The matrices similar to (3.5-6) and (3.5-7) may be established for any other system sensitivity measure, and for higher-order sensitivity functions. Also system model for a closed-loop system can be generated likewise.

3.6. SENSITIVITY INVARIANCE

3.6.1. Introduction

One of the most interesting properties of certain sensitivity measures is the fact that they may be invariant. In other words, there may exist an algebraic or a differential relationship among several presumably independent sensitivity functions, as described subsequently in Examples 3.6-1 & 2. This invariance property, which is not always obvious, must be accounted for in any sensitivity optimization scheme. In particular, the number of sensitivity invariance among various sensitivity measures in circuit theory is large and that is an active area of research. In the following we present two sensitivity invariances for a typical *DCLTI – GS*.

The theory of invariance that is briefly reviewed in conjunction with sensitivity reduction and robustness analysis in Section 6.10 is a separate issue.

3.6.2. Sensitivity Measure Invariants for the Linear System

The following result is a sensitivity invariant which can be used to show the severity of requiring a closed-loop system acts the same as an open-loop system regarding the corresponding total-sensitivity functions.

PROPERTY 3.6-1: Consider the state equation

3.6. Sensitivity Invariance

$$\dot{x}(t) = Ax(t), \quad x(0) = x_o \neq 0, \qquad (3.6\text{-}1)$$

where $x \in R^n$ is the system state vector, and $A = (a_{ij}) \in R^{n \times n}$ is the system coefficient matrix. Then the total-sensitivity function always satisfies

$$\sum_{i,j} a_{ij} \frac{\partial x(t)}{\partial a_{ij}} = t\dot{x}(t), \qquad (3.6\text{-}2)$$

or equivalently the total, normalized-sensitivity function always satisfies

$$\sum_{i,j} S_{a_{ij}}^{x_k(t)} = S_t^{x_k(t)}, \quad k = 1, \ldots, n. \qquad (3.6\text{-}3)$$

PROOF: It is easy to show that (cf., Proposition 3.2-1)

$$\sum_{i,j} a_{ij} \frac{\partial x(t)}{\partial a_{ij}} = \int_0^t \exp[A(t-\tau)]Ax(\tau)d\tau. \qquad (3.6\text{-}4)$$

Substituting for $x(t)$ in (3.6-4), from the solution of (3.6-1), yields.

$$\sum_{i,j} a_{ij} \frac{\partial x(t)}{\partial a_{ij}} = \int_0^t \exp[A(t-\tau)]A \exp(A\tau)d\tau \, x_o. \qquad (3.6\text{-}5)$$

Since A and $\exp(At)$ as well as $A(t-\tau)$ and $A\tau$ commute, thus (3.6-5) becomes

$$\sum_{i,j} a_{ij} \frac{\partial x(t)}{\partial a_{ij}} = \int_0^t A \exp(At)d\tau \, x_o = t \, A \exp(At) \, x_o. \qquad (3.6\text{-}6)$$

Equation (3.6-6) establishes (3.6-2) of the conclusion. Equation (3.6-3) follows from (3.6-2) and the definition of the total, normalized-sensitivity functions. Certainly for a nonsingular A, (3.6-2) does not vanish identically.

Equation (3.6-2) can be Laplace transformed: thus,

$$\sum_{i,j} a_{ij} \frac{\partial x_k(s)}{\partial a_{ij}} = -\frac{d}{ds}[L(\dot{x}(t))] = -\frac{d}{ds}[sx_k(s) - x_k(0)] = -x_k(s) - s\frac{dx_k(s)}{ds}. \qquad (3.6\text{-}7)$$

Dividing both sides of (3.6-7) by $x_k(s)$ and rearranging the terms yields

$$\sum_{i,j} \frac{a_{ij}}{x_k(s)} \frac{\partial x_k(s)}{\partial a_{ij}} + \frac{s}{x_k(s)} \frac{\partial x_k(s)}{\partial s} = -1. \qquad (3.6\text{-}8)$$

That means the total, normalized-sensitivity function of $x_k(s)$ in terms of all the corresponding structural parameters and the complex variable s is invariant.

EXAMPLE 3.6-1 (Sensitivity comparison): Consider a closed-loop *DCLTI – GS* with a linear feedback control law as follows.

$$\dot{x}_c(t) = A_c x_c(t) + B_c u_c(t), \qquad (3.6\text{-}9)$$

$$u_c(t) = -Kx_c(t) + g, \qquad (3.6\text{-}10)$$

where A_c, B_c are system matrices and we assume that $g \in R^m$ is constant ($\dot{g}(t) \equiv 0$). The corresponding set of augmented differential equations for this system is

$$\dot{y}_c(t) \triangleq \begin{Bmatrix} \dot{x}_c(t) \\ \dot{g}(t) \end{Bmatrix} = \begin{bmatrix} A_c - B_c K & B_c \\ 0 & 0 \end{bmatrix} \begin{Bmatrix} x_c(t) \\ g(t) \end{Bmatrix}, \quad y_c(0) \triangleq \begin{Bmatrix} x_c(0) \\ g \end{Bmatrix}. \quad (3.6\text{-}11)$$

Now, consider an open-loop design described by

$$\dot{x}_o(t) = A_o x_o(t) + B_o u_o(t), \quad x_o(0) = x_o. \quad (3.6\text{-}12)$$

Suppose that $u_o(t)$, the open-loop input, is a constant vector ($\dot{u}_o(t) \equiv 0$). Compare the corresponding two system — *TSF's*.

SOLUTION: The (3.6-12) can be exactly rewritten as

$$\dot{y}_o(t) \triangleq \begin{Bmatrix} \dot{x}_o(t) \\ \dot{u}_o(t) \end{Bmatrix} = \begin{bmatrix} A_o & B_o \\ 0 & 0 \end{bmatrix} \begin{Bmatrix} x_o(t) \\ u_o(t) \end{Bmatrix}, \quad y_o(0) \triangleq \begin{Bmatrix} x_o(0) \\ u_o \end{Bmatrix}. \quad (3.6\text{-}13)$$

By Property 3.6-1 the total-sensitivity function of these two sets of homogeneous differential equations in terms of (A_o, B_o) and (A_c, B_c) is invariant.

Thus, if

$$\dot{y}_o(t) \equiv \dot{y}_c(t), \quad (3.6\text{-}14)$$

then from the total-sensitivity function *(TSF)* point of view the two designs are equivalent. A sufficient set of conditions assuming such is

$$\begin{Bmatrix} x_c(0) \\ g \end{Bmatrix} = \begin{Bmatrix} x_o(0) \\ u_o \end{Bmatrix} \Rightarrow \begin{Bmatrix} x_c(0) = x_o(0) \\ g = u_o \end{Bmatrix}, \quad (3.6\text{-}15)$$

and

$$A_c - B_c K = A_o, \quad B_c = B_o. \quad (3.6\text{-}16)$$

For any closed-loop design wherein the plant is characterized by (3.6-9), when utilizing the linear state-feedback control law (3.6-10), an open-loop system with coefficient matrices $(A_c - B_c K)$ and B_c can be determined such that the open-loop and the closed-loop systems have exactly the same total-sensitivity function, provided that the open-loop system has a constant input.[1] Fig. 3.6-1 shows a closed-loop system and an open-loop equivalent in the sense of *TSF*. The above consequence also holds for the case when $g = u_o = 0$. These situations are, of course, very special cases and these requirements are very severe. Thus the main implication of this example is that, in general, we must expect that the equality between the closed-loop and the open-loop designs in terms of the *TSF* or perhaps other sensitivity measures does not exist. In fact, in Section 6.2 we study cases when these two designs are compared with each other, in order to establish the sufficient

---

[1] In Section 6.2 it is shown that this property does not hold for a general linear system.

3.6. Sensitivity Invariance

condition that must be met such that the closed-loop design performs superior when compared with the open-loop design.

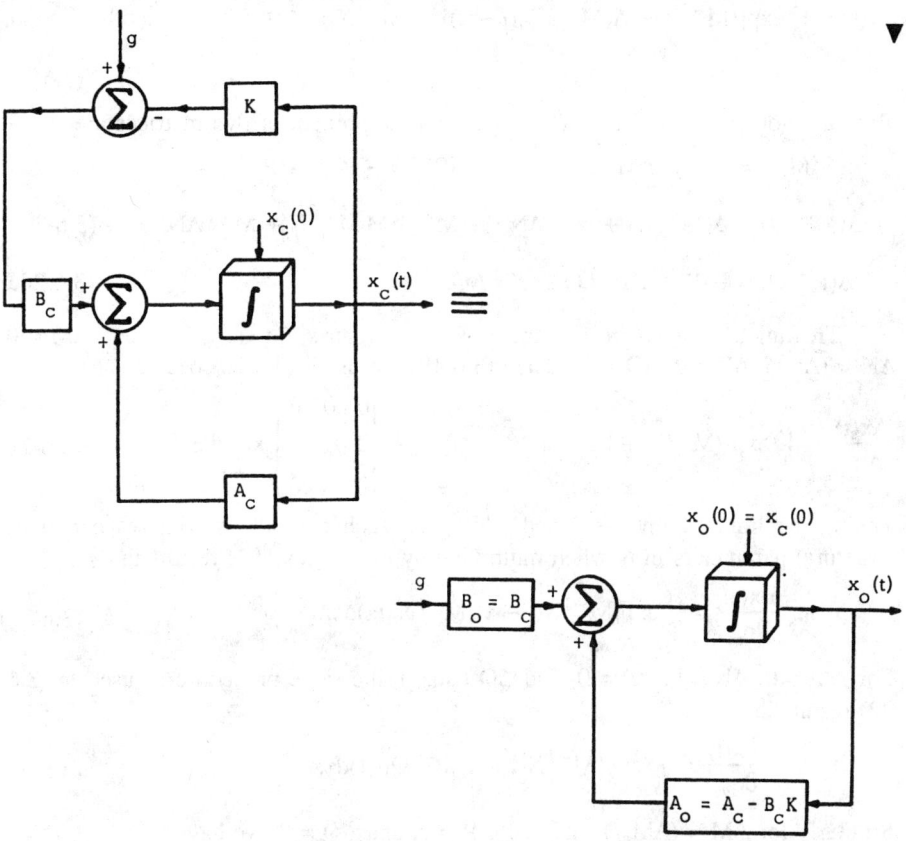

Fig. 3.6-1. An open-loop and a closed-loop equivalent in the sense of *TSF*.

The following theorem demonstrates an important invariance in the linear systems.

THEOREM 3.6-1: Consider the state equation of an open-loop *DCLTIS*,

$$\dot{x}(t) = M^{-1}Nx(t) + P^{-1}Qu(t), \quad x(0) = x_o; \tag{3.6-17}$$

where $M = (m_{kl})$, $N = (n_{rs})$, $P = (p_{ef})$, $Q = (q_{vw})$. Then the total-sensitivity function when $(\alpha_{ij}) = (m_{kl}) \cup (n_{rs}) \cup (p_{ef}) \cup (q_{vw})$ always satisfies

$$\sum_{i,j} \alpha_{ij} \frac{\partial x(t)}{\partial \alpha_{ij}} = 0. \tag{3.6-18}$$

PROOF: From Proposition 3.2-1, if M, N, P, Q and x(t) are replaced by M + ΔM, N + ΔN, P + ΔP, Q + ΔQ, and x(t) + Δx(t), respectively, with Δx(0) = 0, then

$$\Delta x(t) = \int_0^t \exp[(M^{-1}N + \Delta(M^{-1}N))(t - \tau)]\{\Delta(M^{-1}N)x(\tau) + \Delta(P^{-1}Q)u(\tau)\}d\tau.$$

(3.6-19)

The variations in the inverse of a matrix and the product of two matrices are

$$\Delta M^{-1} = -M^{-1}\Delta M M^{-1}, \quad \text{and} \quad \Delta P^{-1} = -P^{-1}\Delta P P^{-1}, \tag{3.6-20a}$$

$$\Delta(M^{-1}N) = \Delta(M^{-1})N + M^{-1}\Delta N = -M^{-1}\Delta M M^{-1}N + M^{-1}\Delta N, \tag{3.6-20b}$$

$$\Delta(P^{-1}Q) = -P^{-1}\Delta P P^{-1}Q + P^{-1}\Delta Q. \tag{3.6-20c}$$

Treating changes in a single system matrix at a time, let $\Delta M = 0$, $\Delta N = (\Delta n_{rs})$, $\Delta P = 0$, $\Delta Q = 0$. Then (3.6-19), on using (3.6-20) becomes

$$\frac{\Delta x(t)}{\Delta n_{rs}} = \int_0^t \exp[(M^{-1}N + M^{-1}\Delta N)(t - \tau)]M^{-1}\begin{bmatrix} 0 & 0 & 0 \\ 0 & 1_{rs} & 0 \\ 0 & 0 & 0 \end{bmatrix} x(\tau)d\tau. \tag{3.6-21}$$

Taking the limit as $\Delta n_{rs} \to 0$ and adding all such derivatives with respect to the structural parameters in N when multiplied by the entries of N results in

$$\sum_{r,s} n_{rs} \frac{\partial x(t)}{\partial n_{rs}} = \int_0^t \exp[M^{-1}N(t - \tau)]M^{-1}Nx(\tau)d\tau. \tag{3.6-22}$$

For $\Delta M = 0$, $\Delta N = 0$, $\Delta P = 0$, and ΔQ (Δq_{vw}), the same procedure as used in (3.6-22) results in

$$\sum_{v,w} q_{vw} \frac{\partial x(t)}{\partial q_{vw}} = \int_0^t \exp[M^{-1}N(t - \tau)]P^{-1}Qu(\tau)d\tau. \tag{3.6-23}$$

Similarly for $\Delta M = (\Delta M_{kl})$, $\Delta N = 0$, $\Delta P = 0$, and $\Delta Q = 0$, we have

$$\sum_{k,l} m_{kl} \frac{\partial x(t)}{\partial m_{kl}} = \int_0^t \exp[M^{-1}N(t - \tau)](-M^{-1}N)x(\tau)d\tau. \tag{3.6-24}$$

Finally, for $\Delta M = 0$, $\Delta N = 0$, $\Delta P = (\Delta p_{ef})$, and $\Delta Q = 0$, we get

$$\sum_{e,f} p_{ef} \frac{\partial x(t)}{\partial p_{e,f}} = \int_0^t \exp[M^{-1}N(t - \tau)](-P^{-1}Q)u(\tau)d\tau. \tag{3.6-25}$$

Adding (3.6-22) to (3.6-25) results in the conclusion. ∎

SPECIAL CASE OF THEOREM 3.6-1: Let M = P, then (3.6-17) becomes

$$\dot{x}(t) = M^{-1}Nx(t) + M^{-1}Qu(t), \quad \text{or} \quad M\dot{x}(t) = Nx(t) + Qu(t), \quad x(0) = x_o. \tag{3.6-26}$$

The property of Theorem 3.6-1 is still true for (3.6-26), i.e.,

3.6. Sensitivity Invariance

$$\sum_{i,j} \alpha_{ij} \frac{\partial x(t)}{\partial \alpha_{ij}} = 0, \quad \text{here } (\alpha_{ij}) = M \cup N \cup Q. \tag{3.6-27}$$

∎

It is also of special interest to note that equation (3.6-26), with M, N, and Q having entries with linear combinations of circuit elements or zero (i.e., no other constant), characterizes a large class of networks expressed in the state-space representation, thus having the property stated in Theorem 3.6-1.

COROLLARY 3.6-1: The second-order, total-sensitivity function of Theorem 3.6-1 is identically zero.

PROOF: Straightforward differentiation of (3.6-18) results in the following:

$$\sum_{i,j} \sum_{m,n} \alpha_{ij} \alpha_{mn} \frac{\partial^2 x(t)}{\partial \alpha_{ij} \partial \alpha_{mn}} = 0. \tag{3.6-28}$$

The results of Theorem 3.6-1 and Corollary 3.6-1 hold in the domain of the integral transform of the system state with respect to the real variable t.

EXAMPLE 3.6-2: Consider a ladder RC-network of Fig. 3.6-2. Form the state-space representation of this network and calculate the total-sensitivity functions of the output voltage, input current, current through R_{n-1}, and the input impedance with respect to R's and C's.

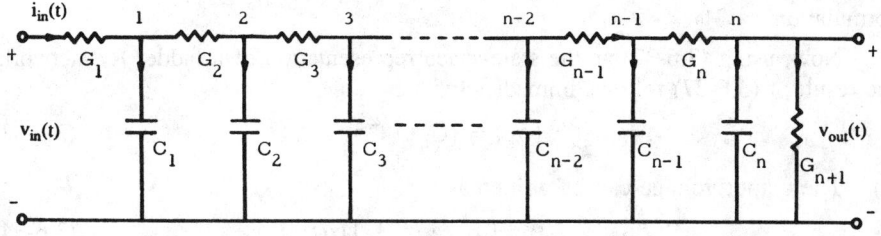

Fig. 3.6-2. Ladder RC-Network.

SOLUTION: Let $x(t) = [x_1(t), \ldots, x_n(t)]^T$ be the voltage across capacitors. Then using KCL at nodes 1, 2, ..., n gives

$$C_1 \frac{dx_1}{dt} = G_1(v_{in} - x_1) + G_2(x_2 - x_1)$$

$$\vdots$$

$$C_n \frac{dx_n}{dt} = G_n(v_{n-1} - x_n) + G_{n+1}(-x_n). \qquad (3.6\text{-}29)$$

An equivalent state-space representation with the following notation

$$C_{nn} \triangleq \text{diag}(C_1, C_2, \cdots, C_n), \quad g \triangleq [G_1, 0, \cdots, 0]^T, \qquad (3.6\text{-}30)$$

and

$$G_{nn} \triangleq \begin{bmatrix} -G_1 - G_2 & G_2 & 0 & 0 & 0 & 0 & 0 \\ G_2 & -G_2 - G_3 & G_3 & 0 & 0 & 0 & 0 \\ 0 & \cdot & \cdot & \cdot & 0 & 0 & \cdot \\ 0 & 0 & 0 & 0 & \cdot & \cdot & 0 \\ 0 & 0 & 0 & 0 & 0 & G_n & -G_n - G_{n+1} \end{bmatrix}, \qquad (3.6\text{-}31)$$

becomes

$$C_{nn} \frac{dx(t)}{dt} = G_{nn} x(t) + gu(t), \quad x(0) = x_o. \qquad (3.6\text{-}32)$$

Generally speaking, much effort is entailed in deriving equations such as (3.6-32) for a large class of networks. However, in this example determination of (3.6-32) is straightforward for the above relatively simple-structured network. Interested readers should consult Kuh and Rohrer [7A. 6], for further insight in this type of formulation.

Now, using (3.6-32) as the state-space representation of a ladder RC-network, the result of (3.6-27) follows immediately. Let

$$\{\alpha_k\} = \{G_i \cup C_j\}, \qquad (3.6\text{-}33)$$

(a) The output voltage can be written as

$$v_{out}(t) = [0, \cdots, 0, 1]x(t). \qquad (3.6\text{-}34)$$

Thus the total-sensitivity function of $v_{out}(t)$ is

$$\sum_k \alpha_k \frac{\partial v_{out}(t)}{\partial \alpha_k} \triangleq \sum_{i=1}^{n+1} G_i \frac{\partial v_{out}(t)}{\partial G_i} + \sum_{j=1}^{n} C_j \frac{\partial v_{out}(t)}{\partial C_j}$$

$$= [0, \cdots, 0, 1] \sum_{\alpha_k} \alpha_k \frac{\partial x(t)}{\partial \alpha_k}. \qquad (3.6\text{-}35)$$

3.6. Sensitivity Invariance

Since C_{mn}^{-1} of (3.6-30) exists, then by (3.6-27) the right-hand side of (3.6-35) is equal to zero, and

$$\sum_k \alpha_k \frac{\partial v_{out}(t)}{\partial \alpha_k} = 0. \tag{3.6-36}$$

(b) The input current can be written as

$$i_{in}(t) = G_1[v_{in}(t) - x_1(t)] = [-G_1, 0, \cdots, 0]x(t) + G_1 v_{in}(t). \tag{3.6-37}$$

Similarly, it can be shown that using (3.6-27) gives

$$\sum_k \alpha_k \frac{\partial i_{in}(t)}{\partial \alpha_k} = i_{in}(t), \quad \text{or}$$

$$\sum_k \frac{\alpha_k}{i_{in}(t)} \cdot \frac{\partial i_{in}(t)}{\partial \alpha_k} = 1. \tag{3.6-38}$$

(c) The current through R_{n-1} can be expressed as

$$i_{n-1}(t) = G_{n-1}[x_{n-2}(t) - x_{n-1}(t)]$$

$$= G_{n-1}[0, \cdots, 0, 1_{n-2}, -1_{n-1}, 0]x(t). \tag{3.6-39}$$

Similarly, it can be shown that

$$\sum_k \frac{\alpha_k}{i_{n-1}} \frac{\partial i_{n-1}}{\partial \alpha_k} = 1. \tag{3.6-40}$$

(d) The input impedance can be written as

$$Z_d(s) = v_{in}(s)/i_{in}(s). \tag{3.6-41}$$

Since the result of Corollary 3.6-1 is also true in the frequency domain, thus the above formulation results in

$$\sum_k \alpha_k \frac{\partial Z_d}{\partial \alpha_k} = v_{in}(s) \sum_k \alpha_k \frac{\partial (1/i_{in}(s))}{\partial \alpha_k}, \quad \text{or}$$

$$\sum_k \alpha_k \frac{\partial Z_d}{\partial d_k} = -Z_d \sum_k \frac{\alpha_k}{i_{in}} \frac{\partial i_{in}}{\partial \alpha_k} = -Z_d. \tag{3.6-42}$$

▼

REMARK 3.6-1: Equation (3.6-35) with $R_i = 1/G_i$ can be written as

$$\sum_{i=1}^{n+1} R_{i+1} \frac{\partial v_{out}(t)}{\partial R_i} = \sum_{j=1}^{n} C_j \frac{\partial v_{out}(t)}{\partial C_j}. \tag{3.6-43}$$

This equation is reported in the literature as a sensitivity invariance property of the ladder RC-network.

3.7. REFERENCES

*[A] Direct Computation and/or Simulation
of Sensitivity Functions*

[1] J.C. Allwright, "Sensitivity function determination for linear feedback systems," *Elect. Lett.*, vol. 6, no. 13, pp. 390-391, June 1970.

[2] S.P. Bingulac, "Simultaneous generation of the second-order sensitivity functions," *IEEE-TAC*, vol. AC-11, no. 3, pp. 563-566, July 1966.

[3] R.N. Biswas and E.S. Kuh, "Multiparameter sensitivity measure for linear systems," *IEEE-TCT*, vol. CT-18, no. 6, pp. 718-719, Nov. 1971.

[4] V. Brajovic, "On-line generation of sensitivity functions using correlation," *IJSS*, vol. 8, no. 12, pp. 1321-1326, Dec. 1977.

[5] J.W. Brewer, "Matrix calculus and sensitivity analysis of linear dynamic systems," *IEEE-TAC*, vol. AC-23, no. 4, pp. 748-751, Aug. 1978.

[6] R.C. Dorf, "System sensitivity in the time-domain," in *Proc. Allerton Conf.*, pp. 46-62, 1965.

[7] M. Eslami and D.P. Brown, "On time domain sensitivity of linear systems," in *Proc. Allerton Conf.*, pp. 382-387, Oct. 1975.

[8] C. Gori-Giorgi and O.M. Grasselli, "Minimal order representations of parameter sensitive impulse response matrices," *RA*, vol. 6, nos. 2-3, pp. 91-107, 1975.

[9] P.V. Kokotovic, "Method of sensitivity points in the investigation and optimization of linear control systems," *ARC*, pp. 1512-1518, Dec. 1964.

[10] P.V. Kokotovic and R.S. Rutman, "Sensitivity matrices and their modelling," *ARC*, vol. 27, no. 6, pp. 1067-1079, June 1966.

[11] W.M. Mazer, "Specification of linear feedback system sensitivity function," *IRE-TAC*, vol. AC-5, pp. 85-93, June 1960.

[12] H.F. Meissinger, "The use of parameter influence coefficients in computer analysis of dynamic systems," in *Proc. Western Joint Computer Conference*, San Francisco, May 1960.

[13] K.S. Miller and F.J. Murray, "A mathematical basis for an error analysis of differential analyzers," *JMP*, vol. 32, nos. 2 and 3, pp. 136-163, July/Oct. 1953.

[14] P.N. Paraskevopoulos, P.G. Sklavounos, and D.A. Karkas, "A new orthogonal series approach to sensitivity analysis," *JFI*, vol. 327, no. 3, pp. 419-433, 1990.

[15] L.A. Pipes, "A perturbation method for the solution of linear matrix differential equations," *JFI*, vol. 283, no. 5, pp. 357-371, May 1967.

[16] J.G. Reid, P.S. Maybeck, R.B. Asher, and J.D. Dillow, "An algebraic representation of parameter sensitivity in linear time-invariant systems," *JFI*, vol. 301, nos. 1 and 2, pp. 123-141, Jan./Feb. 1976.

[17] R.S. Rutman and P.V. Kokotovic, "On the generalization of sensitivity equations," in *Proc. Allerton Conf.*, pp. 27-34, 1965.

[18] L.A. Shirokov, "Compact higher-order sensitivity analyzers for linear dynamic systems," *ARC*, vol. 51, no. 1, pp. 15-22, 1990.

[19] Y. Takahara and M.D. Mesarovic, "Global sensitivity of linear dynamic systems," in *Proc. Joint Auto. Contr. Conf.*, pp. 596-602, 1967.

[20] T.L. Wuu, R.G. Becker, and E. Polak, "On the computation of sensitivity functions of linear time-invariant systems *via* diagonalization," *IEEE-TAC*, vol. AC-31, no. 12, pp. 1141-1143, Dec. 1986.

[B] Sensitivity Functions Generation
"Low-Order Sensitivity Model"

[1] M.F. Aburdene, "Minimal-order systems for generation of sensitivity functions of linear-time invariant systems," *IJC*, vol. 27, no. 1, pp. 139-140, 1978.

[2] D.G. Denery, "Simplification in the computation of the sensitivity functions for constant coefficient linear systems," *IEEE-TAC*, vol. AC-16, pp. 348-350, Aug. 1971.

[3] G. Guardabassi, A. Locatelli, and S. Rinaldi, "On the optimality of the Wilkie-Perkins low-order sensitivity model," *IEEE-TAC*, vol. AC-15, pp. 382-384, June 1970.

[4] N.K. Gupta and R.K. Mehra, "Computational aspects of maximum likelihood estimation and reduction in sensitivity function calculation," *IEEE-TAC*, vol. AC-19, no. 5, pp. 774-783, Dec. 1974.

[5] W. Hafez, K. Loparo, E. Rasmy, and M. Hashish, "Generation of trajectory sensitivities: Minimal-order realization," *IJC*, vol. 49, no. 3, pp. 809-825, 1989.

[6] C.P. Neuman and A.K. Sood, "Sensitivity functions for multi-input linear time-invariant systems," *IJC*, vol 13, no. 6, pp. 1137-1150, 1971.

[7] _____, "Sensitivity functions for single-input linear time-invariant systems," *JFI*, vol. 291, no. 3, pp. 169-179, March 1971.

[8] _____, "Wilkie-Perkins low-order sensitivity model revisited," *IEEE-TAC*, vol. AC-16, no. 1, pp. 96-97, Feb. 1971.

[9] _____, "Sensitivity functions for multi-input linear time-invariant systems – II. Minimal-order models," *IJC*, vol. 15, no. 3, pp. 451-463, 1972.

[10] _____, "Parameter sensitivity in linear systems, a controllability view point," *Proc. IEE*, vo. 119, no. 8, pp. 1217-1219, Aug. 1972.

[11] T. Suzuki, "On simultaneous generation of sensitivity functions," *IEEE-TAC*, vol. AC-15, pp. 493-494, Aug. 1970.

[12] D.F. Wilkie and W.R. Perkins, "Generation of sensitivity functions for linear systems using low-order models," *IEEE-TAC*, vol. AC-14, no. 2, pp. 123-130, April 1969.

[C] General Sensitivity Approaches
with Applications

[1] P.R. Adby and J.R. Baxter, "Frequency-to-time-domain sensitivity matrix," *Elect. Lett.*, vol. 9, no. 19, pp. 455-457, Sept. 1973.

[2] A.T. Bahill, J.R. Latimer, and B.T. Troost, "Sensitivity analysis of linear homeomorphic model for human movement," *IEEE-TSMC*, vol. SMC-10, no. 12, pp. 924-929, Dec. 1980.

[3] H.A. Barker and D.J. Murray-Smith, "Sensitivity relationships for linear feedback systems," *Elect. Lett.*, vol. 3, no. 11, p. 473, Nov. 1967.

[4] R. Bellman, R. Kalaba, and R. Sridhar, "Sensitivity analysis and invariant imbedding," RAND Research Memorandum, no. 4039-PR, March 1964.

[5] I.N. Blinov and A.V. Mozgalevskii, "Selection of parameters for automatic failure detection," *ARC*, vol. 26, no. 10, pp. 1740-1749, Oct. 1965.

[6] V. Brajovic, "On-line generation of sensitivity functions using correlation analysis," *IJSS*, vol. 8, no. 12, pp. 1321-1326, 1977.

[7] G.W. Davis, Jr., "Sensitivity analysis in neural net solutions," *IEEE-TSMC*, vol. SMC-19, no. 5, pp. 1078-1082, Sept./Oct. 1989.

[8] J.K. Fidler and C. Nightingale, "Differential-incremental-sensitivity relationships," *Elect. Lett.*, vol. 8, no. 25, pp. 626-627, Dec. 1972.

[9] A. Ford and P.C. Gardiner, "A new measure of sensitivity for social system simulation models," *IEEE-TSMC*, vol. SMC-9, no. 3, pp. 105-114, March 1979.

[10] L.F. Godbout and D. Jordan, "Sensitivity functions for linear dynamic systems," *IJC*, vol. 34, no. 3, pp. 561-576, 1981.

[11] K. Harada and S. Ueno, "Parameter sensitivity of analog memory with negative feedback," *EEJ*, vol. 89, no. 12, pp. 10-18, 1969.

[12] M.A. Hashish and M.E.M. Rasmy, "A systematic approach to the generation of state sensitivity functions," *IJC*, vol. 35, no. 6, pp. 945-956, 1982.

[13] S. Hosoe and M. Ito, "Error systems in linear servo problems," *EEJ*, vol. 92-C, no. 5, pp. 145-151, 1972.

[14] R. Jain, "Tolerance analysis using fuzzy sets," *IJSS*, vol. 7, no. 12, pp. 1393-1401, 1976.

[15] I. Jaworska, "Sensitivity and tolerance of 2-D linear systems," *FCE*, vol. 8, no. 1, pp. 15-22, 1983.

[16] J.F. Klafin, Jr., and V. Krishnan, "Linear sensitivity analysis applied to a two-loop system with feedback variations," *IJC*, vol. 15, no. 2, pp. 305-317, 1972.

[17] F. Laurent, "Sensitivity properties of linear systems," *Elect. Lett.*, vol. 3, pp. 241-242, June 1967.

[18] C.K. Liu and N.J. Bergman, "Coupling sensitivity in multivariable decoupled systems," in *Proc. Allerton Conf.*, pp. 447-451, 1969.

[19] P.D. McMorran, "Parameter sensitivity and inverse Nyquist method," *Proc. IEE*, vol. 118, no. 6, pp. 802-804, June 1971.

[20] M.D. Mesarovic and Y. Takahara, "On global sensitivity," in *Proc. Allerton Conf.*, pp. 1-7, 1966.

[21] O.A. Mishulina, "Influence of parameter spread on accuracy of automatic systems," *ARC*, vol. 32, no. 9, pp. 1387-1394, 1971.

[22] P.N. Paraskevopoulos and G.T. Kekkeris, "Output sensitivity analysis using orthogonal functions," *IJC*, vol. 40, no. 4, pp. 763-772, 1984.

[23] T.M. Romanowicz, "Sensitivity analysis of linear systems--a structural approach," *IJSS*, vol. 18, no. 1, pp. 91-96, 1987.

[24] J.G. Simes and E.J. Mastascusa, "Minimum sensitivity configurations for numerical integration of linear time invariant systems," in *Proc. Allerton Conf.*, pp. 443-446, 1969.

[25] A.N. Tripathi, "Decoupling multivariable systems by state feedback: Sensitivity considerations," *Proc. IEE*, vol. 119, no. 6, pp. 737-742, June 1972.

[26] J.D. Turner, A. Messac, and J.L. Junkins, "Finite-time matrix convolution integral sensitivity calculations," *JGCD*, vol. 11, no. 5, pp. 473-475, 1988.

[27] K.K. Turski and P.M. Frank, "Sensitivity analysis of pulse-frequency-modulated control using a linearized model," *IJC*, vol. 42, no. 6, pp. 1469-1480, 1985.

[28] C.E. Willbanks, "Sensitivity of premixed compression-initiated supersonic combustion to small perturbations in inlet flow variables," *AIAA J.*, vol. 8, no. 1, pp. 115-119, Jan. 1970.

[D] *Sensitivity Analysis via Signal Flow Graph*

[1] M. Gavrilovic', R. Petrovic', and D.D. Siljak, "Adjoint method in the sensitivity analysis of optimal systems," *JFI*, vol. 276, no. 1, pp. 26-38, July 1963.

[2] W.W. Happ, "Flowgraph techniques for closed systems," *IEEE-TAES*, vol. AES, pp. 252-264, May 1966.

[3] S.Y. Hwang, "On signal flow graph analysis of ladder networks," *IEEE-TCT*, vol. CT-18, no. 5, pp. 563-566, Sept. 1971.

[4] A.Y. Lee, "Signal flow graphs – computer-aided system analysis and sensitivity calculations," *IEEE-TCAS*, vol. CAS-21, no. 2, pp. 209-216, March 1974. Comments by, N.A. Kheir, *ibid.*, vol. CAS-22, no. 5, p. 475, May 1975.

[5] S.J. Mason and H.J. Zimmermann, *Electronic Circuits, Signals, and Systems*. New York: Wiley, 1960.

[6] R. Radzyner, "Adjoint sensitivity models and signal-flowgraph reversal," *Proc. IEEE*, vol. 61, no. 12, pp. 1755-1757, Dec. 1973.

[7] R. Radzyner and W.H. Holmes, "Sensitivity-points models for linear cascaded structure," *Proc. IEEE*, vol. 61, no. 4, pp. 474-476, April 1973.

[8] T.M. Romanowicz, "Structural sensitivity analysis in a certain class of linear systems," *IEEE-TSMC*, vol. SMC-13, no. 3, pp. 413-417, May/June 1983.

[9] R.S. Rutman, "Conditions of zero sensitivity in linear control systems," *ECy*, vol. 7, no. 2, pp. 137-145, March/April 1969.

[10] J. Solymosi and T. Tron, "General interpretation of sensitivity functions," *IJCTA*, vol. 4, pp. 75-80, 1976.

[E] Sensitivity Analysis for Oscillatory Systems

[1] E.G. Bure and E.N. Rozenvasser, "The study of the sensitivity of oscillatory systems," *ARC*, no. 7, pp. 1045-1052, July 1974.

[2] A.N. Gerasimov, "Determining the sensitivity of the oscillation index from the open-loop frequency response," *ARC*, no. 6, pp. 1013-1015, June 1968.

[3] E.O. Kotov, "On the analysis of linear variable-parameter systems I. The case of bounded variation of parameters," *ARC*, vol. 26, no. 12, pp. 2044-2049, 1965.

[4] _____, "Analysis of linear variable-parameter systems II. Rapidly and periodically varying parameters," *ARC*, vol. 28, no. 7, pp. 1010-1016, 1967.

[5] G.L. Madatov and S.S. Chernyavs'ka, "Use of sensitivity theory in the analysis of high-frequency components of motion," *SAC*, vol. 6, no. 4, pp. 5-10, July/Aug. 1973.

[6] E.N. Rozenvasser, "Sensitivity study of oscillatory processes with respect to the excitation frequency," *ARC*, vol. 41, no. 6, pp. 755-759, June 1980.

[7] Y.I. Samoylenko, "Dynamics of the evolution of a system of linear oscillators under parametric resonance control," *SJAIS*, vol. 23, no. 1, pp. 8-18, 1990.

[8] D.D. Siljak and M.R. Stojic, "Sensitivity analysis of self-excited nonlinear oscillations," *IEEE-TAC*, vol. AC-10, no. 4, pp. 413-420, Oct. 1965.

[9] A.A. Tunik and I.B. Yadykin, "Application of the sensitivity-envelope methods to the investigation of a certain class of searchless self-adaptive systems," *ARC*, no. 2, pp. 232-240, Feb. 1975.

[F] Sensitivity Functions Generation Linear Retarded Systems

[1] P. Azema, C. Durante, F. Roubellat, and Y. Sevely, "Etude de la sensibilite des systemes aux variations des retards," *Elect. Lett.*, vol. 3, no. 4, pp. 171-172, April 1967.

[2] S.M. Chiang and Y.P. Shih, "Low sensitivity feedback control of non-linear discrete systems," *IJC*, vol. 14, no. 4, pp. 659-668, 1971.

[3] B. Garland and J.E. Marshall, "Sensitivity considerations of SMITH's method for time-delay systems," *Elect. Lett.*, vol. 10, no. 15, pp. 308-309, July 1974.

[4] _____, "Application of the sensitivity points method to a linear predictor control system," *IJC*, vol. 21, no. 4, pp. 681-688, 1975.

[5] R.E. King, "Sensitivity analysis of a class of differential – difference systems," *IJC*, vol. 5, no. 6, pp. 583-588, 1967.

[6] D.H. Mee, "A sensitivity analysis for control systems having input time delays," *Automatica*, vol. 10, pp. 551-557, 1974.

[G] Sensitivity in the Large Bounded Variations[2]

[1] J.K. Aggarwal, H.T. Banks, and N.H. McClamroch, "Invariance in linear systems," *JMAA*, vol. 29, pp. 498-506, 1970.

[2] S. Barnett, C. Storey, J.B. Cruz, and W.R. Perkins, "Comment on 'Invariance and sensitivity'," *IEEE-TAC*, vol. AC-12, pp. 210-211, April 1967.

[3] J.B. Cruz, Jr., and W.R. Perkins, "On invariance and sensitivity," in *IEEE Int. Conv. Record*, vol. 14, Pt. 7, pp. 159-162, 1966.

[4] _____, "Conditions for signal and parameter invariance in dynamical systems," *IEEE-TAC*, vol. AC-11, pp. 614-615, July 1966.

[5] N.H. McClamroch, J.K. Aggarwal, and L.G. Clark, "On parameter invariance in linear control systems," *IJC*, vol. 5, no. 4, pp. 361-367, 1967.

[6] _____, "Sensitivity of linear control systems to large parameter variations," *Automatica*, vol. 5, pp. 257-263, 1969.

[H] Selected Sensitivity Analyses with Applications in Mechanical Systems

[1] I.Y. Bar-Itzhack, "Modeling of uncertain strapdown heading-sensitive errors in INS [inertial navigation system] error models," *JGCD*, vol. 8, no. 1, pp. 142-144, 1985.

[2] D. Chenais, B. Rousselet, and R. Benedict, "Design sensitivity for arch structures with respect to midsurface shape under static loading," *JOTA*, vol. 58, no. 2, pp. 225-239, 1988.

[3] C.T. Chon, "Sensitivity of total strain energy of a vehicle structure to local joint stiffness," *AIAA J.*, vol. 25. no. 10, pp. 1391-1395, Oct. 1987.

[4] M.C. Delfour and J.-P. Zolesio, "Shape sensitivity analysis *via* min max differentiability," *SIAM-JCO*, vol. 26, no. 4, pp. 834-862, 1988.

[5] Z. Gurdal and R.T. Haftka, "Sensitivity derivatives for static test loading boundary conditions," *AIAA J.*, vol. 23, no. 1, pp. 159-160, 1985.

[6] R.T. Haftka and R.K. Kapania, "Sensitivity of actively damped structures to imperfections and modeling errors," *AIAA J.*, vol. 27. no. 10, pp. 1434-1440, Oct. 1989.

[7] E.J. Haug, "Second-order design sensitivity analysis of structural systems," *AIAA J.*, vol. 19. no. 8, pp. 1087-1088, Aug. 1981.

---

[2] Refer to Remark 6.10-4.

3.7.H. Selected Sensitivity Analyses in Mechanical Systems

[8]   E.J. Haug, K.K. Choi, and V. Komkov, *Design Sensitivity Analysis of Structural Systems.* Orlando, FL: Academic Press, 1986. Reviewed by R.V. Kohn, *SIAM R.,* vol. 30, no. 3, pp. 518-519, 1988.

[9]   J.W. Hou, C. Mei, and Y.X. Xue, "Design sensitivity analysis of beams under nonlinear forced vibrations," *AIAA J.,* vol. 28. no. 6, pp. 1067-1068, June 1990.

[10] J.W. Lim and I. Chopra, "Stability sensitivity analysis of a helicopter rotor," *AIAA J.,* vol. 28. no. 6, pp. 1089-1097, June 1990.

[11] E.C. Mikulcik, "Application of sensitivity analysis to car-trailer stability," *ASME-JDSMC,* vol. 101, pp. 272-274, Sept. 1979.

[12] J.J. Murray and C.P. Neuman, "Linearization and sensitivity models of the Newton-Euler dynamic robot model," *ASME-JDSMC,* vol. 108, pp. 272-276, Sept. 1986.

[13] M.A. Nichols and J.K. Hedrick, "The sensitivity of optimal flight paths to variations in aircraft and atmospheric parameters," in *Proc. Joint Auto. Contr.,* pp. 455-463, 1973.

[14] S.S. Rao, T.-S. Pan, and V.B. Venkayya, "Robustness improvement of actively controlled structures through structural modifications," *AIAA J.,* vol. 28. no. 2, pp. 353-361, Feb. 1990.

[15] B. Rousselet, "Shape design sensitivity of a membrane," *JOTA,* vol. 40, no. 4, pp. 595-623, 1983.

[16] S. Saigal and J.H. Kane, "Design sensitivity analysis of boundary-element substructures," *AIAA J.,* vol. 28, no. 7, pp. 1277-1284, 1990.

[17] G.L. Shaw, "Sensitivity to cross-axis oscillations in a single-axis nuclear gyroscope," *JGCD,* vol. 7, no. 4, pp. 501-502, 1984.

[18] J. Sobieszczanski-Sobieski, "Sensitivity of complex, internally coupled systems," *AIAA J.,* vol. 28, no. 1, pp. 153-160, 1990.

[19] _____, "Higher order sensitivity analysis of complex, coupled systems," *AIAA J.,* vol. 28, no. 4, pp. 756-758, 1990.

[20] J. Sokolowski and J.P. Zolesio, "Shape design sensitivity analysis of plates and plane elastic solids under unilateral constraints," *JOTA,* vol. 54, no. 2, pp. 361-382, 1987.

[21] V.A. Spector and H. Flashner, "Sensitivity of structural models for noncollocated control systems," *ASME-JDSMC,* vol. 111, pp. 646-655, Dec. 1989.

[22] T. Takahashi, "A method of sensitivity analysis and its application to robot manipulators," *CTAT,* vol. 3, no. 1, pp. 73-84, March 1987.

[23] P. Vielsack, "Sensitivity of clamp joints," *DSS,* vol. 2, no. 2, pp. 139-147, 1987.

[24] C.C. Wu and J.S. Arora, "Design sensitivity analysis and optimization of nonlinear structural response using incremental procedure," *AIAA J.,* vol. 25. no. 8, pp. 1118-1125, Aug. 1987.

[25] S.C. Yang and W.L. Garrard, "A low sensitivity, modern control approach to the longitudinal control of automated transit vehicles," in *Proc. Joint Auto. Contr.,* pp. 128-138, 1974.

CHAPTER FOUR

ALGEBRAIC PROPERTIES OF SYSTEM MODELS

TERMINAL CONDITIONS INSENSITIVITY

4.1. INTRODUCTION

In this chapter we establish certain algebraic properties of matrices in a system model *(MSM)*, and provide the requirements for terminal conditions insensitivity *(TCI)* of a linear system with several different system-sensitivity measures. We also review a concise account of the basic concept in guidance theory as pertains to terminal conditions insensitivity.

The system model has been established in Section 3.5, wherein a set of augmented state and state-sensitivity differential equations for a dynamical system has been simultaneously studied. This augmentation is necessary because the solution to state differential equation is a forcing function for the sensitivity differential equations and it must be computed first in order to solve state-sensitivity differential equations. However, the curse of high-dimensionality will soon make the set of augmented differential equations untractable. Since on one hand, the knowledge of certain algebraic properties of coefficient matrices in a system model is essential in order to apply the standard techniques from the optimal linear control theory to the overall system-sensitivity analysis, and on the other hand, it is logical, perhaps, to extract as many algebraic properties of an *MSM* as possible, just by considering those of the generating system (Definition 2.2-1). Thus in this chapter we study the algebraic properties of matrices in a system model with respect to those of the generating system.

Although this area of research is not so active as before, an extensive research on this subject is reported earlier in the literature (cf., § 4.6.A). One interesting report in this category is that given by Sain [A. 12], which connects the existence of Cruz and Perkins sensitivity comparison matrix [6A. 7] to the interesting topic of reproducibility introduced by Brockett and Mesarovic [6A. 4]. "Reproducibility refers to the ability of a system to achieve with its output, something which is desired of it," [6A. 4]. The basic idea of reproducibility, however, is more easily motivated if the system is thought of as an arbitrary transformation which maps the input time function into responses. Let S_u be the input set and S_y be the output set, then a system is said to be reproducible if for any given time-function $y \in S_y$ there is a $u \in S_u$ which generates it. Sain's theorem, states that: for a *DCLTI – GS*, $\dot{x} = Ax + Bu$ with response $z = Cx$, the triple $\{A, B, C\}$ has a classical sensitivity matrix in the sense of Cruz and Perkins if and only if the corresponding system $\{A^T, C^T, B^T\}$ has functionally reproducible homogeneous responses in the sense of Brockett and Mesarovic.

A major difference among references in § 4.6.A is the conditions on the corresponding matrices of generating system. For example, Guardabassi *et al.* [A. 3] assume that in $\dot{x} = Ax + Bu$, A is either in a diagonal or in a Jordan-canonical form, which is perhaps the simplest case to study. However, these sorts of assumptions are often unnecessary as we show in this chapter.

Another subject in this general area is the guidance theory, which is an applied version of terminal conditions insensitivity for linear system.

Therefore the importance of these topics and their basic positions in the overall study of sensitivity theory and the sensitivity-measure optimization; constitutes a separate and an independent research area which is discussed in the remainder of this chapter.

4.2. STABILITY OF SYSTEM MODEL

Consider an open-loop *DCLTI – GS* described by

$$\dot{x}(t; p) = A(p)x(t; p) + B(p)u(t), \quad x(0) = x_o. \quad (4.2\text{-}1)$$

Recall that $x \in R^n$, $u \in R^m$, $p \in R^r$; are system state, system input, and system physical parameters, respectively. Consider a system model for (4.2-1) which contains the trajectory-sensitivity functions of $x(t)$ with respect to p_j, $j = 1, 2, \cdots, r$ (cf., Section 3.5), as follows.

$$\frac{d}{dt} \begin{Bmatrix} x(t) \\ x_1(t) \\ x_2(t) \\ \cdot \\ \cdot \\ \cdot \\ x_r(t) \end{Bmatrix} = \begin{bmatrix} A(p) & 0 & 0 & \cdots & 0 \\ A_1(p) & A(p) & 0 & \cdots & 0 \\ A_2(p) & 0 & A(p) & \cdots & 0 \\ \cdot & \cdot & \cdot & \cdots & \cdot \\ \cdot & \cdot & \cdot & & 0 \\ A_r(p) & 0 & 0 & \cdots & A(p) \end{bmatrix} \begin{Bmatrix} x(t) \\ x_1(t) \\ x_2(t) \\ \cdot \\ \cdot \\ \cdot \\ x_r(t) \end{Bmatrix} + \begin{bmatrix} B(p) \\ B_1(p) \\ \cdot \\ \cdot \\ \cdot \\ B_r(p) \end{bmatrix} u(t)$$

$$\triangleq \tilde{A}_r(p)m(t; p) + \tilde{B}_r(p)u(t), \quad (4.2\text{-}2a)$$

$$m(0, p) = [x^T(0), 0^T, \cdots, 0^T]^T. \quad (4.2\text{-}2b)$$

Here $A_j(p) \triangleq \partial A(p)/\partial p_j$, similarly $B_j(p) \triangleq \partial B(p)/\partial p_j$, $j = 1, 2, ..., r$; provided that certain differentiability conditions for A and B, in terms of p_j's hold.

The following lemma is extensively used in this chapter.

LEMMA 4.2-1 [2C. 28]: Consider an n × n square matrix M, which has been partitioned as below

$$M = \begin{bmatrix} M_1 & M_2 \\ M_3 & M_4 \end{bmatrix}, \qquad (4.2\text{-}3)$$

where M_1 and M_4 are square matrices, M_2 and M_3 are, in general, rectangular matrices. If $\det(M_1) \neq 0$, then

$$\det M = \det(M_1) \det(M_4 - M_3 M_1^{-1} M_2). \qquad (4.2\text{-}4)$$

If $\det(M_4) \neq 0$, then

$$\det M = \det(M_4) \det(M_1 - M_2 M_4^{-1} M_3). \qquad (4.2\text{-}5)$$

∎

Now, consider $\tilde{A}_r(p)$ in (4.2-2). Without any loss of generality, we assume that r = 1. Then

$$\tilde{A}_1(p) \triangleq \begin{bmatrix} A(p) & 0 \\ \dfrac{\partial A(p)}{\partial p} & A(p) \end{bmatrix}, \quad \text{and} \quad \tilde{B}_1(p) \triangleq \begin{bmatrix} B(p) \\ \dfrac{\partial B(p)}{\partial p} \end{bmatrix}. \qquad (4.2\text{-}6)$$

PROPERTY 4.2-1: The set of eigenvalues of $\tilde{A}_1(p)$, in (4.2-6), is the same as that of A(p) with twice the multiplicity.

PROOF: By Lemma 4.2-1

$$\det(\lambda I - \tilde{A}_1) = \det \begin{bmatrix} \lambda I - A & 0 \\ -\dfrac{\partial A}{\partial p} & \lambda I - A \end{bmatrix} = [\det(\lambda I - A)]^2. \qquad (4.2\text{-}7)$$

COMMENT ON PROPERTY 4.2-1: If r > 1, then the above property exists with appropriate increases in the multiplicity for the eigenvalues of $\tilde{A}_r(p)$.

REMARK 4.2-1: Therefore if A, the system matrix of the generating system (4.2-1), is stable then so is $\tilde{A}_r(p)$ of (4.2-2). This remark which is an immediate consequence of Lemma 4.2-1 is first reported by Barnett *et al.* [A. 1].

∎

A considerable effort has been made in [A. 7], to find the minimum polynomial of $\tilde{A}_r(p)$, and to derive the corresponding transformation which results in the Jordan canonical form of $\tilde{A}_r(p)$. Several other properties of $\tilde{A}_r(p)$ with compared to those of A(p) such as cyclicity *etc.*, are included in this reference.

4.2. Stability of System Model

Another pertinent comment is the work by Reid et al. [A. 11], in direct computation of the sensitivity functions which is based on the properties of the minimum polynomial and the eigenvalues of the following matrix

$$\tilde{A}_j(p) \triangleq \begin{bmatrix} A(p) & 0 \\ \frac{\partial A(p)}{\partial p_j} & A(p) \end{bmatrix}, \quad \text{for } j = 1, 2, ..., r. \tag{4.2-8}$$

Clearly

$$\exp[\tilde{A}_j(p)t] = \begin{bmatrix} \exp(At) & 0 \\ \frac{\partial \exp(At)}{\partial p_j} & \exp(At) \end{bmatrix}. \tag{4.2-9}$$

It is also true that

$$\exp(\tilde{A}_j t) = \sum_{i=1}^{2n} \alpha_i(p) \tilde{A}_j^{i-1}, \tag{4.2-10}$$

where $\alpha_i(p)$'s are solved from the set of eigenvalues of $\tilde{A}_j$ with the same procedure as that given in Section 3.3, for a similar computation. Then considering (4.2-9) and (4.2-10), and the fact that

$$\tilde{A}_j^i \triangleq \begin{bmatrix} A^i & 0 \\ \frac{\partial A^i}{\partial p_j} & A^i \end{bmatrix}, \tag{4.2-11}$$

we can show that

$$\frac{\partial \exp(At)}{\partial p_j} = \sum_{i=1}^{2n} \alpha_i(p) \frac{\partial A^i(p)}{\partial p_j}. \tag{4.2-12}$$

Equation (4.2-12) is the basis of an algorithm for direct computation of the sensitivity functions given by Reid et al. [A. 11]. That method, however, requires more computations than the preceding approach given in § 3.3.2.2 which yields the same result.

4.3. CONTROLLABILITY OF SYSTEM MODEL

4.3.1. Introduction

It is of interest to study the controllability of coefficient matrices in a system model in connection with:

(a) Transformation of system model (4.2-2) into its controllability canonical form [2C. 28], in order to generate the minimum number of sensitivity functions which are required for generating any other sensitivity functions for (4.2-1) as an algebraic combination of these minimum number of differential equations.

(b) Terminal conditions insensitivity *(TCI)* of linear systems.

(c) For sensitivity-measure optimization in which a measure of sensitivity function (for example, a quadratic norm of state-sensitivity function) is proposed to be minimized, then the knowledge of these algebraic properties is useful in order to find the system input, such that a given performance index is satisfied. This information is also very useful when applying techniques from the invariance imbedding methodology in order to specify system parameters.

In this section only condition on controllability of system model (4.2-2) and its relationship with the controllability canonical transformation is studied.

4.3.2. Analyses

Recall the *MSM* from (4.2-2), repeated below for convenience, where $\tilde{A}_r(p)$ is an $[n(1+r)] \times [n(1+r)]$ matrix and $\tilde{B}_r(p)$ is an $[n(1+r)] \times m$ matrix, as follows.

$$\tilde{A}_r(p) \triangleq \begin{bmatrix} A(p) & 0 & 0 & \cdots & 0 \\ A_1(p) & A(p) & 0 & \cdots & 0 \\ A_2(p) & 0 & A(p) & \cdots & 0 \\ \cdot & \cdot & \cdot & \cdots & \cdot \\ \cdot & \cdot & \cdot & & \cdot \\ \cdot & \cdot & \cdot & & 0 \\ A_r(p) & 0 & 0 & \cdots & A(p) \end{bmatrix}, \text{ and } \tilde{B}_r(p) \triangleq \begin{bmatrix} B(p) \\ B_1(p) \\ \cdot \\ \cdot \\ \cdot \\ B_r(p) \end{bmatrix}. \quad (4.3\text{-}1)$$

Controllability matrix of the pair $(\tilde{A}_r, \tilde{B}_r)$ is

$$C_{sm} \triangleq (\tilde{B}_r, \tilde{A}_r \tilde{B}_r, \cdots, \tilde{A}_r^{n(1+r)-1} \tilde{B}_r). \quad (4.3\text{-}2)$$

We note that

$$\tilde{A}_r^i \tilde{B}_r \triangleq \begin{bmatrix} A^i B \\ (A^i B)_1 \\ \cdot \\ \cdot \\ \cdot \\ (A^i B)_r \end{bmatrix}; \quad (4.3\text{-}3)$$

where $(A^i B)_j \triangleq \partial (A^i B)/\partial p_j$. Let the controllability matrix of generating system (4.2-1), for the pair (A, B), be C', where

$$C' = (B, AB, \cdots, A^{n-1}B)_{n \times mn}. \quad (4.3\text{-}4)$$

Then we can show that

4.3. Controllability of System Model

$$C_{sm} \triangleq \begin{bmatrix} C' & A^n C' & A^{2n} C' & \cdots & A^{r \times n} C' \\ \dfrac{\partial C'}{\partial p_1} & \dfrac{\partial (A^n C')}{\partial p_1} & \dfrac{\partial (A^{2n} C')}{\partial p_1} & \cdots & \dfrac{\partial (A^{r \times n} C')}{\partial p_1} \\ \vdots & \vdots & \vdots & \vdots & \vdots \\ \dfrac{\partial C'}{\partial p_r} & \dfrac{\partial (A^n C')}{\partial p_r} & \cdots & \cdots & \dfrac{\partial (A^{r \times n} C')}{\partial p_r} \end{bmatrix} \qquad (4.3\text{-}5)$$

A major portion of this study is based on the consequence of the properties of the above $[n(1+r)] \times [mn(1+r)]$ matrix C_{sm}. In fact, this observation of the structure of C_{sm} makes several of its pertinent properties self-describing.

Considering the system model (4.2-2), let $T = (T_1, T_2)$ be a nonsingular transformation matrix such that the columns of T_1 form a basis for the controllable subspace of system (4.2-2); i.e., the basis for C_{sm}, and the columns of T_2 together with those of T_1 form a basis for the whole $n(1+r)$ − dimensional space. Define the transformed state of the system model as

$$m'(t) = T^{-1} m(t). \qquad (4.3\text{-}6)$$

Then the system model (4.2-2) is transformed into its controllability canonical form

$$\dot{m}'(t) = \begin{bmatrix} \tilde{A}'_{11} & \tilde{A}'_{12} \\ 0 & \tilde{A}'_{22} \end{bmatrix} m'(t) + \begin{bmatrix} \tilde{B}'_{11} \\ 0 \end{bmatrix} u(t); \qquad (4.3\text{-}7)$$

where the pair $(\tilde{A}'_{11}, \tilde{B}'_{11})$ is completely controllable. Also, let $m'(t)$ be partitioned accordingly; i.e., $m'(t) = [m_1'^T(t), m_2'^T(t)]^T$. Since the initial condition associated with $m_2'(t)$ is zero, so is its forcing function, we conclude that $m_2'(t) \equiv 0$. Thus the problem of solving system model (4.2-2) reduces to solving the following set of differential equations

$$\dot{m}_1'(t) = \tilde{A}'_{11} m_1'(t) + \tilde{B}'_{11} u(t). \qquad (4.3\text{-}8)$$

Then

$$m(t) = T m'(t) = (T_1, T_2) \begin{Bmatrix} m_1'(t) \\ m_2'(t) \end{Bmatrix} = T_1 m_1'(t). \qquad (4.3\text{-}9)$$

The application of the above technique in generating the minimum number of differential equations in order to solve the whole set of sensitivity functions in a system, is first stated by Gupta and Mehra [A. 9]. In fact, this work of Gupta and Mehra advances the attempts of several other researchers who have noticed the redundancy in the set of differential equations of a system model. Therefore, in the following the rank of C_{sm} for several different cases of matrices A and B are studied.

∎

Because of the importance of the computational aspects of C_{sm}, we establish a computational procedure first, in order to construct C_{sm}. This procedure is based on the special partition of C_{sm} shown in (4.3-5).

(a) To compute $\dfrac{\partial A^i(p)}{\partial p_j}$.

We can show that this quantity for $j = 1, 2, ..., r$ and $i = 1, 2, ...,$ equals

$$\frac{\partial A^i(p)}{\partial p_j} = \frac{\partial A^{i-1}(p)}{\partial p_j} A(p) + A^{i-1} \frac{\partial A(p)}{\partial p_j} . \qquad (4.3\text{-}10)$$

Thus knowing $\dfrac{\partial A(p)}{\partial p_j}$, for all j, all the other $\dfrac{\partial A^i(p)}{\partial p_j}$ can be recursively generated on a computer.

(b) To compute $\dfrac{\partial (A^i B)}{\partial p_j}$.

Upon computing $\dfrac{\partial A^i(p)}{\partial p_j}$, then we can show that by knowing $\dfrac{\partial B(p)}{\partial p_j}$ the $\dfrac{\partial (A^i B)}{\partial p_j}$ can be computed as

$$\frac{\partial (A^i B)}{\partial p_j} = \frac{\partial A^i(p)}{\partial p_j} B(p) + A^i(p) \frac{\partial B(p)}{\partial p_j} , \text{ for all j and i.} \qquad (4.3\text{-}11)$$

An alternative expression for (4.3-11) is

$$\frac{\partial (A^i B)}{\partial p_j} = \frac{\partial A}{\partial p_j}(A^{i-1}B) + A\frac{\partial (A^{i-1}B)}{\partial p_j}. \qquad (4.3\text{-}12)$$

(c) To compute $\dfrac{\partial C'(p)}{\partial p_j}$.

By knowing $\dfrac{\partial (A^i B)}{\partial p_j}$, $i = 0, ..., n$, then $\dfrac{\partial C'(p)}{\partial p_j}$ can be derived as follows.

$$\frac{\partial C'(p)}{\partial p_j} = (\frac{\partial B}{\partial p_j}, \frac{\partial (AB)}{\partial p_j}, \cdots , \frac{\partial (A^{n-1}B)}{\partial p_j}), \text{ for all j.} \qquad (4.3\text{-}13)$$

(d) To compute $\dfrac{\partial (A^k C')}{\partial p_j}$.

4.3. Controllability of System Model

We can show that

$$\frac{\partial(A^k C')}{\partial p_j} = \frac{\partial A^k}{\partial p_j} C' + A^k \frac{\partial C'}{\partial p_j}, \quad \text{for all } j \text{ and } k. \tag{4.3-14}$$

Therefore starting with $\partial A(p)/\partial p_j$ and $\partial B(p)/\partial p_j$ which are evaluated around a given set of nominal values, all the other partitioned matrices can be easily computed on a digital computer and thus C_{sm} can be set up by knowing these partition blocks.

∎

Perhaps it is logical to extract as much properties of C_{sm} as we can, by studying the corresponding generating system. Suppose that the generating system (4.2-1) is derived by a single input. Then the controllability matrix of this system is

$$C \triangleq (b, Ab, \cdots, A^{n-1}b), \tag{4.3-15}$$

and the controllability matrix for the associated system model becomes

$$C'_{sm} \triangleq \begin{bmatrix} C & A^n C & A^{2n} C & \cdots & A^{r \times n} C \\ \dfrac{\partial C}{\partial p_1} & \dfrac{\partial(A^n C)}{\partial p_1} & \dfrac{\partial(A^{2n} C)}{\partial p_1} & \cdots & \dfrac{\partial(A^{r \times n} C)}{\partial p_1} \\ \cdot & \cdot & & & \cdot \\ \cdot & \cdot & & & \cdot \\ \cdot & \cdot & & & \cdot \\ \dfrac{\partial C}{\partial p_r} & \dfrac{\partial(A^n C)}{\partial p_r} & & \cdots & \dfrac{\partial(A^{r \times n} C)}{\partial p_r} \end{bmatrix} \tag{4.3-16}$$

Recall the Cayley-Hamilton Theorem which states that every square matrix satisfies its characteristic equation [2A. 8]. However, the minimum polynomial of certain matrices can be of degree less than that of the characteristic equation. Thus

$$A^n(p) = \sum_{i=0}^{n-1} \alpha_{i0}(p) A^i(p); \tag{4.3-17}$$

where $\alpha_{i0}(p)$'s are the coefficients of characteristic equation of A. Then we can show that

$$A^{n+k}(p) = \sum_{i=0}^{n-1} \alpha_{i0}(p) A^{k+i}(p). \tag{4.3-18}$$

The following two theorems are due to Gupta and Mehra [A. 9], which give insight to the rank of C'_{sm} and C_{sm} and cover most of the attempts made by the preceding researchers in this area. The proofs of these results are based on the applications of (4.3-17) and (4.3-18).

THEOREM 4.3-1 (Gupta and Mehra [A. 9]): For a single-input system, the rank of controllability matrix C'_{sm}, in (4.3-16), is less than or equal to 2n.

THEOREM 4.3-2 (Gupta and Mehra [A. 9]): For a system with m inputs, the rank of C_{sm}, in (4.3-5), cannot exceed $(m+1)n$.

The following theorem is a special case of above theorem.

THEOREM 4.3-3: For a system wherein $A \neq A(p)$, but $B = B(p)$, the rank of C_{sm} cannot exceed mn.

PROOF: Recall C_{sm}, in (4.3-5), and consider

$$\frac{\partial(A^{nk}C')}{\partial p_j} = A^{nk}(\frac{\partial B}{\partial p_j}, A\frac{\partial B}{\partial p_j}, \cdots, A^{n-1}\frac{\partial B}{\partial p_j}). \qquad (4.3\text{-}19)$$

Thus the C_{sm} of (4.3-5) for this case becomes

$$C_{sm} \triangleq \begin{bmatrix} (B, ..., A^{n-1}B) & A^n(B, ..., A^{n-1}B) & \cdots & A^{rn}(B, ..., A^{n-1}B) \\ (\frac{\partial B}{\partial p_1}, ..., A^{n-1}\frac{\partial B}{\partial p_1}) & A^n(\frac{\partial B}{\partial p_1}, ..., A^{n-1}\frac{\partial B}{\partial p_1}) & \cdots & A^{rn}(\frac{\partial B}{\partial p_1}, ..., A^{n-1}\frac{\partial B}{\partial p_1}) \\ \vdots & \vdots & & \vdots \\ (\frac{\partial B}{\partial p_r}, ..., A^{n-1}\frac{\partial B}{\partial p_r}) & A^n(\frac{\partial B}{\partial p_r}, ..., A^{n-1}\frac{\partial B}{\partial p_r}) & \cdots & A^{rn}(\frac{\partial B}{\partial p_r}, ..., A^{n-1}\frac{\partial B}{\partial p_r}) \end{bmatrix}$$

$$(4.3\text{-}20)$$

It is clear, by virtue of $A^n = \sum_{i=0}^{n-1} \alpha_i A^i$, that the rank of C_{sm} does not exceed mn. ∎

CERTAIN SPECIAL CASES: Consider the case wherein the generating system (4.2-1) has only one parameter. Then the corresponding controllability matrix of its system model is

$$C_{sm} \triangleq \begin{bmatrix} C' & A^nC' \\ \frac{\partial C'}{\partial p} & \frac{\partial(A^nC')}{\partial p} \end{bmatrix}. \qquad (4.3\text{-}21)$$

Suppose that the generating system is completely controllable; i.e., there exists a matrix $C \subseteq C'$ such that C^{-1} exists. Then in order to establish the conditions for complete controllability of the corresponding system model the following matrix associated with the columns of C is studied.

4.3. Controllability of System Model

$$C'_{sm} \triangleq \begin{bmatrix} C & A^nC \\ \frac{\partial C}{\partial p} & \frac{\partial A^n C}{\partial p} \end{bmatrix}. \tag{4.3-22}$$

Using Lemma 4.2-1 yields

$$\det C'_{sm} = (\det C)^2 \det[\frac{\partial}{\partial p}(C^{-1}A^nC)]. \tag{4.3-23}$$

PROPERTY 4.3-1: Consider a generating system wherein $A = A(p)$, $B = B(p)$, and $p \in R^1$. If the pair (A, B) is completely controllable (for all p), then for the complete controllability of the corresponding system model, A must be necessarily of maximum rank (for all p). However, this condition is not sufficient.

PROOF: From (4.3-23) if A is not maximum rank for all p, then neither is $C^{-1}A^nC$. Thus $\partial(C^{-1}A^nC)/\partial p$ is not of maximum rank either, and $\det C'_{sm} = 0$. However, by a counterexample it can be shown that $C^{-1}A^nC$ being maximum rank does not imply that $\partial(C^{-1}A^nC)/\partial p$ is of maximum rank. ∎

The matrix A being maximum rank in (4.3-23) as a necessary condition for complete controllability of the corresponding system model is a severe restriction and for a general class of systems A is not maximum rank. However, it is important to notice that in such cases, a closed-loop controller can be designed such that it completely stabilizes the system and results in a corresponding maximum rank A', perhaps in the form of $A' = A - BK$. ∎

Since, by assumption $\det C \neq 0$, for testing the complete controllability it is necessary and sufficient that

$$T(p) = \det[\frac{\partial}{\partial p}(C^{-1}A^nC)] \neq 0. \tag{4.3-24}$$

Here

$$T(p) = \det[\frac{\partial A^n}{\partial p} + A^n(\frac{\partial C}{\partial p}C^{-1}) - (\frac{\partial C}{\partial p}C^{-1})A^n]. \tag{4.3-25}$$

Thus, for the purpose of testing, it is sufficient to evaluate the above $T(p)$ around some nominal values of p by evaluating $\partial A^n(p)/\partial p$ and $(\partial C(p)/\partial p)C^{-1}$ which are described earlier in this section.

THEOREM 4.3-4: Consider a single-input system, where $A = A(p)$ is a diagonal matrix, $b = b(p)$, $p \in R^1$, and the pair (A, b) is completely controllable. Then the corresponding system model is completely controllable if and only if A is diagonal, maximum-rank, and has eigenvalues sensitive to p.

PROOF: The controllability matrix of generating system is

$$C \triangleq (b, Ab, \cdots, A^{n-1}b) = \text{diag}(b_i)\Lambda; \qquad (4.3\text{-}26)$$

where Λ is the Vandermonde matrix

$$\Lambda \triangleq \begin{bmatrix} 1 & \lambda_1 & \lambda_1^2 & , ..., & \lambda_1^{n-1} \\ 1 & \lambda_2 & \lambda_2^2 & , ..., & \lambda_2^{n-1} \\ \cdot \\ \cdot \\ \cdot \\ 1 & \lambda_n & \lambda_n^2 & , ..., & \lambda_n^{n-1} \end{bmatrix} \qquad (4.3\text{-}27)$$

Condition for complete controllability of the corresponding system model is

$$T(p) = \det[\frac{\partial}{\partial p}(C^{-1}A^n C)] = \det[\frac{\partial}{\partial p}(\Lambda^{-1}\text{diag}(\frac{1}{b_i})A^n \text{diag}(b_i)\Lambda)]$$

$$= \det[\frac{\partial}{\partial p}(\Lambda^{-1}A^n \Lambda)] \neq 0. \qquad (4.3\text{-}28)$$

Recall that A is diagonal and the pair (A, b) is completely controllable, therefore $b_i \neq 0$, for all i, (i. e., diag $(1/b_i)$ exists).

Now, using (4.3-25), (4.3-28) becomes

$$T(p) = \det[\frac{\partial A^n}{\partial p} + A^n(\frac{\partial \Lambda}{\partial p}\Lambda^{-1}) - (\frac{\partial \Lambda}{\partial p}\Lambda^{-1})A^n]$$

$$= \det[\frac{\partial A^n}{\partial p} + (A^n\frac{\partial \Lambda}{\partial p})\Lambda^{-1} - \frac{\partial \Lambda}{\partial p}(\Lambda^{-1}A^n)] \neq 0. \qquad (4.3\text{-}29)$$

Suppose that $\partial \lambda_i/\partial p = 0$; i.e., one of the eigenvalues of A is insensitive to the parameter p, then the ith-row and column of $\partial A^n/\partial p$ is zero. So is the ith-row of $\partial \Lambda/\partial p$. Simple matrix bookkeeping shows that the ith-row of $(A^n\frac{\partial \Lambda}{\partial p})\Lambda^{-1} - \frac{\partial \Lambda}{\partial p}(\Lambda^{-1}A^n)$ in (4.3-29), is zero (the parenthesis shows the order of multiplications in order to have the left-hand side matrix with the ith-row zero). Thus if $T(p) = 0$, then $\partial \lambda_i/\partial p = 0$, for some i, and that is a contradiction.

REMARK 4.3-1: Theorem 4.3-4 with slight modification of its proof holds for a system, wherein the pair (A, b) is completely controllable, $A = A(p)$, $b = b(p)$ and $p \in R^1$, and A is of maximum rank, but not diagonal. Furthermore, the above theorem holds for a system, wherein the pair (A, B) is completely controllable from every input (normal system), $A = A(p)$, $b = b(p)$ and $p \in R^1$, and A is of maximum rank, but not diagonal.

4.3. Controllability of System Model

PROPERTY 4.3-2: For the case wherein b = b(p), A ≠ A(p), but A is maximum rank and diagonal, and the pair (A, b) is completely controllable, then the corresponding system model is completely controllable if and only if $\partial b_i(p)/\partial p \neq 0$, for all i.

PROOF: The system model is

$$\dot{m}(t) = \begin{bmatrix} A & 0 \\ 0 & A \end{bmatrix} m(t) + \begin{Bmatrix} b(p) \\ \dfrac{\partial b(p)}{\partial p} \end{Bmatrix} u(t). \qquad (4.3\text{-}30)$$

Since A is diagonal, the system model is completely controllable if and only if $\dfrac{\partial b_i(p)}{\partial p} \neq 0$, for all i = 1, 2, ..., n.

REMARK 4.3-2: To relate the number of independent inputs to the number of independent parameters which can be maneuvered and the related matters is referred to as the structural controllability [A. 2-4]. Recently similar concept regarding the number of allowable varying parameters which do not deteriorate the system stabilizability is reported in [5C. 109].

4.4. TERMINAL CONDITIONS INSENSITIVITY

4.4.1. Introduction

Terminal conditions insensitivity *(TCI)* has several interesting features. For instance, in guidance theory the *TCI* with respect to changes in system parameters plays an important role, since essentially the main goal in guidance is to reach precisely a target by any appropriate means. The mathematical formulation of this problem for the case of linear time-invariant systems shows that the complete controllability of system model is sufficient for the *TCI*. Another version of this problem is to define certain convergence of system trajectory, e.g., convergence in the sense of some form of stability, *per se;* which means to find bounds on system parameters, such that system remains stable. The latter means that the terminal condition goes to a finite limit in spite of variations in the system parameters. Some representative works in this area are reported in § 4.6.B. In particular, it is shown by Kreindler [B. 6], with an example, that an appropriate norm of terminal-error due to parameter variations depends on the location of initial condition and even on the direction of parameter derivations and it does not depend on design being either open-loop or closed-loop.

4.4.2. Formulation and Analysis

Terminal conditions insensitivity of generating system (4.2-1) requires that this system can be steered from any initial condition x(0) to any final condition x(T), such that the variation of x(T) caused by changes in system parameter is zero; i.e., $\Delta x(T) \big|_{p \to p + \Delta p} = 0$. This requirement immediately points out that system (4.2-1) must be necessarily completely controllable, in order that *TCI* can possibly hold. This requirement is not sufficient and in addition we must have $\frac{\partial^i x(t; p)}{\partial p_j^i} \big|_{t=T} = 0$, for all $j = 1, 2, ..., r$ and $i = 1, 2, ...$. However, as a general tendency in the study of sensitivity theory, by appealing to the "small" parameter variation, in order to have *TCI*, we may simply require that $\frac{\partial^i x(t; p)}{\partial p_j^i} \big|_{t=T} = 0$, for $j = 1, 2, ..., r$, and $i = 1$.

The above *TCI* demands that system model (4.2-2) with the initial condition $m(0) = [x^T(0), 0^T, ..., 0^T]^T$ can be steered to a final condition $m(T) = \{x^T(T), [\frac{\partial x(T)}{\partial p_1}]^T, ..., [\frac{\partial x(T)}{\partial p_r}]^T\}^T$, where each of the $[\frac{\partial x(T)}{\partial p_j}]^T$, $j = 1, ..., r$ is zero. This fact implies and is implied by the complete controllability of system model (4.2-2).

Therefore, analysis given in § 4.3.2; i.e., to study the rank of controllability matrix for system model is the same as finding sufficient conditions for the "small" *TCI*. However, certain interpretation or restatement of the previous result can be made. In fact this restatement (which follows immediately) is known as structural uncontrollability [A. 3], wherein the attempt has been made to derive the maximum number of parameters, which can be "desensitized" in the sense of the *TCI* [B. 2].

PROPERTY 4.4-1 *(Parallel to Theorem 4.3-2):* For a multi-input system, if the number of parameters exceeds the number of system inputs, then the system model is not completely controllable.
∎

Similar interpretations for the other theorems can be likewise stated, but are not repeated herein.

In the following the sufficient conditions for *TCI* – and up to the second-order derivative is established. The technique which is established for a system with one parameter includes computational procedures for testing the rank of controllability matrix and it is self-explanatory in regard to its extensions to the higher-order and multi-parameter system matrices.

4.4. Terminal Conditions Insensitivity

First, recall the generating system (4.2-1), repeated below

$$\dot{x}(t) = Ax(t) + Bu(t), \quad x(0) = x_o, \tag{4.4-1}$$

where $x \in R^n$; $u \in R^m$; $p \in R^1$; are the system state, input and parameter vectors, respectively. The first-order and the second-order sensitivity functions of this system are:

$$\frac{d}{dt} \frac{\partial x(t; p)}{\partial p} = A \frac{\partial x(t; p)}{\partial p} + \frac{\partial A(p)}{\partial p} x(t; p) + \frac{\partial B(p)}{\partial p} u(t), \tag{4.4-2}$$

$$\frac{d}{dt} \frac{\partial^2 x(t; p)}{\partial p^2} = A \frac{\partial^2 x(t; p)}{\partial p^2} + 2 \frac{\partial A(p)}{\partial p} \frac{\partial x(t; p)}{\partial p}$$

$$+ \frac{\partial^2 A(p)}{\partial p^2} x(t; p) + \frac{\partial^2 B(p)}{\partial p^2} u(t). \tag{4.4-3}$$

The corresponding augmented set of differential equations (the higher-order system model) is

$$\frac{d}{dt} \left\{ \begin{array}{c} x(t; p) \\ \frac{\partial x(t; p)}{\partial p} \\ \frac{\partial^2 x(t; p)}{\partial p^2} \end{array} \right\} = \left[\begin{array}{ccc} A(p) & 0 & 0 \\ \frac{\partial A(p)}{\partial p} & A(p) & 0 \\ \frac{\partial^2 A(p)}{\partial p^2} & 2\frac{\partial A(p)}{\partial p} & A(p) \end{array} \right] \left\{ \begin{array}{c} x(t; p) \\ \frac{\partial x(t; p)}{\partial p} \\ \frac{\partial^2 x(t; p)}{\partial p^2} \end{array} \right\} + \left[\begin{array}{c} B(p) \\ \frac{\partial B(p)}{\partial p} \\ \frac{\partial^2 B(p)}{\partial p^2} \end{array} \right] u(t)$$

$$\tag{4.4-4a}$$

$$\triangleq \hat{A}(p)m(t; p) + \hat{B}(p)u(t), \tag{4.4-4b}$$

$$m(0; p) = [x_o^T, 0^T, 0^T]^T. \tag{4.4-4c}$$

The higher-order system model (4.4-4) is completely controllable if and only if

$$\hat{C}_{sm} \triangleq (\hat{B}, \hat{A}\hat{B}, \cdots, \hat{A}^{3n-1}\hat{B}) \tag{4.4-5}$$

is of rank equal to 3n. We can show that

$$\hat{A}^i\hat{B} = \left[\begin{array}{c} A^iB \\ (A^iB)_{(1)} \\ (A^iB)_{(2)} \end{array} \right], \tag{4.4-6}$$

where $(A^iB)_{(1)} \triangleq \frac{\partial(A^iB)}{\partial p}$ and $(A^iB)_{(2)} \triangleq \frac{\partial^2(A^iB)}{\partial p^2}$. Recall the controllability matrix of generating system C', from (4.3-4),

$$C' = (B, AB, \cdots, A^{n-1}B). \tag{4.4-7}$$

Then it is clear that $\hat{C}_{sm}$, in (4.4-5) considering (4.4-6) and (4.4-7) can be written as

$$\hat{C}_{sm} \triangleq \begin{bmatrix} C' & A^nC' & A^{2n}C' \\ \dfrac{\partial C'}{\partial p} & \dfrac{\partial(A^nC')}{\partial p} & \dfrac{\partial(A^{2n}C')}{\partial p} \\ \dfrac{\partial^2 C'}{\partial p^2} & \dfrac{\partial^2(A^nC')}{\partial p^2} & \dfrac{\partial^2(A^{2n}C')}{\partial p^2} \end{bmatrix}. \qquad (4.4\text{-}8)$$

$$3n \times 3mn$$

The objective is to study the rank of this matrix. Suppose that the generating system is completely controllable and let $C \subseteq C'$ such that C^{-1} exists and $C \in R^{n \times n}$. Let $\hat{C}'_{sm}$ be the corresponding submatrix of $\hat{C}_{sm}$; i.e.,

$$\hat{C}'_{sm} \triangleq \begin{bmatrix} C & A^nC & A^{2n}C \\ \dfrac{\partial C}{\partial p} & \dfrac{\partial(A^nC)}{\partial p} & \dfrac{\partial(A^{2n}C)}{\partial p} \\ \dfrac{\partial^2 C}{\partial p^2} & \dfrac{\partial^2(A^nC)}{\partial p^2} & \dfrac{\partial^2(A^{2n}C)}{\partial p^2} \end{bmatrix}. \qquad (4.4\text{-}9)$$

$$3n \times 3n$$

To construct this matrix numerically, the procedure is exactly the same as that outlined in § 4.3.2. In order to study the rank of $\hat{C}'_{sm}$, again Lemma 4.2-1 is used for determining the determinant of $\hat{C}'_{sm}$. Here

$$\det \hat{C}'_{sm} = \det C \det \left[\begin{bmatrix} (A^nC)_{(1)} & (A^{2n}C)_{(1)} \\ (A^nC)_{(2)} & (A^{2n}C)_{(2)} \end{bmatrix} - \begin{bmatrix} C_{(1)} \\ C_{(2)} \end{bmatrix} C^{-1}(A^nC, A^{2n}C) \right].$$

$$(4.4\text{-}10)$$

Let $\alpha \triangleq C^{-1}A^nC = (C^{-1}AC)^n$. Then $\alpha^2 = C^{-1}A^{2n}C$. Substituting these quantities into (4.4-10) results in

$$\det \hat{C}'_{sm} = (\det C)^3 \det \begin{bmatrix} \alpha_{(1)} & \alpha^2_{(1)} \\ \alpha_{(2)} + 2C^{-1}C_{(1)}\alpha_{(1)} & \alpha^2_{(2)} + 2C^{-1}C_{(1)}\alpha^2_{(1)} \end{bmatrix}.$$

$$(4.4\text{-}11)$$

Assuming that $\alpha_{(1)}^{-1}$ exists; i.e., the *TCI* holds up to the first-order sensitivity function, then by Lemma 4.2-1, (4.4-11) gives the necessary conditions for establishing the complete *TCI* and up to the second-order sensitivity functions. This derivation is omitted at present. However, it is noted that in the above derivation we assumed that: *(i)* the generating system is completely controllable; and *(ii)* the *TCI* holds up to the first-order sensitivity function.

It is also noted that the controllability canonical transformation, as is described for the first-order sensitivity function, is still a valid approach and it is a practical technique for generating the higher-order sensitivity functions. Also, recall that the extensions of the above procedure to a multi-parameter system has been automatically assumed.

4.5. GUIDANCE THEORY AS PERTAINS TO TERMINAL CONDITIONS INSENSITIVITY

The guidance theory is the goal of the most space control problems. That means to devise techniques for steering a dynamical system from an initial "position" in its state space representation to a certain final "position" therein. For example, the large-scale transportation of lunar materials, for establishing space colonies or manufacturing facilities, to support construction of solar power satellites, or similar large-space systems, requires an extremely sophisticated guidance policy and a highly accurate flight mechanism. Or in view of a complex mission, such as landing of a space craft on a planet that is surrounded by uncertain atmospheres and gravitational fields, the guidance policies are extremely important. These and many other examples which can be cited show the importance of this concept are well known.

The guidance process, in general, may be divided into two parts. In certain guidance problem the target is the main concern and the path is not important. On the other hand, in many other cases the guidance along the entire path is of concern. Perhaps achieving these goals from the point of view of the minimum "cost" is an additional concern for the optimal control theory specialists. But, in general, the guidance theory provides an engineering application for the requirements on the convergence of solution for certain differential equations to some limit points or a boundary trajectory. Based on the these remarks, it is now clear that the theory of sensitivity plays an important role in the framework of guidance theory. This connection between the theory of sensitivity and the objective of guidance theory, is the main reason for bringing these brief remarks to the reader's attention. Typical references related to these matters are given in § 4.6.*C*, and § 3.7.*H*.

4.6. REFERENCES

[A] Algebraic Properties of the Matrices of a System Model

[1] S. Barnett, C. Storey, J.B. Cruz, and W.R. Perkins, "Comment on 'Invariance and sensitivity'," *IEEE-TAC*, vol. AC-12, pp. 210-211, April 1967.

[2] G. Guardabassi, A. Locatelli, and S. Rinaldi, "Structural uncontrollability of sensitivity systems," *Automatica*, vol. 5, no. 3, pp. 297-301, May 1969.

[3] _____, "Further results on controllability and sensitivity," *IEEE-TAC*, vol. AC-15, no. 6, pp. 697-698, Dec. 1970.

[4] _____, "Signal insensitivity, controllability, and observability in linear systems," *IEEE-TAC*, vol. AC-16, no. 3, pp. 277-278, June 1971.

[5] _____, "Structural properties of sensitivity systems," *JFI*, vol. 294, no. 4, pp. 241-248, Oct. 1972.

[6] _____, "Zero-sensitivity in linear systems," *IJC*, vol. 15, no. 4, pp. 785-791, 1972.

[7] _____, "The role of controllability in some sensitivity problems," *IJC*, vol. 19, no. 1, pp. 57-64, 1974.

[8] G. Guardabassi, F. Romeo, and R. Scattolini, "Sensitivity and parameter identifiability in linear systems," *JFI*, vol. 312, nos. 3/4, pp. 167-177, Sept./Oct. 1981.

[9] N.K. Gupta and R.K. Mehra, "Computational aspects of maximum likelihood estimation and reduction in sensitivity function calculations," *IEEE-TAC*, vol. AC-19, no. 6, pp. 774-783, Dec. 1974.

[10] A. Locatelli and S. Rinaldi, "Controllability versus sensitivity in linear discrete systems," *IEEE-TAC*, vol. AC-15, pp. 254-255, April 1970.

[11] J.G. Reid, P.S. Maybeck, R.B. Asher, and J.D. Dillow, "An algebraic representation of parameter sensitivity in linear time-invariant systems," *JFI*, vol. 301, nos. 1 and 2, pp. 123-141, Jan./Feb., 1976.

[12] M.K. Sain, "Functional reproducibility and the existence of the classical sensitivity matrices," *IEEE-TAC*, vol. AC-12, no. 4, p. 458, Aug. 1967.

[13] V.V.S. Sarma and S.N. Singh, "Controllability and sensitivity," *IEEE-TAC*, vol. AC-14, pp. 782-783, Dec. 1969.

[14] E.J.P. G. Schmidt and R.J. Stern, "Invariance theory for infinite dimensional linear control systems," *AMO*, vol. 6, pp. 113-122, 1980.

[B] Terminal Conditions Insensitivity

[1] U. Bertele and G. Guardabassi, "Can the terminal condition of time-invariant linear control systems be made immune against small parameter variations?," *IEEE-TAC*, vol. AC-16, pp. 460-462, 1971.

[2] G. Fronza and G. Guardabassi, "A note on terminal insensitivity," *IEEE-TAC*, vol. AC-17, no. 4, pp. 559-560, Aug. 1972.

[3] G. Guardabassi, A. Locatelli, and S. Rinaldi, "On the insensitivity of the terminal conditions in open-loop control systems," *IEEE-TAC*, vol. AC-13, pp. 442-443, Aug. 1968.

[4] J.M. Holtzman and S. Horing, "The sensitivity of terminal conditions of optimal control systems to parameter variations," *IEEE-TAC*, vol. AC-10, no. 4, pp. 420-426, Oct. 1965.

4.6.B. Terminal Conditions Insensitivity

[5] _____, "Further results on the stability and sensitivity of terminal control systems," in *Proc. Joint Auto. Contr. Conf.*, pp. 76-81, 1966.

[6] E. Kreindler, "On sensitivity of time-optimal systems," *IEEE-TAC*, vol. AC-14, pp. 578-579, Oct. 1969.

[7] O. Rosen, Imanudin, and R. Luus, "Final-state sensitivity for time-optimal control problems," *IJC*, vol. 45, no. 4, pp. 1371-1381, 1987.

[8] Y. Sawaragi, K. Inoue, and K. Asai, "Synthesis of open-loop optimal control with zero sensitive terminal constraints," *Automatica*, vol. 5, pp. 389-394, May 1969.

[C] Certain Issues in Guidance Theory As Pertain to Terminal Conditions Insensitivity

[1] B.R. Barmish, J.A. Fleming, and J.S. Thorp, "A minimax theorem for guaranteed terminal error," *IEEE-TAC*, vol. AC-20, no. 6, pp. 797-799, Dec. 1975.

[2] J.V. Breakwell, J.L. Speyer, and A.E. Bryson, "Optimization and control of nonlinear systems using the second variation," *SIAM J. Contr.*, vol. 1, no. 2, pp. 193-223, 1963.

[3] P.J. Cefola and C.N. Shen, "Vector-matrix second-order sensitivity equation with application to Mars entry," *AIAA J.*, vol. 7, no. 8, pp. 1633-1635, Aug. 1969.

[4] _____, "A parameter dependent spacecraft guidance boundary value problem," *AIAA J.*, vol. 9, no. 10, pp. 1975-1979, Oct. 1971.

[5] J.E. Cochran, Jr., and D.A. Haynes, "Constrained initial guidance algorithm," *JGCD*, vol. 13, no. 2, pp. 193-197, 1990.

[6] T.A. Heppenheimer and D. Kaplan, "Guidance and trajectory considerations in Lunar mass transportation," *AIAA J.*, vol. 5, no. 4, pp. 518-525, April 1977.

[7] R.N. Ingoldby, "Guidance and control system design of the Viking planetary lander," *JGCD*, vol. 1, no. 3, pp. 189-196, 1978.

[8] H.J. Kelley, "Guidance theory and extremal fields," *IRE-TAC*, vol. AC-7, pp. 75-82, Oct. 1962.

[9] _____, "An optimal guidance approximation theory," *IEEE-TAC*, vol. AC-9, pp. 375-380, Oct. 1964.

[10] A. Miele, T. Wang, and W.W. Melvin, "Guidance strategies for near-optimum take-off performance in a windshear," *JOTA*, vol. 50, no. 1, pp. 1-35, 1986.

[11] B.N. Petrov, V.V. Dement'yeva, and M.M. Khrustalev, "Synthesis of an invariant dynamic system with program control," *ECy*, vol. 19, no. 4, pp. 148-151, 1981.

[12] C.N. Shen and P.J. Cefola, "Adaptive control for Mars entry based on sensitivity analysis," *AIAA J.*, vol. 7, no. 6, pp. 1145-1150, June 1969.

[13] J.L. Speyer, "An adaptive terminal guidance scheme based on an exponential cost criterion with application to homing missile guidance," *IEEE-TAC*, vol. AC-21, no. 3, pp. 371-375, June 1976.

[14] G. Singh, P.T. Kabamba, and N.H. McClamroch, "Bang-bang control of flexible spacecraft slewing maneuvers: Guaranteed terminal pointing accuracy," *JGCD*, vol. 13, no. 2, pp. 376-379, 1990.

[15] J.C. Van der Ha, "Long-term evolution of near-geostationary orbits," *JGCD*, vol. 9, no. 3, pp. 363-370, 1986. Comment, with reply, by P. Cefola, *ibid.*, vol. 10, no. 2, pp. 222-223, 1987.

CHAPTER FIVE

STABILITY ROBUSTNESS ANALYSIS

MANEUVERABILITY, AND SENSITIVITY IN THE LARGE

5.1. INTRODUCTION

The primary objective of this chapter is to advance the correlation between the theory of stability and that of sensitivity. Undoubtedly system stability is of paramount concern in design and operation of any dynamical control system. Stability is a gross necessity for any proper functioning of system, while sensitivity more specifically indicates the ability of system to retain required performance characteristics despite changes in operating conditions. Thus the main thrust of following analyses is to describe various issues regarding to the *construction* of a robustly stable dynamical system subject to multiple-parameter variations, because such a construction is the prerequisite to design of any robust, multiple-objective, control system.

Before proceeding further to study issues pertaining to stability robustness analysis *(SRA)*, we recall that there are many different ways to establish system stability. Perhaps the two most widely used methods to conclude stability in dynamical system are stability approaches in the sense of Lyapunov, and stability methods in the sense of bounded-input bounded-output. From our sensitivity point of view, however, since we plan to study mostly the trajectory-sensitivity considerations, it seems appropriate to define system stability, from now on, in the sense of the Lyapunov methods. In general, the concept of stability is very complicated and it is not always clear what is meant by stability, unless otherwise the corresponding definition has been precisely described. For example, by stability we may mean stability with respect to an equilibrium point, or stability with respect to a nonconstant system trajectory, *etc.*. Study in each case depends on the corresponding definition of stability. This chapter aims only at correlating certain notions of stability and sensitivity, also known as maneuverability [B. 106], which is primarily the requirements on multiple-parameter variations that maintain system stability.

To decide whether or not a system achieves what is set out as its stability criterion we must recall a set of definitions and theorems about system stability. An inspiring survey paper of Kalman and Bertram [A. 72] provides a list of these basic

5.1. Introduction

definitions and the history of their development. From this broad set of definitions and theorems we may conclude that stability is synonymous with convergence. This interpretation enables us to utilize the literature on differential equations, where the notion of convergence in different cases has been extensively examined. Among many other survey papers and books on the general theory of stability which are cited in § 5.12.A, that of [A. 151] may serve as another starting point for further research in this field.

To establish the *SRA* for a dynamical system, which is the main goal of this chapter, we need to review the basic conditions that must be met to guarantee the system stability first. Our study is mainly concerned with a linear time-invariant system whose stability can be established in a simpler way than a general nonlinear system.

Thus in this chapter we are concerned mainly with the *SRA* of a linear time-invariant system, where we develop *the largest* "upper bounds," as we define subsequently, for parameter variations in a continuous- and a discrete-time system which assure the system stability. In other words, given an asymptotically stable system, we emphasize the *constructional* aspects of designing the *class* of robustly stable dynamical systems which surrounds the original system. These results are established for systems whose models are given by a set of differential (or difference) equations which are described by matrices or polynomials.

5.1.1. Stability of Linear Time – Invariant System

Since a linear time-invariant system can be characterized in a number of ways, for example, with a rational polynomial, or a rational matrix, to establish stability of a linear time-invariant system we need to investigate the conditions under which a polynomial or a matrix is called *stable*, first, followed by that of *asymptotically stable*. In other words, instead of computing the system transition matrix, which is required when we study the stability behavior of most nonlinear systems, in this case we need only to study the stability behavior of a rational polynomial or a rational matrix.

Polynomial: For a rational polynomial to be called stable it is required that each and every zeros of this polynomial in the complex plane to have a nonpositive real part. The natural connection between stability in a linear time-invariant system and the zeros of a polynomial, makes the study of the behavior in these zeros, as functions of system parameters, one of the earliest research topics in control system theory. Our interests in this regard are to present the principal results on variations in the zeros of a polynomial with respect to variations in system parameters which may affect system stability.

Matrices: We also recall that stability of a rational square matrix S (corresponding to a linear time-invariant system $\dot{x}(t) = Sx(t)$), depends on the eigenvalues of S. The complete stability requires that the real parts of these eigenvalues to be nonpositive. This condition indeed requires that the roots of the corresponding characteristic equation have nonpositive real parts. We recall that the restriction on the sign of the eigenvalues' real-parts for a constant matrix S does not provide a stability criterion for a linear time-varying system $\dot{x}(t) = S(t)x(t)$.

∎

The above two approaches for establishing the stability requirements in polynomials and matrices are indeed complementary of each other. The *asymptotic* stability is when we replace the "nonpositive" with "negative" in each one of the above two cases.

Because the stability in a *DCLTIS* can be studied in at least two different ways (namely, the zeros of a polynomial or the eigenvalues of a matrix), we are also studying their corresponding robustness properties accordingly. Thus in the subsequent sections we briefly review some of the known results for root sensitivity of a polynomial equation, as well as the eigenvalue sensitivity of a matrix. This review is followed by a comprehensive *SRA* based on the second method of Lyapunov. But first we review this method.

One of the most celebrated results on the theory of stability, in particular that of a linear time-invariant system, is the special case of the Lyapunov theorem [A. 8] which is described in the following for both continuous- and discrete-time systems.

COROLLARY 5.1-1 [A. 72]: The equilibrium state $x_e = 0$ of a system described by $\dot{x} = Sx$ is asymptotically stable (a) if and (b) only if given any symmetric, positive-definite matrix Q there exists a symmetric, positive-definite matrix P which is the unique solution of the set of $n(n + 1)/2$ linear equations
(i) $S^T P + PS + Q = 0$, and $\|x\|_P^2$ is a Lyapunov function for this system; or
(ii) $SP + PS^T + Q = 0$, and $\|x\|_P^2$ is similarly a Lyapunov function.

∎

It is well known that P in *(i)* (similarly for (2.6-30)) can be written as follows.

$$P = \int_0^\infty \exp(S^T t) Q \exp(St) dt. \qquad (5.1-1)$$

Similar equation can be stated for *(ii)*. Clearly, the above two P's are numerically different, though they are both symmetric and positive-definite matrices and are shown by the same symbol (cf., Example 5.8-1).

COROLLARY 5.1-2 [A. 73]: All the eigenvalues of $S \in R^{n \times n}$ lie in the unit circle (*convergent* matrix, Shane and Barnett [D. 60]) if and only if for an arbitrary $0 < Q = Q^T \in R^{n \times n}$, the matrix equation

$$S^T P S + Q = P, \qquad (5.1-2)$$

yields a unique solution $0 < P = P^T \in R^{n \times n}$.

∎

5.1. Introduction

To search a closed-form solution for P in (5.1-2), we note that substituting P recursively into itself yields

$$(S^T)^{k+1} P S^{k+1} + \sum_{l=0}^{k} (S^T)^l Q S^l = P. \qquad (5.1-3)$$

For an *asymptotically convergent* S (i.e., a matrix whose norm is less than one), we have $\lim_{k \to \infty} S^k = 0$. Thus for a sufficiently large k and an asymptotically convergent S we have

$$P = \sum_{l=0}^{k=\infty} (S^T)^l Q S^l. \qquad (5.1-4)$$

As a special case, we let Q = I (an identity matrix), then

$$P = \sum_{l=0}^{k=\infty} (S^T)^l S^l. \qquad (5.1-5)$$

Suppose that the above asymptotically convergent S is symmetric, then (5.1-5) becomes

$$P = (I - S^2)^{-1}. \qquad (5.1-6)$$

Also for cases in which the above S is such that $S^T S = S S^T$, we have

$$P = (I - S^T S)^{-1}. \qquad (5.1-7)$$

5.2. ROOT SENSITIVITY AND POLE PLACEMENT

5.2.1. Introduction

Root sensitivity can be studied according to the definition of the corresponding sensitivity measure of *small* or *large*. *First of all,* we note that root-*sensitivity in the large* refers to a case where changes in system parameters yield large variation in the location for a root of a rational polynomial equation in the complex plane. Thus this situation is very much similar to the concept of pole placement and in this sense they are complementary of each other. In a pole placement process we change the location of a system pole according to a stability measure by adjusting certain parameter(s) which is the same as what we call *sensitivity in the large*. Pole placement is a broad area of research as is evident by the large number of papers in that field. However, our interest is from sensitivity point of view, i.e., how much variations for any given poles are acceptable. *Secondly,* the root *small-sensitivity* analysis refers to the study of the rate of change in each corresponding parameter when this change is considered to be infinitesimal. This analysis has also received attentions in the control-theory literature. In general, study of the roots in an equation with respect to changes in its coefficients (more than one parameter) is a very difficult and challenging problem [B. 121] (specially in symbolic form) and [C. 69]. Indeed a major part of this chapter addresses this issue from computational point of view and based on the results which are presented in this book. Therefore we keep the present discussion at the minimum.

In general, root-sensitivity analysis corresponds to various methods of studying the locations for roots of a rational polynomial equation. For example, one major approach to study that problem is the classical root-locus technique. This approach is applied to equations which are of the forms $A(s) + K\, B(s) = 0$, where $A(s)$ and $B(s)$ are two rational polynomials of the complex variable s and K is a real-variable parameter [H. 39 and 21]. The pioneering work of Reza in using root-locus for the SRA should be noted [H. 27]. Although this technique can be conceptually extended to a system with several varying parameters of the form $A(s) + K_1 B_1(s) + K_2 B_2(s) + ... = 0$, the method will soon become cumbersome as the number of parameters exceeds one. Indeed this method which is based on a *point by point* search for finding each stable perturbed system, is very inefficient in practice and that is why researchers have pursued other lines of investigation, in order to come up with a *set by set* search for finding such stable perturbed systems. The root-sensitivity analysis of a polynomial equation which is in a factor form $\prod_{i=1}^{n}(s - s_i) = 0$ has also received attentions in the literature, but that study is only of academic interest, since in most practical situations and for higher-order systems this closed-form expression in terms of system parameters is not easily available.

The Routh and the Hurwitz criteria provide the necessary and sufficient conditions for studying the coefficient changes in a polynomial which brings the zeros of this polynomial to the right-hand side of the complex plane, however, this approach is not very practical for the higher-order systems with unknown varying parameters due to the lack of an appropriate methodology for treating a set of inequalities with multiple-varying parameters. A survey of Routh's criterion is reported in [A. 21]. For another historical prospective refer to [H. 37]. The original work of Morgan [H. 19] also treats root sensitivity of characteristic equation in a linear time-invariant system with respect to its coefficients. This work is also viewed in terms of the eigenvalue sensitivity.

■

In general, the study of zero sensitivity (zeros of the system) is similar to pole sensitivity. For cases in which the numerator polynomial of a transfer function is expressed with respect to system matrices, some results are reported by Morgan in [H. 19].

5.2.2. Certain Formulae On Root Sensitivity

We assume that in general the pole placement does not change system order or (dimension). Also in studying a transfer function, we assume that it does not have a common factor. Indeed we must avoid variations in poles and/or zeros which cause the creation of a common factor in a transfer function, unless otherwise is desired. To this effect several useful equations which relate variation in a zero of a polynomial versus variations in its coefficients are established next.

Consider a monic polynomial $\Delta(s)$ given by

5.2. Root Sensitivity and Pole Placement

$$\Delta(s) \triangleq s^n + a_1 s^{n-1} + \ldots + a_j s^{n-j} + \ldots + a_n. \tag{5.2-1}$$

Suppose that s_o is a zero of this polynomial and $s_o \to s_o + \Delta s_o$, because of only one $a_j \to a_j + \Delta a_j$. Then the objective is to relate this Δa_j to the corresponding Δs_o.

PROPOSITION 5.2-1: Consider the monic polynomial $\Delta(s)$ of (5.2-1). If s_o which is a zero of $\Delta(s)$ becomes $s_o + \Delta s_o$, such that $\Delta s_o \neq -s_o$, and because of $a_j \to a_j + \Delta a_j$. Then

$$\Delta a_j = -\Delta s_o \sum_{k=0}^{n-1} a_k \sum_{i=0}^{n-1-k} (s_o + \Delta s_o)^{(j-1-k)-i} s_o^i,$$

with $a_0 = 1$, and $j = 1, 2, \cdots, n$. \hfill (5.2-2)

PROOF: If $s_o \to s_o + \Delta s_o$ (when $a_j \to a_j + \Delta a_j$), then $\tilde{\Delta}(s_o + \Delta s_o) = 0$ can be written as

$$(s_o + \Delta s_o)^n + a_1(s_o + \Delta s_o)^{n-1} + \ldots + (a_j + \Delta a_j)(s_o + \Delta s_o)^{n-j} \tag{5.2-3}$$

$$+ \ldots + a_{n-1}(s_o + \Delta s_o) + a_n = 0. \tag{5.2-3}$$

From $\Delta(s_o) = 0$, we can show that

$$a_n = -s_o^n - a_1 s_o^{n-1} - \ldots - a_{n-1} s_o. \tag{5.2-4}$$

Substituting a_n from above into (5.2-3) results in

$$[(s_o + \Delta s_o)^n - s_o^n] + a_1[(s_o + \Delta s_o)^{n-1} - s_o^{n-1}] + \ldots + a_j[(s_o + \Delta s_o)^{n-j} - s_o^{n-j}]$$

$$+ \Delta a_j (s_o + \Delta s_o)^{n-j} + \ldots + a_{n-1} \Delta s_o = 0. \tag{5.2-5}$$

Since for every n we have

$$a^n - b^n = (a - b) \sum_{i=0}^{n-1} a^{(n-1)-i} b^i, \tag{5.2-6}$$

equation (5.2-5) on using (5.2-6) becomes

$$\Delta s_o \left[\sum_{k=0}^{n-1} a_k \sum_{i=0}^{n-1-k} (s_o + \Delta s_o)^{(n-1-k)-i} s_o^i \right] + \Delta a_j (s_o + \Delta s_o)^{n-j} = 0. \tag{5.2-7}$$

If $(s_o + \Delta s_o)^{n-j} \neq 0$; i.e., $\Delta s_o \neq -s_o$, then (5.2-7) can be divided by $(s_o + \Delta s_o)^{n-j}$ and the conclusion (5.2-2) follows.

∎

The result of Proposition 5.2-1 has a special merit. Suppose that the difference between a pole and a zero in an appropriate transfer function equals Δs_o. Then (5.2-2) provides a necessary change in Δa_j corresponding to Δs_o, which is causing the occurrence of a common factor that is generally speaking undesirable. Certainly depending on whether zeros or poles of the transfer function are changing, then Δa_j corresponds to either its numerator or its denominator.

REMARKS 5.2-1: (i) From (5.2-2) we conclude that

$$\frac{\partial a_j}{\partial s_o} = -\sum_{k=0}^{n-1} a_k(n-k)s_o^{j-1-k}. \tag{5.2-8}$$

(ii) In certain cases the variations in a_j may have been caused by another parameter which may change several other coefficients of $\Delta(s)$. In that case and unless additional information concerning these coefficient variations is provided we cannot give a useful relationship among them. In other words, the knowledge of only Δs_o is not enough to give information about $\Delta a_j, \ldots, \Delta a_k$ which produce this Δs_o. Clearly this problem has no unique solution and we study that using a different approach in Section 5.7.

(iii) Considering (5.2-3) and in order to solve for an arbitrary Δs_o (i.e., with no restriction on $|\Delta s_o|$) versus Δa_j, (5.2-3) shows that we need to solve a polynomial equation for Δs_o which is of degree n. It is well known that a closed-form analytical solution for $n \geq 5$ is not known. However, for cases of small Δs_o; i.e., when we assume that $|\Delta s_o|^2 \approx 0$, we have

$$\Delta s_o = [-\Delta a_j s_o^{n-j}]/[\Delta a_j(n-j)s_o^{n-j-1} + \sum_{k=0}^{n-1} a_k(n-k)s_o^{n-1-k}],$$

$$j = 1, 2, \cdots, n. \tag{5.2-9}$$

This result can be shown by expanding (5.2-3) as follows.

$$s_o^n + ns_o^{n-1}\Delta s_o + a_1[s_o^{n-1} + (n-1)s_o^{n-2}\Delta s_o] + \cdots + a_j[s_o^{n-j} + (n-j)s_o^{n-j-1}\Delta s_o]$$

$$+ \Delta a_j[s_o^{n-j} + (n-j)s_o^{n-j-1}\Delta s_o] + \cdots + a_{n-1}(s_o + \Delta s_o) + a_n = 0. \tag{5.2-10}$$

Because $\Delta(s_o) = 0$, therefore the conclusion follows.

Equation (5.2-9) provides "small" changes in s_o that is caused by Δa_j which in general may not be "small" We must, however, point out that the resulting Δs_o must be tested to see whether it satisfies the condition of $|\Delta s_o|^2 \approx 0$. Otherwise, we cannot use (5.2-9). On that note, we can show that (5.2-9) is a special case of the next proposition.

PROPOSITION 5.2-2: Consider a polynomial $\Delta(s)$ given by (5.2-1). If any zero of this polynomial such as s_o becomes $s_o + \Delta s_o$, with $|\Delta s_o|^2 \approx 0$, because of $a_j \to a_j + \Delta a_j$ and for all j, then Δs_o becomes

$$\Delta s_o = [-\sum_{j=1}^{n} \Delta a_j s_o^{n-j}]/[\sum_{j=1}^{n-1}(a_j + \Delta a_j)(n-j)s_o^{n-j-1} + ns_o^{n-1}]. \tag{5.2-11}$$

PROOF: Recall that $\Delta(s_o) = 0$. If $s_o \to s_o + \Delta s_o$ such that

$$(s_o + \Delta s_o)^n + (a_1 + \Delta a_1)(s_o + \Delta s_o)^{n-1}$$

$$+ \cdots + (a_{n-1} + \Delta a_{n-1})(s_o + \Delta s_o) + (a_n + \Delta a_n) = 0. \tag{5.2-12}$$

Then because of $|\Delta s_o|^2 \approx 0$, we have

$$s_o^n + ns_o^{n-1}\Delta s_o + (a_1 + \Delta a_1)[s_o^{n-1} + (n-1)s_o^{n-2}\Delta s_o]$$
$$+ \cdots + (a_j + \Delta a_j)[s_o^{n-j} + (n-j)s_o^{n-j-1}\Delta s_o]$$
$$+ \cdots + (a_{n-1} + \Delta a_{n-1})(s_o + \Delta s_o) + a_n + \Delta a_n = 0. \quad (5.2\text{-}13)$$

Collecting the coefficients of Δs_o, and considering that $\Delta(s_o) = 0$, we have

$$\Delta s_o [ns_o^{n-1} + \sum_{j=1}^{n-1} (a_j + \Delta a_j)(n-j)s_o^{n-j-1}] + \sum_{j=1}^{n} \Delta a_j s_o^{n-j} = 0. \quad (5.2\text{-}14)$$

■

In deriving (5.2-11) no constraint has been imposed on the Δa_j's, although their being small is preferable. It seems, however, no such restrictions on the Δa_j's are necessary so long as $|\Delta s_o|$ becomes sufficiently small. Equation (5.2-11) gives more information than a similar equation that is reported by Ryabov and Sachkov [H. 28] concerning with "small" Δa_j's.

5.3. EIGENVALUE AND EIGENVECTOR SENSITIVITY ANALYSIS

Because of intimate relation between stability in a linear time-invariant system described by $\dot{x}(t) = Sx(t)$, and the eigenvalues of S, a large number of papers are devoted to study its corresponding eigenvalue and eigenvector sensitivity functions. A partial list of the pertinent literature is given in § 5.12.I and § 5.12.J.

Many researchers, such as Crossley and Porter [I. 3], have pointed out that the history of eigenvalue sensitivity (in particular large sensitivity) goes back to the original work of Jacobi in the early 1846.

Morgan [H. 19] and [I. 19] also gives many useful equations for deriving the sensitivity functions of roots and eigenvalues in a linear time-invariant system. Additional references are cited in § 5.12.I and § 5.12.J. These results are mostly centered around the small-sensitivity analysis and are mainly for cases whose eigenvalues are distinct.

A generalization of small-sensitivity analysis to the case of an eigenvalue with multiplicity is reported by Paraskevopoulos et al., in [I. 23]. They prove that in a system with distinct eigenvalues λ_i, the change in each eigenvalue in terms of changes in matrix S can be stated as follows.

$$d\lambda_i = \{\text{Tr adj}(\lambda_i I - S)\}_o^{-1} \{\text{adj}(\lambda_i I - S)\}_o * dS, \quad i = 1, \ldots, n, \quad (5.3\text{-}1)$$

where Tr denotes the trace operator (sum of all eigenvalues) and the subscript "o" indicates the nominal value. Here * stands for the inner product of two equidimensional square matrices; i.e., $C*D = \sum_i c^i d_i$, where c^i is the ith-row of C, and d_i is the ith-column of D. It is also well known that for any square matrix S, $d\{\det S\} = \text{adj } S * dS$, and for any such matrix having a repeated eigenvalue λ_r of multiplicity μ the following is true:

$$\{d^k[\operatorname{Tr}\operatorname{adj}(\lambda I - S)]\}_o \begin{cases} = 0 & \text{for } k = 1, \cdots, \mu-2, \\ \neq 0 & \text{for } k = \mu-1. \end{cases} \qquad (5.3\text{-}2)$$

For the case of multiple eigenvalues, the change in λ_r with multiplicity μ is

$$d\lambda_r = [\frac{1}{\mu!}\{d^{\mu-1}\operatorname{Tr}\operatorname{adj}(\lambda I - S)\}_o]^{-1} \sum_{k=1}^{\mu} \frac{1}{k!}\{d^{k-1}\operatorname{adj}(\lambda I - S)\}_o * dS, \qquad (5.3\text{-}3)$$

which reduces to the case of distinct eigenvalues for $\mu = 1$ and coincides with the earlier result. The reader should refer to § 5.12.*I*, for the pertinent literature on eigenvector-sensitivity formulae.

In closing this section we must point out that a number of research activities are reported that describe various sensitivity measures for poles/zeros versus eigenvalues, for instance [*J.* 64]. For additional references on various synthesis methods which are based on the applications of sensitivity measures stemming from the stability considerations (in particular, with respect to poles and/or zeros and eigenvalue-sensitivity functions) refer to § 5.12.*J*.

5.4. STABILITY ROBUSTNESS ANALYSIS: ISSUES AND TECHNIQUES

5.4.1. Introduction

This section is dedicated to the development and discussion of what we call *sensitivity in the large* as pertains to system stability. In general, sensitivity in the large refers to cases in which a set of parameters changes in a manner that the parameter variations are *large* and the objective is to find how *large* these variations can get before a particular qualitative behavior of the system does no longer hold. For instance, in the case of stability as a qualitative behavior of the system, the objective is to find the largest parameter variations whose corresponding perturbed systems remain stable. The very definition of *the largest* in the case of multiple-parameter variations, however, requires special attention, as we elaborate on this matter subsequently. In the early literature on the theory of sensitivity the word *insensitive* was often used to describe a system which was almost invariant in the above sense. In the contemporary control parlance, the word *robust* is being used to describe almost the same circumstance, although robustness, in our opinion, should refer to a situation wherein an actual performance has been met. There are, of course, many other situations (cf., Chapter Six) where we use the adjective robust to describe a system. However, we consider that the true meaning of the robustness is as follows.

To provide condition(s) under which a *class* of systems, preferably *the largest* of its kind, will perform *analogously*.

■

The robustness study and/or analysis is perhaps one of the most fascinating areas of research in system theory and is by no means a simple matter. The

5.4. Stability Robustness Analysis: Issues and Techniques

history of this problem goes back to the early literature on sensitivity theory. In the following we briefly state a few historical approaches as well as results pertaining to the system stability robustness, before we present some exclusive work in this area.

Consider a *DCLTI – GS* which is described by a set of ordinary differential equations *(ODE's)* as follows.

$$\dot{x}(t) = Ax(t) + Bu(t), \quad x(0) = x_0. \tag{5.4-1}$$

Suppose that $A \to A + \Delta A$ and $B \to B + \Delta B$. Then it is expected that $x(t) \to x(t) + \Delta x(t)$, and if (5.4-1) is a closed-loop system, then $u(t) \to u(t) + \Delta u(t)$. Therefore $x(t)$ is no longer what we would have expected had the parameters were not varied. The main thrust of the following investigation is to determine under what conditions (or within which bounds) the system parameters may change such that (5.4-1) remains stable. This determination, preferably in an analytical form, is the main motivation for researchers in this area whose efforts have manifested in one form or another in the literature.

Michael and Merriam [B. 89] consider the robustness problem as follows. Suppose that $u(t)$ is chosen such that it minimizes a quadratic cost given by

$$J = \tfrac{1}{2} \int_0^{t_f} [x^T(t)Rx(t) + u^T(t)Qu(t)]dt. \tag{5.4-2}$$

It is well known that $u^o(t) = -Q^{-1}B^T P(t)x(t)$ minimizes (5.4-2), where $P(t) = P^T(t) > 0$ is the unique solution of the Riccati matrix differential equation

$$-\dot{P}(t) = R + P(t)A + A^T P(t) - P(t)BQ^{-1}B^T P(t), \quad P(t_f) = 0. \tag{5.4-3}$$

The minimum value of $J^o[x_0; u^o]$ is equal to $\tfrac{1}{2}x_0^T P(0)x_0$.

For the case in which A and B are disturbed by the *large* amount, then the previous $u^o(t)$ no longer minimizes the performance index *(PI)* (5.4-2) and in fact this cost has been changed to some new value. Thus a new $u^*(t)$ must be found which is optimal in the sense of minimizing certain other *PI* and produces an asymptotically stable closed-loop system.

Michael and Merriam propose to model the disturbed system as follows.

$$\dot{x}(t) = Ax(t) + Bu(t) + Cv(t), \quad x(0) = x_0, \tag{5.4-4}$$

where the additional term $Cv(t)$ describes the contribution of disturbances in A and/or B. The new *PI* becomes $J^*[x_0; u, v]$, where

$$J^*[x_0; u, v] = \tfrac{1}{2} \int_0^{t_f} [x^T(t)Rx(t) + u^T(t)Qu(t) - v^T(t)Sv(t)]dt. \tag{5.4-5}$$

The optimal control for this auxiliary system is constructed by maximizing J^* with respect to $v(t)$, then minimizing J^* with respect to $u(t)$. That means u^* and v^* are defined by

$$J^*[x_0; u^*, v^*] = \min_u \max_v J^*[x_0; u, v]. \tag{5.4-6}$$

Based on this formulation, Michael and Merriam establish certain bounds on ΔA and ΔB so that the system remains stable. This result is complicated, and because we present a more general result than this one, it is not repeated. Likewise Dergacheva [B. 21] gives certain bounds on these parameter variations. Essentially these two approaches are very similar.

∎

Another major research direction is based on the Lyapunov technique. Upon the publication of [A. 72] a number of researchers began to consider applications of the Lyapunov direct method in design or inverse stability problems. For instance, the very inception of model-reference adaptive control theory in the early 1960's was among the first such applications. The pioneering work of Reza [B. 104] is among the first applications of the Lyapunov method in stability robustness analysis, wherein this method is used to construct a class (or a family) of stable matrices (or systems).

∎

One important aspect of the Lyapunov method has been the actual generation of a Lyapunov function. This difficulty is not so severe in the case of a linear system as is in the case of a nonlinear system. Based on this thought of generating the Lyapunov function for as large a class of [essentially linear] systems as possible, Barnett and Storey [B. 8] and Barnett [B. 7] consider the stability robustness issues from a different point of view. They point out that for a set of linear homogeneous differential equations $\dot{x}(t) = Ax(t)$, where A is asymptotically stable, a square matrix D can be found such that $\dot{y}(t) = (A + D)y(t)$ becomes asymptotically stable in the sense that these two systems have the same Lyapunov function. In other words, they find a class of linear systems, namely $A + D$, that is constructed from A (for a number of different D's), such that the two systems have the same Lyapunov function and therefore they have the same stability behaviors. This result is as follows.

Suppose that A is an asymptotically stable matrix. Then from Corollary 5.1-1, and for an arbitrary $Q = Q^T > 0$, we have $P = P^T > 0$ such that $A^T P + PA + Q = 0$. Now, if a matrix D is chosen such that

$$D = P^{-1}(S - Q_1); \qquad (5.4\text{-}7)$$

where S is an *arbitrary* skew-symmetric matrix, Q_1 is an *arbitrary* non-negative definite matrix and P is given by the preceding Lyapunov equation, then the stability of $\dot{y}(t) = (A + D)y(t)$ is guaranteed [B. 7 & 8], since both $A + D$ and A have the same P.

To interpret the above D as a perturbation of A and thus to use (5.4-7) for the purpose of stability robustness analysis *(SRA)*, has not always been very satisfactory. We note that although this result is useful, D is not necessarily a physical perturbation of neither A nor B, because many arbitrary choices are introduced in order to compute D. In other words, D is a different matrix with different entries, of course, some having no physical meaning. For a typical application of this result the reader may consult [B. 126], which confirms that the engineering applications of this work are limited. However, the above result simply gives a *class* of linear systems which has the same Lyapunov function.

5.4.2. Stability Robustness Analysis Formulation

Research in this area is very active and a number of new methods are proposed in the literature to treat the stability robustness analysis *(SRA)* of a dynamical system, cf., Section 5.12. First we realize that the prerequisite to a meaningful robustness analysis in a dynamical system is to indicate a particular and desirable system behavior, and secondly to specify the precise perturbational mechanism that may not disturb this property. To change the system operating condition as we please, while maintaining its basic operation, is called *maneuverability* [B. 106]. Variations that are attributed to maneuverability are different from those which are attributed to the adaptivity in the sense that the former is not time dependent. (A few papers in § 5.12.F, are inconsistent with this notion.) The current robustness analysis, of course, pertains to system stability and it only concerns with a parameter perturbation mechanism that does not affect this property. But first we make the following assumption.

■

THE FUNDAMENTAL ASSUMPTION: We emphasize that the initial dynamical system is assumed to be in a stable mode and as such, we are only interested to know how we can perturb this system *around this stable-operating mode* without deteriorating its stability. Furthermore, we assume that these large parameter variations will not affect the stabilizability of the underlying system, i.e., no structural unstability is anticipated.

Clearly, we must analyze a meaningful perturbational scheme. To set the record straight we recall the following standard regulator system consisting of two distinct sub-systems which are coupled as follows (cf., Section 2.2).

$$\frac{d}{dt}\begin{Bmatrix} x_1 \\ x_2 \\ x_3 \\ x_4 \end{Bmatrix} = \begin{bmatrix} 0 & 1 & c_{13} & c_{14} \\ \alpha_1 & \beta_1 & c_{23} & c_{24} \\ c_{31} & c_{32} & 0 & 1 \\ c_{41} & c_{42} & \alpha_2 & \beta_2 \end{bmatrix} \begin{Bmatrix} x_1 \\ x_2 \\ x_3 \\ x_4 \end{Bmatrix} + \begin{bmatrix} 0 & 0 \\ \gamma_1 & 0 \\ 0 & 0 \\ 0 & \gamma_2 \end{bmatrix} \begin{Bmatrix} u_1 \\ u_2 \end{Bmatrix}$$

$$\triangleq \begin{bmatrix} A_{11} & C'_{12} \\ C'_{21} & A_{22} \end{bmatrix} x + \begin{bmatrix} B_1 \\ B_2 \end{bmatrix} \begin{Bmatrix} u_1 \\ u_2 \end{Bmatrix} \triangleq Ax + Bu. \qquad (5.4\text{-}8)$$

We assume that u_1 and u_2 are chosen such that $u = Kx$ minimizes a quadratic *PI*. Upon substituting this optimal and stabilizing input in (5.4-8) we have:

$$\dot{x} = (A + BK)x \triangleq S(\theta)x, \qquad (5.4\text{-}9)$$

where

$$S(\theta) \triangleq \begin{bmatrix} 0 & 1 & c_{13} & c_{14} \\ \alpha_1 + \gamma_1 k_{11} & \beta_1 + \gamma_1 k_{12} & c_{23} + \gamma_1 k_{13} & c_{24} + \gamma_1 k_{14} \\ c_{31} & c_{32} & 0 & 1 \\ c_{41} + \gamma_2 k_{21} & c_{42} + \gamma_2 k_{22} & \alpha_2 + \gamma_2 k_{23} & \beta_2 + \gamma_2 k_{24} \end{bmatrix}. \qquad (5.4\text{-}10)$$

Here $\theta \in R^r$ is the generic name for system *parameters,* vector whose nominal value is specified by θ^o.

Clearly and as intended, the above example demonstrates how various system parameters in any given situation propagate into the final system model and thus their variations may affect the overall system stability.

The establishment of the necessary and sufficient conditions for stability robustness properties of linear time-invariant and a class of nonlinear dynamical systems is directly related to that for a set of ordinary differential equations *(ODE's)*, which are characterized in a matrix or a polynomial form. The *SRA* of these systems in turn are directly related to that of a stable matrix $S(\theta)$, where $\theta \in R^r$ is the parameter vector and we assume that when $\theta \to \theta + \Delta\theta$, the multiple large-parameter changes, $\Delta\theta$, will not affect the stabilizability of $S(\theta)$, i.e., no structural problem which deteriorates system stability is predicted. Or similarly, we may formulate the stability requirement of a typical $S(\theta)$ with respect to its characteristic equation and thus the *SRA* for this system may also be concluded from the *SRA* of a stable polynomial $\Delta(s, \theta)$. However, since the *SRA* of any polynomial can be related to that of a special matrix, the main problem that we are investigating in this chapter is as follows.

PROBLEM 5.4-1 *(The Fundamental Question):* Suppose that $S(\theta^o)$, which is an asymptotically stable matrix at a given operating condition $\theta^o \in R^r$, becomes $S + \Delta S$, as a result of $\theta^o \to \theta^o + \Delta\theta$. Then what is *the largest class* of ΔS's which yields asymptotically stable $S + \Delta S$?

■

Clearly the above question implies that we are interchangeably working with two equivalent sets of objects. *One set* consists of points, such as $P \triangleq [\Delta\theta_1, ..., \Delta\theta_r]^T$, in the "parameter variations space" *(PVS)*, which is an Euclidean space and whose origin is at the nominal value or the stable operating point θ^o. This $\theta^o \in R^r$ is also a point in the "parameter space" *(PS)*, whose origin is at zero value for all parameters. Clearly each *PVS* is a translation of the *PS* to the operating point θ^o. *The other set* consists of the perturbation matrices ΔS's which are computed at each of the above points P. The distinction between these two sets, specially for systems with multiple-varying parameters, is required in order to proceed.

■

We must also be very explicit about the structure of our problem. Generally speaking, given $S(\theta)$ with multiple-varying parameters, there are two issues meriting special attention: The *first* one is the definition for the stability robustness measure *(SRM)*, or specifically in our case, the definition for *the largest class* of perturbed systems. The *second* issue is the computational efficiency of generating this *SRM* that deserves serious consideration.

■

Before proceeding further to state our choice for the *SRM* and the rationale behind that choice, we note that in the literature several different *SRM*'s (based on certain matrix norm, or an eigenvalue, or a singular value, *etc.*), are selected and these measures are applied to different numerical examples. In most cases efforts

5.4. Stability Robustness Analysis: Issues and Techniques

are made to classify a system such that it meets all the necessary requirements for a particular *SRM*. As a result of this inconsistency on the definition of the *SRM*, there is always some debate on whose measure should be utilized. However, since most of these choices are initially developed for a particular class of systems and often no mention is made on the computational efforts needed for such system classification, it is a very difficult task to provide a menu (based on the overall computational complexities and/or usefulness of these measures) on how to use any one of these *SRM*'s when we face a new situation. It seems that we really need only one unified approach to define the *SRM* of any given [linear] system, which is to be applicable to every problem. At this juncture we appeal to the following fact that from the physical point of view and when the structure of a system is specified, i.e., when we have, for instance, an asymptotically stable system $\dot{x}(t) = Sx(t)$, $S \in R^{n \times n}$, the true set of points in the associated *PVS*, whose corresponding perturbed matrices are stable, is fixed and is independent from any considerations and/or ways on our parts which may be utilized to find that set. In other words, and no matter which method is used to describe the stability robustness issues, finally the answer as to how far in the *global sense,* we can maneuver a particular parameter in S, *per se,* is the same. The apparent discrepancy is due to the fact that none of the *SRM*'s reported in the literature are global answer to the corresponding Problem 5.4-1, and therefore as long as different measures are defined, different ranges for parameter variations are generated, without any particular advantage of one interval over the other. Perhaps one reason for continuously generating *new* sufficient conditions in the literature is that the globally closed-form solution for system of 5th-order and beyond is not known, and therefore it is very appealing to develop new sufficient conditions. We should explicitly say that, in our opinion however, the real issue is not to advocate or otherwise, any particular measures or approaches for the *SRA,* rather it is to define a measure that is applicable to every problem first, followed by a realistic computational approach to generate that measure.

∎

Motivated by these thoughts we choose the *SRM* of a dynamical system to be *the largest* or *most extended* set of points in the *PVS* (or the *PS*) (which is an Euclidean space), whose corresponding set of perturbed matrices is stable and we call that extended set the *E-Box*. This choice for the *SRM* stems from the *intrinsic* property of a dynamical system regarding its global domain of maneuverability, and thus its construction provides us with a fundamentally invariant property of that system. In most cases and because of the continuous dependence of the solution for a set of *ODE's* on its parameters, this extended set is connected or it is a collection of disjoint connected sets and comes uniquely with each [linear] system. Fig. 5.4-1 depicts a pictorial view of a typical *E-Box* (or a portion of that) for a system with three varying parameters, which this set is equivalent to the "solution" of the first column in the corresponding Routh and the Hurwitz criteria, i.e., the global set of points in the *PS* whose associated perturbed systems are stable. As this figure suggests, in certain cases we may have several directions along which all the parameters may go to infinity, while the system remains stable (cf., Example

5.8-6). To detect an arbitrary infinite directional perturbation for an unknown system, other than those associated with the r axes of the *PVS* and perhaps a few "45° lines" (referring to the condition when $|\Delta\theta_1| = \cdots = |\Delta\theta_r|$), is an open question.

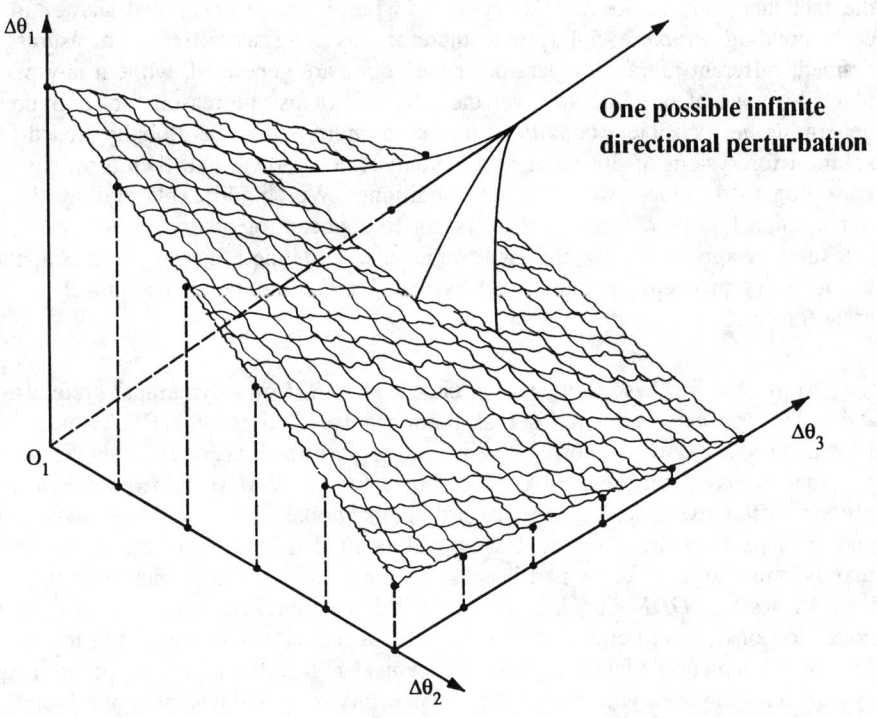

Fig. 5.4-1. A pictorial view of the *largest* or most *extended* set of parameter variations. *The E-Box.*

5.4. Stability Robustness Analysis: Issues and Techniques

Since in most typical applications no analytical approaches can be easily provided to describe this *E-Box*, and from the computational point of view we may need substantial computational-time to generate an *E-Box*, while the existence of such a set cannot be ignored, we approximate the *E-Box* with a set of convex sets (or hyperboxes) that are stacked *under* this *E-Box* in an appropriate way and we call that set of boxes the *SRM* of the system. (The *hyperbox* is an abstraction of a square or rectangle to that in the higher-dimensional Euclidean space, while the *E-Box* is the generalization of *the largest* or *most extended* hyperbox with no geometric abstraction in that space.) The rationale behind this thought is that the allowable-stability region for any dynamical system is a complicated set of points which is independent from our computational methods and it certainly is not a hyperbox. Therefore we cannot dictate a prespecified geometry to the stability region of any system and any school of thought which suggests that must be reevaluated in its entirety. On the other hand, since we cannot analytically develop every *E-Box*, we propose a *probing* (or iterative) technique to check the true stability region of a dynamical system, in order to construct the actual structure of this region as best as possible. It is imperative to use a probing or iterative approach, because every known analytical result is local. Based upon these thoughts we develop Fig. 5.4-2, which shows pictorially our *SRM* that is substantially close to our original *E-Box*.

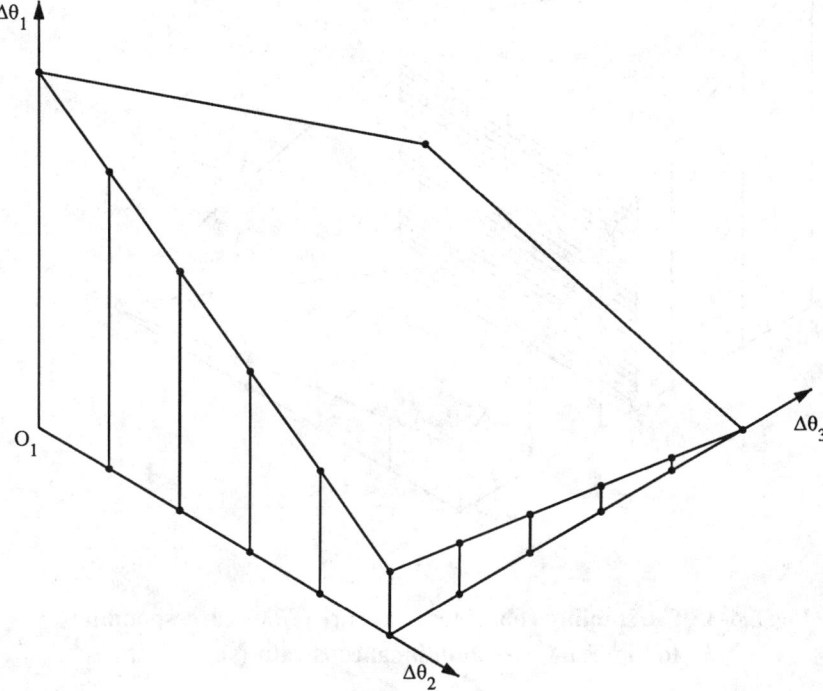

Fig. 5.4-2. The stability robustness measure *(SRM)* corresponding to Fig. 5.4-1.

Finally, we may face a situation that despite some acceptable range of parameter variations in a given system, and from the physical point of view (or perhaps other consideration), there are other constraints which must be met in order to construct the associated *SRM*. For example, in the case of a system corresponding to Fig. 5.4-1 (or Fig. 5.4-2), we may have to meet a physical constraint of $\Delta\theta_2 \leq \Delta\theta_2^{Max}$. In that case Fig. 5.4-2 must be modified to that of Fig. 5.4-3 which shows that a part of the previous *SRM* has been carved to meet the underlying constraint.

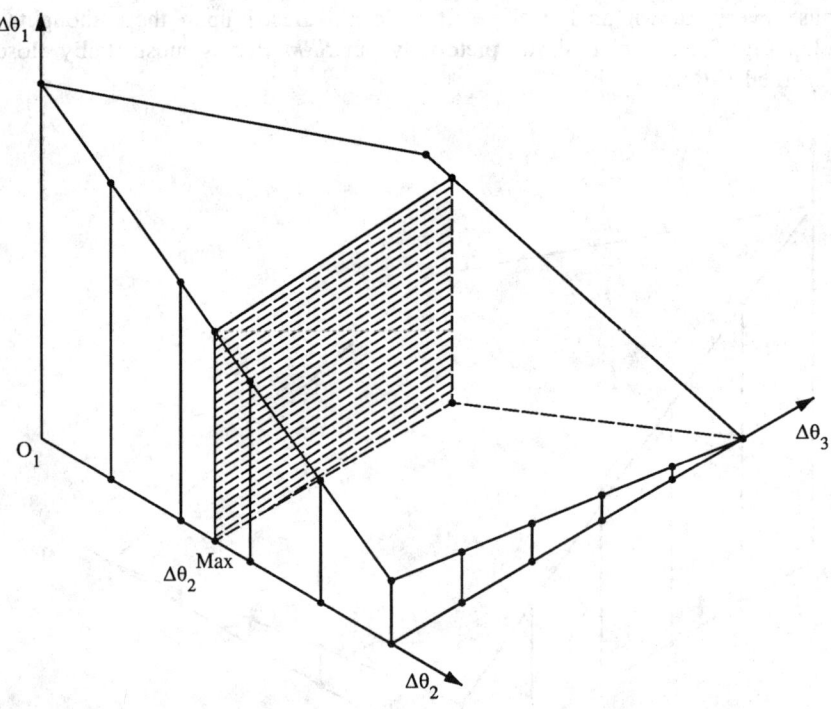

Fig. 5.4-3. The stability robustness measure *(SRM)* corresponding to Fig. 5.4-1 with additional constraint(s).

5.5. CONSTRUCTION OF ROBUST LINEAR DYNAMIC SYSTEMS

5.5.1. Lyapunov Direct Method

Having clarified our *SRM*, we concentrate our efforts on the actual method of generating the corresponding *E-Box* or its best estimate, such that for each point in this set the associated ΔS yields $S + \Delta S$ which is asymptotically stable. Clearly, when the *E-Box* or its best approximation is constructed in the *PVS*, that set entails all the other possible *SRM*'s of the system (including those resulted from optimization methods which are used to stabilize a system), since these *SRM*'s are associated with a subset of the *E-Box*. Furthermore, this measure shows that we should search for a constructional method that generates *set by set* of points in the *PVS* (or correspondingly the *set by set* of perturbed systems), in order to develop Fig. 5.4-2, as *expeditiously* as we possibly can, and indeed this thought has been the underlying idea in our methodology.

To construct our *SRM*, or a typical Fig. 5.4-2, which is perhaps the "best" or "the largest" set of points in the *PVS*, whose corresponding perturbed systems are stable (i.e., for each point $\Delta\theta$ in this set we have $S(\theta^o + \Delta\theta)$ which is stable), we *stack* a number of hyperboxes *under* the *E-Box* in order to reach this set. In other words, we always construct the *E-Box* from *below*. At this juncture we must point out that although the exact *E-Box* is not easily attainable, we have no difficulty to determine the corresponding boundary points of this set on all the r axes and perhaps a few "45° lines". These boundary points are the *guiding points* for constructing the corresponding *E-Box* or its best estimate which we call our *SRM* as we demonstrate these facts in numerical examples of § 5.8.1. Thus within some reasonable computational efforts we can generate the *SRM* that for all practical purposes is more than sufficient for our applications, and from this set we may reach the *E-Box* within some additional computational efforts if we wish to do so, however, the trade-off is how much extra computations we are willing to make.

The main result of this section is described in the next theorem which stems from the direct method of Lyapunov, Corollary 5.1-1. However, we recall that whenever we speak of perturbation we assume that the initial system is sensitive to parameter variations; otherwise there is no merit in further sensitivity analysis. Thus starting with an asymptotically stable matrix $S(\theta)$ we assume that at least some of the eigenvalues of this matrix change as θ changes and as a consequence of that, the system may become unstable when the perturbation policy is being applied. This situation does not necessarily mean that the initial dominant eigenvalues of S change, but always some of these eigenvalues change when θ changes. Recall that we assume the changes in θ do not affect the stabilizability of $S(\theta)$, i.e., no structural problem which deteriorates system stability is predicted. Thus the main thrust of our analyses is to provide a procedure that ultimately answers the fundamental question which is raised in Problem 5.4-1, and this final answer entails the numerical outcomes of all the other relevant results reported in the literature.

THEOREM 5.5-1 [B. 32 and 106]: Consider $\dot{x}(t) = S(\theta)x(t)$, where $S \in R^{n \times n}$ is an asymptotically stable matrix (with $\theta \in R^r$) satisfying the Lyapunov equation $S^T P + PS + Q = 0$, with $Q = Q^T > 0$ and $P = P^T > 0$. Suppose that $\theta \to \theta + \Delta\theta$ or $S \to S + \Delta S$, then $\dot{y}(t) = (S + \Delta S)y(t)$ remains asymptotically stable if

$$\Delta S \; Q^{-1} \; \Delta S^T < \tfrac{1}{4} P^{-1} \; Q \; P^{-1}, \tag{5.5-1}$$

or,

$$\Delta S^T \; P \; Q^{-1} \; P \; \Delta S < \tfrac{1}{4} Q. \tag{5.5-2}$$

If the Lyapunov equation is set as $SP + PS^T + Q = 0$, then the above two inequalities become, respectively, as follows.

$$\Delta S^T \; Q^{-1} \; \Delta S < \tfrac{1}{4} P^{-1} \; Q \; P^{-1}, \tag{5.5-3}$$

and

$$\Delta S \; P \; Q^{-1} \; P \; \Delta S^T < \tfrac{1}{4} Q. \tag{5.5-4}$$

Here by $M < N$, for $0 \le M = M^T \in R^{n \times n}$ and $0 < N = N^T \in R^{n \times n}$, we mean $N - M \ge 0$.

PROOF: The Lyapunov equation $S^T P + PS + Q = 0$ can be written exactly as follows. $(S + \Delta S)P^{-1} + P^{-1}(S + \Delta S)^T + P^{-1} Q P^{-1} - \Delta S P^{-1} - P^{-1} \Delta S^T = 0$. Citing this theorem again, we conclude that $S + \Delta S$ remains a stability matrix if $\tilde{Q} \triangleq P^{-1} Q P^{-1} - \Delta S P^{-1} - P^{-1} \Delta S^T > 0$. If $\tilde{Q} = 0$, then $S + \Delta S$ has purely imaginary eigenvalues, and $S + \Delta S$ becomes completely unstable if $\tilde{Q} < 0$. Denoting by $Q^{\frac{1}{2}}$ the positive-definite symmetric square root of Q, we note that since

$$0 \le (\frac{1}{\sqrt{2}} P^{-1} Q^{\frac{1}{2}} \pm \sqrt{2}\Delta S Q^{-\frac{1}{2}})(\frac{1}{\sqrt{2}} P^{-1} Q^{\frac{1}{2}} \pm \sqrt{2}\Delta S Q^{-\frac{1}{2}})^T$$

$$= \tfrac{1}{2} P^{-1} Q P^{-1} \pm P^{-1} \Delta S^T \pm \Delta S P^{-1} + 2\Delta S Q^{-1} \Delta S^T, \tag{5.5-5}$$

from (5.5-5) we have

$$\pm (P^{-1} \Delta S^T + \Delta S P^{-1}) \le \tfrac{1}{2} P^{-1} Q P^{-1} + 2\Delta S Q^{-1} \Delta S^T. \tag{5.5-6}$$

Hence

$$\tilde{Q} \triangleq P^{-1} Q P^{-1} - P^{-1} \Delta S^T - \Delta S P^{-1} > \tfrac{1}{2} P^{-1} Q P^{-1} - 2\Delta S Q^{-1} \Delta S^T. \tag{5.5-7}$$

The stability of $S + \Delta S$ is guaranteed if $\tilde{Q}$ remains a positive-definite matrix. Thus it is sufficient that the right-hand side of (5.5-7) to be positive definite. That concludes (5.5-1). To reach (5.5-2) we let $\tilde{Q} \triangleq Q - (\Delta S^T P + P \Delta S)$ and we use the following inequality.

$$0 \le (\frac{1}{\sqrt{2}} Q^{\frac{1}{2}} \pm \sqrt{2}\Delta S^T P Q^{-\frac{1}{2}})(\frac{1}{\sqrt{2}} Q^{\frac{1}{2}} \pm \sqrt{2}\Delta S^T P Q^{-\frac{1}{2}})^T. \tag{5.5-8}$$

The other two inequalities can be shown similarly.

5.5. Construction of Robust Linear Dynamic Systems

COMMENT 5.5-1: Clearly the above two Lyapunov equations yield different solutions for P and similarly the above four ΔS's are different. Also in a given problem we do not know *a priori* which one of the (5.5-1) to (5.5-4) yields the "best" ΔS as we show this fact with several numerical examples in § 5.8.1.

COMMENT 5.5-2: It is clear that our main objective in this analysis is to "solve" (5.5-1) to (5.5-4) in terms of ΔS, i.e., to devise a procedure that enables us to construct ΔS in any given problem from any one of the quadratic matrix inequalities given in (5.5-1) to (5.5-4). To that effect we note that depending on the structure of ΔS, we can classify the problem into two major groups, namely: single-parameter variations; and multiple-parameter variations as this classification is described next.

5.5.2. Catalog of Structured Perturbation Matrices

Generally speaking, in each problem, we have some information about the structure of $\Delta S(\Delta \theta)$, $\Delta \theta \in R^r$. In other words, we know how the system parameters are affecting the entries of S or equivalently those of ΔS. It seems that, in general, and no matter how complicated a problem may become, or what the actual r is, we have one of the following possible candidates for ΔS. Here we let $\theta \in R^2$, and assume that $\Delta S \in R^{2 \times 2}$, for the sake of illustration, in order to present the following *catalog* of perturbation matrices.

A: One-parameter, linear perturbation affecting one entry

$$\Delta S \triangleq \begin{bmatrix} 0 & \Delta \theta_1 \\ 0 & 0 \end{bmatrix} = \Delta \theta_1 \begin{bmatrix} 0 & 1 \\ 0 & 0 \end{bmatrix}.$$

B: One-parameter, linear perturbation affecting two or more entries

$$\Delta S \triangleq \begin{bmatrix} \Delta \theta_1 & -\Delta \theta_1 \\ \Delta \theta_1 & -2\Delta \theta_1 \end{bmatrix} = \Delta \theta_1 \begin{bmatrix} 1 & -1 \\ 1 & -2 \end{bmatrix}.$$

C: One-parameter, one nonlinear perturbation affecting one entry

$$\Delta S \triangleq \begin{bmatrix} \Delta \theta_1 + \Delta \theta_1^2 & 0 \\ 0 & 0 \end{bmatrix} = (\Delta \theta_1 + \Delta \theta_1^2) \begin{bmatrix} 1 & 0 \\ 0 & 0 \end{bmatrix}.$$

D: One-parameter, one nonlinear perturbation affecting two or more entries

$$\Delta S \triangleq \begin{bmatrix} \Delta \theta_1 + \Delta \theta_1^2 & 0 \\ \Delta \theta_1 + \Delta \theta_1^2 & 0 \end{bmatrix} = (\Delta \theta_1 + \Delta \theta_1^2) \begin{bmatrix} 1 & 0 \\ 1 & 0 \end{bmatrix}.$$

E: One-parameter, multiple-nonlinear perturbations affecting two or more entries

$$\Delta S \triangleq \begin{bmatrix} \Delta \theta_1 + \Delta \theta_1^2 & \Delta \theta_1 + \Delta \theta_1^2 + \Delta \theta_1^3 \\ \Delta \theta_1 & 0 \end{bmatrix}.$$

F: *Multiple-parameter, linear perturbation affecting two or more entries*

$$\Delta S \triangleq \begin{bmatrix} \Delta\theta_1 & \Delta\theta_2 \\ -2\Delta\theta_2 & \Delta\theta_1 \end{bmatrix} = \Delta\theta_1 \begin{bmatrix} 1 & 0 \\ 0 & 1 \end{bmatrix} + \Delta\theta_2 \begin{bmatrix} 0 & 1 \\ -2 & 0 \end{bmatrix}, \text{ and for polynomials}$$

which are described by phase–variable canonical matrices:

$$\Delta S \triangleq \begin{bmatrix} 0 & 0 \\ -\Delta\theta_2 & -\Delta\theta_1 \end{bmatrix} \text{ (independent variation),}$$

$$\Delta S \triangleq \begin{bmatrix} 0 & 0 \\ -\Delta\theta_2 - \Delta\theta_1 & -\Delta\theta_1 \end{bmatrix} \text{ (dependent variation).}$$

G: *Multiple-parameter, multiple-nonlinear perturbations affecting many entries*

$$\Delta S \triangleq \begin{bmatrix} \Delta\theta_1 + \Delta\theta_2 + \Delta\theta_1\Delta\theta_2 & \Delta\theta_1 + \Delta\theta_1^2 \\ 0 & \Delta\theta_2 + \Delta\theta_2^2 \end{bmatrix}, \text{ or}$$

$$\Delta S \triangleq \begin{bmatrix} 0 & 0 \\ \Delta\theta_1 + \Delta\theta_2 + \Delta\theta_1\Delta\theta_2 & \Delta\theta_2 + \Delta\theta_2^2 \end{bmatrix}, \text{ for polynomials.}$$

Clearly, each and every one of the above cases A to G is computed at a specific operating condition $\theta^o \triangleq [\theta_1^o, \theta_2^o]^T$. Therefore our unknowns, for constructing the corresponding ΔS, are a set of scalar parameter variations $\Delta\theta_i$'s, as shown in the above catalog.

■

In the present context by robustness we mean the "largest" possible variations of θ *around a given stable-operating condition*, whose corresponding perturbed systems remain asymptotically stable. Clearly by variations we mean the construction of a multi-dimensional box (or set) around θ^o in the *PVS*. For future reference the next definition is suggested, although we have already presented its context in § 5.4.2.

DEFINITION 5.5-1, *E-Box:* The "largest" or "most extended" *set* of points in the *PVS*, which is an Euclidean space, that surrounds the initial stable-operating point corresponding to the generating (or original) dynamical system, whose corresponding perturbed systems remain stable is called the *E-Box*.

■

For a pictorial view of a typical *E-Box* and its corresponding "best" estimate which we call the *SRM* we refer to Figs. 5.4-1 & 2.

Recently issues pertaining to the *SRA* in linear time-invariant systems such as (5.4-9) or (5.4-10) are widely examined. As is evident from (5.4-9) or (5.4-10) the *SRA* of this system is a function of the perturbation matrix $\Delta S(\Delta\theta)$, independent

5.5. Construction of Robust Linear Dynamic Systems

from any procedure to formulate its stability determination. Suppose that the perturbation matrix or the perturbation strategy is specified. Then the natural question is: How do we construct the corresponding *E-Box,* or its "best" estimate (which we call the *SRM,* cf., Figs. 5.4-2 & 3)? The answer to this question is at the *heart* of the *SRA* and our subsequent analyses intend to respond to this question. As Fig. 5.4-1 pictorially depicts, a typical *E-Box* cannot be easily derived analytically in its entirety. (Because that in effect means we have developed an algorithm to solve analytically the set of inequalities in the first column of the corresponding Routh and the Hurwitz criteria.) We also note that Fig. 5.4-1 is only one connected set of parameter variations, out of perhaps several such sets, for a given system (a possible *conditional stability*). Furthermore, a given system may have a number of infinite directional perturbations that are not easy to detect. Also there are, of course, other issues in this regard which we describe subsequently.

5.5.3. Matrix Inequalities

We recall that matrix inequality (or ordering) is only valid in the context of a quadratic expression and in the sense of symmetric positive- (semi-) definite matrices. Since our *sole* objective is to construct ΔS from any one of (5.5-1) to (5.5-4), when in general we have $M < N$ for two real-symmetric matrices $M \geq 0$ (which plays the role of unknown) and $N > 0$, our first priority is to provide a tentative procedure to "solve" this type of matrix inequality. Clearly substituting a typical ΔS from any one of its possible candidates into any one of (5.5-1) to (5.5-4) yields an involved algebraic quadratic matrix inequality which must be avoided. Therefore to transform this non-scalar problem into a scalar one, which is more manageable than before, we need to take a few steps. To that effect and in order to solve the above $M < N$, we propose to set the $\lambda_{max}(M)$ ($\triangleq$ the maximum eigenvalue of M) $\leq \lambda_{min}(N)$ ($\triangleq$ the minimum eigenvalue of N) (cf., Comments 5.6-1(*ii*)). A more conservative choice than this one is to set $\text{Tr}(M)$ ($\triangleq$ sum of all eigenvalues of M) $< \lambda_{min}(N)$. Since the above N and M are not often diagonal, we can use the following theorem to diagonalize these matrices simultaneously.

THEOREM 5.5-2 [A. 57]: Let M and K be $n \times n$ Hermitian matrices. If M is positive definite, then there exists an $n \times n$ matrix R for which $R^*MR = I$ and $R^*KR = \text{diag}(\lambda_1, \cdots, \lambda_n)$. (Here * means transpose conjugate.) The numbers λ_j's are real. If K is positive definite, then the λ_j's are generalized eigenvalues satisfying $Kr^j = \lambda_j M r^j$, $r^j \neq 0$, $j = 1, \cdots, n$. If K and M are real, then a real matrix R, with real columns r^j, may be found such that $R^*MR = I$ and $R^*KR = \text{diag}(\lambda_1, \ldots, \lambda_n)$ where λ_j's are the roots of $\det(\lambda M - K) = 0$. ∎

Clearly the matrix inequalities (5.5-1) to (5.5-4) satisfy all the requirements of Theorem 5.5-2 and thus we can apply this theorem to these inequalities, in order to transform them into a set of scalar inequalities which makes the interpretation of (5.5-1) to (5.5-4) very straightforward. Note that all of these suggestions yield conservative "solutions" for the preceding matrix inequalities. We also recall that these inequalities are only sufficient conditions and therefore we only get sufficient solutions as well.

REMARK 5.5-1 [A. 57]: The above R can be computed as follows. Since M is Hermitian, there is a unitary matrix U such that

$$U^* M U = \text{diag}(\mu_1, \mu_2, \ldots, \mu_n). \qquad (5.5\text{-}9)$$

If $M > 0$, then the μ_j's are positive and we can define

$$D = D^* \triangleq \text{diag}(\mu_1^{-\frac{1}{2}}, \mu_2^{-\frac{1}{2}}, \ldots, \mu_1^{-\frac{1}{2}}). \qquad (5.5\text{-}10)$$

Now, we can show that

$$D^*(U^* M U) D = I, \quad \text{and} \quad D^*(U^* K U) D = H, \qquad (5.5\text{-}11)$$

where H is Hermitian. This H can be further diagonalized by a unitary similarity transformation. If $V^* V = I$ and $V^* H V = \text{diag}(\lambda_1, \ldots, \lambda_n)$, then

$$(V^* D^* U^*) M (UDV) = I, \qquad (5.5\text{-}12)$$

and

$$(V^* D^* U^*) K (UDV) = \text{diag}(\lambda_1, \ldots, \lambda_n). \qquad (5.5\text{-}13)$$

Therefore $R \triangleq UDV$, and it is easy to show that the λ_j's are the roots of $\det(\lambda M - K) = 0$.

5.6. COMPUTATIONAL ISSUES

5.6.1. Introduction

We should emphasize that our main objective is to utilize (5.5-1) to (5.5-4) in order to construct the *global* stability robustness measure *(SRM)* for a dynamical system. This utilization involves two issues: *(i)* the choice of Q; and *(ii)* the structure of ΔS. But first we must also point out that none of the above "upper bounds" (5.5-1) to (5.5-4) are global or *the largest* possible perturbation for an asymptotically stable matrix. Not only that, numerical examples can be provided to show that each one of these "upper bounds" yields different quantitative local results when it is applied to different examples. Therefore in a given problem, we cannot select *a priori* any one of these mechanisms as the best perturbation policy, unless we test that first.

On the choices for Q: We note that these "upper bounds" are functions of $Q = Q^T > 0$, and therefore we need to introduce a method that generates *the largest* "upper bound" of ΔS independent from the initial choice of $Q = Q^T > 0$. Since the

5.6. Computational Issues

actual *E-Box* (or its best estimate – *SRM*) of a dynamical system is, of course, independent from any computational or physical measurement methods that are used to generate that, we should somehow be able to eliminate this dependency on Q. A number of researchers have attempted to find the "optimal" Q, in order to generate the "optimal" P, which yields the largest-upper bound *at once*. As it becomes evident from our numerical results, such a search seems fruitless. Therefore we appeal to a different approach for constructing *the largest* "upper bounds" of ΔS, in order to circumvent this dependency on the initial choice of Q. We propose to choose any arbitrary initial $Q = Q^T > 0$, in order to perturb the system. If the system remains stable, then we use the same Q and perturb it again repeatedly. This iterative method, as formulated in Theorem 5.6-1, is an alternative approach to respond to this dependency problem (cf., Comments 5.6-1).

On the structure of ΔS: At this juncture we note that although this issue of dependency on Q can be removed for all practical purposes, we need to be concerned with the fact that we must also study the structure of ΔS, i.e., whether this is a case of single- or a case of multiple-parameter variations. This distinction is extremely important in order to define the meaning of the "upper bounds" which we have repeatedly utilized in the above.

Finally, and upon the initial choices for Q (and ΔS) are made, if the corresponding iterative method for constructing the largest set of points in the *PVS* (a path or a straight line for single-parameter variations and a hyperbox for multiple-parameter variations) can be completed unambiguously (ambiguity may arise in the case of multiple-parameter variations), then we have both the necessary and sufficient conditions for the asymptotic stability of $S + \Delta S$, within our prespecified computational tolerance (cf., § 5.7.2 and § 5.8.2).

5.6.2. Single – Parameter Variations

We recall that our main objective is to "compute" ΔS from any one of (5.5-1) to (5.5-4), which in general is of the form $M < N$ for two real-symmetric matrices $M \geq 0$ (which is unknown) and $N > 0$. Here $\Delta S \triangleq \varepsilon(\theta) \Delta S'$; where $\varepsilon(\theta)$ (or $\varepsilon(\Delta\theta)$ as the case may be) is in effect a single-parameter variation, and $\Delta S'$ is a numerically known matrix that is constructed at a given θ^o, based on the information we may have regarding the parameter variations scheme (cf., cases *A* to *D* of § 5.5.2). (For cases similar to *E* refer to Comment 5.6-2, below.) The construction of *the largest* ΔS in this case can be achieved unambiguously. Here we have only one scalar parameter $\varepsilon(\theta)$ and substituting this ΔS in any one of (5.5-1) to (5.5-4) yields $\varepsilon^2(\theta) M' (\triangleq M) < N$. To "solve" this type of matrix inequality where M' and N are numerically known, as mentioned in § 5.5.3, we propose to set the $\lambda_{max}(M) = \varepsilon^2(\theta) \lambda_{max}(M') \leq \lambda_{min}(N)$, or more conservatively to set $\text{Tr}(M) = \varepsilon^2(\theta) \text{Tr}(M') < \lambda_{min}(N)$. Or we can use Theorem 5.5-2 to diagonalize these matrices simultaneously in order to find $\varepsilon(\theta)$.

As stated before, how to choose Q in order that *the largest* "upper bound" for parameter variations (i.e., the boundary of the *E-Box*), can be achieved from any one of (5.5-1) to (5.5-4) in *the first trial* is not known. The following theorem serves as a computational *tool* to find this *largest*-upper bound by iteratively using a sufficient condition (for ΔS) and an arbitrary initial choice for $Q = Q^T > 0$. By this iterative approach which in effect we are translating the coordinates of the *PVS* along a single line to the new operating point, we reach the largest *directional* perturbation bound for ΔS. Refer to Remark 5.8-7 for additional insights.

THEOREM 5.6-1: Let $Q = Q^T > 0$, $S_1 \equiv S_0$ be an asymptotically stable matrix (each subscript indicates the order of iteration), and $P_1 = P_1^T > 0$ satisfies $S_1^T P_1 + P_1 S_1 + Q = 0$. Let $\Delta S'$ be an $n \times n$ arbitrary matrix. Rewriting the Lyapunov equation as $(S_1 + \varepsilon \Delta S')^T P_1 + P_1(S_1 + \varepsilon \Delta S') + Q - \varepsilon(\Delta S'^T P_1 + P_1 \Delta S') = 0$; then the following can be shown.

(a) There exists a largest $|\varepsilon_1'|$ such that

$$\Pi_1 \triangleq Q - \varepsilon_1(\Delta S'^T P_1 + P_1 \Delta S') > 0, \quad \text{if} \quad |\varepsilon_1| < |\varepsilon_1'|. \tag{5.6-1}$$

(b) Let $S_2 = S_1 + \varepsilon_1 \Delta S'$. If S_2 remains asymptotically stable, then $S_2^T P_2 + P_2 S_2 + Q = 0$; and like (a) there exists a largest ε_2' such that

$$\Pi_2 \triangleq Q - \varepsilon_2(\Delta S'^T P_2 + P_2 \Delta S') > 0, \quad \text{if} \quad |\varepsilon_2| < |\varepsilon_2'|. \tag{5.6-2}$$

(c) Similarly, $\varepsilon_1, \varepsilon_2, \varepsilon_3, \ldots,$ can be found. It can be shown that if $\tilde{\varepsilon}_K \triangleq \sum_{J=1}^{K} \varepsilon_J$, then $S_1 + \tilde{\varepsilon}_\infty \Delta S'$ is not asymptotically stable, but $S_1 + \tilde{\varepsilon}_K \Delta S'$ is asymptotically stable for some finite K when $0 < \tilde{\varepsilon}_K < \tilde{\varepsilon}_\infty$, if $\varepsilon_1 > 0$; and $\tilde{\varepsilon}_\infty < \tilde{\varepsilon}_K < 0$, if $\varepsilon_1 < 0$.

PROOF: (a) It is clear that $\Pi_1 = \Pi_1^T$. Therefore by Theorem 5.5-2, there exists an M_1 such that $M_1^T \Pi_1 M_1 = I - \varepsilon_1 \text{diag}(\lambda_i)$, since $Q = Q^T > 0$, and $\varepsilon_1(\Delta S'^T P_1 + P_1 \Delta S')$ is symmetric. Now, it is seen from $M_1^T \Pi_1 M_1 = I - \varepsilon_1 \text{diag}(\lambda_i)$ how ε_1 can be chosen (term by term) such that $M_1^T \Pi_1 M_1$ remains a positive-definite matrix. Indeed, since in effect this is a single-parameter variation, the existence of the largest ε_1 in each step is guaranteed.

(b) Straightforward proceeding of the result from step (a) proves this part.

(c) To show this step we need to examine the sign of ε_1 first.

Diagonalization of Π_1 may result in both positive and negative perturbation steps because of the sign of $\Delta S'^T P_1 + P_1 \Delta S'$. Since S_1 is a stable matrix (because

5.6. Computational Issues

of our initial assumption on the stability of this matrix), if $\Delta S'$ is a positive- (semi-) definite matrix, and $\varepsilon_1 < 0$, then $S_2 (= S_1 + \varepsilon_1 \Delta S')$ is, generally speaking, more stable than S_1. Similarly, if $\Delta S_1'$ is a negative-definite matrix, and $\varepsilon_1 > 0$, then S_2 is, generally speaking, more stable than S_1. If $\Delta S'$ is sign indifferent, then S_2 must show sensitivity toward the perturbation scheme, because of our fundamental assumption regarding the sensitivity of S_1 with respect to parameter variations, otherwise we have no problem to study. We certainly are not interested in further stabilizing the initial stable matrix. Thus we need to test carefully in order to see how the initial matrix should be perturbed or what is the *troublesome* direction of perturbing S_1. Certainly, there are cases in which both positive- and negative-directional perturbations are possible. These and other special cases must be carefully examined before any applications of these results are being made. We believe that this testing for determining the direction of perturbation is the least difficult issue in this whole process. (It would be very unrealistic to assume that we can select a perturbation matrix for any given stable matrix, and give its troublesome direction for perturbing the initial matrix according to (or in direction of) the given perturbation matrix, and without any testing whatsoever.) A simple eigenvalue testing for determining the troublesome direction of perturbing matrices is a reasonable task.

Proceeding according to the previous steps yields ε_K such that $S_{K+1} = S_K + \varepsilon_K \Delta S'$ remains asymptotically stable or, equivalently, $\Pi_K \triangleq Q - \varepsilon_K (\Delta S'^T P_K + P_K \Delta S') > 0$. This entire process of constructing these perturbation steps is based on the following facts. First of all, this is a single-parameter variation and therefore there exists a largest ε_K in each step. Secondly, each new matrix S_{K+1} has some eigenvalues distributed on the left-hand side of imaginary axis which are closer to this axis than the previous S_K, because of the initial assumption on the sensitivity of S_K. This process of constructing S_K continues and at each iteration the corresponding P_K is also generated. Then for some $K = M$, Π_M is no longer a positive-definite matrix, i.e., an ε_M cannot be found to make $\Pi_M > 0$, because the initial stability matrix S_1 is assumed to be parameter sensitive. As a result of this assumption, at each step the parameter variations bring the eigenvalues of the perturbed matrices closer to the imaginary axis than the preceding one. To avoid computational difficulties we replace imaginary axis with a line parallel to that $-\delta$ units to the left. We can also define a more stringent condition for the distribution of perturbed-matrix eigenvalues if we wish to do so. Therefore after a finite number of iterations some of the eigenvalues of perturbed matrix cross the δ-line and computations stop. Refer to Remark 5.8-7 for additional insights.

The main feature of this process is that each S_{K+1} can be written as $S_{K+1} = S_{K-1} + \varepsilon_{K-1} \Delta S' + \varepsilon_K \Delta S'$. Repeating this substitution results in $S_{K+1} = S_1 + \sum_{J=1}^{K} \varepsilon_J \Delta S'$. This observation completes the proof of this theorem and provides the necessary requirement for the convergent of Algorithm 5.6-1.

COMMENTS 5.6-1 (On the choices for Q): (i) As a consequence of the above theorem, we always choose $Q = I$ (or $2I$) to avoid additional matrix manipulations. We should point out that although the choice for $Q = 2I$ and in a similar analysis is credited to [B. 92], as far as the ΔS is concerned, the (5.5-1) to (5.5-4) give the same result for both $Q = I$ (which we often use) and $Q = 2I$ (which we occasionally use), since Q and P are linearly related.

(ii) Since there is no proof, whether any one of (5.5-1) to (5.5-4) is superior to the other one and for all possible S and ΔS, and since these four "upper bounds" are local (as our numerical examples in § 5.8.1 confirm this fact), a slight improvement of one local-upper bound versus another seems an irrelevant issue, in particular, when it is well-known that the perturbation analysis stemming from the Lyapunov method is not equipped to provide the *global-upper bound*. However, some researchers (while taking the same steps as those in the preceding analysis) have recently claimed that they have developed results which "improve" the *SRM*, of a *DCLTIS*. When these claims are thoroughly scrutinized, it becomes clear that their results are already contained in Theorem 5.5-1. In other words, these results are only algebraic variations of (5.5-1) to (5.5-4) and with no apparent advantages over the original work given in Theorem 5.5-1. For example, in (5.5-2), if we let $Q = 2I$ and $\Delta S = \varepsilon \Delta S'$, i.e., single-parameter variations, or a directional perturbation, where $\Delta S'$ is *any given* numerically known matrix or direction (such as U, a matrix whose entries are all unity, which in essence this is the same as perturbing S *only* along the "45° line", i.e., a single-direction where the magnitudes of all perturbations are equal), then we get

$$M \triangleq \varepsilon^2 \Delta S'^T P P \Delta S' = \varepsilon^2 (P\Delta S')^T(P\Delta S') < I \triangleq N. \quad (5.6\text{-}3)$$

The above inequality in our terminology is interpreted as $\lambda_{max}(M) \leq \lambda_{min}(N)$, which is a conservative-upper bound for the parameter variation that preserves asymptotic stability of $S + \Delta S$. Thus (5.6-3) can be written as follows.

$$\varepsilon^2 \leq \frac{1}{\lambda_{max}[(P\Delta S')^T(P\Delta S')]} \triangleq \frac{1}{\sigma^2_{max}(P\Delta S')}. \quad (5.6\text{-}4)$$

Here $\sigma(\cdot)$ stands for the singular value of $(\cdot)$, but as far as our computation is concerned this change of notation from an eigenvalue [of a symmetric positive- (semi-)definite matrix] to a singular value does not provide any new information let alone improvement. Here the eigenvalues as used are well defined for ordering. There are a few other variations of (5.5-1) to (5.5-4) reported in the literature which we do not pursue to describe them here. We believe that much of the emphases in the literature in regard to finding an "improved-upper bound", if any, is in a wrong direction. Because first of all these results are local and with the advent of computational facilities, supposedly a few percentages of improvement in some *ad hoc* examples do not constitute a global mandate to use any particular formula irrespective of its originality. In Examples 5.8-1 & 5 and 5.11-1 we apply these four

5.6. Computational Issues

inequalities to the corresponding matrices to demonstrate that each inequality gives different relative-upper bound when is applied to a different case, and therefore the issue of "improvement" is not substantiated. We also believe that a conservative-sufficient condition which is used repeatedly is more meaningful approach to use for generating the *SRM* of a dynamical system than constantly changing the perturbational mechanism. Secondly and in the light of Fig. 5.4-1, though pictorially, the real issue in *SRA* is the construction of the corresponding *E-Box*, or its best estimate, which comes with each matrix and its existence is not a matter of choice for us. Thus it seems that we should unify much of these algorithms in order that *expeditiously* construct the *E-Box*. Finally, for both the case of single- and multiple-parameter variations, when we use $Q = 2I$ and maintain ΔS as is, the sufficient condition (5.5-2), *per se,* becomes $\sigma_{\max}(P\Delta S) < 1$ in order to maintain system stability. This condition and for all the four choices can be stated more conservatively than above as follows $\sigma_{\max}(\Delta S) < 1/\sigma_{\max}(P)$.

(iii) Going back to our choice for $Q = I$ or $2I$, we notice that (5.5-1) (or (5.5-3)) has certain advantages over (5.5-2) (or (5.5-4)). In (5.5-1) (or (5.5-3)) the maximum eigenvalue of left-hand side for a given structured perturbation matrix is computed only once, while in (5.5-2) (or (5.5-4)) we must perform the extra matrix multiplications before generating the maximum eigenvalue of its left-hand side. In fact, since (5.5-2) (or (5.5-4)) involves multiplication by P, the formula $\sigma_{\max}(P\Delta S) < 1$, is effective only for directional perturbations. In the subsequent theorem we show that how this requirement can be used to generate *set by set* of ΔS's instead of only directionals for system with multiple-parameter variations. On the other hand, if we use the trace of left-hand side in (5.5-2) (or (5.5-4)), which can be computed easily, in particular, for a sparse ΔS, then the eigenvalue computation is avoided at the expense of increasing the iteration number for our iterative scheme which reaches the *E-Box* or its best estimate that we call the *SRM*.

(iv) Now, suppose that the initial stability matrix S_0 is real and diagonal. If we choose $Q = qI$, where $q > 0$ is an arbitrary scalar number, then for this diagonal S_0 we have $P_1 = -\frac{1}{2}qS_0^{-1}$. Thus from (5.5-1) we have $\Delta S_1^2 \leq S_0^2$ (independent of q); i.e., each entry can be changed at most by its corresponding value, which is (by inspection) the best policy in this case. This observation may suggest that if we choose $Q = qR^{-T}R^{-1}$, where $R^{-1}S_0R$ is a diagonal matrix (also note that depending on which Lyapunov equation is being used we must appropriately change the order of multiplication), then we may get the best-upper bounds for our perturbation matrices. However, this choice of Q has also several shortcomings and it does not give the best results (cf., Example 5.8-2) [B. 23 & 30]. Similarly if the initial stability matrix S_0 is real and diagonal and if we let $Q = \text{diag}(-\lambda_1, \cdots, -\lambda_n)$, then $P = \frac{1}{2}I$ and ΔS becomes its maximum. This observation also gives us an alternative way of generating these upper bounds as is also described in Example 5.8-2. Perhaps if we could have generated the upper and the lower bounds for eigenvalues of P with respect to those of Q, then we could have done better (cf., [A. 128] and the corresponding comments for some minor results on these bounds).

All in all, considering that $Q = I$ is very easy to use, not mentioning that this choice also gives the largest-local-upper bound in each step (this can be shown for our results with an argument similar to that in [B. 92]), we are very content with this choice and we suggest that for all results pertaining to this sort of *SRA's* which are stemming from the direct method of Lyapunov.

COMMENT 5.6-2 (On the structure of perturbation matrices): Referring to cases described in § 5.5.2, for situations similar to the first four cases of A to D the *E-Box* can be generated easily using Theorems 5.5-1 and 5.6-1. Theorem 5.5-1 also strongly holds for the remaining cases of E to G. To apply (5.5-1) to cases similar to that of E, we propose to use $Tr(\Delta S \, \Delta S^T)$, which is a good conservative perturbation step that is computationally attractive, and it can be symbolically generated only once. Then in each iteration step, we can only generate a $\Delta\theta$ that must be *added* (in a proper direction) to the previous θ to complete the construction of the corresponding *E-Box*. From the constructional point of view, we may also consider case E, though single-parameter variations, under the next group of perturbations, by letting each nonlinear term or a group of them as a *new* variable in order to make this case looks like a typical F. This arrangement is more an art than a mathematical procedure. In situation similar to that of F, we need to define the *net* changes for $\Delta\theta_1$ and $\Delta\theta_2$, which are generally different than those generated if $\Delta\theta_1$ and $\Delta\theta_2$ were acting individually. Here changes in θ_1 may expand or it may compress the allowable range of $\Delta\theta_2$ and *vice versa*. This interaction among various parameter variations is unavoidable. Several authors in [B. 115] have suggested various measures (or norms) for stability robustness, in order to study these matters. That means in the case of multiple-parameter variations there is no unique definition for the best perturbational policy, and because of this nonuniqueness, we must carefully define the meaning of the "largest-parameter variations" for these cases which are similar to both F and G. In other words, instead of having a *path* as is the case in single-parameter variation, we now have a *set* of points (or closed paths) in the *PVS*, whose corresponding perturbed systems are stable. We have, of course, defined our stability robustness measure to be *the largest* such set of points in the *PVS* (or the *E-Box*). Therefore our goal is to construct this set for a dynamical system, and when this construction is completed in the *PVS*, naturally the numerical outcome of all the other results that are reported in the literature for a specific matrix norm, *etc.*, are included in this set as well. Finally for the case of dependent-parameter variations where various entries of the perturbed matrix are interdependently and simultaneously changing, we must use an analysis which in essence is similar to that described pictorially in Fig. 5.4-3.

ALGORITHM 5.6-1 (Determination of the Largest Directional Perturbation): We now summarize the process of constructing the largest directional perturbation in flowchart of Fig. 5.6-1. In this manner we can generate ΔS for cases in which the perturbation matrix is defined as a linear function of one parameter variation ε (i.e., $\Delta S = \varepsilon \Delta S'$), whose corresponding $S + \Delta S$ remains asymptotically stable. The

5.6. Computational Issues

stability criterion for avoiding numerical instability is chosen such that $|\text{Re}[\lambda_i(S + \Delta S)]| < 10^{-4}$, where $\lambda_i(\cdot)$ is the eigenvalue of $(\cdot)$. Or we may select other sector (for distribution of λ_i's) to stop our computations.

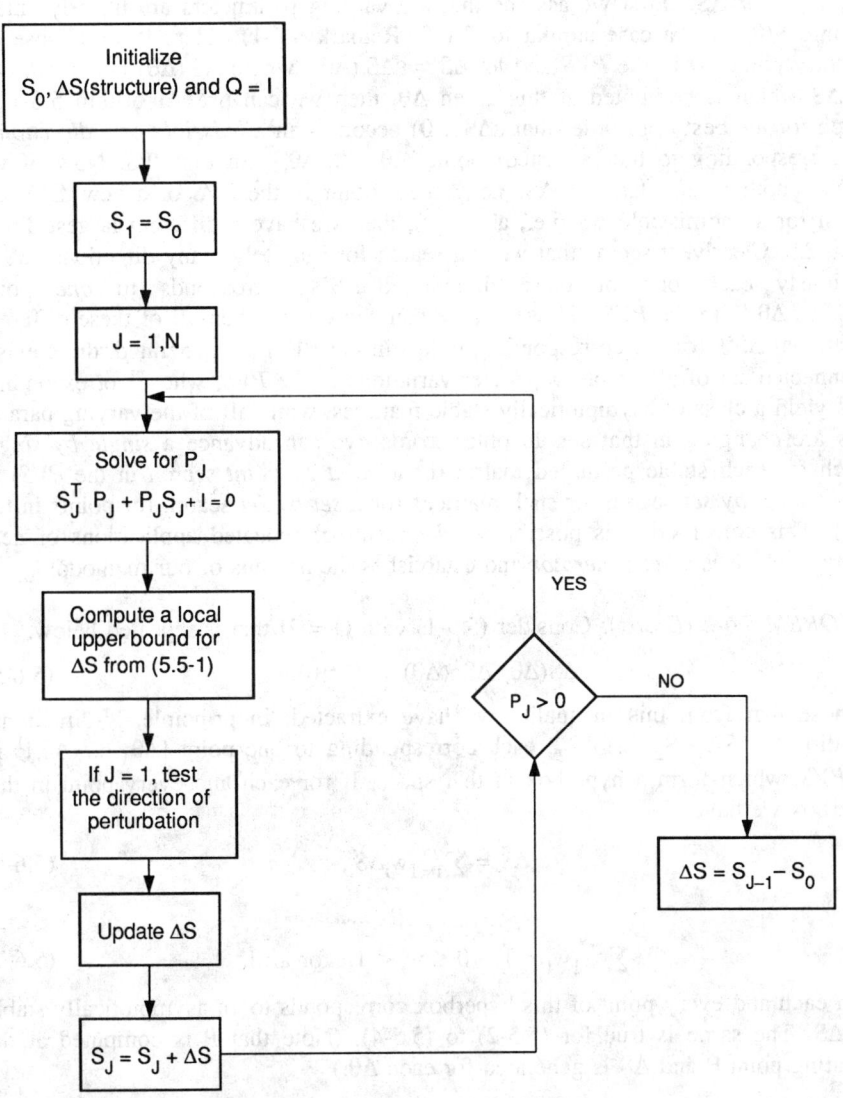

Fig. 5.6-1. A schematic diagram of flow-chart for Algorithm 5.6-1.

5.6.3. Multiple – Parameter Variations

Again we recall that our main objective is to "compute" ΔS from any one of (5.5-1) to (5.5-4), where now ΔS has multiple and perhaps nonlinear varying parameters, i.e., situations which are similar to cases F and G in § 5.5.2. Substituting any one of these ΔS's into any one of (5.5-1) to (5.5-4) yields an involved algebraic quadratic matrix inequality. Thus we must explore any properties of the "solutions" for these inequalities first, in order to simplify the process of solving such an inequality for ΔS. First we assume that the varying parameters are linearly entering into $S(\theta)$, i.e., a case similar to F (cf., Remark 5.6-1). Here if we choose an arbitrary point $\Delta\theta$ in the *PVS* and let $\Delta S = \varepsilon \Delta S'(\Delta\theta)$, where $\Delta S'(\Delta\theta)$ is a *candidate* for ΔS which is computed at this given $\Delta\theta$, then we can use Algorithm 5.6-1 to search for the best ε, in order that $\varepsilon \Delta S'(\Delta\theta)$ becomes an *admissible* or a *directional* ΔS corresponding to the [adjusted] point $[\Delta\theta_1, ..., \Delta\theta_r]^T$ in the *PVS*. Now, if we choose another candidate for ΔS (i.e., a new point in the *PVS* or a new $\Delta S'$) and search for an admissible ΔS (i.e., a new ε), then we have a different largest directional ΔS. Clearly, it seems that we can search for infinitely many directional ΔS's. Obviously each one of these directional ΔS's corresponds to *one* point $[\Delta\theta_1, ..., \Delta\theta_r]^T$ in the *PVS*. However, we can show that when all of these different directional ΔS's (or the corresponding points in the *PVS*) are generated, there exists a connected set of allowable parameter variations in the *PVS*, whose corresponding ΔS's yield a class of asymptotically stable matrices, while all of the varying parameters are changing in that set. In other words, we can advance a *single by single* search for each stable perturbed matrix (or a *point by point* search in the *PVS*) to that of a *set by set* search for such matrices (or a *set by set* search for points in the *PVS*). This construction is possible as the result of repeated applications of next theorem which is a *set generator* and establishes the nucleus of our methodology.

THEOREM 5.6-2 (E-Box): Consider (5.5-1) with $Q = 2I$ that is repeated below.

$$\Delta S(\Delta\theta) \Delta S^T(\Delta\theta) < P^{-2}(\theta). \tag{5.6-5}$$

Suppose that from this inequality we have extracted, in principle, 2^r-directional "solutions": $\Delta S_1, \Delta S_2, ..., \Delta S_{2^r}$, each corresponding to one point $[\Delta\theta_1, ..., \Delta\theta_r]^T$ in the *PVS*, which form a hyperbox in that space. If for each and every point in this hyperbox we have

$$\Delta S = \sum_{i=1}^{2^r} w_i \Delta S_i, \tag{5.6-6}$$

with

$$\sum_{i=1}^{2^r} w_i = 1, \quad 0 \leq w_i < 1, \text{ for all } i. \tag{5.6-7}$$

Then each and every point of this hyperbox corresponds to an asymptotically stable $S + \Delta S$. The same is true for (5.5-2) to (5.5-4). (Note that P is computed at the operating point θ and ΔS is generated for each $\Delta\theta$.)

PROOF: We note that the origin of *PVS* corresponds to the operating condition of

5.6. Computational Issues

the initial dynamical system which is assumed to be asymptotically stable. Here we also assume that the set of allowable-parameter variations for any dynamical system *around a given stable-operating point* is connected and that is a reasonable assumption. To show that $S + \Delta S$ is asymptotically stable we note that

$$\Delta S \Delta S^T = \left(\sum_{i=1}^{2^r} w_i \Delta S_i\right)\left(\sum_{i=1}^{2^r} w_i \Delta S_i\right)^T$$

$$= \sum_{i=1}^{2^r} w_i^2 \Delta S_i \Delta S_i^T + \sum_{k,l=1, k \neq l}^{2^r} (w_k w_l \Delta S_k \Delta S_l^T + w_l w_k \Delta S_l \Delta S_k^T). \quad (5.6\text{-}8)$$

It is also true that we always have

$$(w_k \Delta S_k \pm w_l \Delta S_l)(w_k \Delta S_k \pm w_l \Delta S_l)^T \geq 0. \quad (5.6\text{-}9)$$

Therefore

$$\pm (w_k w_l \Delta S_k \Delta S_l^T + w_l w_k \Delta S_l \Delta S_k^T) \leq w_k^2 \Delta S_k \Delta S_k^T + w_l^2 \Delta S_l \Delta S_l^T. \quad (5.6\text{-}10)$$

Certainly each parenthesis on the left-hand side of (5.6-10) is symmetric and thus we can utilize the matrix inequality. (Since there are a total number of even terms in (5.6-8), we always have a set of inequalities like (5.6-10) in each problem.) We also note that the plus and minus signs in (5.6-10) show that in principle the inequality between symmetric matrices remains valid independent from the sign of each ΔS_i's. Now, using the above set of estimates in (5.6-8), with $\Delta S_i \Delta S_i^T \leq P^{-2}$ for all i, yields

$$\Delta S \Delta S^T \leq 2^r \sum_{i=1}^{2^r} w_i^2 \, \Delta S_i \Delta S_i^T \leq 2^r \sum_{i=1}^{2^r} w_i^2 P^{-2}. \quad (5.6\text{-}11)$$

When we let all the weightings to be equal, then $\sum_{i=1}^{2^r} w_i^2 = (\frac{1}{2})^r$, and (5.6-11) becomes $\Delta S \Delta S^T \leq P^{-2}$, that means ΔS corresponding to the *center* point of the hyperbox yields an asymptotically stable $S + \Delta S$. We now sweep the entire hyperbox by pivoting around this center point and generating more and more asymptotically stable center points inside each sub-hyperboxes. Thus ΔS as constructed above yields an asymptotically stable $S + \Delta S$ for each point in the hyperbox. Similarly for (5.5-3) we can show that the same property holds. With similar reasoning and the following inequality

$$(w_k \Delta S_k \pm w_l \Delta S_l)^T P^2 (w_k \Delta S_k \pm w_l \Delta S_l) \geq 0, \quad (5.6\text{-}12)$$

we may conclude exactly the same property for (5.5-2) and (5.5-4) (with $Q = I$ or $2I$).

REMARK 5.6-1: Clearly (5.6-7) is true for all linear perturbation matrices which are similar to case F (including polynomials). In the case of a nonlinear perturbation matrix we must approximate that matrix, using a Taylor series expansion, with a set of linear perturbation matrices each is computed for *small*-parameter variations, in order to comply with (5.6-7). Thus the results of Theorem 5.6-2 can be made applicable to nonlinear perturbation matrices with some extra computations which depends on the encountered problem (cf., Example 5.10-5).

REMARK 5.6-2: As a result of Theorem 5.6-2 we can construct a hyperbox in the *PVS* whose corresponding ΔS's result in S + ΔS that is asymptotically stable. Then we can show that using an iterative approach similar to that for the single-parameter variations, i.e., Algorithm 5.6-1, we can extend the previous hyperbox, in order to reach the *most extended* such hyperbox which is our *SRM* (or perhaps the *E-Box*). We apply this procedure to several examples in Section 5.8, in order to demonstrate how these constructional methods can be used in practical situations.

COMMENT 5.6-3 (Final thought): In summary, we have an asymptotically stable matrix that is a function of a real-valued parameter vector. We have *somehow* detected a structured perturbational policy for this matrix. Our main objective is to construct the *E-Box* for this matrix, if possible, but without an exhaustive searching procedure which stems from the classical methods. We perform this task by applying the sufficient and local conditions which we have developed, in order to select the best perturbational policy (from the computational point of view) and to detect the direction(s) for applying this policy to a given problem. Because these local-upper bounds are functions of Q and P which in turn P is a function of Q, and we do not know how to choose the optimal Q, if any, we set Q = I, or 2I, and proceed accordingly. For a single-parameter variation Theorem 5.6-1 is used to reach the boundary of the *E-Box* which is a straight line. In the case of multiple-parameter variations, since there are infinitely many paths or directional perturbations in the *PVS* for any given problem, we must judiciously develop a *net* parameter variations scheme to analyze the encountered problem. Development of that scheme has been the main thrust of Theorem 5.6 2, where we introduce a method that generates *set by set* of points in the *PVS* (which is an Euclidean space), whose corresponding perturbed matrices are stable. When appropriate the adjective of *the largest* or *most extended* refers to this set. Applications of these results are given in Section 5.8.

5.7. EXTENSIONS TO HURWITZ POLYNOMIALS

5.7.1. Introduction

Research on the behavior of a rational polynomial $\Delta(s, \theta)$ of a complex variable s with real-valued coefficients or parameters $\theta \in R^r$, when the coefficients or the parameters are changing, is one of the earliest topics in system theory. During the last decade we have witnessed a surge of interest in the *SRA* of these polynomials. Apparently the most widely used approach for the *SRA* of polynomials is the Kharitonov's theorem [C. 58], which is used in an exploratory paper by Barmish [C. 7]. Since the publications of these two papers, a large number of papers have

5.7. Extensions to Hurwitz Polynomials

appeared in the literature which are extending applications of this theorem to various *SRA* of perturbed Hurwitz polynomials. The perturbation mechanism is introduced *via* static changes in coefficients of the corresponding Hurwitz polynomial, which are generally changing as the result of variations in system-physical parameters. The main goal for this sort of *SRA's*, as stated *explicitly* in both [C. 58 and 7], is to show how to generate a hyperbox in the *PS* such that the Hurwitz polynomial remains stable when *all* of its coefficients are changing in that hyperbox. However, it seems *implicitly* that the intentions of both authors may have been to generate *the largest* such hyperbox [or possibly *the E-Box*]. (As mentioned before this *E-Box* can be approximated with a set of convex sets which are stacked appropriately *under* it, and this latter set forms the *SRM* of the corresponding polynomial.) In this section we address issues regarding the *actual* construction (or computation) of this hyperbox or its best estimate for Hurwitz polynomials and we present an alternative method for this construction based on the results which are developed in Sections 5.5 and 5.6.

5.7.2. Brief Review

The following comments are also relevant to our discussions in Sections 5.4 to 5.6. The construction of the *E-Box*, for any linear system whether this system is expressed by a matrix or a polynomial, was one of the many problems that we faced in conjunction with a major sponsored research project on the theory of sensitivity at the University of Wisconsin-Madison, starting in the spring of 1973. Then, Professor Russell developed a novel approach to describe a sufficient condition for perturbing an asymptotically stable matrix, such that this system remains stable. This work was first presented in his 1975 lecture notes [B. 105]. Meanwhile this author developed a modified version of this work and a convergent iterative algorithm that was used to generate, from *the inside,* the *E-Box.* This result was extended to generate the "largest" coefficient variations of stable polynomials and was reported earlier in [B. 23 & 32]. Perhaps a few words on the thought behind *the inside,* mentioned above, is now appropriate, in order to clarify our motivation in developing the procedure that is presented in this book and has led us to our *SRA* for a linear system.

Suppose that we have only a single-input single-output system with one parameter. Then there are cases of these systems whose root-loci are crossing several times the imaginary axis in the complex plane. Therefore each one of these systems is alternatively stable for different (disjoint) sets of parameter variations, i.e., a *conditional stability*. This situation occurs when there are dependencies among various coefficients or entries of a stable matrix. Since our goal is to address the computational issues as well, and it is not always obvious in a given problem whether we have dependencies among various coefficients or not (unless we make some computational efforts that may become very costly to say the least), we are concerned about this sort of discontinuity and the related issues. Thus we are looking for perhaps conservative and sufficient conditions which are to be used as *tools*, in order to avoid the extra computational task of detecting dependencies and to avoid similar jumping and discontinuity in the construction of the stability regions. We are interested in direct construction of these stability regions because the associated computational issues, when we deal with a truly unknown problem whose multiple parameters are varying and we have no prior information about the way that these parameters are changing, are real and they are not trivial. Indeed in a large number of applied problems we do not know the intricacies of the relationships among various varying parameters of a dynamical system, let alone how these parameters can be changed (within an initial region), in order to maintain system stability. Had we known the complete relationship among all of the varying parameters of a given dynamical system *a priori*, then we would have had all of these parameters on a [given] hyperplane (the boundary of the *E-Box*) in the *PS*. But in reality we do not have this information and we cannot make any assumptions regarding the smoothness or any topological properties of a typical such hyperplane other than its existence and the fact that we have only one point on this hyperplane which is the initial [stable] operating point θ^o. Therefore it seems logical that we select θ^o as the starting point, and $\Delta\theta$ as a variable around this θ^o, in a process of constructing the corresponding hyperplane as completely and smoothly as we possibly can. Pictorially speaking, this construction is performed by generating the largest convex set in the *PVS* that approximates the previous hyperplane in the *PS*. When we exhaust this construction then we may perhaps select another operating point which is not connected to the previous set (provided that we have information regarding the conditional stability), and repeat the above procedure in exactly the same way until we complete our constructional task. This idea is in accord with what we would have done in a similar root-locus situation which crosses the imaginary axis repeatedly. However, if somehow in a given system we do not face such a dependency problem, then our construction yields the entire set of parameter variations in the first process. We limit ourselves to one operating point and its corresponding interval of admissible variations at a time (cf., Comment 5.8-1), in order to construct the *E-Box*. This construction is made possible by a convergent algorithm that iteratively uses the same sufficient conditions which are introduced in this chapter and *probes* the *PVS* for additional region of stability. Therefore the final result in the sense of the chosen perturbational policy and from the *computational* point of view is both necessary and sufficient condition for the robust stability, i.e., we can realize the *E-Box*, as sharp as our computational tolerances allow us. Perhaps we should make it clear that by the necessary and sufficient conditions we mean that the region of stability can be constructed as accurately as a prespecified tolerance will allow us, and not in an absolute theoretical sense which may have "sharp" edges. In the following we describe these matters in depth with several numerical examples.

The study of various *SRA* in a dynamical system is a very active research area as is evident by the large number of papers appearing in journals around the world (cf., § 5.12.C, for a partial list). Perhaps the most prominent work in this regard is that stemming from the Kharitonov's theorem [C. 58]. This work was published in

5.7. Extensions to Hurwitz Polynomials

1978 in Russian (English translation appeared in 1979) based on a prior submission in 1975.

THEOREM 5.7-1 [C. 58]: Consider a rational monic-stable polynomial $\Delta(s) = \sum_{j=0}^{n} a_j s^{n-j}$. Let $c \triangleq [a_1, a_2, \ldots, a_n]^T$ and suppose that each coefficient a_j, $j = 1, 2, \ldots, n$, is bounded as follows.

$$a_j^{min} \leq a_j \leq a_j^{max}. \tag{5.7-1}$$

If four monic polynomials which will be constructed in the same manner as $\Delta(s)$ by incorporating the following four sets of coefficients

$$c_1 \triangleq [a_1^{max}, a_2^{max}, a_3^{min}, a_4^{min}, \ldots]^T, \quad c_2 \triangleq [a_1^{min}, a_2^{max}, a_3^{max}, a_4^{min}, \ldots]^T,$$

$$c_3 \triangleq [a_1^{min}, a_2^{min}, a_3^{max}, a_4^{max}, \ldots]^T, \quad c_4 \triangleq [a_1^{max}, a_2^{min}, a_3^{min}, a_4^{max}, \ldots]^T,$$

are stable, then the original polynomial with the set of coefficients $c \triangleq [a_1, a_2, \ldots, a_n]^T$ is stable.

∎

There are a number of different interpretations of this result. But first we recall that, Kharitonov originally started with a set of homogeneous *ODE's* that were expressed in a matrix form such as (5.4-9), and he wanted to know how the entries of the coefficient matrix $S(\theta)$, which he called uncertain elements, may change without deteriorating its stability property [C. 58]. He then translated this problem into that of development of a sufficient condition for a polynomial (that was the characteristic equation of the previous set of *ODE's* with uncertain coefficients) to be Hurwitz as is described by Theorem 5.7-1. This theorem shows that if four additional polynomials which are constructed by incorporating the "largest" possible variations in the coefficients of the original polynomial (or the corresponding variations in the entries of a stable matrix) are Hurwitz, then that is a *sufficient* condition for the original polynomial to be Hurwitz. Naturally, we do not substitute the stability testing of one polynomial to that of four. In other words, the implication of this theorem is that we can construct an asymptotically stable hyperbox (but not the *E-Box*) to surround the original system with four boundary polynomials. The *dimension* of this hyperbox is an indication of the robustness of the original system. Thus in our terminology Theorem 5.7-1 is a *set generator*. Here the *set* refers to a set of points in the *PVS* (which corresponds to the coefficients Δa_1 to Δa_n), or to a set of corresponding perturbed polynomials. Certainly, this is a remarkable theoretical result. Furthermore this theorem is "robust", if we may, with respect to the number of required stability tests, which this concept opens a new school of thought in the *SRA* of systems, based on generating the minimum number of required stability tests when designing various components of a control system. This result says that stability of a family of polynomials that corresponds to a hyperbox in the coefficient space is equivalent to the stability of only four *extreme* polynomials corresponding to the vertexes of that hyperbox.

However, our main concern when using Theorem 5.7-1 to *construct* a class of robust systems is as follows. Given an original stable polynomial, how do we know *a priori* the "largest" possible variations in coefficients of this polynomial? Or equivalently, given an original stable matrix, how do we know the "largest" possible variations in entries of this matrix? Arguments are presented that this information must be sought in physical values of parameters such as the gain of an amplifier or the resistance of a resistor or some other practical considerations. *First of all,* this assertion is true only if we plan to carry out the analysis symbolically, where we can say for sure how coefficients of characteristic equation are changing as elements change. In general, we may have difficulty to trace symbolically changes in system parameters versus that in coefficients of its characteristic equation *per se*. *Secondly,* even if we give a symbolic relationship for any one of these coefficients in terms of system parameters, and even if we figure out how each and every one of these coefficients is changing, based on the extreme variations of its components, we cannot conclude anything whatsoever about stability, unless we make some necessary testings. For example, suppose that an amplifier, whose gain changes between $[k_1', k_2']$, is used in a system that affects coefficient a_j of its characteristic equation. However, since this characteristic equation can also represent many other systems, the allowable range of variations for this particular coefficient and from the stability point of view may be $[k_1'', k_2'']$. Thus the allowable range of variations for this gain and from both stability and physical points of view becomes $[k_1, k_2] \equiv [k_1', k_2'] \cap [k_1'', k_2'']$. To find the previous range of variations for each and every parameters in a system is a nontrivial task. Therefore except in special cases, we cannot provide *a priori* information about the "largest" coefficient variations (for the original polynomial), which will not deteriorate system stability. The fact of the matter is, if we knew the answer to the previous question, then there would be no merit in further investigation and we have accomplished what in essence is the purpose of a *SRA*. Indeed, if there were more than one coefficient that change, then this sort of determination is meaningless, because this method is not applicable to such cases. On the other hand, if these variations are being guessed, even conservatively, and for each choice these "largest" values are tested according to the Kharitonov's theorem, then we face a situation that there is very little information on how to modify *coherently* these guesses about the coefficient variations, in order to meet all the requirements of Theorem 5.7-1.

Therefore, it seems to us that no matter how we look at a stability robustness issue, this sort of analysis and based on testing *alone* may not be the natural way of constructing the *E-Box*. Specifically when we have a numerically defined large-scale system, where the contributions of system parameters have been already propagated into the system numerically, and it is very hard to recover their variations. Furthermore, most control problems (in particular the large-scale control problems) are given in a matrix form, while Kharitonov's result is developed for polynomials. Although theoretically these two settings are equivalent, more computational efforts are required to transform back and forth between them than just one. Therefore we should be alert of when and how to use the Kharitonov's

5.7. Extensions to Hurwitz Polynomials

approach in order to enhance the outcome of a *SRA*. Certainly, when the Kharitonov's theorem is applied to an example (cf., Example 5.8-6), in order to generate these upper and lower bounds (which this is the only application of Kharitonov's Theorem that is of interest to us), we need to develop an elaborate set of algebraic equations to compute these bounds [*C*. 7]. Although the number of boundary polynomials are always four (i.e., these four are independent from the order of the system), the number of algebraic equations that must be generated to determine these bounds, for all the coefficients, increases with the system dimension. Of course, there are now more insights available in this sort of analyses than before, which we can benefit from them when using this method to construct the *SRM* of a polynomial. However, we believe that the following computational method (which is ultimately supported by Theorem 5.6-2) is, in our opinion, more apt for constructing the *E-Box* or its best estimate *(SRM)* for a polynomial than the preceding one.

Finally, we believe that no matter which school of thought we subscribe to, we need to provide a numerical answer for the encountered perturbational problem. The "size" of this numerical answer, together with the efforts that are made to generate that, form the bases for the ultimate judgement on acceptability of the corresponding procedure, when tested on a nontrivial problem.

5.7.3. Alternative Analysis

In the following we present an alternative method to construct *the largest* hyperbox (or preferably the *E-Box*) for the *SRA* of a monic Hurwitz polynomial from that of a stable matrix. The choice of a monic polynomial does not create any lack of generality, since we can always transform a non-monic polynomial into a monic one with normalized coefficients. We must, however, emphasize that the analysis in this section is also based on our fundamental assumption in § 5.4.2, and is in the same spirit as the corresponding Problem 5.4-1. That means we study the perturbational analysis for a polynomial that is in a stable-operating mode corresponding to $\theta^o \in R^r$, and the parameter changes $\Delta\theta$ are referred to this mode. The computational task is to generate the *E-Box* or its best estimate for the encountered polynomial.

First, consider a monic-rational polynomial $\Delta(s, \theta) \triangleq \sum_{j=0}^{n} a_j(\theta) s^{n-j}$, $a_0 \equiv 1$, where $\theta \in R^r$ is a vector of system parameters (and $\Delta\theta$ is its variation), and s is a complex variable. Determination of $\theta \in R^r$ (i.e., a set of independently varying parameters) and how this parameter vector propagates into the coefficients $a_j(\theta)$'s is a nontrivial matter. That is why for simplicity of our analyses we often choose $c \triangleq [a_1, ..., a_n]$ to be the set of varying parameters, and we assume that as $c \to c + \Delta c$ (or $\theta \to \theta + \Delta\theta$), the multiple large parameter changes, Δc (or $\Delta\theta$), will not affect the stabilizability of $\Delta(s, \theta)$. At this juncture we again emphasize that we are simultaneously working with two equivalent sets of objects. *One set* corresponds to points such as $P \triangleq [\Delta a_1, ..., \Delta a_n]^T$ (or $P \triangleq [\Delta \theta_1, ..., \Delta \theta_r]^T$), in the "parameter variations space" *(PVS)*, whose origin is at the nominal value c^o (or θ^o). (This c^o is also a point in the "parameter space" *(PS)*, whose origin is at the zero value for all parameters.) *The other* set corresponds to the perturbed polynomials $\tilde{\Delta}(s, c^o + \Delta c)$ (or $\tilde{\Delta}(s, \theta^o + \Delta \theta)$), which are computed at each of the above points P. The distinction between these two sets, specially for polynomial with multiple-varying parameters, is required in order to proceed.

For the above class of monic polynomials, there corresponds a matrix S in the phase-variable canonical form such that $\det(sI - S(\theta)) = \Delta(s, \theta)$. Thus the problem of studying the largest variations in coefficients of $\Delta(s, \theta)$, as $c \to c + \Delta c$ (or $\theta \to \theta + \Delta\theta$), such that $\tilde{\Delta}(s, c+\Delta c)$ (or $\tilde{\Delta}(s, \theta+\Delta\theta)$) remains stable, becomes that of determining the "allowable range of variations" for the following perturbation matrix which has been studied in Theorem 5.5-1.

$$\Delta S = \begin{bmatrix} & 0 & \\ -\Delta a_n & \cdots & -\Delta a_1 \end{bmatrix}. \tag{5.7-2}$$

We may specialize Theorem 5.5-1 to a monic-stable polynomial by letting S to be in a phase-variable canonical form whose last row represents the coefficients of this polynomial and the corresponding ΔS is at most like (5.7-2).

THEOREM 5.7-2: Consider a monic-stable polynomial $\Delta(s, \theta) = \sum_{j=0}^{n} a_j(\theta) s^{n-j}$. Let $a_j \to a_j + \Delta a_j$. Then there exists a directional-upper bound α_j where

$$\alpha_j^2 = \det Q / [4 c_{ll} \lambda], \quad \text{for each } j = 1, \ldots, n, \tag{5.7-3}$$

such that $\tilde{\Delta}(s, \theta + \Delta\theta) = s^n + a_1 s^{n-1} + \cdots + (a_j + \Delta a_j) s^{n-j} + \cdots + a_n$ remains stable if $(\Delta a_j)^2 < \alpha_j^2$. Here Q is an arbitrary symmetric positive-definite matrix, c_{ll} is the (l, l)-cofactor of Q and $l = n + 1 - j$, $P = P^T > 0$ is the unique solution of $S^T P + PS + Q = 0$ for S in the phase-variable canonical form [i.e., $\Delta(s, \theta) = \det(sI - S(\theta))$, and λ is the largest (or the unique nonzero) root of $\det(\lambda M - K) = 0$, with $M \triangleq P^{-1} Q P^{-1}$, and $K = \text{diag}(0, \cdots, 0, 1_{nn})$.

PROOF: Suppose that the above $S \to S + \Delta S$; then by Theorem 5.5-1, $S + \Delta S$ remains a stability matrix if (5.5-1) holds for the corresponding ΔS. Since ΔS has only one nonzero entry $(-\Delta a_j)$ in the (n, l) entry, where $l = n + 1 - j$, from (5.5-1) we have $\Delta S Q^{-1} \Delta S^T = (c_{ll} (\Delta a_j)^2 / \det Q) K < \frac{1}{4} M$. By simultaneous diagonalization of this inequality the result follows.

COMMENT 5.7-1: In Theorem 5.7-2, we can show that $\lambda = M_{nn}/\det M$, where M_{nn} is the (n, n)-cofactor of M. Therefore, the complete simultaneous diagonalization for determination of λ is not required, and $\alpha_j^2 = \det Q \det M / 4 c_{ll} M_{nn}$.

∎

5.7. Extensions to Hurwitz Polynomials

In many applications, the coefficients of any characteristic equation, are not independent of each other. Therefore, if one parameter changes, then several coefficients *may* change. Next theorem pertains to this situation which shows how to change simultaneously all the coefficients of a stable polynomial.

THEOREM 5.7-3: Consider the monic-stable polynomial $\Delta(s, \theta)$ of Theorem 5.7-2. If all the coefficients of $\Delta(s, \theta)$ change, then this polynomial remains stable subject to the following upper bound,

$$\sum_{j=1}^{n} (\Delta a_j)^2 < \lambda_{\min}(Q) / 4g_{nn}. \qquad (5.7\text{-}4)$$

Here g_{nn} is the (n, n)-entry of $PQ^{-1}P$; P and Q are the same as in Theorem 5.7-2; and $\lambda_{\min}(Q)$ refers to the minimum eigenvalue of Q. (Notice that by asymptotic stability of $S + \Delta S$, $g_{nn} \neq 0$. In other words, care must be exercised to avoid computational instability.)

PROOF: Suppose that $S \to S + \Delta S$, where ΔS has only nonzero elements $-\Delta a_n, \ldots, -\Delta a_1$, respectively, in its last row. Then by (5.5-2), ΔS remains a stability matrix if $\Delta S^T P Q^{-1} P \Delta S < \frac{1}{4}Q$. Assume that $G \triangleq PQ^{-1}P$; then

$$\Delta S^T G \Delta S = g_{nn} \begin{bmatrix} (\Delta a_n)^2 & \Delta a_n \Delta a_{n-1} & \cdots & \Delta a_n \Delta a_1 \\ \Delta a_{n-1} \Delta a_n & (\Delta a_{n-1})^2 & \cdots & \Delta a_{n-1} \Delta a_1 \\ \vdots & \vdots & \cdots & \\ \Delta a_1 \Delta a_n & \Delta a_1 \Delta a_{n-1} & \cdots & (\Delta a_1)^2 \end{bmatrix} \triangleq g_{nn} \Pi < \tfrac{1}{4}Q. \qquad (5.7\text{-}5)$$

In (5.7-5) every element is known *a priori* except Π, which is a nonlinear function of parameter variations. In this situation, however, because the rank of Π is one, its maximum eigenvalue is equal to its trace and we do not need to carry out the complete diagonalization. Thus it is clear that herein a conservative-upper bound becomes $g_{nn} \text{Tr}(\Pi) < \frac{1}{4}\lambda_{\min}(Q)$. ∎

We note that since in Theorem 5.7-3 the underlying system also satisfies Theorem 5.6-2, Theorem 5.7-3 also enables us to construct a set of points in the corresponding *PVS* (which is the space of $[\Delta a_1, \ldots, \Delta a_n]$), whose corresponding set of polynomials is stable (cf., Remark 5.7-2). Thus Theorem 5.7-3 is a *set generator*.

Next theorem which is based on the second half of Theorem 5.5-1 shows how each coefficients in a monic-stable polynomial can be changed without deteriorating its asymptotic stability. This result is particularly interesting for large-scale (or higher-order) polynomials, since it involves no direct eigenvalue computation and it only needs the solution to one Lyapunov equation, in order systematically to change every coefficients of this polynomial, one at a time, while maintaining its overall stability.

THEOREM 5.7-4: Given a monic asymptotically-stable polynomial $\Delta(s) \triangleq \sum_{j=0}^{n} a_j s^{n-j} = \det(sI - S)$ with $S \in R^{n \times n}$ is in a phase-variable canonical form. Suppose that $a_j \to a_j + \Delta a_j$ and thus $\Delta(s) \to \tilde{\Delta}(s)$, such that

$$(\Delta a_j)^2 < 1 / [4 \sum_{i=1}^{n} p_{il}^2], \quad \text{for each } j = 1, \ldots, n. \quad (5.7\text{-}6)$$

Then $\tilde{\Delta}(s)$ remains asymptotically stable. Here p_{il} ($i = 1, \cdots, n$, and $l = n + 1 - j$), is the lth-column of $P = P^T > 0$, with $SP + PS^T + I = 0$.

PROOF: From (5.5-4) and with $Q = I$, we have $\Delta S\, P^2 \Delta S^T < \frac{1}{4} I$. Here ΔS has only one nonzero element at its (n, l) entry which is $-\Delta a_j$. Upon some algebra the previous inequality will have one nonzero element in its (n, n) entry, which is the same as its maximum eigenvalue. Therefore we have

$$(\Delta a_j)^2 \left[p_{1l}^2 + \cdots + p_{nl}^2 \right] < \frac{1}{4}. \quad (5.7\text{-}7)$$

In other words, coefficients can be changed, *one at a time,* according to the inverse Euclidean norm of each corresponding rows of P. No other eigenvalue computation is needed to generate a bank of Hurwitz polynomials surrounding the original Hurwitz polynomial.

COMMENT 5.7-2: In Theorem 5.7-2 if we use inequality (5.5-3) instead of (5.5-1) to find the corresponding upper bound for Δa_j, then we have

$$\frac{(\Delta a_j)^2 c_{nn}}{\det Q} \begin{bmatrix} 0 & 0 & 0 \\ 0 & 1_{ll} & 0 \\ 0 & 0 & 0 \end{bmatrix} \triangleq \frac{(\Delta a_j)^2 c_{nn}}{\det Q} K' < \frac{1}{4} P^{-1} Q P^{-1} \triangleq \frac{1}{4} M'. \quad (5.7\text{-}8)$$

From (5.7-8) we have the following equivalent expression for the result of Theorem 5.7-2.

$$(\Delta a_j)^2 = \det Q / 4 c_{nn} \lambda', \quad \text{for each } j = 1, \ldots, n. \quad (5.7\text{-}9)$$

Here λ' is the unique nonzero root of $\det(\lambda' M' - K') = 0$. (Note that $M' \neq M$, because M' is computed from a different P.) Because K' changes with l, each corresponding λ' is different from the rest. In (5.7-9), if we let $Q = I$, then $\lambda' = \sum_{i=1}^{n} p_{il}^2$ and (5.7-9) becomes the same as (5.7-6).

COMMENT 5.7-3: In general, using (5.5-3) with $Q = I$ and for the same ΔS as in (5.7-2), we have

5.7. Extensions to Hurwitz Polynomials

$$\sum_{j=1}^{n} (\Delta a_j)^2 < \tfrac{1}{4}\, \lambda_{\min}(P^{-2}), \tag{5.7-10}$$

which involves the eigenvalues of P^{-2}. Whether (5.7-4) or (5.7-10) yields the *largest*-upper bound for the total variations in all the coefficients of a Hurwitz polynomial and which one of these algorithms has computational advantages over the other one must be studied. But we must again emphasize that when using results which stem from Theorem 5.5-1, we must select an appropriate Lyapunov function, otherwise the conclusions are inaccurate.

REMARK 5.7-1: Very briefly, Theorem 5.5-1 shows that if an asymptotically stable matrix S is perturbed to $S + \Delta S$, such that ΔS satisfies any one of the inequalities (5.5-1) to (5.5-4) which are developed corresponding to two different Lyapunov functions, then $S + \Delta S$ remains asymptotically stable. In particular, if we let S to be in a phase-variable canonical form (whose characteristic equation represents a stable matrix or a stable polynomial), then ΔS can be at most of the form (5.7-2). What we have shown in this section is that, if we can select the set of $\{\Delta a_j\}_{j=1}^{n}$ in any manner and with no particular *order*, such that (5.7-2) satisfies one of the inequalities (5.5-1) to (5.5-4), then $S + \Delta S$ (or the corresponding perturbed polynomial) remains asymptotically stable. The construction of this bank of Hurwitz polynomials which are robust under coefficient variations is the main thrust of this section and the main consequence of Theorem 5.7-3.

REMARK 5.7-2: In summary, the above results with $Q = 2I$ are reviewed in Table 5.7-1, which shows that for either single- or multiple-parameter variations we have two choices regarding our perturbation policy. Here we cannot say *a priori* which one of these two choices yields the best or the largest perturbation. In each case we test either results before selecting one. This determination can be automated. We also note that in both cases of multiple-parameter variations (where we have one inequality with n variables or a sphere in the *PVS*), any set of possible "solutions" for coefficient variations (or the corresponding points $[\Delta a_1, ..., \Delta a_n]^T$ in the *PVS*), yields an associated "solution" $\Delta S = \Delta S(\Delta a_1, ..., \Delta a_n)$ such that $S + \Delta S$ results in a stable perturbed polynomial. Here we have imbedded the single-parameter variations in the multiple-parameter variations, and therefore there will not be any unacceptable Δa_j. In Column 1 of Table 5.7-1, for instance, each Δa_j can be as large as $1/g_{nn}^{1/2}$. Furthermore, since these multiple-parameter variations are generated from Theorem 5.5-1, with $Q = 2I$, and the corresponding ΔS can be represented as a convex combinations of a set of ΔS_i's (ΔS is of type F according to § 5.5.2), the entire set of points in the *PVS* which connects the corresponding corner points for these "solutions" also results in a set of perturbed polynomials that is stable according to Theorem 5.6-2. That *set by set* generation of points in the *PVS* is the *key* to efficient construction of *the E-Box* for points in the *PVS* whose corresponding set of perturbed polynomials is stable (cf., Examples 5.8-5 & 6). (We must recall that we assume the coefficients of $\Delta(s, \theta)$ (or the entries of $S(\theta)$) can be changed by a large amount without deteriorating its stabilizability.)

Table 5.7-1. Summary of the Coefficient Variations in a Monic Hurwitz Polynomial $\Delta(s) \triangleq \sum_{j=0}^{n} a_j s^{n-j}$.

Column 1	Column 2
$S^T P + PS + 2I = 0$	$SP + PS^T + 2I = 0$
Single-Parameter Variations $(\Delta a_j)^2 < 1/g_{nn}$ g_{nn} is the (n,n) entry of P^2	Single-Parameter Variations $(\Delta a_j)^2 < 1/\sum_{i=1}^{n} p_{il}^2$ $l = n + 1 - j$
Multiple-Parameter Variations $\sum_{j=1}^{n} (\Delta a_j)^2 < 1/g_{nn}$	Multiple-Parameter Variations $\sum_{j=1}^{n} (\Delta a_j)^2 < \lambda_{min}(P^{-2})$

5.8. NUMERICAL EXAMPLES

5.8.1. Introduction

In the following we present a set of tutorial examples, in order to show the applications of certain results developed in this chapter so far. As we proceed, we summarize the pertinent observations, remarks, and comments, which can be generalized to other examples at the end of each solution.

EXAMPLE 5.8-1 (Single-parameter variations or directional perturbation): Consider an asymptotically stable matrix S, whose eigenvalues are $\lambda_1 = -2$ and $\lambda_2 = -1$ as follows [B. 92].

$$S (\equiv S_0) = \begin{bmatrix} -3 & -2 \\ 1 & 0 \end{bmatrix}. \quad (5.8\text{-}1)$$

Let the perturbation scheme be $\Delta S = \varepsilon U$, where U is a unity matrix (a matrix whose entries are all one), that means we are uncertain about every entry of S, but

5.8. Numerical Examples

we are perturbing S *only* along the "45° line" in the *PVS*. The objective is to find *the largest* ε, *vis a vis* the best *directional* perturbation along this line, which is only *one path* in the *PVS* corresponding to the four entries of S.

Incidentally, using classical control methods, we can show that *the largest* ε is 1, where S + U has two eigenvalues $\lambda_1 = 0$ and $\lambda_2 = -1$. (Notice that $\lambda_2 = -1$ is *insensitive* to this perturbation scheme.) Thus if $\varepsilon > 1$, then $S + \varepsilon U$ will have at least one eigenvalue on the right-hand side of complex plane. Thus along this particular line or direction the theoretical $\varepsilon_{max} = 1$. We study this problem based upon the procedures that we have introduced earlier in this chapter.

SOLUTION [B. 24 & 29]: In this situation, which in effect we have only one parameter ε, we must use Algorithm 5.6-1, that primarily depends on the solutions of two corresponding Lyapunov equations. For this and future examples we give these algebraic solutions in Table 5.8-1 for the sake of discussions. In practice this table is generated numerically.

Before proceeding further, and regarding the two Lyapunov equations in Theorem 5.5-1 (or Table 5.8-1), we should point out that although these two Lyapunov equations are related through a linear matrix transformation. That issue is not of concern to us at the moment, because we are searching for a numerical answer and if a transformed state yields that, then so be it.

To determine the upper bound for ε, first we need to select the best perturbation policy by studying the applications of (5.5-1) to (5.5-4) in this example. A corresponding Lyapunov equation for $S^T P + PS + 2I = 0$ yields

$$P = \begin{bmatrix} 1/2 & 1/2 \\ 1/2 & 5/2 \end{bmatrix}. \tag{5.8-2}$$

From this solution the associated local-upper bounds, computed from (5.5-1) and (5.5-2) using ordinary eigenvalues, are as follows. $\varepsilon_{(5.5-1)} = 0.191$ (this upper bound is the same as that of [B. 92]), and $\varepsilon_{(5.5-2)} = 0.2236$.

On the other hand, using (5.5-3) and (5.5-4) requires a solution to the following Lyapunov equation $SP + PS^T + 2I = 0$, which results in

$$P = \begin{bmatrix} 1 & -1 \\ -1 & 2 \end{bmatrix}. \tag{5.8-3}$$

Then from this P we have two new local-upper bounds corresponding to (5.5-3) and (5.5-4) as follows: $\varepsilon_{(5.5-3)} = 0.191$ and $\varepsilon_{(5.5-4)} = 0.707$. Clearly, the latter is superior to the former three local-upper bounds. We must also point out that in this example the *troublesome* direction of perturbation is in the positive direction of ε's. This has been determined by a simple calculation of the perturbed matrix eigenvalues as suggested in [B. 23] and Algorithm 5.6-1 for finding the troublesome direction of a perturbation scheme.

Table 5.8-1. Algebraic Solutions of the Lyapunov Equations.

Column 1	Column 2
$S = \begin{bmatrix} s_{11} & s_{12} \\ s_{21} & s_{22} \end{bmatrix}$	$P = \begin{bmatrix} p_{11} & p_{12} \\ p_{12} & p_{22} \end{bmatrix}$
$S^T P + PS + 2I = 0$	$SP + PS^T + 2I = 0$
$p_{11} = -(s_{21}^2 + s_{22}^2 + \Delta_s)/\Delta$ $p_{12} = (s_{11}s_{21} + s_{12}s_{22})/\Delta$ $p_{22} = -(s_{11}^2 + s_{12}^2 + \Delta_s)/\Delta$ $\Delta_s = s_{11}s_{22} - s_{12}s_{21}, \quad \Delta = (s_{11} + s_{22})\Delta_s$	$p_{11} = -(s_{12}^2 + s_{22}^2 + \Delta_s)/\Delta$ $p_{12} = (s_{11}s_{12} + s_{21}s_{22})/\Delta$ $p_{22} = -(s_{11}^2 + s_{21}^2 + \Delta_s)/\Delta$ $\Delta_s = s_{11}s_{22} - s_{12}s_{21}, \quad \Delta = (s_{11} + s_{22})\Delta_s$

Now, choosing (5.5-4) as our perturbation *mechanism*, then the perturbed S_0 becomes S_1 (where in general $S_J = S_{J-1} + \varepsilon_J U$, and J is the iteration number) as follows.

$$S_1 \triangleq S_0 + \varepsilon_1 U = \begin{bmatrix} -2.293 & -1.293 \\ 1.707 & 0.707 \end{bmatrix}, \quad (5.8\text{-}4a)$$

whose eigenvalues are $\lambda_1 = -0.586$ and $\lambda_2 = -1$. The new corresponding Lyapunov equation, *again* with $Q = 2I$, yields

$$P_1 = 2 \begin{bmatrix} 1.4835 & -2.2445 \\ -2.2445 & 4.7115 \end{bmatrix}. \quad (5.8\text{-}4b)$$

Then from the corresponding (5.5-4), we have

$$\varepsilon_2^2 < 1/53.322, \quad \text{or} \quad \varepsilon_2 = 0.1369. \quad (5.8\text{-}5)$$

This iterative process continues until we can no longer perturb the asymptotically stable matrix S_J. Then *the largest* perturbation step in this example becomes $\varepsilon \triangleq \sum_{J=1}^{6} \varepsilon_J = 0.998 < 1$.

▼

OBSERVATION 5.8-1: We recall that in the above example we have only perturbed S_0 along the "$45°$ line". In Example 5.8-4 we study the case where we perturb all (four) entries of a matrix simultaneously. It is perhaps instructive to apply the method of Example 5.8-4 to the above example.

5.8. Numerical Examples

EXAMPLE 5.8-2 (*On the choices for Q, and polynomials*): Define $\Delta(s) \triangleq s^2 + a_1 s + a_2 \triangleq s^2 + (50 - 10k)s + 200$, where s is a complex variable. Suppose that the nominal value of k is equal to 2, and $k \to k + \Delta k$. Then what is the largest variation of k such that $\tilde{\Delta}(s, k + \Delta k) = 0$ remains asymptotically stable?

SOLUTION [B. 23 & 30]: Certainly we can answer this question by inspection, namely, for asymptotic stability we must have $\Delta k < 3$. We now proceed to derive this answer (partially – only the first iteration) using the results which we have developed in § 5.7.3. We can show that the above $\Delta(s)$, and with k = 2, is the characteristic polynomial of the following matrix.

$$S \equiv S_0 \triangleq \begin{bmatrix} 0 & 1 \\ -a_2 & -a_1 \end{bmatrix} = \begin{bmatrix} 0 & 1 \\ -200 & -30 \end{bmatrix}. \tag{5.8-6}$$

The eigenvalues of this matrix are: $\lambda_1 = -20$ and $\lambda_2 = -10$. Thus this system is asymptotically stable at k = 2 and $S^T P + PS + Q = 0$ for $Q = Q^T > 0$ has a unique solution $P = P^T > 0$. The main purpose of this example is to show the effect of choosing different Q's on the final result, and thus we only emphasize the application of the first part in Theorem 5.5-1 to this example.

CASE I: $Q = I$.
Correspondingly,

$$P = \frac{1}{12000} \begin{bmatrix} 41100 & 30 \\ 30 & 201 \end{bmatrix}. \tag{5.8-7}$$

From (5.7-3) and with $Q = I$, we have $(\Delta a_1)^2 < 1/4\lambda$, where λ is the largest root of $\det(\lambda P^{-2} - K) = 0$. Here $K = \text{diag}(0, 1)$. It is easy to show that $\lambda = 0.000286815$. Thus

$$(\Delta a_1)_I^2 < 1/4\lambda = 871.64, \text{ or } |\Delta a_1|_I < 29.524 \Rightarrow \Delta k_I < 2.9524. \tag{5.8-8}$$

Although the above upper bound is slightly less than $\Delta k = 3$ which is the best result for this example, the $Q = I$ is not the "optimal" choice. To reach the optimal Δk (= 3), we first perturb k by Δk_I, then we repeat the computation for the next iteration. However, the purpose of this example is to compare only the first perturbation step when using different Q's.

CASE II: $Q = R^{-T} Q_1 R^{-1}$, with $Q_1 = I$.
Here R diagonalizes S and this choice is based on Comments 5.6-1(*iv*). In the present example we have

$$R = \begin{bmatrix} 1 & 1 \\ -20 & -10 \end{bmatrix}, \text{ and } R^{-1} = \begin{bmatrix} -1 & -0.1 \\ 2 & 0.1 \end{bmatrix}, \tag{5.8-9a}$$

$$Q \triangleq R^{-T} R^{-1} = \begin{bmatrix} 5 & 0.3 \\ 0.3 & 0.02 \end{bmatrix}, \text{ and } Q^{-1} = \begin{bmatrix} 2 & -30 \\ -30 & 500 \end{bmatrix}, \tag{5.8-9b}$$

and correspondingly, from $S^T P + PS + Q = 0$, we have

$$P = \frac{1}{4000} \begin{bmatrix} 900 & 50 \\ 50 & 3 \end{bmatrix}, \text{ and } P^{-1} Q P^{-1} = 400 \begin{bmatrix} 5 & -90 \\ -90 & 1700 \end{bmatrix}. \tag{5.8-9c}$$

In this case $c_{22} = 5$ and $\det Q = 0.01$. From (5.7-3) we have $(\Delta a_1)^2 < \det Q/4c_{22}\lambda = 1/2000\lambda$. Here λ is the largest root of $\det(\lambda P^{-1}QP^{-1} - K) = 0$, with $K = \text{diag}(0, 1)$. We can show that $\lambda = 10^{-3}/32$. Thus

$$(\Delta a_1)_{II}^2 < 16, \text{ or } |\Delta a_1|_{II} = 4 \Rightarrow \Delta k_{II} < 0.4 < \Delta k_I. \tag{5.8-10}$$

Thus $Q = R^{-T}Q_1R^{-1}$ is not the "optimal" Q either.

CASE III: $Q = \text{diag}(-\lambda_1, -\lambda_2)$.
This choice is based on the following observation that if S is real and diagonal, then this Q yields $P = \frac{1}{2}I$, and certainly the maximum perturbation occurs using (5.5-1). Here we let Q as is and look for P according to a general (or non-modal) form of S. If so, then the corresponding P and (5.5-1) become

$$P = \frac{1}{1200}\begin{pmatrix} 42200 & 60 \\ 60 & 202 \end{pmatrix}, \quad P^{-1} = \frac{6}{42604}\begin{pmatrix} 202 & -60 \\ -60 & 42200 \end{pmatrix}, \tag{5.8-11a}$$

$$P^{-1}QP^{-1} = \begin{pmatrix} 0.0169 & -0.507 \\ -0.507 & 353.206 \end{pmatrix}, \tag{5.8-11b}$$

$$(\Delta a_1)_{III}^2 \begin{pmatrix} 0 & 0 \\ 0 & 0.1 \end{pmatrix} < \frac{1}{4} \begin{pmatrix} 0.0169 & -0.507 \\ -0.507 & 353.206 \end{pmatrix}. \tag{5.8-11c}$$

Using the same approach as before (i.e., a generalized eigenvalue) we have $(\Delta a_1)_{III}^2 < 844.99$, or $|\Delta a_1|_{III} < 29.069$, resulting in $\Delta k_{II} < \Delta k_{III} = 2.9069 < \Delta k_I$. Although this result is very close to (5.8-8), its computation involves more efforts than that required to generate (5.8-8).

CASE IV: Choosing the $P_{optimum} = (RR^T)^{-1}$, [*B*. 122].
Recently it is suggested in [*B*. 122], that if we let Q correspond with the $P_{optimum} = (RR^T)^{-1}$, then we have the optimal stability bound. In this example (5.5-1) becomes

$$P \triangleq R^{-T}R^{-1} = \begin{pmatrix} 5 & 0.3 \\ 0.3 & 0.02 \end{pmatrix}, \text{ and } P^{-1} = \begin{pmatrix} 2 & -30 \\ -30 & 500 \end{pmatrix}, \tag{5.8-12a}$$

$$Q = -S^TP - PS = 2\begin{pmatrix} 60 & 4 \\ 4 & 0.3 \end{pmatrix}, \quad Q^{-1} = \frac{1}{4}\begin{pmatrix} 0.3 & -4 \\ -4 & 60 \end{pmatrix}, \tag{5.8-12b}$$

$$\frac{1}{4}(\Delta a_1)_{IV}^2 \begin{pmatrix} 0 & 0 \\ 0 & 60 \end{pmatrix} < \frac{1}{4}\begin{pmatrix} 60 & -1000 \\ -1000 & 18000 \end{pmatrix}. \tag{5.8-12c}$$

Resulting in

$$(\Delta a_1)_{IV}^2 < 22.2222, \text{ or } |\Delta a_1|_{IV} < 4.714 \Rightarrow \Delta k_{IV} < 0.4714. \tag{5.8-13}$$

This result is clearly smaller than the two out of three upper bounds which we have suggested and is by far smaller than the global answer in this example. Furthermore, we should also point out that we must prove upon choosing this $P_{optimum}$, $Q (= -S^TP_{optimum} - P_{optimum}S)$ becomes a positive-definite matrix.

▼

OBSERVATION 5.8-2: The above two examples serve their purposes by showing the importance of an appropriate choice for Q, and the utilization of an appropriate

5.8. Numerical Examples

Lyapunov equation in Algorithm 5.6-1. In either of these two examples we have used both ordinary and generalized eigenvalues for the corresponding results from Theorem 5.5-1. In *Case I*, we get almost the best result using either of these two eigenvalues. Thus in addition to choosing a proper Q, we should also explore different norms for inequalities (5.5-1) to (5.5-4), before making any attempt to construct a robustly stable structure. In the next example we demonstrate the scope and applications of Theorem 5.6-2, which is the counterpart of Algorithm 5.6-1, for the case of multiple-parameter variations.

EXAMPLE 5.8-3 (Multiple - parameter variations): Consider the following asymptotically stable matrix

$$S(\equiv S_1) = \begin{bmatrix} 0 & 1 & 0 & 0 \\ -1 & -1 & 0 & 0 \\ 0 & 0 & 0 & 1 \\ 1 & 1 & -1 & -4 \end{bmatrix}, \quad (5.8\text{-}14)$$

whose eigenvalues are $\lambda_{1,2} = \frac{1}{2}(-1 \pm j\sqrt{3})$ and $\lambda_{3,4} = -2 \pm \sqrt{3}$. This example (due to Professor D.L. Russell) comes from a fourth-order differential equation characterizing the linear homogeneous part of a mechanical model for an accelerometer. The linearization of system nonlinearities contributes the following perturbation scheme to the linearized model of the system.

$$\Delta S = \Delta\theta_1 \begin{bmatrix} 0 & 0 & 0 & 0 \\ 0 & 1 & 0 & 0 \\ 0 & 0 & 0 & 0 \\ 0 & 0 & 1 & 0 \end{bmatrix} + \Delta\theta_2 \begin{bmatrix} 0 & 0 & 0 & 0 \\ 0 & 0 & -1 & 0 \\ 0 & 0 & 0 & 0 \\ 0 & -1 & 0 & 0 \end{bmatrix}, \quad (5.8\text{-}15)$$

where both $\Delta\theta_1$ and $\Delta\theta_2 > 0$. Our goal is to apply our method to this case and to construct *the largest* set of admissible parameter variations, whose corresponding class of perturbed linear dynamical systems remains [asymptotically] stable.

SOLUTION: Clearly the first step toward solving this or any similar perturbational problem is to define the set of varying parameters. In this case these parameters are $\Delta\theta_1$ and $\Delta\theta_2$. Let $J = 1, 2, ..., N$, be the iteration number.

STEP 0 (To update S): We let $S_J = S_{J-1} \pm \Delta S_{J-1}$ with $S_1 \equiv S_0 \equiv S$ being the initial asymptotically stable matrix and $\Delta S_0 \equiv 0$. (The ($\pm$) means possibly in both directions.)

STEP 1 (To solve the Lyapunov equation): For the current S_J we solve for P_J according to $S_J^T P_J + P_J S_J + 2I = 0$. (At $J = 1$, and in a typical application we may also solve the other Lyapunov equation in Theorem 5.5-1, in order to study which one of the perturbation schemes (5.5-1) to (5.5-4) should be utilized in a given problem. In that case the following steps must be changed accordingly.) At $J = 1$ we have

$$P_1 = \frac{1}{10} \begin{bmatrix} 31 & 11 & 2 & 1 \\ 11 & 24 & 9 & 3 \\ 2 & 9 & 45 & 10 \\ 1 & 3 & 10 & 5 \end{bmatrix}. \quad (5.8\text{-}16)$$

STEP 2 (Directional perturbations): First we search for the largest variations along the "45° line", and all the other axes. Along the "45° line" ($\Delta\theta_1 = \Delta\theta_2 = \varepsilon_{45°}$), the perturbation scheme takes place according to

$$\Delta S = \varepsilon_{45°} \begin{bmatrix} 0 & 0 & 0 & 0 \\ 0 & 1 & -1 & 0 \\ 0 & 0 & 0 & 0 \\ 0 & -1 & 1 & 0 \end{bmatrix}. \quad (5.8\text{-}17)$$

From the physics of this system we know *a priori* that as $\varepsilon_{45°} \to 1$, then the system becomes unstable. Therefore we want to compute the *largest* $\varepsilon_{45°}$ such that $S + \Delta S$ remains asymptotically stable. We have applied the iterative method of Algorithm 5.6-1 (with $Q = I$) to this example (the entire computational procedures are tabulated in [B. 23]) and from this analysis, and after *five* iterations, we have found that for $\varepsilon_{45°} = 0.797481 < 0.8$ (0.8 is the critical value of $\varepsilon_{45°}$ found using the Routh and the Hurwitz criteria) the system remains asymptotically stable. This extreme corner point is shown in the subsequent figures by $C_n^{++} \triangleq (\Delta\theta_1 = \Delta\theta_2 = 0.8)$.

Using the above algorithm we can also show that on the two axes $\Delta\theta_1$ and $\Delta\theta_2$ we have the following two extreme points:

$$A \triangleq (\Delta\theta_1 = 0, \Delta\theta_2 = 2.06), \quad B \triangleq (\Delta\theta_1 = 1.0, \Delta\theta_2 = 0). \quad (5.8\text{-}18)$$

As we continue to study this problem, we realize that perhaps the only question that we need to answer in this part of our solution is the following. Namely, on which path(s) between C_n^{++} to A (or C_n^{++} to B) the system remains stable? The following analysis which is based primarily on Theorem 5.6-2 is intended to answer this question by providing us with a *probing scheme* that allows to search for a closed set of points in the *PVS* corresponding to a class of stable linear perturbed systems. Note that we have described the system with a matrix differential equation and we are considering the stability of matrices along a path in the parameter space or the *PVS*.

STEP 3 (To construct $\Delta S \Delta S^T$): Given any structured ΔS we can easily compute $\Delta S \Delta S^T$ which symbolically remains the same throughout the analyses. For example, for ΔS given in (5.8-15) we have

$$\Delta S \Delta S^T = \begin{bmatrix} 0 & 0 & 0 & 0 \\ 0 & \Delta\theta_1^2 + \Delta\theta_2^2 & 0 & -2\Delta\theta_1\Delta\theta_2 \\ 0 & 0 & 0 & 0 \\ 0 & -2\Delta\theta_1\Delta\theta_2 & 0 & \Delta\theta_1^2 + \Delta\theta_2^2 \end{bmatrix}. \quad (5.8\text{-}19)$$

STEP 4 (To interpret the sufficient condition): Using the sufficient condition $\Delta S \Delta S^T < P_J^{-2}$ we need to establish a set of inequalities which yields sufficient conditions for our parameter variations. Contrary to the case of single-parameter variations, here we do not interpret the previous sufficient condition as, *per se*, $\lambda_{max}(\Delta S \Delta S^T) < \lambda_{min}(P_J^{-2})$ for reasons that soon become apparent. (We study this

5.8. Numerical Examples

inequality in Remark 5.8-3.) Thus we look for alternative conditions on $\Delta\theta_1$ and $\Delta\theta_2$ such that $(P_J^{-2} - \Delta S \Delta S^T) > 0$ holds. At $J = 1$ we have

$$\begin{bmatrix} 0.18423 & -0.175 & 0.021 & -0.032 \\ -0.175 & 0.362 - (\Delta\theta_1^2 + \Delta\theta_2^2) & 0.06 & -0.62 + 2\Delta\theta_1\Delta\theta_2 \\ 0.021 & 0.06 & 0.782 & -3.168 \\ -0.032 & -0.62 + 2\Delta\theta_1\Delta\theta_2 & -3.168 & 14.006 - (\Delta\theta_1^2 + \Delta\theta_2^2) \end{bmatrix} > 0, \quad (5.8\text{-}20)$$

which in turn to satisfy this inequality we must let all of its principal minors to be greater than zero, resulting in the following set of inequalities.

$$0.18423 > 0, \quad (5.8\text{-}21\text{a})$$

$$\Delta\theta_1^2 + \Delta\theta_2^2 < 0.196, \quad (5.8\text{-}21\text{b})$$

$$\Delta\theta_1^2 + \Delta\theta_2^2 < 0.188, \quad (5.8\text{-}21\text{c})$$

$$(\Delta\theta_1^2 - \Delta\theta_2^2)^2 + 0.117 - 1.33(\Delta\theta_1^2 + \Delta\theta_2^2) + 1.316\Delta\theta_1\Delta\theta_2 > 0. \quad (5.8\text{-}21\text{d})$$

STEP 5 (To search for corner points): The final set of the admissible-parameter variations must satisfy the intersection of all these four inequalities, i.e., in this example we must have:

$$\Delta\theta_1^2 + \Delta\theta_2^2 < 0.188, \quad (5.8\text{-}22\text{a})$$

$$(\Delta\theta_1^2 - \Delta\theta_2^2)^2 + 0.117 - 1.33(\Delta\theta_1^2 + \Delta\theta_2^2) + 1.316\Delta\theta_1\Delta\theta_2 > 0. \quad (5.8\text{-}22\text{b})$$

From these inequalities we must extract $2^r = 2^2 = 4$ corner points which satisfy (5.8-22). Obviously and from (5.8-22a), we must search for the admissible-parameter variations which are in a circle of radius 0.43. Certainly, the following points which are directly computed from (5.8-22) are possible candidates for the corner points of the desired set:

$$O_1 \triangleq (\Delta\theta_1 = \Delta\theta_2 = 0), \quad A_1 \triangleq (\Delta\theta_1 = 0, \Delta\theta_2 = 0.295),$$

$$C_1 \triangleq (\Delta\theta_1 = \Delta\theta_2 = 0.295), \quad B_1 \triangleq (\Delta\theta_1 = 0.295, \Delta\theta_2 = 0). \quad (5.8\text{-}23)$$

Substituting any one of these values for parameter variations in (5.8-15) yields the corresponding ΔS's.

STEP 6 (To construct an admissible set): Connecting the above four corner points in (5.8-23), and because of Theorem 5.6-2, we have a set of admissible points in the *PVS* which is shown in Fig. 5.8-1. At this juncture it is imperative to recall that the above four corner points are generated from the *same* P_1, and all the requirements for Theorem 5.6-2 are satisfied, and therefore we are able to connect these four points in order to construct this set.

COMMENT 5.8-1: It is interesting to note that when we let either $\Delta\theta_1$ or $\Delta\theta_2$ in (5.8-22b) to be equal to zero then we get

$$(\Delta\theta_1^2)^2 + 0.117 - 1.33\Delta\theta_1^2 > 0. \tag{5.8-24}$$

This inequality is valid for values of $\Delta\theta_1^2 > 1.2352$ or $0 < \Delta\theta_1^2 < 0.0947$. Note that according to (5.8-18), $\Delta\theta_1 \leq 1$ for the stability requirements. Also note that these conditions suggest that some parameter variations may take place in both positive and negative directions of the parameter variations. Since we are only interested to demonstrate the constructional procedure (and since both $\Delta\theta_1$ and $\Delta\theta_2 > 0$, herein), we are only concentrating on points which are in the *positive* quadrant of the *PVS*. Furthermore, this is an interesting situation that shows there can be a jump between the allowable values for parameter variations, a situation which can easily cause difficulty if it is overlooked. However, since we study these robust constructional schemes starting from an asymptotically stable operating condition and *upward*, we always select points that are immediately connected to the nominal operating condition before perturbing the system to its extreme.

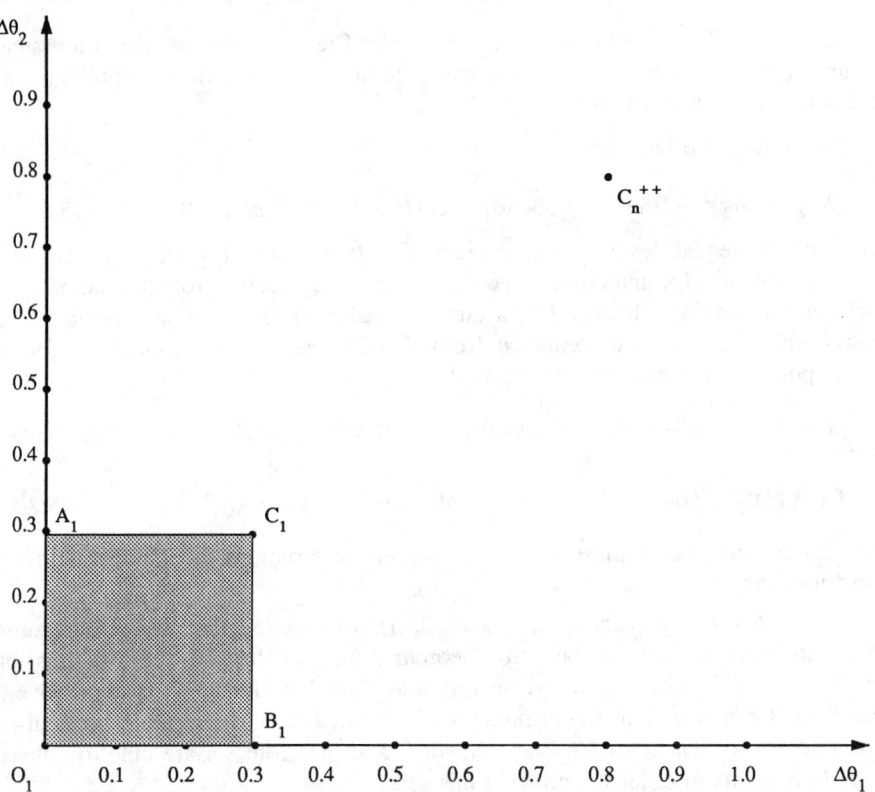

Fig. 5.8-1. Direct construction of an admissible set of parameter variations.

5.8. Numerical Examples

COMMENT 5.8-2: Clearly and as we expect the set of admissible points that is generated from the analytical procedures in Steps 1 to 6 is not the largest possible set. Because in Step 2 of this example, we have already established the fact that at least one other point, say, $C_n^{++} \triangleq (\Delta\theta_1 = \Delta\theta_2 = 0.8)$ results in a stable perturbed system. Thus we now search for the ways of expanding the initial set of admissible points in the *PVS* and that is the purpose of the next step.

Iteration: J+1

STEP 7 *(The construction of the largest or most extended set):* As in the case of single-parameter variations (Algorithm 5.6-1), we now propose to translate the coordinates of the *PVS* to the farthest corner point of the convex set, in order to construct more regions of admissible points in that space. That means we now start with $S_{J+1} = S_J \pm \Delta S_J$ (Step 0), where the corresponding ΔS_J is computed in Step 5, then we go back to Step 1. As we indicate in Comment 5.8-1, some of the parameter variations may indeed have negative signs in a given set of coordinates in the *PVS* and we must be alert of this fact, which is equivalent to reducing the value of one parameter variation, while changing the other one. We must consider this situation very carefully since the implication of this issue is that, as we proceed to translate the coordinates in different directions we must search for points in the preceding set in order to connect the two consecutive convex sets *as smoothly as* we possibly can, and to cover the entire allowable region of admissible points in the *PVS* as smoothly and completely as we can. This method of construction is based on the consequence of Theorem 5.6-2 which requires that the ΔS's corresponding to the 2^r corner points of each convex set in the *PVS* satisfy the *same* sufficient condition. In order to describe this procedure we continue to apply this method to the current example. We summarize our computations in the following and as we proceed we draw each corresponding subsets of admissible parameter variations in Fig. 5.8-2a.

At $J = 2$: From the result in Step 5 we translate the coordinates to point C_1. Therefore now $S_2 = S_1 + \Delta S_1$, where $\Delta S_1 = \Delta S(\Delta\theta_1 = \Delta\theta_2 = 0.295)$. For this S_2 and from Step 1 we have

$$P_2 = \begin{bmatrix} 3.926 & 1.271 & 0.809 & 0.271 \\ 1.271 & 3.487 & 0.599 & 0.266 \\ 0.809 & 0.599 & 5.116 & 1.164 \\ 0.271 & 0.266 & 1.164 & 0.541 \end{bmatrix}. \quad (5.8\text{-}25)$$

The sufficient condition for the asymptotic stability, i.e. $(P_2^{-2} - \Delta S \Delta S^T) > 0$, now becomes

$$\begin{bmatrix} 0.089 & -0.0007 & 0.021 & -0.175 \\ -0.0007 & 0.126 - (\Delta\theta_1^2 + \Delta\theta_2^2) & 0.105 & -0.503 + 2\Delta\theta_1\Delta\theta_2 \\ 0.021 & 0.105 & 0.818 & -3.336 \\ -0.175 & -0.503 + 2\Delta\theta_1\Delta\theta_2 & -3.336 & 14.332 - (\Delta\theta_1^2 + \Delta\theta_2^2) \end{bmatrix} > 0. \quad (5.8\text{-}26)$$

Note that in order to avoid introducing various new symbols or adding new subscripts or superscripts to $\Delta\theta_1$ and $\Delta\theta_2$ we use the same symbols for parameter variations in each new coordinate system or iteration. For (5.8-26) to be true we must let all of its principal minors to be greater than zero. This indeed requires that the new corner points in the *new PVS* satisfy the following two inequalities.

$$\Delta\theta_1^2 + \Delta\theta_2^2 < 0.112, \tag{5.8-27a}$$

$$(\Delta\theta_1^2 - \Delta\theta_2^2)^2 + 0.141 - 0.927(\Delta\theta_1^2 + \Delta\theta_2^2) + 0.303\Delta\theta_1\Delta\theta_2 > 0. \tag{5.8-27b}$$

We again let $\Delta\theta_1 = \Delta\theta_2$. Then we can show that for values such that $|\Delta\theta_1| = |\Delta\theta_2| \leq 0.236$ both of the above two inequalities hold. Therefore we select the following four new corner points which are shown in Fig. 5.8-2a:

$$C_2^{++} \triangleq (\Delta\theta_1 = \Delta\theta_2 = 0.236), \quad C_2^{-+} \triangleq (\Delta\theta_1 = -0.236, \Delta\theta_2 = 0.236),$$

$$C_2^{--} \triangleq (\Delta\theta_1 = \Delta\theta_2 = -0.236), \quad C_2^{+-} \triangleq (\Delta\theta_1 = 0.236, \Delta\theta_2 = -0.236). \tag{5.8-28}$$

At $J = 3$: Continuing in this fashion and from the above result we now translate the coordinates of the *PVS* to point C_2^{++}. Thus $S_3 = S_2 + \Delta S_2$, where $\Delta S_2 = \Delta S(\Delta\theta_1 = \Delta\theta_2 = 0.236)$. For this S_3 and from Step 1 we have P_3. The sufficient condition for the asymptotic stability, i.e. $(P_3^{-2} - \Delta S \Delta S^T) > 0$, yields

$$\begin{bmatrix} 0.053 & -0.016 & 0.005 & -0.146 \\ -0.016 & 0.054 - (\Delta\theta_1^2 + \Delta\theta_2^2) & 0.138 & -0.553 + 2\Delta\theta_1\Delta\theta_2 \\ 0.005 & 0.138 & 0.843 & -3.443 \\ -0.146 & -0.553 + 2\Delta\theta_1\Delta\theta_2 & -3.443 & 14.642 - (\Delta\theta_1^2 + \Delta\theta_2^2) \end{bmatrix} > 0,$$

$$\tag{5.8-29a}$$

$$\Delta\theta_1^2 + \Delta\theta_2^2 < 0.026, \tag{5.8-29b}$$

$$(\Delta\theta_1^2 - \Delta\theta_2^2)^2 + 0.0064 - 0.308(\Delta\theta_1^2 + \Delta\theta_2^2) + 0.117\Delta\theta_1\Delta\theta_2 > 0. \tag{5.8-29c}$$

If we again let $\Delta\theta_1 = \Delta\theta_2$, then we can show that for values such that $|\Delta\theta_1| = |\Delta\theta_2| \leq 0.113$ these two inequalities hold, and we can select the following four new corner points depicted in Fig. 5.8-2a:

$$C_3^{++} \triangleq (\Delta\theta_1 = \Delta\theta_2 = 0.113), \quad C_3^{-+} \triangleq (\Delta\theta_1 = -0.113, \Delta\theta_2 = 0.113),$$

$$C_3^{--} \triangleq (\Delta\theta_1 = \Delta\theta_2 = -0.113), \quad C_3^{+-} \triangleq (\Delta\theta_1 = 0.113, \Delta\theta_2 = -0.113). \tag{5.8-30}$$

We note that the constructional method is progressing along the "45° line" ($\Delta\theta_1 = \Delta\theta_2$) rather quickly. However, we want to know how to fill points on either sides of the previous line. Because, technically speaking, we cannot [yet] connect C_3^{-+} to C_2^{-+} *per se*.

At $J = 4$: Thus we now translate backward the coordinates from O_3 (at C_3^{++}) to O_4 (at C_2^{-+}). That means $S_4 = S_3 + \Delta S_3$, where

5.8. Numerical Examples

$\Delta S_3 = \Delta S(\Delta\theta_1 = -2\times 0.236, \Delta\theta_2 = 0.0)$. For this S_4 and from Step 1 we have P_4 and the sufficient condition for the asymptotic stability is $(P_4^{-2} - \Delta S \Delta S^T) > 0$, which yields

$$\begin{bmatrix} 0.197 & -0.115 & 0.0115 & -0.293 \\ -0.115 & 0.114 - (\Delta\theta_1^2 + \Delta\theta_2^2) & 0.152 & -0.516 + 2\Delta\theta_1\Delta\theta_2 \\ 0.0115 & 0.152 & 0.789 & -3.211 \\ -0.293 & -0.516 + 2\Delta\theta_1\Delta\theta_2 & -3.211 & 14.079 - (\Delta\theta_1^2 + \Delta\theta_2^2) \end{bmatrix} > 0,$$

(5.8-31a)

$$\Delta\theta_1^2 + \Delta\theta_2^2 < 0.01488, \tag{5.8-31b}$$

$$(\Delta\theta_1^2 - \Delta\theta_2^2)^2 + 0.00855 - 0.749(\Delta\theta_1^2 + \Delta\theta_2^2) + 0.176\Delta\theta_1\Delta\theta_2 > 0. \tag{5.8-31c}$$

We can show that the following four new corner points satisfy (5.8-31) (within our computational accuracies), and these points keep our parameter variations in the positive quadrant of the *PVS* as well. These corner points are also shown in Fig. 5.8-2a.

$$C_4^{++} \triangleq (\Delta\theta_1 = 0.059, \Delta\theta_2 = 0.07), \qquad C_4^{-+} \triangleq (\Delta\theta_1 = -0.059, \Delta\theta_2 = 0.07),$$

$$C_4^{--} \triangleq (\Delta\theta_1 = -0.059, \Delta\theta_2 = -0.07), \quad C_4^{+-} \triangleq (\Delta\theta_1 = 0.059, \Delta\theta_2 = -0.07).$$

(5.8-32)

Before we close the solution to this example we offer the following remarks.

REMARK 5.8-1: Now the message is clear. By this iterative approach we can construct the entire set of admissible-parameter variations in the *PVS* such that for each point in this set the corresponding ΔS yields an asymptotically stable $S + \Delta S$. Thus we have a class of asymptotically stable systems surrounding the initial system and indeed that is the true meaning of stability robustness analysis and it is a corresponding response to the associated fundamental question which is raised in Problem 5.4-1. This set, which can be constructed in order to become *the largest* set of admissible-parameter variations or the stability robustness measure *(SRM)* of the system, is shown in Fig. 5.8-2b. Clearly, it seems that by increasing the iteration number and in a few different directions, we can smooth the boundary of the previous set (Fig. 5.8-2b). However there are a few short cuts that we can undertake. For example, the perturbed system corresponding to the nonshaded set between the two sets O_4 and O_1 shown in Figs. 5.8-2a & b, results in an asymptotically stable perturbed system, because of the convexity of perturbation matrices and the fact that this set is surrounded by four other convex stable sets (including the vertical axis). The same can be said for the other boundary sets in that figure. Indeed we may conclude that if we continue the construction along the "45° line", $\Delta\theta_1 = \Delta\theta_2 (= \cdots = \Delta\theta_r)$, in order to reach the point C_n^{++} (here at $\Delta\theta_1 = \Delta\theta_2 = 0.8$), then the entire square box whose farthest corner point is C_n^{++} yields an asymptotically stable matrix provided that each one of the single-parameter variations along the two axes is outside of this box (which is, of course, the case in this example). Because in this square the perturbation matrix remains a convex matrix and that is why it is important to study the type of a perturbation matrix before making any attempt to construct the corresponding *E-Box*. Also since this C_n^{++} is computed as a single-parameter variation, it is unique and it can be evaluated in a number of different ways. Clearly, if the extreme single-parameter variation for each axes is inside of the above square box (or hyperbox), then we need to compress that box accordingly.

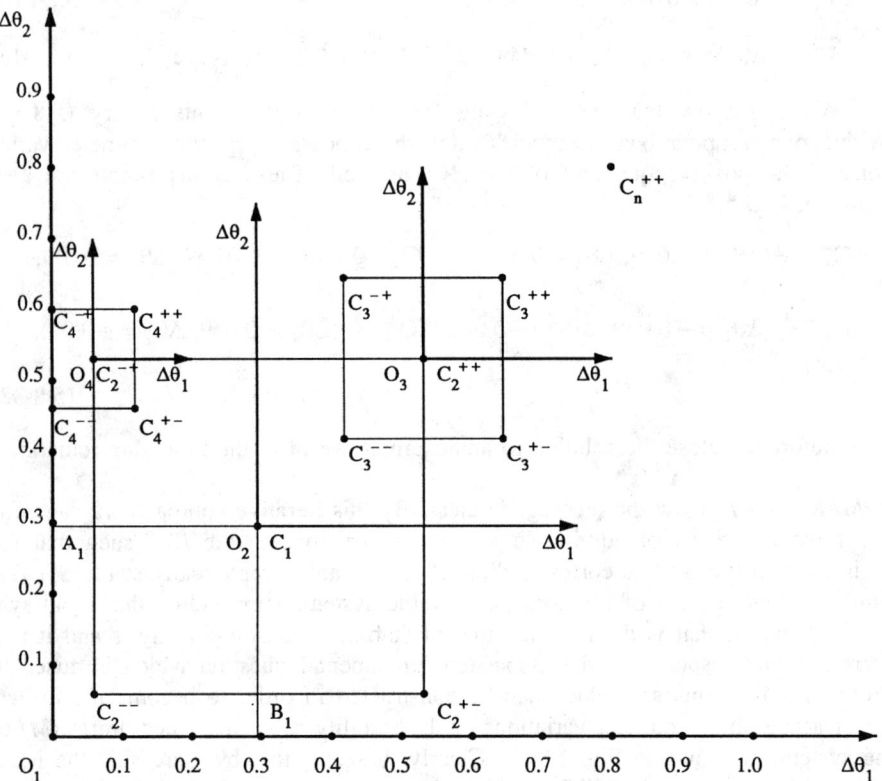

Fig. 5.8-2a. Toward the construction of the *largest* set of parameter variations.

REMARK 5.8-2: As mentioned in Step 2, the real issue in constructing an *E-Box*, or its best estimate *SRM*, is how to connect C_n^{++} to A, or C_n^{++} to B, in (5.8-18). Since to the best of our knowledge, there is no theorem which supports the stability of any perturbed matrix when we connect by a straight line (in order to maintain the convexity) (although that is very conservative) the two points C_n^{++} to A or C_n^{++} to B (in other words, the stability of the perturbed matrix may break down

5.8. Numerical Examples

along this line), we must appeal to this constructional method in order to *expedite* generating the actual *SRM* of the system. However, if we can find the appropriate stable paths between C_n^{++} to A, or C_n^{++} to B, then we can close an area in the *PVS* (considering the two axes and the "45° line") which yield asymptotically stable $(S + \Delta S)$'s. From the construction of this closed path and the convexity of perturbed system inside of this closed path, we can immediately conclude that for all the points inside of this closed path or set, the perturbed systems remain asymptotically stable. From these thoughts we conclude that in general when we have unidirectionally found the extreme points A, B, C_n^{++}, *etc.*, our computational efforts must be centered on finding stable paths (or sets) connecting these extreme points, in order to construct the *SRM* and ultimately the *E-Box*.

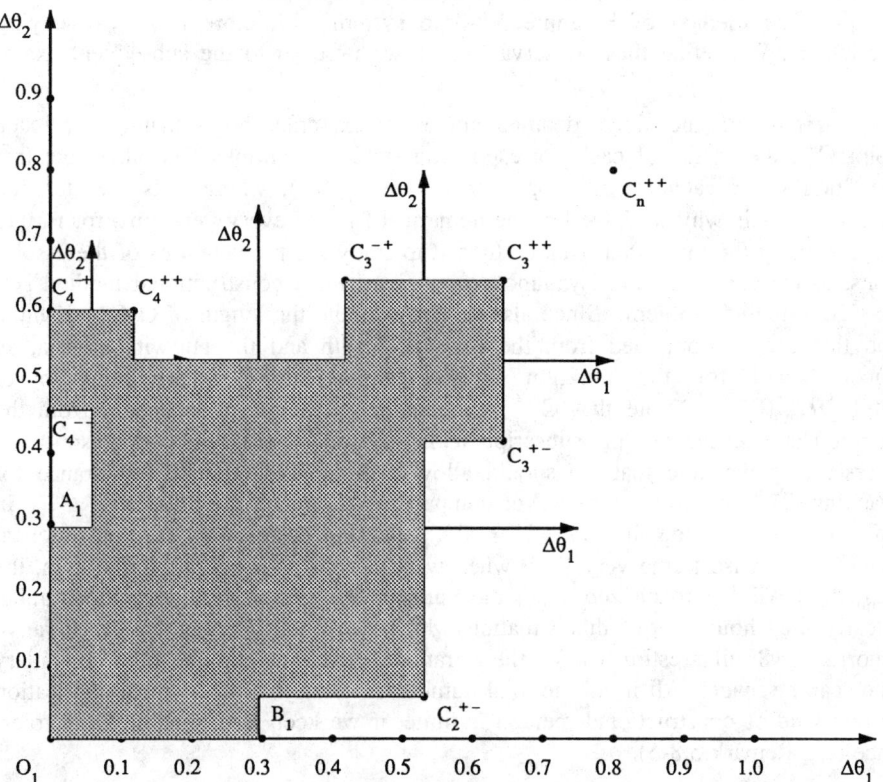

Fig. 5.8-2b. Toward the construction of the *largest* set of parameter variations.

OBSERVATION 5.8-3: It is now clear that when construction of our *SRM* (or ultimately the *E-Box*) in the *PVS* is completed, there are infinitely many paths in this set whose corresponding perturbed systems are stable. Each and every one of these paths corresponds to a different stability robustness measure that is reported in the literature.

OBSERVATION 5.8-4: For a *truly* robust design we must then find the intersection of the preceding set (or Fig. 5.8-2b) *(the E-Box* or its best estimate the *SRM)*, with the set that will be resulted from other design criteria or considerations, in order to get the final set of points in the *PVS* (or *PS*) whose corresponding perturbed systems are *robust* with respect to stability as well as other design objectives (cf., Examples 5.11-1 & 2).

Finally we should point out if we continue the above iterative method in Example 5.8-3 we reach a trapezoidal-like set that represents the largest set of parameter variations or *SRM* for this example. The other two extreme corner points, in addition to the origin at O_1, and C_n^{++}, are at ($\Delta\theta_1 = 1.0$, $\Delta\theta_2 = 0.0$) and ($\Delta\theta_1 = 0.0$, $\Delta\theta_2 = 2.06$) (cf., Fig. 5.11-2).

▼

REMARK 5.8-3: There are a few other observations which we must make in order to apply the method of Example 5.8-3 to system with more than two varying parameters. We refine these observations as we proceed to the subsequent examples.

First of all, the above detailed numerical example shows that as we reach point C_n^{++} the "area" of each corresponding stable set shrinks. We also note that the "radius" of each set is proportional to $\lambda_{min}(P_J^{-2})$, where J is the iteration number. This is why we have left the numerical P_J^{-2}, in every iterations, for further inspection by the interested reader. Indeed to study the propagations of these solutions, to the corresponding Lyapunov equations, in this constructional method is a very informative problem. Since also in this example the length of $O_1C_n^{++}$ is finite and this can be computed from the classical Routh and the Hurwitz criteria, an upper bound for the iteration number along this line is the ratio of $O_1C_n^{++}/\lambda_{min}(P_N^{-2})$. (Note that C_n^{++} is a fixed point and is independent from the iteration number N, thus the subscript 'n' is used only for a symbolic reason.) Of course, here we note that we should allow a reasonable numerical tolerance for reaching C_n^{++}. Since we may never computationally reach the theoretical C_n^{++} (in this example we stop at $0.797481 < 0.8$), and P_N corresponds to this practical value. Otherwise at the very limit when we can no longer perturb the system, the $\lambda_{min}(P_N^{-2})$ will approach zero, i.e., the number of iterations will become infinite. Clearly we should avoid this situation which can occur in any typical iterative algorithm. Similar estimates for the iteration numbers along the other boundary lines can be made. All in all the total number of iterations for a typical application in this kind of constructional method is finite, if we keep a reasonable finite tolerance (cf., Remark 5.8-5).

Secondly, we recall that we open Example 5.8-3 with an assumption that $\Delta\theta_j > 0$ for all j. Otherwise, when we complete our search in the first quadrant of the *PVS*, we should repeat this search in the next quadrant of that space.

Thirdly, we recall that in this example we have only two parameters. In general we have r parameters and we must find all the extreme points first. Then we

5.8. Numerical Examples

must look for ways of connecting these points such that on each closed path (set) the system remains stable. We note that in general when r > 2, then the "45° lines" must be further studied with their corresponding $\{C_n^{++}\}$'s. When we reach each one of the $\{C_n^{++}\}$'s, then we must search for unidirectional extreme points in each remaining sub-spaces of the *PVS*. Recall that we are looking for the global solution of our problem.

COMMENT 5.8-3: To construct inequalities such as (5.8-22a) and (5.8-22b) in a typical application is a nontrivial task involving cumbersome algebra. What we propose as an alternative method for these situations is as follows. We initially make a numerical guess for parameter variations $\Delta\theta_1$ and $\Delta\theta_2$, etc., and substitute these values in ΔS. This substitution makes the parametric ΔS to become a numerically defined perturbation matrix and we call that $\Delta S'$ which is a *candidate* for ΔS. Then we let $\Delta S = \varepsilon \Delta S'$ and look for an ε such that ΔS becomes an *admissible* or a *directional* ΔS. When this ε is found, all the previous guesses are scaled accordingly and we get one directional line or a set of points in the *PVS*. Clearly, we use the same set of $|\Delta\theta_i|$'s (corresponding to ΔS) in different quadrants of the *PVS* and check the corresponding ε_i's. Then using $\varepsilon = \min\{\varepsilon_1, \cdots, \varepsilon_{2^r}\}$ gives us a hyperbox which is stable. We use the farthest corner point of this hyperbox to construct the next hyperbox in the *PVS* whose corresponding perturbed systems are stable.

For example, at $J = 1$, we note that according to the preceding analytical analysis, none of the admissible parameter variations should be outside a circle with radius 0.43 (cf., (5.8-22a)). Suppose that we do not have this information and we assume that $\Delta\theta_1 = \Delta\theta_2 = 0.5$. Then

$$\Delta S \triangleq \varepsilon \Delta S' = \varepsilon \begin{bmatrix} 0 & 0 & 0 & 0 \\ 0 & 0.5 & -0.5 & 0 \\ 0 & 0 & 0 & 0 \\ 0 & -0.5 & 0.5 & 0 \end{bmatrix}. \qquad (5.8\text{-}33)$$

Since we have already computed P_1^{-2}, our sufficient condition becomes

$$\varepsilon^2 \begin{bmatrix} 0 & 0 & 0 & 0 \\ 0 & 0.5 & 0 & -0.5 \\ 0 & 0 & 0 & 0 \\ 0 & -0.5 & 0 & 0.5 \end{bmatrix} < P_1^{-2}. \qquad (5.8\text{-}34)$$

Letting $\varepsilon^2 \text{Tr}(\Delta S' \Delta S'^T) = \varepsilon^2 < \lambda_{\min}(P_1^{-2})$, yields $\varepsilon \leq 0.095$ (this is resulted from a conservative estimate of the minimum eigenvalue), which this in turn yields a point $C_1^* \triangleq (\Delta\theta_1 = \Delta\theta_2 = 0.048)$ (corresponding to C_1). Of course, the above values are very conservative since we have made no attempt to search for an improved ε in this way, and indeed we have only tried to demonstrate the procedure with the minimum computational effort. The remaining steps are similar to the analytical computations.

In practice, we often face a numerically described matrix (corresponding to a *DCLTIS*), that we have no *prior* knowledge about the exact interrelationships among various entries of this matrix. Based on this scenario we want to investigate how to perturb each and/or every entries of this matrix before the system becomes unstable. To address this problem we study the next example which complements the preceding one.

EXAMPLE 5.8-4 *(Directional perturbations/multiple-parameter variations)*: Using the same numerical values as in [Example 5.2 of *B*. 35], consider the stable matrix $S = \begin{pmatrix} -2 & -1 \\ 1 & 0 \end{pmatrix}$ and the corresponding $P = \begin{pmatrix} 1 & 1 \\ 1 & 3 \end{pmatrix}$, which is generated from $S^T P + PS + 2I = 0$. Suppose that we perturb S in three different *directions* according to:

$$\Delta S_\alpha = \varepsilon_\alpha \begin{bmatrix} 0.8 & 0 \\ 0 & 0 \end{bmatrix}, \ \Delta S_\beta = \varepsilon_\beta \begin{bmatrix} 0.2 & -0.5 \\ 0 & 0 \end{bmatrix}, \text{ and } \Delta S_\gamma = \varepsilon_\gamma \begin{bmatrix} -1 & 0.5 \\ 0 & 0 \end{bmatrix}. \quad (5.8\text{-}35)$$

For $\varepsilon_\alpha = \varepsilon_\beta = \varepsilon_\gamma = 1$, (5.8-35) becomes the same as E_1, E_2, and E_3 in [*B*. 35]. Then there are a few important issues which deserve clarifications for the results that are developed in [*B*. 35], as well as future applications of the results developed herein. *(i)* Why are we initially choosing ΔS_α, ΔS_β and ΔS_γ (or E_1, E_2, and E_3 in [*B*. 35]), and what are the significances of these choices, if any? *(ii)* Suppose that we find *the largest* directional perturbations on these lines, then what about the sets of points in the *PVS* which are sandwiched between these lines in the sense that whether their corresponding perturbed systems are stable or not? *(iii)* How do we connect the corner points at the end of these directional perturbations? Finally *(iv)*, how close is the set of points in the *PVS* that we have so far generated to the actual *E-Box* or its best estimate in this example?

SOLUTION [*B*. 29]: Let $J = 1, 2, ..., N$ be the iteration number.

STEP 0 *(To update S)*: We let $S_J = S_{J-1} \pm \Delta S_{J-1}$ with $S_1 \equiv S_0 \equiv S$ the initial asymptotically stable matrix and $\Delta S_0 \equiv 0$.

STEP 1 *(To solve the Lyapunov equation)*: For the current S_J we solve for P_J according to $S_J^T P_J + P_J S_J + 2I = 0$. (At $J = 1$, we may also solve for the other Lyapunov equation in Theorem 5.5-1, in order to choose the best perturbation policy for the underlying problem. In that case the following steps must be changed accordingly).

STEP 2 *(Directional perturbations)*: *First* we must specify r, the total number of independently varying parameters. Since $S_1 \in R^{2 \times 2}$ is given numerically in [*B*. 35], we may speculate that there are $r = 4 \ (= 2 \times 2)$ independently varying parameters. Thus a typical hyperbox in this example has $2^r = 2^4 = 16$ corners. Since we have initially changed s_{11} and s_{12}, in (5.8-35), we may assume that these two parameters are changing first. *Secondly*, we must give directional perturbations along these $r \ (= 4)$ axes of the *PVS* and perhaps a few "45° lines" in each quadrant of this space. This information is needed to initiate the construction. This computation is summarized in Table 5.8-2, and the actual computations are done according to the Algorithm 5.6-1.

5.8. Numerical Examples

Before proceeding further we can show that for the directional perturbations given in (5.8-35) and using Algorithm 5.6-1 we have:

$S_1 + \Delta S_\alpha$ is stable for $\quad \varepsilon_\alpha < 2.5,$ (5.8-36a)

$S_1 + \Delta S_\beta$ is stable for $\quad -2 < \varepsilon_\beta < 10,$ (5.8-36b)

$S_1 + \Delta S_\gamma$ is stable for $\quad -2 < \varepsilon_\gamma < 2.$ (5.8-36c)

In other words, when S_1 is perturbed in the above three directions, then $S_1 + \Delta S_{\alpha \text{ or } \beta \text{ or } \gamma}$ remains stable for a larger directional interval than that proposed in [B. 35]. These directional results are shown in Fig. 5.8-3, and because of this figure we are convinced that the questions which are raised in this example merit further study. For instance, the results that are reported in this book not only do explicitly support the fact that the system remains stable for every points from the origin O_1 (the nominal value) to E_3, *per se*, but also these results show that the system actually remains stable on this direction for as far as ΔS_γ. In other words, the system remains stable for a larger class of perturbation matrices than that resulted from only three single points E_1, E_2, and E_3, in the *PVS*. As we continue to generate these perturbation matrices or their corresponding sets of points in the *PVS*, it becomes very clear that this constructional method which is supported by Theorem 5.6-2 (this theorem requires the convexity of perturbation matrices) will substantially enlarge the initial set of points in the *PVS* relative to a number of methods that are proposed in the literature.

Particularly, in regard to the results reported in [B. 35], we should point out that: *First of all,* as shown in Fig. 5.8-3, for example, for a point in this space which corresponds to $1 < \varepsilon_\alpha = 1.04 < 2.5$, the suggested matrix-measure $\mu_\alpha(M_\alpha = \Delta S_\alpha^T P_1 + P_1 \Delta S_\alpha)$ is equal to 2.0084 which is greater than 2 that is required in [B. 35] for the acceptability of a perturbation matrix, while $S + \Delta S_\alpha (\varepsilon_\alpha = 1.04)$ is stable, similarly for (5.8-36b) and (5.8-36c). *Secondly,* we should point out that the initial P_1 does not correspond to the Lyapunov equation $(S_1 + \Delta S_\alpha)^T P_1 + P_1(S_1 + \Delta S_\alpha) + 2I = 0$ for the *entire* interval of ε_α, similarly for (5.8-36b) and (5.8-36c). Because when $S_1 \to S_1 + \Delta S_{\alpha_1}$, then according to Theorem 5.5-1 (and for $S^T P + PS + 2I = 0$), the perturbation matrices should generally satisfy $\Delta S_{\alpha_j} \Delta S_{\alpha_j}^T < P_J^{-2}$. In other words, the perturbation matrix which is generated from the initial P_1 is within certain (matter of speak) "radius" of the initial or the nominal S_1. If we go beyond this "radius", then this P_1 will not remain valid for the perturbed matrix. But that does not mean that we cannot perturb the initial matrix. On the contrary we can perturb the previous $S + \Delta S_{\alpha_1}$, *per se*, by looking for a new P_2 from the Lyapunov equation for $S + \Delta S_{\alpha_1}$, and subsequently for a new ΔS_{α_2} from this P_2 according to Theorem 5.5-1 such that $S + \Delta S_{\alpha_1} + \Delta S_{\alpha_2}$ remains stable, as this method is used to generate (5.8-36a) to (5.8-36c). This iterative method that seems so true now was first reported in [B. 23]. The thought behind this method was that any physical property of a dynamical system, such as stability, is independent from the method of measurement and thus the system *SRM* must not depend on Q in a Lyapunov setting. We should also mention that if we change Q in $S^T P + PS + Q = 0$ from $2I$ to another $Q = Q^T > 0$, in order to *maintain* P, then that approach will soon breakdown and we cannot use a viable iterative method to construct the perturbation matrices for S which is really the essence of our *SRA*, because in the final analysis we are only interested to develop the corresponding *SRM* of our dynamical system.

STEP 3 (To construct $\Delta S \Delta S^T$): Given any structured ΔS we compute $\Delta S \Delta S^T$ that will remain the same throughout the analyses. (This is why we prefer to work with (5.5-1)).

STEP 4 (To interpret the sufficient condition): Using $\Delta S \Delta S^T < P_J^{-2}$ we need to establish a set of inequalities whose intersection yields a sufficient condition for the parameter variations. This is done by letting all the principal minors of $(P_J^{-2} - \Delta S \Delta S^T)$ to be greater than zero and choosing their intersection. Or more conservatively let $\text{Tr}(\Delta S \Delta S^T) < \lambda_{\min}(P_J^{-2})$.

STEP 5 (To search for the corner points or vertexes): The result from the preceding step provides us with a perturbation policy which is used to find, in principle, 2^r corresponding vertexes.

STEP 6 (To construct an admissible set): Connecting the vertexes found in the previous step and provided that Theorem 5.6-2 is satisfied, gives us a set of admissible parameter variations (or a hyperbox) whose corresponding ΔS's are stable.

STEP 7 (The construction of the most extended set): By translating the coordinates of the *PVS* to the corner vertexes of each admissible stable set (i.e., letting $J \to J + 1$) will enable us to extend the regions or sets of admissible points in that space.

We now continue to apply the above steps to the system of Example 5.8-4.

CASE I: If we only change s_{11} and s_{12}, then a conservative policy to generate sets of points in the *PVS* and the corresponding iterative method for this construction is partially developed and summarized in Table 5.8-3. We note that at each iteration we have translated the coordinates of the *PVS* to the given origin in the left column of Table 5.8-3 and all the values in the right column of that table are relative to the corresponding origin. These results are shown in Figs. 5.8-4a & b.

CASE II: If we now let all four entries of the initial matrix change simultaneously, then the previous analyses hold and the corresponding conservative perturbation policy becomes

$$\text{Tr}(\Delta S \Delta S^T) = \Delta s_{11}^2 + \Delta s_{12}^2 + \Delta s_{21}^2 + \Delta s_{22}^2 < \lambda_{\min}(P_1^{-2}) = 0.085786. \tag{5.8-37}$$

From this set we choose the following 16 corner points and draw the 1/16th of the corresponding hyperbox in the first quadrant of our *PVS* as shown in Fig. 5.8-5:

5.8. Numerical Examples

$$C_1^{++} \triangleq (\Delta s_{11} = \Delta s_{12} = \Delta s_{21} = \Delta s_{22} = 0.1464)$$

...

$$C_1^{--} \triangleq (\Delta s_{11} = \Delta s_{12} = \Delta s_{21} = \Delta s_{22} = -0.1464). \tag{5.8-38}$$

Comparing *appropriately* the first iteration in Fig. 5.8-4a with that of Fig. 5.8-5, reveals that when the number of varying parameters increases, then the corresponding *SRM* decreases. In other words, the common areas or the intersections of the sets for parameter variations as we expect shrinks in order to reflect our conservatism. Here we can certainly apply the same iterative method which we use in *Case I* to enlarge this initial set in *Case II* (cf., Example 5.8-6 for a similar enlargement). We also recall that in *Case II*, we assume that all the four entries of S *may* change independently and simultaneously.

To answer explicitly the questions which are raised in this example, we offer the following: *(i)* The initial choices for ΔS_α, ΔS_β and ΔS_γ really do not constitute any significance in this process and as Fig. 5.8-5b shows we could have chosen none. *(ii)* The sets of points which are sandwiched between these lines are acceptable if they can be included in those sets which we have, or we could have, generated. Here we need more computations to cover more regions, but there are a few short cuts for filling the entire allowable region as described in the previous example. *(iii)* To answer this question we note that in general given two points A and B in the *PVS* whose corresponding stable matrices are S_A and S_B, we cannot say that $S_{A \to B}$ remains stable as we traverse from A to B in the *PVS*. Thus we cannot easily answer this question, because the answer depends on the encountered problem and we cannot draw a straight line (although in many cases that may be a very conservative approach) between each two corner or end points. Indeed, because of this major difficulty, we really need an efficient computational procedure such as Theorem 5.6-2 to generate the *E-Box* or its best estimate *(SRM)*, or a subset of that in any given system. Thus the importance of Theorem 5.6-2 which generates *set by set of* points in the *PVS* now becomes evident. This theorem is responsible for a systematic construction of the *SRM* which enables us to answer accordingly the preceding questions in any given problem. Finally *(iv)* Figs. 5.8-5b & 6 are only a subset of the corresponding *E-Box*'s and naturally here the computations increase as we refine these figures or we continue to *probe* the *SRM* of our system. In any event, not only can we substantially *expedite* the construction of stability region for a given matrix relative to most methods reported in the literature, but also we can do that *systematically* and that direct approach is the main contribution of the method which is used to study this example.

CASE III: In the case of dependent parameters in which there exists a function, say, $g(\Delta s_{11}, ..., \Delta s_{22}) = 0$, we must use the intersection of the previous *E-Box* (or its best estimate) with this hypersurface (cf., Fig. 5.4-3). Since there are infinitely many possibilities regarding these dependencies, this issue must be dealt with individually.

Table 5.8-2. Directional Perturbations and the "45° lines" for

$$S = \begin{pmatrix} -2 & -1 \\ 1 & 0 \end{pmatrix}.$$

Perturbed matrices	The characteristic equations and the directional stability bounds
$\lambda I - S - \Delta S_I = \begin{pmatrix} \lambda+2-\Delta s_{11} & 1 \\ -1 & \lambda \end{pmatrix}$	$\lambda^2 + (2-\Delta s_{11})\lambda + 1 = 0$ is stable if $\Delta s_{11} < 2$
$\lambda I - S - \Delta S_{II} = \begin{pmatrix} \lambda+2 & 1-\Delta s_{12} \\ -1 & \lambda \end{pmatrix}$	$\lambda^2 + 2\lambda + 1 - \Delta s_{12} = 0$ is stable if $\Delta s_{12} < 1$
$\lambda I - S - \Delta S_{III} = \begin{pmatrix} \lambda+2 & 1 \\ -1-\Delta s_{21} & \lambda \end{pmatrix}$	$\lambda^2 + 2\lambda + 1 + \Delta s_{21} = 0$ is stable if $\Delta s_{21} > -1$
$\lambda I - S - \Delta S_{IV} = \begin{pmatrix} \lambda+2 & 1 \\ -1 & \lambda-\Delta s_{22} \end{pmatrix}$	$\lambda^2 + (2-\Delta s_{22})\lambda + 1 - 2\Delta s_{22} = 0$ is stable if $\Delta s_{22} < \frac{1}{2}$
$\lambda I - S - \Delta S_{45° \text{ line}} = \begin{pmatrix} \lambda+2-\varepsilon & 1-\varepsilon \\ -1-\varepsilon & \lambda-\varepsilon \end{pmatrix}$	$\lambda^2 + 2(1-\varepsilon)\lambda + 1 - 2\varepsilon = 0$ is stable if $\varepsilon < \frac{1}{2}$ shown by C_{n4}^{++}
$\lambda I - S - \Delta S_{\text{local } 45° \text{ line}} = \begin{pmatrix} \lambda+2-\delta & 1-\delta \\ -1 & \lambda \end{pmatrix}$	$\lambda^2 + (2-\delta)\lambda + 1 - \delta = 0$ is stable if $\delta < 1$ shown by C_{n2}^{++}

5.8. Numerical Examples

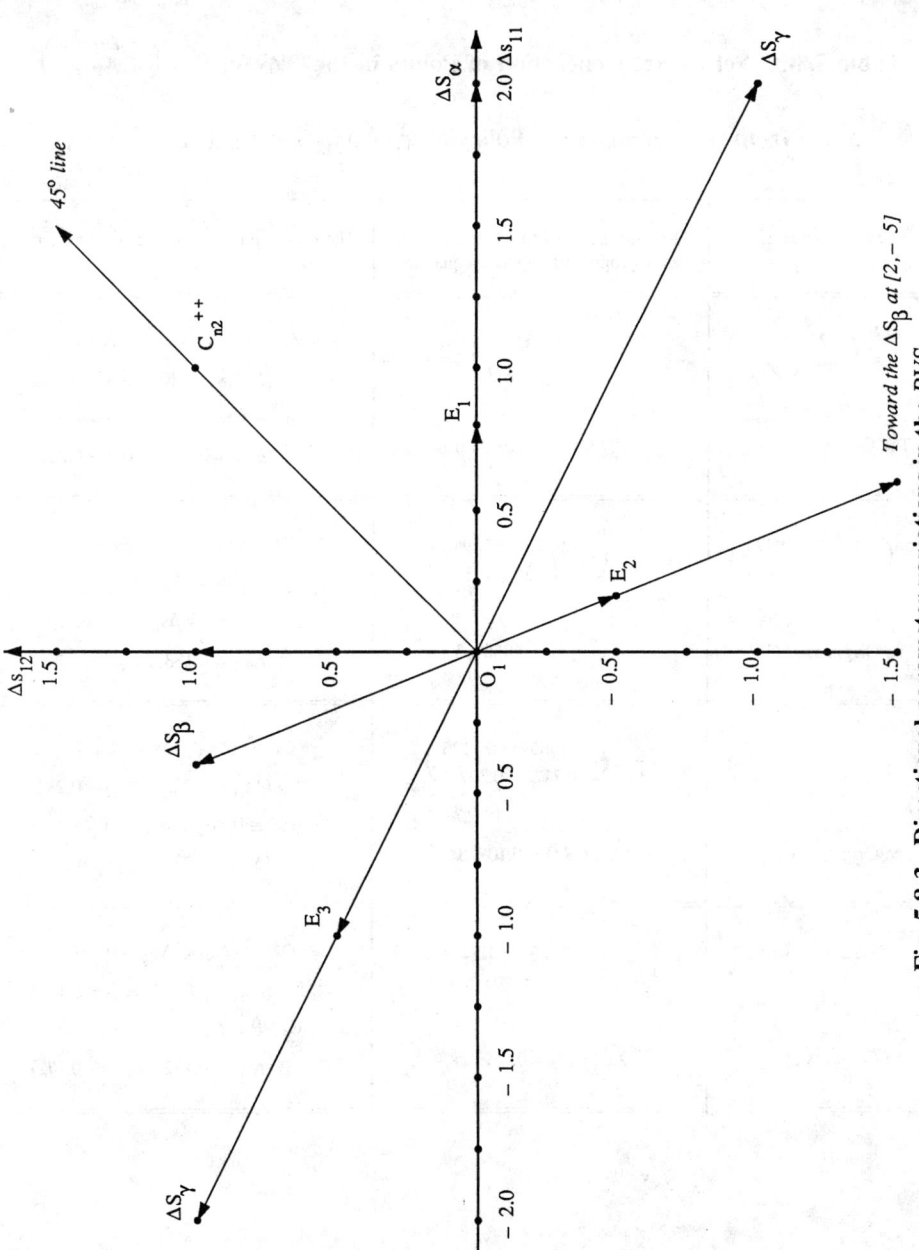

Fig. 5.8-3. Directional parameter variations in the *PVS*.

Table 5.8-3. Set by Set Generation of Points in the *PVS* for $S = \begin{pmatrix} -2 & -1 \\ 1 & 0 \end{pmatrix}$ from the Perturbation Policy $(\Delta s_{11}^2 + \Delta s_{12}^2)_J < \lambda_{\min}(P_J^{-2})$.

Perturbed matrices	The square inverse of the solution to an appropriate Lyapunov equation	The corner points of stable sets in the *PVS*
$S_1 = \begin{pmatrix} -2 & -1 \\ 1 & 0 \end{pmatrix}$ The Origin is at O_1	$P_1^{-2} = \begin{pmatrix} 2.5 & -1 \\ -1 & 0.5 \end{pmatrix}$ $\lambda_{\min}(P_1^{-2}) = 0.085786$	$C_1^{++} \triangleq (\Delta s_{11} = \Delta s_{12} = 0.207)$ $C_1^{-+} \triangleq (\Delta s_{11} = -0.207, \Delta s_{12} = 0.207)$ $C_1^{--} \triangleq (\Delta s_{11} = \Delta s_{12} = -0.207)$ $C_1^{+-} \triangleq (\Delta s_{11} = 0.207, \Delta s_{12} = -0.207)$
$S_2 = \begin{pmatrix} -1.793 & -0.793 \\ 1 & 0 \end{pmatrix}$ The Origin is at C_1^{++}	$P_2^{-2} = \begin{pmatrix} 1.922 & -0.8965 \\ -0.8965 & 0.5 \end{pmatrix}$ $\lambda_{\min}(P_2^{-2}) = 0.066783$	$C_2^{++} \triangleq (\Delta s_{11} = \Delta s_{12} = 0.183)$ $C_2^{-+} \triangleq (\Delta s_{11} = -0.183, \Delta s_{12} = 0.183)$ $C_2^{--} \triangleq (\Delta s_{11} = \Delta s_{12} = -0.183)$ $C_2^{+-} \triangleq (\Delta s_{11} = 0.183, \Delta s_{12} = -0.183)$
$S_3 = \begin{pmatrix} -1.793 & -1.207 \\ 1 & 0 \end{pmatrix}$ The Origin is at C_1^{+-}	$P_3^{-2} = \begin{pmatrix} 1.8565 & -0.7125 \\ -0.7125 & 0.3975 \end{pmatrix}$ $\lambda_{\min}(P_3^{-2}) = 0.10728$	$C_3^{++} \triangleq (\Delta s_{11} = \Delta s_{12} = 0.232)$ $C_3^{-+} \triangleq (\Delta s_{11} = -0.232, \Delta s_{12} = 0.232)$ $C_3^{--} \triangleq (\Delta s_{11} = \Delta s_{12} = -0.232)$ $C_3^{+-} \triangleq (\Delta s_{11} = 0.232, \Delta s_{12} = -0.232)$
$S_4 = \begin{pmatrix} -2.207 & -0.793 \\ 1 & 0 \end{pmatrix}$ The Origin is at C_1^{-+}	$P_4^{-2} = \begin{pmatrix} 3.3125 & -1.3293 \\ -1.3293 & 0.6023 \end{pmatrix}$ $\lambda_{\min}(P_4^{-2}) = 0.059155$	$C_4^{++} \triangleq (\Delta s_{11} = \Delta s_{12} = 0.172)$ $C_4^{-+} \triangleq (\Delta s_{11} = -0.172, \Delta s_{12} = 0.172)$ $C_4^{--} \triangleq (\Delta s_{11} = \Delta s_{12} = -0.172)$ $C_4^{+-} \triangleq (\Delta s_{11} = 0.172, \Delta s_{12} = -0.172)$

5.8. Numerical Examples

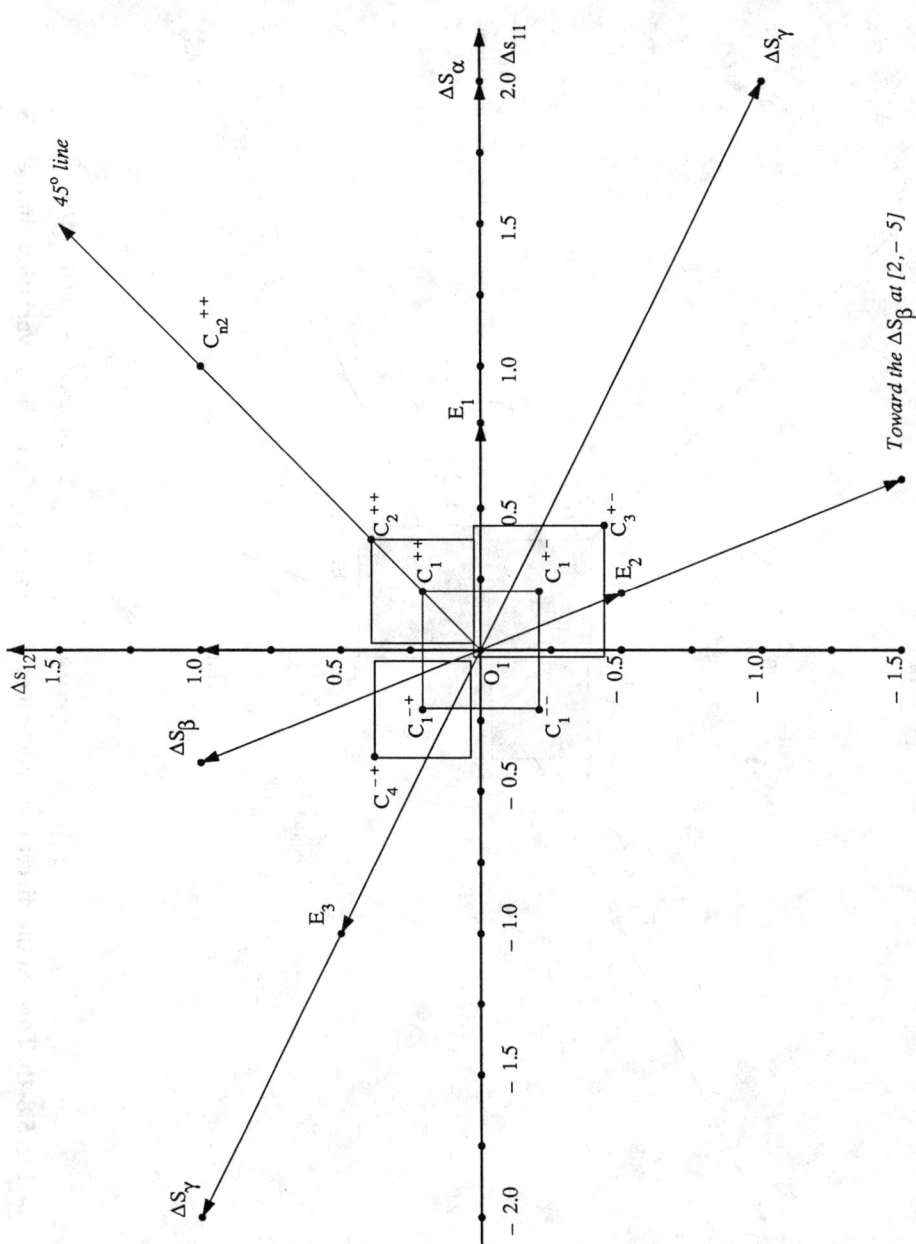

Fig. 5.8-4a. Direct construction of admissible sets of parameter variations in the *PVS*.

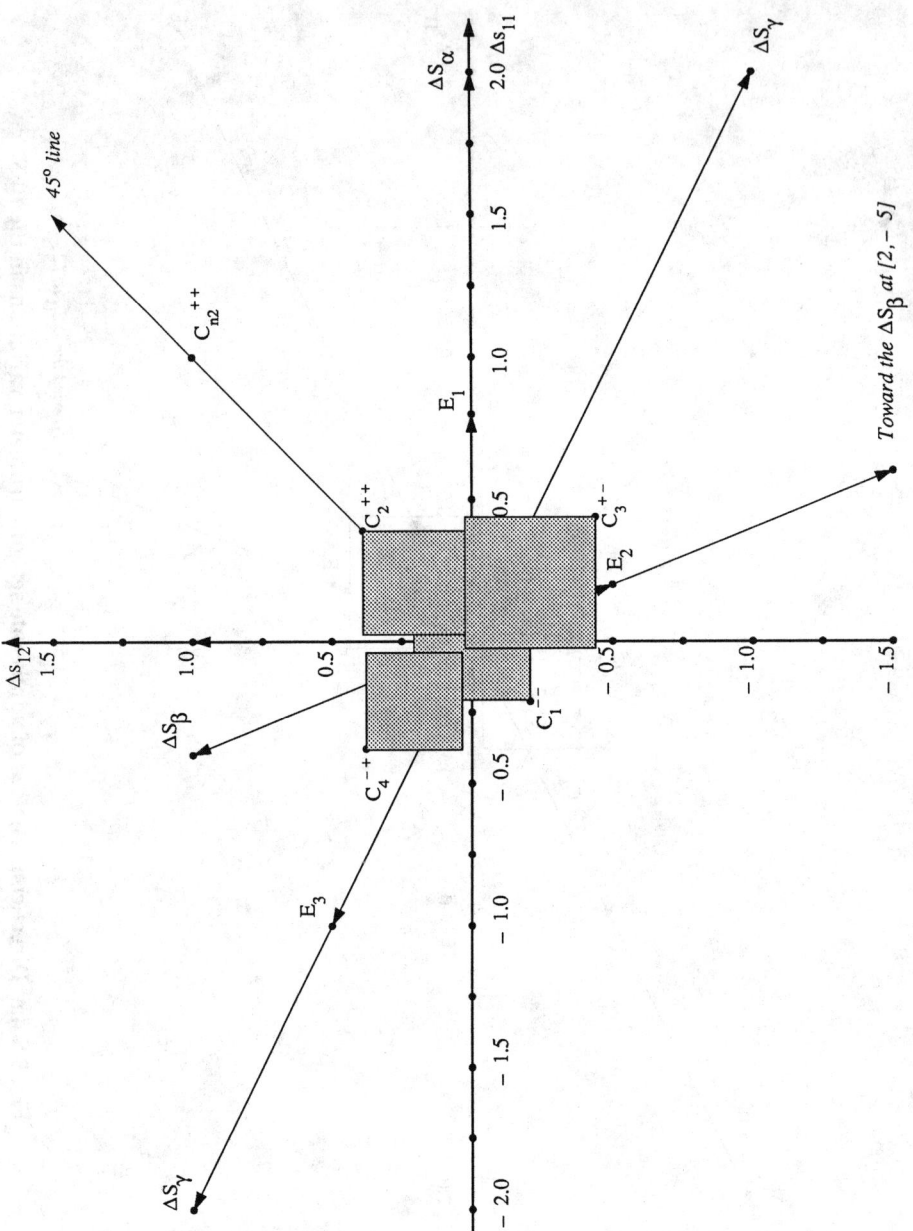

Fig. 5.8-4b. Toward the direct construction of *the largest* set of parameter variations in the *PVS*.

5.8. Numerical Examples

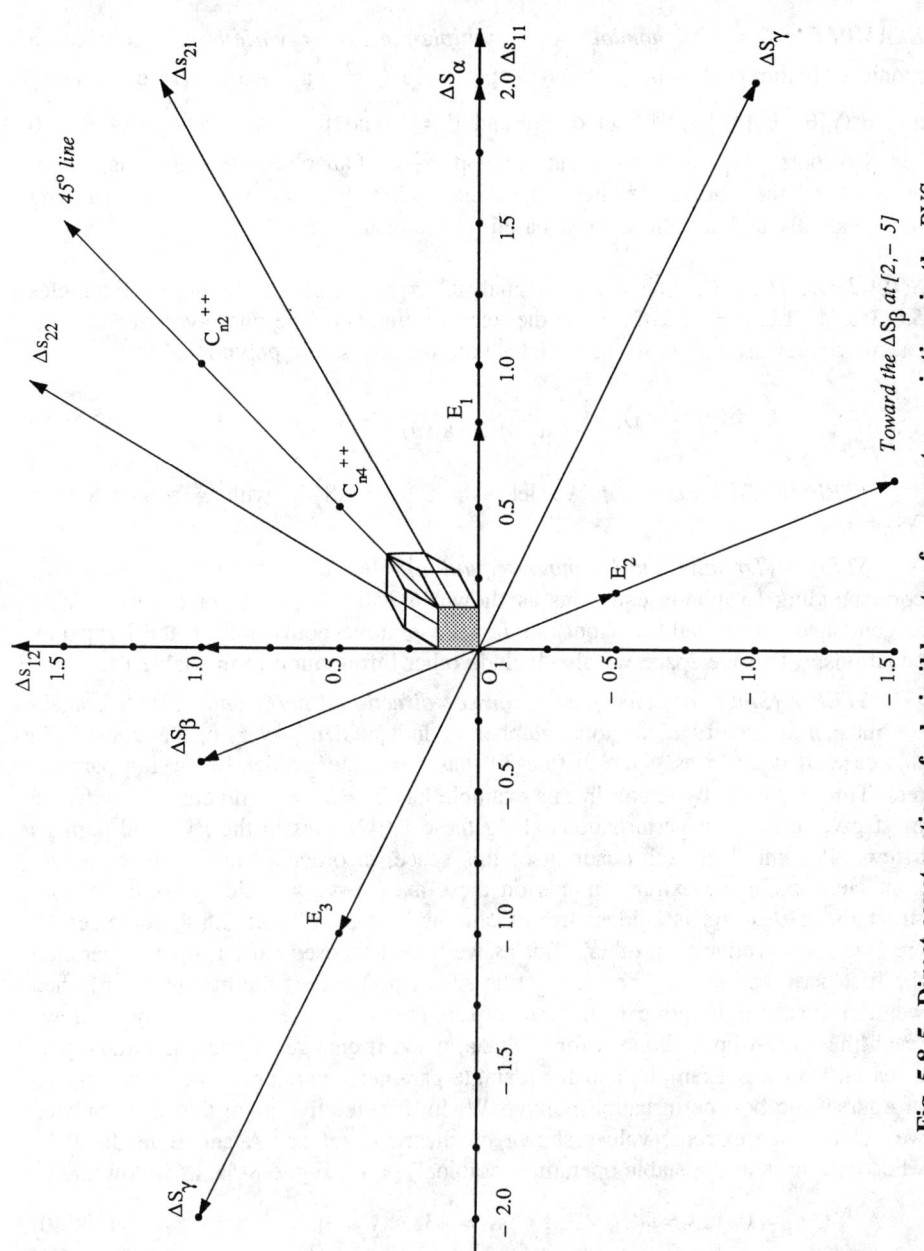

Fig. 5.8-5. Direct construction of admissible sets of parameter variations in the *PVS*.

EXAMPLE 5.8-5 *(Polynomials with multiple-parameter variations):* Consider a monic polynomial $\Delta(s, \theta) \triangleq s^2 + (\alpha + \beta)s + \alpha\beta \triangleq s^2 + a_1 s + a_2$. Here $c \triangleq [a_1, a_2]$ and $\theta \triangleq [\theta_1, \theta_2] \equiv [\alpha, \beta]$. Let $\alpha = 1$ and $\beta = 2$, and for these values $\Delta(s, \theta) = 0$ has two roots at $s_1 = -\alpha = -1$ and $s_2 = -\beta = -2$. Our objective is to construct an estimate of the *E-Box* for this polynomial, when both a_1 and a_1 are changing independently and simultaneously, based on the results of § 5.7.3.

SOLUTION *[C. 35]:* The pertinent analysis is in parallel with that of Examples 5.8-3 & 4. Let $J = 1, 2, ..., N$ be the iteration number. The phase-variable canonical matrix corresponding to the initial asymptotically stable polynomial is

$$S(\theta) (\equiv S_1) \triangleq \begin{pmatrix} 0 & 1 \\ -a_2(\theta) & -a_1(\theta) \end{pmatrix} = \begin{pmatrix} 0 & 1 \\ -2 & -3 \end{pmatrix}. \quad (5.8\text{-}39)$$

STEP 0 *(To update S):* We let $S_J = S_{J-1} \pm \Delta S_{J-1}$ with $S_1 \equiv S_0 \equiv S$, and $\Delta S_0 \equiv 0$.

STEP 1 *(To solve the Lyapunov equation):* For the current S_J we solve the corresponding Lyapunov equations as shown in Table 5.8-4. In practice this table is generated numerically and only at $J = 1$ we solve both cases of the Lyapunov equations. (To save space we also include other information in this table.)

STEP 2 *(Single-parameter variations or directional perturbations):* Here again we must *first* specify r, the total number of independently varying parameters. In this case $S_1 \in R^{2 \times 2}$ as given in (5.8-39) has $r = 2$ independently varying parameters. Thus a typical hyperbox in this example has $2^r = 2^2 = 4$ corners. *Secondly,* we must give directional perturbations along these r (=2) axes in the *PVS* and perhaps a few "45° lines" in each quadrant of this space, in order to initiate the construction. Here again the extreme points on these axes serve as guiding points to construct the *E-Box*. As is evident from Columns 2 & 3 in Table 5.8-4, some results are less conservative than others. But as we have proposed earlier, upon generating the first perturbation step, i.e., $(\Delta a_j)_1$ (the subscript 1 means the first iteration), then we must use our iterative method to repeat the above step accordingly until we reach the largest-upper bounds for each Δa_j when it changes *alone* (directional perturbation). In this example, and for a single-parameter variation, Theorem 5.7-2 is, of course, the best perturbation policy. When this iterative algorithm is completed we get the two extreme values shown by the two vertexes A and B in the *PVS*, whose origin is at the stable-operating (nominal) point, Fig. 5.8-6a, as follows.

$$A \triangleq (\Delta a_1 = 0, \Delta a_2 = -2), \quad B \triangleq (\Delta a_1 = -3, \Delta a_2 = 0). \quad (5.8\text{-}40)$$

These numbers can also be confirmed from the classical stability analysis. Since we can show that when $\alpha \to \alpha + \Delta\alpha$ or $\beta \to \beta + \Delta\beta$, the system remains stable if $-1 < \Delta\alpha$ or $-2 < \Delta\beta$. These values in turn yield $-3 < \Delta a_1 \ (= \Delta\alpha + \Delta\beta)$ or $-2 < \Delta a_2 \ (= \alpha\Delta\beta + \beta\Delta\alpha + \Delta\alpha\Delta\beta)$.

5.8. Numerical Examples

As we continue to study this problem, we realize that perhaps the question which we must again answer in this type of situation is, how to connect these extreme points, in order that: first to close the set in the *PVS*, and secondly to generate the most extended set in this space such that for each point in that set the corresponding perturbed polynomial remains stable.

Table 5.8-4. Applications of Results from § 5.7.3 to Example 5.8-5.

Column 1	Column 2	Column 3	Column 4
$S^T P + PS + 2I = 0$	Theorem 5.7-2		Theorem 5.7-3
$P = \begin{bmatrix} \dfrac{a_1^2 + a_2(1+a_2)}{a_1 a_2} & \dfrac{1}{a_2} \\ \dfrac{1}{a_2} & \dfrac{a_2+1}{a_1 a_2} \end{bmatrix}$	$(\Delta a_1)^2 = 2$ $(\Delta a_2)^2 = 2$		$\sum\limits_{j=1}^{2}(\Delta a_j)^2 = 2$
$SP + PS^T + 2I = 0$	Theorem 5.7-4	Comment 5.7-2	Comment 5.7-3
$P = \begin{bmatrix} \dfrac{a_1^2 + a_2 + 1}{a_1 a_2} & -1 \\ -1 & \dfrac{a_2+1}{a_1} \end{bmatrix}$	$(\Delta a_1)^2 = \dfrac{1}{2}$ $(\Delta a_2)^2 = \dfrac{1}{5}$	$(\Delta a_1)^2 = \dfrac{1}{2}$ $(\Delta a_2)^2 = \dfrac{1}{5}$	$\sum\limits_{j=1}^{2}(\Delta a_j)^2 = 0.146$

STEP 3 (To interpret the sufficient condition for multiple-parameter variations): In general, we may use, for instance, $\Delta S \Delta S^T < P_J^{-2}$, in order to establish a set of inequalities whose intersection yields a sufficient condition for the multiple-parameter variations. In the current situation, however, we may use either Theorem 5.7-3 or Comment 5.7-3, which in effect these are the consequences of the preceding thought. Table 5.8-4 reveals that in this example the best policy for multiple-parameter variations is Theorem 5.7-3 which at the first iteration yields

$$(\Delta a_1)^2 + (\Delta a_2)^2 \leq 2. \tag{5.8-41}$$

STEP 4 (To search for the vertexes): The result from the preceding step provides us with a perturbation policy which can be used to find, in principle, 2^r corresponding vertexes. Clearly (5.8-41) is a circle in the *PVS*, but that fact does not really affect the remaining analyses. To satisfy (5.8-41) we can show that the following $2^r = 2^2 = 4$ vertexes are acceptable choices:

$$O_1 \triangleq (\Delta a_1 = \Delta a_2 = 0), \qquad A_1 \triangleq (\Delta a_1 = 0, \Delta a_2 = -1),$$

$$C_1^{--} \triangleq (\Delta a_1 = \Delta a_2 = -1), \quad B_1 \triangleq (\Delta a_1 = -1, \Delta a_2 = 0). \tag{5.8-42}$$

STEP 5 (To construct an admissible set): Connecting the vertexes found in the previous step gives us a set of admissible-parameter variations (a hyperbox), whose corresponding ΔS's are stable (Theorem 5.6-2). To determine the troublesome direction(s) for perturbing this system, we construct ΔS at the above four vertexes. Then we look into the eigenvalues of our perturbed system at these points. From distribution of these eigenvalues we can determine the troublesome direction(s) of our perturbation policy. In this example, we have already determined the single-parameter variations on each axes. Therefore from (5.8-40) we note that the troublesome direction of perturbation is in the third quadrant of the *PVS*, i.e., $S_{J+1} = S_J - \Delta S_J$. (This fact is used in choosing (5.8-42).) We also note that the two extreme points A and B (in (5.8-40)) are outside of the above set. Therefore as we expect, the above set of admissible points is not the largest possible set. Thus we now search for ways of extending this initial set of admissible points.

STEP 6 (The construction of the most extended set): By translating the coordinates of the *PVS* to the farthest (troublesome) corner vertexes of each admissible-stable set (i.e., $J \to J+1$) enables us to extend the regions or sets of admissible points in that space. That means we now start with $S_{J+1} = S_J - \Delta S_J$, where ΔS_J is computed from (5.8-42) (here at C_1^{--}). Certainly, some of the parameter variations may indeed have both positive and negative signs in each set of coordinates in the *PVS*, and we must be alert of this fact. This situation is equivalent to reducing the value of one parameter variation, while changing the other one. (For example, at the first iteration, the mirror images of points in (5.8 42) with respect to the other three quadrants in Fig. 5.8-6a, also satisfy (5.8-41). But we have only concentrated in the third quadrant of that space.) We must always consider this situation very carefully, since the implication of this issue is that, as we proceed to translate the coordinates in certain direction we must search for points in the previous set, in order to connect the two consecutive convex sets *as smoothly as* we possibly can and to cover the entire allowable region of admissible points in the *PVS* as completely as we can. This method of construction is based upon the consequence of Theorem 5.6-2. We continue to apply this method to Example 5.8-5 and summarize our computations in Fig. 5.8-6a.

At J = 2: We translate the coordinates to the corner point C_1^{--}, while maintaining the *same* notation. Therefore now the new $S_2 = S_1 - \Delta S_1$, and the corresponding P_2 are:

$$S_2 = \begin{bmatrix} 0 & 1 \\ -1 & -2 \end{bmatrix} \text{ and } P_2 = \begin{bmatrix} 3\frac{1}{2} & \frac{1}{2} \\ \frac{1}{2} & \frac{1}{2} \end{bmatrix}. \tag{5.8-43}$$

Here again using Theorem 5.7-3 in the *new* coordinate system results in

5.8. Numerical Examples

$$(\Delta a_1)^2 + (\Delta a_2)^2 \leq \tfrac{1}{2}. \tag{5.8-44}$$

We now select the following *new* vertexes (along the "$\pm 45°$ lines" in the new coordinate system), Fig. 5.8-6a.

$$C_2^{++} \triangleq (\Delta a_1 = \Delta a_2 = 0.5), \quad C_2^{-+} \triangleq (\Delta a_1 = -0.5, \Delta a_2 = 0.5),$$

$$C_2^{--} \triangleq (\Delta a_1 = \Delta a_2 = -0.5), \quad C_2^{+-} \triangleq (\Delta a_1 = 0.5, \Delta a_2 = -0.5). \tag{5.8-45}$$

Clearly, using this iterative approach we can construct the entire set of admissible parameter variations such that for each point in that set the corresponding ΔS yields an asymptotically stable $S \pm \Delta S$ or $\tilde{\Delta}(s, \theta)$. Thus we have a class of asymptotically stable polynomials surrounding the initial polynomial and indeed that is the true meaning of the stability robustness analysis. This set, which can be constructed in order to become *the largest* set of its kind, is shown in Fig. 5.8-6b. As this figure shows by increasing the iteration number and in a few different directions we can smooth the boundary of the set in that figure. At this juncture we must, however, point out that the main purpose of this study is to find an *optimal* (in the sense of robust stability) path which connects the extreme points (here A to B). In this example, we must search along the "$-45°$ line" for a new extreme point, in order to complete the construction. Here we note that the corner point C_2^{--} has already passed the direct line connecting the extreme point A to B, which we always attempt to choose despite of its being very conservative, and therefore we may claim that the extra computational effort has paid off. We can continue to enlarge the previous set.

▼

EXAMPLE 5.8-6 *(A comparison with the method stemming from the Kharitonov's Theorem):* Consider a strictly Hurwitz monic polynomial

$$\Delta(s, \theta) \triangleq s^4 + a_1 s^3 + a_2 s^2 + a_3 s + a_4 = s^4 + 5s^3 + 8s^2 + 8s + 3. \tag{5.8-46}$$

According to Barmish [C. 7] and based upon Theorem 5.7-1, we can perturb all four coefficients a_1 to a_4 by $\varepsilon_{max} = 1.81$, while the polynomial remains stable. In other words, when each and every coefficients is changing by ± 1.81, then the corresponding set of polynomials remains stable. Choosing the positive direction for this perturbation, we draw this hyperbox in the *PVS* (whose origin O_1 is at the nominal values of a_1 to a_4), Fig. 5.8-7. In an elaborate setting, Barmish uses Theorem 5.7-1, in order to conclude this result and in so doing he uses four sets of conditions (similar to four Routh-Hurwitz tables) for each of the four boundary polynomials to be stable. (We note that each of these conditions requires four tests of its own. In fact the saving in this approach begins with polynomials of order more than four.) From these conditions the above upper and lower bounds for a_1 to a_4 are found. This method of analysis, for systems which we have no prior information about their upper and lower bounds, is purely algebraic. Furthermore the final result has shown to be often conservative. Therefore the main thrust of the following unified analysis (unified in the sense that is applicable to both

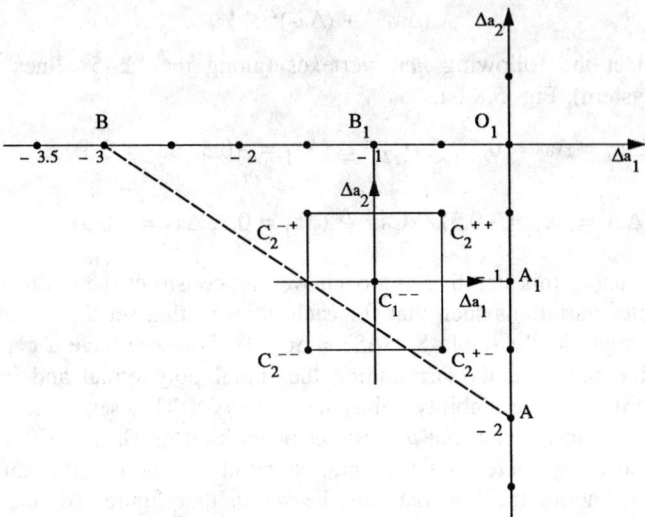

Fig. 5.8-6a. Toward the direct construction of the largest admissible set of parameter variations.

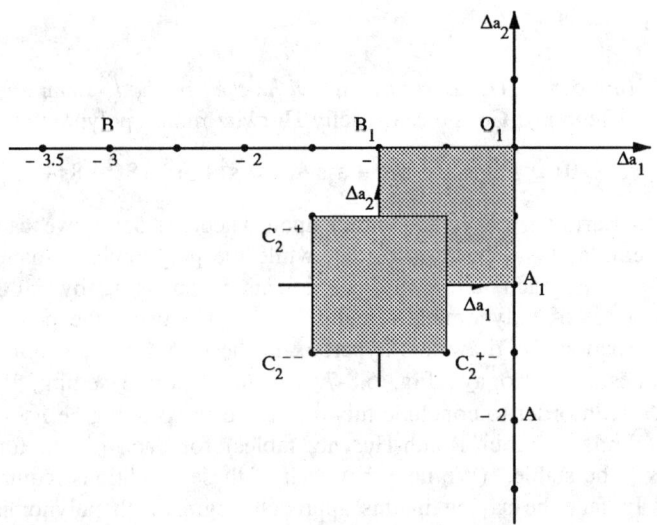

Fig. 5.8-6b. Toward the direct construction of the largest admissible set of parameter variations.

5.8. Numerical Examples

matrices and polynomials) is to show that the same method which is used to analyze Examples 5.8-3 to 5 can be used as an alternative method to construct this hyperbox, and/or to *enlarge* this ε_{max}, if possible.

In order to demonstrate as succinctly as possible the application of our method for constructing a hyperbox (or perhaps *the E-Box*), for a typical Hurwitz monic polynomial and as in Example 5.8-5, we apply the information from Table 5.7-1 to the fourth-order polynomial $\Delta(s, \theta)$, given in (5.8-46). For this analysis we need the solutions to two corresponding Lyapunov equations, whose *algebraic* solutions are summarized in Table 5.8-5, which in practice this table is replaced by a numerical algorithm.

Table 5.8-5. Two Algebraic Solutions for the Lyapunov Equations Corresponding to a Fourth-Order Monic Polynomial.

Column 1	Column 2
$S = \begin{bmatrix} 0 & 1 & 0 & 0 \\ 0 & 0 & 1 & 0 \\ 0 & 0 & 0 & 1 \\ -a_4 & -a_3 & -a_2 & -a_1 \end{bmatrix}$	$P = \begin{bmatrix} p_{11} & p_{12} & p_{13} & p_{14} \\ p_{12} & p_{22} & p_{23} & p_{24} \\ p_{13} & p_{23} & p_{33} & p_{34} \\ p_{14} & p_{24} & p_{34} & p_{44} \end{bmatrix}$
$S^T P + PS + 2I = 0$	$SP + PS^T + 2I = 0$
$\begin{bmatrix} a_3 & -a_4 a_1 \\ a_1 & a_3 - a_2 a_1 \end{bmatrix} \begin{Bmatrix} p_{24} \\ p_{44} \end{Bmatrix} = \begin{Bmatrix} 1 - a_4 + (a_2/a_4) \\ -1 - a_2 + (1/a_4) \end{Bmatrix}$	$\begin{bmatrix} a_1 a_4 & a_3 - a_1 a_2 \\ a_3 & -a_1 \end{bmatrix} \begin{Bmatrix} p_{13} \\ p_{24} \end{Bmatrix} = \begin{Bmatrix} 1 + a_2 - a_4 + a_1^2 + a_1 a_3 \\ 1 - a_2 - a_4 \end{Bmatrix}$
$p_{14} = 1/a_4, \quad p_{12} = a_3 p_{24} - 1, \quad p_{34} = a_1 p_{44} - 1$	$p_{12} = -1, \quad p_{23} = -1, \quad p_{34} = -1$
$p_{23} = a_2 a_1 p_{44} - a_2 - 1, \quad p_{11} = a_4 p_{24} + (a_3/a_4)$	$p_{14} = 1, \quad p_{22} = -p_{13}, \quad p_{33} = -p_{24}$
$p_{13} = a_4 p_{44} + (a_1/a_4), \quad p_{33} = (a_1^2 + a_2) p_{44} - p_{24} - a_1$	$p_{11} = (a_3 - a_1 - a_2 p_{13} + p_{24})/a_4$
$p_{22} = a_2 p_{24} + (a_3 a_1 - a_4) p_{44} - a_3 - (a_1/a_4)$	$p_{44} = a_4 p_{13} - a_2 p_{24} - a_1 - a_3$

Furthermore, to speed up the computations and since we have already demonstrated our method in the preceding examples, we start from the Barmish's results, Fig. 5.8-7, which is drawn for the sake of simplicity only in the first quadrant of the *PVS*.

By the method of construction the entire hyperbox in Fig. 5.8-7 yields a set of strictly Hurwitz polynomials. Now, we consider O_2, the farthest corner along the "45° line", as the new origin of the *PVS*, and translate the coordinates of this space from O_1 to O_2. We keep the symbolic coefficients the same as before in order to use Tables 5.7-1 and 5.8-5. Thus

$$\tilde{\Delta}(s, \theta) \triangleq s^4 + a_1 s^3 + a_2 s^2 + a_3 s + a_4 = s^4 + 6.81s^3 + 9.81s^2 + 9.81s + 4.81. \tag{5.8-47}$$

The solutions to two corresponding Lyapunov equations for (5.8-47), are given in Table 5.8-6.

Table 5.8-6. The Solutions for Two Lyapunov Equations Corresponding to the Fourth-Order Monic Polynomial (5.8-47).

Column 1	Column 2
$S = \begin{bmatrix} 0 & 1 & 0 & 0 \\ 0 & 0 & 1 & 0 \\ 0 & 0 & 0 & 1 \\ -4.81 & -9.81 & -9.81 & -6.81 \end{bmatrix}$ $a_1 = 6.81, \ a_2 = a_3 = 9.81, \ a_4 = 4.81$	$P = \begin{bmatrix} p_{11} & p_{12} & p_{13} & p_{14} \\ p_{12} & p_{22} & p_{23} & p_{24} \\ p_{13} & p_{23} & p_{33} & p_{34} \\ p_{14} & p_{24} & p_{34} & p_{44} \end{bmatrix}$
$S^T P + PS + 2I = 0$	$SP + PS^T + 2I = 0$
$P = \begin{bmatrix} 5.565 & 6.191 & 2.729 & 0.208 \\ 6.191 & 12.89 & 7.428 & 0.733 \\ 2.729 & 7.428 & 7.796 & 0.859 \\ 0.208 & 0.733 & 0.859 & 0.273 \end{bmatrix}$	$P = \begin{bmatrix} 9.261 & -1 & -4.725 & 1 \\ -1 & 4.725 & -1 & -4.806 \\ -4.725 & -1 & 4.806 & -1 \\ 1 & -4.806 & -1 & 7.8 \end{bmatrix}$

Using the numerical results from Table 5.8-6 in Table 5.7-1 gives us various ways of perturbing $\tilde{\Delta}(s, \theta)$ in (5.8-47). Indeed from these numerical results we conclude that we can further perturb this polynomial in several different directions. For example, choosing Column 1 of Table 5.7-1 (incidentally in this example this is the best policy to perturb $\tilde{\Delta}(s, \theta)$), we note that we can perturb each coefficients of (5.8-47) independently by $1/g_{nn}^{1/2} = \pm 0.8473$. But if we change (or we allow to change) all four of them simultaneously, then we have ± 0.42365. These numbers immediately show that the initial hyperbox of Fig. 5.8-7 can be enlarged. Of course, at this stage we cannot [yet] claim that we have enlarged the entire

5.8. Numerical Examples

hyperbox of Fig. 5.8-7 by ± 0.42365. Instead as in the preceding example, we have only built a new hyperbox centered at the farthest corner of the preceding hyperbox (at point O_2) as shown in Fig. 5.8-7.

The procedure to enlarge the initial hyperbox of Fig. 5.8-7 is now clear. We must translate the coordinates of the *PVS* to the other corners of each hyperbox and to perturb repeatedly at those points, because the current analytical results *alone* do not provide the detailed largest set of allowable-parameter variations. The method that we are proposing is iterative and each hyperbox must again be constructed such that it *overlaps* with the preceding one as shown in Fig. 5.8-7 (for example, the two hyperboxes that are centered at O_3 and O_4). This *unidirectional* method of construction in each quadrant of the *PVS* guarantees that the final set of points covers those points in this space which results in stable perturbed polynomials. Here again we conclude that in order to construct the *E-Box* for the *SRA* of a typical dynamical system, we need to study the perturbation along each axes first. This analysis for the initial $\Delta(s, \theta)$ is summarized in Table 5.8-7, which confirms the fact that the results in [C. 7] is conservative. Then we should look into the ways (Table 5.7-1) of extending the initial hyperbox of Fig. 5.8-7, in order to reach the *E-Box*. That extension is possible only as the result of an iterative procedure such as the one which is presented in this chapter, i.e., Theorem 5.6-2, because it is a well-known fact that a closed-form solution for higher-order polynomials (≥ 5) is not known. When the construction is completed in one quadrant of the *PVS*, then we must repeat the same computational steps in the next quadrant of that space. We invite the reader to study this matter further.

Table 5.8-7. The Directional Perturbations of $\Delta(s, \theta) = s^4 + 5s^3 + 8s^2 + 8s + 3$.

$\tilde{\Delta}_{45° \text{ line}}(s) = s^4 + (5+\delta)s^3 + (8+\delta)s^2 + (8+\delta)s + (3+\delta)$	$-3 < \delta < \infty$
$\tilde{\Delta}_1(s) = s^4 + (5+\delta)s^3 + 8s^2 + 8s + 3$	$-3.948 < \delta < 15.28$
$\tilde{\Delta}_2(s) = s^4 + 5s^3 + (8+\delta)s^2 + 8s + 3$	$-4.525 < \delta < \infty$
$\tilde{\Delta}_3(s) = s^4 + 5s^3 + 8s^2 + (8+\delta)s + 3$	$-6.027 < \delta < 30.02$
$\tilde{\Delta}_4(s) = s^4 + 5s^3 + 8s^2 + 8s + (3+\delta)$	$-3 < \delta < 48.2$

208 Chapter Five: Stability Robustness Analysis, Maneuverability

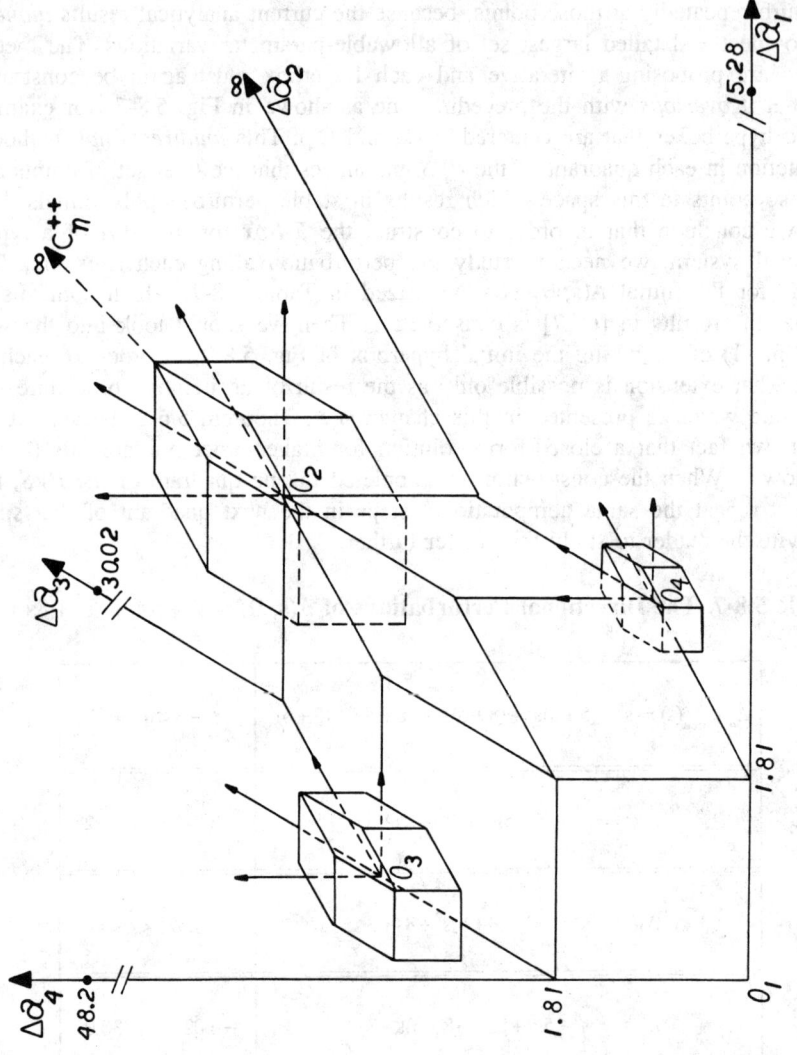

Fig. 5.8-7. Direct construction of admissible sets of parameter variations in the *PVS*.

5.8.2. Concluding Remarks on Numerical Examples

The following remarks should be noted when we apply the results of this chapter to other problems, in particular the subsequent examples in this chapter.

REMARK 5.8-5 (On the iterative or probing procedure): First of all, and from the physical point of view, the stability robustness measure *(SRM)* of a dynamical system, i.e., *the largest* set of points in the *PVS* whose corresponding set of perturbed systems (polynomials or matrices) is stable, is independent from any procedure which is used for its determination. Also as the preceding analysis in Example 5.8-6 confirms this set is far different from a hyperbox that is reported in [C. 7], although the associated method in this reference is perhaps the most widely used method for determination of the *SRM* in polynomials. We have also studied the fourth-order example given in [C. 48], and have concluded that we can substantially enlarge the outcome of that example as well. We invite the reader to apply the procedure of Example 5.8-6 to the numerical example in [C. 48]. All in all we are convinced that the current analytical results in spite of their overwhelming appeal do not provide the detailed largest set of parameter variations with all the intricacies among various (more than three) varying parameters. We are convinced and many numerical examples will confirm that, what we need is an efficient computational algorithm to detect the hidden corners and to keep tracking them for the iterative method, in order to construct a typical *E-Box*. Clearly, if we use this iterative approach, then we can probe the entire set of points in the *PVS* whose corresponding ΔS's yield a set of asymptotically stable systems [$S \pm \Delta S$ or $\tilde{\Delta}(s)$]. As shown in Examples 5.8-3 to 6, by increasing the iteration number and in a few different directions, we can smooth the boundary of each corresponding *SRM* and construct the associated *E-Box*. Of course, the trade-off is how much computations we are willing to make, in order to have the *E-Box*. Since this procedure can be automated, this construction is manageable, however, if that construction is not of the highest priority, then certainly we may be content with what we have as long as we know where we stand. For instance, the choice of subscript 'max' in $\varepsilon_{\max} = 1.81$ of Example 5.8-6 should not be read in the global sense, because that construction is not completed.

REMARK 5.8-6 (On the computational issues): We should point out that this probing method which searches for *sets of points,* instead of each single point, in the *PVS,* can be easily automated. We have extensively used our proposed method for matrices and polynomials of order 50. Note that this is an off-line advisory computation, and how fast or how slow is not an issue. In general, we need a reliable algorithm to compute solutions for the two Lyapunov equations and an algorithm to evaluate eigenvalues of various matrices for testing their signs, *etc.,* and a set of subroutines to perform different matrix algebra.

REMARK 5.8-7 *(On the iteration number):* In the case of a single-parameter variation we can use a number of classical methods to compute the associated largest perturbation, which for the moment we assume is finite. Therefore if we take the ratio of this largest perturbation over the "radius" of the last set (which is proportional to the norm of P^{-2}), then that ratio gives us a practical-upper bound for a directional iteration number. When this "radius" goes to zero (or requiring more accuracy), then the iteration number will go to infinity (or increases substantially), which we must avoid this situation by letting a practical (tolerance) value for the "radius", typically 10^{-4}. However, if the perturbations are infinite (for instance, in Example 5.8-6 along the Δa_2-axis and the "45° line", cf., Remark 5.8-9), then these computations will require infinite iterations, which must be avoided by giving an upper limit for the iteration number, say $N = 100$, and studying the final result. In these situations when we detect an infinite directional perturbation, we can easily switch to a number of classical methods to determine these directional perturbations. Similarly, we can estimate the length of the other extreme boundary lines of the hyperbox, in order to get an estimate of the total iterations required to cover those boundaries. Using the Hadamard's inequality (cf., § 2.5.8*(iv)*), we can estimate the "volume" of the extreme *SRM* and divide that by the "volume" of the "smallest" hyperbox generated at one of the farthest corner points in order to reach a practical-upper bound for the number of iterations needed to fill the *SRM*. Certainly, this is truly an exaggerated estimate of the iteration number, but if we maintain a reasonable computational tolerance, as is required in any such numerical computation, then our iteration number will remain finite and the iterative method will converge within a pre-specified finite tolerance. The final set of points (or the allowable-parameter variations) in the *PVS* may become the largest set that we can construct in order to approximate the global and perhaps unattainable set of points in the *PS* (i.e., the *E-Box*), if we are willing to make the extra, though straightforward, computations. We should again recall that our objective or the nature of our problem is numerical determination of *the largest* stability region for a truly unknown and possibly of higher-dimensional system. The issue of iteration and convergence should be viewed in this context of finding a global answer for this type of system, which we have no *a priori* knowledge about its global stability region. Therefore we cannot analytically say in how many exact iterations we will get to the *E-Box,* unless we use a plethora of methodologies to estimate different roots of various polynomials and a host of other tests to classify various encountered problems. The very issues which we are trying to avoid, in order to maintain the methodology accessible to everyone. What our method tells us is where to probe, and if we do not want to get too far in a given direction, then we can stop. So the iteration number, or the probing method, should be looked at from this angle; and a possible lack of reaching the global answer for the stability region does not have the same detrimental effect that is associated to the undesirable consequences relating to the lack of convergence in an algorithmic sense.

5.8. Numerical Examples 211

REMARK 5.8-8 (On the propagations of P's): Upon choosing the iterative policy, it is also very instructive to study the propagations of P's in the subsequent "smaller" sets. Note that in all these computations we have kept Q = 2I. This study shows that using the Lyapunov method we definitely need an iterative procedure. Because the "radius" of each set is proportional to the norm of P^{-2} which gets smaller as we get further away from the origin of the *PVS* (because in most cases we cannot indefinitely perturb the system unless we are in the stabilizing direction). Loosely speaking, each P's can only be "stretched" up to a certain point and beyond that we cannot use this P to perturb the system, while the stability robustness measure of dynamical system as described earlier is independent from this P or *any* other computational *tools* which is used to generate that. This observation may disappoint those researchers who are seeking an "optimal" P for the "largest-upper bound" of the stability robustness measure.

REMARK 5.8-9 (On the directional perturbations): As Fig. 5.8-7 or Table 5.8-7 of Example 5.8-6 depicts, it is evident that the perturbations along the "$45°$ line" and the Δa_2-axis extend to infinity. Here we can show by classical methods that indeed every points on the plane (Δa_2, "$45°$ line") corresponds to one stable perturbed system. On the other hand, the allowable single-perturbations along the Δa_1-axis, Δa_3-axis, and Δa_4-axis are bounded, which suggests that in this example the *E-Box* looks like a funnel. Clearly, and in many given situations, there may exist a number of other directional perturbations which extend to infinity. How to detect these directions in any given situation is a very interesting problem. Indeed these *peakings* are very important phenomena.

REMARK 5.8-10 (Other applications): Since the preceding analyses are based on the Lyapunov technique, there are a number of other applications for these results in quadratic optimization and covariance analysis which can be cited.

5.9. CONSTRUCTION OF ROBUSTLY STABLE LARGE – SCALE SYSTEMS

5.9.1. Introduction

First we must point out that by a large-scale dynamical system we mean a system that is both constructed by coupling several dynamic sub-systems [A. 151], as well as a system that has a large number of parameters and/or variables, and thus it can be very computationally involved. These large-scale dynamical systems are modeled by sets of *ODE's,* that are characterized in either matrices or polynomials. The analyses that are presented in the preceding sections can be applied to this class of systems very straightforwardly. However, when analyzing the parameter variations of any dynamical system the knowledge about the structure of that system is always very helpful. For example, in certain class of large-scale systems there are some hidden properties, regarding the parameter variations, which knowing that will simplify the *SRA* and otherwise the analytical complications will be overwhelming. We now present some results that are referred to as robust structures, whose corresponding solutions to the appropriate Lyapunov equations can also be generated with the minimum computational effort. This section closely follows [B. 31].

LEMMA 5.9-1: Given an asymptotically stable matrix $S_0 \in R^{n \times n}$ such that $S_0^T S_0 = S_0 S_0^T$ and $S_0 + S_0^T = -2\nu I$, where $\nu > 0$ is a constant, and I is an identity matrix, then S_0 is robustly stable with respect to the individual linear perturbations affecting one or several of its off-diagonal entries.

PROOF: Since S_0 is asymptotically stable,

$$S_0^T P_0 + P_0 S_0 + Q = 0 \quad \rightarrow \quad P_0 = \int_0^\infty e^{S_0^T \tau} Q\, e^{S_0 \tau} d\tau. \tag{5.9-1}$$

Letting Q = I, and because $S_0^T S_0 = S_0 S_0^T$, we have

$$P_0 = \int_0^\infty e^{(S_0^T + S_0)\tau} d\tau = \int_0^\infty e^{-2\nu I \tau} d\tau = I/2\nu. \tag{5.9-2}$$

Let ΔS_1 be the allowable perturbation for the off-diagonal entries of S_0, i.e., $S_1 \triangleq S_0 + \Delta S_1$ remains asymptotically stable. Since $S_0 + S_0^T = -2\nu I$ and ν is constant, we have $\Delta S_1 = -\Delta S_1^T$. By Theorem 5.5-1, with Q = I, and in order that S_1 remains stable, it is sufficient to have $\Delta S_1 \Delta S_1^T < \frac{1}{4} P_0^{-2} = \nu^2 I$. Suppose that ΔS_1 has one nonzero element ε_1 at its (i, j) entry, then we have

$$\varepsilon_1^2 \operatorname{diag}(0, \ldots, 0, 1_{ii}, 0, \ldots, 0) < \nu^2 I. \tag{5.9-3}$$

If we choose $|\varepsilon_1| = \nu$, then S_1 remains asymptotically stable. We must utilize Algorithm 5.6-1, here. But first, we note that $S_1 + S_1^T = -2\nu I$, which is true immediately, and $S_1 S_1^T = S_1^T S_1$ because $S_1 S_1^T - S_1^T S_1 = 0$. Therefore if we now perturb S_1 such that $S_2 \triangleq S_1 + \Delta S_2$ remains asymptotically stable, then using (5.5-1) yields $\Delta S_2 = \Delta S_1$. We can show that $\Delta S_N = \Delta S_{N-1} = \cdots = \Delta S_1$. Therefore for $\Delta S = \sum_{J=0} \Delta S_J$, with J, the iteration number, is being arbitrarily large, we have $S_0 \pm \Delta S$ that is asymptotically stable and it remains in the correspondingly very large *E-Box*. When ΔS_1 has more than one nonzero element, then (5.9-3) becomes $\varepsilon_1^2 M' < \nu^2 I$. Here M' is a constant matrix and we can choose any particular norm from both sides of the previous inequality, in order to find the largest ε_1. In effect this situation is the same as the case corresponding to (5.9-3) except that the new ε_1 is scaled according to the norm of M'. Similarly, we can show that herein $\Delta S_N = \Delta S_{N-1} = \cdots = \Delta S_2$ are scaled by the same amount as ΔS_1, and the final result is the same as before.

EXAMPLE 5.9-1: Determine the *SRA* of the following matrix $S \in R^{3 \times 3}$.

5.9. Construction of Robustly Stable Large – Scale Systems

$$S \triangleq \begin{bmatrix} -v & \omega_1 & \omega_2 \\ -\omega_1 & -v & \omega_3 \\ -\omega_2 & -\omega_3 & -v \end{bmatrix}, \quad v > 0. \tag{5.9-4}$$

SOLUTION: Clearly $S + S^T = -2vI$ and $S^T S = SS^T$. Letting $Q = I$, it is easy to show that $P = P^T = I/2v > 0$ satisfies the Lyapunov equation and thus as expected S is asymptotically stable. Furthermore this matrix satisfies Lemma 5.9-1 and we may perturb arbitrarily each and every ω_i *individually* as we please without deteriorating the asymptotic stability of S. Here the choice of this 3×3 matrix is only of academic interest. However, its first 2×2 minor (or similar structure) has received attentions because it represents one pair of complex poles in a system. For instance, recently in [B. 49] and upon an elaborate analysis a similar result is shown for a 2×2 matrix. Note that herein we do not restrict the sign of ω_i. We can easily show that this property can be generalized to a corresponding large-scale matrix.
▼

LEMMA 5.9-2: Given a stable matrix $S = \text{diag}(S_{11}, \ldots, S_{mm}) \in R^{n \times n}$ such that each and every $S_{ii} \in R^{n_i \times n_i}$, with $n = \sum_{i=1}^{m} n_i$, satisfies Lemma 5.9-1. Then S is robustly stable with respect to linear variations in every off-diagonal entries of each S_{ii} independently. Furthermore, if $n_i = 2$, for all i, then S is robustly stable with respect to a uniform shift in all its off-diagonal entries.

PROOF: Since each S_{ii} satisfies Lemma 5.9-1 and the perturbation is taking place in the off-diagonal entries of each S_{ii} independently, $\Delta S_1 = (\Delta S_{ii})$, with ΔS_{ii} having the same properties as that of ΔS_1 in Lemma 5.9-1. The only difference is that here with $Q = I$ we have $P_0 = \text{diag}((I/2v_1), \ldots, (I/2v_m))$. Furthermore, if $n_i = 2$, and $\omega_i \to \omega_i + \varepsilon$, for all $i = 1, \ldots, m$, then we can show that $\varepsilon^2 I < v^2 I$, where $v \triangleq \min(v_1, \ldots, v_m)$. We can repeat this perturbation as often as we please.

5.9.2. Special Coupled Systems

Consider an asymptotically stable coupled system $S \triangleq S_0 + C$, such that

$$S_0 \triangleq \begin{bmatrix} S_{11} & 0 \\ 0 & S_{22} \end{bmatrix}, \text{ and } C \triangleq \begin{bmatrix} 0 & C_{12} \\ C_{21} & 0 \end{bmatrix}. \tag{5.9-5}$$

Here we assume that S_0 has the same structure as that given in Lemma 5.9-2. The coupling matrix C is to be designed such that S remains in the *E-Box*. Also each square matrices S_{ii} may have different dimension. The coupling parameters or disturbances are considered to be one of the following cases:

(i) $C_{12} = C_{21}$; (ii) $C_{12} = C_{21}^T$; (iii) $C_{12} = -C_{21}^T$; and (iv) $C_{12} \neq C_{21}$.

In all these four cases a sufficient condition for ΔS in order that $S_0 + \Delta S$

(herein $S_0 + C$) remains in the E-Box is that stated in Theorem 5.5-1, particularly (5.5-1), with $Q = I$. With S_0 as given in (5.9-5) and $Q = I$, the Lyapunov equation (5.9-1) yields $P_0 = \text{diag}((I/2v_1), (I/2v_2))$. Using S_0 (as the asymptotically stable matrix), with C (as its disturbance matrix), given in (5.9-5), we now present the following results. From (5.5-1), with $Q = I$, we have

$$\Delta S \Delta S^T \triangleq \begin{bmatrix} 0 & C_{12} \\ C_{21} & 0 \end{bmatrix} \begin{bmatrix} 0 & C_{21}^T \\ C_{12}^T & 0 \end{bmatrix} = \begin{bmatrix} C_{12}C_{12}^T & 0 \\ 0 & C_{21}C_{21}^T \end{bmatrix} \leq \begin{bmatrix} v_1^2 I & 0 \\ 0 & v_2^2 I \end{bmatrix}. \quad (5.9\text{-}6)$$

From this inequality we have

$$C_{12}C_{12}^T \leq v_1^2 I \quad \text{and} \quad C_{21}C_{21}^T \leq v_2^2 I. \quad (5.9\text{-}7)$$

A quick conservative estimate of (5.9-7) yields

$$\text{Tr}(C_{12}C_{12}^T) = v_1^2 \quad \text{and} \quad \text{Tr}(C_{21}C_{21}^T) = v_2^2. \quad (5.9\text{-}8)$$

In case (i), $C_{12} = C_{21}$, then (5.9-8) becomes

$$\text{Tr}(C_{12}C_{12}^T) = \text{Tr}(C_{21}C_{21}^T) = v^2 \triangleq \min(v_1^2, v_2^2). \quad (5.9\text{-}9)$$

In case (ii), $C_{12} = C_{21}^T$, then (5.9-8) becomes

$$\text{Tr}(C_{12}C_{21}) = \text{Tr}(C_{21}C_{12}) = v^2 \triangleq \min(v_1^2, v_2^2). \quad (5.9\text{-}10)$$

In case (iii), $C_{12} = -C_{21}^T$, then (5.9-8) becomes

$$\text{Tr}(-C_{12}C_{21}) = \text{Tr}(-C_{21}C_{12}) = v^2 \triangleq \min(v_1^2, v_2^2). \quad (5.9\text{-}11)$$

All the above are conservative estimates for the *first* perturbation step in Algorithm 5.6-1. During the next iteration, naturally the solution to the Lyapunov equation will not remain diagonal or so easily computed. Nevertheless, this example shows how quickly a sufficient condition for asymptotic stability of a large-scale system can be developed with the minimum computational effort.

The conclusion for case (iv), where $C_{12} \neq C_{21}$, is already stated in (5.9-8).

EXAMPLE 5.9-2: Find ε such that $S = S_0 + \varepsilon C$ remains asymptotically stable matrix, where

$$S_0 = \begin{bmatrix} -\frac{1}{2} & 1 & 0 & 0 \\ -1 & -\frac{1}{2} & 0 & 0 \\ 0 & 0 & -1 & 2 \\ 0 & 0 & -2 & -1 \end{bmatrix}, \text{ and } C = \begin{bmatrix} 0 & 0 & 0 & 1 \\ 0 & 0 & 1 & 2 \\ 0 & 1 & 0 & 0 \\ 1 & 2 & 0 & 0 \end{bmatrix}. \quad (5.9\text{-}12)$$

SOLUTION: From $S_0^T P_0 + P_0 S_0 + I = 0$ we have $P_0 = \text{diag}((I/2v_1), (I/2v_2))$. Here we must apply (5.9-9) to this example, resulting in

$$\text{Tr}\left[\varepsilon_1^2 C_{12} C_{12}^T\right] = \text{Tr}\left[\varepsilon_1^2 \begin{bmatrix} 1 & 2 \\ 2 & 5 \end{bmatrix}\right] = 6\varepsilon_1^2 \leq \frac{1}{4}. \quad (5.9\text{-}13)$$

Therefore $|\varepsilon_1| = 0.2041$. This conservative estimate result in a new

5.9. Construction of Robustly Stable Large – Scale Systems

$S_1 \triangleq S_0 + |\varepsilon_1| C$. We can show that a proper choice is $\varepsilon_1 = 0.2041$. This is a quick way of generating *an* asymptotically stable S. Using the maximum eigenvalue instead of the trace, we can actually show that the corresponding (5.5-1) yields $\varepsilon_1 = 0.2068$.

In this example using Algorithm 5.6-1 and after 22 iterations, however, we reach $\varepsilon = \sum_{J=1}^{22} \varepsilon_J = 0.797$, such that $S_0 + \varepsilon C$ remains in the *E-Box*, and that is the best we can do. It is to be noted that only P_0 in this process is diagonal. The remaining matrix solutions for the corresponding Lyapunov equations are not diagonal matrices and these must be computed in each perturbation steps in order to complete the construction of this robustly stable S.

▼

EXAMPLE 5.9-3: Consider a special case of (5.9-5) wherein [B. 49]

$$C_{ij} = \begin{bmatrix} \alpha_{ij} & \beta_{ij} \\ -\beta_{ij} & \alpha_{ij} \end{bmatrix}. \tag{5.9-14}$$

Determine a perturbation policy for this system, in order to study its stability robustness property.

SOLUTION: Using (5.9-7) yields

$$C_{12} C_{12}^T = \begin{bmatrix} \alpha_{12}^2 + \beta_{12}^2 & 0 \\ 0 & \alpha_{12}^2 + \beta_{12}^2 \end{bmatrix} \leq v_1^2 I, \tag{5.9-15a}$$

and

$$C_{21} C_{21}^T = \begin{bmatrix} \alpha_{21}^2 + \beta_{21}^2 & 0 \\ 0 & \alpha_{21}^2 + \beta_{21}^2 \end{bmatrix} \leq v_2^2 I. \tag{5.9-15b}$$

Here the robustness is guaranteed if $\det(C_{12}) \leq v_1^2$ and $\det(C_{21}) \leq v_2^2$. Each of these inequalities may also be used at this iteration as a "set generator", allowing both α_{ij} and β_{ij} to change simultaneously. Here again and after the first iteration, the special diagonal solutions for the corresponding Lyapunov equations, which are used to conclude these results, must be replaced with new solutions which are not diagonal matrices. Here Theorem 5.6-2 can be applied to construct the *E-Box* when both α_{ij} and β_{ij} are allowed to change simultaneously. In that case v_1 and v_2 must be replaced with appropriate conservative estimates for the eigenvalues of P_J^{-2}. If in addition to (5.9-14), we assume that $C_{ij} = C_{ji}$, then we can show that a sufficient condition for $S_0 + C$ to remain in the *E-Box* is that $\det(C_{12}) = \det(C_{21}) = v^2 \triangleq \min(v_1^2, v_2^2)$. This property also holds for $C_{ji} = C_{ij}^T$ and $C_{ij} = -C_{ji}^T$ as well.

▼

The special structured matrices shown in the previous examples have applications in modeling of beams and other mechanical and/or aerospace systems. Furthermore, the above two lemmas play an important role in relating the frequency- and time-domain robustness analyses.

5.10. CONSTRUCTION OF ROBUSTLY STABLE (CONVERGENT) DISCRETE − TIME SYSTEMS

5.10.1. Introduction

Consider a set of linear homogeneous difference equations $x(l+1, \theta) = S(\theta)x(l, \theta)$, where $x(l) \in R^n(l)$ is the system state, $\theta \in R^r$ is the generic name for system parameters and $S(\theta) \in R^{n \times n}(\theta)$ is the system coefficient matrix. These equations appear naturally in descriptions of many physical phenomena when observations or measurements are made discretely in time. Furthermore, difference (discrete-time) equations are generated in order to study discretization methods for differential (continuous-time) equations. In general, there exist many similarities between the two versions of representing a dynamical system. However, there are properties in the discrete-time problems which do not correspond to their counterparts in the continuous-time situations as these issues are reviewed extensively in [D. 43]. In this section we are only concerned with the stability (which for discrete-time systems this property is also called the convergent [D. 60]) behavior of $S(\theta)$ and its corresponding computational aspects, when θ changes. The changes in θ and thus in $S(\theta)$ may have been the results of the following possible inaccuracies: *(i)* in system parameters (including those of design parameters); and *(ii)* in system model. In the latter case inaccuracies may have been also resulted from approximating the model or linearizing the nonlinear model of system with $\Delta S(\Delta \theta) x(l)$ ($\Delta \theta$ is the variation of θ with respect to a nominal condition θ^o). In all these cases the main problem becomes how to determine the stability (or the convergent) behavior of system when $\theta \to \theta + \Delta\theta$, or $S \to S + \Delta S$, with ΔS having *multiple-varying parameters*. Here again we bear in mind that this situation means we are simultaneously monitoring two sets of objects. One set consists of point such as $[\Delta\theta_1, ..., \Delta\theta_r]^T$ in the *PVS*, whose origin is at θ^o, and the other set consists of convergent matrices $S(\theta^o + \Delta\theta)$ which corresponds to each point $\Delta\theta$ in the previous set. (We also assume that the entries of $S(\theta)$ can be changed by a large amount without affecting its stabilizability.)

Much of the thoughts and motivations for analyses in this section are in parallel with their counterparts in Sections 5.4 to 5.7, specially in regard to the fundamental assumption that our original system is sensitive to parameter variations. In particular, issues pertaining to the *SRA's* for a convergent matrix, with *multiple-varying parameters* (such as the catalog of perturbation matrices, and the issues pertaining to computational matters), are in the same spirit as their counterparts for a continuous-time system. Since this is an important topic, however, we briefly present our main results for *SRA* of discrete-time systems in the following. But first we note that the notion of *the largest* or *most extended* set of points in the *PVS* whose corresponding perturbed systems are convergent can also be applied to this case. In other words, we may pictorially depict an *E-Box* for a convergent system in the *same manner* as shown in Fig. 5.4-1. Then this presentation suggests that we may also define the "largest" set which approximates Fig. 5.4-1 to be called the "convergent robustness measure" *(CRM)* (i.e., a figure which is *similar* to Fig. 5.4-

5.10. Construction of Robustly Stable Discrete – Time Systems

2). The same goes for a system with constrained perturbation, i.e., a figure which is similar to Fig. 5.4-3. Although this generalization seems very straightforward, we bear in mind that the two *E-Box's*, which are generated for a continuous-time system and its discrete-time version need *not* necessarily be exactly the same. The discrepancy is because that the discretization of a continuous-time system is not a unique process. The same thought propagates into Figs. 5.4-2 & 3.

Having stated these facts, the method that is presented in this section has the same two distinct differences with most of the results published in this area. *First of all,* we define the "convergent robustness measure" *(CRM),* to be *the largest* (or perhaps the global) region or set of points in the *PVS,* whose corresponding perturbed matrices remain convergent, i.e., their eigenvalues are in a unit circle. Here we do not restrict ourselves to any particular scalar measure such as an eigenvalue or a singular value, *etc.,* because the robustness property resulting from these measures can be concluded from a set of points in the *PVS* which is, of course, a subset of the *CRM.* We call *the largest* or *most extended* such *CRM* as before the *E-Box.* When the *E-Box* is found, then this set will contain all the other possible results which are generally local.

Secondly, we put our emphases on the *direct* construction of this *CRM* (or the *E-Box*) for a convergent system that we have no prior information about its region of convergent. The rationale behind a direct computational approach is that the computational issues which we face, when studying a truly unknown problem whose multiple parameters are varying and we have no prior information about the way that the system parameters are changing, as mentioned before, are real and they are not trivial. Indeed in a large number of applied problems, as in the case of continuous-time systems, we do not know the intricacies of the relations among various varying parameters in a dynamical system, let alone how these parameters can be changed (within an initial region), in order to maintain system stability. Again this construction is performed by generating a set of convex sets (or hyperboxes) in the *PVS* that are stacked *under* the *E-Box* in the *PS,* in order to approximate that set the best.

In summary, to emphasize the computationally efficient methods of constructing the *entire* set of points in the *PVS* for a convergent matrix, with single- or multiple-parameter variations, such that each point of this set corresponds to a convergent perturbed matrix, is the main thrust of this section. In this regard we extend the region of convergence by an iterative procedure that *probes* this region and yields *the largest* such region which is called the *CRM* of the system. Finally, with additional computations we reach the corresponding *E-Box*. In passing, we should point out that the extensions of the Kharitonov's theorem to discrete-time system has been actively studied by a number of researchers (cf., § 5.12.D) including [D. 39]. This school of thought is primarily based on determination of the minimum number of required testings for *SRA.* We maintain our course, however, in the same manner as the preceding sections, in order to emphasize the breadth of its applications and scope of its uniformities.

5.10.2. Discrete Lyapunov Method

The analysis in this section stems from discrete-time version of the Lyapunov equation – Corollary 5.1-2. This analysis closely follows [D. 14 to 17].

THEOREM 5.10-1: For a parameter perturbation $\Delta\theta$, $S \rightarrow S + \Delta S$, then $S + \Delta S$ remains a convergent matrix if

$$\Delta S^T (P + 2P\, S\, Q^{-1}\, S^T\, P)\, \Delta S < \tfrac{1}{2} Q. \qquad (5.10\text{-}1)$$

Here $0 < Q = Q^T \in R^{n \times n}$ and $0 < P = P^T \in R^{n \times n}$ satisfy (5.1-2). Also by $M < N$ for two matrices $0 \le M = M^T \in R^{n \times n}$ and $0 < N = N^T \in R^{n \times n}$ we mean $N - M > 0$.

PROOF: Equation (5.1-2) can be exactly rewritten as follows.

$$(S + \Delta S)^T\, P(S + \Delta S) - P = -Q + \Delta S^T\, P\, S + S^T\, P\, \Delta S + \Delta S^T\, P\, \Delta S. \qquad (5.10\text{-}2)$$

Citing the Lyapunov equation again, we note that $S + \Delta S$ remains a convergent matrix if

$$\tilde{Q} \triangleq Q - \Delta S^T\, P\, S - S^T\, P\, \Delta S - \Delta S^T\, P\, \Delta S > 0. \qquad (5.10\text{-}3)$$

To put (5.10-3), which is stronger than (5.10-1) and often as useful as (5.10-1), in the desired form of (5.10-1), we use the following inequality.

$$0 \le \left(\tfrac{1}{\sqrt{2}} Q^{1/2} \pm \sqrt{2}\, \Delta S^T\, P\, S\, Q^{-1/2}\right) \left(\tfrac{1}{\sqrt{2}} Q^{1/2} \pm \sqrt{2}\, \Delta S^T\, P\, S\, Q^{-1/2}\right)^T$$

$$= \tfrac{1}{2} Q \pm \Delta S^T\, P\, S \pm S^T\, P\, \Delta S + 2\, \Delta S^T\, P\, S\, Q^{-1}\, S^T\, P\, \Delta S. \qquad (5.10\text{-}4)$$

Here $Q^{1/2}$ is the symmetric positive-definite square root of Q. From (5.10-4) we have

$$\tfrac{1}{2} Q - \Delta S^T\, P\, S - S^T\, P\, \Delta S \ge -2\, \Delta S^T\, P\, S\, Q^{-1}\, S^T\, P\, \Delta S, \qquad (5.10\text{-}5)$$

hence

$$\tilde{Q} \ge \tfrac{1}{2} Q - 2\, \Delta S^T\, P\, S\, Q^{-1}\, S^T\, P\, \Delta S - \Delta S^T\, P\, \Delta S. \qquad (5.10\text{-}6)$$

For the convergent property of $S + \Delta S$, it is sufficient to have the right-hand side of (5.10-6) a positive-definite matrix. ∎

From (5.1-6) and for cases such that $S = S^T$ and $Q = I$, we have $P = (I - S^2)^{-1}$. Substituting this P into (5.10-1) yields

$$\Delta S^T\, P\, (I + S^2)\, P\, \Delta S < \tfrac{1}{2} I. \qquad (5.10\text{-}7)$$

Similarly from (5.1-7) for cases such that $SS^T = S^T S$, $Q = I$, we have $P = (I - SS^T)^{-1}$, and substituting this P into (5.10-1) results in

$$\Delta S^T\, P\, (I + S\, S^T)\, P\, \Delta S < \tfrac{1}{2} I. \qquad (5.10\text{-}8)$$

5.10. Construction of Robustly Stable Discrete – Time Systems

REMARK 5.10-1: To "compute" ΔS from (5.10-1) such that $S + \Delta S$ remains a convergent matrix in a typical application is the main issue in the remaining part of this study. This computation is an old question, that goes back to several decades ago. For example, in [*D.* 60] a linear matrix equation was solved in order to compute ΔS in a similar setting as that of (5.10-3). This solution was found based on the fact that the *entire* class of perturbed systems should have the same P (different Q's) as the original system. Because to generate P (or equivalently to generate a Lyapunov function) for a number of systems was the research focal in the late 1960's and the early 1970's. In today's parlance this property can be rephrased in terms of a 'family of perturbed (or interval) matrices which has the same P as the unperturbed matrix and therefore the entire family of matrices is convergent.' Perhaps a brief mention of this work should be of interest to some reader. Using their original notation, Shane and Barnett show that if A (the same as our S) $\to$ A + B (B is the same as our ΔS) such that $B = (S - \frac{1}{2}P)^{-1}(SY + \frac{1}{2}PY + PA)$, with $A^TPA + Q = P$, and Y is *any* real symmetric solution of $YPY + A^TPY + YPA + 2Q_0 = 0$, where Q_0 is *any* symmetric matrix for which $Q + 2Q_0 > 0$, and S is *an arbitrary* real skew-symmetric matrix. Then A + B is convergent. Again the proof rests on the fact that A + B and A have the same P (with different Q), and therefore they both have the same Lyapunov function ($V = x^TPx$). The procedure that is suggested in [*D.* 60] does not find engineering applications, because of the way that B is generated, which in most cases does not directly correspond to the actual or physical perturbation of the system, due to many arbitrary choices made in this work. Note that as in the case of continuous-time systems, instead of generating B (which we call ΔS), we have provided a sufficient condition that represents the actual mechanism for change or the perturbation policy.

Our *sole* objective in this study is to "compute" ΔS from (5.10-1). To avoid repetition, we also recall that all the relevant discussions in § 5.5.3, on matrix inequalities and the diagonalization theorem, directly apply to this study. As shown in Theorem 5.10-3, we may utilize (5.10-1) as a "set generator", in order to construct ΔS. Also in order to "generate" ΔS, we recall that this solution depends on the structure of ΔS, which has the same catalog as in § 5.5.2. We can generally divide this catalog into two groups: *(i)* single-parameter variations; and *(ii)* multiple-parameter variations. Numerical examples have shown that (5.10-1) does not provide the global or *the largest* possible perturbation for an asymptotically convergent matrix (a matrix whose norm is less than one). Clearly this "local-upper bound" is only a sufficient condition and it is a function of $Q = Q^T > 0$, and it seems that one way to rectify this dependency on the initial choice of Q, is to use an iterative or probing method in order to circumvent this situation, just as we have proposed in the continuous-time case. We now study these matters as succinctly as possible.

5.10.3. Computational Issues

5.10.3.1. Single – Parameter Variations

To apply (5.10-1) to any structured perturbation matrices which are similar to those detailed in § 5.5.2, and corresponding to the single-parameter variations, we face the same question that we address in § 5.6.2, namely: given $\Delta S = \varepsilon \Delta S'$ (where $\Delta S'$ is numerically known and is a *candidate* for ΔS, and we want that $\varepsilon \Delta S'$ becomes an *admissible* ΔS), for what "value" of $Q = Q^T > 0$, we have the best result? By the best result we mean a value for $\Delta \theta$, that yields $S + \Delta S$ such that all its eigenvalues are inside the unit circle in the z-plane, and any *additional* changes in $\Delta \theta$ will push these eigenvalues towards and subsequently outside of this circle. For computational purposes we may choose a circle that its radius δ is slightly less than one to avoid computational instability. This circle is called the δ-circle. Since we have in effect only one single parameter ε, this situation is also called *the largest directional* ΔS whose determination is independent from any computational procedure and its final value is fixed. We propose the following theorem in order to generate this *largest*-upper bound for $\Delta \theta$ or for that matter ΔS, such that $S + \Delta S$ remains an asymptotically convergent matrix. The procedure that is supported by this theorem is independent from the initial choice for $Q = Q^T > 0$.

THEOREM 5.10-2: Let $Q = Q^T > 0$, $S_1 \equiv S_0$ be an asymptotically convergent matrix (each subscript J refers to the iteration number), and $P_1 = P_1^T > 0$ satisfies $S_1^T P_1 S_1 + Q = P_1$. Let $\Delta S'$ be an arbitrary $n \times n$ numerical matrix and rewrite the previous equation as follows.

$$(S_1 + \varepsilon \Delta S')^T P_1 (S_1 + \varepsilon \Delta S') + Q = P_1 + \varepsilon (\Delta S'^T P_1 S_1 + S_1^T P_1 \Delta S') + \varepsilon^2 \Delta S'^T P_1 \Delta S'.$$

(5.10-9)

Then the following can be shown.

(a) There is a largest $|\varepsilon_1'| < 1$ such that

$$\Pi_1 \triangleq Q - \varepsilon_1 (\Delta S'^T P_1 S_1 + S_1^T P_1 \Delta S') - \varepsilon_1^2 \Delta S'^T P_1 \Delta S' > 0, \text{ if } |\varepsilon_1| < |\varepsilon_1'| < 1.$$

(5.10-10)

(b) Let $S_2 \triangleq S_1 + \varepsilon_1 \Delta S'$. If S_2 remains asymptotically convergent, then $S_2^T P_2 S_2 + Q = P_2$; and like (a), an $|\varepsilon_2'| < 1$ can be found such that

$$\Pi_2 \triangleq Q - \varepsilon_2 (\Delta S'^T P_2 S_2 + S_2^T P_2 \Delta S') - \varepsilon_2^2 \Delta S'^T P_2 \Delta S' > 0, \text{ if } |\varepsilon_2| < |\varepsilon_2'| < 1.$$

(5.10-11)

5.10. Construction of Robustly Stable Discrete – Time Systems

(c) Similarly $\varepsilon_1, \varepsilon_2, \varepsilon_3, \ldots,$ can be generated. It can be shown that if $\tilde{\varepsilon}_K = \sum_{J=1}^{K} \varepsilon_J,$ then $S_1 + \tilde{\varepsilon}_\infty \Delta S'$ is not asymptotically convergent, but $S_1 + \tilde{\varepsilon}_K \Delta S'$ is asymptotically convergent if $|\tilde{\varepsilon}_K| < |\tilde{\varepsilon}_\infty|$.

PROOF: (a) From $|\varepsilon'_1| < 1$ and Theorem 5.10-1, we have

$$\varepsilon_1^2 \Delta S'^T (P + 2P S Q^{-1} S^T P) \Delta S' < \tfrac{1}{2} Q. \tag{5.10-12}$$

From (5.10-12), we have $-\varepsilon_1^2 \Delta S'^T P \Delta S' > -\tfrac{1}{2} Q + \gamma(\varepsilon_1) Q$, where $\gamma(\varepsilon_1) \triangleq (2\varepsilon_1^2 \Delta S'^T P S Q^{-1} S^T P \Delta S') Q^{-1}$. If we let $\varepsilon_1 = 1$ (which results in a conservative-upper bound), then we can compute $\gamma(\varepsilon_1 = 1) \triangleq \gamma'$ very easily. Thus

$$\Pi_1 > \Pi'_1 \triangleq (\tfrac{1}{2} I + \gamma') Q - \varepsilon_1 (S_1^T P_1 \Delta S' + \Delta S'^T P_1 S_1). \tag{5.10-13}$$

From (5.10-13), and Theorem 5.5-2, we can simultaneously diagonalize the symmetric positive-definite matrix $(\tfrac{1}{2} I + \gamma')Q$ and the symmetric matrix $(S_1^T P_1 \Delta S' + \Delta S'^T P_1 S_1)$, and to find ε_1 such that $\Pi'_1 > 0$. We must realize that the possibility of having both $\pm \varepsilon_1$ in certain problem exists and this situation must be carefully analyzed.

(b) Straightforward repetition of the previous step.

(c) Continuing similarly as in step (a), we can show that by the appropriate choice of ε_K a sequence of matrices can be established such that at each stage $S_{K+1} = S_K + \varepsilon_K \Delta S'$ remains asymptotically convergent; i.e.,

$$\Pi_K \triangleq Q - \varepsilon_K (\Delta S'^T P_K S_K + S_K^T P_K \Delta S') - \varepsilon_K^2 \Delta S'^T P_K \Delta S' > 0. \tag{5.10-14}$$

Each S_{K+1} has some eigenvalues distributed radially closer to the δ-circle than those of S_K. As K increases, say $K = M$, and as we generate Π_M and S_{M+1}, we reach a limiting case wherein S_{M+1} is no longer a convergent matrix or Π_M is no longer a positive-definite matrix. Recall that by the fundamental assumption our initial convergent matrix is sensitive to parameter variations. The key point is that S_{K+1} can be generated by K-times repetition of the previous step, or

$$S_{K+1} = S_1 + \sum_{J=1}^{K} \varepsilon_K \Delta S' \triangleq S_1 + \tilde{\varepsilon}_K \Delta S'. \tag{5.10-15}$$

Thus the largest perturbation matrix is $S_{K+1} - S_1$. This provides the proof of this theorem and the requirement for the convergence of the next algorithm. ∎

Thus based on the consequence of Theorem 5.10-2, we let $Q = I$ or $2I$. ∎

ALGORITHM 5.10-1: The flow-chart of Fig. 5.10-1, shows the basic concept in computation of *the largest* $\Delta\theta$, for a desired single-parameter variation, structurally perturbed matrix that can be presented as $\varepsilon \Delta S'$ such that $S_1 + \varepsilon \Delta S'$ remains asymptotically convergent. Here we select the δ-circle, such that $|\delta| = 0.9999$. In most cases we specify the maximum number of iterations, say $N = 100$, and we study the final result when we reach this limiting value.

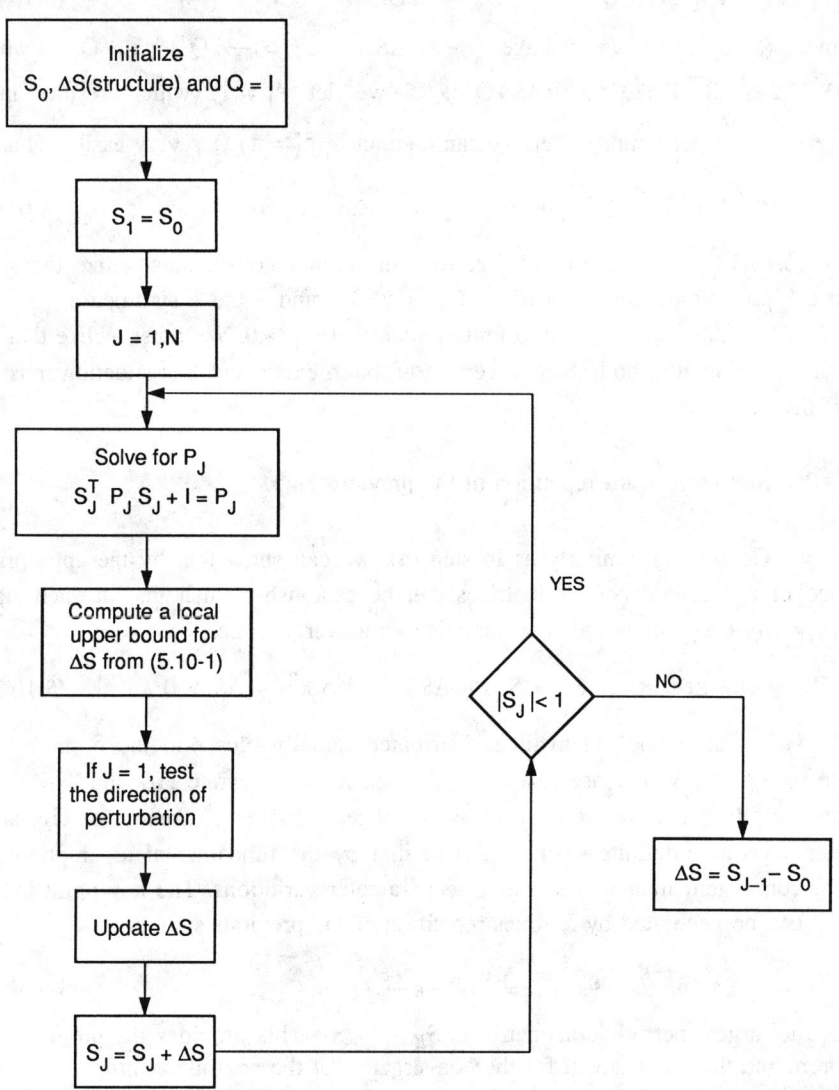

Fig. 5.10-1. A schematic diagram of flow-chart for Algorithm 5.10-1.

5.10.3.2. Multiple – Parameter Variations

The main thrust of this section is to find out how we can judiciously benefit from the properties of (5.10-1) and ΔS, in order to simplify the process of solving a quadratic matrix inequality such as (5.10-1), particularly when ΔS has multiple-varying parameters. Although the following discussion is almost *identical* with that of § 5.6.3, the underlying message is that contrary to many other methods which are reported in the literature, the development of the *SRM* (for the continuous-time system) and the *CRM* (for the discrete-time system) in this book can be concluded from one unified analytical approach and that is really the strength of this methodology.

As in the case of continuous-time systems we want to avoid *direct* algebraic solution for ΔS in (5.10-1), when this ΔS is of type F or G (or possibly E) given in § 5.5.2. To do so we apply the constructional procedure that is suggested for the case of single-parameter variations (i.e., when we let $\Delta S = \varepsilon \Delta S'$, where $\Delta S'$ is known and we search for the best ε), to F or G, in order to generate *an* admissible ΔS. This ΔS that corresponds to one point $[\Delta\theta_1, ..., \Delta\theta_r]^T$ in the *PVS* may be *an* acceptable perturbation matrix. However, if we choose another candidate for ΔS (i.e., a new $\Delta S'$) and search for a new admissible ΔS (i.e., a new ε), then we have a completely different directional ΔS. Obviously each one of these directional ΔS's corresponds to *one* point $[\Delta\theta_1, ..., \Delta\theta_r]^T$ in the *PVS*. Having generated these different directional ΔS's (or the corresponding points in the *PVS*), then we can show that there exists a connected set of allowable parameter variations in the *PVS* whose corresponding ΔS's yield a class of asymptotically convergent matrices when all the varying parameters are changing in that set. Since there are infinitely many possible such ΔS's, it is essential to have a procedure to construct this set from the information which stems from a finite number of admissible ΔS's. That construction is indeed possible as the consequence of next theorem, which in effect advances a *single by single* (or directional) search for a convergent perturbed matrix (or a *point by point* search in the *PVS*) to that of a *set by set* search for such matrices (or a *set by set* search for points in the *PVS*), and is the main result of this section which is essentially identical with its continuous-time counterpart.

THEOREM 5.10-3 (E-Box): Consider (5.10-1) with $Q = 2I$ repeated below.

$$\Delta S^T(\Delta\theta)(P + P\,S\,S^T\,P)\,\Delta S(\Delta\theta) \triangleq \Delta S^T(\Delta\theta)X\Delta S(\Delta\theta) < I. \qquad (5.10\text{-}16)$$

Suppose that from this inequality we have extracted, in principle, 2^r-directional "solutions": $\Delta S_1, \Delta S_2, ..., \Delta S_{2^r}$; each corresponding to one point $[\Delta\theta_1, ..., \Delta\theta_r]^T$ in the *PVS*, which form a hyperbox in that space. If for each and every point in this hyperbox we have

$$\Delta S = \sum_{i=1}^{2^r} w_i \Delta S_i, \text{ with } \quad \sum_{i=1}^{2^r} w_i = 1, \quad 0 \le w_i < 1, \text{ for all i.} \qquad (5.10\text{-}17)$$

Then each and every point of this hyperbox corresponds to an asymptotically convergent $S + \Delta S$.

PROOF: Here we assume that the set of allowable-parameter variations of a dynamical system *around a given convergent-operating point* (the origin of the *PVS*) is connected. Since the proof is line by line the same as that of Theorem 5.6-2, we only state the differences. To show that $S + \Delta S$ is asymptotically convergent we note that

$$\Delta S^T X \Delta S = \left(\sum_{i=1}^{2^r} w_i \Delta S_i^T\right) X \left(\sum_{i=1}^{2^r} w_i \Delta S_i\right)$$

$$= \sum_{i=1}^{2^r} w_i^2 \Delta S_i^T X \Delta S_i + \sum_{k,l=1, k \neq l}^{2^r} (w_k w_l \Delta S_k^T X \Delta S_l + w_l w_k \Delta S_l^T X \Delta S_k).$$

(5.10-18)

It is also true that we always have

$$(w_k \Delta S_k \pm w_l \Delta S_l)^T X (w_k \Delta S_k \pm w_l \Delta S_l) \geq 0. \qquad (5.10\text{-}19)$$

Therefore

$$\pm (w_k w_l \Delta S_k^T X \Delta S_l + w_l w_k \Delta S_l^T X \Delta S_k) \leq w_k^2 \Delta S_k^T X \Delta S_k + w_l^2 \Delta S_l^T X \Delta S_l. \quad (5.10\text{-}20)$$

Using the above set of estimates in (5.10-18), with $\Delta S_i^T(\Delta \theta) X \Delta S_i(\Delta \theta) < I$ for all i, yields

$$\Delta S^T X \Delta S \leq 2^r \sum_{i=1}^{2^r} w_i^2 \Delta S_i^T X \Delta S_i \leq 2^r \sum_{i=1}^{2^r} w_i^2 I. \qquad (5.10\text{-}21)$$

The rest of the argument is exactly the same as in Theorem 5.6-2. Thus ΔS as constructed above yields asymptotically convergent $S + \Delta S$ for each point in that hyperbox.

REMARK 5.10-2: Clearly the above convexity condition is true for linear perturbation matrices (cases similar to *F*, in § 5.5.2, including polynomials). The situations for the nonlinear perturbation matrices that are similar to case *G*, in § 5.5.2, require the same treatment which we describe in Remark 5.6-1. Furthermore, the set that is to be constructed as a result of Theorem 5.10-3 is not the "largest" set as numerical examples show, and here again we must devise an iterative method to extend this set. Example 5.10-5 shows one such application.

5.10.4. Extensions to Polynomials

We now specialize (5.10-1) to the case where S is in a phase-variable canonical form. This representation is used to generate the allowable perturbations for the coefficients in a monic-convergent polynomial $\Delta(z) = \det(zI - S)$. The following derivation is in parallel with that of § 5.7.3.

THEOREM 5.10-4: Consider a monic-convergent polynomial $\Delta(z, \theta) = \sum_{j=0}^{n} b_j(\theta) z^{n-j}$. Let $\theta \to \theta + \Delta\theta$ such that $b_j \to b_j + \Delta b_j$. Then there exists an upper bound β_l, where $\beta_l^2 = 1/2 y_{nn} \lambda$, $l = n + 1 - j$, such that

5.10. Construction of Robustly Stable Discrete – Time Systems

$\tilde{\Delta}(z, \theta+\Delta\theta) = z^n + b_1 z^{n-1} + \cdots + (b_j + \Delta b_j) z^{n-j} + \cdots + b_n$, remains convergent if

$$(\Delta b_j)^2 \le \beta_l^2. \qquad (5.10\text{-}22)$$

Here y_{nn} is the (n, n)-entry of $(P + 2PSQ^{-1}S^T P)$, $Q = Q^T > 0$, and $P = P^T > 0$, satisfies (5.1-2) for S in a phase-variable canonical form such that $\Delta(z, \theta) = \det(zI - S(\theta))$; and λ is the largest (or the unique nonzero) root of $\det(\lambda Q - K)$, with $K = \text{diag}(0, \ldots, 0, 1_{ll}, 0, \ldots, 0)$. (Notice that by the asymptotic convergent property of $\tilde{\Delta}(z, \theta+\Delta\theta)$, $y_{nn} \ne 0$.)

PROOF: Let $S \to S + \Delta S$, then $S + \Delta S$ remains a convergent matrix if $\Delta S^T (P + 2PSQ^{-1}S^T P) \Delta S \triangleq \Delta S^T Y \Delta S < \tfrac{1}{2}Q$. If $b_j \to b_j + \Delta b_j$, then ΔS has only one nonzero entry $-\Delta b_j$ in its (n, l) entry. Carrying out the multiplication on the left-hand side of the previous inequality yields $(\Delta b_j)^2 y_{nn} K < \tfrac{1}{2}Q$. By simultaneous diagonalization of K and Q we have $(\Delta b_j)^2 y_{nn} \text{diag}(\lambda_1, \ldots, \lambda_n) < \tfrac{1}{2} I$. ∎

For cases in which all the coefficients are functions of one parameter and the changes in this parameter make $b_j \to b_j + \Delta b_j$, for all j, or when we allow all the coefficients to change simultaneously, we present the following theorem.

THEOREM 5.10-5: Consider the monic-convergent polynomial $\Delta(z, \theta)$ of Theorem 5.10-4. If a perturbation causes all the coefficients of $\Delta(z, \theta)$ to change, then subject to the following upper bound this polynomial remains convergent.

$$\sum_{j=1}^{n} (\Delta b_j)^2 < \lambda_{\min}(Q) /2 y_{nn}. \qquad (5.10\text{-}23)$$

Here y_{nn} and Q are the same as in Theorem 5.10-4, and $\lambda_{\min}(Q)$ refers to the minimum eigenvalue of Q.

PROOF: Suppose that $S \to S + \Delta S$ with ΔS having nonzero entries $-\Delta b_j$, $j = n, \ldots, 1$, at its last row. Then by Theorem 5.10-1, $S + \Delta S$ remains an asymptotically convergent matrix if (5.10-1) holds. As in Theorem 5.10-4, we let $Y \triangleq P + 2PSQ^{-1}S^T P$, then

$$\Delta S^T Y \Delta S \triangleq y_{nn} \Pi < \tfrac{1}{2}Q, \qquad (5.10\text{-}24)$$

where Π is similar to that in (5.7-5) for b_j's. In (5.10-24) every element is known except matrix Π, which in general is a nonlinear function of the encountered perturbation – the quantity that we are seeking. Because the rank of Π is one, its maximum eigenvalue is equal to its trace. Thus we have $y_{nn} \text{Tr}(\Pi) < \tfrac{1}{2}\lambda_{\min}(Q)$. ∎

Clearly if for this polynomial, the encountered perturbation is a vector, i.e., multiple parameters variations, then (5.10-24) should be used as a *set generator*. Because in effect the applications of Theorem 5.10-3 yields the same perturbation policy as that of (5.10-24) for changing all the coefficients in a polynomial. We demonstrate this case in Example 5.10-4.

5.10.5. Numerical Examples

In this section we present several simple numerical examples to emphasize certain applications of the results developed in this section. These examples should be considered in conjunction with those presented in § 5.8.1.

EXAMPLE 5.10-1 (*Single-parameter variations / directional perturbations*): Consider a symmetric asymptotic convergent matrix S_0 whose eigenvalues are $\lambda_1 = 0.64$ and $\lambda_2 = -0.39$ as follows.

$$S_1 \equiv S_0 = \frac{1}{4}\begin{bmatrix} 1 & 2 \\ 2 & 0 \end{bmatrix}. \qquad (5.10\text{-}25)$$

This choice enables us to use the closed-form solution for the corresponding Lyapunov equation with $Q = I$ and as follows.

$$P_1 = (I - S_1^2)^{-1} = \frac{1}{8}\begin{bmatrix} 12 & 2 \\ 2 & 11 \end{bmatrix}. \qquad (5.10\text{-}26)$$

Let the perturbation scheme be according to one of the following three directional matrices:

$$\Delta S_\alpha = \begin{bmatrix} 1 & 0 \\ 0 & 0 \end{bmatrix}, \quad \Delta S_\beta = \begin{bmatrix} 0 & 1 \\ 1 & 0 \end{bmatrix}, \text{ and } \Delta S_\gamma = \begin{bmatrix} 1 & 1 \\ 1 & 1 \end{bmatrix}. \qquad (5.10\text{-}27)$$

The objective is to compute *the largest* $|\varepsilon|$, such that $S_1 + |\varepsilon| \Delta S_{\alpha \text{ or } \beta \text{ or } \gamma}$ remains a convergent matrix.

SOLUTION [D. 14]: The perturbation matrix for the above three cases must satisfy (5.10-1). For ΔS_α, we have

$$\frac{\varepsilon_1^2}{1024}\begin{bmatrix} 1 & 0 \\ 0 & 0 \end{bmatrix}\begin{bmatrix} 12 & 2 \\ 2 & 11 \end{bmatrix}\begin{bmatrix} 21 & 2 \\ 2 & 20 \end{bmatrix}\begin{bmatrix} 12 & 2 \\ 2 & 11 \end{bmatrix}\begin{bmatrix} 1 & 0 \\ 0 & 0 \end{bmatrix} < \frac{I}{2}, \qquad (5.10\text{-}28)$$

which yields $|\varepsilon_1| = 0.4$. If $\varepsilon_1 = 0.4$, then

$$S_2^+ = S_1 + 0.4\,\Delta S_\alpha = \begin{bmatrix} 0.65 & 0.5 \\ 0.5 & 0 \end{bmatrix}, \qquad (5.10\text{-}29)$$

whose eigenvalues are $\lambda_1 = 0.92$ and $\lambda_2 = -0.27$. On the other hand, if we choose $\varepsilon_1 = -0.4$, then

$$S_2^- = S_1 - 0.4\,\Delta S_\alpha = \begin{bmatrix} -0.15 & 0.5 \\ 0.5 & 0 \end{bmatrix}, \qquad (5.10\text{-}30)$$

whose eigenvalues are $\lambda_1 = 0.43$ and $\lambda_2 = -0.58$.

Comparing the above two pairs of eigenvalues with the original pair of $\lambda_1 = 0.64$ and $\lambda_2 = -0.39$, we notice that the perturbations in both positive and negative directions of ε are acceptable. We continue our studies by choosing $\varepsilon_1 = 0.4$ first, and thus

5.10. Construction of Robustly Stable Discrete – Time Systems

$$S_2^+ = \frac{1}{20} \begin{bmatrix} 13 & 10 \\ 10 & 0 \end{bmatrix} \rightarrow (S_2^+)^2 = \frac{1}{400} \begin{bmatrix} 269 & 130 \\ 130 & 100 \end{bmatrix}. \quad (5.10\text{-}31)$$

Then

$$P_2 = (I - (S_2^+)^2)^{-1} = \frac{1}{28} \begin{pmatrix} 150 & 65 \\ 65 & 131/2 \end{pmatrix}. \quad (5.10\text{-}32)$$

For this S_2^+, we construct $S_3^+ = S_2^+ + \varepsilon_2 \Delta S_\alpha$, where using the corresponding (5.10-1) yields.

$$\varepsilon_2^2 \Delta S_\alpha^T P_2 (I + (S_2^+)^2) P_2 \Delta S_\alpha < \tfrac{1}{2} I. \quad (5.10\text{-}33)$$

Substituting for $(S_2^+)^2$ and P_2 from (5.10-31) and (5.10-32) into (5.10-33) results in $\varepsilon_2 = 0.089$. By Algorithm 5.10-1 and proceeding in that manner, we reach $\varepsilon_\alpha = \sum_{J=1} \varepsilon_J \le 0.5$, which is *the largest* ε_α for this structured (or directional) perturbation and in the positive direction of ε_α. Resulting in $\lambda_1 = 1$ and $\lambda_2 = -0.25$. We now go back and choose $\varepsilon_1 = -0.4$ to generate *the largest* ε_α and in the negative direction which becomes $\varepsilon_\alpha \ge -1.0$. Resulting in $\lambda_1 = 0.25$ and $\lambda_2 = -1.0$. Therefore *the super largest* interval in this problem is $-1.0 \le \varepsilon_\alpha \le 0.5$, where in this interval both eigenvalues of the perturbed system are in the unit circle.

Similarly, for $S_2 = S_1 + \varepsilon_1 \Delta S_\beta$, we can use (5.10-1) to generate $\varepsilon_1^2 = 0.12339$, which yields $\varepsilon_1 = 0.3513$. We choose the positive sign for ε_1 and proceed according to Algorithm 5.10-1, in order to generate *the largest* $\varepsilon_\beta = \sum_{J=1} \varepsilon_J \le 0.366$, which in this case is in the positive direction.

Finally, for $S_2 = S_1 + \varepsilon_1 \Delta S_\gamma$, i.e., the case that we are uncertain about every elements of S_0, but we are perturbing that *only* along the "$45°$ line", we can show that (5.10-7) yields $\varepsilon_1 = 0.176$. Proceeding in this manner results in $\varepsilon_\gamma = \sum_{J=1} \varepsilon_J \le 0.1818$.

▼

It is interesting to note that the first local-upper bound in each of the above three cases is very close to the global-upper bound (around the operating point) that is generated from Algorithm 5.10-1. Furthermore, we may look at this example in the same context as in Example 5.8-4, in terms of the questions which are raised in that example for the corresponding directional perturbations.

■

EXAMPLE 5.10-2 (*Directional perturbations and polynomials*): For a nonsymmetric (asymptotic) convergent matrix $S_0 \triangleq \begin{pmatrix} 0.0 & 1.0 \\ 0.125 & 0.25 \end{pmatrix}$ whose eigenvalues are $\lambda_1 = 0.5$ and $\lambda_2 = -0.25$, suppose that we perturb this matrix according to $\Delta S_J \triangleq \varepsilon_J \begin{pmatrix} 0.0 & 0.0 \\ 1.0 & 0.0 \end{pmatrix}$, where $J = 1, 2, \ldots$, is the iteration number. Then our objective is to find the largest $\varepsilon \triangleq \sum_{J=1} \varepsilon_J$, such that $(S_0 + \Delta S) \triangleq (S_0 + \Delta S_1 + \Delta S_2 + \cdots)$ remains a convergent matrix. (Note that this example is also applicable to convergent polynomials.)

SOLUTION [D. 14]: From classical control theory the largest ε for this system is 0.625, resulting in $\lambda_1 = 1$ and $\lambda_2 = -0.75$ for the eigenvalues of the perturbed system. Therefore the perturbed system becomes unstable for $\varepsilon > 0.625$. To find ε according to Algorithm 5.10-1 we have ($S_1 \equiv S_0$ and $Q = I$)

$$P_1 = \begin{pmatrix} 1.0346 & 0.0788 \\ 0.0788 & 2.2123 \end{pmatrix} \rightarrow P_1 + 2P_1 S_1 S_1^T P_1 = \begin{pmatrix} 3.2579 & 1.4167 \\ 1.4167 & 3.1638 \end{pmatrix}. \quad (5.10\text{-}34)$$

The perturbation matrix must satisfy (5.10-1) and in this case we have

$$\varepsilon_1^2 \begin{pmatrix} 3.1638 & 0.0 \\ 0.0 & 0.0 \end{pmatrix} < \tfrac{1}{2} I \rightarrow \varepsilon_1 \leq 0.3975. \quad (5.10\text{-}35)$$

Letting $S_2 \triangleq S_1 + \Delta S_1$, we search for ΔS_2 such that $S_2 + \Delta S_2$ remains a convergent matrix. Following the above steps we reach $\varepsilon_2 \leq 0.1601$. Continuing in this fashion we get $\varepsilon_3 \leq 0.0475$, and $\varepsilon_4 \leq 0.014$, and finally $\varepsilon_5 \leq 0.0058$. Thus the largest perturbation step is $\varepsilon \triangleq \varepsilon_1 + \varepsilon_2 + \varepsilon_3 + \varepsilon_4 + \varepsilon_5 = 0.6249 < 0.625$.

▼

REMARK 5.10-3: Recently in [D. 40], it is shown that a sufficient condition for parameter variation of the type studied in Theorem 5.10-1 (the first iteration), is that $\sigma_{max}(\Delta S_1) < -\sigma_{max}(S_1) + \sqrt{[\sigma_{max}(S_1)]^2 + \sigma_{min}(Q) / \sigma_{max}(P_1)}$, where $\sigma(\cdot)$ stands for a singular value of $(\cdot)$. Applying, for example, the previous inequality to Example 5.10-2 yields $\varepsilon_1^2 < -1.063432 + \sqrt{[1.063432]^2 + 1/4.917} = 0.09167$ and thus $\varepsilon_1 = 0.3028$, which is smaller than the $\varepsilon_1 = 0.3975$ generated from Algorithm 5.10-1.

REMARK 5.10-4: Recently in [D. 80], several alternative *CRM's* are introduced which are based on a scaled-Lyapunov equation $(1 + \alpha)S^T P S - P = -I$. In particular, it is shown that if we let $\Delta S = \varepsilon U$, where U is a unity matrix, then for $\varepsilon < \{(1+\alpha^{-1})\sigma_{max}(U|P|U)\}^{-1/2}$, $S + \Delta S$ remains asymptotically convergent. This result is valid for a general $\Delta S = \varepsilon \Delta S'$. Here σ stands for singular value and $0 < \alpha < \infty$ is chosen in order that all the eigenvalues of the asymptotically convergent matrix *remain* in a disk with a radius $(1 + \alpha)^{-1/2}$, and P is the solution of the above scaled-Lyapunov equation. In our application of this result, however, we note that as we change α and compute each corresponding ε_α's and subsequently the eigenvalues of the perturbed matrix $S + \varepsilon_\alpha U$, these eigenvalues move outside the disk with radius $(1 + \alpha)^{-1/2}$.

First of all, since we can always let $Q = I$ and assume that $P + 2PSS^T P \leq \tfrac{1}{2}(1 + \alpha^{-1})P$, for some scalar $\alpha > 0$, and rewrite (5.10-1) as $(1 + \alpha^{-1})\Delta S^T P \Delta S < I$, in essence this result is essentially similar to that of Theorem 5.10-1. *Secondly,* Theorem 5.10-1 for the proposed perturbational scheme of $\Delta S = \varepsilon U$ (U is a unity matrix), with $Q = I$, yields

5.10. Construction of Robustly Stable Discrete – Time Systems

$$\varepsilon^2 U(P + 2P\,S\,S^T\,P)\,U < \tfrac{1}{2}I. \qquad (5.10\text{-}36)$$

Here ε is a scalar and using the well-defined maximum and minimum eigenvalues, (5.10-36) reads as

$$\varepsilon^2 \lambda_{max}[U(P + 2P\,S\,S^T\,P)U] = \varepsilon^2 [n\sum_{i,j=1}^{n}(P + 2P\,S\,S^T\,P)_{i,j}] < \lambda_{min}(\tfrac{1}{2}I) = \tfrac{1}{2}.$$

$$(5.10\text{-}37)$$

In (5.10-37) we use the fact that given any $A \triangleq (A_{i,j}) \in R^{n \times n}$ and a unity matrix U, we always have $\lambda_{max}(UAU) = n\sum_{i,j=1}^{n} A_{i,j}$. Thus similarly the upper bound given in [D. 80] can be written as $\varepsilon < \{(1+\alpha^{-1})\,\sigma_{max}(U|P'|U)\}^{-\frac{1}{2}} = \{n(1+\alpha^{-1})\sum_{i,j=1}^{n}|P'_{i,j}|\}^{-\frac{1}{2}}$, or as $\tfrac{1}{2}\varepsilon^2 \{n(1+\alpha^{-1})\sum_{i,j=1}^{n}|P'_{i,j}|\} < \tfrac{1}{2}$. Note that here we use P' in order to emphasize that this upper bound and (5.10-37) use different Lyapunov equations. In order to make a comparison between these two upper bounds, we note that if we assume for the sake of discussion that $P = P'$, and since $\sum_{i,j=1}^{n}(P_{i,j}) < \sum_{i,j=1}^{n}[(P + 2P\,S\,S^T\,P)_{i,j}]$, then we may conclude that the result in [D. 80] has improved (5.10-37). However, multiplying the left-hand side of the previous inequality by $\tfrac{1}{2}(1 + \alpha^{-1})$ and depending on the choice of α the preceding result *may* reverse its direction, i.e., unless α is chosen properly the results in [D. 80] may be smaller than (5.10-1).

EXAMPLE 5.10-3 [D. 80]: Consider an asymptotically convergent matrix $S(\equiv S_1) = \begin{pmatrix} 0.5 & 1.0 \\ 0.0 & 0.0 \end{pmatrix}$ whose eigenvalues are $\lambda_1 = 0.5$ and $\lambda_2 = 0.0$. Suppose that we perturb this matrix along the "45° line" according to $\Delta S = \varepsilon U$. Then what is the *largest* $\varepsilon > 0$ such that $S + \Delta S$ remains convergent?

SOLUTION: The characteristic equation of the perturbed system is

$$\Delta(z) = (zI - S - \Delta S) = \det\begin{pmatrix} z - 0.5 - \varepsilon & -1 - \varepsilon \\ -\varepsilon & z - \varepsilon \end{pmatrix}$$

$$= z^2 - (2\varepsilon + 0.5)z - 0.5\varepsilon = 0. \qquad (5.10\text{-}38)$$

Using the classical stability test we conclude that for $-1 < \varepsilon < 0.2$ the system remains convergent. In other words, the *CRM* (or the *E-Box*) is $[-1, 0.2]$. We call $\varepsilon_{max} = 0.2$.

To apply Theorem 5.10-1 (or (5.10-37)) to this situation, we compute P from $S^T P S + I = P$, followed by $U(P + 2PSS^T P)U$. Here

$$P = \tfrac{1}{3}\begin{pmatrix} 4 & 2 \\ 2 & 7 \end{pmatrix}, \quad U(P + 2PSS^T P)U = \tfrac{135}{9}\begin{pmatrix} 1 & 1 \\ 1 & 1 \end{pmatrix}. \qquad (5.10\text{-}39)$$

Thus the corresponding (5.10-37) becomes

$$\varepsilon^2 \frac{135}{9} \begin{pmatrix} 1 & 1 \\ 1 & 1 \end{pmatrix} < \begin{pmatrix} \frac{1}{2} & 0 \\ 0 & \frac{1}{2} \end{pmatrix}. \qquad (5.10\text{-}40)$$

From (5.10-40) we conclude that for $\varepsilon^2 < 9/540$ the system remains convergent, resulting in $\varepsilon < 0.1291$, which is about 64.5% of the $\varepsilon_{max} = 0.2$. Now, using the upper bound given in [D. 80], and because we want to compare under *equal* condition with Theorem 5.10-1, we let $\alpha = 0.00001$, which yields a radius of $(1 + \alpha)^{-\frac{1}{2}} = 0.999995$, and for this α the $(1 + \alpha)S^T PS + I = P$ gives almost the same P as (5.10-39) resulting in

$$\varepsilon < \{(1+\alpha^{-1})\sigma_{max}(U|P|U)\}^{-\frac{1}{2}} = \{100001\, \sigma_{max}\begin{pmatrix} 5 & 5 \\ 5 & 5 \end{pmatrix}\}^{-\frac{1}{2}} < 0.001.$$

$$(5.10\text{-}41)$$

This measure is about ½% of ε_{max} and is by far smaller than the preceding result. Obviously this study confirms that we should choose a proper α for developing the *CRM* of our system. Since no clear procedure is suggested in [D. 80], in order to find an "optimal" α, and the heuristic method of numerical search, which is implied in [D. 80], may not be an efficient way as discussed next, we choose this α according to Lemma 5.10-1.

But first, we note that there is an important issue in [D. 80] which in our opinion has not received proper attention. Earlier we point out that the dependency on Q (in Lyapunov equation *and* relative to the determination of the *global* solution) can be eliminated by an iterative method. Thus we let Q = I (or 2I) and repeatedly and accordingly proceed until we reach the global answer. In [D. 80], however, it is proposed to search *numerically* for an *optimal* α in order to *improve* the upper bound. But the real question is what does this search entail? The heuristic method that is proposed for this search is not efficient, because any numerical change in α and for the purpose of finding its optimal value requires that a new P be computed which in turn a new ε must be found. Then as we study these new ε's (or any other comparable upper bounds), we save those which are "improved" and we drop the lesser ones. At the end of this process we call the best ε (which is not, of course, the global ε_{max} in the classical sense) the upper bound for the encountered problem. The main difference between the proposed method in this paper and that presented herein is that all these computational efforts for finding the supposedly non-improved upper bounds will be discarded at the end. For instance, it is proposed that, in order to "improve" the upper bound, one should use, what we call a *[circular]* iterative search, which means several values are tried until one improved-upper bound is found. Only one out of many such choices for α is utilized and the remaining choices, with all their necessary

5.10. Construction of Robustly Stable Discrete – Time Systems

computations, are discarded, while the final result is also a local-upper bound and often with no apparent advantage over the known results.

We suggest, instead, to benefit from all these intermediate computations by using a *[directional]* iterative method which is the key to tackle these types of problems that have no globally explicit analytical solutions. For instance, we set $\varepsilon_1 (\equiv \varepsilon) = 0.1291$ as the first perturbation step toward computing ε_{max} in Example 5.10-3, then we can set $S_2 = S_1 + \varepsilon_1 U$ and perturb S_2 such that $S_2 + \varepsilon_2 U$ remains convergent. Here $S_2 = \begin{pmatrix} 0.6291 & 1.1291 \\ 0.1291 & 0.1291 \end{pmatrix}$ and from $S_2^T P_2 S_2 + I = P_2$, we have approximately $P_2 = \begin{pmatrix} 2.415 & 2.564 \\ 2.564 & 4.84 \end{pmatrix}$, $U(P_2 + 2P_2 S_2 S_2^T P_2)U = 132.271U$, resulting in $\varepsilon_2^2 132.271U < \frac{1}{2}I$ which yields $\varepsilon_2 = 0.043$. Augmenting this perturbation step with the previous one results in $\varepsilon = \varepsilon_1 + \varepsilon_2 = 0.1721$, which is about 86% of ε_{max} and this is just the *second* iteration. If we continue this well-defined path we reach the global answer very quickly. Having stated our case, we now propose the following method to search for an "optimal" α.

LEMMA 5.10-1: Consider a perturbation policy of the type $\Delta S = \varepsilon \Delta S'$ for a convergent matrix S, where $\Delta S'$ is a numerically known matrix and ε changes according to $\varepsilon^2 < 1/[(1 + \alpha^{-1}) \|\Delta S'^T P \Delta S'\|]$, $P = P^T > 0$ is the solution of $(1 + \alpha)S^T PS + I = P$ with $0 < \alpha < \infty$. Then the least-upper bound (l.u.b.) of ε^2 is maximized at $\alpha + 1 = 1/\|S\|$.

PROOF: It is easy to show that $\|P\| \leq 1/[1 - (1 + \alpha)\|S\|^2]$, and $\|\Delta S'^T P \Delta S'\| \leq \|\Delta S'\|^2/[1 - (1 + \alpha)\|S\|^2]$, which yields $\varepsilon^2 \leq [1 - (1 + \alpha)\|S\|^2] / [\|\Delta S'\|^2(1 + \alpha^{-1})]$ as its l.u.b.. Here $d\varepsilon^2/d\alpha = 0$ results in $1 + \alpha = 1/\|S\|$ and straightforward differentiation shows that for this α, the $d^2\varepsilon^2/d\alpha^2 < 0$, and thus this α corresponds to a maximum.

∎

Applying this "optimal" α to Example 5.10-3, we note that since different matrix norms are defined in the literature, we get different α's and different "upper bounds" corresponding to each appropriate norm of S. In this example using the maximum eigenvalue or spectral radius for $\|S\|$ gives $\alpha = 1$, which this in turn yields $\varepsilon = 0.1507$ which is better than $\varepsilon_1 = 0.1291$, but it is less than the $\varepsilon_1 + \varepsilon_2 = 0.1721$ and certainly this is less than $\varepsilon_{max} = 0.2$ as expected. In other words, no *global* answer can be provided using the Lyapunov technique *alone* in spite of this recent new parameterization. Thus the iterative method presented in this book is the only viable approach for finding the *global* solution for *CRM's*.

▼

REMARK 5.10-5: As stated in the introduction of this section (§ 5.10.1), there are many cases where we have a numerically defined asymptotically convergent matrix and we want to construct a set (or a family) of matrices *around* this initial matrix in order that this entire family of matrices remains asymptotically convergent. To give a point of view of what we mean by a numerically defined matrix, consider an example by Mori and Kokame [D. 51]. According to this reference the following set of matrices are convergent

$$\left\{ \begin{matrix} [-0.5, 0.5] & [0, 0.6] \\ [-0.25, 0.75] & 0 \end{matrix} \right\}. \quad (5.10\text{-}42)$$

If we consider the following matrix from the above set

$$\begin{bmatrix} 0.0 & 0.3 \\ 0.5 & 0 \end{bmatrix}, \quad (5.10\text{-}43)$$

then we propose the following question that suppose (5.10-43) is given and the task is the inverse problem of generating (5.10-42). That is the kind of problem which we want to address in the following example that also demonstrates the scope and applicability of Theorem 5.10-3. Incidentally, (5.10-42) also confirms our earlier notion about the nature of the multiple-parameters variations in a matrix and the fact that we need to define the corresponding "largest" variations very carefully. Here (5.10-42) represents a box (set). However, if we apply our method to this example (with three variables), we get a connected set which is not in the same geometrical shape as in (5.10-42), but we may stack a number of these boxes under it, which finally it may become "larger" than (5.10-42). We invite the reader to study this problem.

EXAMPLE 5.10-4 (Multiple-parameter variations): Consider an asymptotically convergent matrix of Example 5.10-2 whose eigenvalues are $\lambda_1 = 0.5$ and $\lambda_2 = -0.25$ as follows.

$$S \; (\equiv S_1) \triangleq \begin{bmatrix} 0 & 1 \\ \theta_2 & \theta_1 \end{bmatrix} = \begin{bmatrix} 0.0 & 1.0 \\ 0.125 & 0.25 \end{bmatrix}. \quad (5.10\text{-}44)$$

Our objective is to study the ways of constructing *the largest* set of r (= 2) independently varying parameters ΔS such that $S + \Delta S$ remains convergent, when

$$\Delta S = \begin{pmatrix} 0 & 0 \\ \Delta\theta_2 & \Delta\theta_1 \end{pmatrix}. \quad (5.10\text{-}45)$$

Or equivalently determine *the largest* set of points in the PVS (i.e., for $\Delta\theta_1$, $\Delta\theta_2$), whose corresponding perturbed matrices are convergent.

SOLUTION [D. 17]: Let $J = 1, 2, ..., N$ be the iteration number. The perturbation matrices must satisfy (5.10-16). Now, we take the following steps to complete the solution to this problem.

5.10. Construction of Robustly Stable Discrete – Time Systems

STEP 0 (To update S): We let $S_J = S_{J-1} \pm \Delta S_{J-1}$, with $S_1 \equiv S_0 \equiv S$ being the initial asymptotically convergent matrix and $\Delta S_0 \equiv 0$.

STEP 1 (To solve the Lyapunov equation): For the current S_J and $Q = 2I$ we solve $S_J^T P_J S_J + 2I = P_J$ for P_J. At $J = 1$ we have

$$P_1 = \begin{bmatrix} 2.069 & 0.158 \\ 0.158 & 4.425 \end{bmatrix}. \tag{5.10-46}$$

STEP 2 (Directional perturbations): We perturb the system iteratively along all the r axes and in a few "45° lines". We compute the corresponding largest directional perturbations along each one of these axes according to Algorithm 5.10-1. These results are summarized in Table 5.10-1.

Table 5.10-1. Single-Parameter Variations in Example 5.10-4.

$\Delta S_{\Delta\theta_2} = \Delta\theta_2 \begin{pmatrix} 0 & 0 \\ 1 & 0 \end{pmatrix}$	$S + \Delta S_{\Delta\theta_2}$ is convergent if $-1.125 < \Delta\theta_2 < 0.625$
$\Delta S_{\Delta\theta_1} = \Delta\theta_1 \begin{pmatrix} 0 & 0 \\ 0 & 1 \end{pmatrix}$	$S + \Delta S_{\Delta\theta_1}$ is convergent if $-1.125 < \Delta\theta_1 < 0.625$
$\Delta S_{45^o} = \varepsilon_{45^o} \begin{pmatrix} 0 & 0 \\ 1 & 1 \end{pmatrix}$	$S + \Delta S_{45^o}$ is convergent if $-1.125 < \varepsilon_{45^o} < 0.3125$

STEP 3 (To interpret the sufficient conditions): We need to study $\Delta S^T X \Delta S < I$, in order to establish a set of inequalities whose intersection yields the set of sufficient conditions for all parameter variations. Since in most cases ΔS has many zero entries, in each problem we simplify this condition before we apply the iterative method. Here we have

$$\begin{bmatrix} 0 & \Delta\theta_2 \\ 0 & \Delta\theta_1 \end{bmatrix} \begin{bmatrix} x_{11} & x_{12} \\ x_{21} & x_{22} \end{bmatrix} \begin{bmatrix} 0 & 0 \\ \Delta\theta_2 & \Delta\theta_1 \end{bmatrix} = x_{22} \begin{bmatrix} \Delta\theta_2^2 & \Delta\theta_2 \Delta\theta_1 \\ \Delta\theta_1 \Delta\theta_2 & \Delta\theta_1^2 \end{bmatrix} < I. \tag{5.10-47}$$

The sufficient condition for the asymptotic convergent of (5.10-47), which comes by letting $\text{Tr}(\Delta S^T X \Delta S) < 1$, is

$$\Delta\theta_1^2 + \Delta\theta_2^2 < 1/x_{22}. \tag{5.10-48}$$

STEP 4 (To compute $X \triangleq P + P S S^T P$): With the current S_J and P_J we compute X. In this problem we only need x_{22}. At $J = 1$, $x_{22} = 6.329$.

STEP 5 (To search for corner points): The set of the admissible-parameter variations must satisfy the sufficient condition (5.10-48). At $J = 1$,

$$\Delta\theta_1^2 + \Delta\theta_2^2 < 0.158. \tag{5.10-49}$$

From (5.10-49) we must extract $2^r = 2^2 = 4$ sets of corner points for each corresponding hyperbox. Clearly the following points which we directly compute from (5.10-49) are possible candidates for the corner points of the desired set:

$$C_1^{++} \triangleq (\Delta\theta_1 = \Delta\theta_2 = 0.281), \quad C_1^{-+} \triangleq (\Delta\theta_1 = -0.281, \Delta\theta_2 = 0.281),$$

$$C_1^{--} \triangleq (\Delta\theta_1 = \Delta\theta_2 = -0.281), \quad C_1^{+-} \triangleq (\Delta\theta_1 = 0.281, \Delta\theta_2 = -0.281).$$

$$\tag{5.10-50}$$

Substituting any one of these values for parameter variations in (5.10-45) yields the corresponding ΔS, which is a "solution" to (5.10-1) in this example.

STEP 6 (To construct the admissible set): Connecting the above $2^r = 2^2 = 4$ corner points in (5.10-50) yields a set (or a hyperbox) of admissible points in the *PVS*, shown in Fig. 5.10-1a. Every point of this set yields a ΔS such that $S + \Delta S$ remains convergent.

COMMENT 5.10-1: To find (5.10-48) in a typical application may require some efforts. We propose instead to make a numerical guess for each parameter variations $\Delta\theta_1$ and $\Delta\theta_2$, etc., in order to have a *candidate* for ΔS which we call $\Delta S'$. Then we let $\Delta S = \varepsilon \Delta S'$ and look for an ε, such that ΔS becomes an *admissible* perturbation matrix corresponding to one point in the *PVS*. As is discussed in Comment 5.8-3, we use the same set of $|\Delta\theta_i|$'s (corresponding to this ΔS) in different quadrants of the *PVS* in order to find new ε_i's. Using $\varepsilon = \min\{\varepsilon_1, \cdots, \varepsilon_{2^r}\}$ gives us a hyperbox which is asymptotically convergent. We use the farthest corner point of this hyperbox to build a convex set with $2^r (= 4)$ corners around it. If the requirement of Theorem 5.10-3 is satisfied in this box, then the entire box is admissible. Otherwise we adjust the *dimension* of the convex set accordingly.

For example, at $J = 1$, and according to (5.10-49) the set of the admissible parameter variations is in a circle with radius 0.3975. But we let $\Delta S = \varepsilon \Delta S' (\Delta\theta_1 = \Delta\theta_2 = 0.6)$ and $Q = 2I$, in order to use (5.10-1) (with S_1 and P_1) to compute ε as follows.

$$\varepsilon^2 \begin{bmatrix} 0 & 0.6 \\ 0 & 0.6 \end{bmatrix} \begin{bmatrix} 6.515 & 2.835 \\ 2.835 & 6.329 \end{bmatrix} \begin{bmatrix} 0.0 & 0.0 \\ 0.6 & 0.6 \end{bmatrix} \triangleq M < I. \tag{5.10-51}$$

Using $\text{Tr}(M) < 1$ results in $|\varepsilon| < 0.468$ and thus $|\Delta\theta_1| = |\Delta\theta_2| = 0.281$, which is the same as that of the analytical result. However, in general, the above procedure yields a more conservative-upper bound than that resulted from the analytical approach. But this approach is more amenable for computer automation than the analytical method.

5.10. Construction of Robustly Stable Discrete – Time Systems

COMMENT 5.10-2: From Table 5.10-1 we know that at least one other point which we call $C_n^{++} \triangleq (\Delta\theta_1 = \Delta\theta_2 = \varepsilon_{45°} = 0.3125)$ results in a convergent perturbed matrix. Therefore we now search for iterative ways of expanding the initial set of admissible points in the *PVS*, in order to reach point C_n^{++}.

Iteration: J+1

STEP 7 (Construction of the largest set): We now propose to translate the coordinates of the *PVS* to the farthest corner of the current convex set, in order to construct additional regions of admissible points in that space. That means we now start with $S_{J+1} = S_J \pm \Delta S_J$ (Step 0), where $\Delta S_J = \Delta S(\Delta\theta_1, \Delta\theta_2, ..., $ at $O_{J+1})$ and we invoke Steps 1 to 6 to S_{J+1}. In this manner we continue our construction until we find the largest set in a given problem. Of course, we must translate the coordinates in several different directions in order to connect the consecutive convex sets *as smoothly as* we can such that the entire allowable region of admissible points in the *PVS* is covered. We continue to apply this *probing* method to the current example, and we draw Fig. 5.10-1a, in order to depict partially these results.

At $J = 2$: From the result in Step 5 we translate the coordinates to point C_1^{++} (which is not shown in Fig. 5.10-1a) with $S_2 = S_1 + \Delta S_1$, here $\Delta S_1 = \Delta S(\Delta\theta_1 = \Delta\theta_2 = 0.281)$. Using the same symbols for parameter variations in each iteration, and for this S_2 and the corresponding P_2, the sufficient condition for the asymptotically convergent matrix becomes

$$\Delta\theta_1^2 + \Delta\theta_2^2 < 0.00175. \tag{5.10-52}$$

From (5.10-52) we select the following four new corner ponts, shown partially in Fig. 5.10-1a:

$$C_2^{++} \triangleq (\Delta\theta_1 = \Delta\theta_2 = 0.0296), \quad C_2^{-+} \triangleq (\Delta\theta_1 = -0.0296, \Delta\theta_2 = 0.0296),$$

$$C_2^{--} \triangleq (\Delta\theta_1 = \Delta\theta_2 = -0.0296), \quad C_2^{+-} \triangleq (\Delta\theta_1 = 0.0296, \Delta\theta_2 = -0.0296).$$

$$\tag{5.10-53}$$

At $J = 3$: Continuing in this fashion we now translate the coordinates of the *PVS* to point C_2^{++} which is not shown in Fig. 5.10-1a, but it is almost over C_n^{++}. Therefore now $S_3 = S_2 + \Delta S_2$, where $\Delta S_2 = \Delta S(\Delta\theta_1 = \Delta\theta_2 = 0.0296)$. Here the sufficient condition becomes

$$\Delta\theta_1^2 + \Delta\theta_2^2 < 0.000006267. \tag{5.10-54}$$

Resulting in the following four new corners which are so close to each other that we could not draw them on Fig. 5.10-1a.

$$C_3^{++} \triangleq (\Delta\theta_1 = \Delta\theta_2 = 0.00177), \quad C_3^{-+} \triangleq (\Delta\theta_1 = -0.00177, \Delta\theta_2 = 0.00177),$$

$$C_3^{--} \triangleq (\Delta\theta_1 = \Delta\theta_2 = -0.00177), \quad C_3^{+-} \triangleq (\Delta\theta_1 = 0.00177, \Delta\theta_2 = -0.00117).$$

(5.10-55)

At J = 4: We now translate backward the coordinates from O_3 (at C_2^{++} which is close to C_n^{++}) to O_4. Here $\Delta S_3 = \Delta S(\Delta\theta_1 = -0.0296 - 0.281, \Delta\theta_2 = -0.0296)$, $S_4 = S_3 + \Delta S_3$, and the sufficient condition for the asymptotic convergent is

$$\Delta\theta_1^2 + \Delta\theta_2^2 < 0.04508. \quad (5.10\text{-}56)$$

We can show that the following four new corners (shown in Fig. 5.10-1a) also satisfy (5.10-56):

$$C_4^{++} \triangleq (\Delta\theta_1 = \Delta\theta_2 = 0.15), \quad C_4^{-+} \triangleq (\Delta\theta_1 = -0.15, \Delta\theta_2 = 0.15),$$

$$C_4^{--} \triangleq (\Delta\theta_1 = \Delta\theta_2 = -0.15), \quad C_4^{+-} \triangleq (\Delta\theta_1 = 0.15, \Delta\theta_2 = -0.15). \quad (5.10\text{-}57)$$

▼

EXAMPLE 5.10-5 *(Linear systems with multiple nonlinear parameter variations):* Consider the system which is asymptotically convergent at its operating condition $\theta^o \triangleq \frac{1}{4}[2, \sqrt{2}]^T$ as follows.

$$S(\equiv S_1) \triangleq \begin{bmatrix} 0 & 1 \\ \theta_2^2 & \theta_1^2 \end{bmatrix} \equiv \begin{bmatrix} 0 & 1 \\ 0.125 & 0.25 \end{bmatrix}. \quad (5.10\text{-}58)$$

Our objective is to construct *the largest* set of r (= 2) independently varying parameters $\Delta\theta_1$ and $\Delta\theta_2$ from their operating condition $\theta^o \triangleq \frac{1}{4}[2, \sqrt{2}]^T$ such that $S + \Delta S$ remains convergent, where the perturbation matrix ΔS at each operating condition θ^o is a type G perturbation matrix (cf., § 5.5.2).

$$\Delta S \bigg|_{\theta^o} \triangleq \begin{bmatrix} 0 & 0 \\ 2\theta_2^o \Delta\theta_2 + (\Delta\theta_2)^2 & 2\theta_1^o \Delta\theta_1 + (\Delta\theta_1)^2 \end{bmatrix}. \quad (5.10\text{-}59)$$

SOLUTION: Clearly at any given θ^o, (5.10-59) does not meet the convexity requirement of Theorem 5.10-3 and we must seek other approach to construct its CRM. But first, we explore any special property of the original system, if we may. In fact, because of the way in which nonlinearity has appeared in (5.10-58), and as suggested in Comment 5.6-2, we may group the (2,1) entry as one parameter and

5.10. Construction of Robustly Stable Discrete – Time Systems

the (2,2) entry as another. For instance, we let $\theta_2^2 = \alpha$ and $\theta_1^2 = \beta$ with the corresponding initial values given in (5.10-58). Next, (5.10-58) becomes $\begin{bmatrix} 0 & 1 \\ \alpha & \beta \end{bmatrix}$. Perturbing this matrix in the space of (α, β) results in the corresponding *CRM*. Indeed this measure is already established in Example 5.10-4 and is shown in Fig. 5.10-2b, but to interpret that *CRM* in terms of θ_1 and θ_2 may not be so easy as it looks. Nevertheless, this is a thought that has to be explored. On the other hand, we can always apply Theorem 2 to (5.10-58) and its corresponding ΔS, but we face a complicated algebraic problem. Therefore, in the following instead of an *ad hoc* procedure, we review a systematic method which works and its complexity depends on the encounter problem.

As Remark 5.10-2 (or Remark 5.6-1) suggests, we propose to expand (5.10-59) by its corresponding Taylor series expansion as follows.

$$\Delta S(\Delta\theta) = \Delta S(\Delta\theta)\Big|_{\Delta\theta^o} + \sum_{j=1}^{r} \frac{\partial \Delta S}{\partial \Delta\theta_j}\Big|_{\Delta\theta^o} \frac{\Delta\theta_j}{1!} + \left[\frac{\partial^2 \Delta S}{\partial(\Delta\theta_1)^2} \frac{(\Delta\theta_1)^2}{2!}\right.$$

$$+ \frac{\partial^2 \Delta S}{\partial \Delta\theta_1 \partial \Delta\theta_2} \frac{(\Delta\theta_1)(\Delta\theta_2)}{2!} + \cdots + \frac{\partial^2 \Delta S}{\partial(\Delta\theta_r)^2} \frac{(\Delta\theta_r)^2}{2!}\bigg]\bigg|_{\Delta\theta^o} + \cdots , \quad (5.10\text{-}60)$$

where by working in the *PVS*, $\Delta\theta^o \equiv 0$. Applying (5.10-60) to (5.10-59) results in

$$\Delta S = \begin{bmatrix} 0 & 0 \\ 0 & 2\theta_1^o \end{bmatrix} \Delta\theta_1 + \begin{bmatrix} 0 & 0 \\ 2\theta_2^o & 0 \end{bmatrix} \Delta\theta_2$$

$$+ \begin{bmatrix} 0 & 0 \\ 0 & 2 \end{bmatrix} \frac{(\Delta\theta_1)^2}{2} + \begin{bmatrix} 0 & 0 \\ 2 & 0 \end{bmatrix} \frac{(\Delta\theta_2)^2}{2}. \quad (5.10\text{-}61)$$

As expected in this case, all the remaining higher-order terms are zero. In order that (5.10-61) meets the convexity requirement, we must be able to justify dropping the higher-order (second-order) perturbations. For instance, if we let $\Delta\theta_1 < 0.3$ and $\Delta\theta_2 < 0.3$, resulting in $(\Delta\theta_1)^2 < 0.09$, and $(\Delta\theta_2)^2 < 0.09$, then we can approximate ΔS as follows.

$$\Delta S \approx \begin{bmatrix} 0 & 0 \\ 0 & 2\theta_1^o \end{bmatrix} \Delta\theta_1 + \begin{bmatrix} 0 & 0 \\ 2\theta_2^o & 0 \end{bmatrix} \Delta\theta_2, \quad (5.10\text{-}62)$$

where the initial $\theta^o \triangleq \frac{1}{4}[2, \sqrt{2}]^T$ remains as a running variable for each newly operating condition which is selected at the farthest vertex of a new generated set. Clearly, to approximate (5.10-61) with (5.10-62) in order to meet the convexity

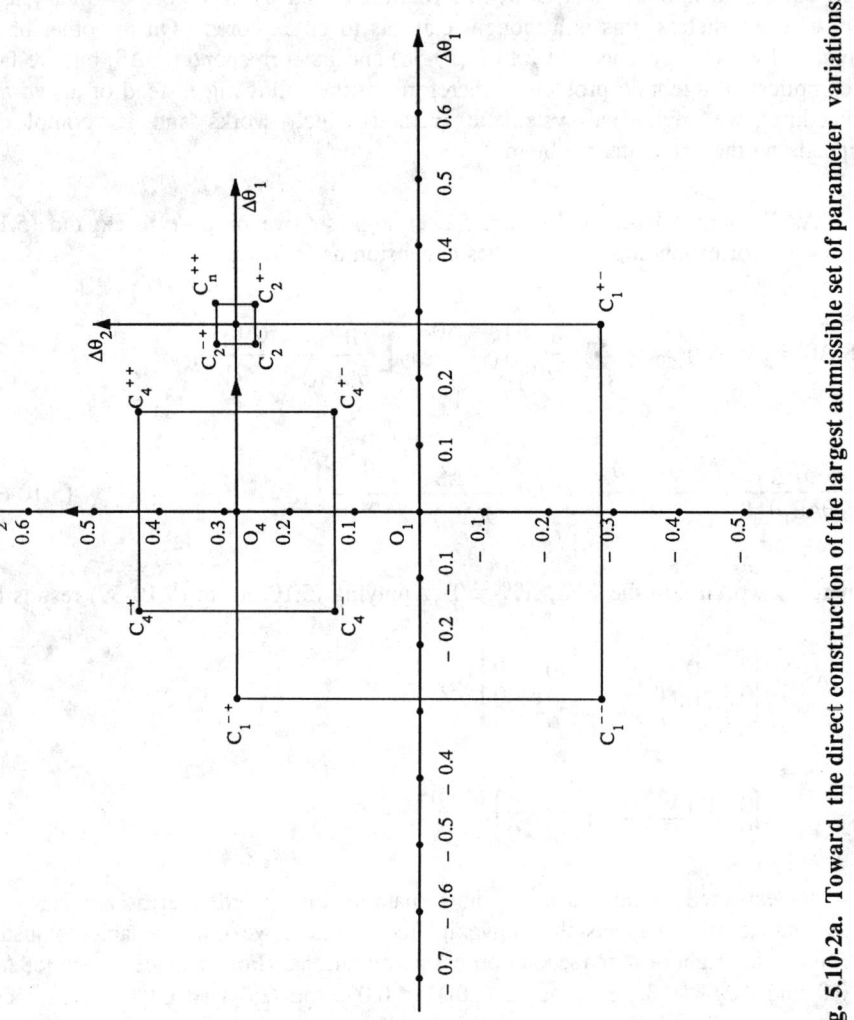

Fig. 5.10-2a. Toward the direct construction of the largest admissible set of parameter variations.

5.10. Construction of Robustly Stable Discrete-Time Systems

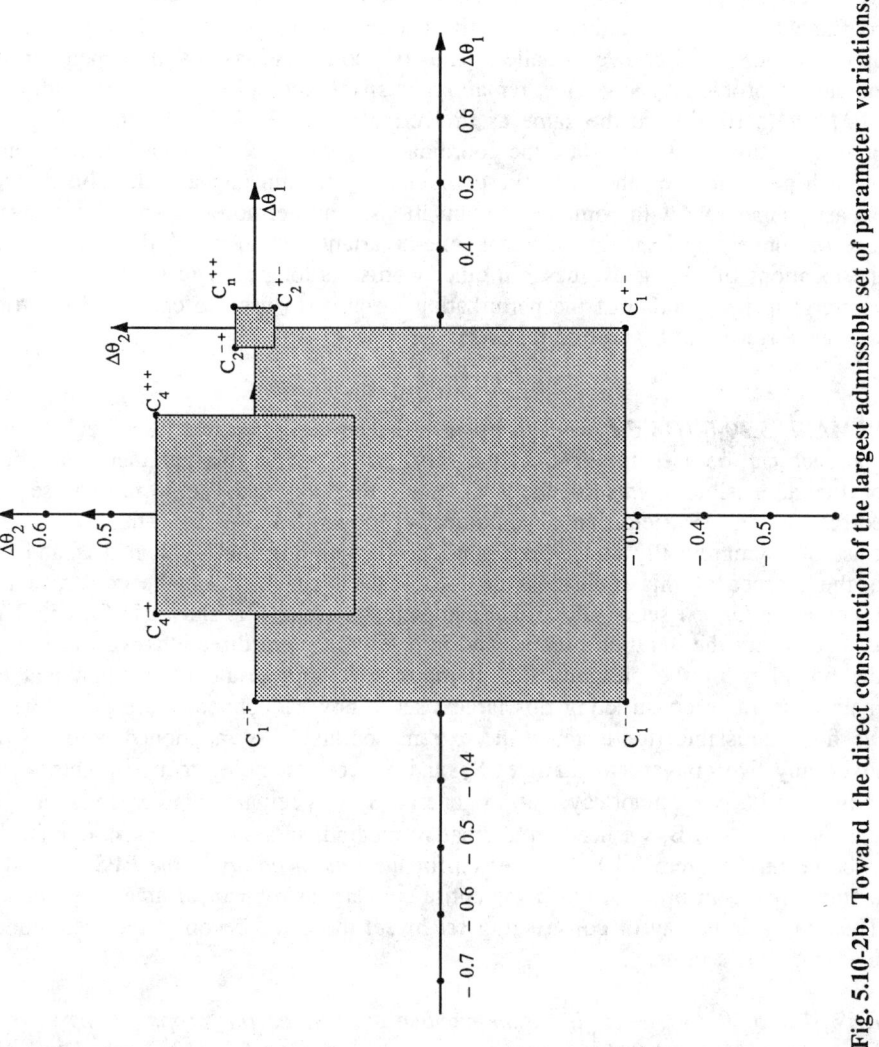

Fig. 5.10-2b. Toward the direct construction of the largest admissible set of parameter variations.

requirement of Theorem 3, one must consider the norms of the coefficient matrices as well as the magnitude of parameter variations. When these conditions are met, by using Theorem 5.10-3 (Example 5.10-4) we can construct the corresponding *CRM*. However, in each iteration, *only* that portion of this *CRM* is acceptable where $\Delta\theta_1 < 0.3$ and $\Delta\theta_2 < 0.3$. A limitation that we inherently face. If the coefficient matrices of higher-order perturbations had larger norms, then the chosen upper bound would be even smaller. Thus our computations heavily depend on the encounter problem. Now, the remaining steps to complete the construction of *CRM* for (5.10-62) are the same as in Example 5.10-4. We note that using this procedure, and as we translate the coordinates of the *PVS* to various corner points of each generated set, the symbolic computations remain unchanged. This method is easily amenable with computer computations. Furthermore we are taking advantage of convergent behavior of linear time-invariant system which depends only on the locations of its eigenvalues. In other words, as long as S in (5.10-58) remains convergent, irrespective of the perturbation mechanism used to changed its entries, we can perturb that S and find its *CRM*.

▼

REMARK 5.10-6 *(On the procedure)*: It is indeed clear that this iteratively probing approach can be used to construct the *entire* set of admissible parameter variations or the admissible points in the *PVS*, such that for each point in that set the corresponding ΔS yields an asymptotically convergent $S + \Delta S$. Thus we have a class of asymptotically convergent matrices surrounding the initial matrix and that is the essence of this robustness analysis. This set, which can be constructed to become *the largest* set of admissible parameter variations, is shown in Fig. 5.10-1b. By increasing the iteration number and in a few different directions we can smooth the boundary of the previous set. Perhaps it is appropriate to mention that our main concern in constructing this largest set is how to connect those points which we find admissible (for instance the extreme points). As mentioned before, since given any two convergent matrices S_A and S_B corresponding to two points A and B in the *PVS*, we cannot say that, in general, $S_{A \to B}$ remains convergent as a point moves from A to B, we need an alternative method, such as the present one (which is based on Theorem 5.10-3), to search for the *sets of points* in the *PVS*, instead of a single point, in order to cover the entire convergent robustness area very quickly. This expeditious way of constructing *set by set* the *CRM* becomes the main contribution of this section.

REMARK 5.10-7 *(On the iteration number as well as the propagations of* P's*)*: Here again, to avoid the repetition we refer to Remarks 5.8-7 & 8 which are also true here (the "radius" herein is proportional to the inverse norm of X).

Finally, we note that Examples 5.10-4 & 5, which are in a phase-variable canonical forms, also represent monic polynomials. Thus the above procedure shows how this method of construction can be applied to polynomials.

5.11. CERTAIN EXTENSIONS TO NONLINEAR SYSTEMS

5.11.1. Continuous – Time Nonlinear Systems

The results developed in this chapter can be used to generate the necessary and sufficient condition for absolute stability of a special set of nonlinear differential (or difference) equations. This topic is also known as absolute stability or sector problem. The importance of this topic is emphasized in Aizerman [A. 3], Gaiduk [A. 58], Narendra and Taylor [A. 102], and Pyatnytsky [A. 115], among many other additional references in § 5.12.A, and § 5.12.F. The following analysis closely follows [F. 44]. The underlying issue is as follows.

Consider a class of continuous-time nonlinear (feedback) systems described by

$$\dot{x}(t) = Sx(t) + bf(\sigma), \quad \sigma(t) = c^T x(t); \quad (5.11\text{-}1)$$

where $x(t) \in R^n(t)$ is the state vector, $f(\sigma)$ is a nonlinear scalar function of σ, and b and c are constant vectors. Here S is assumed to be an asymptotically stable matrix. Replacing $f(\sigma)$ by a linear function $k\sigma$, where k is a constant scalar value, results in a system which is asymptotically stable for all $k \in [k_1, k_2]$. In the literature this classical control problem is referred to as absolute stability of a nonlinear dynamical system or the sector problem. The importance and engineering applicability of this concept is emphasized in the above references among many others in § 5.12.A, § 5.12.B, and § 5.12.F. The main issue is to find *the largest* interval $[k_1, k_2]$ such that the nonlinear system (5.11-1) remains asymptotically stable for [any] initial disturbances. "A vast literature has been devoted to this problem and many powerful methods are available for its solution. However, these solutions yield only sufficient conditions, and despite the efforts of many scientists over a period of more than 20 years, the necessary and sufficient conditions have not been found as yet [A. 3]." In the following *a* necessary and sufficient condition for determination of *the largest* interval $[k_1, k_2]$ for (5.11-1), when S is asymptotically stable, is proposed such that (5.11-1) remains asymptotically stable. If S is not stable, then we must first stabilize S, provided that this stabilization is possible.

REMARK 5.11-1: It is well known that any linearization of the model of a nonlinear dynamical system results in a local model for that system. Thus the notion of *a* necessary *and* sufficient condition which yields information on the stability behavior of original nonlinear system, when a particular linearization method is utilized, is inherently local, unless we are absolutely convinced that the above nonlinear system is globally "boxed" in a sector that is characterized by two straight lines in which we *may* provide additional information about the global stability behavior of the system. There are, however, nontrivial and potentially important counterexamples to this sort of linearization procedures. An excellent exposition of this critical matter is reported in [A. 158, pp. 177-182], which shows by an example that unexpected periodic solutions may exist in a class of nonlinear feedback systems such as (5.11-1). We also recall that in (5.11-1) the nonlinearity appears as a scalar function with one parameter. The stability issue of a nonlinear system with multiple nonlinearities and several varying parameters becomes very complicated. As we briefly mention this issue subsequently, however, we can manage this situation using Theorem 5.6-2. In spite of these subtleties, linearization methods when constructed prudently work in a number of engineering problems and that is the reason behind our analysis.

∎

Suppose that in (5.11-1) we can approximate $f(\sigma)$ at a given operating point (entailing the entire state of the system) by $k\sigma$ such that $k_1' < f(\sigma)/\sigma < k_2'$. Then substituting for $f(\sigma)$, $k\sigma$, in (5.11-1) results in

$$\dot{x}(t) = (S + kbc^T)x(t) \triangleq S(k)x(t). \tag{5.11-2}$$

Since we have assumed that S is asymptotically stable, we can show that for some $k_1'' < k < k_2''$, $S(k)$ remains asymptotically stable.

PROBLEM 5.11-1: Find the largest interval $[k_1, k_2] \equiv [k_1', k_2'] \cap [k_1'', k_2'']$.

This intersection implies that we have accounted for the contributions of nonlinearity as well as the linear homogeneous part of the system, in our overall absolute stability analysis of this nonlinear system. In other words, the *SRA* of only linear part without any consideration of the nonlinearity and *vice versa* is not acceptable for the absolute stability analysis. We further elaborate on this matter and the choice for our system operating point in Example 5.11-1.

SOLUTION: Since $\Delta S \triangleq kbc^T$, with $Q = I$ and $S^T P + PS + I = 0$. Then using the sufficient condition (5.5-1) yields

$$k^2[(bc^T)(bc^T)^T] < \tfrac{1}{4}P^{-2}. \tag{5.11-3}$$

Because the rank of the left-hand side of (5.11-3) is one, thus its maximum eigenvalue is equal to its trace which we set that equal to the minimum eigenvalue of the right-hand side of (5.11-3) for a conservative-upper bound, i.e.,

$$k^2 = \lambda_{\min}(P^{-2}) / 4\operatorname{Tr}[(bc^T)(bc^T)^T]. \tag{5.11-4}$$

Using the above upper bound iteratively (cf., Algorithm 5.6-1) we can construct the largest interval $[k_1'', k_2'']$ such that for $k \in [k_1'', k_2'']$, $S + kbc^T$ remains asymptotically stable. On the other hand in any given problem, if the following is true

$$k_1' < f(\sigma)/\sigma < k_2', \quad \text{at a given operating point } x. \tag{5.11-5}$$

Then *the largest* interval becomes as follows.

$$[k_1, k_2] \equiv [k_1', k_2'] \cap [k_1'', k_2'']. \tag{5.11-6}$$

EXAMPLE 5.11-1 (Quadratic Duffing's equation): Consider a nonlinear system described by

$$\ddot{x}(t) + \dot{x}(t) + \alpha x(t) + x^2(t) = 0, \tag{5.11-7}$$

5.11. Certain Extensions to Nonlinear Systems

which is a slightly modified version of an example given in [A. 105, pp. 587-588]. Here we have adopted the parameter α ($> \frac{1}{4}$), in order that we demonstrate a different situation later. To analyze this second-order differential equation (or system) we must specify *both* $x(0)$ and $\dot{x}(0)$. A phase-plane analysis of this system reveals that there are two singular points, namely $S_1 \triangleq (x = 0, \dot{x} = 0)$ (a stable focus) and $S_2 \triangleq (x = -\alpha, \dot{x} = 0)$ (a saddle point). A schematic diagram for the phase-plane portrait of this system, Fig. 5.11-1, shows that depending on the system initial condition, we have two separate stability behaviors for this system. In one connected area in the phase plane the trajectories approach to the origin as $t \to \infty$, and in the remaining area of that plane the trajectories diverge to infinity as $t \to \infty$.

We now study the stability properties of this system using our method.

STEP 0 (To present S and ΔS): We let $x = x_1$ and $\dot{x} = x_2$, in order to transform (5.11-7) into the following state-space form.

$$\frac{d}{dt}\begin{Bmatrix} x_1(t) \\ x_2(t) \end{Bmatrix} = \begin{bmatrix} 0 & 1 \\ -\alpha & -1 \end{bmatrix}\begin{Bmatrix} x_1(t) \\ x_2(t) \end{Bmatrix} + \begin{bmatrix} 0 \\ -1 \end{bmatrix}x_1^2(t)$$

$$\triangleq Sx(t) + b\sigma^2(t), \quad \sigma(t) = [1, 0]x(t) \triangleq c^T x(t). \tag{5.11-8}$$

Here S is an asymptotically stable matrix for all $\alpha > \frac{1}{4}$. If we let $\sigma^2 \triangleq f(\sigma) = k\sigma$, then our linearized system becomes $\dot{x}(t) = (S + \Delta S)x(t)$, where $\Delta S \triangleq kbc^T$ is

$$\Delta S \triangleq k\begin{Bmatrix} 0 \\ -1 \end{Bmatrix}[1, 0] = k\begin{bmatrix} 0 & 0 \\ -1 & 0 \end{bmatrix}. \tag{5.11-9}$$

STEP 1 (To study the perturbational steps): We apply Theorem 5.5-1 to S in (5.11-8) and ΔS in (5.11-9), in order to compute the best perturbation step for ΔS. (Or equivalently to use (5.11-3).) Since we do not know *a priori* which one of (5.5-1) to (5.5-4) will yield the best result, we compute all four of them in order to choose the best one. This computation is summarized in Table 5.11-1. In this table we have used the ordinary eigenvalues of both sides of inequalities (5.5-2) and (5.5-4). But we have left (5.5-1) and (5.5-3) untackled due to their complications. From Table 5.11-1, it is clear that $k^2_{(5.5-2)} > k^2_{(5.5-4)}$ for all α. (Here the subscript $(\cdot)$ refers to inequality $(\cdot)$.) But we cannot say much about $k^2_{(5.5-1)}$ and $k^2_{(5.5-3)}$ in their present forms. Therefore we use the nominal value of $\alpha = 2$, resulting in $k^2_{(5.5-1)} = 0.0764$, $k^2_{(5.5-2)} = 0.4$, $k^2_{(5.5-3)} = 0.0764$, $k^2_{(5.5-4)} = 0.2$. (We have used the minimum ordinary eigenvalue of the right-hand sides of (5.5-1) and (5.5-3).) Thus we choose (5.5-2) as our perturbation policy.

Table 5.11-1. Applications of Theorem 5.5-1 to Example 5.11-1.

$S^T P + PS + I = 0$	Inequality (5.5-1)	Inequality (5.5-2)
$P = \begin{bmatrix} \dfrac{\alpha^2 + \alpha + 1}{2\alpha} & \dfrac{1}{2\alpha} \\ \dfrac{1}{2\alpha} & \dfrac{\alpha + 1}{2\alpha} \end{bmatrix}$	$\begin{bmatrix} 0 & 0 \\ 0 & k^2 \end{bmatrix} < \dfrac{1}{\alpha^2 + 2\alpha + 2} \times \begin{bmatrix} 1 & -1 \\ -1 & \dfrac{(\alpha^2 + \alpha + 1)^2 + 1}{\alpha^2 + 2\alpha + 2} \end{bmatrix}$	$k^2 < \dfrac{\alpha^2}{\alpha^2 + 2\alpha + 2}$
$SP + PS^T + I = 0$	Inequality (5.5-3)	Inequality (5.5-4)
$P = \begin{bmatrix} \dfrac{\alpha + 2}{2\alpha} & -\dfrac{1}{2} \\ -\dfrac{1}{2} & \dfrac{\alpha + 1}{2\alpha} \end{bmatrix}$	$\begin{bmatrix} k^2 & 0 \\ 0 & 0 \end{bmatrix} < \dfrac{1}{\alpha^2 + 2\alpha + 2} \times \begin{bmatrix} \alpha^2 & \alpha \\ \alpha & 2 \end{bmatrix}$	$k^2 < \dfrac{\alpha^2}{2(\alpha^2 + 2\alpha + 2)}$

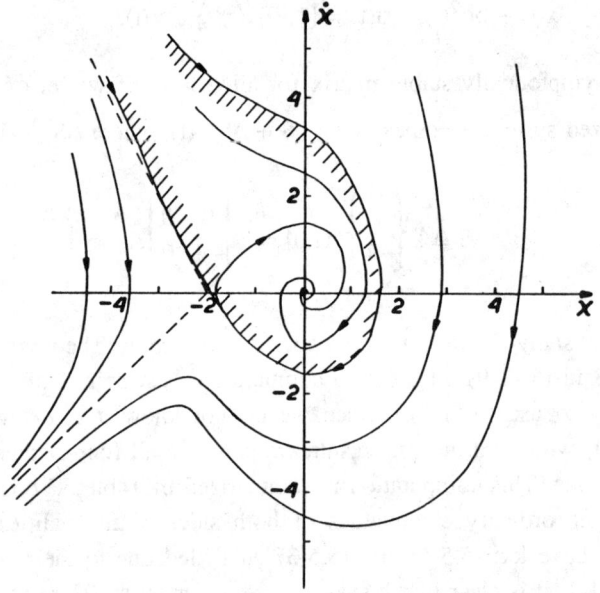

Fig. 5.11-1. A Schematic diagram of the phase-plane portraits of (5.11-7) with $\alpha = 2$.

5.11. Certain Extensions to Nonlinear Systems

STEP 2 (To choose the direction for the perturbation step): By studying the eigenvalues of perturbed system we can determine the direction for our perturbation step. In this example, the proper direction for applying the perturbation step is when k < 0, because for k > 0 the eigenvalues of perturbed system move further to the left-hand side of the complex plane. (In certain cases both positive and negative perturbations may occur).

STEP 3 (To apply the iterative procedure): By selecting the appropriate perturbation step ΔS (here ΔS corresponds to $k_{(5.5-2)}$), and using Theorem 5.6-1, with $Q = I$, we reach *the largest* ΔS. This is done by letting $S_{J+1} = S_J + \Delta S_J$ (where J is the order of iteration), and searching for the corresponding ΔS_{J+1} from Steps 2 and 3. Here *the largest* ΔS corresponds to $k = -\alpha(= -2)$, i. e., $[k_1'', k_2''] \equiv [-2, \infty)$.

STEP 4 (To linearize the nonlinear function): Clearly the results from Step 3, in general, lead us to the direction of an appropriate operating condition. In this example, we can show that for $x_1 \in [-1, 1]$, $-1 < f(\sigma)/\sigma < 1$, i. e., $[k_1', k_2'] \equiv [-1, 1]$.

STEP 5 (The final answer): The solution to this example is therefore as follows.

$$[k_1, k_2] \equiv [-2, \infty) \cap [-1, 1] = [-1, 1]. \qquad (5.11\text{-}10)$$

▼

COMMENT 5.11-1: As mentioned before, we must select an operating point that entails the entire state of the system when we undertake a linearization method. That means to specify the range of only a portion of the state (here x_1) is not sufficient to conclude asymptotic stability of (5.11-1). In this example, because of the special type of nonlinearity, in Step 4 we have selected an operating point considering only x_1. We must, however, investigate by an appropriate means an allowable interval for x_2, since x_2 has not explicitly appeared in $f(\sigma)$. (Fig. 5.11-1 approximately indicates this interval.) Based on that information and $x_1 \in [-1, 1]$, we have a set of initial conditions such that (5.11-1) is asymptotically stable for [any] initial disturbance from this set. Fig. 5.11-1, though approximate, is intended to convey this message. For example, if the initial condition is chosen at $x(0) = 0$ ($\in [-1, 1]$), and $\dot{x}(0) = 5$, *per se*, then the corresponding trajectories will diverge as $t \to \infty$. Furthermore, and needless to say that, this entire analysis is a substitution for the complete solution of the nonlinear differential equation. Therefore in a typical application we must also be concerned about the computational cost, as this balancing act has been our main message throughout this book.

REMARK 5.11-2: In Example 5.11-1, had we selected $\alpha = \frac{1}{2}$ instead of 2, then $[k_1'', k_2''] \equiv [-\frac{1}{2}, \infty)$ and (5.11-10) would have been $[-\frac{1}{2}, 1]$, i. e., the final solution to our problem as we have expected, must account for both linear homogeneous as well as the nonlinearity parts of the system.

REMARK 5.11-3: The salient feature of the above Problem 5.11-1 is that the overall system has only one varying parameter which has resulted from either the linearization scheme or perhaps an actual system parameter. When b in (5.11-1) is a matrix and/or ΔS is a function of more than one parameter which are changing simultaneously, the procedure for constructing the "largest" interval $[k_1, k_2]$ or sector in principle is the same as above, however, this construction must now be completed based upon the results developed in Theorem 5.6-2 as described next. ■

An accurate mathematical model for a class of engineering problems is a set of nonlinear ordinary differential equations, which is characterized as follows.

$$\dot{x}(t) = Ax(t) + Bf(\sigma(t)), \quad \sigma(t) = c^T x(t), \qquad (5.11\text{-}11)$$

where $x(t) \in R^n(t)$ is the state vector, $f(\sigma(t)) \in R^m(t)$ is a vector of system nonlinearities or the vector of modeling errors; A and B are system coefficient matrices, and c is a constant vector. Often the direct analysis of (5.11-11) is avoided and the trend is to replace (5.11-11) with its linearized version such that it represents the original system for as wide an operating range as possible. In particular, when we seek information about stability behavior of (5.11-11), subject to the large-parameter variations, we may conclude that property from its equivalent class of linear systems provided that certain conditions are met (cf., Remark 5.11-2). Here we seek the best (as we further elaborate) class of linear systems whose stability characteristics are similar to (5.11-11), and thus these linear models can represent the stability robustness properties of (5.11-11).

The procedure to develop this class of equivalent linear systems is to replace $f(\sigma(t))$ with a linear term $Kx(t)$, $K \in R^{m \times n}$, and subsequently (5.11-11) becomes a standard set of linear *ODE's* as follows.

$$\dot{x} = (A + BK)x \triangleq S(\theta)x. \qquad (5.11\text{-}12)$$

Here $\theta \in R^r$ is a generic name for the vector of system *parameters* and we note that A, B, and K are matrices. Another implication of this representation is that the construction of the largest class of robust linear systems will encompass the variations in system parameters A and B as well as the design or the linearization parameters K corresponding to a nonlinear system with multiple nonlinearities. This $\theta \in R^r$ also includes those parameters that represent the physical constraints on the system.

The problem that we face now is an extension of Problem 5.11-1 to nonlinear system with multiple large-parameter variations and system with multiple nonlinearities. In this regard we seek *a* necessary and sufficient condition for determination of *the largest* or *most extended* set (instead of interval) of parameter variations, χ, for (5.11-11), whose corresponding set of perturbed systems is stable. The meaning and/or definition of the preceding set are examined subsequently. In this regard we propose the following problem.

PROBLEM 5.11-2: Find *the largest set* $\chi \equiv \chi_{linearization} \cap \chi_{perturbation}$.

This intersection, as in Problem 5.11-1, implies that we must consider the contribution and/or validity of linearization, and the contribution of linear

5.11. Certain Extensions to Nonlinear Systems

homogeneous part of the system, in order to determine the overall stability robustness measure *(SRM)* for the nonlinear system.

A Conceptual Solution: The conceptual solution for Problem 5.11-2 consists of generating two sets. The first set $\chi_{linearization}$ depends on the encountered nonlinear problem and this set has to be treated in each situation accordingly. This set may also represent the physical constraints on various system parameters, or an intersection of these constraints and the system nonlinearities. The second set $\chi_{perturbation}$ depends on the stability robustness properties of the corresponding linear systems, which for generating this set we have a general procedure (supported by Theorem 5.6-2) and that is applicable to all such problems.

In regard to the nonlinearity, as in Problem 5.11-1, we recall that any linearization of the model of a nonlinear dynamical system results in a local model for that system. Thus, our analysis must guarantee that the nonlinear system is globally 'boxed' in a closed set, otherwise we cannot conclude that the linearization of the nonlinear system yields information on the global stability of that system. Therefore when linearizing a nonlinear system entailing its entire state, care must be exercised to find a closed stable path in the *PVS,* which boxes the nonlinear system at the operating point. If this task cannot be performed then our conclusion regarding the *equivalence* of stability properties between the nonlinear and linear systems at this point is unfounded.

EXAMPLE 5.11-2: Consider the asymptotically stable matrix of Example 5.8-3, where both $\Delta\theta_1$ and $\Delta\theta_2 > 0$. Our goal is to apply our method to this case and to construct *the largest* set of admissible parameter variations, whose corresponding class of perturbed dynamical systems remains stable (or asymptotically stable).

SOLUTION: The solution to this problem requires that two sets of points in the *PVS* be generated first, before we find their intersection.

I - $\chi_{linearization}$: This set is constructed from the information which we have regarding system nonlinearities. For instance, in this example we know *a priori* that the system becomes unstable as $\Delta\theta_1$ and $\Delta\theta_2 \to 1$. Furthermore, we know that in this case the linearization is valid only for a square in the *PVS* where $\Delta\theta_j \le 1$, j = 1, 2. Thus $\chi_{linearization}$ is a set of points inside of the box $\Delta\theta_j \le 1$, j = 1, 2, Fig. 5.11-2.

We note that this set can also represent a typical set of physical constraints on system parameters.

II - $\chi_{perturbation}$: This set which corresponds to the overall homogeneous part of our system (i.e., (5.11-12)), is constructed according to the steps taken in Example 5.8-3, and is shown in Fig. 5.11-2.

Thus the solution to Example 5.11-2 is intersection of these two sets as shown in Fig. 5.11-2. Note that in this case we may choose only the perfect shaded square (whose corner is at C_n^{++}), although the two additional dashed sub-sets on the top and on the side of this square can be included in this intersection as well.

Finally, the method of solution for Problem 5.11-2 can be used to incorporate the physical constraints on the overall system parameter variations by replacing $\chi_{linearization}$ with $\chi_{constraints}$, or $\chi_{linearization} \cap \chi_{constraints}$.

▼

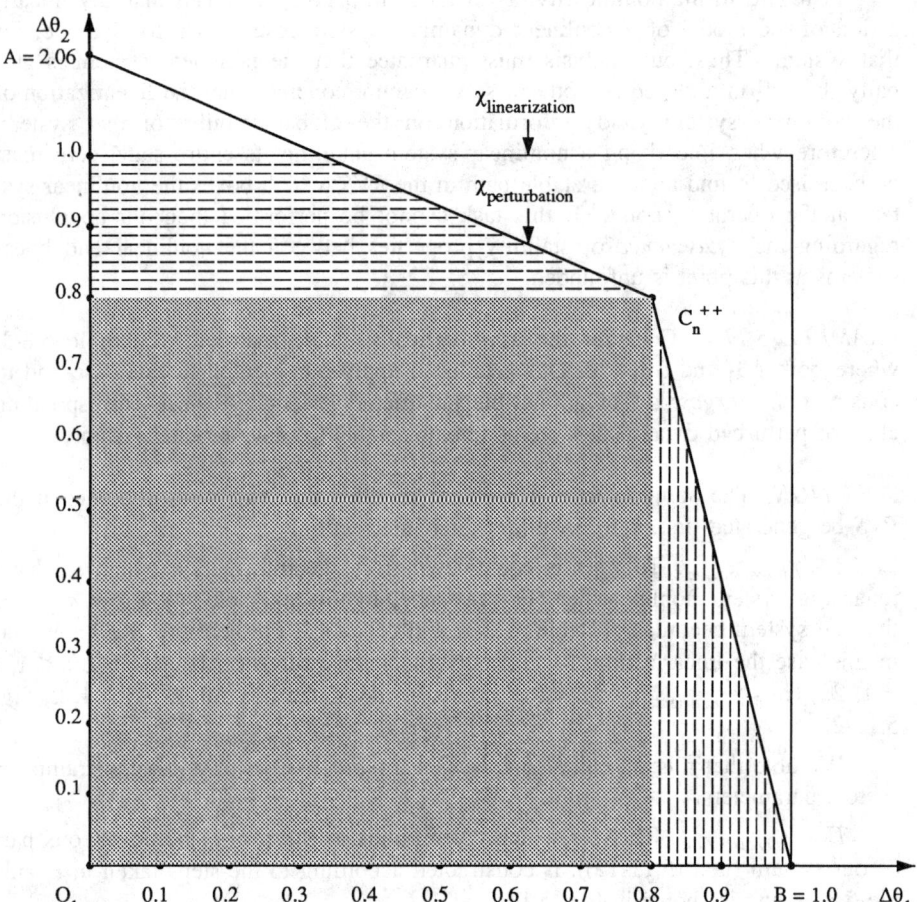

Fig. 5.11-2. Direct construction of *the largest* set of parameter variations.

5.11. Certain Extensions to Nonlinear Systems

We now revisit Example 5.11-1 with a new set of parameters in order to describe a different aspect of the method for solving Problem 5.11-2.

EXAMPLE 5.11-3 (Quadratic Duffing's equation): Consider the following classical nonlinear system described by

$$\ddot{x}(t) + \theta_1 \dot{x}(t) + \theta_2 x(t) + x^2(t) = 0. \tag{5.11-13}$$

Here $\theta \triangleq [\theta_1, \theta_2]^T$ is the vector of physical parameters with the nominal value of $\theta^o = [1, 2]^T$. Thus the main difference between this case and (5.11-7) is the introduction of another independently varying parameter. To analyze this second-order differential equation (system) we must specify *both* $x(0)$ and $\dot{x}(0)$. The preliminary phase-plane analysis of case is exactly the same as before, except that the saddle point is at $S_2 \triangleq (x = -\theta_2, \dot{x} = 0)$.

Our objective is to study the stability properties of this system when *both* θ_1 and θ_2 change subject to a *physical* constraint of $|\Delta \theta_1| < 1$.

SOLUTION: Here, we must generate three sets of points in the *PVS* (considering the physical constraint), followed by finding their intersection. But first, we take the next standard step.

STEP 0 (To present S and ΔS): We let $x = x_1$ and $\dot{x} = x_2$, in order to transform (5.11-13) into the following state-space form.

$$\frac{d}{dt}\begin{Bmatrix} x_1(t) \\ x_2(t) \end{Bmatrix} = \begin{bmatrix} 0 & 1 \\ -\theta_2 & -\theta_1 \end{bmatrix} \begin{Bmatrix} x_1(t) \\ x_2(t) \end{Bmatrix} + \begin{Bmatrix} 0 \\ -1 \end{Bmatrix} x_1^2(t)$$

$$\triangleq Sx(t) + b\sigma^2(t), \quad \sigma(t) = [1, 0]x(t) \triangleq c^T x(t). \tag{5.11-14}$$

Here S is an asymptotically stable matrix around its nominal value of $\theta^o \triangleq [1, 2]^T$. If we let $\sigma^2 \triangleq f(\sigma) = k\sigma$, then our linearized system becomes $\dot{x}(t) = (S + \Delta S)x(t)$, where

$$\Delta S \triangleq kbc^T \triangleq k\begin{Bmatrix} 0 \\ -1 \end{Bmatrix}[1, 0] = k\begin{bmatrix} 0 & 0 \\ -1 & 0 \end{bmatrix}. \tag{5.11-15}$$

In order that this *special* perturbation matrix be acceptable from the stability point of view, it must be within the allowable range of perturbation matrices for S. In general, however,

$$\Delta S \triangleq \begin{bmatrix} 0 & 0 \\ \Delta\theta_2 & \Delta\theta_1 \end{bmatrix}. \tag{5.11-16}$$

$I - \chi_{perturbation}$: This set corresponds to the overall homogeneous part of the system and is the largest set of ΔS's such that $S + \Delta S$ remains asymptotically stable. Next, this set is constructed according to the following steps, and subsequently we return to (5.11-15) when we construct $\chi_{linearization}$. We note that in this *special* case the linearization does not introduce an independently varying parameters in ΔS. Let $J = 1, 2, ..., N$, be the iteration number.

STEP 1 (To update S): We let $S_J = S_{J-1} \pm \Delta S_{J-1}$ with $S_1 \equiv S_0 \equiv S$ being the initial asymptotically stable matrix and $\Delta S_0 \equiv 0$. (The $\pm$ means in both directions.)

STEP 2 (To solve the Lyapunov equation): For the current S_J we solve for P_J according to $S_J^T P_J + P_J S_J + 2I = 0$. Because of this rather simple problem, P_J is generated symbolically herein, but usually this computation is done numerically.

$$P_J = \frac{1}{\theta_1 \theta_2} \begin{bmatrix} \theta_1^2 + \theta_2^2 + \theta_2 & \theta_1 \\ \theta_1 & \theta_2 + 1 \end{bmatrix}. \qquad (5.11\text{-}17)$$

STEP 3 (Directional perturbations): First we search for the largest variations from the operating condition $\theta^o = [1, 2]^T$ along the r $(= 2)$ axes of $\Delta\theta_1$ and $\Delta\theta_2$ and the "45° line". This analysis is summarized below.

Table 5.11-2. Applications of Theorem 1 to Example 5.11-3.

$\Delta S_{\Delta\theta_2} = \Delta\theta_2 \begin{pmatrix} 0 & 0 \\ 1 & 0 \end{pmatrix}$	$S + \Delta S_{\Delta\theta_2}$ is stable if $\Delta\theta_2 < 2.0$, shown by point B_n in Fig. 5.11–3.
$\Delta S_{\Delta\theta_1} = \Delta\theta_1 \begin{pmatrix} 0 & 0 \\ 0 & 1 \end{pmatrix}$	$S + \Delta S_{\Delta\theta_1}$ is stable if $\Delta\theta_1 < 1.0$, shown by point A_n in Fig. 5.11–3.
$\Delta S_{\varepsilon_{45°}} = \varepsilon_{45°} \begin{pmatrix} 0 & 0 \\ 1 & 1 \end{pmatrix}$	$S + \Delta S_{\varepsilon_{45°}}$ is stable if $\varepsilon_{45°} < 1.0$ shown by point C_n^{++} in Fig. 5.11–3.

STEP 4 (To construct $\Delta S \Delta S^T$): Here,

$$\Delta S \Delta S^T = \begin{bmatrix} 0 & 0 \\ 0 & \Delta\theta_1^2 + \Delta\theta_2^2 \end{bmatrix}. \qquad (5.11\text{-}18)$$

STEP 5 (To interpret the sufficient condition): Using the sufficient condition $\Delta S \Delta S^T < P_J^{-2}$ we establish a set of inequalities which yield sufficient conditions for parameter variations. As before, we let $\text{Tr}(\Delta S \Delta S^T) < \lambda_{\min}(P_J^{-2})$. Here, because of this simple case, at each iteration we have

5.11. Certain Extensions to Nonlinear Systems

$$\Delta\theta_1^2 + \Delta\theta_2^2 < \frac{\theta_1^2}{2[\theta_1^2 + (\theta_2 + 1)^2]}\{(\theta_1^2 + \theta_2^2 + 1) - \sqrt{(\theta_1^2 + \theta_2^2 + 1)^2 - 4\theta_2^2}\}.$$

(5.11-19)

Recall that each P_J is generated in terms of θ, while ΔS is in terms of $\Delta\theta$.

STEP 6 (To search for vertices): The final set of the admissible-parameter variations must satisfy (5.11-19). At $J = 1$, we have:

$$\Delta\theta_1^2 + \Delta\theta_2^2 < 0.076393, \qquad (5.11\text{-}20)$$

From this inequality we extract $2^r = 2^2 = 4$ points. The following points computed directly from (5.11-20) are possible candidates for the vertices of the desired set:

$$C_1^{++} \triangleq (\Delta\theta_1 = \Delta\theta_2 = 0.1954), \quad C_1^{-+} \triangleq (\Delta\theta_1 = -0.1954, \Delta\theta_2 = 0.1954),$$

$$C_1^{--} \triangleq (\Delta\theta_1 = \Delta\theta_2 = -0.1954), \quad C_1^{+-} \triangleq (\Delta\theta_1 = 0.1954, \Delta\theta_2 = -0.1954).$$

(5.11-21)

Substituting any one of these values for parameter variations in (5.11-16) yields the corresponding ΔS's.

STEP 7 (To construct an admissible set): Connecting the above four vertices in (5.11-21) yields a set of admissible points for parameter variations which is shown by a shaded square in Fig. 5.11-3. At this juncture it is imperative to recall that the four vertices in (5.11-21) satisfy all the requirements of Theorem 5.6-2, since they are generated from the *same* P_1. Indeed, the entire boundary of this set has the same Lyapunov function, however, this property does not hold for the largest set of admissible parameter variations in this example.

COMMENT 5.11-2: As before, the set of admissible points that is generated in Step 7 is not the corresponding largest set. Going through the same steps, as in the previous example, we can complete the construction of the final set shown in Figs. 5.11-3 & 4. In these figures, point $[\Delta\theta_1, \Delta\theta_2]^T = [1, 2]^T$ is shown by D which by our assumption is stable (its corresponding system eigenvalues are at the origin of the complex plane.)

Now, the construction of set $\chi_{perturbation}$ is completed.

$II - \chi_{linearization}$: This set is constructed from the information which we have regarding our system nonlinearities. For instance, as the results of Fig. 5.11-3, which in general leads us to the proper choice of an appropriate operating condition for linearization, and Fig. 5.11-1, we note that in this example when $x_1 \in [-1, 1]$, $-1 < f(\sigma)/\sigma < 1$, we have $\chi_{linearization} \triangleq [k_1', k_2'] \equiv [-1, 1]$. On the other hand, and returning to (5.11-15), we note that as far as the stability is concerned, this *special* perturbation matrix which corresponds to $\Delta\theta_2$ is valid for all $[k_1'', k_2''] \equiv [-2, \infty)$ and is compatible with $\chi_{perturbation}$. Thus $\chi_{linearization}$ along the

$\Delta\theta_2$-axis is $[-1, 1]$. Often this aspect of our solution requires an elaborate analysis of its own.

III − $\chi_{constraints}$: Since we have already stipulated that our physical constraint is of the form $|\Delta\theta_1| < 1$, we now have all the three sets needed to complete the solution to this example.

The Final Answer: The solution to Example 5.11-3 is therefore constructed by intersecting the preceding three sets as shown by the shaded square in Fig. 5.11-4. Each point on the boundary of this square corresponds to a stable perturbed system. It is interesting to note that the corresponding Lyapunov equation for this boundary set is not the same. In fact there is a set of such functions.

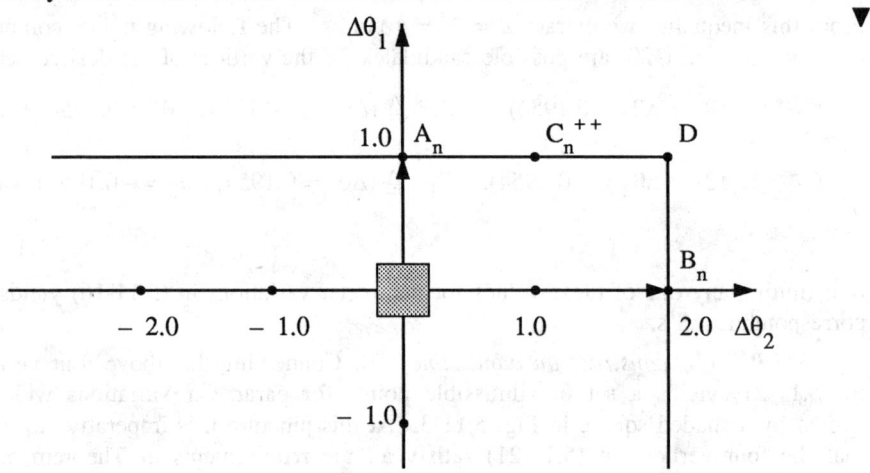

Fig. 5.11-3. Direct construction of *the largest* set of parameter variations.

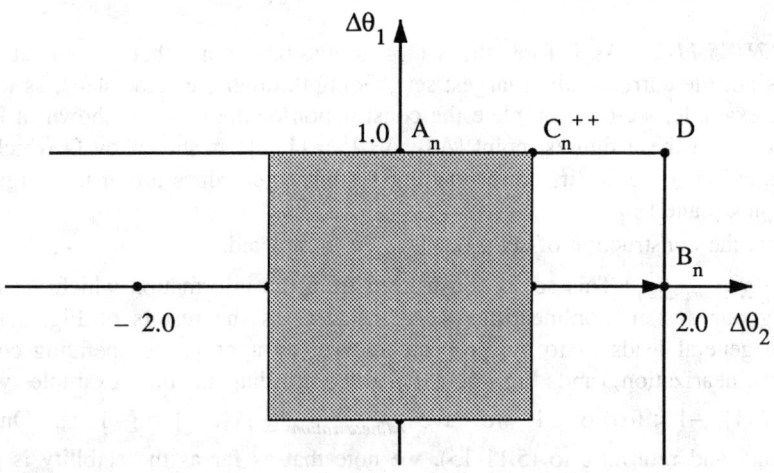

Fig. 5.11-4. The final answer for the admissible set of parameter variations.

5.11.2. Discrete – Time Nonlinear Systems

Now, we consider a set of nonlinear difference-equations described by

$$x(l + 1) = Sx(l) + bf(\sigma), \quad \sigma(l) = c^T x(l), \tag{5.11-22}$$

which is the discrete-time counterpart of (5.11-1) and as before $x(l) \in R^n(l)$ is state vector; S is an $n \times n$ convergent matrix; $f(\sigma)$ is a nonlinear scalar function of σ; and b and c are constant vectors. Replacing $f(\sigma)$ by a linear function $\eta\sigma$ (η is a scalar), results in

$$x(l + 1) = (S + \eta bc^T)x(l) \triangleq S(\eta)x(l). \tag{5.11-23}$$

This linearized system is asymptotically convergent for all time and for η in an interval $[\eta_1'', \eta_2'']$, while the nonlinear function $f(\sigma)$ is satisfying $\eta_1' < f(\sigma)/\sigma < \eta_2'$ at a given operating point considering the entire state of the system.

PROBLEM 5.11-3: Find the largest interval $[\eta_1, \eta_2] \equiv [\eta_1', \eta_2'] \cap [\eta_1'', \eta_2'']$.

Here again to solve the above problem we appeal to the discrete-time version of both the Lyapunov equation (Corollary 5.1-2) and Theorems 5.10-1 & 2.

Indeed for the above S, ηbc^T is a perturbation matrix and therefore $S + \Delta S$ remains asymptotically convergent if

$$(\eta bc^T)^T X (\eta bc^T) < \tfrac{1}{2} Q, \tag{5.11-24}$$

here $X \triangleq P + 2PSQ^{-1}S^TP$ and $S^T PS + Q = P$. Since the rank of $(bc^T)^T X (bc^T)$ is one, we may use the following conservative-upper bound

$$\eta^2 < \lambda_{min}(Q) / 2\text{Tr}[(bc^T)^T X(bc^T)]. \tag{5.11-25}$$

Here $\lambda_{min}(Q)$ is the minimum eigenvalue of Q. Or

$$\eta^2 < 1/2\lambda, \tag{5.11-26}$$

where λ is the unique nonzero generalized eigenvalue of $(bc^T)^T X(bc^T)$ with respect to Q. Using (5.11-25) and/or (5.11-26) and the same iterative procedure as before we can reach the largest $[\eta_1'', \eta_2'']$. On the other hand, the nonlinearity in each problem will determine $[\eta_1', \eta_2']$. Having found these two sets, then the solution to Problem 5.11-3 follows immediately.

REMARK 5.11-4: Finally, for the case wherein we have multiple nonlinearities, i.e., $x(l + 1) = Sx(l) + Bf(\sigma)$, then the linearized version of this problem becomes $x(l + 1) = (S + \Delta S)x(l)$, with ΔS having multiple-varying parameters and the system is having multiple nonlinearities. This problem is now manageable in the context of Theorem 5.10-3. Here we must seek *a* necessary and sufficient condition for determination of *the largest* or *most extended* set (instead of interval) for parameter variations, χ, for the above nonlinear state equation, whose corresponding set of perturbed systems is convergent. In this regard we propose the following problem.

PROBLEM 5.11-4: Find *the largest set* $\chi \equiv \chi_{linearization} \cap \chi_{perturbation}$.

This intersection, as in Problem 5.11-3, requires that we consider the contribution and/or validity of linearization, and the contribution of linear homogeneous part of the system, in order to determine the overall convergent robustness measure *(CRM)* of the nonlinear system. The conceptual solution for Problem 5.11-4 also consists of generating two sets. The necessary steps to complete this solution are exactly the same as in Problem 5.11-2, supported by Theorem 5.10-3 for generating $\chi_{perturbation}$.

5.11.3. Final Remarks

One of the most interesting features for the Lyapunov equation is that it applies equally well to both deterministic and stochastic linear control problems. Many problems in linear estimation theory and/or LQG are dual of linear optimization problems. It is a well-known fact that any time we solve a Lyapunov equation we either are optimizing a process or we are estimating some state equations – in addition to the stability conclusion which stems from this equation. Because of this versatility we believe that the results which are developed in this chapter with respect to the construction of a robustly stable matrix with multiple-varying parameters based on the Lyapunov equation can be easily used in both linear deterministic and certain linear stochastic control problems. Although it may seem that at times we are repetitive in the latter sections of this chapter, the fact of the matter is that the *methodology* which is presented here provides a coherent and harmonious approach to tackle a wide range of problems and it also provides a forum to discuss many other applications that are beyond the scope of this book.

Having completed the preceding analyses, we are indeed in the position of designing a robust system in its true meaning, namely a system which *performs* robustly. This designing task involves two phases. In the first phase when we choose the structure of the system, we construct the largest set of parameter variations from the stability robustness analysis point of view (i.e., we determine the corresponding stability robustness measure) based upon the results developed in this chapter. Next, and in the second phase of this process, we intersect the previous set of allowable parameter variations with a set of varying parameters: stemming from either the analysis and/or other design considerations in order to complete the task of designing a robustly stable multiple-objective system. In Chapter Six, we look into several such methods for designing systems which are optimal with respect to certain measures of sensitivity. The augmentation of these issues is the final aspect of this robust design.

5.12. REFERENCES

[A] Theory of Stability: Survey Papers and General References

[1] E.H. Abed, "Strong D-stability," *SCL*, vol. 7, pp. 207-212, 1986.

[2] J.K. Aggarwal, *Notes on Nonlinear Systems*. New York: Van Nostrand Reinhold, 1972. Reviewed by E.F. Infante, *IEEE-TAC*, vol. AC-18, no. 6, p. 693, 1973.

[3] M.A. Aizerman, "Some unsolved problems in theory of automatic control and fuzzy proofs," *IEEE-TAC*, vol. AC-22, no. 1, pp. 116-118, Feb. 1977.

[4] V.P. Alekperov and V.S. Khabarov, "Extension of parametric stability region with the aid of nonlinear controls," *ARC*, vol. 26, no. 11, pp. 1959-1965, Nov. 1965.

[5] B.D.O. Anderson, "A System theory criterion for positive real matrices," *SIAM J. Contr.*, vol. 5, no. 2, pp. 171-182, 1967.

[6] B.D.O. Anderson, "A simplified viewpoint of hyperstability," *IEEE-TAC*, vol. AC-13, pp. 292-294, June 1968.

[7] B.D.O. Anderson, R.R. Bitmead, C.R. Johnson, Jr., P.V. Kokotovic, R.L. Kosut, I.M.Y. Mareels, L. Praly, and B.D. Riedle, *Stability of Adaptive Systems: Passivity and Averaging Analysis*. Cambridge, MA: MIT Press, 1986. Reviewed by J. Baumeister, *SIAM R.*, vol. 30, no. 2, pp. 323-324, 1988, and by P.C. Parks, *Automatica*, vol. 25, no. 1, pp. 155-156, 1989.

[8] Y.N. Andreev, "Algebraic methods of state space in linear object control theory (survey of foreign literature)," *ARC*, vol. 38, pp. 305-342, March 1977.

[9] M. Araki, "Stability of large-scale nonlinear systems – quadratic-order theory of composite-system method using M-matrices," *IEEE-TAC*, vol. AC-23, no. 2, pp. 129-142, April 1978.

[10] M. Artzrouni, "On the local stability of nonautonomous difference equations in R^n," *JMAA*, vol. 122, pp. 519-537, 1987.

[11] D.P. Atherton, *Nonlinear Control Engineering, describing function analysis and design*. New York: Van Nostrand Reinhold, 1975 (student edition 1982). Reviewed by G.M. Siouris, *IEEE-TSMC*, vol. SMC-7, pp. 567-568, 1977, and by D.R. Towill, *ibid.*, p. 678.

[12] F. Attarzadeh, "Relative stability test for continuous and sampled-data control systems using the generalised sign matrix," *Proc. IEE*, vol. 129, Pt. D, no. 5, pp. 189-192, 1982.

[13] S. Baizaev and E. Mukhamadiev, "Tests for instability of stationary modes in nonlinear systems," *ARC*, vol. 48, no. 4, pp. 425-429, April 1987.

[14] A.T. Barabanov and Y.F. Starozhilov, "Investigation of the stability of solutions of continuous discrete systems by Lyapunov's second method," *SJAIS*, vol. 21, no. 6, pp. 35-41, 1988.

[15] N.E. Barabanov, "A dichotomy of nonlinear control systems that satisfy a differential constraint," *ARC*, vol. 43, no. 2, pp. 137-139, Feb. 1982.

[16] _____, "Stability, instability, and dichotomy of control systems with gradient nonlinearities," *ARC*, vol. 43, no. 4, pp. 425-428, April 1982.

[17] _____, "Stability and instability of multivariable control systems with a nonunique equilibrium state," *ARC*, vol. 45, no. 6, pp. 685-692, June 1984.

[18] _____, "Stabilization of nonstationary linear systems with uncertainty in the coefficients," *ARC*, vol. 51, no. 10, pp. 1326-1332, 1990.

[19] E.A. Barbashin, *Introduction to the theory of stability*. The Netherlands: Wolters-Noordhoff Pub., 1970. [Translation editor, T. Lukes.]

[20] S. Barnett, "Matrices, polynomials, and linear time-invariant systems," *IEEE-TAC*, vol. AC-18, no. 1, pp. 1-10, Feb. 1973.

[21] S. Barnett and D.D. Siljak, "Routh's algorithm: A centennial survey," *SIAM R.*, vol. 19, no. 3, pp. 472-489, July 1977.

[22] S. Barnett and C. Storey, *Matrix Methods in Stability Theory*. New York: Barnes and Nobel, 1971.

[23] D.S. Bernstein, "Robust static and dynamic output-feedback stabilization: Deterministic and stochastic perspectives," *IEEE-TAC*, vol. AC-32, no. 12, pp. 1076-1084, Dec. 1987.

[24] N.P. Bhatia and G.P. Szego, *Dynamical Systems: Stability Theory and Applications*. New York: Springer – Verlag, 1967. Reviewed by K.R. Meyer, *IEEE-TAC*, vol. AC-13, no. 4, p. 460, 1968.

[25] _____, *Stability Theory of Dynamical Systems*. New York: Springer – Verlag, 1970. Reviewed by R.J. Sacker, *SIAM R.*, vol. 16, no. 1, pp. 106-107, 1974.

[26] S.P. Bhattacharyya, L.H. Keel, and J.W. Howze, "Stabilizability conditions using linear programming," *IEEE-TAC*, vol. AC-33, pp. 460-463, May 1988.

[27] G. Birkhoff and G.C. Rota, *Ordinary Differential Equations*. Lexington, MA: Balisdell Publishing Company, 1969.

[28] M. Bisiacco, "Stabilization theory for single-input/single-output two-dimensional systems," *CSSP*, vol. 6, no. 1, pp. 77-93, 1987.

[29] R.W. Brockett, "The status of stability theory for deterministic systems," *IEEE-TAC*, vol. AC-11, no. 3, pp. 596-606, 1966.

[30] C.F. Chen and C.H. Hsiao, "Stability tests for singular cases of discrete systems," *CSSP*, vol. 8, no. 2, pp. 123-132, 1989.

[31] C.T. Chen, "Stability of linear multivariable feedback systems," *Proc. IEEE*, vol. 56, pp. 821-828, May 1968.

[32] _____, *Linear System Theory and Design*. New York: Holt, Rinehart and Winston, 1984. The 1970 edition is reviewed by A.S. Morse, *IEEE-TAC*, vol. AC-17, no. 5, pp. 748-749, 1972, and the 1984 edition is reviewed by S. Barnett, *Automatica*, vol. 22, no. 3, pp. 385-386, 1986.

[33] M.O. Cesar and G. Zampieri, "On Liapunov stability for $\ddot{x} + xf(x) = 0$, $\ddot{y} + yw(x) = 0$," *DSS*, vol. 3, nos. 3 & 4, pp. 177-185, 1988.

[34] P.S.M. Chin, "Extensions to the intrinsic method and its application to stability problems," *IJC*, vol. 48, no. 5, pp. 2139-2146, 1988.

[35] J.E. Cochran, Jr., and C.-S. Ho, "Stability of aircraft motion in critical cases," *JGCD*, vol. 6, no. 4, pp. 272-279, 1983.

[36] E.A. Coddington and N. Levinson, *Theory of Ordinary Differential Equations*. New York: McGraw-Hill, 1955. (Reprint, R.E. Krieger, 1984).

[37] J. Cronin, "A criterion for asymptotic stability," *JMAA*, vol. 74, pp. 247-269, 1980.

[38] J.B. Cruz, Jr., J.S. Freudenberg, and D.P. Looze, "A relationship between sensitivity and stability of multivariable feedback systems," *IEEE-TAC*, vol. AC-26, pp. 66-74, Feb. 1981.

[39] F.M. Dannan and S. Elaydi, "Lipchitz stability of nonlinear systems of differential equations," *JMAA*, vol. 113, pp. 562-577, 1986.

[40] _____, "Lipchitz stability of nonlinear systems of differential equations. II. Liapunov functions," *JMAA*, vol. 143, pp. 517-529, 1989.

[41] B.S. Darkhovskii, "Conditions of structural stability of a locally optimal stabilization system," *ARC*, vol. 49, no. 5, pp. 580-588, 1988.

[42] R.R. Del Rio Castillo, "Instability of absolutely continuous spectrum of ordinary differential operators under local perturbations," *JMAA*, vol. 142, pp. 591-604, 1989.

[43] J.W. Demmel, "A counterexample for two conjectures about stability," *IEEE-TAC*, vol. AC-32, pp. 340-342, April 1987. [cf., [6P. 74].]

5.12.A. Theory of Stability: Survey Papers and General References

[44] C.A. Desoer, *Notes for a Second Course on Linear Systems*. New York: Van Nostrand Reinhold, 1970. Reviewed by R. Liu, *IEEE-TAC*, vol. AC-17, no. 2, pp. 279-280, 1972.

[45] C.A. Desoer and M. Vidyasagar, *Feedback Systems: Input-Output Properties*. New York: Academic Press, 1975. Reviewed by J.C. Willems, *IEEE-TAC*, vol. AC-21, no. 2, pp. 300-301, 1976.

[46] S.N. Diligenskii, "Analysis of the stability of nonlinear dynamical systems," *ARC*, vol. 45, no. 6, pp. 699-707, June 1984.

[47] V. Dolezal, "Some practical stability criteria for semistate equations," *CSSP*, vol. 6, no. 3, pp. 335-345, 1987.

[48] _____, "Some results on robust stability of general input-output systems," *CSSP*, vol. 9, no. 3, pp. 343-364, 1990.

[49] _____, "Sensitivity and robust stability of general input-output systems," *IEEE-TAC*, vol. AC-36, no. 5, pp. 539-550, 1991.

[50] _____, "Sensitivity and robust stability of some control systems containing a nonlinear plant," *IEEE-TAC*, vol. AC-37, no. 12, pp. 1949-1956, 1992.

[51] W. Eckhaus, *Studies in Non-Linear Stability Theory*. New York: Springer – Verlag, 1965. Reviewed by L. Weiss, *IEEE-TAC*, vol. AC-12 no. 2, pp. 232-233, 1967.

[52] L.E. Faibusovich, "Stabilization of infinite-dimensional linear dynamical systems by the Kalman-Letov method," *ARC*, vol. 46, no. 2, pp. 182-188, Feb. 1985.

[53] A.F. Filippov, "Stability conditions in homogeneous systems with arbitrary regime switching," *ARC*, vol. 41, no. 8, pp. 1078-1085, Aug. 1980.

[54] _____, *Differential Equations with Discontinuous Righthand Sides*. Dordrecht, the Netherlands: Kluwer, 1988. Reviewed by S.R. Bernfeld, *SIAM R.*, vol. 32, no. 2, pp. 312-315, 1990.

[55] N.Y. Foo, "Stability preservation under homomorphisms," *IEEE-TSMC*, vol. SMC-7, no. 10, pp. 750-754, Oct. 1977.

[56] L.N.F. Franca and J.C.F. de Oliveira, "Stability of some mechanical systems with dry friction," *DSS*, vol. 5, no. 3, pp. 129-136, 1990.

[57] J.N. Franklin, *Matrix Theory*. Englewood Cliffs, NJ: Prentice-Hall, 1968.

[58] A.I. Gaiduk, "Absolute stability of control systems with several nonlinearities," *ARC*, vol. 37, no. 6, pp. 815-821, Nov. 1976.

[59] A.K. Gelig, "Averaging method in the theory of stability of nonlinear sampled-data systems," *ARC*, vol. 44, no. 5, pp. 591-600, May 1983.

[60] R. Genesio, M. Tartaglia, and A. Vicino, "On the estimation of asymptotic stability regions: State of the art and new proposals," *IEEE-TAC*, vol. AC-30, pp. 747-755, Aug. 1985.

[61] S.P. Gordon, "A stability theory for perturbed difference equations," *SIAM J. Contr.*, vol. 10, no. 4, pp. 671-678, 1972.

[62] W. Hahn, *Theory and Application of Liapunov's Direct Method*. [Translated from German by H.H. Hosenthien and S.H. Lehnigh.] Englewood Cliffs, NJ: Prentice-Hall 1963.

[63] _____, *Stability of Motion*. [Translated from German by A.P. Baartz.] New York: Springer – Verlag, 1967. Reviewed by E.F. Infante, *IEEE-TAC*, vol. AC-14, no. 1, pp. 118-119, 1969, and by R.F. Datko, *SIAM R.*, vol. 15, no. 1, pp. 228-230, 1973.

[64] J.K. Hale, *Ordinary Differential Equations*. New York: Wiley, 1969. (Reprint, R.E. Krieger, 1980). The 1969 edition is reviewed by K.R. Meyer, *SIAM R.*, vol. 14, no. 2, pp. 348-350, 1972.

[65] J.K. Hale, E.F. Infante, and F.-S. P. Tsen, "Stability in linear delay equations," *JMAA*, vol. 105, pp. 533-555, 1985.

[66] H. Hermes, "On the synthesis of a stabilizing feedback control via Lie algebraic methods," *SIAM-JCO*, vol. 18, no. 4, pp. 352-361, 1980.

[67] D. Hertz, E.I. Jury, and E. Zeheb, "Stability independent and dependent of delay for delay differential systems," *JFI*, vol. 318, no. 3, pp. 143-150, Sept. 1984.

[68] J.M. Holtzman, *Nonlinear System Theory, a functional analysis approach.* Englewood Cliffs, NJ: Prentice-Hall, 1970. Reviewed by R.M. DeSantis, *IEEE-TAC*, vol. AC-17, no. 4, pp. 584-585, 1972, and by J.G. Lankford and V.E. Mablekos, *IEEE-TSMC*, vol. SMC-3, no. 1, p. 116, 1973, and by B.A. Fleishman, *SIAM R.*, vol. 18, no. 2, pp. 308-310, 1976.

[69] M. Jamshidi, *Large-Scale Systems: Modeling and Control.* New York: North Holland, 1983. Reviewed by M.F. Hassan, *Automatica*, vol. 21, no. 2, pp. 219-220, 1985.

[70] E.I. Jury, *Inners and Stability of Dynamic Systems.* New York: Wiley, 1974. Reviewed by P.C. Parks, *IEEE-TAC*, vol. AC-21, no. 5, pp. 809-810, 1976.

[71] R.E. Kalman, "Lyapunov functions for the problem of Lure in automatic control," *Proc. Nat. Acad. Sci., U.S.*, vol. 49, pp. 201-205, Feb. 1963.

[72] R.E. Kalman and J.E. Bertram, "Control system analysis and design via the 'second method' of Lyapunov, I. Continuous-time systems," *ASME-JBE*, vol. 82, no. 2, pp. 371-393, 1960.

[73] _____, "Control system analysis and design via the 'second method' of Lyapunov, II. Discrete-time systems," *ibid*, pp. 394-400, June 1960.

[74] N. Kalouptsidis and J. Tsinias, "Stability improvement of nonlinear systems by feedback," *IEEE-TAC*, vol. AC-29, no. 4, pp. 364-367, April 1984.

[75] U. Kirchgraber, "A note on Liapunov's center theorem," *JMAA*, vol. 73, pp. 568-570, 1980.

[76] A.V. Knyazev, "Analysis of feedback control systems by the averaging method," *ARC*, vol. 51, no. 11, pp. 1487-1494, 1990.

[77] D.E. Koditschek and K.S. Narendra, "The stability of second-order quadratic differential equations," *IEEE-TAC*, vol. AC-27, no. 4, pp. 783-798, Aug. 1982.

[78] M.A. Krasnosel'skii and A.V. Pokrovskii, "Absolute stability of systems with discrete time," *ARC*, vol. 39, no. 2, pp. 189-199, Feb. 1978.

[79] V.I. Krutin, "Stabilizing the output signals of linear control systems in a finite time," *ARC*, vol. 49, no. 10, pp. 1300-1303, 1988.

[80] W.E. Kwon and A.E. Pearson, "A note on the algebraic matrix Riccati equation," *IEEE-TAC*, vol. AC-22, no. 1, pp. 143-144, Feb. 1977. Further corrections, by M.M. Fahmy and A.A.R. Hanafy, *ibid.*, vol. AC-26, no. 2, p. 619, April 1981.

[81] J.P. LaSalle and S. Lefschetz, *Stability by Lyapunov's Direct Method with Applications.* New York: Academic Press, 1961.

[82] S. Lefschetz, *Stability of Nonlinear Control Systems.* New York: Academic Press, 1965. Reviewed by R.W. Brockett, *IEEE-TAC*, vol. AC-11, no. 3, p. 634, 1966.

[83] N. Levan, "Approximate stabilizability via the algebraic Riccati equation," *SIAM-JCO*, vol. 23, no. 1, pp. 153-160, Jan. 1985.

[84] A.M. Liapunov, *Stability of Motion.* [With a contribution by V.A. Pliss and an introduction by V.P. Basov.] [Translated from Russian by F. Abramovici and M. Shimshoni.] New York: Academic Press, 1966.

[85] A.V. Lipatov, "Stability of stationary system with one nonlinear unit I. Fundamental theorems," *ARC*, vol. 43, no. 6, pp. 737-746, June 1982.

[86] _____, "Stability of a stationary system with one nonlinear unit II. Geometrical criterion," *ARC*, vol. 43, no. 7, pp. 865-871, July 1982.

[87] _____, "Stability of discrete stationary system with one nonlinear element," *ARC*, vol. 45, no. 10, pp. 1162-1170, Sept. 1984.

5.12.A. Theory of Stability: Survey Papers and General References

[88] A.I. Lur'e, *Certain (Some) Nonlinear Problems in the Theory of Automatic Control [in Russian]*. Moscow: Gostekhizdat, 1951. [English Translation: H.M.S.O., London, 1957.]

[89] M.M. Lychak, "A new approach to investigation of the stability of nonlinear dynamic systems," *SAC*, vol. 13, no. 1, pp. 30-36, 1980.

[90] R. Marino and P. Tomei, "Dynamic output feedback linearization and global stabilization," *SCL*, vol. 17, pp. 115-121, 1991.

[91] J.L. Massera, "Contributions to stability theory," *Ann. Math.*, vol. 64, pp. 182-206, 1956.

[92] B.G. Mertzios and M.A. Christodoulou, "Decoupling and data sensitivity in singular systems," *Proc. IEE*, vol. 135, Pt. D, no. 2, pp. 106-110, March 1988.

[93] A.N. Michel, "On the status of stability of interconnected systems," *IEEE-TAC*, vol. AC-28, no. 6, pp. 639-653, June 1983.

[94] R.K. Miller and A.N. Michel, "Asymptotic stability of systems: Results involving the system topology," *SIAM-JCO*, vol. 18, no. 2, pp. 181-190, 1980.

[95] N. Minorsky, *Theory of Nonlinear Control Systems*. New York: McGraw-Hill, 1969. Reviewed by E.F. Infante, *IEEE-TAC*, vol. AC-16, no. 5, pp. 524-525, 1971.

[96] I.V. Miroshnik, "Stabilization of motion over a manifold," *SJAIS*, vol. 19, no. 4, pp. 69-72, 1986.

[97] H. Miyagi and K. Yamashita, "Stability studies of control systems using non-Lure type Lyapunov function," *IEEE-TAC*, vol. AC-31, no. 10, pp. 970-972, Oct. 1986.

[98] ———, "A generalized Lyapunov function for power systems with product-type nonlinearities," *EEJ*, vol. 108, no. 4, pp. 67-74, 1988.

[99] J.C. Moore, "A new concept of stability, M_o-stability," *JMAA*, vol. 112, pp. 1-13, 1985.

[100] M.V. Morozov, "An algorithm of analysis of the stability of linear periodic systems and its computer realization," *ARC*, vol. 51, no. 4, pp. 444-451, 1990.

[101] S. Mossaheb, "The relationship between various L^p stabilities of time-varying feedback systems," *IJC*, vol. 38, no. 6, pp. 1199-1212, 1983.

[102] K.S. Narendra and J.H. Taylor, *Frequency Domain Criteria for Absolute Stability*. New York: Academic Press, 1973. Reviewed by B.A. Asner, *SIAM R.*, vol. 17, no. 1, pp. 177-178, 1975, and by C.A. Desoer, *IEEE-TAC*, vol. AC-20, no. 4, p. 583, 1975, and by M. Vidyasagar, *JFI*, vol. 300, no. 4, pp. 312-314, 1975, and by Y.V. Venkatesh, *ARC*, vol. 38, no. 9, pp. 1428-1429, Sept. 1977.

[103] R.A. Nesbit, "Several applications of the direct method of Liapunov," in *Control and Dynamic Systems advances in theory and applications*, vol. 2 (C.T. Leondes, Ed.). New York: Academic Press, pp. 269-311, 1965.

[104] J.W. Nieuwenhuis, "About positive invariance and asymptotic stability," *AMO*, vol. 12, pp. 81-87, 1984.

[105] K. Ogata, *Modern Control Engineering*. Englewood Cliff, NJ: Prentice-Hall, 1970. Reviewed by S.R. Liberty, *IEEE-TAC*, vol. AC-17, no. 3, p. 419, 1972, and by R.P. Batni, *JFI*, vol. 293, no. 1, pp. 70-71, 1972.

[106] Y. Ohta, H. Imanishi, L. Gong, and H. Haneda, "Computer generated Lyapunov functions for a class of nonlinear systems," *IEEE-TCAS*, Pt. I, vol. 40, no. 5, pp. 343-354, 1993.

[107] V.N. Orlov, "Conditions for instability in systems of interacting subsystems and their applications to chemical kinetics problems," *ARC*, vol. 41, pp. 1349-1356, Oct. 1980.

[108] L. Pandolfi, "A Lyapunov theorem for semigroups of operators," *SCL*, vol. 15, pp. 147-151, 1990.

[109] T.K.C. Peng, "Invariance and stability for bounded uncertain systems," *SIAM J. Contr.*, vol. 10, no. 4, pp. 679-690, Nov. 1972.

[110] S.C. Persek, "Asymptotic stability on slow time scales for periodic systems," *SIAM-JAM*, vol. 41, no. 1, pp. 16-28, 1981.

[111] V.M. Popov, "Absolute stability of nonlinear systems of automatic control," *ARC*, vol. 22, no. 8, pp. 857-875, 1961.

[112] _____, "The solution of a new stability problem for controlled systems," *ARC*, vol. 24, pp. 1-23, 1963.

[113] _____, *Hyperstability of Control Systems*. New York: Springer – Verlag, 1970.

[114] S. Pradeep and S.K. Shrivastava, "Stability of dynamical systems: An overview," *JGCD*, vol. 13, no. 3, pp. 385-393, 1990.

[115] E.S. Pyatnytsky, "New research on absolute stability of automatic control systems," (Review), *ARC*, vol. 29, no. 6, pp. 855-881, June 1968.

[116] W.T. Reid, *Sturmian Theory for Ordinary Differential Equations*. New York: Springer – Verlag, 1980. Reviewed by A. Lazer, *SIAM R.*, vol. 24, no. 4, pp. 485-489, 1982.

[117] R. Rebarber, "Semigroup generation and stabilization by A^p-bounded feedback perturbations," *SCL*, vol. 14, pp. 333-340, 1990.

[118] F. Reza, *Stability of Linear Dynamic Systems*. Teheran, Iran: Teheran University Press, 1974.

[119] V.P. Reztsov and L.M. Salikov, "Lyapunov functions for linear nonstationary sampled-data systems," *ARC*, vol. 39, no. 4, pp. 624-626, 1978.

[120] E. Roxin, "Stability in general control systems," *JDE*, vol. 1, no. 2, pp. 115-150, 1965.

[121] D.L. Russell, "Linear stabilization of the linear oscillator in Hilbert space," *JMAA*, vol. 25, pp. 663-675, 1969.

[122] R. Saeks, Ed., *Special Issue on Large Scale Systems*. IEEE-TAC, vol. AC-28, no. 6, June 1983.

[123] M.G. Safonov, *Stability and Robustness of Multivariable Feedback Systems*. Cambridge, MA: MIT Press, 1980. Reviewed by M. Vidyasagar, *IEEE-TAC*, vol. AC-29, no. 1, pp. 93-94, 1984.

[124] I.W. Sandberg, "Discrete-time p-powers and stability," *CSSP*, vol. 9, no. 4, pp. 435-448, 1990.

[125] D.G. Schultz, "The generation of Liapunov functions," in *Control and Dynamic Systems advances in theory and applications*, vol. 2 (C.T. Leondes, Ed.). New York: Academic Press, pp. 1-64, 1965.

[126] P. Seibert and R. Suarez, "Global stabilization of nonlinear cascade systems," *SCL*, vol. 14, pp. 347-352, 1990.

[127] _____, "Global stabilization of a certain class of nonlinear systems," *SCL*, vol. 16, pp. 17-23, 1991.

[128] E.Y. Shapiro, "On the Lyapunov matrix equation," *IEEE-TAC*, vol. AC-19, no. 5, pp. 594-596, Oct. 1974. Comments, by J.J. Montemayer and B.F. Womack, *ibid.*, vol. AC-20, no. 6, pp. 814-815, Dec. 1975. [Also refer to [80].]

[129] A.G. Shevelev and A.V. Polukhin, "On absolute stability of equilibrium position of a class of nonlinear systems," *SAC*, vol. 10, no. 2, pp. 38-43, March/April 1977.

[130] S.K. Shrivastava and S. Pradeep, "Stability of multidimensional linear time-varying systems," *JGCD*, vol. 8, no. 5, pp. 579-583, 1985. Comment, with reply, by P.C. Muller, *ibid.*, vol. 10, no. 1, pp. 127-128, 1987.

[131] D.D. Siljak, "Absolute stability and parameter sensitivity," *IJC*, vol. 8, no. 3, pp. 279-283, 1968.

[132] D.D. Siljak and M.B. Vukcevic, "Large-scale systems: Stability, complexity, reliability," *JFI*, vol. 301, nos. 1-2, pp. 49-69, Jan./Feb. 1976.

5.12.A. Theory of Stability: Survey Papers and General References

[133] P.S. Simeonov and D.D. Bainov, "Stability under persistent disturbances for systems with impulse effect," *JMAA*, vol. 109, pp. 546-563, 1985.

[134] _____, "Stability with respect to part of the variables in systems with impulse effect," *JMAA*, vol. 117, pp. 247-263, 1986.

[135] _____, "Stability with respect to part of the variables in systems with impulse effect," *JMAA*, vol. 124, pp. 547-560, 1987.

[136] _____, "Stability of the solutions of singularly perturbed systems with impulse effect," *JMAA*, vol. 136, pp. 575-588, 1988.

[137] _____, "Exponential stability of the solutions of singularly perturbed systems with impulse effect," *JMAA*, vol. 151, pp. 462-487, 1990.

[138] R.A. Smith, "Bounds for quadratic Lyapunov functions," *JMAA*, vol. 12, pp. 425-435, 1965.

[139] E.M. Solnechnyy, "Stability of one- and two-channel invariant systems with parameter deviations," *SAC*, vol. 4, no. 1, pp. 10-24, 1971.

[140] K.-P. Sondergold, "A generalization of the Routh-Hurwitz stability criteria and an application to a problem in robust controller design," *IEEE-TAC*, vol. AC-28, pp. 965-970, Oct. 1983.

[141] E.D. Sontag, "Smooth stabilization implies coprime factorization," *IEEE-TAC*, vol. AC-34, no. 4, pp. 435-443, April 1989.

[142] E.D. Sontag and A. Bacciotti, Eds., "Special issue on smooth and continuous stabilizability," *CTAT*, vol. 6, no. 4, 1990.

[143] R.J. Stern, "A note on positively invariant cones," *AMO*, vol. 9, pp. 67-72, 1982.

[144] T.-J. Tarn, X. Zeng, and J.R. Zavgren, Jr., "Feedback stabilization of linear dynamic systems with multirate sampled output," *SIAM-JCO*, vol. 26, no. 4, pp. 812-833, 1988.

[145] J. Tsinias, "A theorem on global stabilization of nonlinear systems by linear feedback," *SCL*, vol. 17, pp. 357-362, 1991.

[146] _____, "A local stabilization theorem for interconnected systems," *SCL*, vol. 18, pp. 429-434, 1992.

[147] A. Vannelli and M. Vidyasagar, "Maximal Lyapunov functions and domains of attraction for autonomous nonlinear systems," *Automatica*, vol. 21, no. 1, pp. 69-80, 1985.

[148] V.M. Vaplyushkin, S.I. Vdovin, and E.F. Sabaev, "Stability in the large of a class of control systems," *ARC*, vol. 40, no. 7, pp. 949-955, July 1979.

[149] Y.V. Venkatesh, "Riesz-Thorin theorem and l_p-stability of nonlinear time-varying discrete systems," *JMAA*, vol. 135, pp. 627-643, 1988.

[150] M. Vidyasagar, *Nonlinear Systems Analysis*. Englewood Cliffs, NJ: Prentice-Hall, 1978. Reviewed by L.G. Clark, *IEEE-TAC*, vol. AC-25, no. 3, pp. 613-614, 1980, and by C.A. Desoer, *IEEE-TSMC*, vol. SMC-10, no. 8, pp. 537-538, 1980.

[151] _____, "New directions of research in nonlinear system theory," *Proc. IEEE*, vol. 74, no. 8, pp. 1060-1091, Aug. 1986.

[152] _____, "A state-space interpretation of simultaneous stabilization," *IEEE-TAC*, vol. AC-33, no. 5, pp. 506-509, May 1988.

[153] A.A. Voronov, "Absolutely stable systems with a differentiable nondecreasing nonlinearity," *ARC*, vol. 39, no. 7, pp. 947-958, July 1978.

[154] _____, "Present state and problems of stability theory," (Survey) *ARC*, vol. 43, no. 5, pp. 573-592, May 1982.

[155] M.B. Vukcevic, "On decentralized hierarchic stabilization of a large-scale system with constraints," *CSSP*, vol. 6, no. 4, pp. 391-419, 1987. Comment by the author, *ibid.*, vol. 7, no. 3, pp. 409-411, 1988.

[156] J.T. Wen, "Stability of a class of interconnected evolution systems," *IEEE-TAC*, vol. AC-37, no. 3, pp. 342-347, 1992.

[157] J.C. Willems, "Stability, instability, invertibility and causality," *SIAM J. Contr.*, vol. 7, no. 4, pp. 645-671, Nov. 1969.

[158] _____, *The Analysis of Feedback Systems*. Cambridge, MA: MIT Press, 1970. Reviewed by R. Saeks, *IEEE-TAC*, vol. AC-17, no. 5, pp. 745-746, 1972.

[159] _____, "Mechanisms for the stability and instability in feedback systems," *Proc. IEEE*, vol. 64, pp. 24-35, Jan. 1976.

[160] J.L. Willems, *Stability Theory of Dynamical Systems*. London: Thomas Nelson, 1970. Reviewed by G. Zames, *IEEE-TAC*, vol. AC-18, no. 1, pp. 83-84, 1973.

[161] G. Zames, "On the input-output stability of time-varying nonlinear feedback systems – Part I: Conditions derived using concepts of loop gain, conicity, and positivity," *IEEE-TAC*, vol. AC-11, no. 2, pp. 228-238, April 1966.

[162] _____, "On the input-output stability of time-varying nonlinear feedback systems – Part II: Conditions involving circles in the frequency plane and sector nonlinearities," *IEEE-TAC*, vol. AC-11, no. 3, pp. 465-476, July 1966.

[163] B.G. Zaslavskii, "Positive stabilizability of control processes," *ARC*, vol. 51, no. 3, pp. 291-294, 1990.

[164] W. Zhan, T.J. Tarn, and A. Isidori, "A canonical dynamic extension for noninteraction with stability for affine nonlinear square systems," *SCL*, vol. 17, pp. 177-184, 1991.

[165] V.P. Zhukov, "Method of sources for the study of stability of nonlinear systems," *ARC*, vol. 40, no. 3, pp. 330-335, March 1979.

[166] _____, "Stability study of one class of nonlinear systems," *ARC*, vol. 42, no. 1, pp. 4-8, Jan. 1981.

[B] Stability Robustness Analysis
Maneuverability, and Sensitivity in the Large
Group I: Matrices and Polynomials

[1] A.-A. A. Abdul-Wahab, "Robustness measure bounds for Lyapunov-type state-feedback systems," *Proc. IEE*, vol. 137, Pt. D, no. 5, pp. 337-340, 1990.

[2] B. Aulbach, "On linearly perturbed linear systems," *JMAA*, vol. 112, pp. 317-327, 1985.

[3] R.I. Badr, M.F. Hassan, J. Bernussou, and A.Y. Bilal, "Stability and performance robustness for multivariable linear systems," *Automatica*, vol. 25, no. 6, pp. 935-942, 1989.

[4] N.E. Barabanov, "Stabilization of nonstationary linear systems with uncertainty in the coefficients," *ARC*, vol. 51, no. 10, pp. 1326-1332, 1990.

[5] B.R. Barmish, "Robust solutions of perturbed linear equations," *IEEE-TAC*, vol. Ac-22, no. 1, pp. 123-124, Feb. 1977.

[6] B.R. Barmish, M. Fu, and S. Saleh, "Stability of a polytope of matrices: Counterexamples," *IEEE-TAC*, vol. AC-33, no. 6, pp. 569-572, June 1988. Comments, with reply, by Y.C. Soh and Y.K. Foo, *ibid.*, vol. AC-34, no. 12, pp. 1324-1326, Dec. 1989.

[7] S. Barnett, "Insensitivity of control systems," *IJC*, vol. 10, no. 6, pp. 665-675, 1969.

[8] S. Barnett and C. Storey, "Some results on the sensitivity and synthesis of asymptotically stable linear and nonlinear systems," *Automatica*, vol. 4, pp. 187-194, 1968.

[9] A. Belhouari, E. Tissir, and A. Hmamed, "Stability of interval matrix polynomial in continuous and discrete cases," *SCL*, vol. 18, pp. 183-189, 1992.

[10] S. Bialas, "A necessary and sufficient condition for the stability of interval matrices," *IJC*, vol. 37, no. 4, pp. 717-722, 1983. Comments by W.C. Karl, J.P. Greschak, and G.C. Verghese, *ibid.*, vol. 39, no. 4, pp. 849-851, 1984. Counter example by B.R. Barmish and C.V. Hollot, *ibid.*, no. 5, pp. 1103-1104, 1984.

[11] S.S.L. Chang and T.K.C. Peng, "Adaptive guaranteed cost control of systems with uncertain parameters," *IEEE-TAC*, vol. AC-17, pp. 474-483, 1972.

[12] J. Chen, "Sufficient conditions on stability of interval matrices: Connections and new results," *IEEE-TAC*, vol. AC-37, no. 4, pp. 541-544, 1992.

[13] W.L. Chen and J.S. Gibson, "A Lyapunov robustness bound for linear systems with periodic uncertainties," *Automatica*, vol. 27, no. 3, pp. 545-547, 1991.

[14] J.-H. Chou, "Stability robustness of linear state space models with structured perturbations," *SCL*, vol. 15, pp. 207-210, 1990.

[15] J.-H. Chou and B.-S. Chen, "New approach for the stability analysis of interval matrices," *CTAT*, vol. 6, no. 4, pp. 725-730, 1990.

[16] J.-H. Chou and I.-R. Horng, "On sufficient conditions for the stability of interval matrices," *JFI*, vol. 326, no. 1, pp. 19-25, 1989.

[17] L.G. Chouinard, J.P. Dauer, and G. Leitmann, "Properties of matrices used in uncertain linear control systems," *SIAM-JCO*, vol. 23, no. 3, pp. 381-389, May 1985.

[18] J.D. Cobb, "The minimal dimension of stable faces required to guarantee stability of a matrix polytope: *D*-stability," *IEEE-TAC*, vol. AC-35, no. 4, pp. 469-473, April 1990.

[19] J.D. Cobb and C.L. DeMarco, "The minimal dimension of stable faces required to guarantee stability of a matrix polytope," *IEEE-TAC*, vol. AC-34, no. 9, pp. 990-992, Sept. 1989.

[20] X. Daoyi, "Simple criteria for stability of interval matrices," *IJC*, vol. 41, no. 1, pp. 289-295, 1985.

[21] E.I. Dergacheva, "Stabilizing the steady-state displacements in a control system under continuous perturbations," *ARC*, no. 2, pp. 182-192, Feb. 1967.

[22] A. Dickman, "On robustness of multivariable linear feedback systems in state-space representation," *IEEE-TAC*, vol. AC-32, no. 5, pp. 407-410, May 1987.

[23] M. Eslami, "Sensitivity analysis and synthesis in automatic control systems," Ph.D. Dissertation, Univ. Wisconsin-Madison, Madison, May 1978.

[24] _____, "Robust stability analysis of large scale systems with certain structural perturbation matrices," in *Proc. 27th IEEE Conf. on Decision and Contr.*, pp. 2209-2214, Dec. 1988.

[25] _____, "A unified approach to the stability robustness analysis of systems with multiple varying parameters," The University of Illinois at Chicago, Dept. EECS, Tech. Report UIC-EECS-90-3, 38 pp., April 1990.

[26] _____, "On the theory of stability robustness of dynamic systems," in *Proc. Circuits, Systems, and Information [A Tribute to Professor F. Reza, UCLA, CA, May 1990]* (M. Jamshidi, M. Ahmadi, and M. Nahvi, Eds.). Albuquerque, NM: TSI Press, 1991, pp. 65-68.

[27] _____, "On stability robustness analysis of dynamic systems with multiple large parameter variations and multiple nonlinearities," in *Proc. Allerton Conf.*, pp. 107-116, Oct. 1991.

[28] _____, "Numerical case studies for robustly stable dynamic systems," in *Proc. 30th IEEE Conf. on Decision and Contr.*, pp. 3037-3042, Dec. 1991.

[29] _____, "Set by set generation of robust stable matrices," *IEEE-TAC*, vol. AC-38, no. 4, pp. 646-651, April 1993.

[30] _____, "On optimization of a stability bound," *SIAM R.*, vol. 35, no. 4, pp. 625-630, Dec. 1993.

[31] _____, "On construction of robustly stable large-scale dynamic systems," *IJC*, vol. 57, no. 6, 1499-1508, June 1993.

[32] M. Eslami and D.L. Russell, "On stability with large parameter variations: Stemming from the direct method of Lyapunov," *IEEE-TAC*, vol. AC-25, no. 6, pp. 1231-1234, Dec. 1980.

[33] E. Fabian and W.M. Wonham, "Decoupling and data sensitivity," *IEEE-TAC*, vol. AC-20, no. 3, pp. 338-344, June 1975.

[34] _____, "Decoupling and disturbance rejection," *IEEE-TAC*, vol. AC-20, no. 3, pp. 399-401, June 1975.

[35] Y.K. Foo and Y.C. Soh, "Stability analysis of a family of matrices," *IEEE-TAC*, vol. AC-35, no. 11, pp. 1257-1259, Nov. 1990. Comments by J.-H. Kim and Z. Bien, *IEEE-TAC*, vol. AC-37, no. 9, pp. 1470-1472, 1992.

[36] _____, "Robustness analysis of matrices with highly structured uncertainties," *IEEE-TAC*, vol. AC-37, no. 12, pp. 1974-1976, 1992.

[37] M. Fu and B.R. Barmish, "Maximal unidirectional perturbation bounds for stability of polynomials and matrices," *SCL*, vol. 11, pp. 173-179, 1988.

[38] A.R. Galimidi and B.R. Barmish, "The constrained Lyapunov problem and its applications to robust output feedback stabilization," *IEEE-TAC*, vol. AC-31, no. 5, pp. 410-419, May 1986.

[39] Z. Gao and P.J. Antsaklis, "Explicit asymmetric bounds for robust stability of continuous and discrete-time systems," *IEEE-TAC*, vol. AC-38, no. 2, pp. 332-335, 1993.

[40] F.G. Garashchenko, "Directional stability and construction of optimal Lyapunov functions and stability sets," *SJAIS*, vol. 19, no. 5, pp. 35-38, 1986.

[41] F. Garofalo, G. Celentano, and L. Glielmo, "Stability robustness of interval matrices *via* Lyapunov quadratic forms," *IEEE-TAC*, vol. AC-38, no. 2, pp. 281-284, 1993.

[42] R. Genesio, M. Tartaglia, and A. Vicino, "On the estimation of asymptotic stability regions: State of the art and new proposals," *IEEE-TAC*, vol. AC-30, no. 8, pp. 747-755, Aug. 1985.

[43] J.C. Geromel and A.O.E. Santo, "On the robustness of linear continuous-time dynamic systems," *IEEE-TAC*, vol. AC-31, no. 12, pp. 1136-1138, Dec. 1986.

[44] K. Gu and Y.H. Chen, "Linear control guaranteeing stability of uncertain systems *via* orthogonal decomposition," *Automatica*, vol. 27, no. 5, pp. 873-876, 1991.

[45] K. Gu, Y.H. Chen, M.A. Zohdy, and N.K. Loh, "Quadratic stabilizability of uncertain systems: A two level optimization setup," *Automatica*, vol. 27, no. 1, pp. 161-165, 1991.

[46] K. Gu and N.K. Loh, "Direct computation of stability bound for systems with polytopic uncertainties," *IEEE-TAC*, vol. AC-38, no. 2, pp. 363-366, 1993.

[47] K. Gu, M.A. Zohdy, and N.K. Loh, "Necessary and sufficient conditions of quadratic stability of uncertain linear systems," *IEEE-TAC*, vol. AC-35, no. 5, pp. 601-604, May 1990.

[48] G.W. Harrison, "Stability of linear systems with uncertain parameters," *IJSS*, vol. 9, no. 9, pp. 1043-1053, 1978.

[49] D.C. Hayland and D.S. Bernstein, "The majorant Lyapunov equation: A nonnegative matrix equation for robust stability and performance of large scale systems," *IEEE-TAC*, vol. AC-32, no. 11, pp. 1005-1013, Nov. 1987.

[50] D.C. Hayland and E.G. Collins, Jr., "An M-matrix and majorant approach to robust stability and performance analysis for systems with structured uncertainty," *IEEE-TAC*, vol. AC-34, no. 7, pp. 699-710, July 1989.

[51] J.A. Heinen, "Sufficient conditions for stability of interval matrices," *IJC*, vol. 39, no. 6, pp. 1323-1328, 1984.

[52] G. Hewer and C. Kenney, "The sensitivity of the stable Lyapunov equation," *SIAM-JCO*, vol. 26, no. 2, pp. 321-344, 1988.

[53] D. Hinrichsen and A.J. Pritchard, "Stability radii of linear systems," *SCL*, vol. 7, pp. 1-10, 1986.

5.12.B. Group I: Matrices and Polynomials

[54] _____, "Stability radius for structural perturbations and the algebraic Riccati equation," *SCL,* vol. 8, pp. 105-113, 1986.

[55] _____, "A note on some differences between real and complex stability radii," *SCL,* vol. 14, pp. 401-408, 1990.

[56] C.V. Hollot and A.R. Galimidi, "Stablizing uncertain systems: Recovering full state feedback performance *via* an observer," *IEEE-TAC,* vol. AC-31, no. 11, pp. 1050-1053, Nov. 1986.

[57] H.P. Horisberger and P.R. Belanger, "Regulators for linear time-invariant plants with uncertain parameters," *IEEE-TAC,* vol. AC-21, no. 5, pp. 705-7-8, Oct. 1976.

[58] Y.S. Hung and D.J.N. Limebeer, "Robust stability of additively perturbed interconnected systems," *IEEE-TAC,* vol. AC-29, no. 12, pp. 1069-1075, Dec. 1984.

[59] O. Huseyin, M.E. Sezer, and D.D. Siljak, "Robust decentralised control using output feedback," *Proc. IEE,* vol. 129, Pt. D, no. 6, pp. 310-314, Nov. 1982.

[60] F. Jabbari and W.E. Schmitendorf, "A noniterative method for the design of linear robust controllers," *IEEE-TAC,* vol. AC-35, no. 8, pp. 954-957, Aug. 1990.

[61] C.-L. Jiang, "Sufficient condition for the asymptotic stability of interval matrices," *IJC,* vol. 46, no. 5, pp. 1803-1810, 1987. Comments by M. Mansour, *ibid.,* vol. 47, no. 6, pp. 1973-1974, 1988, and by E.H. Abed and A.L. Tits, *ibid.,* p. 1975, and by Z.H. Luo, *ibid.,* p. 1977, and by C.B. Soh, *ibid.,* vol. 49, no. 5, p. 1817, 1989.

[62] _____, "Robust stability in linear state space model," *IJC,* vol. 48, no. 2, pp. 813-816, 1988. Comments by W.C. Karl and G.V. Verghese, *ibid.,* vol. 49, no. 3, p. 1093, 1989, and by C.B. Soh, *ibid.,* vol. 49, no. 5, p. 1815, 1989.

[63] S.M. Joshi and N.J. Groom, "Stability bounds for the control of large space structures," *JGCD,* vol. 2, no. 4, pp. 349-351, 1979.

[64] Y.-T. Juang, Z.-C. Hong, and Y.-T. Wang, "Robustness of pole-assignment in a specified region," *IEEE-TAC,* vol. AC-34, no. 7, pp. 758-760, July 1989.

[65] Y.-T. Juang, T.-S. Kuo, and S.-L. Tung, "Stability analysis of continuous and discrete interval systems," *CTAT,* vol. 6, no. 2, pp. 221-235, 1990.

[66] Y.-T. Juang and C.-S. Shao, "Stability analysis of dynamic interval systems," *IJC,* vol. 49, no. 4, pp. 1401-1408, 1989.

[67] V.A. Kamenetskiy and Y.S. Pyatnitskiy, "An iterative method of Lyapunov function construction for differential inclusions," *SCL,* vol. 8, pp. 445-451, 1987.

[68] L.H. Keel, S.P. Bhattacharyya, and J.W. Howze, "Robust control with structured perturbations," *IEEE-TAC,* vol. AC-33, no. 1, pp. 68-78, Jan. 1988.

[69] C. Kenney and G. Hewer, "The sensitivity of the algebraic and differential Riccati equations," *SIAM-JCO,* vol. 28, no. 1, pp. 50-69, Jan. 1990.

[70] H. Kiendl, "Totale stabilitat von linearen regelungssystemen bei ungenau bekannten parametern der regelstrecke," *Automatisierungstechnik,* vol. 33, pp. 379-386, 1985. [This entry is suggested by Prof. Mansour of ETH, Zurich.]

[71] H. Kokame and T. Mori, "An exact quadratic stability condition of uncertain linear systems," *IEEE-TAC,* vol. AC-38, no. 2, p. 280, 1993.

[72] O.I. Kosmidou, "Robust stability and performance of systems with structured and bounded uncertainties: An extension of the guaranteed cost control approach," *IJC,* vol. 52, no. 3, pp. 627-640, 1990.

[73] O.I. Kosmidou and P. Bertrand, "Robust-controller design for systems with large parameter variations," *IJC,* vol. 45, no. 3, pp. 927-938, 1987.

[74] V.M. Kuntsevich, "Synthesis of control systems under conditions of uncertainty (robustness of closed control systems)," *SJAIS,* vol. 23, no. 1, pp. 1-7, 1990.

[75] M.A. Leal and J.S. Gibson, "A first-order Lyapunov robustness method for linear systems with uncertain parameters," *IEEE-TAC*, vol. AC-35, no. 9, pp. 1068-1070, Sept. 1990.

[76] I. Lewkowicz, "When are the complex and the real stability radii equal?" *IEEE-TAC*, vol. AC-37, no. 6, pp. 880-883, 1992.

[77] I. Lewkowicz and R. Sivan, "Maximal stability robustness for state equations," *IEEE-TAC*, vol AC-33, no. 3, pp. 297-300, March 1988.

[78] W. Lian, "On the construction of generalized Lyapunov's function," *Sc. S*, vol. 24, no. 4, pp. 686-704, April 1981.

[79] X.-X. Liao and J.-L. Qian, "Some new results for stability of interval matrices," *CTAT*, vol. 4, no. 2, pp. 265-275, June 1988.

[80] C.-L. Lin, F.-B. Hsiao, and B.-S. Chen, "Stabilization of large structural systems under mode truncation, parameter perturbations and actuator saturations," *IJSS*, vol. 21, no. 8, pp. 1423-1440, 1990.

[81] M. Mansour, "Simplified sufficient conditions for the asymptotic stability of interval matrices," *IJC*, vol. 50, no. 1, pp. 443-444, 1989.

[82] J.M. Martin, "State-space measures for stability robustness," *IEEE-TAC*, vol. AC-32, no. 6, pp. 509-512, June 1987.

[83] J.M. Martin and G.A. Hewer, "Smallest destablizing perturbations for linear systems," *IJC*, vol. 45, no. 5, pp. 1495-1504, 1987. Comments by C.B. Soh, *ibid.*, vol. 49, no. 5, pp. 1813-1814, 1989.

[84] A.G. Mazko, "The Lyapunov matrix equation for a certain class of regions bounded by algebraic curves," *SAC*, vol. 13, no. 3, pp. 37-42, 1980.

[85] _____, "Generalization of Lyapunov's theorem for the class of regions bounded by algebraic curves," *SAC*, vol. 14, no. 1, pp. 93-96, 1982.

[86] A.M. Meilakhs, "Design of stable control systems subject to parametric perturbation," *ARC*, vol. 39, no. 10, pp. 1409-1418, Oct. 1978.

[87] _____, "Stabilization of linear control systems subjected to continuous perturbations," *ARC*, vol. 41, no. 11, pp. 1500-1505, Nov. 1980.

[88] _____, "Stabilization of linear control systems," *ARC*, vol. 43, no. 6, pp. 747-751, June 1982.

[89] G.J. Michael and C.W. Merriam, III, "Stability of parametrically disturbed linear optimal systems," *JMAA*, vol. 28, pp. 294-302, 1969.

[90] T. Mori and H. Kokame, "Stabilization of perturbed systems *via* linear optimal regulator," *IJC*, vol. 47, no. 1, pp. 363-372, 1988.

[91] D.H. Owens and A. Choatai, "Robust stability of multivariable feedback systems with respect to linear and nonlinear feedback perturbations," *IEEE-TAC*, vol. AC-27, no. 1, pp. 254-256, Feb. 1982.

[92] R.V. Patel and M. Toda, "Quantitative measures of robustness for multivariable systems," in *Proc. Joint Auto. Contr. Conf.*, paper TP8-A, 1980.

[93] R.V. Patel, M. Toda, and B. Sridhar, "Robustness of linear quadratic state feedback designs in the presence of system uncertainty," *IEEE-TAC*, vol. AC-22, no. 6, pp. 945-949, Dec. 1977.

[94] I.R. Petersen, "A Riccati equation approach to the design of stablizing controllers and observers for a class of uncertain linear systems," *IEEE-TAC*, vol. AC-30, no. 9, pp. 904-907, Sept. 1985.

[95] _____, "Quadratic stabilizability of uncertain linear systems containing both constant and time-varying uncertain parameters," *JOTA*, vol. 57, no. 3, pp. 439-461, 1988.

5.12.B. Group I: Matrices and Polynomials

[96] _____, "Stabilization of an uncertain linear system in which uncertain parameters enter into the input matrix," *SIAM-JCO*, vol. 26, no. 6, pp. 1257-1264, 1988.

[97] _____, "A class of stability regions for which a Kharitonov-like theorem holds," *IEEE-TAC*, vol. AC-34, no. 10, pp. 1111-1115, Oct. 1989.

[98] I.R. Petersen and C.V. Hollot, "A Riccati equation approach to the stabilization of uncertain linear systems," *Automatica*, vol. 22, pp. 397-411, July 1986.

[99] D. Petkovski, "Robustness of decentralised control systems subject to sensor perturbations," *Proc. IEE*, vol. 132, Pt. D, no. 2, pp. 53-60, March 1985.

[100] _____, "Decentralised design of robust controllers," *Proc. IEE*, vol. 134, Pt. D, no. 5, pp. 317-326, Sept. 1987.

[101] D.B. Petkovski, "Stability analysis of interval matrices: Improved bounds," *IJC*, vol. 48, no. 6, pp. 2265-2273, 1988. Counter examples by Y.-T. Juang and J.-D. Chen, *ibid.*, vol. 51, no. 2, pp. 497-498, 1990. Comments by M. Buslowicz, *ibid.*, no. 6, pp. 1485-1486, 1990.

[102] _____, "Improved time-domain stability robustness measures for linear regulators," *JGCD*, vol. 12, no. 4, pp. 595-598, 1989.

[103] R.X. Qian and C.L. DeMarco, "An approach to robust stability of matrix polytopes through copositive homogeneous polynomials," *IEEE-TAC*, vol. AC-37, no. 6, pp. 848-852, 1992.

[104] F. Reza, "Derived stable matrices," *Proc. IEEE*, vol. 55, no. 6, p. 1112, June 1967.

[105] D.L. Russell, "Topics in ordinary differential equations," The University of Wisconsin-Madison, Dept. Mathematics, unpublished lecture notes, Spring 1975.

[106] _____, *Mathematics of Finite-Dimensional Control System, theory and design*. New York: Marcel Dekker, 1979. Reviewed by H. Sagan, *SIAM R.*, vol. 23, no. 1, pp. 115-116, 1981, and by M. Eslami, *IEEE-TSMC* vol. SMC-14, no. 1, pp. 166-168, 1984.

[107] W.E. Schmitendorf, "Designing stabilizing controllers for uncertain systems using the Riccati equation Approach," *IEEE-TAC*, vol. AC-33, no. 4, pp. 376-379, April 1988.

[108] M.E. Sezer and D.D. Siljak, "A note on robust stability bounds," *IEEE-TAC*, vol. AC-34, no. 11, pp. 1212-1215, Nov. 1989.

[109] B.A. Shane and S. Barnett, "Insensitivity of constant linear systems to finite variations in parameters," in *Recent Mathematical Developments in Control* (D.J. Bell, Ed.). New York: Academic Press, pp. 393-404, 1973.

[110] _____, "Insensitivity of asymptotically stable matrices to variations in submatrices," *IJSS*, vol. 6, no. 7, pp. 621-632, 1975.

[111] A.V. Shabalin, "A matrix scheme of stability analysis," *ARC*, vol. 51, no. 5, pp. 609-614, 1990.

[112] V.N. Shchennikov, "Stability under continuous disturbances," *ARC*, vol. 46, no. 2, pp. 197-200, Feb. 1985.

[113] Z.-C. Shi and W.-B. Gao, "A necessary and sufficient condition for the positive-definiteness of interval symmetric matrices," *IJC*, vol. 43, no. 1, pp. 325-328, 1986. Comments by C.-L. Jiang, *ibid.*, vol. 49, no. 4, p. 1439, 1989.

[114] _____, "Stability of interval parameter matrices," *IJC*, vol. 45, no. 3, pp. 1093-1101, 1987.

[115] D.D. Siljak, "Parameter space methods for robust control design: A guided tour," *IEEE-TAC*, vol. AC-34, no. 7, pp. 674-688, July 1989. [Also Santa Clara University, Tech. Rep. No. EECS-031588, Jan. 1988.]

[116] C.B. Soh, "Characterizations of eigenvalues of interval matrices," *CTAT*, vol. 6, no. 1, pp. 123-131, 1990.

[117] C.B. Soh and C.K. Chan, "Destabilizing perturbations for linear systems," *IJC*, vol. 50, no. 5, pp. 2101-2104, 1989.

[118] Y.C. Soh and R.J. Evans, "Stability analysis of interval matrices--continuous and discrete systems," *IJC*, vol. 47, no. 1, pp. 25-32, 1988.

[119] E.D. Sontag, "Parametric stabilization is easy," *SCL*, vol. 4, pp. 181-188, 1984.

[120] M.K. Sundareshan and P.C.K. Huang, "On the design of a decentralized observation scheme for large-scale system," *IEEE-TAC*, vol. AC-29, no. 3, pp. 274-276, March 1984.

[121] O. Taussky, "On the variation of the characteristic roots of a finite matrix under various changes of its elements," in *Recent Advances in Matrix Theory* (Proc. of Conf. in Madison, Oct. 1963, H. Schneider, Ed.), Milwaukee, WI: University of Wisconsin Press, pp. 125-138, 1964.

[122] L.M. van Woerkom, "Optimization of a stability bound," *SIAM R.*, vol. 32, no. 2, p. 300, June 1990. Solutions, by P.Th.L.M. Hanau, and by M. Eslami, *ibid.*, vol. 33, no. 2, pp. 277-281, June 1991.

[123] A. Vicino, A. Tesi, and M. Milanese, "Computation of nonconservative stability perturbation bounds for systems with nonlinearly correlated uncertainties," *IEEE-TAC*, vol. AC-35, no. 7, pp. 835-841, July 1990.

[124] S.-D. Wang, T.-S. Kuo, Y.-H. Lin, C.-F. Hsu, and Y.-T. Juang, "Robust control design for linear systems with uncertain parameters," *IJC*, vol. 46, no. 5, pp. 1557-1567, 1987.

[125] K. Wei, "Quadratic stabilizability of linear systems with structural independent time-varying uncertainties," *IEEE-TAC*, vol. AC-35, no. 3, pp. 268-277, March 1990.

[126] P.K. Wong and M. Athans, "Closed-loop structural stability for linear-quadratic optimal systems," *IEEE-TAC*, vol. AC-22, no. 1, pp. 94-99, Feb. 1977.

[127] L.X. Xin, "Stability of interval matrices," *IJC*, vol. 45, no. 1, pp. 203-210, 1987. Comment by M. Mansour, *ibid.*, vol. 46, no. 5, p. 1845, 1987.

[128] _____, "Necessary and sufficient conditions for stability of a class of interval matrices," *IJC*, vol. 45, no. 1, pp. 211-214, 1987.

[129] H.-H. Yeh, S.S. Banda, and D.B. Ridgely, "Stability robustness measures utilizing structural information," *IJC*, vol. 41, no. 2, pp. 365-387, 1985.

[130] D.-Z. Zheng, "A method for determining the parameter stability regions of linear control systems," *IEEE-TAC*, vol. AC-29, no. 2, pp. 183-186, Feb. 1984.

[131] C.S. Zhou and J.L. Deng, "The stability of the grey linear system," *IJC*, vol. 43, no. 1, pp. 313-320, 1986.

[132] K. Zhou and P.P. Khargonekar, "Stability robustness bounds for linear state-space models with structured uncertainty," *IEEE-TAC*, vol. AC-32, no. 7, pp. 621-623, July 1987.

[133] _____, "On the stabilization of uncertain linear systems *via* bound invariant Lyapunov functions," *SIAM-JCO*, vol. 26, no. 6, pp. 1265-1273, 1988.

[C] Stability Robustness Analysis Maneuverability, and Sensitivity in the Large Group II: Polynomials

[1] J.J. Anagnost, C.A. Desoer, and R.J. Minnichelli, "Graphical stability robustness tests for linear time-invariant systems: Generalization of Kharitonov's stability theorem," in *Proc. 27th IEEE Conf. on Decision and Contr.*, 1988.

[2] B.D.O. Anderson, E.I. Jury, and M. Mansour, "On robust Hurwitz polynomials," *IEEE-TAC*, vol. AC-32, no. 10, pp. 909-913, Oct. 1987.

[3] M.B. Argoun, "Frequency domain conditions for the stability of perturbed polynomials," *IEEE-TAC*, vol. AC-32, no. 10, pp. 913-916, Oct. 1987.

5.12.C. Group II: Polynomials

[4] _____, "Stability of a Hurwitz polynomial under coefficient perturbations: Necessary and sufficient conditions," *IJC*, vol. 45, no. 2, pp. 739-744, 1987. Comment by S.-J. Xu and C.-X. Shao, *ibid.*, vol. 49, no. 1, pp. 379-381, 1989. Correction by C.B. Soh, *ibid.*, vol. 51, no. 5, pp. 1151-1154, 1990.

[5] S.F. Babak, B.G. Il'yasov, B.N. Petrov, and I.Y. Yusupov, "Structural instability of systems," *ARC*, vol. 41, no. 8, pp. 1060-1065, Aug. 1980.

[6] B.R. Barmish, "Robust solutions of perturbed linear equations," *IEEE-TAC*, vol. AC-22, no. 1, pp. 123-124, Feb. 1977.

[7] _____, "Invariance of the strict Hurwitz property for polynomials with perturbed coefficients," *IEEE-TAC*, vol. AC-29, no. 10, pp. 935-936, Oct. 1984.

[8] _____, "A generalization of Kharitonov's four-polynomial concept for robust stability problems with linearly dependent coefficient perturbations," *IEEE-TAC*, vol. AC-34, no. 2, pp. 157-165, Feb. 1989. Comments by J.-Y. Juang, *ibid.*, vol. AC-35, no. 8, pp. 987-988, Aug. 1990.

[9] B.R. Barmish, C.V. Hollot, F.J. Kraus, and R. Tempo, "Extreme point results for robust stabilization of interval plants with first order compensation," *IEEE-TAC*, vol. AC-37, no. 6, pp. 707-714, 1992. Comments, with reply, by K.K. Yen and S.F. Zhou, *IEEE-TAC*, vol. AC-38, no. 2, p. 384, 1993.

[10] B.R. Barmish, P.P. Khargonekar, Z.C. Shi, and R. Tempo, "Robustness margin need not be a continuous function of the problem data," *SCL*, vol. 15, pp. 91-98, 1990. [cf., [49].]

[11] B.R. Barmish and Z. Shi, "Robust stability of a class of polynomials with coefficients depending multilinearly on perturbations," *IEEE-TAC*, vol. AC-35, no. 9, pp. 1040-1044, Sept. 1990.

[12] B.R. Barmish and R. Tempo, "The robust root locus," *Automatica*, vol. 26, no. 2, pp. 283-292, 1990.

[13] _____, "On the spectral set for a family of polynomials," *IEEE-TAC*, vol. AC-36, no. 1, pp. 111-115, 1991.

[14] B.R. Barmish, R. Tempo, C.V. Hollot, and H.I. Kang, "An extreme point result for robust stability of a diamond of polynomials," *IEEE-TAC*, vol. AC-37, no. 9, pp. 1460-1462, 1992.

[15] A.C. Bartlett, "Vertex results for the steady state analysis of uncertain systems," *IEEE-TAC*, vol. AC-37, no. 11, pp. 1758-1762, 1992.

[16] A.C. Bartlett, C.V. Hollot, and H. Lin, "Root locations of an entire polytope of polynomials: It suffices to check the edges," *MCSS*, vol. 1, pp. 61-71, 1988.

[17] S. Bialas and J. Garloff, "Stability of polynomials under coefficient perturbation," *IEEE-TAC*, vol. AC-30, no. 3, pp. 310-312, March 1985.

[18] R.M. Biernacki, "Sensitivities of stability constraints and their applications," *IEEE-TAC*, vol. AC-31, no. 7, pp. 639-642, July 1986.

[19] Y.M. Borushko and V.M. Vartanyan, "Automated synthesis of automated control systems using regions of specified quality," *SJAIS*, vol. 21, no. 3, pp. 64-66, 1988.

[20] N.K. Bose, E.I. Jury, and E. Zeheb, "On robust Hurwitz and Schur polynomials," *IEEE-TAC*, vol. AC-33, no. 12, pp. 1166-1169, Dec. 1988.

[21] N.K. Bose and K.D. Kim, "Stability of a complex polynomial set with coefficients in a diamond and generalizations," *IEEE-TCAS*, vol. CAS-36, no. 9, pp. 1168-1174, 1989. Counterexample and correction by H.I. Kang, B.R. Barmish, R. Tempo, and C.V. Hollot, *ibid.*, vol. CAS-38, no. 11, pp. 1370-1373, 1991. Authors' Reply, *ibid.*, pp. 1397-1400.

[22] N.K. Bose and Y.Q. Shi, "A simple general proof of Kharitonov's generalized stability criterion," *IEEE-TCAS*, vol. CAS-34, pp. 1233-1237, Oct. 1987.

[23] H. Bouguerra, B.C. Chang, H.H. Yeh, and S.S. Banda, "Fast stability checking for the convex combination of stable polynomials," *IEEE-TAC*, vol. AC-35, no. 5, pp. 586-588, May 1990.

[24] A. Cavallo, G. Celentano, and G. De Maria, "Robust stability analysis of polynomials with linearly dependent coefficient perturbations," *IEEE-TAC*, vol. AC-36, no. 3, pp. 380-384, 1991.

[25] H. Chapellat and S.P. Bhattacharyya, "A generalization of Kharitonov's theorem: Robust stability of interval plants," *IEEE-TAC*, vol. AC-34, no. 3, pp. 306-311, March 1989.

[26] _____, "An alternative proof of Kharitonov's theorem," *IEEE-TAC*, vol. AC-34, no. 4, pp. 448-450, April 1989.

[27] H. Chapellat, M.A. Dahleh, and S.P. Bhattacharyya, "Robust stability under structured and unstructured perturbations," *IEEE-TAC*, vol. AC-35, no. 10, pp. 1100-1108, Oct. 1990.

[28] H. Chapellat, M. Dahleh, and S.P. Bhattacharyya, "Robust stability manifolds for multilinear interval systems," *IEEE-TAC*, vol. AC-38, no. 2, pp. 314-318, 1993.

[29] C.-L. Chen and N. Munro, "Calculation of the largest generalized stability hypersphere in the robust stability problem for the maximum setting-time and minimum damping-ratio cases," *IEEE-TAC*, vol. AC-36, no. 6, pp. 756-759, 1991.

[30] T.E. Djaferis and C.V. Hollot, "The stability of a family of polynomials can be deduced from a finite number $O(k^3)$ of frequency checks," *IEEE-TAC*, vol. AC-34, no. 9, pp. 982-986, Sept. 1989.

[31] _____, "Parameter partitioning *via* shaping conditions for the stability of families of polynomials," *IEEE-TAC*, vol. AC-34, no. 11, pp. 1205-1209, Nov. 1989.

[32] _____, "A Routh-like test for the stability of families of polynomials with linear uncertainty," *SCL*, vol. 13, pp. 23-29, 1989.

[33] M. Eslami, "Sensitivity analysis and synthesis in automatic control systems," Ph.D. Dissertation, Univ. Wisconsin-Madison, Madison, May 1978.

[34] _____, "On robust Hurwitz polynomials," in *Proc. the 1989 IEEE Int. Symp. on Circuits and Systems*, pp. 1776-1779, May 1989.

[35] _____, "On construction of a bank of robust Hurwitz polynomials," *IEEE-TCAS*, Part I, vol. CAS-40, pp. 493-496, July 1993.

[36] L. Faxun, "Simple criteria for stability interval polynomials," *IJC*, vol. 50, no. 1, pp. 339-347, 1989.

[37] Y.K. Foo and Y.C. Soh, "Root clustering of interval polynomials in the left-sector," *SCL*, vol. 13, pp. 239-245, 1989.

[38] _____, "Damping margins of interval polynomials," *IEEE-TAC*, vol. AC-35, no. 4, pp. 477-479, April 1990.

[39] _____, "A generalization of strong Kharitonov theorems to polytopes of polynomials," *IEEE-TAC*, vol. AC-35, no. 8, pp. 936-939, Aug. 1990.

[40] _____, "Stability of a family of polynomials with coefficients bounded in a diamond," *IEEE-TAC*, vol. AC-36, no. 12, pp. 1501-1502, 1991.

[41] _____, "Characterization of zero locations of polytopes of real polynomials," *IEEE-TAC*, vol. AC-37, no. 8, pp. 1227-1230, 1992.

[42] _____, "Strong Kharitonov theorems for low-order polynomials," *IEEE-TAC*, vol. AC-37, no. 11, pp. 1816-1820, 1992.

[43] _____, "Strict positive realness of a family of polytopic plants," *IEEE-TAC*, vol. AC-38, no. 2, pp. 287-289, 1993.

[44] M. Fu and B.R. Barmish, "Polytopes of polynomials with zeros in a prescribed set," *IEEE-TAC*, vol. AC-34, no. 5, pp. 544-546, May 1989.

[45] M.Y. Fu, "Robustness bounds of Hurwitz and Schur polynomials," *JOTA*, vol. 62, no. 3, pp. 405-417, 1989.

5.12.C. Group II: Polynomials

[46] J. Garloff and N.K. Bose, "Boundary implications for stability properties: Present status," in *The Role of Interval Methods in Scientific Computing* (R.E. Moore, Ed.). New York: Academic, pp. 391-402, 1988.

[47] E.P. Gil'bo, I.B. Chelpanov, and G.L. Shevlyakov, "Robust approximation of functions in case of uncertainty," *ARC*, vol. 40, no. 10, pp. 522-529, April 1979.

[48] J.P. Guiver and N.K. Bose, "Strictly Hurwitz property invariance of quartics under coefficient perturbation," *IEEE-TAC*, vol. AC-28, no. 1, pp. 106-107, Jan. 1983.

[49] S. Gutman, "Data and uncertain parameters in robust analysis," *IEEE-TAC*, vol. AC-37, no. 8, pp. 1238-1239, 1992.

[50] _____, "Root clustering for convex combination of complex polynomials," *IEEE-TAC*, vol. AC-37, no. 10, pp. 1520-1522, 1992.

[51] C.V. Hollot, "Kharitonov-like results in the space of Markov parameters," *IEEE-TAC*, vol. AC-34, no. 5, pp. 536-538, May 1989.

[52] C.V. Hollot and A.R. Galimidi, "Stablizing uncertain systems: Recovering full state feedback performance *via* an observer," *IEEE-TAC*, vol. AC-31, no. 11, pp. 1050-1053, Nov. 1986.

[53] C.V. Hollot, D.P. Looze, and A.C. Bartlett, "Parametric uncertainty and unmodeled dynamics: Analysis *via* parameter space methods," *Automatica*, vol. 26, no. 2, pp. 269-282, 1990.

[54] K. Horiguchi, S. Hayashi, and N. Hamada, "New sets of stability criteria as a generalization of Markov's stability theorem," *IJC*, vol. 43, no. 5, pp. 1581-1591, 1986.

[55] Y.-T. Juang, "Strictly Hurwitz property analysis for polynomials with uncertain coefficients," *CTAT*, vol. 5, no. 2, pp. 215-226, June 1989.

[56] Y.-T. Juang, T.-S. Kuo, C.-F. Hsu, and S.-D. Wang, "Root-locus approach to the stability analysis of interval matrices," *IJC*, vol. 46, no. 3, pp. 817-822, 1987. Counter example by H. Qing-Long and W. Qi-Bin, *ibid.*, vol. 51, no. 2, pp. 499-500, 1990.

[57] E.I. Jury and A. Katbab, "A note on Kharitonov-type results in the space of Markov parameters," *IEEE-TAC*, vol. AC-37, no. 1, pp. 155-158, 1992.

[58] V.L. Kharitonov, "Asymptotic stability of an equilibrium position of a family of systems of linear differential equations," *Differential Equations*, vol. 14, pp. 1483-1485, Nov. 1978 (English translation published in 1979).

[59] K.D. Kim and N.K. Bose, "Invariance of strict Hurwitz property for bivariate polynomials under coefficient perturbations," *IEEE-TAC*, vol. AC-33, no. 12, pp. 1172-1174, Dec. 1988.

[60] J. Kogan, "How near is a stable polynomial to an unstable polynomial?" *IEEE-TCAS*, Part I, vol. CAS-39, no. 8, pp. 676-680, 1992.

[61] H. Kokame and T. Mori, "A Kharitonov-like theorem for interval polynomial matrices," *SCL*, vol. 16, pp. 107-116, 1991.

[62] F.J. Kraus, B.D.O. Anderson, and M. Mansour, "Robust stability of polynomials with multilinear parameter dependence," *IJC*, vol. 50, no. 5, pp. 1745-1762, 1989.

[63] P. Lancaster and M. Tismenetsky, "Some extensions and modifications of classical stability tests for polynomials," *IJC*, vol. 38, no. 2, pp. 369-380, 1983.

[64] Y. Li, K.M. Nagpal, and E.B. Lee, "Stability analysis of polynomials with coefficients in disks," *IEEE-TAC*, vol. AC-37, no. 4, pp. 509-513, 1992.

[65] H. Lin and C.V. Hollot, "Results on positive pairs of polynomials and their application to the construction of stability domains," *IJC*, vol. 46, no. 1, pp. 153-159, 1987.

[66] H. Lin, C.V. Hollot, and A.C. Bartlett, "Stability of families of polynomials: Geometric considerations in coefficient space," *IJC*, vol. 45, no. 2, pp. 649-660, 1987.

[67] S.H. Lin, I.K. Fong, Y.T. Juang, T.S. Kuo, and C.F. Hsu, "Stability of perturbed polynomials based on the argument principle and Nyquist criterion," *IJC*, vol. 50, no. 1, pp. 55-63, 1989.

[68] M. Mansour, E.I. Jury, and L.F. Chaparro, "Estimation of the margin of stability for linear continuous and discrete systems," *IJC*, vol. 30, no. 1, pp. 49-69, 1979.

[69] M. Marden, *Geometry of Polynomials*. Providence, Rhode Island: American Mathematical Society (Math. Surveys no. 3), 2nd edition 1966.

[70] J. Maroulas and S. Barnett, "Polynomials with respect to a general basis. I. Theory," *JMAA*, vol. 72, pp. 177-194, 1979. II. Applications, *ibid.*, pp. 599-614.

[71] R.J. Minnichelli, J.J. Anagnost, and C.A. Desoer, "An elementary proof of Kharitonov's stability theorem with extensions," *IEEE-TAC*, vol. AC-34, no. 9, pp. 995-998, Sept. 1989.

[72] T. Mori and H. Kokame, "Aperiodicity conditions for polynomials with uncertain coefficient parameters," *IJC*, vol. 51, no. 5, pp. 1147-1150, 1990.

[73] P. Padmanabhan and C.V. Hollot, "Complete instability of a box of polynomials," *IEEE-TAC*, vol. AC-37, no. 8, pp. 1230-1233, 1992.

[74] E.R. Panier, M.K.H. Fan, and A.L. Tits, "On the robust stability of polynomials with no cross-coupling between the perturbations in the coefficients of even and odd powers," *SCL*, vol. 12, pp. 291-299, 1989.

[75] B.T. Polyak and Y.Z. Tsypkin, "Frequency criteria of robust stability and aperiodicity of linear systems," *ARC*, vol. 51, no. 9, pp. 1192-1201, 1990.

[76] L.R. Pujara and N. Shanbhag, "Some stability theorems for polygons of polynomials," *IEEE-TAC*, vol. AC-37, no. 11, pp. 1845-1849, 1992.

[77] L. Qiu and E.J. Davison, "A simple procedure for the exact stability robustness computation of polynomials with affine coefficient perturbations," *SCL*, vol. 13, pp. 413-420, 1989.

[78] A. Rantzer, "Hurwitz testing sets for parallel polytopes of polynomials," *SCL*, vol. 15, pp. 99-104, 1990.

[79] _____, "Stability conditions for polytopes of polynomials," *IEEE-TAC*, vol. AC-37, no. 1, pp. 79-89, 1992.

[80] F. Reza, "RLC canonic forms," *J. Applied Physics*, vol. 25, no. 3, pp. 297-301, March 1954.

[81] H. Rotstein, R. Sanchez Pena, J. Bandoni, A. Desages, and J. Romagnoli, "Robust characteristic polynomial assignment," *Automatica*, vol. 27, no. 4, pp. 711-715, 1991.

[82] B.O. Ryabov and G.P. Sachkov, "Sensitivity of accumulated disturbances in linear systems," *SAC*, vol. 2, no. 5, pp. 6-13, 1969.

[83] W.E. Schmitendorf and C.V. Hollot, "Simultaneous stabilization *via* linear feedback control," *IEEE-TAC*, vol. AC-34, no. 9, pp. 1001-1005, Sept. 1989.

[84] B.A. Shane and S. Barnett, "Sensitivity of stable linear systems," *IEEE-TAC*, vol. AC-17, no. 1, pp. 148-150, Feb. 1972.

[85] J. Shaw and S. Jayasuriya, "Robust stability of an interval plant with respect to a convex region in the complex plane," *IEEE-TAC*, vol. AC-38, no. 2, pp. 284-287, 1993.

[86] Y.Q. Shi, "Robust (strictly) positive interval rational functions," *IEEE-TCAS*, vol. CAS-38, no. 5, pp. 552-554, 1991.

[87] Y.Q. Shi, K.K. Yen, and C.M. Chen, "Two necessary conditions for a complex polynomial to be strictly Hurwitz and their applications in robust stability analysis," *IEEE-TAC*, vol. AC-38, no. 1, pp. 125-128, 1993.

[88] Y.Q. Shi and S.F. Zhou, "Stability of a set of multivariate complex polynomials with coefficients varying in diamond domain," *IEEE-TCAS*, Part I, vol. CAS-39, no. 8, pp. 683-688, 1992.

[89] A. Sideris and B.R. Barmish, "An edge theorem for polytopes of polynomials which can drop in degree," *SCL*, vol. 13, pp. 233-238, 1989.

5.12.C. Group II: Polynomials

[90] C.B. Soh, "Invariance of the aperiodic property for polynomials with perturbed coefficients," *IEEE-TAC*, vol. AC-35, no. 5, pp. 616-618, May 1990. Comments by L. Saydy, A.L. Tits, and E.H. Abed, *IEEE-TAC*, vol. AC-37, no. 5, p. 697, 1992.

[91] _____, "On a root distribution criterion for interval polynomials," *IEEE-TAC*, vol. AC-37, no. 12, pp. 1977-1978, 1992.

[92] C.B. Soh and C.S. Berger, "Damping margins of polynomials with perturbed coefficients," *IEEE-TAC*, vol. AC-33, no. 5, pp. 509-511, May 1988.

[93] _____, "Strict aperiodic property of polynomials with perturbed coefficients," *IEEE-TAC*, vol. AC-34, no. 5, pp. 546-548, May 1989.

[94] C.B. Soh, C.S. Berger, and K.P. Dabke, "On the stability properties of polynomials with perturbed coefficients," *IEEE-TAC*, vol. AC-30, no. 10, pp. 1033-1036, Oct 1985. Addendum, *ibid.*, vol. AC-32, no. 3, pp. 239-240, March 1987.

[95] Y.C. Soh, "Strict Hurwitz property of polynomials under coefficient perturbations," *IEEE-TAC*, vol. AC-34, no. 6, pp. 629-632, June 1989.

[96] _____, "Stability of an entire polytope of polynomials," *IJC*, vol. 49, no. 3, pp. 993-999, 1989. Comments by Y.Q. Shi, *ibid.*, vol. 51, no. 2, pp. 495-496, 1990.

[97] _____, "Zero locations of an entire family of polytope polynomials," *IJC*, vol. 49, no. 6, pp. 1851-1859, 1989.

[98] Y.C. Soh and Y.K. Foo, "Generalized edge theorem," *SCL*, vol. 12, pp. 219-224, 1989.

[99] _____, "A note on the edge theorem," *SCL*, vol. 15, pp. 41-43, 1990.

[100] _____, "Kharitonov regions: It suffices to check a subset of vertex polynomials," *IEEE-TAC*, vol. AC-36, no. 9, pp. 1102-1105, 1991.

[101] Y.C. Soh, Y.K. Foo, and C.B. Soh, "Perturbed polynomials with zeros in the left-sector," *SCL*, vol. 12, pp. 415-419, 1989.

[102] E.M. Solnechnyi, "Robustness of a linear system which depends on a scalar parameter," *ARC*, vol. 51, no. 11, pp. 1510-1517, 1990.

[103] A. Tesi and A. Vicino, "A new fast algorithm for robust stability analysis of control systems with linearly dependent parametric uncertainties," *SCL*, vol. 13, pp. 321-329, 1989.

[104] _____, "Frequency response of interval plant-controller families," *SCL*, vol. 18, pp. 347-354, 1992.

[105] A. Thowsen, "The Routh-Hurwitz method for stability determination of linear differential-difference systems," *IJC*, vol. 33, no. 5, pp. 991-995, 1981. Comments by E.I. Jury and M. Mansour, *ibid.*, vol. 36, no. 5, p. 903, 1982.

[106] V.A. Tkachenko, "Analysis of the degree of stability of linear control systems," *ARC*, vol. 41, no. 3, pp. 316-324, March 1980.

[107] Y.Z. Tsypkin and B.T. Polyak, "Frequency domain criteria for l^p-robust stability of continuous linear systems," *IEEE-TAC*, vol. AC-36, no. 12, pp. 1464-1469, 1991.

[108] A. Vicino, "Maximal polytopic stability domains in parameter space for uncertain systems," *IJC*, vol. 49, no. 1, pp. 351-361, 1989.

[109] K. Wei, "The solution of a transcendental problem and its applications in simultaneous stabilization problems," *IEEE-TAC*, vol. AC-37, no. 9, pp. 1305-1315, 1992.

[110] K. Wei and B.R. Barmish, "Making a polynomial Hurwitz-invariant by choice of feedback gains," *IJC*, vol. 50, no. 4, pp. 1025-1038, 1989.

[111] K.S. Yeung and S.S. Wang, "A simple proof of Kharitonov's theorem," *IEEE-TAC*, vol. AC-32, no. 9, pp. 822-823, Sept. 1987.

[112] V. Zakian, "Computation of the abscissa of stability by repeated use of the Routh test," *IEEE-TAC*, vol. AC-24, no. 4, pp. 604-607, Aug. 1979.

[113] E. Zeheb, "Necessary and sufficient conditions for root clustering of a polytope of polynomials in a simply connected domain," *IEEE-TAC*, vol. AC-34, no. 9, pp. 986-990, 1989.

[114] _____, "On the largest modulus of polynomial zeros," *IEEE-TCAS*, vol. CAS-38, no. 3, pp. 333-337, 1991.

[D] Stability Robustness Analysis
Maneuverability, and Sensitivity in the Large
Group III: Discrete-Time Systems

[1] A.-A. A. Abdul-Wahab, "Lyapunov bounds for root clustering in the presence of system uncertainty," *IJSS*, vol. 21, no. 12, pp. 2603-2611, 1990.

[2] A. Aziz and Q.G. Mohammad, "On the zeros of a certain class of polynomials and related analytic functions," *JMAA*, vol. 75, pp. 495-502, 1980.

[3] E.-W. Bai and S. Dasgupta, "Robust control design of sampled systems," *IJSS*, vol. 21, no. 5, pp. 985-992, 1990.

[4] B.R. Barmish and J. Sankaran, "The propagation of parametric uncertainty *via* polytopes," *IEEE-TAC*, vol. AC-24, no. 2, pp. 346-349, April 1979.

[5] A.C. Bartlett and C.V. Hollot, "A necessary and sufficient condition for Schur invariance and generalized stability of polytopes of polynomials," *IEEE-TAC*, vol. AC-33, pp. 575-578, June 1988.

[6] D.S. Bernstein and C.V. Hollot, "Robust stability for sampled-data control systems," *SCL*, vol. 13, pp. 217-226, 1989.

[7] N.K. Bose, E.I. Jury, and E. Zeheb, "On robust Hurwitz and Schur polynomials," *IEEE-TAC*, vol. AC-33, no. 12, pp. 1166-1169, Dec. 1988.

[8] N.K. Bose and E. Zeheb, "Kharitonov's theorem and stability test of multidimensional digital filters," *Proc. IEE*, vol. 133, Pt. G, no. 4, pp. 187-190, Aug. 1986.

[9] H. Chapellat, S.P. Bhattacharyya, and M. Dahleh, "Robust stability of a family of disc polynomials," *IJC*, vol. 51, no. 6, pp. 1353-1362, 1990.

[10] K.P. Dabke, "A simple criterion for stability of linear discrete systems," *IJC*, vol. 37, no. 3, pp. 657-659, 1983. Comments by J.A. Heinen, *ibid.*, vol. 42, no. 1, pp. 269-270, 1985, and by T. Mori, *ibid.*, vol. 43, no. 2, pp. 737-739, 1986.

[11] J.F. Delansky and N.K. Bose, "Schur stability and stability domain construction," *IJC*, vol. 49, no. 4, pp. 1175-1183, 1989.

[12] M. Eslami, "Sensitivity analysis and synthesis in automatic control systems," Ph.D. Dissertation, Univ. Wisconsin-Madison, Madison, May 1978.

[13] _____, "On sensitivity of nonlinear digital systems," in *Proc. Allerton Conf.*, pp. 424-432, Oct. 1980.

[14] _____, "Computer-aided determination of stability robustness measure of linear discrete-time systems," *Automatica*, vol. 26, no. 3, pp. 623-627, May 1990.

[15] _____, "On stability robustness analysis for discrete-time dynamic systems," in *Proc. 31st IEEE Conf. on Decision and Contr.*, pp. 1569-1574, Dec. 1992.

[16] _____, "On stability robustness analysis in discrete-time systems," in *Proc. Symp. on Fundamentals of Discrete-Time Systems [In honor of Professor E.I. Jury, Chicago, June 1992]* (M. Jamshidi, *et al.*, Eds.). Albuquerque, NM: TSI Press, 1993, pp. 243-255.

[17] _____, "Stability robustness analysis of discrete-time systems with multiple large parameter variations," *Automatica*, vol. 30, no. 2, pp. 357-361, 1994.

5.12.D. Group III: Discrete-Time Systems

[18] M. Farsi, K. Warwick, and M. Guilandoust, "Stable reduced-order models for discrete-time systems," *Proc. IEE*, vol. 133, Pt. D, no. 3, pp. 137-141, May 1986.

[19] M. Fu, A.W. Olbrot, and M.P. Polis, "The edge theorem and graphical tests for robust stability of neutral time-delay systems," *Automatica*, vol. 27, no. 4, pp. 739-741, 1991.

[20] V.I. Gostev and V.I. Polivanov, "The stability of automatic control systems with digital regulations of two kinds," *SJAIS*, vol. 22, no. 5, pp. 63-66, 1989.

[21] _____, "Parametric synthesis of digital regulators in discrete-continuous systems by numerical method of unconditional optimization," *SJAIS*, vol. 23, no. 1, pp. 75-80, 1990.

[22] N.K. Govil, "On the maximum modulus of polynomials," *JMAA*, vol. 112, pp. 253-258, 1985.

[23] D.C. Hayland and E.G. Collins, Jr., "Some majorant robustness results for discrete-time systems," *Automatica*, vol. 27, no. 1, pp. 167-172, 1991.

[24] J.A. Heinen, "A simple and direct proof of a bound on the zeros of a polynomial," *IJC*, vol. 42, no. 1, pp. 269-270, 1985.

[25] C.V. Hollot and A.C. Bartlett, "Some discrete-time counterparts to Kharitonov's stability criterion for uncertain systems," *IEEE-TAC*, vol. AC-31, no. 4, pp. 355-356, April 1986.

[26] T. Ishihara, "Robust stability bounds for a class of discrete-time regulators with computation delays," *Automatica*, vol. 24, no. 5, pp. 697-700, 1988.

[27] C.-L. Jiang, "Sufficient and necessary condition for the asymptotic stability of discrete linear interval systems," *IJC*, vol. 47, no. 5, pp. 1563-1565, 1988. Comments by W.C. Karl and G.C. Verghese, *ibid.*, vol. 48, no. 5, pp. 2159-2160, 1988, and by W.M. Grimm and U. Piechottka, *ibid.*, vol. 49, no. 3, pp. 1099-1102, 1989, and by A. Rachid, *ibid.*, vol. 49, no. 3, p. 1105, 1989. Counter examples by S.R. Kolla and J.B. Farison, *ibid.*, vol. 48, no. 4, pp. 1751-1752, 1988, and by K.-Y. Zhao, *ibid.*, vol. 49, no. 5, pp. 1811-1812, 1989, and by Y.Q. Shi and H. Zhang, *ibid.*, vol. 49, no. 3, pp. 1095-1097, 1989, and by C.B. Soh, *ibid.*, vol. 49, no. 3, pp. 1107-1108, 1989.

[28] Y.-T. Juang, T.-S. Kuo, and C.-F. Hsu, "Stability robustness analysis of digital control systems in state-space models," *IJC*, vol. 46, no. 5, pp. 1547-1556, 1987.

[29] Y.-T. Juang, S.-L. Tung, and T.-C. Ho, "Sufficient condition for asymptotic stability of discrete interval systems," *IJC*, vol. 49, no. 5, pp. 1799-1803, 1989.

[30] E.I. Jury, "Robustness of a discrete system," (Survey) *ARC*, vol. 51, no. 5, pp. 571-592, 1990.

[31] E.I. Jury and T. Pavlidis, "Stability and aperiodicity constraints for systems design," *IEEE-TCT*, vol. CT-10, pp. 137-141, 1963.

[32] E.W. Kamen and P.P. Khargonekar, "On the control of linear systems whose coefficients are functions of parameters," *IEEE-TAC*, vol. AC-29, no. 1, pp. 25-33, Jan. 1984.

[33] A. Kanellakis, S.G. Tzafestas, and N. Theodorou, "Computation of the stability margin of two-dimensional discrete systems," *IEEE-TAC*, vol. AC-37, no. 6, pp. 824-827, 1992.

[34] A. Katbab and E.I. Jury, "Robust Schur-stability of control systems with interval plants," *IJC*, vol. 51, no. 6, pp. 1343-1352, 1990.

[35] _____, "On the stability of a general polynomial in the space of Markov parameters," *SCL*, vol. 14, pp. 419-430, 1990.

[36] _____, "Robust Schur stability of a complex-coefficient polynomials set with coefficients in a diamond," *JFI*, vol. 327, no. 5, pp. 687-698, 1990.

[37] _____, "A note on two methods related to stability robustness of polynomials in a sector (relative stability)," *IEEE-TAC*, vol. AC-38, no. 2, pp. 380-383, 1993.

[38] A. Katbab, E.I. Jury, and M. Mansour, "On robust Schur property of discrete-time polynomials," *IEEE-TCAS*, Part I, vol. CAS-39, no. 6, pp. 467-470, 1992.

[39] V.L. Kharitonov, "Stability of a family of difference systems," *ARC*, vol. 51, no. 3, pp. 309-315, 1990.

[40] S.R. Kolla and J.B. Farison, "Improved robust stability bounds for discrete-time linear regulators with computational delays," *Automatica*, vol. 26, no. 3, pp. 619-621, May 1990.

[41] _____, "Improved stability robustness bounds using state transformation for linear discrete systems," *Automatica*, vol. 26, no. 5, pp. 933-935, 1990.

[42] F.J. Kraus, B.D.O. Anderson, and M. Mansour, "Robust Schur polynomial stability and Kharitonov's theorem," *IJC*, vol. 47, no. 5, pp. 1213-1225, 1988.

[43] V. Lakshmikantham and D. Trigiante, *Theory of Difference Equations: Numerical Methods and Applications*. New York: Academic Press, 1988. Reviewed by J.C. Butcher, *SIAM R.*, vol. 31, no. 4, pp. 689-690, 1989.

[44] T.-H.S. Li and J.-H. Li, "Stabilization bound of discrete two-time-scale systems," *SCL*, vol. 18, pp. 479-489, 1992.

[45] X.-Y. Liu and M. Mansour, "Stability test and stability conditions for delay differential systems," *IJC*, vol. 39, no. 6, pp. 1229-1242, 1984. Comments by M. Buslowicz, *ibid.*, vol. 45, no. 2, p. 745, 1987.

[46] M.S. Mahmoud and A.A. Bahnasawi, "Some properties of absolutely stable discrete systems," *CTAT*, vol. 4, no. 2, pp. 243-250, June 1988.

[47] _____, "Asymptotic stability for a class of linear discrete systems with bounded uncertainties," *IEEE-TAC*, vol. AC-33, pp. 572-575, June 1988. Comments, by J.B. Farison and S.R. Kolla, *ibid.*, vol. AC-35, pp. 382-383, March 1990, and Comments, with reply, by M. Corless, *ibid.*, p. 384.

[48] M. Mansour, E.I. Jury, and L.F. Chaparro, "Estimation of the margin of stability for linear continuous and discrete systems," *IJC*, vol. 30, no. 1, pp. 49-69, 1979.

[49] T. Mori, "Criteria for asymptotic stability of linear time-delay systems," *IEEE-TAC*, vol. AC-30, no. 2, pp. 158-161, Feb. 1985.

[50] _____, "On the relationship between the spectral radius and stability radius for discrete systems," *IEEE-TAC*, vol. AC-35, no. 7, p. 835, Feb. 1982.

[51] T. Mori and H. Kokame, "Convergence property of interval matrices and interval polynomials," *IJC*, vol. 45, no. 2, pp. 481-484, 1987. Comments by C.B. Soh, *ibid.*, vol. 51, no. 1, pp. 445-446, 1989.

[52] L.T. Movchan, "Exponential stability of discrete control systems," *SAC*, vol. 13, no. 3, pp. 71-73, 1980.

[53] X. Niu, J.A. De Abreu-Garcia, and E. Yaz, "Improved bounds for linear discrete-time systems with structured perturbations," *IEEE-TAC*, vol. AC-37, no. 8, pp. 1170-1173, 1992. Correction, *ibid.*, vol. AC-38, no. 5, p. 832, 1993.

[54] O. Ocali and M. Sezer, "Robust sampled-data control," *IEEE-TAC*, vol. AC-37, no. 10, pp. 1591-1597, 1992.

[55] A. Rachid, "Robustness of pole assignment in a specified region for perturbed systems," *IJSS*, vol. 21, no. 3, pp. 579-585, 1990.

[56] Z.V. Rekasius, "Stability of digital control with computer interruptions," *IEEE-TAC*, vol. AC-31, no. 4, pp. 356-359, April 1986.

[57] M.E. Sezer and D.D. Siljak, "Robust stability of discrete systems," *IJC*, vol. 48, no. 5, pp. 2055-2063, 1988.

[58] B. Shafai, K. Perev, D. Cowley, and Y. Chehab, "A necessary and sufficient condition for the stability of nonnegative interval discrete systems," *IEEE-TAC*, vol. AC-36, no. 6, pp. 742-746, 1991. Comments by T. Mori and H. Kokame, *IEEE-TAC*, vol. AC-37, no. 11, pp. 1853-1854, 1992, and by J. Chen, *IEEE-TAC*, vol. AC-38, no. 1, p. 189, 1993.

[59] U. Shaked, "Guaranteed stability margins for the discrete-time linear quadratic optimal regulator," *IEEE-TAC*, vol. AC-31, no. 2, pp. 162-165, Feb. 1986.

5.12.D. Group III: Discrete-Time Systems

[60] B.A. Shane and S. Barnett, "Sensitivity of convergent matrices and polynomials," *JMAA*, vol. 43, pp. 114-122, 1973.

[61] C.B. Soh, "Necessary and sufficient conditions for stability of symmetric interval matrices," *IJC*, vol. 51, no. 1, pp. 243-248, 1990.

[62] _____, "Robust stability of discrete-time systems using delta operators," *IEEE-TAC*, vol. AC-36, no. 3, pp. 377-380, 1991.

[63] C.B. Soh, C.S. Berger, and K.P. Dabke, "On the stability properties of polynomials with perturbed coefficients," *IEEE-TAC*, vol. AC-30, no. 10, pp. 1033-1036, Oct 1985. Addendum, *ibid.*, vol. AC-32, no. 3, pp. 239-240, March 1987.

[64] _____, "Stability of linear discrete-time systems: Geometric considerations," *IJC*, vol. 49, no. 1, pp. 15-23, 1989.

[65] C.B. Soh, Y.H. Lim, and T.C. Lim, "Damping ratio of linear uncertain discrete-time systems," *Proc. IEE*, vol. 136, Pt. D, no. 6, pp. 298-300, Nov. 1989.

[66] T.-J. Su and C.-G. Huang, "Robust stability of delay dependence for linear uncertain systems," *IEEE-TAC*, vol. AC-37, no. 10, pp. 1656-1659, 1992.

[67] T.-J. Su, T.-S. Kuo, and Y.-Y. Sun, "Robust stability of linear discrete systems with saturating actuators," *IJSS*, vol. 21, no. 7, pp. 1273-1279, 1990.

[68] _____, "Robust stability of linear perturbed discrete large-scale systems with saturating actuators," *IJSS*, vol. 21, no. 11, pp. 2263-2272, 1990.

[69] H.-K. Sung and S. Hara, "Properties of sensitivity and complementary sensitivity functions in single-input single-output digital control systems," *IJC*, vol. 48, no. 6, pp. 2429-2439, 1988.

[70] M. Sznaier, "Norm based robust control of state-constrained discrete-time linear systems," *IEEE-TAC*, vol. AC-37, no. 7, pp. 1057-1062, 1992.

[71] T. Taniguchi, "On the estimate of solutions of perturbed linear difference equations," *JMAA*, vol. 149, pp. 599-610, 1990.

[72] J. Tokarzewski, "Sufficient stabilizability conditions for multirate sampled-data systems," *IJC*, vol. 39, no. 2, pp. 257-277, 1984.

[73] Y.Z. Tsypkin, "Frequency criteria of linear discrete system modality," *SJAIS*, vol. 23, no. 3, pp. 1-7, 1990.

[74] _____, "Stabilizing nonlinear discrete systems under nonparametric uncertainty conditions," *SJAIS*, vol. 24, no. 4, pp. 1-5, 1991.

[75] Y.Z. Tsypkin and B.T. Polyak, "Frequency criteria of robust modality in linear discrete systems," *SJAIS*, vol. 23, no. 4, pp. 1-7, 1990.

[76] V.I. Vorotnikov, "Stability and stabilization of motion with respect to some of the variables for linear delayed systems," *ARC*, vol. 41, no. 8, pp. 1068-1077, Aug. 1980.

[77] D. Xu, "On the bounds for the roots of a polynomials," *IEEE-TAC*, vol. AC-32, no. 4, pp. 336-337, April 1987.

[78] J.-H. Xu, R.E. Skelton, and G. Zhu, "Upper and lower covariance bounds for perturbed linear systems," *IEEE-TAC*, vol. AC-35, no. 8, pp. 944-949, Aug. 1990.

[79] E. Yaz, "Constant feedback stabilization of discrete-time systems with random-coefficients," *IJSS*, vol. 17, no. 5, pp. 819-827, 1986.

[80] _____, "Deterministic and stochastic robustness measures for discrete systems," *IEEE-TAC*, vol. AC-33, no. 10, pp. 952-955, Oct. 1988.

[81] E. Yaz and X. Niu, "Stability robustness of linear discrete-time systems in the presence of uncertainty," *IJC*, vol. 50, no. 1, pp. 173-182, 1989.

[82] C.S. Zhou and J.L. Deng, "Stability analysis of gray discrete-time systems," *IEEE-TAC*, vol. AC-34, no. 2, pp. 173-175, Feb. 1989.

[E] Stability Robustness Analysis
Maneuverability, and Sensitivity in the Large
Group IV: The Classical Frequency-Domain / Servomechanism

[1] J. Ackermann, H.Z. Hu, and D. Kaesbauer, "Robustness analysis: A case study," *IEEE-TAC*, vol. AC-35, no. 3, pp. 352-356, March 1990.

[2] M.B. Argoun, "Frequency domain conditions for the stability of perturbed polynomials," *IEEE-TAC*, vol. AC-32, no. 10, pp. 913-916, Oct. 1987.

[3] R.J. Benhabib, R.P. Iwens, and R.L. Jackson, "Stability of large space structure control systems using positivity concepts," *JGCD*, vol. 4, no. 5, pp. 487-494, 1981.

[4] A. Bhaya and C.A. Desoer, "Robust stability under additive perturbations," *IEEE-TAC*, vol. AC-30, no. 2, pp. 1233-1234, Dec. 1985.

[5] R.M. Biernacki, "Sensitivities of stability constraints and their applications," *IEEE-TAC*, vol. AC-31, no. 7, pp. 639-642, July 1986.

[6] R.M. Biernacki, H. Hwang, and S.P. Bhattacharyya, "Robust stability with structured real parameter perturbations," *IEEE-TAC*, vol. AC-32, no. 6, pp. 495-506, June 1987.

[7] H. Chapellat and S.P. Bhattacharyya, "Some recent results in structured robust stability: An overview," *IJACSP*, vol. 2, pp. 311-325, 1988.

[8] B.-S. Chen, "Controller synthesis of optimal sensitivity: Multivariable case," *Proc. IEE*, vol. 131, Pt. D, no. 1, pp. 47-51, Jan. 1984.

[9] B.-S. Chen and C.-C. Chiang, "Robust stabilizer synthesis for feedback systems containing time-varying nonlinear perturbations," *IEEE-TAC*, vol. AC-31, no. 8, pp. 768-771, Aug. 1986.

[10] B.-S. Chen, S.S. Wang, and H.-C. Lu, "Minimal sensitivity perfect model matching control," *IEEE-TAC*, vol. AC-34, no. 12, pp. 1279-1283, Dec. 1989.

[11] M.-J. Chen and C.A. Desoer, "Algebraic theory for robust stability of interconnected systems: Necessary and sufficient conditions," *IEEE-TAC*, vol. AC-29, no. 6, pp. 511-519, June 1984.

[12] R.W. Daniel, "Frequency-response design of robust optimal controllers," *Proc. IEE*, vol. 129, Pt. D, no. 6, pp. 257-262, Nov. 1982.

[13] E.J. Davison, "The robust decentralized control of a general servomechanism problem," *IEEE-TAC*, vol. AC-21, no. 1, pp. 14-24, Feb. 1976. Erratum, *ibid.*, vol. AC-21, no. 4, p. 631, Aug. 1976.

[14] _____, "The robust control of a servomechanism problem for linear time-invariant multivariable systems," *IEEE-TAC*, vol. AC-21, no. 1, pp. 25-34, Feb. 1976. Addendum, *ibid.*, vol. AC-22, no. 2, p. 283, April 1977.

[15] _____, "Multivariable tuning regulators: The feedforward and robust control of a general servomechanism problem," *IEEE-TAC*, vo. AC-21, no. 1, pp. 35-47, Feb. 1976. Errata, *ibid.*, vol. AC-21, no. 4, p. 631, Aug. 1976, vol. AC-22, no. 2, p. 286, April 1977, and vol. AC-24, no. 5, p. 809, Oct. 1979.

[16] _____, "The robust decentralized servomechanism problem with extra stablizing control agents," *IEEE-TAC*, vol. AC-22, no. 2, pp. 256-258, April 1977.

[17] _____, "Decentralized robust control of unknown systems using tuning regulators," *IEEE-TAC*, vol. AC-23, no. 2, pp. 276-289, April 1978.

[18] R.R.E. de Gaston and M.G. Safonov, "Exact calculation of the multiloop stability margin," *IEEE-TAC*, vol AC-33, no. 2, pp. 156-171, Feb. 1988.

[19] C.A. Desoer and Y.T. Wang, "On the minimum order of a robust servocompensator," *IEEE-TAC*, vol. AC-23, no. 1, pp. 70-73, Feb. 1978.

[20] R. Doraiswami, "Robust decentralised control of a servomechanism problem for a class of nonlinear systems," *Proc. IEE*, vol. 128, Pt. D, no. 2, pp. 33-40, March 1981.

5.12.E. Group IV: The Classical Frequency-Domain / Servomechanism

[21] _____, "Robust control strategy for a linear time-invariant multivariable sampled-data servomechanism problem," *Proc. IEE*, vol. 129, Pt. D, no. 6, pp. 283-292, Nov. 1982.

[22] _____, "Robust control of a class of nonlinear servomechanism problem," *Proc. IEE*, vol. 130, Pt. D, no. 2, pp. 63-71, March 1983.

[23] J. Doyle, "Analysis of feedback systems with structured uncertainties," *Proc. IEE*, vol. 129, Pt. D, no. 6, pp. 242-250, Nov. 1982.

[24] P.M.G. Ferreira, "Robustness and partial robustness in the servo problem," *IJC*, vol. 32, no. 5, pp. 891-898, 1980.

[25] Y.K. Foo, "Stability and performance robustness analysis with highly structured uncertainties: Necessary and sufficient test," *IEEE-TAC*, vol. AC-35, no. 3, pp. 350-352, March 1990.

[26] Y.K. Foo and I. Postlethwaite, "Extensions of small-μ test for robust stability," *IEEE-TAC*, vol. AC-33, no. 2, pp. 172-176, Feb. 1988.

[27] B.A. Francis, "On robustness of stability of feedback systems," *IEEE-TAC*, vol. AC-25, no. 4, pp. 817-819, Aug. 1980.

[28] _____, "Stabilizability of linear systems with uncertain gains," *SCL*, vol. 7, pp. 195-198, 1986.

[29] J.S. Freudenberg, D.P. Looze, and J.B. Cruz, Jr., "Robustness analysis using singular value sensitivities," *IJC*, vol. 35, no. 1, pp. 95-116, 1982.

[30] M. Fujita and E. Shimemura, "Integrity conditions for a class of robust servo systems," *EEJ*, vol. 109, no. 5, pp. 65-74, 1989.

[31] R.A. Haddad and J.G. Truxal, "Sensitivity and stability in multiloop systems," *IEEE-TAC*, vol. AC-9, pp. 548-551, Oct. 1964.

[32] D.J. Hill, "A generalization of the small-gain theorem for nonlinear feedback systems," *Automatica*, vol. 27, no. 6, pp. 1043-1045, 1991.

[33] I. Horowitz, "Quantitative synthesis of uncertain non-linear feedback systems with non-minimum phase inputs," *IJSS*, vol. 12, no. 1, pp. 55-76, 1981.

[34] _____, "Quantitative feedback theory," *Proc. IEE*, vol. 129, Pt. D, no. 6, pp. 215-226, Nov. 1982.

[35] I. Horowitz and M. Breiner, "Quantitative synthesis of feedback systems with uncertain nonlinear multivariable plants," *IJSS*, vol. 12, no. 5, pp. 539-563, 1981.

[36] I. Horowitz and M. Sidi, "Practical design of feedback systems with uncertain multivariable plants," *IJSS*, vol. 11, no. 7, pp. 851-875, 1980.

[37] I. Horowitz and T.-S. Wang, "Quantitative synthesis of multiple-loop feedback systems with large uncertainty," *IJSS*, vol. 10, no. 11, pp. 1235-1268, 1979.

[38] Q. Huang and R. Liu, "A necessary and sufficient condition for stability of a perturbed system," *IEEE-TAC*, vol. AC-32, pp. 337-340, April 1987.

[39] Y.S. Hung, "Robust stability: Parameter-dependent perturbations," *IJC*, vol. 38, no. 1, pp. 87-105, 1983.

[40] Y.S. Hung and D.J.N. Limebeer, "Robust stability of additively perturbed interconnected systems," *IEEE-TAC*, vol. AC-29, no. 12, pp. 1069-1075, Dec. 1984.

[41] A. Iftar and U. Ozguner, "Modeling of uncertain dynamic for robust controller design in state space," *Automatica*, vol. 27, no. 1, pp. 141-146, 1991.

[42] J.C. Kantor and R.P. Andres, "Characterization of 'allowable perturbations' for robust stability," *IEEE-TAC*, vol. AC-28, no. 1, pp. 107-109, Jan. 1983.

[43] P.P. Khargonekar, T.T. Georgiou, and A.M. Pascoal, "On the robust stabilizability of linear time-invariant plants with unstructured uncertainty," *IEEE-TAC*, vol. AC-32, no. 3, pp. 201-207, March 1987.

[44] H. Kimura, "Robust stabilizability for a class of transfer functions," *IEEE-TAC*, vol. AC-29, no. 9, pp. 788-793, Sept. 1984.

[45] B. Kouvaritakis and H. Latchman, "Necessary and sufficient stability conditions for the case of dependent additive perturbations," *IJC*, vol. 43, no. 6, pp. 1615-1629, 1986.

[46] B. Kouvaritakis and I. Postlethwaite, "Principal gains and phases: Insensitivity robustness measures for assessing the closed-loop stability property," *Proc. IEE*, vol. 129, Pt. D, pp. 233-241, Nov. 1982.

[47] H. Kwakernaak, "A condition for robust stabilizability," *SCL*, vol. 2, pp. 1-5, 1982.

[48] D.J.N. Limebeer and Y.S. Hung, "Robust stability of interconnected systems," *IEEE-TAC*, vol. AC-28, no. 6, pp. 710-716, June 1983.

[49] J. Lunze, "Robustness tests for feedback control systems using multidimensional uncertainty bounds," *SCL*, vol. 4, pp. 85-89, 1984.

[50] _____, "Robustness analysis of control systems by means of a structured uncertainty description," *FCE*, vol. 10, no. 4, pp. 201-213, 1985.

[51] J.B. Moore and M. Tomizuka, "On the class of all stabilizing regulators," *IEEE-TAC*, vol. AC-34, no. 10, pp. 1115-1120, Oct. 1989.

[52] M. Morari, "Robust stability of systems with integral control," *IEEE-TAC*, vol. AC-30, no. 6, pp. 574-577, June 1985.

[53] V. Mukhopadhyay, "Stability robustness improvement using constrained optimization techniques," *JGCD*, vol. 10, no. 2, pp. 172-177, 1987.

[54] O.D.I. Nwokah, "On nonsingular value based design of controller for robust stability," *Proc. IEE*, vol. 133, Pt. D, no. 2, pp. 57-64, March 1986.

[55] D.H. Owens, "The numerical range: A tool for robust stability studies?" *SCL*, vol. 5, pp. 153-158, 1984.

[56] D.H. Owens and A. Choatai, "Robust stability of multivariable feedback systems with respect to linear and nonlinear feedback perturbations," *IEEE-TAC*, vol. AC-27, no. 1, pp. 254-256, Feb. 1982.

[57] A.N. Penchuk, P.D. Hattis, and E.T. Kubiak, "A frequency domain stability analysis of a phase plane control system," *JGCD*, vol. 8, no. 1, pp. 50-55, 1985.

[58] M. Saeki, "A method of robust stability analysis with highly structured uncertainties," *IEEE-TAC*, vol. AC-31, no. 10, pp. 935-940, Oct. 1986.

[59] M.G. Safonov, "Stability margins of diagonally perturbed multivariable feedback systems," *Proc. IEE*, vol. 129, Pt. D, no. 6, pp. 251-256, Nov. 1982.

[60] M.G. Safonov and M. Athans, "Gain and phase margins for multiloop LQG regulators," *IEEE-TAC*, vol. AC-22, pp. 173-179, April 1977.

[61] M.G. Safonov and B.S. Chen, "Multivariable stability-margin optimisation with decoupling and output regulation," *Proc. IEE*, vol. 129, Pt. D, no. 6, pp. 276-282, Nov. 1982.

[62] M.G. Safonov and G. Wyetzner, "Computer-aided stability analysis renders Popov criterion obsolete," *IEEE-TAC*, vol. AC-32, no. 12, pp. 1128-1131, Dec. 1987.

[63] R.S. Sanchez Pena and A. Sideris, "Robustness with real parametric and structured complex uncertainty," *IJC*, vol. 52, no. 3, pp. 753-765, 1990.

[64] A. Sideris, "Elimination of frequency search from robustness tests," *IEEE-TAC*, vol. AC-37, no. 10, pp. 1635-1640, 1992.

[65] A. Sideris and R.S. Sanchez Pena, "Fast computation of the multivariable stability margin for real interrelated uncertain parameters," *IEEE-TAC*, vol. AC-34, no. 12, pp. 1272-1276, Dec. 1989.

[66] C.B. Soh, C.S. Berger, and K.P. Dabke, "On the stability properties of polynomials with perturbed coefficients," *IEEE-TAC*, vol. AC-30, no. 10, pp. 1033-1036, Oct 1985. Addendum, *ibid.*, vol. AC-32, no. 3, pp. 239-240, March 1987.

[67] H.J. Sussmann, "Limitations on the stabilizability of globally minimum phase systems," *IEEE-TAC*, vol. AC-35, no. 1, pp. 117-119, Jan. 1990.

[68] J.A. Tekawy, M.G. Safonov, and R.Y. Chiang, "Convexity property of the one-sided multivariable stability margin," *IEEE-TAC*, vol. AC-37, no. 4, pp. 496-498, 1992.

[69] M. Verma and E. Jonckheere, "L^∞–compensation with mixed sensitivity as a broadband matching problem," *SCL*, vol. 4, pp. 125-129, 1984.

[70] M.S. Verma, "Robust stabilization of linear time-invariant systems," *IEEE-TAC*, vol. AC-34, no. 8, pp. 870-875, Aug. 1989.

[71] M. Vidyasagar and H. Kimura, "Robust controllers for uncertain linear multivariable systems," *Automatica*, vol. 22, no. 1, pp. 85-94, 1986.

[72] K. Wei and B.R. Barmish, "An iterative design procedure for simultaneous stabilization of MIMO systems," *Automatica*, vol. 24, no. 5, pp. 643-652, 1988.

[73] B. Wie and A.E. Bryson, Jr., "Multivariable control robustness examples: A classical approach," *JGCD*, vol. 10, no. 1, pp. 118-120, 1987.

[74] D. Williamson, "Structural state space sensitivity in linear systems," *SCL*, vol. 7, pp. 301-307, 1986.

[75] W.-Y. Yan, B.D.O. Anderson, and R.R. Bitmead, "The combined sensitivity and phase margin problem," *Automatica*, vol. 28, no. 2, pp. 417-421, 1992.

[76] O. Yaniv, "Arbitrarily small sensitivity in multiple-input-output uncertain feedback systems," *Automatica*, vol. 27, no. 3, pp. 565-568, 1991.

[77] H.-H. Yeh, S.S. Banda, and D.B. Ridgely, "Stability robustness measures utilizing structural information," *IJC*, vol. 41, no. 2, pp. 365-387, 1985.

[78] H.-H. Yeh, D.B. Ridgely, and S.S. Banda, "Nonconservative evaluation of uniform stability margins of multivariable feedback systems," *JGCD*, vol. 8, no. 2, pp. 167-174, 1985.

[79] L.F. Yeung and G.F. Bryant, "Robust stability of diagonally dominant systems," *Proc. IEE*, vol. 131, Pt. D, no. 6, pp. 253-260, Nov. 1984.

[80] D.-N. Zhang, M. Saeki, and K. Ando, "Stability margin calculation of systems with structured time-delay uncertainties," *IEEE-TAC*, vol. AC-37, no. 6, pp. 865-868, 1992.

[81] K. Zhou and G. Gu, "Robust stability of multivariable systems with both real parametric and norm bounded uncertainties," *IEEE-TAC*, vol. AC-37, no. 10, pp. 1533-1537, 1992.

[F] Stability Robustness Analysis
Maneuverability, and Sensitivity in the Large
Group V: Time-varying, Nonlinear, and Delayed Systems
The Concept of Absolute Stability

[1] G. Ahmadi and A.M. Morshedi, "On the stability of stochastic chemical systems," *JFI*, vol. 306, no. 1, pp. 77-86, 1978.

[2] E.Y. Aivazyan, "A frequency criterion of stability of nonlinear sampled-data systems," *ARC*, vol. 49, no. 3, pp. 279-283, 1988.

[3] M.A. Aizerman and F.R. Gantmacher, *Absolute Stability of Regulator Systems*. San Francisco: Holden-Day, Inc. 1964. [Translated from Russian by E. Polak.] Reviewed by R.W. Brockett, *IEEE-TAC*, vol. AC-11, no. 3, p. 634, 1966, and by R.E. Kalman, *SIAM R.*, vol. 12, no. 1, pp. 161-162, 1970.

[4] V.V. Andrusevich, "Absolute instability of monotonic nonlinear systems," *ARC*, vol. 47, no. 7, pp. 904-911, July 1986.

[5] D.P. Atherton and S. Shankar, "Some observations on the Aizerman conjecture," *Proc. IEE*, vol. 126, no. 12, pp. 1305-1306, 1979.

[6] A. Bacciotti, "Potentially global stablizability," *IEEE-TAC*, vol. AC-31, no. 10, pp. 974-976, Oct. 1986.

[7] _____, "Further remarks on potentially global stablizability," *IEEE-TAC*, vol. AC-34, no. 6, pp. 637-638, June 1989.

[8] R. Balasubramanian, "Stability of limit cycles in feedback systems containing a relay," *Proc. IEE*, vol. 128, Pt. D, no. 1, pp. 24-29, Jan. 1981.

[9] N.E. Barabanov, "The Lyapunov indicator of discrete inclusions," I, II, III, *ARC*, vol. 49, no. 2, pp. 152-157, no. 3, pp. 283-287, and no. 5, pp. 558-565, 1988.

[10] _____, "Absolute stability of sampled-data control systems," *ARC*, vol. 49, no. 8, pp. 981-988, 1988.

[11] I. Bar-Kana, "On the Lur'e problem and stability of nonlinear controllers," *JFI*, vol. 325, no. 6, pp. 687-693, 1988.

[12] A.I. Barkin, "L_{2p}-Stability and absolute stability of nonlinear systems," *ARC*, vol. 44, no. 10, pp. 1290-1295, Oct. 1983.

[13] A.I. Barkin and A.L. Zelentsovskii, "Absolute stability criterion for nonlinear control systems," *ARC*, vol. 42, no. 7, pp. 853-857, July 1981.

[14] _____, "A new criterion for absolute stability: Non-linear transformation technique," *IJSS*, vol. 14, no. 10, pp. 1217-1228, 1983.

[15] _____, "Method of power transformations for analysis of stability of nonlinear control systems," *SCL*, vol. 3, pp. 303-310, 1983.

[16] B.R. Barmish and Z. Shi, "Robust stability of perturbed systems with time delays," *Automatica*, vol. 25, no. 3, pp. 371-381, 1989.

[17] S. Barnett and C. Storey, "Some results on the sensitivity and synthesis of asymptotically stable linear and nonlinear systems," *Automatica*, vol. 4, pp. 187-194, 1968.

[18] D.S. Bernstein, "Robust stability and performance *via* fixed-order dynamic compensator," *SIAM-JCO*, vol. 27, no. 2, pp. 389-406, March 1989.

[19] A.M. Bloch, "Stabilizability of nonholonomic control systems," *Automatica*, vol. 28, no. 2, pp. 431-435, 1992.

[20] A.M. Bloch, M. Reyhanoglu, and N.H. McClamroch, "Control and stabilization of nonholonomic dynamic systems," *IEEE-TAC*, vol. AC-37, no. 11, pp. 1746-1757, 1992.

[21] N.A. Bobylev, "Method of stability analysis of gradient systems," *ARC*, vol. 41, no. 8, pp. 1065-1067, Aug. 1980.

[22] H. Bourles, Y. Joannic, and O. Mercier, "ρ-stability and robustness: Discrete-time case," *IJC*, vol. 52, no. 5, pp. 1217-1239, 1990.

[23] M. Buslowicz, "Simple stability criterion for a class of delay differential systems," *IJSS*, vol. 18, no. 5, pp. 993-995, 1987.

[24] H. Chapellat, M. Dahleh, and S.P. Bhattacharyya, "On robust nonlinear stability of interval control systems," *IEEE-TAC*, vol. AC-36, no. 1, pp. 59-67, 1991.

[25] B. Charlet, "Stability and robustness for nonlinear systems decoupled and linearized by feedback," *SCL*, vol. 8, pp. 367-374, 1987.

[26] J. Chegancas and C. Burgat, "Polyhedral cones associated to M-matrices and stability of time varying discrete time systems," *JMAA*, vol. 118, pp. 88-96, 1986.

[27] Y.N. Chekhovoy, "Stability of discrete automatic systems containing nonlinear and quasilinear blocks," *SAC*, vol. 17, no. 2, pp. 86-89, 1984.

[28] _____, "Stability of nonlinear dynamic systems with quasilinear blocks," *SAC*, vol. 17, no. 6, pp. 84-86, 1984.

[29] B.-S. Chen and C.-C. Chiang, "Robust stabilizer synthesis for feedback systems containing time-varying nonlinear perturbations," *IEEE-TAC*, vol. AC-31, no. 8, pp. 768-771, Aug. 1986.

[30] B.-S. Chen and C.H. Lo, "Necessary and sufficient conditions for robust stabilization of an observer-based compensating system suffering nonlinear time-varying perturbation," *IEEE-TAC*, vol. AC-34, no. 8, pp. 899-900, Aug. 1989.

[31] B.-S. Chen and S.-S. Wang, "The stability of feedback control with nonlinear saturating actuator: Time domain approach," *IEEE-TAC*, vol. AC-33, no. 1, pp. 483-487, May 1988.

[32] G. Chen and R.J.P. de Figueiredo, "On robust stabilization of nonlinear control systems," *SCL*, vol. 12, pp. 373-379, 1989.

[33] E. Cheres, S. Gutman, and Z. Palmor, "Stabilization of uncertain dynamic systems including state delay," *IEEE-TAC*, vol. AC-34, no. 11, pp. 1199-1203, Nov. 1989.

[34] E. Cheres, Z. Palmor, and S. Gutman, "Quantitative measures of robustness for systems including delayed perturbations," *IEEE-TAC*, vol. AC-34, no. 11, pp. 1203-1205, Nov. 1989.

[35] H.-D. Chiang, M.W. Hirsch, and F.F. Wu, "Stability region of nonlinear autonomous dynamical systems," *IEEE-TAC*, vol. AC-33, no. 1, pp. 16-27, Jan. 1988.

[36] H.-D. Chiang and J.S. Thorp, "Stability regions of nonlinear dynamical systems: A constructive methodology," *IEEE-TAC*, vol. AC-34, no. 12, pp. 1229-1241, Dec. 1989.

[37] S.Y. Chimishkyan, "Normal approximation, numerical range, and boundary criteria for absolute stability of nonlinear multiply connected automatic control systems," *SJAIS*, vol. 22, no. 6, pp. 33-36, 1989.

[38] P.L. Chow and K.L. Chiou, "Asymptotic stability of randomly perturbed linear periodic systems," *SIAM-JAM*, vol. 40, no. 2, pp. 315-326, 1981.

[39] K.L. Cooke and J.M. Ferreira, "Stability conditions for linear retarded functional difference equations," *JMAA*, vol. 96, pp. 480-504, 1983.

[40] M. Corless and G. Leitmann, "Adaptive control of systems containing uncertain functions and unknown functions with uncertain bounds," *JOTA*, vol. 41, no. 1, pp. 155-168, 1983.

[41] J.J. da Cruz and J.C. Geromel, "Frequency-domain approach to the absolute stability analysis of discrete-time linear-quadratic regulators," *Proc. IEE*, vol. 137, Pt. D, no. 2, pp. 104-106, 1990.

[42] E.I. Dergacheva, "Stabilizing the steady-state displacements in a control system under continuous perturbations," *ARC*, no. 2, pp. 182-192, Feb. 1967.

[43] M. Eslami, "On sensitivity of nonlinear digital systems," in *Proc. Allerton Conf.*, pp. 424-432, Oct. 1980.

[44] _____, "Robust nonlinear dynamic systems – The old question with a new answer," *SCL*, vol. 14, no. 5, pp. 445-451, June 1990.

[45] _____, "On stability robustness analysis of dynamic systems with multiple large parameter variations and multiple nonlinearities," in *Proc. Allerton Conf.*, pp. 107-116, Oct. 1991.

[46] M. Fu, A.W. Olbrot, and M.P. Polis, "Robust stability for time-delay systems: The edge theorem and graphical test," *IEEE-TAC*, vol. AC-34, no. 8, pp. 813-820, Aug. 1989.

[47] F.G. Garashchenko and L.A. Pantaliyenko, "Solution of sensitivity theory problems by the practical stability method," *SJAIS*, vol. 21, no. 4, pp. 65-69, 1988.

[48] A.K. Gelig and N.I. L'yanova, "A circle criterion of stability of nonlinear sampled-data systems with a new type of quadratic constraints," *ARC*, vol. 49, no. 6, pp. 714-719, 1988.

[49] R. Genesio, M. Tartaglia, and A. Vicino, "On the estimation of asymptotic stability regions: State of the art and new proposals," *IEEE-TAC*, vol. AC-30, no. 8, pp. 747-755, Aug. 1985.

[50] M.I. Gil', "Stability in the large of nonstationary systems," *ECy*, vol. 19, no. 6, pp. 125-130, 1981.

[51] _____, "A class of absolutely stable systems," *ARC*, vol. 44, no. 10, pp. 1295-1300, Oct. 1983.

[52] _____, "Absolute stability of nonlinear nonstationary systems with distributed parameters," *ARC*, vol. 46, no. 6, pp. 685-693, June 1985.

[53] W.L. Green and E.W. Kamen, "Stabilizability of linear systems over a commutative normed algebra with applications to spatially-distributed and parameter-dependent systems," *SIAM-JCO*, vol. 23, no. 1, pp. 1-18, 1985.

[54] L.T. Grujic, "Solutions for the Lurie-Postnikov and Aizerman problems," *IJSS*, vol. 9, no. 12, pp. 1359-1372, 1978.

[55] _____, "On absolute stability and the Aizerman conjecture," *Automatica*, vol. 17, no. 2, pp. 335-349, 1981.

[56] _____, "Lyapunov-like solutions for stability problems of the most general stationary Lurie-Postnikov systems," *IJSS*, vol. 12, no. 7, pp. 813-833, 1981.

[57] L.T. Grujic and D. Petkovski, "On robustness of Lurie systems with a single nonlinearity," *CTAT*, vol. 2, no. 4, pp. 627-632, Dec. 1986.

[58] _____, "On robustness of Lurie systems with multiple non-linearities," *Automatica*, vol. 23, no. 3, pp. 327-334, 1987.

[59] _____, "Robust absolutely stable Lurie systems," *IJC*, vol. 46, no. 1, pp. 357-368, 1987.

[60] M. Hached, S.M. Medani-Esfahani, and S.H. Zak, "Stabilization of uncertain systems subject to hard bounds on control with application to a robot manipulator," *IEEE-JRA*, vol. 4, no. 3, pp. 310-323, June 1988. Correction, by M. Arshad and Z. Bahri, *ibid.*, vol. 5, pp. 394-395, June 1989. Errata, *ibid.*, p. 396.

[61] A. Hmamed, "On the stability of time-delay systems: New results," *IJC*, vol. 43, no. 1, pp. 321-324, 1986.

[62] C.V. Hollot, "Bounds invariant Lyapunov functions: A means for enlarging the class of stablizable uncertain systems," *IJC*, vol. 46, no. 1, pp. 161-184, 1987.

[63] S. Hui and S.H. Zak, "Robust control synthesis for uncertain/nonlinear dynamical systems," *Automatica*, vol. 28, no. 2, pp. 289-298, 1992.

[64] K. Ichikawa and R. Ortega, "On stabilization of nonlinear systems with enlarged domain of attraction," *Automatica*, vol. 28, no. 3, pp. 623-626, 1992.

[65] S.M. Joshi, "Stability of multiloop LQ regulators with nonlinearities – Part I: Region of attraction," *IEEE-TAC*, vol. AC-31, no. 4, pp. 364-367, April 1986.

[66] _____, "Stability of multiloop LQ regulators with nonlinearities – Part II: Region of ultimate boundedness," *IEEE-TAC*, vol. AC-31, no. 4, pp. 367-370, April 1986.

[67] _____, "Stability regions for multiloop LQ-regulated systems with state estimators," *IEEE-TAC*, vol. AC-31, no. 12, pp. 1151-1153, Dec. 1986.

[68] V.A. Kamenetskii, "Absolute stability and absolute instability of control systems with several nonlinear nonstationary elements," *ARC*, vol. 44, no. 12, pp. 1543-1552, Dec. 1983.

[69] Y.S. Kan and A.I. Kibzun, "Stabilization of a dynamic system which is under the action of undetermined and random disturbances," *ARC*, vol. 51, no. 12, pp. 1665-1673, 1990.

[70] E. Kaszkurewicz and L. Hsu, "Stability of Nonlinear systems: A structural approach," *Automatica*, vol. 15, pp. 609-614, 1979.

[71] P.P. Khargonekar, A.M. Pascoal, and R. Ravi, "Strong, simultaneous, and reliable stabilization of finite-dimensional linear time-varying plants," *IEEE-TAC*, vol. AC-33, no. 12, pp. 1158-1161, Dec. 1988.

5.12.F. Group V: Time-varying, Nonlinear, and Delayed Systems

[72] P.P. Khargonekar and K. Poolla, "Robust stabilization of distributed systems," *Automatica*, vol. 22, no. 1, pp. 77-84, 1986.

[73] K. Khorasani, "Robust stabilization of non-linear systems with unmodelled dynamics," *IJC*, vol. 50, no. 3, pp. 827-844, 1989.

[74] D. Kinderlehrer, "Estimates for the solution and its stability in Signorini's problem," *AMO*, vol. 8, pp. 159-188, 1982.

[75] A.I. Korshunov, "Analysis of stability in the large of linearizable sampled-data systems with the aid of two Lyapunov functions," *ARC*, vol. 51, no. 5, pp. 647-657, 1990.

[76] _____, "Robustness of pulse-width systems," *ARC*, vol. 51, no. 12, pp. 1702-1709, 1990.

[77] V.S. Kozyakin, "Algebraic unsolvability of problem of absolute stability of desynchronized systems," *ARC*, vol. 51, no. 6, pp. 754-759, 1990.

[78] _____, "Stability of phase-frequency desynchronized systems under a perturbation of the switching instants of the components," *ARC*, vol. 51, no. 8, pp. 1034-1040, 1990.

[79] _____, "Absolute stability of systems with asynchronous sampled-data elements," *ARC*, vol. 51, no. 10, pp. 1349-1355, 1990.

[80] I.B. Krasil'nikov and F.R. Sadykov, "Frequency-domain analysis of input-output stability of nonlinear multidimensional systems," *ARC*, vol. 49, no. 12, pp. 1575-1583, 1988.

[81] M.A. Krasnosel'skii and A.V. Pokrovskii, "Absolute stability of systems with discrete time," *ARC*, vol. 39, no. 2, pp. 189-199, 1978.

[82] V.P. Kuznetsov and E.P. Kukareko, "Regions of asymptotic stability of linear continuous nonstationary systems," *ARC*, vol. 41, no. 3, pp. 309-312, March 1980.

[83] G.A. Leonov, "Necessary frequency conditions for the absolute stability of nonstationary systems," *ARC*, vol. 42, no. 1, pp. 9-13, Jan. 1981.

[84] G. Leugering, "Control and stabilization of a flexible robot arm," *DSS*, vol. 5, no. 1, pp. 37-46, 1990.

[85] I. Lewkowicz and R. Sivan, "Stability robustness of almost linear state equations," *IEEE-TAC*, vol. AC-38, no. 2, pp. 262-266, 1993.

[86] M.R. Liberzon, "New results on absolute stability of nonstationary controlled systems," (Survey) *ARC*, vol. 40, no. 8, pp. 1124-1140, Aug. 1979.

[87] _____, "Absolute stability of a class of servomechanisms," *ARC*, vol. 40, no. 12, pp. 1729-1732, Dec. 1979.

[88] A.V. Lipatov, F.R. Sadykov, and G.Y. Soloveichik, "Graphical methods for stability analysis of continuous systems with one nonlinearity of various classes," *ARC*, vol. 46, no. 3, pp. 300-306, March 1985.

[89] K. Liu and F.L. Lewis, "An improved result on the stability analysis of nonlinear systems," *IEEE-TAC*, vol. AC-37, no. 9, pp. 1425-1431, 1992.

[90] C.C.H. Ma and M. Vidyasagar, "Parametric conditions for stability of reduced-order linear time-varying control systems," *Automatica*, vol. 23, no. 5, pp. 625-634, 1987.

[91] V.A. Mayorov, Y.G. Borisenko, O.B. Kerber, V.V. Pavlov, and O.S. Yakovlev, "The problem of synthesizing a nonlinear control law," *SJAIS*, vol. 21, no. 2, pp. 41-49, 1988.

[92] S.M. Medani-Esfahani, M. Hached, and S.H. Zak, "Estimation of sliding mode domains of uncertain variable structure systems with bounded controllers," *IEEE-TAC*, vol. AC-35, no. 4, pp. 446-449, April 1990.

[93] A.M. Meilakhs, "Nonmodal approach to stabilization of nonlinear control systems," *ARC*, vol. 51, no. 4, pp. 440-443, 1990.

[94] A.N. Michel, N.R. Sarabudla, and R.K. Miller, "Stability analysis of complex dynamical systems, some computational methods," *CSSP*, vol. 1, no. 2, pp. 171-202, 1982.

[95] B.J. Min, C. Slivinsky, and R.G. Hoft, "Absolute stability analysis of PWM systems," *IEEE-TAC*, vol. AC-22, no. 3, pp. 447-452, June 1977.

[96] H. Miyagi and K. Yamashita, "A generalized Lyapunov function for power systems with product-type nonlinearities," *EEJ*, vol. 108, no. 4, pp. 67-74, 1984.

[97] _____, "Robust stability of Lure systems with multiple nonlinearities," *IEEE-TAC*, vol. AC-37, no. 6, pp. 883-886, 1992.

[98] A.P. Molchanov, "Absolute stability of sampled-data systems with several nonstationary nonlinear elements," *ARC*, vol. 40, no. 3, pp. 355-362, March 1979.

[99] _____, "Criterion of absolute stability for sampled-data systems with a nonstationary nonlinearity. I," *ARC*, vol. 44, no. 5, pp. 607-616, May 1983.

[100] _____, "A criterion of absolute stability of sampled-data systems with a nonstationary nonlinearity. II," *ARC*, vol. 44, no. 6, pp. 719-729, June 1983.

[101] A.P. Molchanov and E.S. Pyatnitskii, "Absolute instability of nonlinear nonstationary systems," Part I, *ARC*, vol. 43, no. 1, pp. 13-20, Jan. 1982, Part II, *ARC*, vol. 43, no. 2, pp. 147-157, Feb. 1982, Part III, *ARC*, vol. 43, no. 3, pp. 289-299, March 1982.

[102] _____, "Lyapunov functions that specify necessary and sufficient conditions of absolute stability of nonlinear nonstationary control systems," Part I, *ARC*, vol. 47, no. 3, pp. 344-354, March 1986, Part II, *ARC*, vol. 47, no. 4, pp. 443-451, April 1986, Part III, *ARC*, vol. 47, no. 5, pp. 620-630, May 1986.

[103] A.P. Molchanov and Y.S. Pyatnitskiy, "Criteria of asymptotic stability of differential and difference inclusions encountered in control theory," *SCL*, vol. 13, pp. 59-64, 1989.

[104] T. Mori, N. Fukuma, and M. Kuwahara, "A stability criterion for linear time-varying systems," *IJC*, vol. 34, no. 3, pp. 585-591, 1981. Comments by R.I. Silva-Madriz, *ibid.*, vol. 39, no. 2, pp. 413-416, 1984.

[105] V.M. Mutter, "Analytical conditions for absolute stability in a certain class of nonlinear systems," *ECy*, vol. 13, no. 1, pp. 145-153, 1975.

[106] A.W. Olbrot, "Robust stabilization of uncertain systems by periodic feedback," *IJC*, vol. 45, no. 3, pp. 747-758, 1987.

[107] V.I. Opoitsev, "Unbounded regions of asymptotic stability," *ARC*, vol. 42, no. 3, pp. 275-282, March 1981.

[108] D.H. Owens and A. Choatai, "Robust stability of multivariable feedback systems with respect to linear and nonlinear feedback perturbations," *IEEE-TAC*, vol. AC-27, no. 1, pp. 254-256, Feb. 1982.

[109] D.H. Owens and A. Raya, "Robust stability of Smith predictor controllers for time-delay systems," *Proc. IEE*, vol. 129, Pt. D, no. 6, pp. 298-304, Nov. 1982.

[110] P.V. Pakshin, "An algebraic criterion of absolute stochastic stability of nonlinear discrete-time systems with multiplicative white noise," *ARC*, vol. 49, no. 6, pp. 749-756, 1988.

[111] L. Pandolfi, "Some observations on the asymptotic behaviour of the solutions of the equation $\dot{x}(t) = A(t)x(\lambda t) + B(t)x(t)$ $\lambda > 0$," *JMAA*, vol. 67, pp. 483-489, 1979.

[112] I.R. Petersen, "Notions of stabilizability and controllability for a class of uncertain systems," *IJC*, vol. 46, no. 2, pp. 409-422, 1987.

[113] _____, "A stabilization algorithm for a class of uncertain linear systems," *SCL*, vol. 8, pp. 351-357, 1987. Correction by J.M. Collado and P.I. Petersen, *ibid.*, vol. 11, p. 83, 1988.

[114] I.R. Petersen and B.R. Barmish, "Control effort considerations in the stabilization of uncertain dynamical systems," *SCL*, vol. 9, pp. 417-422, 1987.

[115] J.-B. Pomet and L. Praly, "A result on robust boundedness," *SCL*, vol. 10, pp. 83-92, 1988.

[116] E.S. Pyatnitskii and V.I. Skorodinskii, "A criterion of absolute stability of nonlinear sampled-data control systems in the form of numerical procedures," *ARC*, vol. 47, no. 9, pp. 1190-1198, 1986.

5.12.F. Group V: Time-varying, Nonlinear, and Delayed Systems

[117] M.R.M. Rao and S. Sivasundaram, "Asymptotic stability for equations with unbounded delay," *JMAA,* vol. 131, pp. 97-105, 1988.

[118] L.B. Rapoport, "Absolute stability of control systems with several nonlinear stationary elements," *ARC,* vol. 48, no. 5, pp. 623-630, May 1987.

[119] _____, "Limit of absolute stability of nonlinear nonstationary systems and its connection with the construction of invariant functions," *ARC,* vol. 51, no. 10, pp. 1368-1375, 1990.

[120] T.A. Sabaeva, "Stability of multidimensional variable structure systems," *ARC,* vol. 44, no. 12, pp. 1552-1558, Dec. 1983.

[121] F.M.A. Salam, "Feedback stabilization of nonlinear pendulum under uncertainty: A robustness issue," *SCL,* vol. 7, pp. 199-206, 1986.

[122] T. Sasagawa, "Stability and stabilization of random parameter linear systems," *EEJ,* vol. 102, no. 6, pp. 146-154, 1982.

[123] A.V. Savkin, "An absolute stability criterion for nonlinear control systems with a periodically nonstationary linear part," *ARC,* vol. 51, no. 8, pp. 1047-1051, 1990.

[124] R.H. Seacat, L.F. Judd, and D.R. Frannin, "A method of specifying parameter changes to insure absolute stability of nonlinear systems," in *Proc. Midwest Symp. Circuit Theory,* vol. 11, pp. 273-280, 1968.

[125] K. Shamsa and H. Flashner, "A class of discrete-time stabilizing controllers for flexible mechanical systems," *ASME-JDSMC,* vol. 112, pp. 55-61, March 1990.

[126] R. Shoureshi, M.E. Momot, and M.D. Roesler, "Robust control for manipulators with uncertain dynamics," *Automatica,* vol. 26, no. 2, pp. 353-359, 1990.

[127] V.I. Skorodinskii, "Absolute stability and absolute instability of control systems with two nonlinear nonstationary elements. I," *ARC,* vol. 42, no. 9, pp. 1149-1157, Sept. 1981.

[128] _____, "Absolute stability and absolute instability of control with two nonlinear nonstationary elements. II," *ARC,* vol. 43, no. 6, pp. 775-781, June 1982.

[129] G.V. Smirnov, "Weak asymptotic stability of differential inclusions. I and II," *ARC,* vol. 51, no. 7, pp. 901-908, and no. 8, pp. 1052-1058, 1990.

[130] V.I. Sokolov, "Asymptotic stability of linear nonstationary control systems," *ARC,* vol. 42, no. 8, pp. 1012-1018, 1981.

[131] E.M. Solnechnyi, "Sufficient conditions for stability of nonautonomous linear system and estimation of domains of attraction of nonlinear systems," *ARC,* vol. 46, no. 1, pp. 34-40, Jan. 1985.

[132] T. Taniguchi, "Stability theorems of perturbed linear ordinary differential equations," *JMAA,* vol. 149, pp. 583-598, 1990.

[133] A. Tesi and A. Vicino, "Robust absolute stability of Lur's control systems in parameter space," *Automatica,* vol. 27, no. 1, pp. 147-151, 1991.

[134] A. Tesi, A. Vicino, and G. Zappa, "Clockwise property of the Nyquist plot with implications for absolute stability," *Automatica,* vol. 28, no. 1, pp. 71-80, 1992.

[135] J. Tsinias, "Sufficient Lyapunov-like conditions for stabilization," *MCSS,* vol. 2, pp. 343-357, 1989.

[136] D. Valsamis, D.H. Owens, and J.O. Gray, "Stability and performance bounds for the linearisation-based analysis and design of multivariable nonlinear systems," *Proc. IEE,* vol. 135, Pt. D, no. 2, pp. 79-89, March 1988.

[137] A. Vannelli and M. Vidyasagar, "Maximal Lyapunov functions and domains of attraction for autonomous nonlinear systems," *Automatica,* vol. 21, no. 1, pp. 69-80, 1985.

[138] V.M. Vaplyushkin, S.I. Vdovin, and E.F. Sabaev, "Stability in the large of a class of control systems," *ARC,* vol. 40, no. 7, pp. 949-955, July 1979.

[139] A.A. Voronov, "Absolutely stable systems with a differentiable nondecreasing nonlinearity," *ARC*, vol. 39, no. 7, pp. 947-958, 1978.

[140] S.-S. Wang, B.-S. Chen, and T.-P. Lin, "Robust stability of uncertain time-delay systems," *IJC*, vol. 46, no. 3, pp. 963-976, 1987.

[141] J.L. Willems and J.C. Willems, "Robust stabilization of uncertain systems," *SIAM-JCO*, vol. 21, no. 3, pp. 352-374, May 1983.

[142] D.-Y. Xu, "Robust stability analysis of uncertain neutral delay-differential systems via difference inequality," *CTAT*, vol. 5, no. 3, pp. 301-313, Sept. 1989.

[143] V.A. Yakubovich, "Absolute stability of nonlinear distributed-parameter systems. I," *ARC*, vol. 44, no. 6, pp. 729-736, June 1983.

[144] _____, "Dichotomy and absolute stability of nonlinear systems with periodically nonstationary linear part," *SCL*, vol. 11, pp. 221-228, 1988.

[145] R.T. Yanushevskiy, "Synthesis of a certain class of absolutely stable nonlinear control systems," *ECy*, vol. 17, no. 2, pp. 128-132, 1979.

[146] T. Yoneyama and J. Sugie, "On the stability region of scalar delay-differential equations," *JMAA*, vol. 134, pp. 408-425, 1988.

[147] Y. Yu, "On stablizing uncertain linear delay systems," *JOTA*, vol. 41, no. 3, pp. 503-508, 1983.

[148] V.V. Zaitsev, "Lyapunov's second method for estimating the reachability regions of nonlinear systems," *ARC*, vol. 51, no. 6, pp. 746-754, 1990.

[149] K. Zhou and P.P. Khargonekar, "Robust stabilization of linear systems with norm-bounded time-varying uncertainty," *SCL*, vol. 10, pp. 17-20, 1988.

[150] V.P. Zhukov, "Necessary and sufficient conditions of instability of nonlinear autonomous dynamic systems," *ARC*, vol. 51, no. 12, pp. 1652-1657, 1990.

[G] Stability Robustness Analysis
Maneuverability, and Sensitivity in the Large
Group VI: Stabilization Using the H^∞ and Other Optimization Approaches

[*] Also consult Section *[H]* of the references in Chapter Six, since almost all of those entries are relevant to this section.

[1] A.G. Aleksandrov, "Finite-frequency criteria for stability of systems with undetermined parameters," *ARC*, vol. 49, no. 7, pp. 838-847, 1988.

[2] J. Amillo and F.A. Mata, "Robust stabilization of systems with multiple real pole uncertainties," *IEEE-TAC*, vol. AC-36, no. 6, pp. 749-752, 1991.

[3] H.-H. Chou, B.-S. Chen, and Y.-P. Lin, "Robust pole placement: A frequency-domain approach," *IJSS*, vol. 21, no. 2, pp. 317-333, 1990.

[4] M.A. Dahleh, "BIBO stability robustness in the presence of coprime factor perturbations," *IEEE-TAC*, vol. AC-37, no. 3, pp. 352-355, 1992.

[5] R. Devanathan, "Robust stabilization of a SISO system with uncertainty in time delay," *IEEE-TAC*, vol. AC-37, no. 11, pp. 1820-1823, 1992.

[6] T.T. Georgiou, A.M. Pascoal, and P.P. Khargonekar, "On the robust stabilizability of uncertain linear time-invariant plants using nonlinear time-varying controllers," *Automatica*, vol. 23, no. 5, pp. 617-624, 1987.

[7] K. Glover, "Robust stabilization of linear multivariable systems: Relations to approximation," *IJC*, vol. 43, no. 3, pp. 741-766, 1986.

5.12.G. Group VI: Stabilization Using Optimization Approaches

[8] J.-P. Guerin and X.C. Zhong, "H^∞ robust stabilizability," *SCL*, vol. 9, pp. 289-293, 1987.

[9] E.M. Kasenally and D.J.N. Limebeer, "Closed formulae for a parametric mixed sensitivity problem," *SCL*, vol. 12, pp. 1-7, 1989.

[10] B. Kouvaritakis, R.F. Harrison, and H.A. Latchman, "Stability and dynamic performance bounds for nonlinear multivariable systems," *Proc. IEE*, vol. 134, Pt. D, no. 2, pp. 115-123, March 1987.

[11] V.B. Larin, "Parametrization of the set of stabilizing regulators in a standard synthesis problem," *SJAIS*, vol. 23, no. 2, pp. 21-26, 1990.

[12] C.C.H. Ma, "Stability robustness of repetitive control systems with zero phase compensation," *ASME-JDSMC*, vol. 112, pp. 320-324, Sept. 1990.

[13] S.D. O'Young and B.A. Francis, "Optimal performance and robust stabilization," *Automatica*, vol. 22, no. 2, pp. 171-183, 1986.

[14] I.R. Petersen, "Disturbance attenuation and H^∞ optimization: A design method based on the algebraic Riccati equation," *IEEE-TAC*, vol. AC-32, no. 5, pp. 427-429, May 1987.

[15] I.R. Petersen and C. Hollot, "High gain observers applied to problems in the stabilization of uncertain linear systems, disturbance attenuation and H^∞ optimization," *IJACSP*, vol. 2, pp. 347-369, 1988.

[16] E. Polak and T.L. Wuu, "On the design of stablizing compensators *via* semiinfinite optimization," *IEEE-TAC*, vol. AC-34, no. 2, pp. 196-200, Feb. 1989.

[17] I. Postlethwaite, S.D. O'Young, and D.-W. Gu, "Stable H – an H^∞ control-system design package," *TIMC*, vol. 10, no. 2, pp. 103-109, 1988.

[18] D.D. Siljak, *Large-Scale Dynamic Systems, stability and structure*. New York: North-Holland, 1978. Reviewed by V. Lakshmikantham, *IEEE-TAC*, vol. AC-26, no. 4, pp. 976-977, 1981.

[19] X.-L. Tan, D.D. Siljak, and M. Ikeda, "Reliable stabilization *via* factorization methods," *IEEE-TAC*, vol. AC-37, no. 11, pp. 1786-1791, 1992.

[20] S. Townley, "Stability radius optimization: A geometric approach," *SCL*, vol. 14, pp. 199-207, 1990.

[21] D.J. Walker, "Robust stabilizability of discrete-time systems with normalized stable factor perturbation," *IJC*, vol. 52, no. 2, pp. 441-455, 1990.

[22] W. Yan, R.R. Bitmead, and B.D.O. Anderson, "Design problems for sensitivity and complementary sensitivity," *SCL*, vol. 15, pp. 137-146, 1990.

[H] Root Sensitivity and Pole Placement

[1] M.M. Bayoumi and T.L. Duffield, "Output feedback decoupling and Placement in linear time-invariant systems," *IEEE-TAC*, vol. AC-22, no. 1, pp. 142-143, Feb. 1977.

[2] F.G. Boese and W.J. Luther, "A note on a classical bound for the moduli of all zeros of a polynomial," *IEEE-TAC*, vol. AC-34, pp. 998-1001, Sept. 1989.

[3] F.M. Brasch, Jr., and J.B. Pearson, "Pole placement using dynamic compensators," *IEEE-TAC*, vol. AC-15, no. 1, pp. 34-43, Feb. 1970.

[4] H.-H. Chou, B.-S. Chen, and Y.-P. Lin, "Robust pole placement: A frequency-domain approach," *IJSS*, vol. 21, no. 2, pp. 317-333, 1990.

[5] S. Elangovan and G. Srinivasan, "New algorithm for the optimal for the stabilization of linear systems using pole-assignment techniques," *Proc. IEE*, vol. 122, no. 11, pp. 1316-1319, Nov. 1975.

[6] E.G. Gilbert, "Conditions for minimizing the norm sensitivity of characteristic roots," *IEEE-TAC*, vol. AC-29, no. 7, pp. 658-661, July 1984.

[7] V. Gourishankar and K. Ramar, "Pole assignment with minimum eigenvalue sensitivity to plant parameter variations," *IJC*, vol. 23, no. 4, pp. 493-504, 1976.

[8] K. Hasegawa, "Root sensitivity of linear feedback control systems," *EEJ*, vol. 88, no. 4, pp. 10-20, 1968.

[9] Y.-T. Juang, "Robust stability and robust pole assignment of linear systems with structured uncertainty," *IEEE-TAC*, vol. AC-36, no. 5, pp. 635-637, 1991.

[10] Y.-T. Juang and K.-H. Chen, "Robust pole-assignment of linear dynamic systems," *CTAT*, vol. 5, no. 1, pp. 67-74, March 1989.

[11] Y.-T. Juang, Z.-C. Hong, and Y.-T. Wang, "Lyapunov approach to robust pole-assignment analysis," *IJC*, vol. 49, no. 3, pp. 921-927, 1989.

[12] _____, "Robustness of pole-assignment in a specified region," *IEEE-TAC*, vol. AC-34, no. 7, pp. 758-760, July 1989.

[13] J. Kautsky and N.K. Nichols, "Robust pole assignment in systems subject to structured perturbations," *SCL*, vol. 15, pp. 373-380, 1990.

[14] H. Kimura, "A further result on the problem of pole assignment by output feedback," *IEEE-TAC*, vol. AC-22, no. 3, pp. 458-463, June 1977.

[15] P.V. Kokotovic and D.D. Siljak, "The sensitivity problem in continuous and sampled-data linear control systems by generalized Mitrovic method," *IEEE Trans. on Applications and Industry*, vol. 83, pp. 321-324, 1964.

[16] P. Kudva and V. Gourishankar, "Adaptive scheme for pole assignment in linear systems," *Proc. IEE*, vol. 123, no. 3, pp. 267-270, March 1976.

[17] Y.S. Lee, "Sensitivity analysis and synthesis of linear feedback systems," in *Proc. Joint Auto. Contr. Conf.*, pp. 581-586, 1967.

[18] T. Mita, "On zeros and responses of linear regulators and linear observers," *IEEE-TAC*, vol. AC-22, no. 3, pp. 423-428, June 1977.

[19] B.S. Morgan, Jr., "Sensitivity analysis and synthesis of multivariable systems," *IEEE-TAC*, vol. AC-11, no. 3, pp. 506-512, July 1966.

[20] J.B. Nail, J.R. Mitchell, and W.L. McDaniel, "Determination of pole sensitivities by Danilevskii's method," *AIAA J.*, vol. 15, no. 10, pp. 1525-1527, Oct. 1977.

[21] A. Papoulis, "Displacement of the zeros of the impedance z(s) due to an incremental variation in the network elements," *Proc. IRE*, vol. 43, pp. 79-82, Jan. 1955.

[22] C. Pierre, "Root sensitivity to parameter uncertainties: A statistical approach," *IJC*, vol. 49, no. 2, pp. 521-532, 1989.

[23] I. Postlethwaite and Y.K. Foo, "Robustness with simultaneous pole and zero movement across the jω-axis," *Automatica*, vol. 21, no. 4, pp. 433-443, 1985.

[24] I. Postlethwaite, S.D. O'Young, and D.-W. Gu, "Stable H – an H^∞ control-system design package," *TIMC*, vol. 10, no. 2, pp. 103-109, 1988.

[25] K. Ramar and V. Gourishankar, "Pole assignment with optimality and minimum eigenvalue sensitivity," *Proc. IEE*, vol. 122, no. 12, pp. 1437-1438, 1975.

[26] M.J. Remec, "On the upper and lower bounds of excursions of critical-frequencies of a perturbed network," in *Proc. Allerton Conf.*, pp. 502-508, 1963.

[27] F. Reza, "Some mathematical properties of root loci for control system design," *American Inst. of Electrical Engineers, Communication and Electronics*, March 1956.

[28] B.O. Ryabov and G.P. Sachkov, "Sensitivity of accumulated disturbances in linear systems," *SAC*, vol. 14, no. 5, pp. 6-13, 1969.

[29] A. Sambandan and S.L. Hakkapakki, "New results on optimal pole assignment," *CCy*, vol. 17, no. 4, pp. 291-307, 1988.

[30] L. Shaw, "Pole placement: Stability and sensitivity of dynamic compensators," *IEEE-TAC*, vol. AC-16, no. 2, p. 210, April 1971.

[31] R.E. Skelton and D.A. Wagie, "Minimal root sensitivity in linear systems," *JGCD*, vol. 7, no. 5, pp. 570-574, 1985.

[32] Y.C. Soh, "Robust pole-placement for uncertain interval systems," *Proc. IEE*, vol. 136, Pt. D, no. 6, pp. 301-306, Nov. 1989.

[33] Y.C. Soh, R.J. Evans, I.R. Petersen, and R.E. Betz, "Robust pole assignment," *Automatica*, vol. 23, no. 5, pp. 601-610, 1987.

[34] R.L. Stapleford and D.T. McRuer, "Sensitivity of multiloop flight control system roots to open-loop parameter variations," in *Proc. Joint Auto. Contr. Conf.*, pp. 399-407, 1966.

[35] M.R. Stojic, M.M. Fedenia, and R.M. Stojic, "Sensitivities of the prescribed pole spectrum in a closed-loop control system," *Automatica*, vol. 24, no. 2, pp. 257-260, 1988.

[36] O. Taussky, "On the variation of the characteristic roots of a finite matrix under various changes of its elements," in *Recent Advances in Matrix Theory* (Proc. of Conf. in Madison, Oct. 1963, H. Schneider, Ed.), Milwaukee, WI: University of Wisconsin Press, pp. 125-138, 1964.

[37] J.S.H. Tsai and S.S. Chen, "Root distribution of a polynomial in subregions of the complex plane," *IEEE-TAC*, vol. AC-38, no. 1, pp. 173-178, 1993. [This entry has an excellent survey on the very early development of this concept.]

[38] A.B. Turgeon, "Methode systematique de compensation des systemes asservis," *Automatisme*, vol. XVIII, no. 1, pp. 16-22, 1973.

[39] H. Ur, "Root locus properties and sensitivity relations in control systems," *IRE-TAC*, vol. AC-5, pp. 57-65, Jan. 1960.

[40] E.S. Vattuone and R.C. Dorf, "Root sensitivity as a design criterion," in *Proc. Asimolar Conf.*, pp. 287-291, 1968.

[41] E. Zeheb, "Necessary and sufficient conditions for root clustering of a polytope of polynomials in a simply connected domain," *IEEE-TAC*, vol. AC-34, no. 9, pp. 986-990, Sept. 1989.

[42] M.S. Zilovic, L.M. Roytman, P.L. Combettes, and M.N.S. Swamy, "A bound for the zeros of polynomials," *IEEE-TCAS*, Part I, vol. CAS-39, no. 6, pp. 476-478, 1992.

[I] Eigenvalue and Eigenvector Sensitivity

[1] T.M. Bakri and G.K.F. Lee, "A parametric approach to the discrete robust eigenstructure assignment problem," *CAC*, vol. 16, pp. 39-43, 1988.

[2] H.H. Chi and C.-T. Chen, "Sensitivity studies of analog simulations," in *Proc. Allerton Conf.*, pp. 845-854, 1969.

[3] T.R. Crossley and B. Porter, "Eigenvalue and Eigenvector sensitivities in linear systems theory," *IJC*, vol. 10, no. 2, pp. 163-170, 1969.

[4] _____, "High-order eigenproblem sensitivity methods: Theory and application to the design of linear dynamical systems," *IJC*, vol. 10, no. 3, pp. 315-329, 1969.

[5] R.L. Dailey, "Eigenvector derivatives with repeated eigenvalues," *AIAA J.*, vol. 27, no. 4, pp. 486-491, 1989. Comment by W.C. Mills-Curran, *ibid.*, vol. 28, no. 10, p. 1846, 1990.

[6] S.M. DeCaro and D.J. Inman, "Stable eigenvalue placement by constrained optimization," *DSS*, vol. 5, no. 4, pp. 191-199, 1990.

[7] C. de Oliveira, "Eigenvalue sensitivity of some transfer function realizations," *IEEE-TAC*, vol. AC-20, no. 2, pp. 276-278, April 1975.

[8] C.A. Desoer, "Perturbation of eigenvalues and eigenvectors of a network," in *Proc. Allerton Conf.*, pp. 8-11, 1967.

[9] N.D. Francis, "Sensitivity analysis in multivariable-system design," *Elect. Lett.*, vol. 3, no. 9, pp. 404-405, Sept. 1967.

[10] J.A. Gibson and M.A. Hamilton-Jenkins, "Modal sensitivity to discretization," *IJSS*, vol. 13, no. 1, pp. 85-91, 1982.

[11] J.-N. Juang, P. Ghaemmaghami, and K.B. Lim, "Eigenvalue and eigenvector derivatives of a nondefective matrix," *JGCD*, vol. 12, no. 4, pp. 480-486, 1989.

[12] J.-N. Juang, K.B. Lim, and J.L. Junkins, "Robust eigensystem assignment for flexible structures," *JGCD*, vol. 12, no. 3, pp. 381-387, 1989.

[13] G. Klein and B.C. Moore, "Eigenvalue-generalized eigenvector assignment with state feedback," *IEEE-TAC*, vol. AC-22, no. 1, pp. 140-141, Feb. 1977.

[14] K.B. Lim and J.-N. Juang, "Eigenvector derivatives of repeated eigenvalues using singular value decomposition," *JGCD*, vol. 12, no. 2, pp. 282-283, 1989.

[15] K.B. Lim and J.L. Junkins, "Robustness optimization of structural and controller parameters," *JGCD*, vol. 12, no. 1, pp. 89-96, 1989.

[16] P.E. Mantey, "Eigenvalue sensitivity and state-variable selection," *IEEE-TAC*, vol. AC-13, pp. 263-269, June 1968.

[17] A.G. Mazko, "Geberalization of Lyapunov's theorem for regions bounded by algebraic and transcendental curves," *SJAIS*, vol. 18, no. 3, pp. 52-58, 1985.

[18] B.C. Moore, "On the flexibility offered by state feedback in multivariable systems beyond closed-loop eigenvalue assignment," *IEEE-TAC*, vol. AC-21, no. 5, pp. 689-692, Oct. 1976.

[19] B.S. Morgan, "Computational procedure for the sensitivity of an eigenvalue," *Elect. Lett.*, vol. 2, p. 197, 1966.

[20] H. Nicholson, "Sensitivity properties of linear systems," *Elect. Lett.*, vol. 3, no. 6, pp. 241-242, June 1967.

[21] _____, "Eigenvalue and state-transition sensitivity of linear systems," *Proc. IEE*, vol. 114, no. 12, pp. 1991-1995, Dec. 1967.

[22] T.J. Owens, "Parametric output feedback control with left eigenstructure insensitivity," *IJSS*, vol. 21, no. 8, pp. 1603-1630, 1990.

[23] P.N. Paraskevopoulos, C.A. Tsonis, and S.G. Tzafestas, "Eigenvalue sensitivity of linear time-invariant control systems with repeated eigenvalues," *IEEE-TAC*, vol. AC-19, no. 5, pp. 610-612, Oct. 1974.

[24] B. Porter, "Assignment of closed-loop eigenvalues by direct method of Liapunov," *IJC*, vol. 10, no. 2, pp. 153-157, 1969.

[25] _____, "Eigenvalue sensitivity of modal control systems to loop-gain variations," *IJC*, vol. 10, no. 2, pp. 159-162, 1969.

[26] L. Qiu and E.J. Davison, "The stability robustness of generalized eigenvalues," *IEEE-TAC*, vol. AC-37, no. 6, pp. 886-891, 1992.

[27] D.C. Reddy, "Sensitivity of an eigenvalue of a multivariable control system," *Elect. Lett.*, vol. 2, p. 446, 1966.

[28] _____, "Eigenfunction sensitivity – an approach to the solution of the tolerance problem," in *Proc. Allerton Conf.*, pp. 12-18, 1967.

[29] _____, "Eigenfunction sensitivity and the parameter variation problem," *IJC*, vol. 9, no. 5, pp. 561-568, 1969.

[30] S.E. Reyer and J.A. Heinen, "Eigenvalue sensitivity of digital filters based on the expanded state model," *CAC*, vol. 9, pp. 53-55, 1981.

[31] J. Rohn, "Stability of interval matrices: The real eigenvalue case," *IEEE-TAC*, vol. AC-37, no. 10, pp. 1604-1605, 1992.

[32] H.H. Rosenbrock, "Sensitivity of an eigenvalue to changes in the matrix," *Elect. Lett.*, vol. 1, p. 278, 1965.

[33] C.A. Sandridge and R.T. Haftka, "Accuracy of eigenvalue derivatives from reduced-order structural models," *JGCD*, vol. 12, no. 6, pp. 822-829, 1989.

[34] A.L. Tits and L. Saydy, "On robust eigenvalue configuration," *IEEE-TCAS*, vol. CAS-38, no. 1, pp. 138-139, 1991.

[35] S.G. Tzafestas, "Eigenvalue controller design of reduced sensitivity," *Proc. IEEE*, vol. 63, no. 7, pp. 1080-1081, July 1975.

[36] _____, "Sensitivity reduction in feedback eigenvalue controller design," *Proc. IEEE*, vol. 63, no. 12, pp. 1723-1725, Dec. 1975.

[37] S.G. Tzafestas and P.N. Paraskevopoulos, "Sensitivity reduction in modal control systems," *JFI*, vol. 298, no. 1, pp. 29-43, July 1974.

[38] S.-S. Wang and W.-G. Lin, "On the analysis of eigenvalue assignment robustness," *IEEE-TAC*, vol. AC-37, no. 10, pp. 1561-1564, 1992.

[39] H.M. Zein El-Din, R.H. Alden, and P.C. Chakravarti, "Second-order eigenvalue sensitivities applied to multivariable control systems," *Proc. IEEE*, vol. 65, no. 2, pp. 277-278, Feb. 1977.

[J] Synthesis Considering Certain Sensitivity Equations Resulted From System Stability

[1] J. Ackermann, "Parameter space design of robust control systems," *IEEE-TAC*, vol. AC-25, pp. 1058-1072, 1980.

[2] _____, "Robustness against sensor failures," *Automatica*, vol. 20, no. 2, pp. 211-215,, 1984.

[3] J. Ackermann and H.Z. Hu, "Robustness of sampled-data control systems with uncertain physical plant parameters," *Automatica*, vol. 27, no. 4, pp. 705-710, 1991.

[4] J. Ackermann, D. Kaesbauer, and R. Muench, "Robust gamma-stability analysis in a plant parameter space," *Automatica*, vol. 27, no. 1, pp. 75-85, 1991.

[5] V.G. Andronov and Y.G. Belousov, "The stability of polynomial optimization problems with linear constraints," *ECy*, vol. 23, no. 4, pp. 97-106, 1985.

[6] P.R. Apkarian, "Structured stability robustness improvement by eigenspace techniques: A hybrid methodology," *JGCD*, vol. 12, no. 2, pp. 162-168, 1989.

[7] A. Arbel and N.K. Gupta, "Robust colocated control for large flexible space structures," *JGCD*, vol. 4, no. 5, pp. 480-486, 1981.

[8] M.J. Ashworth and D.R. Towill, "Sensitivity reduction in ship-manoeuvring performance *via* nonlinear compensation," *Proc. IEE*, vol. 129, Pt. D, no. 6, pp. 227-232, Nov. 1982.

[9] G.M. Bakan, V.V. Volosov, and A.S. Kalita, "Improvement of the quality of functioning of local systems for stabilization of automatic control systems for continuous technological processes," *SJAIS*, vol. 19, no. 1, pp. 85-88, 1986.

[10] A.V. Balakrishnan, "Compensator design for stability enhancement with collocated controllers: Explicit solutions," *IEEE-TAC*, vol. AC-38, no. 3, pp. 505-507, 1993.

[11] M.J. Balas, "Discrete-time stability of continuous-time controller designs for large space structures," *JGCD*, vol. 5, no. 5, pp. 541-543, 1982.

[12] J. Bandoni, H. Rotstein, A. Desages, and J. Romagnoli, "Necessary and sufficient conditions for robust stabilization with fixed order conditions," *CTAT*, vol. 6, no. 1, pp. 75-91, 1990.

[13] K.C. Bhattacharyya, "Stability of a precision attitude determination scheme," *JGCD*, vol. 3, no. 6, pp. 586-587, 1980.

[14] R.K. Cavin III and S.P. Bhattacharyya, "Robust and well-conditioned eigenstructure assignment *via* Sylvester's equation," *JOCAM*, vol. 4, pp. 205-212, 1983.

[15] C.-H. Chang and K.-W. Han, "Gain margins and phase margins for control systems with adjustable parameters," *JGCD*, vol. 13, no. 3, pp. 404-408, 1990.

[16] T.R. Crossley and B. Porter, "High-order eigen-problem sensitivity methods: Theory and application to the design of linear dynamical systems," *IJC*, vol. 10, no. 3, pp. 315-329, 1969.

[17] J.G. Crouch, W.J. Anderson, and D.T. Greenwood, "Eigenvalue errors in the method of weighted residuals," *AIAA J.*, vol. 8, no. 11, pp. 2048-2054, Nov. 1970.

[18] P.L. Dandeno and P. Kundur, "Practical application of eigenvalue techniques in the analysis of power system dynamic stability problems," *CEEJ*, vol. 1, no. 1, pp. 35-46, 1976.

[19] C.A. Desoer, "Perturbations of eigenvalues and eigenvectors of a network," in *Proc. Allerton Conf.*, pp. 8-11, 1967.

[20] J.B. Edwards, "Stability of accelerating repetitive systems," *Proc. IEE*, vol. 130, Pt. D, no. 4, pp. 183-187, July 1983.

[21] L. El Ghaoui, "Fast computation of the largest stability radius for a two-parameter linear system," *IEEE-TAC*, vol. AC-37, no. 7, pp. 1033-1037, 1992.

[22] M.M. Elmetwally and N.D. Rao, "Sensitivity analysis using second method of Lyapunov," *Elect. Lett.*, vol. 7, no. 20, pp. 622-624, Oct. 1971.

[23] A.M. Foss, "Criterion to assess stability of a 'lowest wins' control strategy," *Proc. IEE*, vol. 128, Pt. D, no. 1, pp. 1-8, Jan. 1981.

[24] S.N. Franklin and J. Ackermann, "Robust flight control: A design example," *JGCD*, vol. 4, no. 6, pp. 597-605, 1981.

[25] M. Gopal and J.G. Ghodekar, "Sensitivity reduced design of discrete linear regulator with prescribed degree of stability," *Proc. IEE*, vol. 130, Pt. D, no. 5, pp. 261-266, Sept. 1983.

[26] M. Gopal and P. Pratapachandran, "Sensitivity-reduced optimal discrete linear regulator with prescribed closed-loop eigenvalues," *Proc. IEE*, vol. 132, Pt. D, no. 1, pp. 18-24, Jan. 1985.

[27] T.L. Greenlee, "An eigenvalue sensitivity study of a high temperature gas-cooled reactor system," *IEEE-TAC*, vol. AC-21, no. 3, pp. 422-423, June 1976.

[28] W.M. Greenlee, "Perturbation of eigenvalues with an engineering applications," *JMAA*, vol. 27, pp. 1-20, 1969.

[29] V.V. Grigor'ev, "Synthesis of control functions for variable-parameter systems," *ARC*, vol. 44, no. 2, pp. 189-194, Feb. 1983.

[30] M.J. Grimble, "Design of optimal output regulators using multivariable root loci," *Proc. IEE*, vol. 128, Pt. D, no. 2, pp. 41-49, March 1981.

[31] V. Guruprasada Rau, "Eigenvalue sensitivity method for optimal stabilisation of linear systems," *Elect. Lett.*, vol. 14, no. 3, pp. 59-60, Feb. 1978.

[32] K. Harada and T. Fukao, "On the stability of a 'reaction-diffusion' model in ecological systems," *EEJ*, vol. 98, no. 6, pp. 121-126, 1978.

[33] A. Haraldsdottir, P.T. Kabamba, and A.G. Ulsoy, "Sensitivity reduction by state derivative feedback," *ASME-JDSMC*, vol. 110, pp. 84-93, March 1988.

[34] A. Herrera-Vaillard, J. Paduano, and D. Downing, "Sensitivity analysis of automatic flight control systems using singular-value concepts," *JGCD*, vol. 9, no. 6, pp. 621-626, 1986.

5.12.J. Synthesis Considering Certain Sensitivity Equations

[35] R.D. Hill and M.E. Halpern, "Minimum overshoot design for SISO discrete-time systems," *IEEE-TAC*, vol. AC-38, no. 1, pp. 155-158, 1993.

[36] J.W. Howze and R.K. Cavin, III, "Regulator design with modal insensitivity," *IEEE-TAC*, vol. AC-24, no. 3, pp. 466-469, June 1979. Errata, *ibid.*, vol. AC-26, no. 4, p. 975, Aug. 1981.

[37] J.W. Howze and D. Ohm, "Control system synthesis with response insensitivity," *Automatica*, vol. 19, no. 4, pp. 365-371, 1983.

[38] P.C. Hughes and D.J. McTavish, "Stability of discrete-time attitude control systems," *DSS*, vol. 3, nos. 1 & 2, pp. 1-24, 1988.

[39] Y.-T. Juang and T.-C. Yu, "Robust pole-assignment for structured perturbation systems," *CTAT*, vol. 6, no. 3, pp. 451-461, 1990.

[40] Y. Kamiya, "Multivariable servo-system design technique using sensitivity as a design parameter," *Proc. IEE*, vol. 134, Pt. D, no. 5, pp. 333-337, Sept. 1987.

[41] N.V. Kondrat'eva and G.A. Leonov, "Dynamic stability of synchronous machines with strong control of excitation," *ARC*, vol. 51, no. 6, pp. 768-778, 1990.

[42] V.A. Korneyev, A.A. Melikyan, and I.N. Titovskiy, "A minimax formulation of the stabilization of the glide path of an aircraft in the presence of wind perturbations," *ECy*, vol. 23, no. 5, pp. 74-81, 1985.

[43] R.L. Kosut, H. Salzwedel, and A. Emami-Naeini, "Robust control of flexible spacecraft," *JGCD*, vol. 6, no. 2, pp. 104-111, 1983.

[44] V.V. Kozorez, "Parametric stabilization of superconducting dynamic systems," *SJAIS*, vol. 22, no. 5, pp. 67-71, 1989.

[45] P.D. Krut'ko, "Synthesis of discrete controls from Lyapunov functions," *ECy*, vol. 22, no. 3, pp. 95-102, 1984.

[46] A.V. Kuntsevich and V.M. Kuntsevich, "The 'robust stability' tool system for the analysis of the robust stability of dynamic systems," *SJAIS*, vol. 23, no. 6, pp. 1-6, 1990.

[47] Y.P. Ladikov and M.T. Paguta, "Stability of a thermonuclear reaction with large $-\beta$," *SAC*, vol. 14, no. 5, pp. 61-67, 1981.

[48] V.B. Larin, "Parametrization of the set of stabilizing regulators in a standard synthesis problem," *SJAIS*, vol. 23, no. 2, pp. 21-26, 1990.

[49] A. Lepschy, G.A. Mian, and U. Viaro, "Stability preservation and computational aspects of a newly proposed reduction method," *IEEE-TAC*, vol. AC-33, no. 3, pp. 307-310, March 1988.

[50] T. Li and R.W. Longman, "Stability relationships between gyrostats with free, constant-speed, and speed-controlled rotors," *JGCD*, vol. 5, no. 6, pp. 545-552, 1982.

[51] J.W. Lim and I. Chopra, "Stability sensitivity analysis of a helicopter rotor," *AIAA J.*, vol. 28, no. 6, pp. 1089-1097, June 1990.

[52] M.S. Lukich and D.L. Mingori, "Attitude stability of dual-spin spacecraft with unsymmetrical bodies," *JGCD*, vol. 8, no. 1, pp. 110-117, 1985.

[53] P.E. Mantey, "Eigenvalue sensitivity and state-variable selection," *IEEE-TAC*, vol. AC-13, pp. 263-269, June 1968.

[54] T. Meressi, D. Chen, and B. Paden, "Application of Kharitonov's theorem to mechanical systems," *IEEE-TAC*, vol. AC-38, no. 3, pp. 488-491, 1993.

[55] L.H. Meyer, F.Q. Raines, T.J. Tarn, and S.K. Gupta, "Optimal coordination of aggregate stabilization policy and price controls: A sensitivity analysis," in *Proc. IFAC*, vol. 3 (of the 6th congress), pp. 62.4.1-62.4.8, 1975.

[56] Y. Murotsu and F. Ohba, "Design of control systems considering accuracy requirements," *IEEE-TAC*, vol. AC-17, no. 4, pp. 575-576, Aug. 1972.

[57] F.W. Nesline, Jr., and P. Zarchan, "Digital homing guidance – stability vs performance tradeoffs," *JGCD*, vol. 8, no. 2, pp. 255-261, 1985.

[58] T. Okada, M. Kihara, and M. Ikeda, "Robust control system design synthesis with observers," *JGCD*, vol. 13, no. 2, pp. 337-342, 1990.

[59] T.J. Owens, "Parametric output feedback control with response insensitivity," *IJC*, vol. 48, no. 3, pp. 1213-1239, 1988.

[60] T.J. Owens and J. O'Reilly, "Parametric state-feedback control with response insensitivity," *IJC*, vol. 45, no. 3, pp. 791-809, 1987.

[61] _____, "Parameter state-feedback control for arbitrary eigenvalue assignment with minimum sensitivity," *Proc. IEE*, vol. 136, Pt. D, no. 6, pp. 307-313, Nov. 1989.

[62] J.D. Paduano and D.R. Downing, "Sensitivity analysis of digital flight control systems using singular-value concepts," *JGCD*, vol. 12, no. 3, pp. 297-303, 1989.

[63] L.R. Pujara, "On the parameterization of stabilizing controllers for SISO control systems with applications to aircraft flying qualities," *CTAT*, vol. 6, no. 1, pp. 143-152, 1990.

[64] K. Ramar and V. Gourishankar, "Pole assignment with optimality and minimum eigenvalue sensitivity," *Proc. IEE*, vol. 122, no. 12, pp. 1437-1438, Dec. 1975.

[65] D.C. Reddy, "Eigenfunction sensitivity – an approach to the solution of the tolerance problem," in *Proc. Allerton Conf.*, pp. 12-18, 1967.

[66] E.P. Ryan, G. Leitmann, and M. Corless, "Practical stabilizability of uncertain dynamical systems: Applications to robotic tracking," *JOTA*, vol. 47, no. 2, pp. 235-252, 1985.

[67] Y.I. Samoilenko, "Negative sensitivity of dynamic systems and its utilization for the stabilization of unstable processes," *ARC*, vol. 39, no. 5, pp. 632-639, 1978.

[68] S.L. Shah, D.G. Fisher, and D.E. Seborg, "Eigenvalue invariance to system parameter variations by eigenvector assignment," *IJC*, vol. 26, no. 6, pp. 871-881, 1977.

[69] D.D. Siljak and A. Burzio, "Minimization of sensitivity with stability constraints in linear control systems," in *Proc. Allerton Conf.*, pp. 230-240, 1965.

[70] R.A. Singer, "Selecting state variables to minimize eigenvalue sensitivity of multivariable systems," *Automatica*, vol. 5, pp. 85-93, 1969.

[71] R.E. Skelton and D.A. Wagie, "Minimal root sensitivity in linear systems," *JGCD*, vol. 7, no. 5, pp. 570-574, 1985.

[72] H.M. Soliman and A. Bahgat, "Minimum sensitivity pole placer," *DSS*, vol. 3, nos. 1 & 2, pp. 51-56, 1988.

[73] E.M. Solnechnyi, "Stability and robustness of linear systems with respect to various classes of signals," *ARC*, vol. 40, no. 9, pp. 1278-1286, 1979.

[74] B.V. Ulanov, "Stabilization of nonstationary dynamic plants with unknown parameters without measuring the derivatives of the controlled variable," *ARC*, vol. 51, no. 7, pp. 908-913, 1990.

[75] E.I. Veremei and V.M. Korchanov, "Multiobjective stabilization of a certain class of dynamic systems," *ARC*, vol. 49, no. 9, pp. 1210-1219, 1988.

[76] A. Vicino, A. Tesi, and M. Milanese, "Computation of nonconservative stability perturbation bounds for systems with nonlinearly correlated uncertainties," *IEEE-TAC*, vol. AC-35, no. 7, pp. 835-841, July 1990.

[77] V.V. Zdor, "Definition and assignment of tolerances on the parameters of control systems," *ECy*, vol. 15, pp. 80-87, Nov./Dec. 1977.

[78] M.A. Zohdy, A.A. Abdul-Wahab, N.K. Loh, and J. Liu, "On robust parametric dynamic output feedback," *ASME-JDSMC*, vol. 112, pp. 507-512, Sept. 1990.

CHAPTER SIX

SENSITIVITY REDUCTION AND ROBUSTNESS

6.1. INTRODUCTION

In Chapters Two to Five we devote our attentions primarily to issues regarding parameter-sensitivity analysis. In spite of different *uncertainties* which may exist in a dynamical system, we expect that our controller will ultimately maintain *qualitatively* and/or *quantitatively* the overall system operation as desired. Or it will maintain this operation within a prespecified *class* of such operations which this is really the *key* issue in any robustness methodology. Therefore we must incorporate in control synthesis algorithm any information regarding possible *discrepancies* in system "output", in order to generate new controllers that can stand against the consequences of some possible uncertainties. This incorporation is by no means a simple task, and before we let our expectations exceed our means, we examine some of our earlier assumptions and review the class of problems which currently can be analyzed.

Starting with two generic names: uncertainty and discrepancy, we note that the latter is meant to be the consequence of the former. To discover the nature of uncertainty in any dynamical system is a complicated matter and often is not clear how and when this effect has started. To that end, it is a common practice to approximate system that is subject to uncertainty within certain class of problems which are mathematically manageable. To elucidate this thought as succinctly as possible we refer to Table 6.1-1, where in reality, we only consider the linear time-invariant system, however, most of the following general ideas can be extended to other dynamical systems; and some of the following solutions and techniques in this chapter can be extended to certain class of nonlinear systems. Therefore for the first reading of this table, we concentrate on the classifications of system uncertainty rather than plant itself. These models can be conceptually extended to other systems of interests in different disciplines.

Referring to Table 6.1-1, in *Case 1*, it is assumed that a statistically known signal (indeed the White noise) is acting on the plant (or system) that is otherwise *fixed*. (Here under certain restrictive conditions *small*-parameter variations may be allowed.) This condition is the main premise of standard linear optimal control theory and its solution for feedback controller in the sense of minimizing a

Table 6.1-1. Meeting Uncertainty.

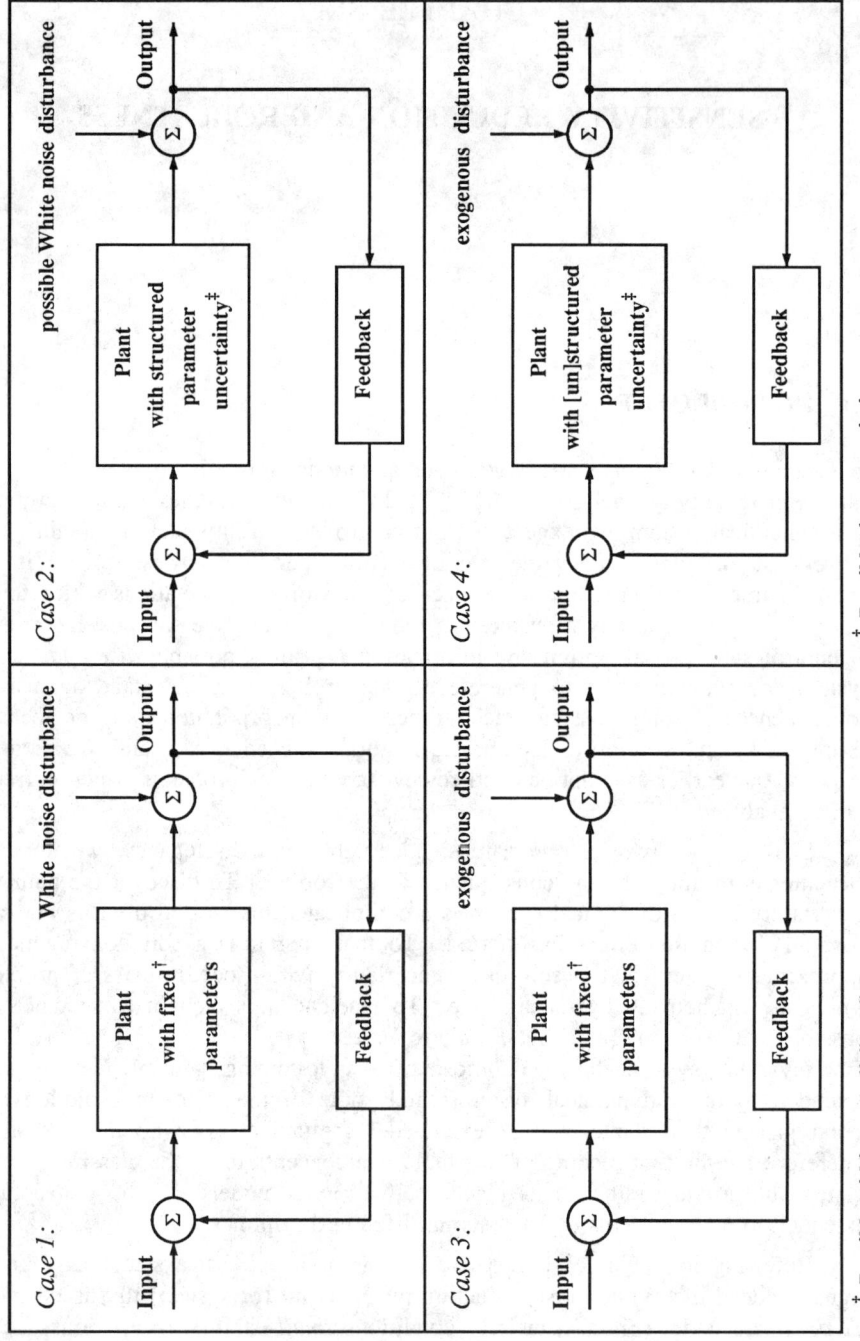

† Possibly within a *small* – "ball".
‡ Possibly *large* – variations.

6.1. Introduction

quadratic performance index is well known [2C. 28, 34 and 35]. So is its robustness property or the lack of it. Thus this case serves only as a point of reference for the subsequent developments of *Cases 2* to *4* in Table 6.1-1 which are the main objectives of our study.

In *Case 2*, uncertainty in system dynamics is partly due to: *(i)* various approximational schemes which is inherent in system identification with all the corresponding noise *etc.;* and *(ii)* some further parameter variations mainly in plant and possibly in controllers, resulted from a list of possible stimuli as described in Section 1.1. However, by and large it is assumed that in this situation the prominent source of uncertainty is that resulted from [large] parameter variations. We again reiterate that in spite of terms such as "parameter variations," we mean *static* changes, i.e., we do not associate neither time-dependency nor any statistical properties to these variations whatsoever. This type of variations is also the same for the remaining two cases in Table 6.1-1. That means we consider parameters to be constant, or they change very "slowly" with respect to time and that is the fundamental assumption in sensitivity theory. The forthcoming analyses are performed within the *static*-parameter variations and based on some possible *projected worst* scenario. Methodologies that are resulted from this school of thought often yield a system which is overdesigned, due to insufficient information about the mechanism of parameter variations. In other words, certain anticipated parameter variations may be weighted more heavily than the actual changes and that exaggeration is always costly, although the anticipated changes may never materialize. On the other hand, to rectify some of these inherent difficulties regarding the static-parameter variations we may turn to a sophisticated control-design scheme (stemming from methodologies in adaptive or stochastic adaptive control theory), but such an utilization soon becomes prohibitive and costly requiring sophisticated hardware. Here by maintaining the static-parameter variations we are trying to make the best out of as little *hardware* as possible, and that is really the essence of this assumption.

Thus in *Case 2*, we mainly consider a dynamical system whose source of uncertainty is its parameter variations resulting in discrepancy in its "output". Thus our task in minimizing the unwanted effects of this discrepancy is to model uncertainty by *a* parameter-sensitivity measure and proceed to study the resulting discrepancy and ways of reducing its undesirable effects. We recall that parameter-sensitivity measures are not unique, however, they are mainly classified into two groups of *small*- and *large*-sensitivity measures.

Again referring to *Case 2*, there are two major approaches to utilize sensitivity theory in design methodologies. The *first* approach is concerned with the qualitative behavior of a particular system property. This approach does not involve direct manipulation of any sensitivity measure, that means, only a sufficient condition which guarantees, for example, system A becomes less sensitive than system B is developed without actually computing any sensitivity functions for these two systems. For instance, it is a well-known fact that the main reason for utilizing a feedback control law is its usefulness in cancelling, to some extent, the effects of disturbances and parameter variations in system output. Therefore to expect that a closed-loop designed system may behave more efficiently than an open-loop designed system, stimulated the idea of comparing these two in the sense of a sensitivity measure. Example 3.6-1, however, shows that, in general, the equivalence with respect to [certain] sensitivity measure between an open-loop and a closed-loop designs does not exist, therefore we should seek conditions which guarantee the superiority of one design versus another well in advance (cf., Sections 6.2 and 6.3). This type of design may serve adequately for a number of problems. In this classical approach no sensitivity optimization is performed, although quite a few attempts were made. However, due to certain technicality the optimization approach has not been successful until recently and in the context of the H^∞-sensitivity optimization as described in the following.

Secondly, and again referring to *Case 2,* there are many situations wherein the previous [comparison] approach *alone* is not enough to satisfy design requirements. Here, again we incorporate an appropriate parametric-sensitivity function in design process. However, since there is no unique best sensitivity measure to represent all kinds of systems, we use trajectory-sensitivity functions as system-sensitivity measure in order to continue our analysis. There are a number of other choices for these parametric-sensitivity measures which are used by different authors in § 6.12.*F*. Upon choosing this measure, a desired functional of various system quantities are selected as the performance index. By developing a set of new approaches, we proceed to extremize the corresponding performance index which involves our system-sensitivity measure. In the final analysis we have a number of algorithms that yield optimally sensitive systems.

It is, however, very instructive that given any dynamical system subject to parametric uncertainty alone, its sensitivity be evaluated in terms of the above two general design approaches as well as different sensitivity measures, before a final decision on how to reduce its sensitivity will be made. For a case study in which several different measures of system sensitivity are reviewed, we refer to a comprehensive research reported in [1*B*. 15]. Although this is a 1970's research project, its premise is very current. Here it is shown that among all possible and [then] known system-sensitivity measures, which are applied to *a given theme example* (the model of a C5-Galaxy airplane), the trajectory- (or response-variation) sensitivity measure results in the best performance. This conclusion, of course, does not necessarily hold in every applications.

Before closing this case we note that, in the literature various names are attributed to systems that exhibit certain degrees of improvement against parameter variations. For instance, in the early stage of sensitivity theory, the word *insensitive* was often used to describe a system that has improved with respect to *a* specific *small*-sensitivity measure. In other words, insensitivity means to construct a class of systems, although *small,* which behaves similarly. We note that an insensitive system does not necessarily mean a zero-sensitive system. Nowadays, the word *robust* is used to define a class of systems that can tolerate *large*-parameter variations, or *large*-external disturbances, without deteriorating its performance. The robustness property, when exists, is a remarkable system property, but tests are needed to assure whether the correct assumption regarding system variations is met or not, before a system is called robust.

6.1. Introduction

Sections 6.2 to 6.8 provide procedures to design controllers for systems subject to both *small-* and *large-*parameter variations. In these cases we assume that almost no White noise disturbance exists except the assumptions on initial conditions. These results can be easily extended to both nonlinear systems with quadratic performance and systems subject to White noise disturbances.

In *Case 3* of Table 6.1-1, we consider the situation when an exogenous signal is acting on the system. The primary approach to design controller that can handle this signal (which is no longer the White noise), is the newly developed technique in the context of H^∞-sensitivity optimization methodologies. In the preceding case when we discuss the classical sensitivity-comparison scheme we point out that no optimization is performed. Zames has formulated this sensitivity-reduction problem with feedback as an optimization problem in his 1981-pioneering work [*H2.* 66]. In Section 6.9, we present the novel solution provided by Doyle, Glover, Khargonekar and Francis [*H3.* 6] for this type of problems when the underlying system is represented by its state-space model. This result considers primarily a system of type *Case 3,* i.e., a plant with fixed parameters and a bounded exogenous disturbance.

As another point of view to consider *Case 3,* we note that during the last several decades some activities are reported in the Russian literature regarding *the theory of invariance.* The theory of invariance (or synonymously: The theory of compensation for *external* disturbances) is a branch of control theory which discusses various existing techniques for compensating the effects of *external* disturbances on plant operation. The existing theory mostly covers cases wherein an additive disturbance is entering into the system and the objective is to reduce its effects on system operation. For the case of parameter variations, however, the existing theory must be modified in order to reduce the effects of these parameter changes (that should be called *internal* disturbances) in system operations, although parameter changes do not always occur because of *external* disturbances. Here the dynamical system is represented by a set of *ODE's* and the *external* input is modeled as an additional-input vector for this system, i.e., the *external* input is modeled as some sort of additive noise in the system. Therefore the main emphasis of invariance theory is to study the effects of these input components in system transient response. Thus from this point of view, it is clear that this theory is a synthesis procedure for control system which compensates for variations in its *forced* components (disturbance). Also there is a difference between invariance in terms of forced components and system initial conditions. Development of any technique in the theory of invariance requires that certain method of analysis (in the form of incremental changes) for system *output* (with respect to any given element) be established first; followed by the synthesis procedure to design system such that these variations become in some defined manner acceptable. If these objectives are perceptively analyzed, then the relation between sensitivity and invariance theories becomes evident. In the above first step, derivation of incremental changes in system output due to incremental changes in any given system quantity becomes the same as the main thrust of sensitivity theory. The next step in above involves imposition of conditions or bounds on incremental changes in certain *chosen* quantities such that the overall system performance becomes acceptable according to a given criterion. This concept in effect is parallel to that of robust system. Thus we may consider our design stemming from theory of invariance complements that resulted from robustness analysis.

In invariance theory the time variations of disturbances are not so important as are the maximum of their absolute values (bounded L_∞-norms) and their corresponding effects on system operation. Therefore this theory should not be confused with other branches of control theory, where parameter variations are modeled by time-varying probability density functions. To achieve absolute (i.e., zero) invariance of a system *quantity* (or for all practical purposes within a small quantity $\varepsilon = 0^+$), with respect to a given varying system element, is the goal of the design algorithm, however, under practical circumstances system variables specially control variables may change and in order to keep them within specified bounds we must perform additional optimization tasks. Thus when the system is not absolutely invariant and deviations of controlled variables are unavoidable, these unwanted responses must be eliminated if the actual control capabilities are limited. This situation is a typical problem in optimal control and/or calculus of variations. Therefore in synthesis of high-performance control systems the principles of both sensitivity or invariance theory and optimal control theory must be jointly considered. Some introductory aspects of these issues are reviewed in Section 6.10.

In *Case 4,* of Table 6.1-1, we consider a situation when both plant variations (possibly unstructured) and an external signal acting on the system exist. This situation is perhaps the most challenging of all, and the underlying problem soon becomes highly nonlinear and the preceding two cases need to be reexamined before applying to this case. There are alternative methods to handle this situation. One such method which considers both types of uncertainties is stemming from the Lyapunov design technique. This Lyapunov method discussed in Section 6.11 is intended to look at these situations. Section 6.11 ends with additional comments on disturbance rejection.

In summary, this chapter provides a number of choices for final design of a highly reliable controller which meets many challenging obstacles. It is instructive, however, to study differences among various cases in Table 6.1-1 before proceeding further. We meet systems similar to: *Case 2* in Sections 6.2 to 6.8; *Case 3* in Sections 6.9 and 6.10; and *Case 4* in Sections 6.11 and 6.10 from our point of view. Our final thought is presented in Remark 6.11-2.

6.2. TRAJECTORY – SENSITIVITY COMPARISON AND CERTAIN FEEDBACK PROPERTIES OF LINEAR SYSTEMS

6.2.1. Introduction

In this section we present the first of a series of methods in order to provide sufficient conditions which guarantee to affect and indeed to lower the sensitivity

6.2. Trajectory – Sensitivity Comparison, Feedback Properties

performance (by that we mean the actual numerical value of a chosen sensitivity measure) in the underlying control design. Since lowering is a relative term, here we compare the sensitivity performance of a closed-loop design with that of a nominally equivalent open-loop design. This comparison allows us to seek sufficient conditions such that the closed-loop design has a lower sensitivity performance than its corresponding nominally equivalent open-loop design. This "improvement," however, does not imply that the final design stemming from this approach cannot be further improved upon. Indeed there are many cases which the final outcome from this sort of improvement is not completely satisfactory and we must seek other approach to lower the sensitivity performance. Thus we should explicitly point out that the sensitivity measure which is resulted from this study is only an aspect of this theory and it is not the final chapter on the difficult choice of selecting an appropriate sensitivity measure for a dynamical system.

It is a well-known fact that the main reason for utilizing a feedback control law is the usefulness of a closed-loop design in cancelling, to some extent, the effects of disturbances and system-parameter variations in system output. It is also well known (cf., Proposition 3.2-2 and the subsequent discussions) that the response or output of an open-loop design is very susceptible to disturbances and system-parameter variations. On the other hand, from Example 3.6-1 we know that in general the equivalence in terms of sensitivity performance (the actual value of a sensitivity measure) between these two open-loop and closed-loop designs only exist under a very severe and an unattainable requirement. Since these two systems exhibit different sensitivity performance, we should seek conditions which guarantee the superiority (i.e., lower sensitivity performance) of one design versus another one, well in advance and without direct computations of trajectory-sensitivity measures. Eliminating this computational difficulty was another motivating factor for the development of this concept. Furthermore, the closed-loop design is always more expensive and requiring more hardware than the corresponding nominally equivalent open-loop design. Therefore when we are making the extra efforts to build a feedback controller, we must get as much improvement from that design as possible. Based upon these remarks and the fact that a closed-loop designed system may behave more efficiently than an open-loop designed system, the idea of comparing these two designs in the sense of *a* sensitivity measure was evolved and became an important aspect of this theory and indeed is one of the most widely used sensitivity measures for all dynamical systems.

Historically speaking, Dorato [B. 2] in 1963, and Cruz and Perkins [A. 7] in the early 1964, proposed a new treatment for the analysis of system-sensitivity measure. Both reports consider two equal dimension systems: (a) an open-loop design, and (b) a closed-loop design. Here the term closed-loop refers to the case where system input is implemented as a function of system state, otherwise it is called open-loop or in some cases semi-closed loop [A. 8]. If the above two systems produce an equal state when we apply two equal inputs, then they are called nominally-equivalent systems (cf., Definition 2.2-9). Because of this flexible definition, the concept of sensitivity comparison can be generalized to a number of other criteria, such as: terminal conditions comparison of two designs or output comparison of two designs (with respect to certain external disturbances), *etc.*. Dorato suggested that we compare the variations in the quadratic-performance indices of open-loop and closed-loop designs, while Cruz and Perkins suggested to compare the variations in the trajectories of these two nominally-equivalent systems. In this section we concentrate on the latter and we postpone discussions on the former topic to the next section. As we develop this measure in the following, it turns out that the sensitivity-comparison matrix defined for the trajectory of a multi-input, multi-output, multi-parameter system becomes a natural generalization of the normalized-sensitivity function (or Bode's sensitivity function, Definition 2.2-20) defined for a system with one input, one output and a single parameter. Due to this special appeal of the trajectory-sensitivity comparison concept, this area of research has been continuously the focus many researchers since its introduction and is the main sensitivity measure for the H^∞-optimization techniques.

In general, the main reason for developing a trajectory-sensitivity comparison criterion is to have a set of sufficient conditions such that a closed-loop design will definitely function less sensitively than its corresponding nominally equivalent open-loop design, and without direct computations of various trajectory-sensitivity functions. However, since an extensive effort has been already made in Sections 3.2 and 3.3 to derive several different trajectory-sensitivity functions, in the following two ways of comparing these sensitivity functions are considered: *(i)* with respect to plant parameters, assuming all the controllers are invariant, which is similar to the original work of Cruz and Perkins; and *(ii)* with respect to all the parameters of plant and controllers. However, in both cases we soon realize that the use of these sensitivity functions has only a catalyst effect, in order to generate the final criterion. With this thought in mind, the sensitivity comparison with respect to both cases *(i)* and *(ii)* are presented. In particular, we note that since a closed-loop system has more controller components than an open-loop system, it is natural to expect that these controllers exhibit more sensitivity than those of an equivalent open-loop system and that is the reason for emphasizing case *(ii)*. Certain pertinent references are given in § 6.12.*A*. and § 6.12.*B*.

6.2.2. Analyses and Algorithms

Consider two nominally equivalent open-loop (shown by o) and closed-loop (shown by c) systems described by

$$\dot{x}_o(t) = Ax_o(t) + Bu_o(t), \qquad (6.2\text{-}1)$$

$$\dot{x}_c(t) = Ax_c(t) + Bu_c(t). \qquad (6.2\text{-}2)$$

The main task ahead of us is to relate an appropriate sensitivity measure of each of these two designs, in order to establish a set of sufficient conditions which guarantees the superiority of one design versus another one.

6.2. Trajectory – Sensitivity Comparison, Feedback Properties

PROPOSITION 6.2-1: For the two nominally equivalent systems described by (6.2-1) and (6.2-2) we have

$$\Delta x_c(t) = \Delta x_o(t) + \int_0^t \exp[(A + \Delta A)(t - \tau)](B + \Delta B)\Delta u_c(\tau)d\tau; \qquad (6.2\text{-}3)$$

where $\Delta x_c(t)$ and $\Delta x_o(t)$ are the closed-loop and the open-loop system-trajectory variations, respectively. Recall that the two nominally equivalent systems are defined by $x_c(t) \equiv x_o(t) \equiv x(t)$ and $u_c(t) \equiv u_o(t) \equiv u(t)$.

PROOF: Suppose that $A \rightarrow A + \Delta A$, $B \rightarrow B + \Delta B$, then $x_c(t) \rightarrow x_c(t) + \Delta x_c(t)$, and $u_c(t) \rightarrow u_c(t) + \Delta u_c(t)$, where $\Delta x_c(0) = 0$. From Propositions 3.2-1 & 4 we have

$$\Delta x_o(t) = \int_0^t \exp[(A + \Delta A)(t - \tau)]\{\Delta A x_o(\tau) + \Delta B u_o(\tau)\}d\tau, \qquad (6.2\text{-}4)$$

$$\Delta x_c(t) = \int_0^t \exp[(A + \Delta A)(t - \tau)]\{\Delta A x_c(\tau) + \Delta B u_c(\tau)$$
$$+ (B + \Delta B)\Delta u_c(\tau)\}d\tau. \qquad (6.2\text{-}5)$$

By the linearity of the integral operator and (6.2-4), we can write (6.2-5) in the form of (6.2-3). ∎

From (6.2-5) it is evident that in order to evaluate $\Delta x_c(t)$, $\Delta u_c(t)$ must be known in advance; i.e., changes in $u_c(t)$ caused by changes in system parameters.

PROPOSITION 6.2-2: Let $H \triangleq (\eta_{ij}) = (a_{kl}) \cup (b_{pq})$ and $f_{ul}(t) = \sum_{i,j} \eta_{ij}\dfrac{\partial u_c(t)}{\partial \eta_{ij}}$, then we can show that for the two nominally-equivalent systems described by (6.2-1) and (6.2-2) we have

$$f_{c1}(t) = f_{o1}(t) + \int_0^t \exp[A(t - \tau)]Bf_{ul}(\tau)d\tau. \qquad (6.2\text{-}6)$$

Here $f_{c1}(t)$ and $f_{o1}(t)$ are the two corresponding system – *TSF's*.

PROOF: Because the corresponding system – *TSF's* for these two designs are (cf., (3.2-43) and (3.2-65)), respectively,

$$f_{o1}(t) = \int_0^t \exp[A(t - \tau)]\{Ax(\tau) + Bu(\tau)\}d\tau, \qquad (6.2\text{-}7)$$

and

$$f_{c1}(t) \triangleq \sum_{i,j} \eta_{ij}\frac{\partial x(t)}{\partial \eta_{ij}} = \int_0^t \exp[A(t - \tau)]\{Ax(\tau) + Bu(\tau) + Bf_{ul}(\tau)\}d\tau; \qquad (6.2\text{-}8)$$

therefore the conclusion (6.2-6) follows immediately. ∎

Equation (6.2-6) is a general integral equation for an arbitrary input function $u(t)$ and its *TSF*. Therefore at this stage due to the lack of sufficient information about $f_{ul}(t)$ we cannot make any definitive assertion. However, it is worth mentioning that $f_{o1}(t) = f_{c1}(t)$, if and only if $Bf_{ul}(t) \equiv 0$. The latter does not occur for a system whose B is of the maximum rank. Thus in a single input phase-variable canonical form, *per se,* no choice of input-vector will produce a nominally equivalent open-loop system with the same sensitivity performance as that of a closed-loop system. In fact, it is shown that the closed-loop system is always less sensitive than its nominally equivalent open-loop system [A. 8].

■

We can specialize the above results using the following general linear feedback control law

$$u(t) = -Kx(t) + Gy(t), \qquad (6.2\text{-}9)$$

where G and K are gain matrices of appropriate dimensions, and $y(t) \in R^v$ is a component of the input vector which is not disturbed by plant parameters.

CASE I: Let $y(t) = 0$ in (6.2-9), and choose $H \triangleq (\eta_{ij}) = A \cup B$, then on the one hand, the system – *TSF* for the closed-loop system of (6.2-2) and with respect to plant parameters A and B (with $u_c(t) = -Kx_c(t)$) becomes

$$f_{c1}(t) \triangleq \sum_{i,j} \eta_{ij} \frac{\partial x_c(t)}{\partial \eta_{ij}} = \int_0^t \exp[(A - BK)(t - \tau)](A - BK)x_c(\tau)d\tau. \qquad (6.2\text{-}10)$$

On the other hand, because $u_c(t) = -Kx_c(t)$, with $H \triangleq (\eta_{ij}) = A \cup B$ (i.e., we assume that K is fixed), $f_{u1}(t) = -Kf_{c1}(t)$ and substituting this $f_{u1}(t)$ into (6.2-6) results in

$$f_{c1}(t) = f_{o1}(t) - \int_0^t \exp[A(t - \tau)]BKf_{c1}(\tau)d\tau. \qquad (6.2\text{-}11)$$

Of course, the above two equations for $f_{c1}(t)$ are equivalent. Since $f_{o1}(t)$, given in (6.2-7), does not vanish identically (cf., Proposition 3.2-2), it is clear from (6.2-6) that $f_{c1}(t)$ does not vanish either. In other words, no linear input can be found to make the right-hand side of (6.2-8) identically zero. Thus this system – *TSF* of the closed-loop system is *again* a reasonable qualitative sensitivity measure for our system.

Now, taking the Laplace transform of both sides of (6.2-11) yields

$$f_{c1}(s) = [I + (sI - A)^{-1}BK]^{-1}f_{o1}(s) \triangleq S_c(s)f_{o1}(s). \qquad (6.2\text{-}12)$$

Here the matrix $S_c(s)$, which relates *a* sensitivity-measure of a closed-loop design to that of an open-loop design, is called *the sensitivity comparison matrix* and is similar to that widely used in the literature. The $S_c(s)$ resembles Bode's sensitivity function defined for the transmission between input and output of a single-loop single-parameter system, and this likeness can be interpreted as the generalization of Bode's sensitivity function, if one so chooses.

In order that the previous closed-loop system behaves less sensitively than a nominally equivalent open-loop system we must compare certain norms of these two sensitivity measures. For example, given $Q = Q^T > 0$, we may choose the following norm

6.2. Trajectory – Sensitivity Comparison, Feedback Properties

$$\int_0^t f_{c1}^T(\tau) Q f_{c1}(\tau) d\tau < \int_0^t f_{o1}^T(\tau) Q f_{o1}(\tau) d\tau, \tag{6.2-13}$$

as a way of establishing a sufficient condition for the superiority of the closed-loop design versus the open-loop design. Here on using the Parseval's identity and with $t = \infty$, (6.2-13) is equivalent to

$$\int_{-\infty}^{+\infty} f_{c1}^T(-j\omega) Q f_{c1}(j\omega) d\omega < \int_{-\infty}^{+\infty} f_{o1}^T(-j\omega) Q f_{o1}(j\omega) d\omega. \tag{6.2-14}$$

On substituting (6.2-12) (with $s = j\omega$) for $f_{c1}(j\omega)$ into (6.2-14), results in

$$\int_{-\infty}^{+\infty} f_{o1}^T(-j\omega)[S_c^T(-j\omega) Q S_c(j\omega) - Q] f_{o1}(j\omega) d\omega < 0. \tag{6.2-15}$$

ALGORITHM 6.2-1 (Sensitivity Comparison – Plant Parameters): The previous inequality holds if and only if the integrand is a negative-definite matrix for a large band of frequencies, i.e.,

$$S_c^T(-j\omega) Q S_c(j\omega) - Q < 0, \tag{6.2-16}$$

where

$$S_c(s) \triangleq [I + (sI - A)^{-1} BK]^{-1}. \tag{6.2-17}$$

In other words, if we select the controller gain K in (6.2-17) such that the corresponding $S_c(j\omega)$ makes (6.2-16) hold for a large band of frequencies, then we have accomplished the design objective. Also it should be noted that the system – TSF does not explicitly appear in this algorithm.

∎

Note that in a typical (6.2-16), as pointed out by Kreindler [A. 22], $Q = I$ is *not* necessarily the best choice.

CASE II: Consider the system described by (6.2-2) and (6.2-9) with $y(t) = 0$. Let $H' \triangleq (\eta'_{ij}) = A \cup B \cup K$, then the system – TSF in terms of this H' is given by (3.2-74) which is repeated below for convenience

$$\dot{f}'_{c1} = (A - BK)f'_{c1} + (A - 2BK)x_c(t), \quad f'_{c1}(0) = 0. \tag{6.2-18}$$

On substituting $u_c(t) = -Kx_c(t)$ into (6.2-2), $x_c(t)$ satisfies the following differential equation

$$\dot{x}_c(t) = (A - BK)x_c(t), \quad x_c(0) = x_o \neq 0. \tag{6.2-19}$$

Also on using $u_c(t) = -Kx_c(t)$ with $H' \triangleq [\eta'_{ij}] = A \cup B \cup K$, $f_{u1} \triangleq \sum_{i,j} \eta'_{ij} \frac{\partial u_c(t)}{\partial \eta'_{ij}}$

$= -Kf'_{c1}(t) - Kx_c(t)$, and substituting this $f_{u1}(t)$ into (6.2-8), results in

$$f'_{c1}(t) = f'_{o1}(t) - \int_0^t \exp[A(t - \tau)] BK\{f'_{c1}(\tau) + x_c(\tau)\} d\tau. \tag{6.2-20}$$

Equation (6.2-20) (the same as (6.2-18)) can be written as

$$f'_{c1}(t) = \int_0^t \exp[(A - BK)(t - \tau)](A - 2BK)x(\tau) d\tau, \tag{6.2-21}$$

with

$$f'_{o1}(t) = \int_0^t \exp[A(t - \tau)]\{Ax(\tau) + Bu(\tau)\}d\tau. \tag{6.2-22}$$

Taking the Laplace transform from both sides of (6.2-21) and (6.2-22) yields

$$f'_{c1}(s) = [sI - (A - BK)]^{-1}(A - 2BK)x(s), \tag{6.2-23}$$

and

$$f'_{o1}(s) = (sI - A)^{-1}(A - BK)x(s). \tag{6.2-24}$$

If K is chosen such that $A - BK$ is completely stable (i.e., all of the eigenvalues of $A - BK$ have negative real parts), then $(A - BK)^{-1}$ exists and

$$f'_{c1}(s) = (sI - A + BK)^{-1}(A - 2BK)(A - BK)^{-1}(sI - A)f'_{o1}(s)$$

$$= [I + (sI - A)^{-1}BK]^{-1}[I - (sI - A)^{-1}BK(A - BK)^{-1}(sI - A)]f'_{o1}(s)$$

$$\triangleq S_c(s)W(s, K)f'_{o1}(s)$$

$$\triangleq S_{ac}(s)f'_{o1}(s). \tag{6.2-25}$$

We call the $S_{ac}(s) \triangleq S_c(s)W(s, K)$ an *adaptive sensitivity comparison matrix*, which is equal to the previous nonadaptive sensitivity comparison matrix $S_c(s)$, multiplied or weighted by $W(s, K)$ that is the contribution of the non-fixed controller gain K.

ALGORITHM 6.2-2 (Sensitivity Comparison – Plant and Controllers Parameters): The sufficient condition for a closed-loop design behaves less sensitively than a nominally equivalent open-loop system, under the same set of conditions that led to Algorithm 6.2-1, and for a large band of frequencies is as follows.

$$S_{ac}^T(-j\omega) Q S_{ac}(j\omega) - Q < 0. \tag{6.2-26}$$

Condition (6.2-26) provides certain design guidance for choosing the controller gain K such that $A - BK$ becomes completely stable while (6.2-26) holds. The implication of this statement is: Among all possible choices for stabilizing matrix A such that $A - BK$ becomes completely stable (if the pair (A, B) is completely controllable and $x(t)$ is reconstructible), choose the one which satisfies (6.2-26) over the widest range of frequencies.

■

Since both open-loop and closed-loop designs under the above two cases exhibit nonzero trajectory-sensitivity measures, it is important to compare these designs with respect to the previous or any other equivalent sensitivity measures, in order to select the best possible design within this frame of reference. Each one of the preceding two algorithms provides a procedure to guarantee a lower-sensitivity performance for the corresponding closed-loop design versus that of the open-loop design. But this sensitivity measure has also room for improvement. Indeed the optimization of the H_∞–norm of this measure is the topic of an extensive research area as we discuss in Section 6.9.

6.2. Trajectory – Sensitivity Comparison, Feedback Properties

REMARK 6.2-1: A sensitivity comparison matrix can be developed for comparing the effect of any particular physical change in two separate designs. The following case is one such application.

Consider the closed-loop system-trajectory variation given in (6.2-3). Suppose that $u_c(t) = -Kx_c(t)$. Then (6.2-3), on substituting for $\Delta u_c(t) = -K\Delta x_c(t)$ and considering (6.2-4), becomes

$$\Delta x_c(t) = \Delta x_o(t) - \int_0^t \exp[(A + \Delta A)(t-\tau)](B + \Delta B)K\Delta x_c(\tau)d\tau. \qquad (6.2-27)$$

Taking the Laplace transform from both sides of (6.2-27) yields

$$\Delta x_c(s) = [I + (sI - A - \Delta A)^{-1}(B + \Delta B)K]^{-1}\Delta x_o(s)$$

$$\triangleq S_c'(s)x_o(s), \qquad (6.2-28)$$

where $S_c'(s)$ is the sensitivity comparison matrix *similar* to that given first by Cruz and Perkins [A. 7]. Establishing a criterion such that

$$\int_{-\infty}^{+\infty} \Delta x_c^T(-j\omega) Q \Delta x_c(j\omega)d\omega < \int_{-\infty}^{+\infty} \Delta x_o^T(-j\omega) Q \Delta x_o(j\omega)d\omega, \qquad (6.2-29)$$

results in a similar condition on $S_c'(j\omega)$ as that of Algorithm 6.2-1. Here the closed-loop system exhibits less sensitivity performance than the corresponding nominally equivalent open-loop system if we choose K such that over a large band of frequencies

$$S_c'^T(-j\omega) Q S_c'(j\omega) - Q < 0. \qquad (6.2-30)$$

COMMENT 6.2-1: As is evident in (6.2-28), and this has been the case in the original work of Cruz and Perkins [A. 7], the terms ΔA and ΔB are appearing explicitly in this expression, which reflects the fact that this measure of sensitivity is for *large*-parameter variations. However, most researchers have considered both ΔA and ΔB to be *small* or almost zero. This assumption is acceptable in its own place, wherein we consider only the *small*-parameter variations. If both ΔA and ΔB are set equal to zero, then $S_c' \to S_c$, and in effect we are considering the comparison with respect to the system – TSF. Under this situation it is inappropriate to claim that the ultimate design, whose sensitivity measure is based on the total unnormalized *small*-sensitivity functions, is *robust* in the sense of the *large*-parametric sensitivity measure. These are important issues that we should be fully aware of, in order to make a meaningful use of these results. ∎

We must also point out that since the introduction of this sensitivity measure, many researchers have generalized this concept of comparison to the abstract mathematical spaces, in order to expand this analysis to a broader class of systems such as nonlinear and distributed parameter systems. Here we may cite W.A. Porter as one of the most prominent contributors to this aspect of sensitivity comparison problems [A. 35 & 36]. For additional information on these matters we refer to § 6.12.*L* to § 6.12.*O*.

6.3. PERFORMANCE – INDEX SENSITIVITY COMPARISON

Conventionally every system is designed according to a given criterion, and presumably the system operation is fulfilling its *optimality* index. But realistically because of uncertainty about plant parameters, system measurements, also the inherent uncertainties which exist in the choice for a proper mathematical model of the system: both for the controlled system and the controllers, the system performance does not always follow its prespecified pattern of operation. This known fact is the main concern of sensitivity analysis, and in particular; the variations in system performance index *(PI)* with respect to plant and/or controller parameter variations. However, the major concern about sensitivity of *PI* with respect to the parameter variations is to avoid the inverse optimal problem (as cited in [2D. 5], and we do not plan to discuss this issue any further). Also the sensitivity analysis of a general and perhaps a nonlinear integral *PI,* is not the purpose of the following discussion. Even the sensitivity of a quadratic *PI,* as used frequently in the optimal control theory is deferred to Section 6.4. What is discussed in this section is only a concise account of the *PI*–sensitivity comparison which is given only for the sake of its historical merits.

Historically speaking, Dorato [B. 2] was the first to bring to the attentions of researchers the design potential in the *PI*–sensitivity with respect to plant parameters. Dorato developed certain qualitative *PI*–sensitivity functions, and proposed comparing the *PI*–sensitivity function of an open-loop design with that of a closed-loop design, in order to make sure that the latter is *better* than the former (as it was believed to be so). In other words, this comparison became a design criterion; i.e., to find out precisely when a closed-loop system is less sensitive than its nominally equivalent open-loop system in the sense of certain functional (generally speaking an Euclidean norm) of the corresponding *PI*–sensitivity functions. Surprisingly, in the process of using Dorato's suggestion for a linear system with a quadratic *PI,* Pagurek [B. 8] realized that indeed these two quantities are equivalent. Later Witsenhausen [B. 14] proved that Pagurek's result holds for a more abstract system than that stated in [B. 9]. Also the Pagurek-Witsenhausen's result evolves as a special case of *PI*–expansion in an optimal adaptive control given by Werner and Cruz [C. 31]. Among many reports, the paper by Courtin and Rootenberg [B. 1] shows that the sensitivity of the performance indices (integral of a quadratic norm) in two nominally equivalent designs, as stated by Dorato and under certain conditions, are indeed equal with respect to the first-order sensitivity function. In this short paper, analysis of *small* sensitivity of *PI* in the optimal control problem with initial and final conditions, free end time, with terminal penalty appearing in the *PI* is performed. From this result of [B. 1], it can be easily seen that the *PI*–sensitivity cannot be disassociated from the system trajectory sensitivity. Also, "when based on performance index consideration, the concept of 'insensitivity' may turn out to be an illusion [B. 1]." For further details and a stronger definition of equally-sensitive system the interested reader may consult Courtin and Rootenberg [B. 1].

We note that the equality of the *PI*–sensitivity functions for an open-loop and a closed-loop system (which is, indeed, the consequence of *PI* being stationary with respect to variations in input) does not hold for the higher-order sensitivity functions [*B*. 14]. In general *sensitivity in the large* (Definition 2.2-17) of an open-loop design is greater than or equal to that in a closed-loop design as an example cited by Sinha and Atluri [*B*. 11] shows. It is also to be emphasized that a closed-loop design is less sensitive than an open-loop design based on the system trajectory and not on the system – *PI*.

6.4. OPTIMALITY – INDEX SENSITIVITY:

ANALYSIS AND SYNTHESIS

6.4.1. Introduction

The perfect knowledge of system model is an essential requirement for an *optimal* control of a dynamical system. In fact this requirement becomes more stringent than before, for practical cases in which controllers are generated using state-feedback. In general, the system state must be measured and reconstructed for implementation of the control law. On the other hand, the performance index *(PI)* of an *optimal* control system, which is a function of system parameters, is affected by various uncertainties that may exist in a dynamical system resulting in discrepancies in its output. Therefore the overall discrepancies, will cause a substantial deviation for the system – *PI*. Hence, it is natural to study sensitivity of the *PI* in an *optimal* design with respect to parameters of controlled system and/or controllers [1*B*. 27]. However, it is clear that no specific equation can be given in this matter, unless the *PI* is selected first.

We note that there are several phases in studying the *PI*–variation in a general *optimal* control design. For example, in many practical problems, especially in an open-loop design, it is not always possible to have the exact input function which is required in order to satisfy a given *PI*. Therefore it is very realistic in this type of problems to study the *PI* – or more desirable than that to study system trajectory-variation with respect to system-input variations. However, in order to maintain the mathematical sophistication at the current level (which avoids the use of functional analysis) this kind of sensitivity analysis is not considered herein. But, for a simpler version than that requiring advanced mathematics, the interested reader may refer to Belanger [*F*. 1], and Howard and Rekasius [*F*. 39].

In the following several sensitivity equations for a quadratic – *PI* are derived, in order to illustrate the procedure and its potential applications in quadratic *optimal* control theory. Thus from now on by *optimal* we mean optimality in the sense of a quadratic – *PI*.

6.4.2. PI – Small – Sensitivity Analysis

In general, the *PI* – small-sensitivity analysis is the same as small-sensitivity analysis of any general function. However, the importance and usefulness of this subject makes it necessary to study that independently. In the following an integral of a quadratic expression of system-state and system-input vectors is chosen as our *PI*. Then *small*-sensitivity functions of this *PI* with respect to the plant parameters are developed. In the next section, this analysis is extended to generate an upper bound for *PI*, with respect to the *largest* perturbations of system parameters. The following procedures are simple, and are dependent on the applications of some trace differentiation rules.

Consider a *DCLTI* – *GS* described by

$$\dot{x}(t) = Ax(t) + Bu(t), \quad x(0) = x_o, \qquad (6.4-1)$$

with the following *PI*

$$J = \min \int_0^\infty [x^T(t)W_0 x(t) + u^T(t)Wu(t)]dt, \qquad (6.4-2)$$

where $W_0 = W_0^T \geq 0$, and $W = W^T > 0$. It is well known that the optimal input is $u^o(t) = -W^{-1}B^T P x(t) \triangleq -K^o x(t)$, where $P = P^T > 0$ is the unique solution of the Riccati algebraic-matrix equation $A^T P + PA + W_0 - PBW^{-1}B^T P = 0$. The optimal (minimal) cost is $J^o = x_o^T P x_o$. Substituting $K^o = W^{-1}B^T P$ into this Riccati algebraic equation (and upon some manipulations) results in the following Lyapunov equation for P,

$$(A - BK^o)^T P + P(A - BK^o) + W_0 + K^{oT}WK^o = 0, \qquad (6.4-3)$$

which is equivalent to

$$P = \int_0^\infty \exp[(A - BK^o)^T t](W_0 + K^{oT}WK^o)\exp[(A - BK^o)t]\,dt. \qquad (6.4-4)$$

The results that are presented in several of the remaining sections in this chapter are extensively based upon the consequence of next Lemma. It is worth mentioning that the following rather straightforward lemma, which is reported in [*C.* 10], is very instrumental in providing an alternative method for solving a number of *now* classical quadratic-optimization problems. So much so that at times we may look very repetitive in certain remaining sections of this chapter. Indeed, as a consequence of this result, the impact and the importance of the Lyapunov approach in linear optimization theory becomes very clear. This idea of having a unified method for analysis of a number of control problems has been our main thought in § 5.11.3.

LEMMA 6.4-1 [C. 10]: Given an asymptotically stable matrix H and $Q = Q^T > 0$, then there exists a unique $P = P^T > 0$ such that $H^T P + PH + Q = 0$. If $H \to H + \Delta H$, where $H + \Delta H$ remains asymptotically stable, and $Q \to Q + \Delta Q$ with $Q + \Delta Q$ and $\Delta H^T P + P\Delta H + \Delta Q$ remain symmetric positive- (semi-) definite matrices, then $P \to P + \Delta P$; where

6.4. Optimality – Index Sensitivity: Analysis and Synthesis

$$\Delta P = \int_0^\infty \exp[(H + \Delta H)^T t] \, (\Delta H^T P + P\Delta H + \Delta Q) \exp[(H + \Delta H)t] \, dt. \tag{6.4-5}$$

PROOF: Let $H \to H + \Delta H$ and $Q \to Q + \Delta Q$, as above. Then $P \to P + \Delta P$; where $(H + \Delta H)^T (P + \Delta P) + (P + \Delta P)(H + \Delta H) + (Q + \Delta Q) = 0$. This equation, considering $H^T P + PH + Q = 0$, simplifies to

$$(H + \Delta H)^T \Delta P + \Delta P (H + \Delta H) + \Delta H^T P + P\Delta H + \Delta Q = 0, \tag{6.4-6}$$

which (6.4-6) is equivalent to (6.4-5). If ΔH and ΔQ are sufficiently "small" which is of interest to us, then the positive- (semi-) definiteness of various matrices are met.

REMARK 6.4-1 [2D. 1]: Recall the following trace operations, provided that the matrix products are well defined: *(i)* $\mathrm{Tr}(AB) = \mathrm{Tr}(BA)$ (or $\mathrm{Tr}(B^T A^T)$); *(ii)* $\frac{\partial}{\partial A} \mathrm{Tr}(AB) = B^T$; *(iii)* $\frac{\partial}{\partial A} \mathrm{Tr}(BA^T) = B$; *(iv)* $\frac{\partial}{\partial \alpha} \mathrm{Tr}(A\alpha B) = A^T B^T$; and *(v)* $\frac{\partial}{\partial \alpha} \mathrm{Tr}(A\alpha^T B) = BA$.

THEOREM 6.4-1: For an optimal system described by (6.4-1) and (6.4-2), the following relationships are true

$$\frac{\partial J}{\partial A} = 2PE, \quad \text{and} \quad \frac{\partial J}{\partial B} = -2PEK^{oT}; \tag{6.4-7}$$

where conventionally $\frac{\partial J}{\partial A} = (\frac{\partial J}{\partial a_{ij}}) \in R^{n \times n}$,

$$E \triangleq \int_0^\infty \exp(Ht) X \exp(H^T t) dt, \tag{6.4-8}$$

and $H \triangleq A - BK^o$, with $X = x(0)x^T(0)$. In order that the performance index does not depend directly on the initial condition, which is not generally available, a statistical average or intensity of this initial condition, namely $\mathrm{cov}(x(0), x^T(0))$ (cov stands for covariance), is considered hereafter instead of X; i.e., we let $X = \mathrm{cov}(x(0), x^T(0))$.

PROOF: We can show that the optimal cost is $J^o = x_0^T P x_0 = \mathrm{Tr}(XP)$, where P is given by (6.4-4). If $H \to H + \Delta H$, then $P \to P + \Delta P$, where, by Lemma 6.4-1, ΔP is given by (6.4-5). Correspondingly, if $P \to P + \Delta P$, then $J \to J + \Delta J$; where $\Delta J = \mathrm{Tr}(X \Delta P)$. Substituting for ΔP from (6.4-5) and with some trace operations, we can show that

$$\Delta J = \mathrm{Tr}\{\int_0^\infty \exp[(H + \Delta H)t] X \exp[(H + \Delta H)^T t] \, dt \, (\Delta H^T P + P\Delta H + \Delta Q)\}$$

$$\triangleq \mathrm{Tr}\{\tilde{E}(\Delta H^T P + P\Delta H + \Delta Q)\}; \tag{6.4-9}$$

where $\tilde{E} \triangleq \int_0^\infty \exp[(H + \Delta H)t] X \exp[(H + \Delta H)^T t] \, dt$.

Consider the case where $\Delta H = \Delta A$, then $\Delta Q = 0$, and from (6.4-9), with the aid of Remark 6.4-1, it can be deduced that

$$\frac{\partial J}{\partial A} = \lim_{\Delta A \to 0} (\frac{\Delta J}{\Delta A}) = \lim_{\Delta A \to 0} \{Tr[\tilde{E}(\Delta A^T P + P\Delta A)] / \Delta A\} = 2PE. \quad (6.4\text{-}10)$$

Similarly, if $\Delta H = -\Delta BK^o$, then $\Delta Q = 0$, and from (6.4-9) and as above, it can be deduced that

$$\frac{\partial J}{\partial B} = \lim_{\Delta B \to 0} (\frac{\Delta J}{\Delta B})$$

$$= \lim_{\Delta B \to 0} \{Tr[\tilde{E}(-K^{oT}\Delta B^{TP} - P\Delta BK^o)] / \Delta B\} = -2PEK^{oT}. \quad (6.4\text{-}11)$$

REMARK 6.4-2: In the case of *small*-sensitivity function of J with respect to the weighting matrices the same procedure as above goes through in order to find, first ΔJ then $\partial J/\partial W$. As an example, consider the case where $W \to W + \Delta W$, then $\Delta Q = -K^{oT}\Delta WK^o$ (recall that $\Delta W^{-1} = -W^{-1}\Delta W W^{-1}$ and $K^o = W^{-1}B^TP$), and $\Delta H = BW^{-1}\Delta W K^o$, and $\Delta J = Tr\{\tilde{E}(\Delta H^T P + P\Delta H + \Delta Q)\} = Tr\{\tilde{E}K^{oT}\Delta WK^o\}$. Therefore

$$\frac{\partial J}{\partial W} = \lim_{\Delta W \to 0} (\frac{\Delta J}{\Delta W}) = \lim_{\Delta W \to 0} \{Tr[\tilde{E}K^{oT}\Delta WK^o] / \Delta W\} = K^o EK^{oT}. \quad (6.4\text{-}12)$$

REMARK 6.4-3: It is certainly possible that both A and B become functions of the same parameter. Then the small variation of J with respect to changes in this parameter is $\delta J = (\frac{\partial J}{\partial A}\delta A^T + \frac{\partial J}{\partial B}\delta B^T)$. The extension of this idea to more parameters is now straightforward. Similar results to that given above, but with different algebra and terminology, are also reported in [C. 12 and 6].

6.4.3. PI – Large – Sensitivity Analysis

As always the extension of *small*-sensitivity analysis to that of *large* is of special interest. One recent such extension is reported in [C. 5]. Recall the optimal cost $J^o = Tr(XP)$, where P is given by (6.4-4) and suppose that $P \to P + \Delta P$, because $H \triangleq (A - BK^o) \to H + \Delta H$, provided that $H + \Delta H$ remains stable. Then $J^o \to J^o + \Delta J$, where $\Delta J = Tr(X\Delta P)$, and ΔP is given by (6.4-5). We must point out that several system-sensitivity measures are defined considering this ΔJ. For example, Rohrer and Sobral [C. 23] consider $S = |\Delta J|/J^o$, as the system-sensitivity measure. Salmon [C. 26] gives a different point of view for this ΔJ, by advancing theorems pertinent to design of a system, based upon minimizing the maximum deviation of *PI* caused by parameter variations. However, the crucial question is how to determine the *largest*-upper bound of J^o which we call J_M. Clearly, the cause of this problem must be first studied thoroughly before an analytical procedure can be provided to tackle its computation. As mentioned above this variation is due to $H \to H + \Delta H$. Since ΔH can be one of the following choices: ΔA; $-\Delta BK$; $\Delta A - \Delta BK$; $-B\Delta K$; $\Delta A - B\Delta K$; $\Delta A - B\Delta K$; or $\Delta A - B\Delta K - \Delta BK$,

6.4. Optimality – Index Sensitivity: Analysis and Synthesis

and indeed any one of these variations can be either single- or multiple-parameter variations, the corresponding analysis must be performed accordingly.

But first we establish the following facts. From $J^o = \text{Tr}(XP)$ with P given by (6.4-4), we conclude that $J^o = J(p^o)$, where $p \in R^r$ is the vector of system physical parameters and p^o is its nominal value. Therefore if $J^o = J(p^o)$ is explicitly stated, or if it could be derived, then it seems that a simple differentiation of J^o with respect to p_j, $j = 1, 2, ..., r$, gives information about how to find the maximum variations in J^o with respect to these parameters. Suppose that $p \in R^1$, and $\partial J^o/\partial p = 0$, then solving this equation for p results in p^* and correspondingly $J^* = J(p^*)$. If $\partial^2 J^o/\partial p^2|_{p=p^*}$ has a definitive sign, then the maximum variation in J^o is $|J^* - J^o| = \Delta J$ corresponding to $\Delta p = |p^o - p^*|$. However, this expectation of analytically generating ΔJ is unrealistic, because we cannot easily derive $J = J(p)$. To give a point of view, consider an example by Platzman and Athans [C. 21] (Example 6.4-1), which shows an expression for J in terms of a single-parameter variation. This derivation soon becomes very complicated for most typical applications. However, no matter how we look at this problem, it seems that we should relate ΔJ to that of ΔH directly and numerically, because the cause of variations in ΔJ must be sought in H and this must definitely be such that the overall system stability is maintained. Therefore it seems that in the context of our stability robustness analysis *(SRA)*, in Chapter Five, we must show that indeed an algorithm can be developed which provides *the largest*-upper bound for J^o – say – J_M, caused by the *largest-acceptable*-parameter variations in H. Thus the heuristic approach suggested in [C. 21] to study a special example is modified to a unified technique for studying this type of problems.

LEMMA 6.4-2: Suppose that M and N are two, n × n, stable matrices, such that $S \triangleq N - M$ is a positive-definite matrix. If R_1 and R_2 are solutions of the following two Lyapunov equations $M^T R_1 + R_1 M + Q = 0$, and $N^T R_2 + R_2 N + Q = 0$, with $Q = Q^T > 0$. Then $J_2 \triangleq x_o^T R_2 x_o > J_1 \triangleq x_o^T R_1 x_o$.

PROOF: By assumption $S \triangleq N - M$ is a positive-definite matrix. Thus substituting for $N = M + S$ into the Lyapunov equation corresponding to R_2 results in

$$M^T R_2 + R_2 M + Q + S^T R_2 + R_2 S = 0. \qquad (6.4\text{-}13)$$

Subtracting $M^T R_1 + R_1 M + Q = 0$ from (6.4-13), yields

$$M^T(R_2 - R_1) + (R_2 - R_1)M + S^T R_2 + R_2 S = 0. \qquad (6.4\text{-}14)$$

Since $R_2 S + S^T R_2$ is symmetric and positive-definite matrix, (6.4-14) implies that $R_2 - R_1 > 0$. Therefore $J_2 - J_1 = x_o^T(R_2 - R_1)x_o > 0$, which implies $J_2 > J_1$.

REMARK 6.4-4: Recall that P is the unique symmetric positive-definite solution of (6.4-3), with $H \triangleq A - BK^o$. It is well known that this optimal gain K^o always guarantees the asymptotic stability of H. Therefore if $H \to H \pm \Delta H$ such that this perturbed H remains asymptotically stable, then $P \to P \pm \Delta P$, and correspondingly $J \to J^o \pm \Delta J$. Suppose that the troublesome direction of parameter variation is in the positive direction, i.e., $J \to J + \Delta J$. If we assume that ΔH is a single-parameter perturbation matrix. The method to generate the *largest* ΔH is discussed in Algorithm 5.6-1, where for such an upper bound for ΔH, some of the eigenvalues of $H + \Delta H$ on the left-hand side of the imaginary axis in the complex plane are closer to this axis than the dominant eigenvalues of H. Since H is an asymptotically stable matrix, if we assume that ΔH is a positive- (semi-) definite matrix, then this perturbation leads to $\Delta J \geq 0$ (cf., Lemma 6.4-2). However, it is possible that ΔH does not have a definitive sign but $H + \Delta H$ remains stable. Here the term *largest* perturbation of H is that *value* of ΔH such that some or all the eigenvalues of $H + \Delta H$ become closer to the imaginary axis (approaching from the left) than those of H. This issue has been the main thrust of the discussions in Section 5.5. However, the first question that must be raised is what will happen to the remaining eigenvalues of $H + \Delta H$, when compared to the remaining eigenvalues of H? The answer is that the eigenvalues of $H + \Delta H$ may move in all directions on the left-hand side of the imaginary axis in the complex plane, but the dominant eigenvalues of $H + \Delta H$ are closer to the imaginary axis, approaching from the left, than those of H, based on our fundamental assumption on sensitivity of original system. On the other hand, the eigenvalues of the solution to the Lyapunov equation

$$(H + \Delta H)^T(P + \Delta P) + (P + \Delta P)(H + \Delta H) + Q = 0, \qquad (6.4\text{-}15)$$

which is given by the eigenvalues of the following matrix

$$P + \Delta P = \int_0^\infty \exp[(H + \Delta H)^T t] Q \exp[(H + \Delta H) t] dt \qquad (6.4\text{-}16)$$

are more or less (with some abuse of terminology) the mirror image of the eigenvalues of $H + \Delta H$ with respect to the imaginary axis. That means as the eigenvalues of perturbed matrix $H + \Delta H$ reaches the imaginary axis from the *left*, the corresponding solution of (6.4-16) has eigenvalues which are closer to the imaginary axis from the *right* than the solution corresponding to unperturbed system. Perhaps the situation for the remaining (or more stable) eigenvalues of $H + \Delta H$ versus those of the corresponding P are not so apparent as the dominant eigenvalues. Therefore as the *largest* ΔH is being applied, $P \to P + \Delta P$ such that $x_o^T(P + \Delta P)x_o > x_o^T P x_o$, which results in the largest ΔJ.

The content of Remark 6.4-4 concludes the assertion which has been made therein, and completes the proof of the following proposition.

PROPOSITION 6.4-1: Consider the optimal control system described by (6.4-1) and (6.4-2). The optimal gain $K^o = W^{-1}B^T P$, with P given by (6.4-4), results in $H \triangleq A - BK^o$ which is an asymptotically stable matrix, with the optimal cost $J^o = x_o^T P x_o$. Then the variation in J^o, caused by ΔH, is increasing in the direction of the *largest* ΔH which is given by Algorithm 5.6-1. The *largest* J – say J_M, occurs when $H \to H + \Delta H$ with ΔH being the *largest* perturbation of H.

6.4. Optimality – Index Sensitivity: Analysis and Synthesis

PROOF: Refer to Remark 6.4-5. We must note that $J_M(H + \Delta H) < \infty$, but $J_M(H + \Delta H + \varepsilon I) = \infty$, for an arbitrary *small*-troublesome ε. ∎

Based upon the result of Proposition 6.4-1 we have the following algorithm.

ALGORITHM 6.4-1 (Single – Large – Parameter Variation of PI): Given $\dot{x}(t) = Ax(t) + Bu(t)$, with $E(x_o) = 0$ and $X = \text{cov}(x(0), x^T(0))$ (E and cov mean the statistical expectation and covariance, respectively), and the following performance index $J = \min \int_0^\infty \{x^T(t)W_0 x(t) + u^T(t)Wu(t)\}dt$; the optimal input is $u^o(t) \triangleq -K^o x = -W^{-1}B^T P x(t)$, and the single-parameter $H \triangleq A - BK^o$ is an asymptotically stable matrix, with the optimal cost $J^o = \text{Tr}(XP)$ and P is given by $H^T P + PH + (W_0 + K^{oT}WK^o) = 0$. Based on the result of Proposition 6.4-1, the *troublesome* direction of parameter variation in ΔH which causes the eigenvalues of $H \pm \Delta H$ to become closer to the left-hand side of the imaginary axis than those of H, gives the *largest* ΔJ. This result is the direct consequence of Algorithm 5.6-1. ∎

The following example is given by Platzman and Athans [C. 21], and is reported here for comparison with the results developed in this section.

EXAMPLE 6.4-1: Consider the system described by $\dot{x}(t) = (A + \alpha_0 bc^T)x(t)$, and $J(\alpha) = E\{\int_0^\infty x^T(t)x(t)dt\}$, where $E(x_o) = 0$, $\text{cov}(x(0), x^T(0)) = I$, and the nominal values of parameters are $A = \begin{pmatrix} -1 & 1 \\ -\frac{1}{2} & 0 \end{pmatrix}$, $b = \begin{Bmatrix} 1 \\ -\frac{1}{2} \end{Bmatrix}$, $c^T = [1, 0]$, and $\alpha_0 = 0$. Find a feasible range of variations for $\Delta\alpha$ such that $J(\alpha_0 + \Delta\alpha)$ remains bounded.

SOLUTION: According to [C. 21], $J(\alpha_0 + \Delta\alpha) = J(\Delta\alpha) = (3.25 - 0.5\Delta\alpha + 1.25\Delta\alpha^2)/(1 - \Delta\alpha^2)$, therefore $J(\Delta\alpha)$ exists for $-1 < \Delta\alpha < 1$. Certainly, because of this rather simple situation we are able analytically to give a feasible range of variations for $\Delta\alpha$. But, in general, we can avoid this direct analytical (symbolic) approach, by using Algorithms 6.4-1 and 5.6-1 in order to come up with the proper numerical answer to this type of problems as explained next.

Considering the results of Theorem 5.5-1, if $A \to A + \Delta A$ for $\Delta A = \Delta\alpha bc^T$, then to maintain the stability of $A + \Delta A$ we can use the following sufficient condition $\Delta A^T Q^{-1} \Delta A < \frac{1}{4} P^{-1} Q P^{-1}$, where $AP + PA^T + Q = 0$, for a symmetric positive-definite matrix Q, such as $Q = I$. Here we have

$$\Delta\alpha^2 \begin{bmatrix} 1 & -\frac{1}{2} \\ 0 & 0 \end{bmatrix} \begin{bmatrix} 1 & 0 \\ -\frac{1}{2} & 0 \end{bmatrix} < \frac{1}{4} P^{-2}. \quad (6.4\text{-}17)$$

Or

$$\Delta\alpha^2 \begin{bmatrix} 1.25 & 0 \\ 0 & 0 \end{bmatrix} < \frac{1}{4} \begin{bmatrix} \frac{20}{13} & -\frac{16}{13} \\ -\frac{16}{13} & \frac{16}{13} \end{bmatrix}. \quad (6.4\text{-}18)$$

From (6.4-18), and by simultaneous diagonalization of the above inequality, we have $\Delta\alpha^2 < 1/4\lambda$, where λ is the largest root of $\det(\lambda P^{-2} - K) = 0$, with $K = \mathrm{diag}(1.25, 0)$. Here $\lambda = 65/16$ and $|\Delta\alpha| < 0.248$; i.e., $-0.248 < \Delta\alpha < 0.248$. Of course, this result is the first iteration in Algorithm 5.6-1 and without an explicit analytical derivation. This interval can be enlarged using this algorithm. The corresponding *largest* ΔJ can be computed when the *largest* $\Delta\alpha$ and the correspondingly the *largest* $\Delta A = \Delta\alpha bc^T$ is generated. Thus as this example suggests, we can compute the allowable range of parameter variations for a single-parameter quadratic *PI* with the help of Algorithms 6.4-1 and 5.6-1 and without the need for an explicit analytical derivation. We invite the reader to finish this computation.

▼

In the case of multiple-parameter variations, the situation is different from the above, requiring special attention to the meaning of "the largest perturbation," first. As we have studied this situation in § 5.6.3, when two or more parameters change, then we have a set of allowable-parameter variations in the *PVS* that can be constructed according to Theorem 5.6-2. The construction of this set is the prerequisite to answering the current issue. If we map the boundary of this set into J(p), then that set provides us with a mechanism to generate the corresponding set of allowable-parameter variations for ΔJ. The computational procedure is exactly the same as that suggested in Examples 5.8-3 & 4 for asymptotically stable matrices.

6.4.4. PI – Small – Sensitivity Synthesis

In the sequel an algorithm for deriving the optimal controller gain K, which minimizes a combined-quadratic cost; consisting of the plant – *PI* and the sensitivity of this *PI* with respect to *small*-parameter variations is developed. The present derivation is also concluded from the main procedure (i.e., Lemma 6.4-1) which is developed in this book primarily for trajectory-sensitivity minimization and is presented subsequently.

Consider a *DCLTI – GS* described by

$$\dot{x}(t) = Ax(t) + Bu(t), \quad x(0) = x_o, \quad (6.4\text{-}19a)$$

$$z(t) = Cx(t). \quad (6.4\text{-}19b)$$

In general, x(0) is not exactly known. Thus in the following analysis we assume that $E(x(0)) = 0$, and $\mathrm{cov}(x(0), x^T(0)) = X$, with the plant-quadratic cost denoted as follows.

$$J_p = \int_0^\infty [z^T(t)W_0 z(t) + u^T(t)Wu(t)]dt, \quad (6.4\text{-}20)$$

where $W_0 = W_0^T \geq 0$ and $W = W^T > 0$ are the weighting matrices of appropriate dimensions. A sensitivity cost for this system is considered as follows.

6.4. Optimality – Index Sensitivity: Analysis and Synthesis

$$J_s = \tfrac{1}{2} \operatorname{Tr}(\frac{\partial J_p}{\partial \eta_{ij}} \delta \eta_{ij}^T); \qquad (6.4\text{-}21)$$

where $H \triangleq (\eta_{ij}) = A \cup B$. From Theorem 6.4-1, and for $K^o \equiv K_p \triangleq W^{-1} B^T P$, we have

$$\frac{\partial J_p}{\partial A} = 2PE, \quad \text{and} \quad \frac{\partial J_p}{\partial B} = -2PEK_p^T; \qquad (6.4\text{-}22)$$

with

$$E \triangleq \int_0^\infty \exp(H_p t) X \exp(H_p^T t)\, dt, \qquad (6.4\text{-}23)$$

and $P = P^T > 0$ is the unique solution of

$$H_p^T P + P H_p + C^T W_0 C + K_p^T W K_p = 0. \qquad (6.4\text{-}24)$$

Here $H_p = A - B K_p$ with $u(t) = -K_p x(t)$. Using these equations in (6.4-20) yields

$$J_p = \operatorname{Tr}(X P). \qquad (6.4\text{-}25)$$

The objective of PI – sensitivity reduction is: To choose a new controller gain matrix K^o, such that $u(t) = -K^o x(t)$ minimizes the following combined-quadratic cost

$$J = J_p + \alpha J_s, \qquad (6.4\text{-}26)$$

where $\alpha > 0$ is an arbitrary constant, useful in computational algorithm. Clearly, the K_p which minimizes J_p (cf., (6.4-20) or (6.4-25)) is no longer the optimal controller gain matrix in the sense of minimizing (6.4-26).

PROBLEM 6.4-1 (PI – Small – Sensitivity Minimization): Compute the optimal controller gain K^o such that $u(t) = -K^o x(t)$ minimizes the combined-quadratic cost (6.4-26) subject to (6.4-19).

SOLUTION: A necessary condition for $u(t) = -K^o x(t)$ to minimize the performance index J given in (6.4-26) is that K satisfies the following matrix equation

$$\frac{\partial J}{\partial K} = 0. \qquad (6.4\text{-}27)$$

To compute this derivative is the main thrust of the following analysis. Substituting for J_p (from (6.4-25)) and for J_s (from (6.4-21) to (6.4-24)) into (6.4-26) yields

$$J = \operatorname{Tr}(X P + \alpha P E \delta A^T - \alpha P E K^T \delta B^T). \qquad (6.4\text{-}28)$$

Note that K_p is no longer the acceptable controller gain. Using certain trace operations (cf., Remark 6.4-1), and substituting for E, from (6.4-23), into (6.4-28), results in

$$J = \operatorname{Tr}\{X P + \alpha X \int_0^\infty \exp(H^T t)(\delta A - \delta B\, K)^T P \exp(Ht) dt\} \triangleq \operatorname{Tr}\{X(P + \alpha Q)\}; \qquad (6.4\text{-}29)$$

where Q is the solution of the following equation

$$H^TQ + QH + (\delta A - \delta B\,K)^TP = 0. \tag{6.4-30}$$

In order to derive $\partial J/\partial K$, the following approach is advanced. Suppose that $K \to K + \delta K$, such that δK is "small," then $P \to P + \delta P$, $Q \to Q + \delta Q$ and correspondingly $J \to J + \delta J$; where

$$\delta J = \text{Tr}\{X(\delta P + \alpha \delta Q)\}. \tag{6.4-31}$$

From (6.4-24) and (6.4-30) (with K_p is being replaced by K), if $K \to K + \delta K$, with δK being "small," then $P \to P + \delta P$ and $Q \to Q + \delta Q$; where (cf., Lemma 6.4-1),

$$\delta P = \int_0^\infty \exp(H^Tt)[\delta K^T(WK - B^TP) + (K^TW - PB)\delta K]\exp(Ht)dt, \tag{6.4-32}$$

$$\delta Q = \int_0^\infty \exp(H^Tt)[(\delta A - \delta BK)^T\delta P - \delta K^T(\delta B^TP + B^TQ)$$

$$- QB\delta K]\exp(Ht)dt. \tag{6.4-33}$$

Substituting for δP and δQ, from (6.4-32) and (6.4-33) into (6.4-31) results in

$$\delta J = \text{Tr}\{X\int_0^\infty \exp(H^Tt)[\delta K^T(WK - B^TP - \alpha(\delta B^TP + B^TQ))$$

$$+ (K^TW - PB - \alpha QB)\delta K + \alpha(\delta A - \delta BK)^T\delta P]\exp(Ht)dt\}$$

$$= \text{Tr}\{[\delta K^T(WK - B^TP - \alpha(\delta B^TP + B^TQ)) + (K^TW - PB - \alpha QB)\delta K$$

$$+ \alpha(\delta A - \delta BK)^T\delta P]E\}. \tag{6.4-34}$$

Here E is the same as in (6.4-23), and substituting for δP, from (6.4-32), into (6.4-34), yields

$$\delta J = \text{Tr}\{[\delta K^T(WK - B^TP - \alpha(\delta B^TP + B^TQ)) + (K^TW - PB - \alpha QB)\delta K]E$$

$$+ \alpha[\delta K^T(WK - B^TP) + (K^TW - PB)\delta K]E'\}; \tag{6.4-35}$$

where

$$E' \triangleq \int_0^\infty \exp(Ht)E(\delta A - \delta BK)^T\exp(H^Tt)\,dt. \tag{6.4-36}$$

Equation (6.4-35), on using some trace operations, further simplifies to

$$\delta J = \text{Tr}\{\delta K^T[(WK - B^TP - \alpha(\delta B^TP + B^TQ))E + \alpha(WK - B^TP)E']$$

$$+ [E(K^TW - PB - \alpha QB) + \alpha E'(K^TW - PB)]\delta K\}. \tag{6.4-37}$$

From (6.4-37), we can show that (cf., Remark 6.4-1),

$$\frac{\partial J}{\partial K} = \lim_{\delta K \to 0}\frac{\delta J}{\delta K} = [2WK - 2B^TP - \alpha B^T(Q + Q^T) - \alpha \delta B^TP]E$$

$$+ \alpha(WK - B^TP)(E' + E'^T). \tag{6.4-38}$$

6.4. Optimality – Index Sensitivity: Analysis and Synthesis

The optimality of K requires that necessarily $\partial J/\partial K = 0$. Thus the optimal controller gain must satisfy the following matrix equation

$$WK\overline{E} - B^T P \overline{E} - \tfrac{1}{2}\alpha[B^T(Q + Q^T) + \delta B^T P] E = 0; \quad (6.4\text{-}39a)$$

where

$$\overline{E} \triangleq E + \tfrac{1}{2}\alpha(E' + E'^T). \quad (6.4\text{-}39b)$$

From the above equation and recalling that $W = W^T > 0$, the optimal controller gain becomes

$$K^o = W^{-1}\{B^T P + \tfrac{1}{2}\alpha[B^T(Q + Q^T) + \delta B^T P]E\overline{E}^\dagger\}; \quad (6.4\text{-}40)$$

where $\overline{E}^\dagger$ is a generalized inverse matrix of $\overline{E} = \overline{E}^T$ (given in (6.4-39b)), which is used only for a symbolic reason and does not play any other role. The P and Q are described by (6.4-24) and (6.4-30).

▼

The optimal controller gain given in (6.4-40) and the other relevant equations are quadratic function of K^o. The following computational procedure is proposed to solve these equations iteratively.

ALGORITHM 6.4-2 (Optimal Controller Gain: PI – Small – Sensitivity Reduction):
The following steps are proposed in order to compute the optimal controller gain K^o given in (6.4-40). Here we use one such sub-optimal solution to iterate in the policy space in order to seek the optimal solution. Let $J = 1, 2, ..., N$ be the iteration number.

STEP 1 (To initialize): Given A and B, select weighting matrices W_0, W and choose δA and δB in the troublesome direction for parameter variations in the *PI*. *Comment:* This task of choosing δA and δB is perhaps the most difficult step in this optimization process. Nevertheless this is a problem dependent issue and it must be studied case by case.

STEP 2 (To compute the initial K): Set $J = 1$ and $\alpha = 0$. From (6.4-26) the *PI* becomes the same as J_p and from (6.4-40) $K_1 = W^{-1} B^T P$, with P given by $A^T P + PA + C^T W_0 C - PBW^{-1}B^T P = 0$.

STEP 3 (To update P, Q, E, and E'): With the current K, compute a new P and Q, from (6.4-24) and (6.4-30), respectively. Also with this K, compute E and E', from (6.4-23) and (6.4-36), respectively. Then with the actual α compute $\overline{E}$ from (6.4-39b).

STEP 4 (To update the controller gain): Substitute the current P, Q, E, $\overline{E}$ into (6.4-40) to generate a new optimal controller gain K_J.

STEP 5 (To test for the optimal controller gain): If $|(K_{i,j})_J - (K_{i,j})_{J-1}| < 10^{-8} * |(K_{i,j})_{J-1}|$, then stop, $K_J \equiv K^o$ is the final optimal solution. Otherwise go back to Step 3 and set $J \to J + 1$.

■

6.5. TRAJECTORY – SENSITIVITY OPTIMIZATION

6.5.1. Introduction and Review of the Existing Methods

In this section trajectory-sensitivity optimization refers to the minimization of a quadratic-sensitivity function that is integrated over an infinite horizon. The procedure for this minimization, however, is exactly the same as that for a trajectory-sensitivity maximization (cf., Remark 6.5-2): useful in parameter specification (cf., Definition 2.2-4). (This infinite-horizon optimization is the prerequisite for deriving a solution of a finite-horizon optimization.) Trajectory-sensitivity optimization has received attentions from many researchers, who wish to design the so-called insensitive-control systems; wherein *a* system-trajectory sensitivity measure is minimized. The main drawback of the earlier work in this area is the direct attempt to apply results from linear optimal control theory (or a regulator problem) to this sort of minimization problems [D. 41]. As is discussed subsequently, that formulation is not completely correct, therefore the resulting algorithm is not accurate. Hence, in the following, a concise review of the earlier results pertaining to trajectory-sensitivity minimization is reviewed for the sake of historical reasons.

Consider a single-parameter ($p \in R^1$) open-loop $DCLTI - GS$ described by

$$\dot{x}(t; p) = A(p)x(t; p) + B(p)u(t), \quad x(0; p) = x_o. \quad (6.5\text{-}1)$$

Then the corresponding system model for (6.5-1) is

$$\frac{d}{dt} \begin{Bmatrix} x(t; p) \\ \frac{\partial x(t; p)}{\partial p} \end{Bmatrix} = \begin{bmatrix} A(p) & 0 \\ \frac{\partial A(p)}{\partial p} & A(p) \end{bmatrix} \begin{Bmatrix} x(t; p) \\ \frac{\partial x(t; p)}{\partial p} \end{Bmatrix} + \begin{bmatrix} B(p) \\ \frac{\partial B(p)}{\partial p} \end{bmatrix} u(t)$$

$$\triangleq \tilde{A}_1 m_1(t; p) + \tilde{B}_1(p) u(t), \quad (6.5\text{-}2a)$$

$$m_1(0; p) = [x_o^T, 0^T]^T. \quad (6.5\text{-}2b)$$

Minimization of the following *PI*, subject to (6.5-2), which results in an explicit form for the input function, has received researchers' attentions. However, the main difficulty arises when the input function resulting from this method is to be implemented.

$$J_1 = \min\{ \int_{t_0}^{t_f} [m_1^T(t)Q_1 m_1(t) + u^T(t)Wu(t)]dt + m_1^T(t_f)Sm(t_f)\}; \quad (6.5\text{-}3)$$

where $Q_1 = Q_1^T \triangleq \text{diag}(W_0, W_1) \geq 0$, $W = W^T > 0$, and $S = S^T \geq 0$.

The open-loop implementation of input vector that is generated from (6.5-3) can be accomplished without any elaboration. On the other hand, the problem arises when its closed-loop implementation is being sought, to that end certain difficulties arise which are explained subsequently. If the input is implemented in a feedback form, then the sensitivity function satisfies the following differential equation

6.5. Trajectory – Sensitivity Optimization

$$\frac{d}{dt}\frac{\partial x(t; p)}{\partial p} = A(p)\frac{\partial x(t; p)}{\partial p} + \frac{\partial A(p)}{\partial p}x(t; p) + \frac{\partial B(p)}{\partial p}u(t; p) + B(p)\frac{\partial u(t; p)}{\partial p},$$

$$\left.\frac{\partial x(t; p)}{\partial p}\right|_{t=0} = 0. \tag{6.5-4}$$

It is evident from (6.5-4) that, in general, the lack of information about $\partial u(t; p)/\partial p$, prevents further application of (6.5-4). One assumption that is made often for the case of linear input is to neglect $\partial^2 x(t; p)/\partial p^2$, and to proceed accordingly, although this approximation is seldom valid. On the other hand, if this assumption were not made, then it is required that the next higher-order sensitivity function vanish. In Section 6.6 another difficulty of this problem is explained.

Considering (6.5-3), the earlier researchers have speculated that the *optimal* input should be of the following form

$$u(t; p) = -[K_1, K_2] m_1(t; p). \tag{6.5-5}$$

If this input, with the assumption that $\partial^2 x(t; p)/\partial p^2 = 0$, is substituted into (6.5-4), then (6.5-4) results in an equation for $\partial x(t; p)/\partial p$ which is an approximation to this quantity. To avoid the confusion between the exact $\partial x(t; p)/\partial p$ and its approximate value (approximate in the sense of letting $\partial^2 x/\partial p^2 = 0$), the latter is shown by $\rho(t; p)$. Kreindler has described well the above procedure in [D. 32]. The paper by Fleming and Newmann [D. 16], presents an advancement of this procedure. There are, however, a number of problems with this type of formulations which are summarized in Remark 6.5-1.

REMARK 6.5-1: Consider a *DCLTI – GS* with multiple-varying parameters $p \in R^r$. Perhaps the main shortcomings of the earlier results in trajectory-sensitivity minimization area are: *(i)* the erroneous assumption that $\partial^2 x/\partial p_j^2 = 0$, where p_j, $j=1, 2, \ldots, r$, is system-physical parameter and $x(t)$ is system-state vector [D. 49]; *(ii)* the direct attempt to apply results from linear optimal control theory (or a regulator problem) for the minimization of an integral of a quadratic-sensitivity function [D. 41]; *(iii)* the requirement on the convergence of the corresponding algorithm has not been established [D. 32 & 33]; *(iv)* the computational burden of solving the pertinent Riccati equation is always excessive even for lower-dimensional system with a few parameters [D. 32 & 33 and 52]; *(v)* the inclusion of the above mentioned approximations (in particular, the approximate closed-loop sensitivity function) causes the entire computational process to be nonexact; and finally *(vi)* the inclusion of the above approximations and the resultant computational inaccuracies lead to additional design difficulties in the already expensive process of reconstructing system-trajectory sensitivity functions that is needed for the proposed closed-loop control policy. Thus the ultimate $x(t; p)$ is not *the optimal trajectory* as it is expected to be, and because of these shortcomings, this approach is no longer of any use.

In spirit of these discussions we present, an alternative approach to derive directly an optimal controller gain for a quadratic trajectory-sensitivity minimization. Subsequently, we present an algorithm to generate a solution for this optimization problem.

6.5.2. Alternative Approach to Generate an Optimal Controller Gain for Trajectory – Sensitivity Minimization

A new method for deriving an optimal controller gain matrix K (such that $u = -Kx$) is developed in [D. 10] that results in a minimum combined – quadratic cost; consisting of standard regulator cost and system-sensitivity cost denoted as an appropriate weighted norm of system-trajectory sensitivity functions. This algorithm is developed to rectify most of the shortcomings that are mentioned in Remark 6.5-1. The unique feature of this new approach is that the optimal input does not have any component of system-sensitivity measure. This general procedure that is developed for a multidimensional and multi-parameter system is specialized to a single-parameter case with more details for the sake of illustration. Similar results are also reported in [D. 31, 52, and 53 & 54].

PROBLEM 6.5-1 (Quadratic-Sensitivity Optimization): Consider the following DCLTI – GS, described by

$$\dot{x}(t; p) = A(p)x(t; p) + B(p)u(t), \quad x(0) = x_o \neq 0, \quad (6.5\text{-}6)$$

with the following plant-quadratic cost

$$J_p = \int_0^\infty \{x^T(t)W_0 x(t) + u^T(t)Wu(t)\}dt; \quad (6.5\text{-}7)$$

where $W_0 = W_0^T \geq 0$, and $W = W^T > 0$. Let the J_{sr} be the system-sensitivity cost as follows.

$$J_{sr} = \int_0^\infty \sum_{j=1}^r \frac{\partial x^T}{\partial p_j} W_j \frac{\partial x}{\partial p_j} dt; \quad (6.5\text{-}8)$$

where $W_j = W_j^T \geq 0$, $j = 1, 2, ..., r$. Then the objective is: To choose the optimal controller gain matrix K^o, such that $u(t; p) = -K^o x(t; p)$ minimizes the following combined cost

$$J_r' = J_p + \alpha J_{sr}. \quad (6.5\text{-}9)$$

Here $\alpha > 0$ is a constant which is useful in computational algorithm.

SOLUTION: The heuristic approach for solving the above problem is to derive $\partial J_r'/\partial K$, and to solve $\partial J_r'/\partial K = 0$, for the optimal controller gain K^o. Indeed this derivation that yields a necessary condition for this optimization problem is the main purpose of the following analysis.

Substituting $u(t; p) = -Kx(t; p)$ into (6.5-6) yields

$$\dot{x}(t; p) = [A(p) - B(p)K]x(t; p). \quad (6.5\text{-}10)$$

Equation (6.5-10) is used to generate all system-trajectory sensitivity functions; i.e.,

6.5. Trajectory – Sensitivity Optimization

$\partial x/\partial p_j$, $j = 1, 2, \cdots, r$. Augmenting these differential equations results in the following system model

$$\dot{m}_r(t; p) = \tilde{A}_r(p) m_r(t; p); \qquad (6.5\text{-}11)$$

where $m_r(t; p) \triangleq [x^T(t; p), \dfrac{\partial x^T}{\partial p_1}, \cdots, \dfrac{\partial x^T}{\partial p_r}]^T$, $m_r(0; p) \triangleq [x_0^T, 0^T, \cdots, 0^T]^T$,

$$\tilde{A}_r(p) \triangleq \begin{bmatrix} A - BK & 0 & \cdots & 0 \\ \dfrac{\partial(A-BK)}{\partial p_1} & A - BK & \cdots & 0 \\ \vdots & \vdots & \ddots & \vdots \\ \dfrac{\partial(A-BK)}{\partial p_r} & 0 & 0 & A - BK \end{bmatrix}, \qquad (6.5\text{-}12a)$$

with

$$\dfrac{\partial(A-BK)}{\partial p_j} = \dfrac{\partial A}{\partial p_j} - \dfrac{\partial B}{\partial p_j} K, \quad j = 1, 2, \cdots, r. \qquad (6.5\text{-}12b)$$

Considering the system model (6.5-11), J_r', given in (6.5-9), can be written as follows.

$$J_r' = \min \int_0^\infty \{ x^T(t)(W_0 + K^T W K) x(t) + \alpha \sum_{j=1}^r \dfrac{\partial x^T}{\partial p_j} W_j \dfrac{\partial x}{\partial p_j} \} dt, \qquad (6.5\text{-}13)$$

which further simplifies to

$$J_r' = \min \{ m_r^T(0; p) \int_0^\infty \exp(\tilde{A}_r^T t) \tilde{W}_r \exp(\tilde{A}_r t) \, dt \, m_r(0; p) \}$$

$$\triangleq \min \text{Tr}\{ m_r^T(0; p) \tilde{P}_r m_r(0; p) \}. \qquad (6.5\text{-}14)$$

Here, $\tilde{P}_r = \tilde{P}_r^T > 0$ is the unique solution of the following Lyapunov equation

$$\tilde{A}_r^T \tilde{P}_r + \tilde{P}_r \tilde{A}_r + \tilde{W}_r = 0, \qquad (6.5\text{-}15)$$

with

$$\tilde{W}_r \triangleq \begin{bmatrix} W_0 + K^T W K & 0 & \cdots & 0 \\ 0 & \alpha W_1 & \cdots & 0 \\ \vdots & \vdots & \ddots & \vdots \\ 0 & 0 & \cdots & \alpha W_r \end{bmatrix}. \qquad (6.5\text{-}16)$$

In order that the performance index does not directly depend on the initial condition, a statistical average of this performance index is considered hereafter.

Therefore the new performance index J_r becomes

$$J_r = \min \text{Tr}(\tilde{P}_r M_r), \qquad (6.5\text{-}17)$$

where $M_r \triangleq \text{cov}(m_r(0; p), m_r^T(0; p))$.

We now search for a controller gain that minimizes J_r given in (6.5-17) which necessarily has to satisfy $\partial J_r/\partial K = 0$. To derive $\partial J_r/\partial K$, the following approach is advanced. If $K \to K + \delta K$, with δK being "small," then $\tilde{A}_r \to \tilde{A}_r + \delta \tilde{A}_r$, where $\delta \tilde{A}_r$ is "small" and $\delta \tilde{A}_r \to 0$ as $\delta K \to 0$. Also as $K \to K + \delta K$, with δK being "small," $\tilde{W}_r \to \tilde{W}_r + \delta \tilde{W}_r$, and $\tilde{P}_r \to \tilde{P}_r + \delta \tilde{P}_r$ such that $\tilde{P}_r + \delta \tilde{P}_r$ remains a positive-definite matrix, and (cf., Lemma 6.4-1)

$$\delta \tilde{P}_r = \int_0^\infty \exp(\tilde{A}_r^T t)(\delta \tilde{A}_r^T \tilde{P}_r + \tilde{P}_r \delta \tilde{A}_r + \delta \tilde{W}_r) \exp(\tilde{A}_r t)\,dt. \qquad (6.5\text{-}18)$$

Correspondingly, as $K \to K + \delta K$, $J_r \to J_r + \delta J_r$, where $\delta J_r = \text{Tr}(\delta \tilde{P}_r M_r)$. Using certain trace operations this δJ_r, with $\delta \tilde{P}_r$ from (6.5-18), becomes

$$\delta J_r = \text{Tr}(\Pi_r E_r), \qquad (6.5\text{-}19)$$

where

$$\Pi_r \triangleq \delta \tilde{A}_r^T \tilde{P}_r + \tilde{P}_r \delta \tilde{A}_r + \delta \tilde{W}_r \in R^{n(1+r) \times n(1+r)}, \qquad (6.5\text{-}20)$$

$$E_r \triangleq \int_0^\infty \exp(\tilde{A}_r t) M_r \exp(\tilde{A}_r^T t)\,dt \in R^{n(1+r) \times n(1+r)}, \qquad (6.5\text{-}21)$$

From (6.5-12) and (6.5-16), $\delta \tilde{A}_r$ and $\delta \tilde{W}_r$ are as follows.

$$\delta \tilde{A}_r \triangleq \begin{bmatrix} -B\delta K & 0 & \cdots & 0 \\ -\dfrac{\partial(B\delta K)}{\partial p_1} & -B\delta K & \cdots & 0 \\ \vdots & \vdots & & \vdots \\ -\dfrac{\partial(B\delta K)}{\partial p_r} & 0 & \cdots & -B\delta K \end{bmatrix}, \qquad (6.5\text{-}22)$$

and

$$\delta \tilde{W}_r \triangleq \begin{bmatrix} \delta K^T W K + K^T W \delta K & 0 \\ 0 & 0 \end{bmatrix}. \qquad (6.5\text{-}23)$$

Substituting (6.5-22) and (6.5-23) into (6.5-20), and partitioning $\tilde{P}_r$ accordingly, yields

6.5. Trajectory – Sensitivity Optimization

$$\pi_{ij} \triangleq \delta K^T \phi_{ij}^T + \psi_{ij} \delta K, \quad (6.5\text{-}24)$$

where $\pi_{ij} = \pi_{ji}^T$ are the $n \times n$ block partitions of Π_r; i.e.,

$$\Pi_r \triangleq \begin{bmatrix} \pi_{11} & \pi_{12} & \cdots & \pi_{1(1+r)} \\ \pi_{21} & \pi_{22} & \cdots & \cdot \\ \cdot & & & \cdot \\ \cdot & & & \cdot \\ \cdot & & & \cdot \\ \pi_{(1+r)1} & \cdot & \cdots & \pi_{(1+r)(1+r)} \end{bmatrix}, \quad (6.5\text{-}25)$$

and ϕ_{ij}, ψ_{ij} are explicit functions of K, B, $\partial B/\partial p_j$, $j = 1, 2, \cdots, r$, W and P_{ij} $(i, j = 1, \cdots, (1+r))$. In particular, it can be shown that

$$\phi_{11}^T = WK - B^T P_{11} - \sum_{j=1}^{r} \frac{\partial B^T}{\partial p_j} P_{1(1+j)}^T, \quad (6.5\text{-}26)$$

$$\psi_{11} = K^T W - P_{11}B - \sum_{j=1}^{r} P_{1(1+j)} \frac{\partial B}{\partial p_j}. \quad (6.5\text{-}27)$$

If E_r is also partitioned similar to (6.5-25), then it can be shown that (6.5-19) with $E_{ij} = E_{ji}^T$ is equivalent to

$$\delta J_r = \text{Tr} \sum_{i,j=1}^{1+r} \pi_{ij} E_{ji}. \quad (6.5\text{-}28)$$

Substituting (6.5-24) into (6.5-28) and using $\partial \text{Tr}(AB)/\partial A = B^T$ and $\partial \text{Tr}(A^T B)/\partial A = A$ (cf., Remark 6.4-1), results in

$$\delta J_r = \text{Tr} \sum_{i,j=1}^{1+r} (\delta K^T \phi_{ij}^T + \psi_{ij} \delta K) E_{ji} = \text{Tr} \sum_{i,j=1}^{1+r} (\delta K^T \phi_{ij}^T E_{ij}^T + E_{ji} \psi_{ij} \delta K). \quad (6.5\text{-}29)$$

Finally

$$\frac{\partial J_r}{\partial K} = \lim_{\delta K \to 0} \frac{\partial J_r}{\partial K} = \sum_{i,j=1}^{1+r} (\phi_{ij}^T E_{ij}^T + \psi_{ij}^T E_{ij})$$

$$= (\phi_{11}^T + \psi_{11}^T) E_{11} + \sum_{\substack{i,j=1 \\ (i,j) \neq (1,1)}}^{1+r} (\phi_{ij}^T E_{ij}^T + \psi_{ij}^T E_{ij}). \quad (6.5\text{-}30)$$

The optimality of K requires that necessarily $\partial J_r/\partial K = 0$. Thus the optimal controller gain must satisfy the following matrix equation

$$(\phi_{11}^T + \psi_{11}^T) E_{11} + \sum_{\substack{i,j=1 \\ (i,j) \neq (1,1)}}^{1+r} (\phi_{ij}^T E_{ij}^T + \psi_{ij}^T E_{ij}) = 0. \quad (6.5\text{-}31)$$

Substituting for ϕ_{11}^T and ψ_{11}^T, from (6.5-26) and (6.5-27), into (6.5-31) results in

$$2[WK - B^T P_{11} - \sum_{j=1}^{r} \frac{\partial B^T}{\partial p_j} P_{1(1+j)}^T]E_{11} + \sum_{\substack{i,j=1 \\ (i,j) \neq (1,1)}}^{1+r} (\phi_{ij}^T E_{ij}^T + \psi_{ij}^T E_{ij}) = 0. \qquad (6.5\text{-}32)$$

Therefore, and because of $W = W^T > 0$, the optimal controller gain necessarily satisfy the following matrix equation

$$K^o = W^{-1}[B^T P_{11} + \sum_{j=1}^{r} \frac{\partial B^T}{\partial p_j} P_{1(1+j)}^T - \tfrac{1}{2} \sum_{\substack{i,j=1 \\ (i,j) \neq (1,1)}}^{1+r} (\phi_{ij}^T E_{ij}^T + \psi_{ij}^T E_{ij}) E_{11}^\dagger], \qquad (6.5\text{-}33)$$

where $E_{11}^\dagger$ is a generalized inverse matrix of $E_{11} = E_{11}^T$, and it is used only for a symbolic reason and does not play any other role.

▼

The above derivations of optimal controller gain have been carried out systematically for a general multi-parameter system. Here derivations of ϕ_{ij} and ψ_{ij} $(i, j = 1, \ldots, (1 + r))$, are purely algebraic manipulations that cannot be significantly simplified at this stage. Therefore this procedure has to be appropriately carried out in each specific problem and is not discussed any further.

To compute numerically this optimal controller gain K^o, we note that (6.5-33) is a function of K, B, $\partial B/\partial p_j$ $(j = 1, \cdots, r)$; and ϕ_{ij}, E_{ij}, ψ_{ij}, and P_{ij} $(i, j = 1, \cdots, (1 + r))$. Therefore a special computational procedure is needed to cope with this problem. To address this issue and without any loss of generality we choose a single-parameter system to describe the procedure. For the case of a multi-parameter system the similar algorithm, with appropriate modifications, can be established.

SUB-PROBLEM 6.5-1 (Single-Parameter Case): Suppose that in Problem 6.5-1, $r = 1$ and we let $p_1 \triangleq p$, then from (6.5-12) we have

$$\tilde{A}_1(p) \triangleq \begin{bmatrix} A - BK & 0 \\ \dfrac{\partial(A - BK)}{\partial p} & A - BK \end{bmatrix}, \qquad (6.5\text{-}34)$$

where

$$\frac{\partial}{\partial p}(A - BK) = \frac{\partial A}{\partial p} - \frac{\partial B}{\partial p} K. \qquad (6.5\text{-}35)$$

Similarly, from (6.5-15), (6.5-16), (6.5-22) and (6.5-23), we have, respectively,

$$\tilde{A}_1^T \tilde{P}_1 + \tilde{P}_1 \tilde{A}_1 + \tilde{W}_1 = 0, \qquad (6.5\text{-}36)$$

6.5. Trajectory – Sensitivity Optimization

$$\tilde{W}_1 \triangleq \begin{bmatrix} W_0 + K^T W K & 0 \\ 0 & \alpha W_1 \end{bmatrix}, \qquad (6.5\text{-}37)$$

$$\delta \tilde{A}_1(p) = \begin{bmatrix} -B\delta K & 0 \\ -\dfrac{\partial B}{\partial p}\delta K & -B\delta K \end{bmatrix}, \qquad (6.5\text{-}38)$$

$$\delta \tilde{W}_1 \triangleq \begin{bmatrix} \delta K^T W K + K^T W \delta K & 0 \\ 0 & 0 \end{bmatrix}. \qquad (6.5\text{-}39)$$

Using the above $\delta \tilde{A}_1(p)$ and $\delta \tilde{W}_1$, we have the corresponding (6.5-20) for this single-parameter case which is

$$\Pi_1 \triangleq \begin{bmatrix} \pi_{11} & \pi_{12} \\ \pi_{21} & \pi_{22} \end{bmatrix}. \qquad (6.5\text{-}40)$$

Here

$$\pi_{11} \triangleq \delta K^T \phi_{11}^T + \psi_{11} \delta K$$

$$\triangleq \delta K^T (WK - B^T P_{11} - \frac{\partial B^T}{\partial p} P_{12}^T) + (K^T W - P_{11}B - P_{12} \frac{\partial B}{\partial p})\delta K, \qquad (6.5\text{-}41)$$

$$\pi_{21}^T = \pi_{12} \triangleq \delta K^T (-B^T P_{12} - \frac{\partial B^T}{\partial p} P_{22}) - P_{12} B \delta K, \qquad (6.5\text{-}42)$$

$$\pi_{22} \triangleq -\delta K^T B^T P_{22} - P_{22} B \delta K. \qquad (6.5\text{-}43)$$

Similarly from (6.5-21) it is seen that

$$E_1 \triangleq \begin{bmatrix} E_{11} & E_{12} \\ E_{12}^T & E_{22} \end{bmatrix} \triangleq \int_0^\infty \exp(\tilde{A}_1 t) M_1 \exp(\tilde{A}_1^T t) dt, \qquad (6.5\text{-}44)$$

where without any loss of generality it is assumed that

$$M_1 = \begin{bmatrix} X & 0 \\ 0 & 0 \end{bmatrix}, \qquad (6.5\text{-}45)$$

with $X = \text{cov}(x(0), x^T(0))$.

From (6.5-28) and considering (6.5-41) to (6.5-44), it is seen that

$$\delta J_1 = \text{Tr}(\pi_{11} E_{11} + \pi_{12} E_{12}^T + \pi_{12}^T E_{12} + \pi_{22} E_{22}). \qquad (6.5\text{-}46)$$

Substituting for π_{ij} in above and using certain trace operations yields

$$\delta J_1 = \text{Tr}\{\delta K^T [(WK - B^T P_{11} - \frac{\partial B^T}{\partial p} P_{12}^T) E_{11} - (B^T P_{12} + \frac{\partial B^T}{\partial p} P_{22}) E_{12}^T$$

$$- B^T P_{12}^T E_{12} - B^T P_{22} E_{22}] + [E_{11}(K^T W - P_{11}B - P_{12}\frac{\partial B}{\partial p})$$

$$- E_{12}^T P_{12} B - E_{12}(P_{12}^T B + P_{22}\frac{\partial B}{\partial p}) - E_{22} P_{22} B]\delta K\}. \tag{6.5-47}$$

Or equivalently, we have

$$\frac{\partial J_1}{\partial K} = 2[(WK - B^T P_{11} - \frac{\partial B^T}{\partial p} P_{12}^T)E_{11}$$

$$- (B^T P_{12} + \frac{\partial B^T}{\partial p} P_{22})E_{12}^T - B^T P_{12}^T E_{12} - B^T P_{22} E_{22}]. \tag{6.5-48}$$

Recalling that $W = W^T > 0$, the optimal controller gain then satisfies

$$K^o = W^{-1}\{B^T P_{11} + \frac{\partial B^T}{\partial p} P_{12}^T$$

$$+ [(B^T P_{12} + \frac{\partial B^T}{\partial p} P_{22})E_{12}^T + B^T P_{12}^T E_{12} + B^T P_{22} E_{22}]E_{11}^\dagger\}, \tag{6.5-49}$$

where $E_{11}^\dagger$ is a generalized inverse matrix of $E_{11} = E_{11}^T$.

To compute K^o, E_{ij} and P_{ij} (i, j = 1, 2) must be determined first; recall that A, B, $\partial A/\partial p$, $\partial B/\partial p$, X, α, W_0, W and W_1 are known. From (6.5-36) with (6.5-34) and (6.5-37) it is seen that P_1 satisfies

$$\begin{bmatrix} A - BK & 0 \\ \frac{\partial(A - BK)}{\partial p} & A - BK \end{bmatrix}^T \begin{bmatrix} P_{11} & P_{12} \\ P_{12}^T & P_{22} \end{bmatrix} + \begin{bmatrix} P_{11} & P_{12} \\ P_{12}^T & P_{22} \end{bmatrix} \begin{bmatrix} A - BK & 0 \\ \frac{\partial(A - BK)}{\partial p} & A - BK \end{bmatrix}$$

$$+ \begin{bmatrix} W_0 + K^T WK & 0 \\ 0 & \alpha W_1 \end{bmatrix} = 0; \tag{6.5-50}$$

where $\partial(A-BK)/\partial p$ is given by (6.5-35). The above $2n \times 2n$ linear-matrix equation is equivalent to the following three sets of $n \times n$ linear-matrix equations

$$(A - BK)^T P_{22} + P_{22}(A - BK) + \alpha W_1 = 0, \tag{6.5-51}$$

$$(A - BK)^T P_{12} + P_{12}(A - BK) + \frac{\partial(A - BK)^T}{\partial p} P_{22} = 0, \tag{6.5-52}$$

$$(A - BK)^T P_{11} + P_{11}(A - BK) + W_0 + K^T WK$$

$$+ \frac{\partial(A - BK)^T}{\partial p} P_{12}^T + P_{12}\frac{\partial(A - BK)}{\partial p} = 0. \tag{6.5-53}$$

From (6.5-44) with (6.5-34) and (6.5-45) it is seen that E_1 satisfies

6.5. Trajectory – Sensitivity Optimization

$$\left[\begin{matrix} A-BK & 0 \\ \frac{\partial(A-BK)}{\partial p} & A-BK \end{matrix} \right] \left[\begin{matrix} E_{11} & E_{12} \\ E_{12}^T & E_{22} \end{matrix} \right] + \left[\begin{matrix} E_{11} & E_{12} \\ E_{12}^T & E_{22} \end{matrix} \right] \left[\begin{matrix} A-BK & 0 \\ \frac{\partial(A-BK)}{\partial p} & A-BK \end{matrix} \right]^T$$

$$+ \begin{bmatrix} X & 0 \\ 0 & 0 \end{bmatrix} = 0. \tag{6.5-54}$$

This 2n × 2n linear-matrix equation is equivalent to the following three sets of n × n linear-matrix equations

$$(A - BK)E_{11} + E_{11}(A - BK)^T + X = 0, \tag{6.5-55}$$

$$(A - BK)E_{12} + E_{12}(A - BK)^T + E_{11}\frac{\partial(A-BK)^T}{\partial p} = 0, \tag{6.5-56}$$

$$(A - BK)E_{22} + E_{22}(A - BK)^T + \frac{\partial(A-BK)}{\partial p}E_{12} + E_{12}^T\frac{\partial(A-BK)^T}{\partial p} = 0. \tag{6.5-57}$$

▼

ALGORITHM 6.5-1 (Trajectory-Sensitivity Minimization): To compute the optimal controller gain K^o, given in (6.5-49) we propose the following iterative procedure. Let $J = 1, 2, ..., N$ be the iteration number.

STEP 1 (To initialize and to compute a starting controller gain): Given A, B, $\partial A/\partial p$, $\partial B/\partial p$, X, α, W_0, W and W_1, set $\alpha = 0$, then $J_1 = J_p$ and $K_1 = W^{-1}B^TP_{11}$, with $A^TP_{11} + P_{11}A + W_0 - P_{11}BW^{-1}B^TP_{11} = 0$.

STEP 2 (To generate new $\tilde{P}_1$ and E_1): With the current K and the actual α solve (6.5-51) to (6.5-53) and (6.5-55) to (6.5-57), for $\tilde{P}_1$ and E_1, respectively.

STEP 3 (To update the controller gain): Compute a new K^o from (6.5-49) with the current values of P_{ij} and E_{ij} which are generated in Step 2.

STEP 4 (To test for the final solution): If $|(K_{i,j})_J - (K_{i,j})_{J-1}| < 10^{-8} * |(K_{i,j})_{J-1}|$, then stop, K_J is the final optimal solution. Otherwise go to Step 2 and set $J \to J + 1$.

∎

REMARK 6.5-2: Consider a *DCLTI – GS* described as follows.

$$\dot{x}(t, \theta) = A(\theta)x(t, \theta) + B(\theta)u(t), \tag{6.5-58}$$

where $\theta \in R^r$ is its vector of unknown but identifiable (or in our terminology, specifiable) parameters. The *optimal* input $\{u(t), 0 \le t \le T\}$, from the optimization point of view, refers to an input that maximizes a suitable norm of the information matrix M subject to suitable constraints [G. 18]. The information matrix is shown to be as follows [G. 22].

$$M = \int_0^T \left(\frac{\partial x}{\partial \theta}\right)^T W_s \left(\frac{\partial x}{\partial \theta}\right) dt, \qquad (6.5\text{-}59)$$

where $\partial x / \partial \theta$ is given by (2.4-2) and $W_s = W_s^T \geq 0$ is a weighting matrix.

The M^{-1} gives the Cramer-Rao lower bound for the covariance of an unbiased estimator of θ, which suggests that a proper way of choosing the input is to minimize a norm of this matrix. However, the direct minimization of, say $\text{Tr}(M^{-1})$ is very involved computationally, while the maximization of $\text{Tr}(M)$ or perhaps $\text{Tr}(WM)$, where W is another weighting matrix, results in a more manageable problem than the previous one. This maximization means to increase the state sensitivity (or eventually the output sensitivity) with respect to system parameters, if we wish to identify these parameters [G. 18]. Of course, the maximization of $\text{Tr}(WM)$ is equivalent to the minimization of $\text{Tr}(-WM)$, which is similar to the sensitivity cost J_{sr} given in (6.5-8). Thus the analysis in this section is very much applicable to this class of problems.

6.6. SENSITIVITY OPTIMIZATION WITH ADAPTIVE CONTROLLER

6.6.1. Introduction and Problem Statement

In Section 6.5 we introduce a method to develop an optimal controller gain which minimizes a combined-quadratic cost under the assumption that the controller gain is "constant" with respect to system-physical parameters. In other words, as in (6.5-12b) we let $\partial K/\partial p_j = 0$, for all j. In this section we study the same optimization problem with the assumption that $K = K(p)$ and thus (6.5-12b) must be modified accordingly to incorporate the nonzero term $\partial K/\partial p_j$. This choice for $K = K(p)$ is reasonable, although this choice opens a number of new questions. But first, we must point out that the implication of $K = K(p)$ is that either this controller is designed (hardware) adaptively, or it simply is computed with certain consideration for possible contributions of initial uncertainty about system-parameter-nominal values. Clearly, when this optimal controller becomes parameter independent, then the result of this section simplifies to that of the previous section.

PROBLEM 6.6-1 (Quadratic-Sensitivity Optimization with Adaptive Controller):
Consider the Problem 6.5-1 with $K = K(p)$ and thus

$$\frac{\partial (A - BK)}{\partial p_j} = \frac{\partial A}{\partial p_j} - \frac{\partial B}{\partial p_j} K - B \frac{\partial K}{\partial p_j}, \ j = 1, 2, \cdots, r. \qquad (6.6\text{-}1)$$

Find an optimal $K^{o\,a}$ such that $u(t; p) = -K^{o\,a}(p)x(t; p)$ minimizes the combined quadratic cost given by (6.5-9).

SOLUTION: The procedure to solve this problem is exactly the same as that of Problem 6.5-1, except in (6.5-12a) and (6.5-12b). The new (6.5-12b) is given in (6.6-1), which this equation results in a new (6.5-15), (6.5-16), (6.5-21) and, of

6.6. Sensitivity Optimization with Adaptive Controller

course, a new optimal controller gain $K^{o\,a}$. The final $K^{o\,a}$ is similar to (6.5-33) but its components must be computed considering (6.6-1). To compute this optimal adaptive controller gain numerically, however, we face an obvious question that how do we generate *a priori* $\partial K/\partial p_j$, for all j?

Without any loss of generality, and for the sake of simplicity, a single-parameter system is considered in the following. For this single-parameter case an answer to the preceding question and an algorithm for the numerical computations of optimal controller gain is proposed. For the case of multi-parameter system a similar algorithm, with appropriate modifications, can be established. This section closely follows [D. 14 & 15].

SUB-PROBLEM 6.6-1 (Single-Parameter Case): Suppose that in Problem 6.6-1, r = 1 and we let $p_1 \triangleq p$. Then the list of one to one corresponding equations between this case and Sub-Problem 6.5-1 is given next using the superscript "a" to distinguish this adaptive case from the previous case.

$$A_1^a(p) \triangleq \begin{bmatrix} A-BK & 0 \\ \dfrac{\partial(A-BK)}{\partial p} & A-BK \end{bmatrix}, \quad (6.6\text{-}2)$$

with $\partial(A-BK)/\partial p$ given by (6.6-1). Similarly we have

$$A_1^{aT} \tilde{P}_1^a + \tilde{P}_1^a A_1^a + \tilde{W}_1^a = 0, \quad (6.6\text{-}3)$$

$$\tilde{W}_1^a \triangleq \begin{bmatrix} W_0 + K^T W K & 0 \\ 0 & \alpha W_1 \end{bmatrix}, \quad (6.6\text{-}4)$$

$$\delta A_1^a(p) = \begin{bmatrix} -B\delta K & 0 \\ -\dfrac{\partial B}{\partial p}\delta K & -B\delta K \end{bmatrix}, \quad (6.6\text{-}5)$$

$$\delta \tilde{W}_1^a \triangleq \begin{bmatrix} \delta K^T W K + K^T W \delta K & 0 \\ 0 & 0 \end{bmatrix}. \quad (6.6\text{-}6)$$

With the above δA_1^a and $\delta \tilde{W}_1^a$ we have a parameter perturbation matrix which is

$$\Pi_1^a \triangleq \begin{bmatrix} \pi_{11}^a & \pi_{12}^a \\ \pi_{21}^a & \pi_{22}^a \end{bmatrix}. \quad (6.6\text{-}7)$$

Here

$$\pi_{11}^a \triangleq \delta K^T \phi_{11}^{aT} + \psi_{11}^a \delta K$$

$$\triangleq \delta K^T (WK - B^T P_{11}^a - \dfrac{\partial B^T}{\partial p} P_{12}^{aT}) + (K^T W - P_{11}^a B - P_{12}^a \dfrac{\partial B}{\partial p})\delta K, \quad (6.6\text{-}8)$$

$$\pi_{21}^{aT} = \pi_{12}^a$$

$$\triangleq \delta K^T(-B^T P_{12}^a - \frac{\partial B^T}{\partial p} P_{22}^a) - P_{12}^a B \, \delta K, \qquad (6.6\text{-}9)$$

$$\pi_{22}^a \triangleq -\delta K^T B^T P_{22}^a - P_{22}^a B \, \delta K. \qquad (6.6\text{-}10)$$

Similarly it is seen that

$$E_1^a \triangleq \begin{bmatrix} E_{11}^a & E_{12}^a \\ E_{12}^{aT} & E_{22}^a \end{bmatrix} \triangleq \int_0^\infty \exp(\tilde{A}_1^a t) M_1 \exp(\tilde{A}_1^a t)^T dt, \qquad (6.6\text{-}11)$$

where without any loss of generality it is assumed that

$$M_1 = \begin{bmatrix} X & 0 \\ 0 & 0 \end{bmatrix}, \qquad (6.6\text{-}12)$$

with $X = \text{cov}(x(0), x^T(0))$.

Computing the first variation of the cost yields

$$\delta J_1^a = \text{Tr}(\pi_{11}^a E_{11}^a + \pi_{12}^a E_{12}^{aT} + \pi_{12}^{aT} E_{12}^a + \pi_{22}^a E_{22}^a). \qquad (6.6\text{-}13)$$

Here (6.6-13) on substituting for π_{ij} and using certain trace operations, becomes

$$\delta J_1^a = \text{Tr}\{\delta K^T[(WK - B^T P_{11}^a - \frac{\partial B^T}{\partial p} P_{12}^{aT}) E_{11}^a - (B^T P_{12}^a + \frac{\partial B^T}{\partial p} P_{22}^a) E_{12}^{aT}$$

$$- B^T P_{12}^{aT} E_{12}^a - B^T P_{22}^a E_{22}^a] + [E_{11}^a (K^T W - P_{11}^a B - P_{12}^a \frac{\partial B}{\partial p})$$

$$- E_{12}^{aT} P_{12}^a B - E_{12}^a (P_{12}^{aT} B + P_{22}^a \frac{\partial B}{\partial p}) - E_{22}^a P_{22}^a B] \delta K\}. \qquad (6.6\text{-}14)$$

Thus

$$\frac{\partial J_1^a}{\partial K} = 2[(WK - B^T P_{11}^a - \frac{\partial B^T}{\partial p} P_{12}^{aT}) E_{11}^a$$

$$- (B^T P_{12}^a + \frac{\partial B^T}{\partial p} P_{22}^a) E_{12}^{aT} - B^T P_{12}^{aT} E_{12}^a - B^T P_{22}^a E_{22}^a]. \qquad (6.6\text{-}15)$$

Since $W = W^T > 0$, and the optimal adaptive controller gain must satisfy $\partial J_1^a / \partial K = 0$, we have

$$K^{o\,a} = W^{-1}\{B^T P_{11}^a + \frac{\partial B^T}{\partial p} P_{12}^{aT} \qquad (6.6\text{-}16)$$

$$+ [(B^T P_{12}^a + \frac{\partial B^T}{\partial p} P_{22}^a) E_{12}^{aT} + B^T P_{12}^{aT} E_{12}^a + B^T P_{22}^a E_{22}^a] E_{11}^{a\dagger}\}, \qquad (6.6\text{-}16)$$

where $E_{11}^{a\dagger}$ is a generalized inverse matrix of $E_{11}^a = E_{11}^{aT}$.

6.6. Sensitivity Optimization with Adaptive Controller

To compute $K^{o\,a}$ numerically, P_{ij}^a and E_{ij}^a ($i, j = 1, 2$) must be determined first; (recall that A, B, $\partial A/\partial p$, $\partial B/\partial p$, X, α, W_0, W and W_1 are known). We can show that P_{ij} must satisfy the following set of three $n \times n$ linear-matrix equations.

$$(A - BK)^T P_{22}^a + P_{22}^a(A - BK) + \alpha W_1 = 0, \qquad (6.6\text{-}17)$$

$$(A - BK)^T P_{12}^a + P_{12}^a(A - BK) + \frac{\partial (A - BK)^T}{\partial p} P_{22}^a = 0, \qquad (6.6\text{-}18)$$

$$(A - BK)^T P_{12}^a + P_{11}^a(A - BK) + W_0 + K^T W K$$
$$+ \frac{\partial (A - BK)^T}{\partial p} P_{12}^{aT} + P_{12}^a \frac{\partial (A - BK)}{\partial p} = 0. \qquad (6.6\text{-}19)$$

Similarly E_{ij} must satisfy the following set of three $n \times n$ linear-matrix equations.

$$0 = (A - BK)E_{11}^a + E_{11}^a(A - BK)^T + X, \qquad (6.6\text{-}20)$$

$$0 = (A - BK)E_{12}^a + E_{12}^a(A - BK)^T + E_{11}^a \frac{\partial (A - BK)^T}{\partial p}, \qquad (6.6\text{-}21)$$

$$0 = (A - BK)E_{22}^a + E_{22}^a(A - BK)^T + \frac{\partial (A - BK)}{\partial p} E_{12}^a + E_{12}^{aT} \frac{\partial (A - BK)^T}{\partial p}. \qquad (6.6\text{-}22)$$

Given the optimal adaptive controller gain, indicated in (6.6-16) and supported by the solution of (6.6-17) to (6.6-22), it might appear that the optimal controller gain has been established. This is not quite the case. One problem remains, the expression $\partial K/\partial p$ in the above is yet to be determined. Addressing this problem the following procedure is proposed.

The initial $\partial K/\partial p$ is approximated as follows. Set $\alpha = 0$, then from (6.6-17) $P_{22}^a = 0$, and from (6.6-18) $P_{12}^a = 0$, and from (6.6-19) we have

$$(A - BK)^T P_{11}^a + P_{11}^a(A - BK) + W_0 + K^T W K = 0. \qquad (6.6\text{-}23)$$

Correspondingly, the initial K becomes $K = W^{-1} B^T P_{11}^a$, and the initial $\frac{\partial K}{\partial p}$ is

$$\frac{\partial K}{\partial p} = W^{-1}(\frac{\partial B^T}{\partial p} P_{11}^a + B^T \frac{\partial P_{11}^a}{\partial p}). \qquad (6.6\text{-}24)$$

Here the initial P_{11}^a and $\partial P_{11}^a/\partial p$ are given, respectively, by

$$A^T P_{11}^a + P_{11}^a A + W_0 - P_{11}^a B W^{-1} B^T P_{11}^a = 0, \qquad (6.6\text{-}25)$$

$$(A - BK)^T \frac{\partial P_{11}^a}{\partial p} + \frac{\partial P_{11}^a}{\partial p}(A - BK) + \frac{\partial A^T}{\partial p} P_{11}^a$$
$$+ P_{11}^a \frac{\partial A}{\partial p} - P_{11}^a \frac{\partial (BW^{-1}B^T)}{\partial p} P_{11}^a = 0. \qquad (6.6\text{-}26)$$

To complete the computations of a sub-optimal $K^{o\,a}$, we assume that $\partial^2 K/\partial p^2 = 0$ and proceed as follows.

$$\frac{\partial^2(A-BK)}{\partial p^2} = \frac{\partial^2 A}{\partial p^2} - \frac{\partial^2 B}{\partial p^2}K - 2\frac{\partial B}{\partial p}\frac{\partial K}{\partial p}, \tag{6.6-27}$$

$$(A-BK)^T\frac{\partial P_{22}^a}{\partial p} + \frac{\partial P_{22}^a}{\partial p}(A-BK) + \frac{\partial(A-BK)^T}{\partial p}P_{22}^a + P_{22}^a\frac{\partial(A-BK)}{\partial p} = 0, \tag{6.6-28}$$

$$(A-BK)^T\frac{\partial P_{12}^a}{\partial p} + \frac{\partial P_{12}^a}{\partial p}(A-BK) + \frac{\partial(A-BK)^T}{\partial p}P_{12}^a + P_{12}^a\frac{\partial(A-BK)}{\partial p}$$
$$+ \frac{\partial^2(A-BK)^T}{\partial p^2}P_{22}^a + \frac{\partial(A-BK)^T}{\partial p}\frac{\partial P_{22}^a}{\partial p} = 0, \tag{6.6-29}$$

$$(A-BK)^T\frac{\partial P_{11}^a}{\partial p} + \frac{\partial P_{11}^a}{\partial p}(A-BK) + \frac{\partial(A-BK)^T}{\partial p}(P_{11}^a + \frac{\partial P_{12}^{aT}}{\partial p}) + (P_{11}^a + \frac{\partial P_{12}^a}{\partial p})\frac{\partial(A-BK)}{\partial p}$$
$$+ \frac{\partial}{\partial p}(K^T W K) + \frac{\partial^2(A-BK)^T}{\partial p^2}P_{12}^{aT} + P_{12}^a\frac{\partial^2(A-BK)}{\partial p^2} = 0, \tag{6.6-30}$$

$$(A-BK)\frac{\partial E_{11}^a}{\partial p} + \frac{\partial E_{11}^a}{\partial p}(A-BK)^T + \frac{\partial(A-BK)}{\partial p}E_{11}^a + E_{11}^a\frac{\partial(A-BK)^T}{\partial p} = 0, \tag{6.6-31}$$

$$(A-BK)\frac{\partial E_{12}^a}{\partial p} + \frac{\partial E_{12}^a}{\partial p}(A-BK)^T + \frac{\partial(A-BK)}{\partial p}E_{12}^a + (E_{12}^a + \frac{\partial E_{11}^a}{\partial p})\frac{\partial(A-BK)^T}{\partial p}$$
$$+ E_{11}^a\frac{\partial^2(A-BK)^T}{\partial p^2} = 0, \tag{6.6-32}$$

$$(A-BK)\frac{\partial E_{22}^a}{\partial p} + \frac{\partial E_{22}^a}{\partial p}(A-BK)^T + \frac{\partial(A-BK)}{\partial p}(E_{22}^a + \frac{\partial E_{12}^a}{\partial p})$$
$$+ (E_{22}^a + \frac{\partial E_{12}^{aT}}{\partial p})\frac{\partial(A-BK)^T}{\partial p} + \frac{\partial^2(A-BK)}{\partial p^2}E_{12}^a + E_{12}^{aT}\frac{\partial^2(A-BK)^T}{\partial p^2} = 0, \tag{6.6-33}$$

and finally

$$\frac{\partial K^{o\,a}}{\partial p} = W^{-1}\left\{\frac{\partial B^T}{\partial p}P_{11}^a + B^T\frac{\partial P_{11}^a}{\partial p} + \frac{\partial^2 B^T}{\partial p^2}P_{12}^{aT} + \frac{\partial B^T}{\partial p}\frac{\partial P_{12}^{aT}}{\partial p}\right.$$
$$\left. + \frac{\partial}{\partial p}\{[(B^T P_{12}^a + \frac{\partial B^T}{\partial p}P_{22}^a)E_{12}^{aT} + B^T P_{12}^{aT}E_{12}^a + B^T P_{22}^a E_{22}^a]E_{11}^{a\dagger}\}\right\}. \tag{6.6-34}$$

Here $E_{11}^{a\dagger}$ is a generalized inverse of $E_{11}^a = E_{11}^{aT}$.

Clearly, when we reach the end of the first iteration (for computation of an up-dated $\partial K/\partial p$) we need values such as $\partial P_{ij}^a/\partial p$ and $\partial E_{ij}^a/\partial p$ (i, j = 1, 2). These in

6.6. Sensitivity Optimization with Adaptive Controller

turn require the knowledge of $\partial^2 K/\partial p^2$, which we set equal to zero. Otherwise, if we desire to compute the optimal adaptive controller gain as accurately as possible, then we need to compute the higher order $\partial^i K/\partial p^i$, $i \geq 2$, for which the computational cost increases exponentially. This difficulty as severe as it seems is manageable because in most cases the controllers are linear functions of system parameters, therefore to set the higher-order derivatives of these gains equal to zero is not unreasonable.

▼

6.7. SENSITIVITY – MEASURE OPTIMIZATION FOR SYSTEMS WITH LARGE PARAMETER VARIATIONS

6.7.1. Introduction

The apparent discrepancy between the nominal and actual trajectories of any system has concerned many researchers. Irrespective of what might have been the cause of this discrepancy from engineering point of view this situation must be manageable. In particular, when the difference between the nominal (model) and actual (real-operating) trajectory becomes *large* due to large-parameter variations, the conventional *small*-sensitivity analyses and syntheses of Section 6.5 are no longer useful. Because in that section we are concerned with *small*-parameter variations which means $|\Delta p| \ll 1$ or $|\Delta p|^2 \approx 0$, and the sensitivity measure is the first derivative of a particular signal with respect to the varying parameter p. This derivative is sufficient for evaluating the new value of that signal and thus this first derivative is a valid system-sensitivity measure when $|\Delta p|^2 \approx 0$. On the other hand, if $|\Delta p|^2 \neq 0$, then we cannot benefit from the previous analysis which is based on a Taylor-series expansion, because we need to evaluate the higher-order derivatives of signal with respect to parameters and that approach soon becomes prohibitive and costly to say the least. In other words, when the parameter variations are *large* resulting in the difference between the nominal (model) and actual (real-operating) trajectories becomes *large,* we must use directly this difference as our system sensitivity measure. This measure is called system *large-sensitivity measure*. The rationale behind this choice is that the first-order derivatives do not provide sufficient information about the large changes which are taking place in the system and thus the minimization of these derivatives (*small-sensitivity measures*) is a useless task. Perhaps, the main shortcoming of minimizing *small-sensitivity measure* of a system is that this approach does not account for the inevitable interactions among various parameters which are changing simultaneously. The purpose of this section is to provide synthesis techniques which result in the optimal controller gains in spite of *large-parameter variations* for system of *Case 2* in Table 6.1-1.

Since the criterion to achieve the above goal is to have the nominal and an actual system trajectories close to each other, it is clear that to accomplish this task some functional of this difference must be minimized. This difference which is considered as the sensitivity measure of the system, expresses a degree of tracking between the nominal and actual system trajectories. However, when we are studying this new system-sensitivity measure and its minimization, it may seem to some researchers that we are studying a "tracking problem" instead of a parametric-sensitivity minimization problem. Indeed we strongly believe that such problem is inherently a tracking problem, because whenever we minimize a given system-sensitivity measure, in effect, we expect that the real-operating system tracks an ideal system and in that sense every sensitivity minimization problem becomes a "distance" minimization or a tracking problem. Another advantage of the large-sensitivity measure is that the entire analysis can be performed from a lower-dimensional matrix equation than is required in the small-sensitivity analysis regardless of the number of varying parameters; and this approach to a large extent accounts for interactions among various parameters. Thus the *large-sensitivity measure* is a sound choice and its minimization has engineering applications.

The algorithms in this section are based on quadratic minimization of the projected "effects" for some "worst"-parameter variations that may happen on the actual system trajectory. Here we anticipate a set of parameter variations and we design the controller gains based on that choice (set); or equivalently, we can use these algorithms to generate some sort of boundaries for acceptable-parameter variations in the sense of minimizing a quadratic − *PI*. Indeed, to construct this class of systems that are optimal in this quadratic sense is the ultimate goal of any system synthesis which incorporates such large-parameter variations schemes in order to generate optimal controller gains. These algorithms may also be used to design adaptive controllers by intermittent adaptation, which is a very desirable features that assures robustness in the system operation.

The following analysis is based on the derivation of an optimal controller gain so as a combined-quadratic cost; consisting of the standard regulator cost and an appropriate norm of the weighted system-response variation due to the system *large*-parameter variations, is minimized. This optimal controller gain, depending on the original open-loop system, may have either one component corresponding to trajectory feedback or two components corresponding to both trajectory and the model feedback. Furthermore, this result directly provides certain parameter-variations bounds, or a hyperbox in the *PVS*, such that the resulting system remains "optimal" in that hyperbox (of course, with respect to a class of quadratic − *PI's*). A numerical algorithm for computations of optimal controller gains as well as a numerical example to compare the system performance under these designs methodologies are also included. These results by and large are more applicable to engineering problems than those resulted from optimization of small-sensitivity measures which have been extensively studied during the last three decades. This section closely follows [*D*. 10 to 13]. Similar results are also reported in [*D*. 34, 35 and 57].

6.7.2. Problem Statement

Consider a real-operating system whose state and the corresponding real-response are described, respectively, by

$$\dot{\xi}(t) = (A + \Delta A)\xi(t) + (B + \Delta B)u(t), \quad \xi(0) = \xi_o, \quad (6.7\text{-}1)$$

$$\zeta(t) = (H + \Delta H)\xi(t) + (F + \Delta F)u(t). \quad (6.7\text{-}2)$$

Suppose that this real-operating system is *a* perturbed system corresponding to *the* nominal system state and the corresponding nominal response described, respectively, by

$$\dot{x}(t) = Ax(t) + Bu(t), \quad x(0) = x_o, \quad (6.7\text{-}3)$$

$$z(t) = Hx(t) + Fu(t). \quad (6.7\text{-}4)$$

We assume that $x \in R^n$; $\xi \in R^n$; $z \in R^\omega$; $\zeta \in R^\omega$; $u \in R^m$; and A, B, H, and F are matrices of the appropriate dimensions. The perturbation matrices ΔA, ΔB, ΔH, and ΔF are generated with respect to the nominal values of system matrices A, B, H, and F, respectively. These perturbations are subsequently constrained numerically.

Augmenting (6.7-1) and (6.7-3), results in

$$\dot{\eta}(t) \triangleq \begin{Bmatrix} \dot{\xi}(t) \\ \dot{x}(t) \end{Bmatrix} = \begin{bmatrix} A + \Delta A & 0 \\ 0 & A \end{bmatrix} \begin{Bmatrix} \xi(t) \\ x(t) \end{Bmatrix} + \begin{bmatrix} B + \Delta B \\ B \end{bmatrix} u(t), \quad (6.7\text{-}5a)$$

$$\eta(0) = \begin{Bmatrix} \xi(0) \\ x(0) \end{Bmatrix}. \quad (6.7\text{-}5b)$$

The objective as "sensitivity-measure optimization", is to minimize the following combined-quadratic cost subject to (6.7-5).

$$\tilde{J}_l = \min \int_0^\infty \{[\zeta(t) - z(t)]^T W_s[\zeta(t) - z(t)] + z^T(t)W_0 z(t) + u^T(t)Wu(t)\}dt; \quad (6.7\text{-}6)$$

where the subscript "*l*" stands for the *large*-parameter variations; and the weighting matrices $W_s = W_s^T \geq 0$, $W_0 = W_0^T \geq 0$, and $W = W^T > 0$ are of appropriate dimensions. In (6.7-6), $\zeta(t) - z(t)$ (instead of $\frac{\partial \eta}{\partial p_j}$ in Section 6.5) represents the tracking error or the *large*-sensitivity measure of the system.

Before simplifying the above problem, we note that in reality, the designer has some expectation about possible inaccuracy of system model or some possible worst-parameter variations. Thus the implication of the present study is that the designer uses this information either to anticipate the worst that can happen to the system, or to incorporate the most pessimistic situation in regard to parameter variations when designing the optimal controller gain.

The combined-quadratic cost given in (6.7-6), on using (6.7-2) and (6.7-4), can be rewritten as

$$\tilde{J}_l = \min\int_0^\infty \{[(H + \Delta H)\xi(t) + (F + \Delta F)u(t) - Hx(t) - Fu(t)]^T W_s[(H + \Delta H)\xi(t) +$$

$$(F+\Delta F)u(t) - Hx(t) - Fu(t)] + [Hx(t)+Fu(t)]^T W_0[Hx(t)+Fu(t)] + u^T(t)Wu(t)\}dt$$

$$\triangleq \min\int_0^\infty \{[\xi^T, x^T, u^T]W_s'[\xi^T, x^T, u^T]^T + [x^T, u^T]W_0'[x^T, u^T]^T + u^T Wu\}dt; \quad (6.7\text{-}7)$$

where the dependence on t is implied, and

$$W_s' \triangleq \begin{bmatrix} (H + \Delta H)^T \\ -H^T \\ \Delta F^T \end{bmatrix} W_s \begin{bmatrix} H + \Delta H, & -H, & \Delta F \end{bmatrix}$$

$$= \begin{bmatrix} (H + \Delta H)^T W_s(H + \Delta H) & -(H + \Delta H)^T W_s H & (H + \Delta H)^T W_s \Delta F \\ -H^T W_s(H + \Delta H) & H^T W_s H & -H^T W_s \Delta F \\ \Delta F^T W_s(H + \Delta H) & -\Delta F^T W_s H & \Delta F^T W_s \Delta F \end{bmatrix}, \quad (6.7\text{-}8)$$

with

$$W_0' \triangleq \begin{bmatrix} H^T \\ F^T \end{bmatrix} W_0(H, F) = \begin{bmatrix} H^T W_0 H & H^T W_0 F \\ F^T W_0 H & F^T W_0 F \end{bmatrix}. \quad (6.7\text{-}9)$$

Imbedding matrices in (6.7-7) results in

$$\tilde{J}_l = \min\int_0^\infty [\xi^T, x^T, u^T] Q[\xi^T, x^T, u^T]^T dt; \quad (6.7\text{-}10)$$

where

$$Q \triangleq \begin{bmatrix} (H + \Delta H)^T W_s(H + \Delta H) & -(H + \Delta H)^T W_s H & (H + \Delta H)^T W_s \Delta F \\ -H^T W_s(H + \Delta H) & H^T(W_0 + W_s)H & H^T(W_0 F - W_s \Delta F) \\ \Delta F^T W_s(H + \Delta H) & (F^T W_0 - \Delta F^T W_s)H & \Delta F^T W_s \Delta F + F^T W_0 F + W \end{bmatrix}.$$

$$(6.7\text{-}11)$$

We recall that the objective is to minimize $\tilde{J}_l$ given in (6.7-10) subject to the differential equation (6.7-5). In general, and in regard to choosing u(t) in (6.7-5), we have two ways to consider this optimization problem. Namely, to have trajectory feedback or to have both trajectory and model feedback.

6.7.3. Optimal Controller Gain with Trajectory Feedback

Suppose that the system input is chosen from the real-operating trajectory as

$$u(t) = R\xi(t). \quad (6.7\text{-}12)$$

Because in reality the trajectory which is available for feedback is $\xi(t)$ and not the nominal trajectory which is x(t) and it must be reconstructed. However, this choice

6.7. Sensitivity – Measure Optimization, Large Parameter Variations

of u(t) is valid if A is a stable matrix. Otherwise, the implication of this requirement is that in order to have a bounded cost, A must be first stabilized. Then the differential equation (6.7-5), on using (6.7-12), becomes

$$\dot{\eta}(t) = \begin{bmatrix} A' + B'R & 0 \\ BR & A \end{bmatrix} \eta(t) \triangleq A_l \, \eta(t); \qquad (6.7\text{-}13)$$

where $A' \triangleq A + \Delta A$, and $B' \triangleq B + \Delta B$. Correspondingly, the combined-quadratic cost $\tilde{J}_l$ becomes

$$\tilde{J}_l = \min \int_0^\infty \eta^T(t) \begin{bmatrix} I & 0 & R^T \\ 0 & I & 0 \end{bmatrix} Q \begin{bmatrix} I & 0 \\ 0 & I \\ R & 0 \end{bmatrix} \eta(t) dt; \qquad (6.7\text{-}14)$$

where I is an identity matrix. On substituting the solution of (6.7-13) for $\eta(t)$ into (6.7-14), and using a statistical average of the initial condition $\eta(0)$, results in the following cost.

$$J_l = \min \{\eta^T(0) \int_0^\infty \exp(A_l^T t) W_l \exp(A_l t) dt \, \eta(0)\}$$

$$= \min \mathrm{Tr} \{ \int_0^\infty \exp(A_l^T t) W_l \exp(A_l t) \, dt \, [\eta(0)\eta^T(0)] \}$$

$$\triangleq \min \mathrm{Tr}(P_l H), \qquad (6.7\text{-}15)$$

where

$$W_l \triangleq \begin{bmatrix} W_{l11} & W_{l12} \\ W_{l12}^T & W_{l22} \end{bmatrix} \triangleq \begin{bmatrix} I & 0 & R^T \\ 0 & I & 0 \end{bmatrix} Q \begin{bmatrix} I & 0 \\ 0 & I \\ R & 0 \end{bmatrix}$$

$$= \begin{bmatrix} Q_{11} + R^T Q_{13}^T + Q_{13}R + R^T Q_{33} R & Q_{12} + R^T Q_{23}^T \\ Q_{12}^T + Q_{23} R & Q_{22} \end{bmatrix}, \qquad (6.7\text{-}16)$$

and $P_l = P_l^T > 0$ is the unique solution of the following Lyapunov equation

$$A_l^T P_l + P_l A_l + W_l = 0, \qquad (6.7\text{-}17)$$

with

$$H \triangleq \mathrm{cov}(\eta(0), \eta^T(0)). \qquad (6.7\text{-}18)$$

PROBLEM 6.7-1 *(Large-Sensitivity Minimization with Trajectory Feedback):* Compute the optimal controller gain R, such that $u(t) = R\xi(t)$ minimizes the quadratic cost (6.7-15) subject to the differential equation (6.7-13).

SOLUTION: The necessary condition to minimize J_l given in (6.7-15) is to have $\partial J_l / \partial R = 0$. To compute this quantity we propose the following procedure.

Let $R \to R + \delta R$, with δR being "small". Then $A_l \to A_l + \delta A_l$ such that $\delta A_l \to 0$ as $\delta R \to 0$ and $A_l + \delta A_l$ remains a stability matrix. Also, as $R \to R + \delta R$, with δR being "small", then $P_l \to P_l + \delta P_l$; where (cf., Lemma 6.4-1),

$$\delta P_l = \int_0^\infty \exp(A_l^T t)(\delta A_l^T P_l + P_l \delta A_l + \delta W_l)\exp(A_l t)dt. \qquad (6.7\text{-}19)$$

Correspondingly, from (6.7-15), if $P_l \to P_l + \delta P_l$, then $J_l \to J_l + \delta J_l$, where

$$\delta J_l = \text{Tr}(\delta P_l H). \qquad (6.7\text{-}20)$$

Substituting (6.7-19) into (6.7-20), and using some trace operations, results in

$$\delta J_l = \text{Tr}(\Pi_l E_l); \qquad (6.7\text{-}21)$$

where

$$\Pi_l \triangleq (\delta A_l^T P_l + P_l \delta A_l + \delta W_l), \qquad (6.7\text{-}22)$$

$$E_l \triangleq \int_0^\infty \exp(A_l t) H \exp(A_l^T t) dt. \qquad (6.7\text{-}23)$$

From (6.7-13) it is seen that if $R \to R + \delta R$, then

$$\delta A_l \triangleq \begin{bmatrix} B'\delta R & 0 \\ B\delta R & 0 \end{bmatrix}. \qquad (6.7\text{-}24)$$

From (6.7-16) it is clear that if $R \to R + \delta R$, then

$$\delta W_l \triangleq \begin{bmatrix} \delta R^T(Q_{13}^T + Q_{33}R) + (R^T Q_{33} + Q_{13})\delta R & \delta R^T Q_{23}^T \\ Q_{23}\delta R & 0 \end{bmatrix}. \qquad (6.7\text{-}25)$$

Therefore, substituting (6.7-24) and (6.7-25) into (6.7-22), with P_l being partitioned accordingly, we can show that

$$\Pi_l \triangleq \begin{bmatrix} \pi_{l11} & \pi_{l12} \\ \pi_{l12}^T & \pi_{l22} \end{bmatrix}; \qquad (6.7\text{-}26)$$

where $\pi_{l22} = 0$ and

$$\pi_{l11} \triangleq \delta R^T(B'^T P_{l11} + B^T P_{l12}^T + Q_{13}^T + Q_{33}R)$$

$$+ (P_{l11}B' + P_{l12}B + Q_{13} + R^T Q_{33})\delta R, \qquad (6.7\text{-}27)$$

$$\pi_{l12} \triangleq \delta R^T(B^T P_{l22} + B'^T P_{l12} + Q_{23}^T). \qquad (6.7\text{-}28)$$

Also, E_l of (6.7-23) can be similarly partitioned. Then (6.7-21), on using this partitioned E_l with (6.7-27) and (6.7-28), becomes

$$\delta J_l = \text{Tr}(\pi_{l11}E_{l11} + \pi_{l12}E_{l12}^T + \pi_{l12}^T E_{l12}). \qquad (6.7\text{-}29)$$

6.7. Sensitivity – Measure Optimization, Large Parameter Variations

Substituting for π_{lij} into (6.7-29) and using some trace operations results in

$$\delta J_l = \text{Tr}\{\delta R^T[(B'^T P_{l11} + B^T P_{l12}^T + Q_{13}^T + Q_{33}R)E_{l11}$$
$$+ (B^T P_{l22} + B'^T P_{l12} + Q_{23}^T)E_{l12}^T] + [E_{l12}(P_{l22}B + P_{l12}B' + Q_{23})$$
$$+ E_{l11}(P_{l11}B' + P_{l12}B + Q_{13} + R^T Q_{33})]\delta R\}. \tag{6.7-30}$$

Using $\dfrac{\partial \text{Tr}(AB)}{\partial A} = B^T$ and $\dfrac{\partial \text{Tr}(A^T B)}{\partial A} = B$, we can show that

$$\frac{\partial J_l}{\partial R} = \lim_{\delta R \to 0} \frac{\delta J_l}{\delta R} = 2[(B'^T P_{l11} + B^T P_{l12}^T + Q_{13}^T + Q_{33}R)E_{l11}$$
$$+ (B^T P_{l22} + B'^T P_{l12} + Q_{23}^T)E_{l12}^T]. \tag{6.7-31}$$

The optimality of R requires that $\dfrac{\partial J_l}{\partial R} = 0$, i.e.,

$$(B'^T P_{l11} + B^T P_{l12}^T + Q_{13}^T + Q_{33}R)E_{l11} + (B^T P_{l22} + B'^T P_{l12} + Q_{23}^T)E_{l12}^T = 0. \tag{6.7-32}$$

Thus the optimal controller gain must necessarily satisfy the following equation

$$R^o = -Q_{33}^{-1}\{B'^T P_{l11} + B^T P_{l12}^T + Q_{13}^T + (B^T P_{l22} + B'^T P_{l12} + Q_{23}^T)E_{l12}^T E_{l11}^\dagger\}, \tag{6.7-33}$$

where $E_{l11}^\dagger$ is a generalized inverse matrix of $E_{l11} = E_{l11}^T$ (which is used only for a symbolic reason and does not play any other role), and $Q_{33} = Q_{33}^T > 0$. To prove the sufficiency requirement for this minimization we need to develop $\dfrac{\partial^2 J_l}{\partial R^2}$ to test its sign. This derivation is purely algebraic and is not reported herein. ▼

ALGORITHM 6.7-1 (Optimal Controller Gain with Trajectory Feedback): To compute the above R^o numerically, we propose the following steps. Let $J = 1, 2, ..., N$ be the iteration number.

STEP 1.0 (To initialize): Given, B, H, F, A', B', H', F', H (or X), W, W_0, and W_1, we invoke the following steps to compute R^o.

STEP 1.1 (Generating P_l): From (6.7-17), the P_l can be determined from the following $2n \times 2n$ linear-matrix equation

$$\begin{bmatrix} A' + B'R & 0 \\ BR & A \end{bmatrix}^T \begin{bmatrix} P_{l11} & P_{l12} \\ P_{l12}^T & P_{l22} \end{bmatrix} + \begin{bmatrix} P_{l11} & P_{l12} \\ P_{l12}^T & P_{l22} \end{bmatrix} \begin{bmatrix} A' + B'R & 0 \\ BR & A \end{bmatrix}$$
$$+ \begin{bmatrix} Q_{11} + R^T Q_{13}^T + Q_{13}R + R^T Q_{33}R & Q_{12} + R^T Q_{23}^T \\ Q_{12}^T + Q_{23}R & Q_{22} \end{bmatrix} = 0. \tag{6.7-34}$$

As before, we replace the above $2n \times 2n$ linear-matrix equation with the following three sets of n × n linear-matrix equations:

$$A^T P_{l22} + P_{l22}A + Q_{22} = 0, \tag{6.7-35}$$

$$(A' + B'R)^T P_{l12} + P_{l12}A + Q_{12} + R^T(B^T P_{l22} + Q_{23}^T) = 0, \tag{6.7-36}$$

$$(A' + B'R)^T P_{l11} + P_{l11}(A' + B'R) + Q_{11}$$
$$+ R^T(Q_{13}^T + B^T P_{l12}^T) + (Q_{13} + P_{l12}B)R + R^T Q_{33}R = 0. \tag{6.7-37}$$

By assumption on A being a stability matrix and the fact that the ultimate R must stabilize $A' + B'R$ (actually we can prove this fact easily), all of the above three equations can be easily computed with minor modifications of the algorithm for solving a linear-matrix equation.

STEP 1.2 (Generating E_l): Similarly E_l of (6.7-23) can be determined as follows.

$$A_l E_l + E_l A_l^T + H = 0, \tag{6.7-38}$$

where the above 2n × 2n linear-matrix equation is equivalent to three sets of n × n linear-matrix equations as follows.

$$(A' + B'R)E_{l11} + E_{l11}(A' + B'R)^T + X = 0, \tag{6.7-39}$$

$$(A' + B'R)E_{l12} + E_{l12}A^T + E_{l11}R^T B^T = 0, \tag{6.7-40}$$

$$AE_{l22} + E_{l22}A^T + BRE_{l12} + E_{l12}^T R^T B^T + X = 0. \tag{6.7-41}$$

STEP 1.3 (Construction of a truncated optimal gain): First we choose R to be

$$R = -Q_{33}^{-1}(B'^T P_{l11} + B^T P_{l12}^T + Q_{13}^T). \tag{6.7-42}$$

That means we ignore the contributions of the remaining terms in R^o which are introduced because of sensitivity reduction algorithm.

STEP 1.4 (Computation of the initial P_l): The above choice for R does not change (6.7-35), but substituting this R into (6.7-36) results in

$$[A' - B'Q_{33}^{-1}(B'^T P_{l11} + Q_{13}^T)]^T P_{l12} + P_{l12}[A - BQ_{33}^{-1}(B^T P_{l22} + Q_{23}^T)]$$
$$+ Q_{12} - (Q_{13} + P_{l11}B')Q_{33}^{-1}(B^T P_{l22} + Q_{23}^T) - P_{l12}BQ_{33}^{-1}B'^T P_{l12} = 0. \tag{6.7-43}$$

Similarly, substituting (6.7-42) into (6.7-37) yields

$$A_*^T P_{l11} + P_{l11}A_* + Q_* - P_{l11}B'Q_{33}^{-1}B'^T P_{l11} = 0; \tag{6.7-44}$$

where

$$A_* \triangleq A' - B'Q_{33}^{-1}(Q_{13}^T + B^T P_{l12}^T), \tag{6.7-45}$$

and

6.7. Sensitivity – Measure Optimization, Large Parameter Variations

$$Q_* \triangleq Q_{11} - (Q_{13} + P_{l12}B)Q_{33}^{-1}(B^T P_{l12}^T + Q_{13}^T). \quad (6.7\text{-}46)$$

It is clear that equations (6.7-43) and (6.7-44) are two coupled quadratic-matrix equations which generally need some additional iterative techniques for constructing their solutions. The following procedure is proposed.

In (6.7-45) and (6.7-46), it is assumed that $P_{l12} = 0$ (i.e., we consider only the trajectory-norm minimization as the starting point). Then P_{l11} becomes

$$(A' - B'Q_{33}^{-1}Q_{13}^T)^T P_{l11} + P_{l11}(A' - B'Q_{33}^{-1}Q_{13}^T)$$

$$+ Q_{11} - Q_{13}Q_{33}^{-1}Q_{13}^T - P_{l11}B'Q_{33}^{-1}B'^T P_{l11} = 0. \quad (6.7\text{-}47)$$

Equation (6.7-47) can be routinely solved for P_{l11} if $Q_{11} - Q_{13}Q_{33}^{-1}Q_{13}^T \geq 0$ (otherwise we need to develop a special routine to solve this equation for P_{l11}). Substituting this P_{l11} into (6.7-43) results in the initial P_{l12}. Then this P_{l12} is substituted into (6.7-45) and (6.7-46) in order to compute a new P_{l11} from (6.7-44). This process continues until the "best" of P_{l11} and P_{l12} are found according to the next comment.

COMMENT 6.7-1: Generally speaking by the "best" P_{l11} and P_{l12}, we mean that at certain iteration number J we have

$$\left| [(P_{l11})_{i,j}]_J - [(P_{l11})_{i,j}]_{J-1} \right| \leq \text{Tol.*} \left| [(P_{l11})_{i,j}]_{J-1} \right|, \quad (6.7\text{-}48)$$

and

$$\left| [(P_{l12})_{i,j}]_J - [(P_{l12})_{i,j}]_{J-1} \right| \leq \text{Tol.*} \left| [(P_{l12})_{i,j}]_{J-1} \right|. \quad (6.7\text{-}49)$$

Here the Tol. is a prespecified tolerance or the rate of convergence for this iterative scheme. Furthermore, the final P_l which we have iteratively computed using the above partitions, must be tested for its positive definiteness.

STEP 1.5 (Computation of the initial truncated optimal gain): Substituting the above P_l into (6.7-42), results in a truncated optimal controller gain R which serves as the initial "value" for R.

STEP 1.6 (Computation of the optimal gain R^o): Substituting the above initial "value" for R into (6.7-35) to (6.7-37) and (6.7-39) to (6.7-40) results in new P_{lij} and E_{lij} (i, j = 1, 2). Correspondingly, substituting all of these new "values" for P_{lij} and E_{lij} (i, j = 1, 2), into the optimal controller gain (6.7-33) yields a new "value" for that controller gain. This process continues until the new optimal controller gain converges (within the prespecified rate of convergence) to a fixed value. In general, we use the optimal cost (6.7-15) as the figure of merit for stopping the iterative algorithm. ∎

REMARKS 6.7-1: (i) In the process of computing the above optimal controller gain we assume that in (6.7-11) $Q = Q^T \geq 0$. Furthermore, to compute P_{l11} of (6.7-44) we assume that $Q_* = Q_*^T \geq 0$, where Q_* is given by (6.7-46) with the following components

$$Q_{11} \triangleq (H + \Delta H)^T W_s (H + \Delta H), \tag{6.7-50}$$

$$Q_{13} \triangleq (H + \Delta H)^T W_s \Delta F, \tag{6.7-51}$$

$$Q_{33} \triangleq \Delta F^T W_s \Delta F + F^T W_0 F + W. \tag{6.7-52}$$

Also we must recall that by assumption: A is asymptotically stable, and $Q_{11} - Q_{13} Q_{33}^{-1} Q_{13}^T \geq 0$ for P_{l11} to be a positive- (semi-) definite matrix. First of all, the assumption that A is asymptotically stable means we do not need to use an initial feedback to stabilize A. Secondly, the positive (semi-) definitiness of Q and Q_* as well as $Q_{11} - Q_{13} Q_{33}^{-1} Q_{13}^T \geq$ provide us with certain measure of the *largest* perturbations for matrices A, B, H, and F such that an optimal solution exists. Should any one of these inequalities fail, the perturbation ought to be changed accordingly, or pictorially the optimality holds in a *smaller* hyperbox than before. Admittedly there is no unique way of attaining the previous set of inequalities, but it seems reasonable to have such a constraint. After all, if there were no constraints or bounds on the system perturbations, implying that the system is absolutely 'insensitive' with respect to the parameter variations, then the results were too good to be true. On the other hand, this non-uniqueness of the *large-parameter variations*, is what we should expect from such robustness analysis, since these parameter variations may compensate for each other, as the result of interactions among them. That means physically as well as mathematically speaking, there are many paths between any two points in the *PVS,* therefore when more than one parameter change, then some of these changes may vary the "troublesome" variational interval of the other parameters and may result in a smaller (or a larger) such interval.

(ii) We note that in general, the inverse of the previous algorithm is of interest. Here we choose certain worst cases that may occur in a system operation and design the optimal gain for that worse situation. Certainly we must seek a set of allowable-parameter variations for our final choice (set). We thus generate a class of systems which are optimal in the sense of a quadratic cost, and that construction is the true meaning of this robust design.

(iii) We must show that the initial $A' + B'R$ is indeed a stability matrix. Because if (6.7-44) holds, then (6.7-44) results in a solution $P_{l11} > 0$. By the inverse optimal problem this P_{l11} results in a gain matrix R_* such that for the pair (A_*, B'), $A_* + B'R_*$ is asymptotically stable, where $R_* = -Q_{33}^{-1} B'^T P_{l11}$, and A_* is given by (6.7-45). Thus

$$A_* + B'R_* = A' - B'Q_{33}^{-1}(Q_{13}^T + B^T P_{l12}^T) - B'Q_{33}^{-1} B'^T P_{l11}$$

$$= A' - B'Q_{33}^{-1}(B'^T P_{l11} + B^T P_{12}^T + Q_{13}^T), \tag{6.7-53}$$

is asymptotically stable. A quick look at (6.7-53) reveals that indeed $A_* + B'R_*$ equals $A' + B'R$.

6.7. Sensitivity – Measure Optimization, Large Parameter Variations

(iv) The structure of R^o in (6.7-33) is as follows

$$R^o = - Q_{33}^{-1}(B'^T P_{l11} + \rho), \qquad (6.7\text{-}54)$$

where ρ is the remaining terms of (6.7-33). To prove that this R^o ultimately stabilizes $A' + B'R^o$ the procedure is exactly the same as that used for truncated R in *(iii)*, above.

(v) The present study considers an extremely pessimistic circumstance in regard to the parameter perturbations and the single choice of $u(t) = R\xi(t)$, i.e., no feedback of the system model is used. Thus it seems that too much has been asked, and indeed the optimal controller gain in (6.7-33) may not always be needed. In other words, the truncated R, given in (6.7-42), might as well be sufficient for most practical purposes.

6.7.4. Optimal Controller Gain with Both Trajectory and Model Feedback

Now, suppose that the input $u(t)$ in (6.7-5) is chosen as a linear function of both trajectory and model states:

$$u(t) \triangleq (R_1, R_2) \begin{Bmatrix} \xi(t) \\ x(t) \end{Bmatrix} = R_1 \xi(t) + R_2 x(t), \qquad (6.7\text{-}55)$$

where "trajectory" refers to the real-operating state of the system, i.e., $\xi(t)$, and "model" refers to the nominal state of the system, i.e., $x(t)$, which must be reconstructed. We must emphasize that to reconstruct the state of the ideal model (or any sensitivity model) for feedback purposes is an expensive proposition. Nevertheless this case of large-sensitivity model is more manageable than the corresponding case of small-sensitivity measures whose high dimensionality soon becomes very problematic.

Considering (6.7-55), the augmented differential equation (6.7-5) is

$$\frac{d}{dt} \begin{Bmatrix} \xi(t) \\ x(t) \end{Bmatrix} = \begin{bmatrix} A' + B'R_1 & B'R_2 \\ BR_1 & A + BR_2 \end{bmatrix} \begin{Bmatrix} \xi(t) \\ x(t) \end{Bmatrix} \triangleq A'_l \eta'(t), \qquad (6.7\text{-}56)$$

with $\eta'(0) \equiv \eta(0)$. Here the new combined-quadratic cost is

$$J'_l = \min \int_0^\infty \eta'^T(t) W'_l \eta'(t)\, dt, \qquad (6.7\text{-}57)$$

where

$$W'_l \triangleq \begin{bmatrix} W'_{l11} & W'_{l12} \\ W'^T_{l12} & W'_{l22} \end{bmatrix} = \begin{bmatrix} I & 0 & R_1^T \\ 0 & I & R_2^T \end{bmatrix} Q \begin{bmatrix} I & 0 \\ 0 & I \\ R_1 & R_2 \end{bmatrix}, \qquad (6.7\text{-}58)$$

or equivalently

$$W'_{l11} \triangleq Q_{11} + R_1^T Q_{13}^T + Q_{13} R_1 + R_1^T Q_{33} R_1, \qquad (6.7\text{-}59)$$

$$W'_{l12} \triangleq Q_{12} + R_1^T Q_{23}^T + Q_{13} R_2 + R_1^T Q_{33} R_2, \qquad (6.7\text{-}60)$$

$$W'_{l22} \triangleq Q_{22} + R_2^T Q_{23}^T + Q_{23} R_2 + R_2^T Q_{33} R_2. \qquad (6.7\text{-}61)$$

As in the previous case, J'_l on substituting for $\eta'(t)$ from (6.7-56) becomes

$$J'_l = \min\{\eta^T(0) \int_0^\infty \exp(A_l'^T t) W'_l \exp(A_l' t) dt\, \eta(0)\} \triangleq \min \operatorname{Tr}(P'_l H); \qquad (6.7\text{-}62)$$

where H is the same as in (6.7-18), and $P'_l = {P'_l}^T > 0$ is the unique solution of

$$A_l'^T P'_l + P'_l A'_l + W'_l = 0. \qquad (6.7\text{-}63)$$

PROBLEM 6.7-2 *(Large-Sensitivity Minimization with Trajectory and Model Feedback):* Compute the optimal controller gains R_1 and R_2 of (6.7-55) such that the cost function (6.7-62) is minimized subject to (6.7-56)

SOLUTION: The necessary condition to minimize J'_l of (6.7-62) is that of the previous problem, namely $\partial J'_l/\partial R_i = 0$, $i = 1, 2$. To generate these derivatives we propose the following procedures.

Let $R_i \to R_i + \delta R_i$, $i = 1, 2$, with δR_i being "small". Then $A'_l \to A'_l + \delta A'_l$, such that $\delta A'_l \to 0$ as $\delta R_i \to 0$, $i = 1, 2$. Also, as $R_i \to R_i + \delta R_i$, $i = 1, 2$, with δR_i being "small", $P'_l \to P'_l + \delta P'_l$, $J'_l \to J'_l + \delta J'_l$; where (cf., Lemma 6.4-1),

$$\delta P'_l = \int_0^\infty \exp(A_l'^T t)(\delta A_l'^T P'_l + P'_l \delta A'_l + \delta W'_l) \exp(A'_l t) dt, \qquad (6.7\text{-}64)$$

and

$$\delta J'_l = \operatorname{Tr}(\delta P'_l H). \qquad (6.7\text{-}65)$$

Substituting for $\delta P'_l$ from (6.7-64) into (6.7-65) and using some trace operations results in

$$\delta J'_l = \operatorname{Tr}(\Pi'_l E'_l); \qquad (6.7\text{-}66)$$

where

$$\Pi'_l \triangleq \delta A_l'^T P'_l + P'_l \delta A'_l + \delta W'_l, \qquad (6.7\text{-}67)$$

$$E'_l \triangleq \int_0^\infty \exp(A'_l t) H \exp(A_l'^T t) dt. \qquad (6.7\text{-}68)$$

From (6.7-56), if $R_i \to R_i + \delta R_i$, $i = 1, 2$, then $A'_l \to A'_l + \delta A'_l$, where

$$\delta A'_l \triangleq \begin{bmatrix} B' \delta R_1 & B' \delta R_2 \\ B \delta R_1 & B \delta R_2 \end{bmatrix}, \qquad (6.7\text{-}69)$$

and from (6.7-58), if $R_i \to R_i + \delta R_i$, $i = 1, 2$, then $W'_l \to W'_l + \delta W'_l$. Here the

6.7. Sensitivity – Measure Optimization, Large Parameter Variations

partitions of $\delta W_l'$ are:

$$\delta W_{l11}' \triangleq \delta R_1^T(Q_{13}^T + Q_{33}R_1) + (R_1^TQ_{33} + Q_{13})\delta R_1, \qquad (6.7\text{-}70)$$

$$\delta W_{l12}' \triangleq \delta R_1^T(Q_{23}^T + Q_{33}R_2) + (R_1^TQ_{33} + Q_{13})\delta R_2, \qquad (6.7\text{-}71)$$

$$\delta W_{l22}' \triangleq \delta R_2^T(Q_{23}^T + Q_{33}R_2) + (R_2^TQ_{33} + Q_{23})\delta R_2. \qquad (6.7\text{-}72)$$

Partitioning P_l' accordingly, then the partitions of (6.7-67), on using (6.7-69) and (6.7-70) to (6.7-72), become

$$\pi_{l11}' \triangleq \delta R_1^T(B'^TP_{l11}' + B^TP_{l12}'^T + Q_{13}^T + Q_{33}R_1)$$

$$+ (P_{l11}'B' + P_{l12}'B + Q_{13} + R_1^TQ_{33})\delta R_1, \qquad (6.7\text{-}73)$$

$$\pi_{l21}'^T = \pi_{l12}' \triangleq \delta R_1^T(B'^TP_{l12}' + B^TP_{l22}' + Q_{23}^T + Q_{33}R_2)$$

$$+ (P_{l12}'B + P_{l11}'B' + Q_{13} + R_1^TQ_{33})\delta R_2, \qquad (6.7\text{-}74)$$

$$\pi_{l22}' \triangleq \delta R_2^T(B^TP_{l22}' + B'^TP_{l12}' + Q_{23}^T + Q_{33}R_2)$$

$$+ (P_{l22}'B + P_{l12}'^TB' + Q_{23} + R_2^TQ_{33})\delta R_2. \qquad (6.7\text{-}75)$$

Partitioning E_l' the same way as Π_l', and substituting these partitions into (6.7-66), yields

$$\delta J_l' = \text{Tr}(\pi_{l11}'E_{l11}' + \pi_{l12}'E_{l12}'^T + \pi_{l12}'^TE_{l12}' + \pi_{l22}'E_{l22}'). \qquad (6.7\text{-}76)$$

Substituting for π_{lij}' (i, j = 1, 2), from (6.7-73) to (6.7-75), into (6.7-76) and using some trace operations results in

$$\delta J_l' = \text{Tr}\{\delta R_1^T[(B'^TP_{l11}' + B^TP_{l12}'^T + Q_{13}^T + Q_{33}R_1)E_{l11}'$$

$$+ (B'^TP_{l12}' + B^TP_{l22}' + Q_{23}^T + Q_{33}R_2)E_{l12}'^T] + [E_{l11}'(P_{l11}'B' + P_{l12}'B + Q_{13}$$

$$+ R_1^TQ_{33}) + E_{l12}'(P_{l12}'^TB' + P_{l22}'B + Q_{23} + R_2^TQ_{33})]\delta R_1$$

$$+ \delta R_2^T[(B'^TP_{l11}' + B^TP_{l12}'^T + Q_{13}^T + Q_{33}R_1)E_{l12}' + (B^TP_{l22}' + B'^TP_{l12}' + Q_{23}^T$$

$$+ Q_{33}R_2)E_{l22}'] + [E_{l12}'^T(P_{l12}'B + P_{l11}'B' + Q_{13} + R_1^TQ_{33})$$

$$+ E_{l22}'(P_{l22}'B + P_{l12}'^TB' + Q_{23} + R_2^TQ_{33})]\delta R_2\}. \qquad (6.7\text{-}77)$$

Using $\dfrac{\partial \text{Tr}(AB)}{\partial A} = B^T$ and $\dfrac{\partial \text{Tr}(A^TB)}{\partial A} = B$, then from (6.7-77) we can show that:

$$\frac{\partial J'_l}{\partial R_1} = \lim_{\delta R_1 \to 0 \text{ and } \delta R_2 \to 0} \frac{\delta J'_l}{\delta R_1} = 2[(B'^T P'_{l11} + B^T P'^T_{l12} + Q^T_{13} + Q_{33}R_1)E'_{l11}$$

$$+ (B'^T P'_{l12} + B^T P'_{l22} + Q^T_{23} + Q_{33}R_2)E'^T_{l12}], \tag{6.7-78}$$

$$\frac{\partial J'_l}{\partial R_2} = \lim_{\delta R_2 \to 0 \text{ and } \delta R_1 \to 0} \frac{\delta J'_l}{\delta R_2} = 2[(B'^T P'_{l11} + B^T P'^T_{l12} + Q^T_{13} + Q_{33}R_1)E'_{l12}$$

$$+ (B^T P'_{l22} + B'^T P'_{l12} + Q^T_{23} + Q_{33}R_2)E'_{l22}]. \tag{6.7-79}$$

The optimality of R_1 and R_2 requires that $\dfrac{\partial J'_l}{\partial R_i} = 0$, i = 1, 2. Thus

$$(B'^T P'_{l11} + B^T P'^T_{l12} + Q^T_{13} + Q_{33}R_1)E'_{l11}$$

$$+ (B'^T P'_{l12} + B^T P'_{l22} + Q^T_{23} + Q_{33}R_2)E'^T_{l12} = 0, \tag{6.7-80}$$

and

$$(B'^T P'_{l11} + B^T P'^T_{l12} + Q^T_{13} + Q_{33}R_1)E'_{l12}$$

$$+ (B^T P'_{l22} + B'^T P'_{l12} + Q^T_{23} + Q_{33}R_2)E'_{l22} = 0. \tag{6.7-81}$$

From the above two equations we have the optimal controller gains as follows.

$$R^o_1 = - Q^{-1}_{33}[B'^T P'_{l11} + B^T P'^T_{l12} + Q^T_{13}$$

$$+ (B'^T P'_{l12} + B^T P'_{l22} + Q^T_{23} + Q_{33}R_2)E'^T_{l12}E'^{\dagger}_{l11}], \tag{6.7-82}$$

and

$$R^o_2 = - Q^{-1}_{33}[B^T P'_{l22} + B'^T P'_{l12} + Q^T_{23}$$

$$+ (B'^T P'_{l11} + B^T P'^T_{l12} + Q^T_{13} + Q_{33}R_1)E'_{l12}E'^{\dagger}_{l22}]. \tag{6.7-83}$$

Here the two symmetric matrices $E'^{\dagger}_{l11}$ and $E'^{\dagger}_{l22}$ are generalized inverse matrices of E'_{l11} and E'_{l22}, respectively (which are used only for symbolic reasons and do not play any other roles), and $Q_{33} = Q^T_{33} > 0$.

▼

ALGORITHM 6.7-2 (Optimal Controller Gains with Trajectory and Model Feedback): To compute R^o_1 and R^o_2 from (6.7-82) and (6.7-83), we propose that the following steps to be taken. Let $J = 1, 2, ..., N$ be the iteration number.

STEP 2.0 (To initialize): Given A, B, H, F, A', B', H', F', H (or X), W, W_0, and W_1, we invoke the following steps to compute R^o_1 and R^o_2.

STEP 2.1 (Generating P'_l): Equations resulting in P'_{lij} (i, j = 1, 2) must be derived from (6.7-63) which are the following three sets of n × n linear-matrix equations.

6.7. Sensitivity – Measure Optimization, Large Parameter Variations

$$(A' + B'R_1)^T P'_{l11} + P'_{l11}(A' + B'R_1) + Q_{11} + R_1^T(B^T P'^T_{l12} + Q_{13}^T)$$

$$+ (Q_{13} + P'_{l12}B)R_1 + R_1^T Q_{33} R_1 = 0, \tag{6.7-84}$$

$$(A + BR_2)^T P'_{l22} + P'_{l22}(A + BR_2) + Q_{22} + R_2^T(B'^T P'_{l12} + Q_{23}^T)$$

$$+ (Q_{23} + P'^T_{l12}B')R_2 + R_2^T Q_{33} R_2 = 0, \tag{6.7-85}$$

and finally the coupling equation

$$(A' + B'R_1)^T P'_{l12} + P'_{l12}(A + BR_2) + Q_{12} + R_1^T(B^T P'_{l22} + Q_{23}^T)$$

$$+ (Q_{13} + P'_{l11}B')R_2 + R_1^T Q_{33} R_2 = 0. \tag{6.7-86}$$

STEP 2.2 *(Generating E'_l)*: From (6.7-68), the partitions of E'_l can be derived as follows.

$$(A' + B'R_1)E'_{l11} + E'_{l11}(A' + B'R_1)^T + B'R_2 E'^T_{l12} + E'_{l12} R_2^T B'^T + X = 0, \tag{6.7-87}$$

$$(A + BR_2)E'_{l22} + E'_{l22}(A + BR_2)^T + BR_1 E'_{l12} + E'^T_{l12} R_1^T B^T + X = 0, \tag{6.7-88}$$

$$(A' + B'R_1)E'_{l12} + E'_{l12}(A + BR_2)^T + B'R_2 E'_{l22} + E'_{l11} R_1^T B^T = 0. \tag{6.7-89}$$

The above six equations can be solved in the same manner as the previous sets of such coupled-matrix equations. Namely first, by letting the coupling (or $l12$) terms to vanish in order to compute the starting ($l11$) and ($l22$) terms. Then these values are used to generate a new coupling term, and this process continues until the "best" P'_l and E'_l are computed according to the contents of Comment 6.7-1.

STEP 2.3 *(Construction of truncated optimal gains)*: To start computation of the optimal controller gains, we let

$$R_1 = - Q_{33}^{-1}(B'^T P'_{l11} + Q_{13}^T + B^T P'^T_{l12}), \tag{6.7-90}$$

$$R_2 = - Q_{33}^{-1}(B^T P'_{l22} + Q_{23}^T + B'^T P'_{l12}). \tag{6.7-91}$$

STEP 2.4 *(Computation of the initial P'_l)*: Substituting the above initial truncated controller gains into (6.7-84) to (6.7-86) results in

$$[A' - B'Q_{33}^{-1}(Q_{13}^T + B^T P'^T_{l12})]^T P'_{l11} + P'_{l11}[A' - B'Q_{33}^{-1}(Q_{13}^T + B^T P'^T_{l12})]$$

$$+ [Q_{11} - (Q_{13} + P'_{l12}B)Q_{33}^{-1}(B^T P'^T_{l12} + Q_{13}^T)] - P'_{l11}B'Q_{33}^{-1}B'^T P'_{l11} = 0, \tag{6.7-92}$$

$$[A - BQ_{33}^{-1}(Q_{23}^T + B'^T P'_{l12})]^T P'_{l22} + P'_{l22}[A - BQ_{33}^{-1}(Q_{23}^T + B'^T P'_{l12})]$$

$$+ [Q_{22} - (Q_{23} + P'^T_{l12}B')Q_{33}^{-1}(B'^T P'_{l12} + Q_{23}^T)] - P'_{l22}BQ_{33}^{-1}B^T P'_{l22} = 0, \tag{6.7-93}$$

and

$$[A' - B'Q_{33}^{-1}(B'^T P'_{l11} + Q_{13}^T)]^T P'_{l12} + P'_{l12}[A - BQ_{33}^{-1}(B^T P'_{l22} + Q_{23}^T)]$$

$$+[Q_{12} - (Q_{13} + P'_{l11}B')Q_{33}^{-1}(B^T P'_{l22} + Q_{23}^T)] - P'_{l12}BQ_{33}^{-1}B'^T P'_{l12} = 0. \quad (6.7\text{-}94)$$

The above three equations are three quadratic-matrix equations, and an iterative scheme may be devised to generate P'_{lij} (i, j = 1, 2). First we assume that $P'_{l12} = 0$. Then, from (6.7-92) and (6.7-93), it is clear that the "starting" P'_{l11} and P'_{l22} must satisfy the following two equations.

$$(A' - B'Q_{33}^{-1}Q_{13}^T)^T P'_{l11} + P'_{l11}(A' - B'Q_{33}^{-1}Q_{13}^T)$$

$$+ Q_{11} - Q_{13}Q_{33}^{-1}Q_{13}^T - P'_{l11}B'Q_{33}^{-1}B'^T P'_{l11} = 0, \quad (6.7\text{-}95)$$

and

$$(A - BQ_{33}^{-1}Q_{23}^T)^T P'_{l22} + P'_{l22}(A - BQ_{33}^{-1}Q_{23}^T)$$

$$+ Q_{22} - Q_{23}Q_{33}^{-1}Q_{23}^T - P'_{l22}BQ_{33}^{-1}B^T P'_{l22} = 0. \quad (6.7\text{-}96)$$

Substituting these starting "values" of P'_{l11} and P'_{l22} into (6.7-94), results in a starting "value" for P'_{l12}. The process continues until P'_{lij} (i, j = 1, 2), are computed within some degree of accuracy as that described in Comment 6.7-1.

STEP 2.5 (Computation of initial truncated optimal gains): Correspondingly, substituting the "best" numerical P'_{lij} (i, j = 1, 2), into R_1 and R_2, which are given by (6.7-90) and (6.7-91), respectively, yields the starting or initial truncated "values" for R_1 and R_2.

STEP 2.6 (Computation of the optimal controller gains): Having determined these starting "values" for R_1 and R_2, the remaining process of numerically computing the optimal controller gains is the same as that of Algorithm 6.7-1. Namely, we substitute these values in (6.7-84) to (6.7-89) to generate new P'_{lij} and E'_{lij} (i, j = 1, 2) and subsequently we substitute these values in (6.7-82) and (6.7-83) to develop new values for R_1, and R_2. Note that for R_1 and R_2 on the right-hand sides of (6.7-82) and (6.7-83) we use the previous values of these quantities. These equations are quadratic and their convergence to the final values are generally quadratically fast.

∎

6.7.5. Numerical Example

In this section we present a tutorial example to show the applicability of the preceding results when they are utilized in the same situation.

EXAMPLE 6.7-1: Consider a real-operating system and its corresponding nominal system which is initialized with the following set of parameters.

6.7. Sensitivity – Measure Optimization, Large Parameter Variations

$$A = -1, \quad \Delta A = -1, \quad A' \triangleq A + \Delta A = -2,$$

$$B = 2, \quad \Delta B = 1, \quad B' \triangleq B + \Delta B = 3,$$

$$H = 1, \quad \Delta H = 1, \quad H' \triangleq H + \Delta H = 2,$$

$$F = 1, \quad \Delta F = 1, \quad F' \triangleq F + \Delta F = 2,$$

$$W_s = 1, \quad W_0 = 1, \quad W = 1, \quad X = 1. \tag{6.7-97}$$

Apply Algorithms 6.7-1 & 2 to this system and compare the outcomes of each one of these algorithms when designing optimal controller gains.

SOLUTION: From the above values and (6.7-11) we have the following $Q = Q^T > 0$.

$$Q \triangleq \begin{bmatrix} 4 & -2 & 2 \\ -2 & 2 & 0 \\ 2 & 0 & 3 \end{bmatrix}. \tag{6.7-98}$$

Clearly, $Q_{11} = 4 > 0$ and $Q_{11} - Q_{13} Q_{33}^{-1} Q_{13}^T = 8/3 > 0$. Let $J = 1, 2, ..., N$ be the iteration number.

First we utilize Algorithm 6.7-1 with the above numerical values to generate the optimal controller gain R^o of (6.7-33) as follows.

$$R^o = -\frac{1}{3}[3P_{l11} + 2P_{l12} + 2 + (2P_{l22} + 3P_{l12})E_{l12}^T E_{l11}^\dagger]. \tag{6.7-99}$$

STEP 1.1: From (6.7-35) to (6.7-37) we have

$$P_{l22} = 1, \quad P_{l12} = -2/3, \quad P_{l11} = (9R^2 + 4R + 12)/(12 - 18R).$$

STEP 1.2: From (6.7-39) to (6.7-41) we conclude that

$$E_{l11} = 1/(4 - 6R), \quad E_{l12} = 2RE_{l11}/(3 - 3R), \quad E_{l22} = 0.5 + 2RE_{l12}.$$

STEP 1.3: From (6.7-42) and Step 1.1 we have

$$R = -\frac{1}{3}(3P_{l11} + 2P_{l12} + 2) = -P_{l11} - 0.222.$$

STEP 1.4: Technically speaking, we must let $P_{l12} = 0$ and from (6.7-47) we must generate P_{l11}. Then this P_{l11} must be substituted in (6.7-43) to generate a new P_{l12} until the "best" initial P_{l11} and P_{l12} are computed according to the Comment 6.7-1. However, in this example $P_{l12} \equiv -2/3$, therefore we use this value in (6.7-44) to (6.7-46) to compute the initial $P_{l11} = 0.551$.

STEP 1.5: Substituting $P_{l11} = 0.551$ in R of Step 1.3 yields the starting $R = -0.773$.

STEP 1.6: Computation of the new P_l and E_l and thus R yields at the first iteration $R^o = -0.773$. In other words, the truncated R in Step 1.5 is indeed the optimal-controller gain because in (6.7-99), we have $2P'_{l22} + 3P'_{l12} \equiv 0$. Finally, the corresponding optimal cost for this example becomes $J = \text{Tr}(P_l H) = 1.551$.

Secondly, we apply Algorithm 6.7-2 to the above example.

STEP 2.1: From (6.7-84) to (6.7-86) we have

$$2(3R_1 - 2)P'_{l11} + 4 + 4R_1 + 3R_1^2 + 4R_1 P'_{l12} = 0,$$

$$2(2R_2 - 1)P'_{l22} + 2 + 3R_2^2 + 6R_2 P'_{l12} = 0,$$

$$(3R_1 + 2R_2 - 3)P'_{l12} - 2 + 2R_2 + 3R_1 R_2 + 3R_2 P'_{l11} + 2R_1 P'_{l22} = 0.$$

STEP 2.2: From (84) to (6.7-89) we have

$$2(3R_1 - 2)E'_{l11} + 6R_2 E'_{l12} + 1 = 0,$$

$$2(2R_2 - 1)E'_{l22} + 4R_1 E'_{l12} + 1 = 0,$$

$$(3R_1 + 2R_2 - 3)E'_{l12} + 3R_2 E'_{l22} + 2R_1 E'_{l11} = 0.$$

STEP 2.3: From (6.7-90) to (6.7-91) we have the following truncated optimal gains.

$$R_1 = -\frac{1}{3}(3P'_{l11} + 2P'_{l12} + 2), \text{ and}$$

$$R_2 = -\frac{1}{3}(2P'_{l22} + 3P'_{l12}).$$

STEP 2.4: From (6.7-95) and (6.7-96) (with $P'_{l12} = 0.0$) we have $P'_{l11} = 0.3$ and $P'_{l22} = 0.686$. Substituting these values in (6.7-95) yields $P'_{l12} = -0.59$. Now substituting this P'_{l12} into (6.7-91) and (6.7-94) yields $P'_{l11} = 0.524$ and $P'_{l22} = 0.88$. Continuing in this manner results in the following starting P'_l which is, of course, a positive-definite matrix.

$$P'_l = \begin{bmatrix} 0.524 & -0.63 \\ -0.63 & 0.88 \end{bmatrix}. \tag{6.7-100}$$

STEP 2.5: The initial truncated optimal gains are now become

$$R_1 = -0.771 \quad \text{and} \quad R_2 = 0.043.$$

STEP 2.6: Using the above starting R_1 and R_2 in Step 2.1 and Step 2.2 yield

$$P'_l = \begin{bmatrix} 0.551 & -0.667 \\ -0.667 & 1.008 \end{bmatrix}, \text{ and } E'_l = \begin{bmatrix} 0.115 & -0.02 \\ -0.02 & 0.58 \end{bmatrix},$$

$$R_1^o = -0.765 \quad \text{and} \quad R_2^o = -0.0016. \tag{6.7-101}$$

These values correspond to a new P'_l as follows.

6.7. Sensitivity – Measure Optimization, Large Parameter Variations

$$P_l' = \begin{bmatrix} 0.552 & -0.668 \\ -0.668 & 1.000 \end{bmatrix}. \qquad (6.7\text{-}102)$$

Because the P_l' of (6.7-102) is very close to that of (6.7-101) and the optimal cost of (6.7-62) evaluated at this iteration does not change substantially from its previous value, we stop at this iteration. In this case the optimal cost of (6.7-62) becomes $J = \text{Tr}(P_l H) = 1.552$.

Thirdly, we only consider the nominal system as described by (6.7-3) and (6.7-4), i. e., all the coefficient perturbations are assumed to be zero. Here we can show that the quadratic cost of (6.7-6) associated with this nominal system yields

$$R_{nominal} = -0.618 \quad \text{and} \quad J_{nominal} = 0.118. \qquad (6.7\text{-}103)$$

REMARK 6.7-2 (On perturbation matrices): Perhaps we should make it clear that the main implication of these perturbation matrices is that we are uncertain about the actual values of the system parameters. Therefore we want to design our controllers with this uncertainty in mind that if needs arise, then we can incorporate the necessary changes in our controllers, in order to stand for a range of possible "worst"-case parameter variations. This is in contrast with fixed controllers which cannot be changed. We have also accounted for possible interactions among various parameter variations by using the *large*-sensitivity measure. Clearly, if we know exactly the value of each parameters and they are fixed, then we do not need to incorporate this analysis or any other sensitivity-reduction schemes whatsoever. But in reality we do not know the exact value of each parameter and thus we appeal to this sort of analysis in order to build enough capability in our controllers which in turn enable them to withstand a range of possible values for system parameter variations. In this sort of analysis in which we have stretched the linear time-invariant property of our system to its limit, any further restriction on the nature of these assumptions or on the "dynamics" of these parameter variations will put this system outside the realm of deterministic linear time-invariant system. Indeed, this formulation is already on the periphery of linear time-invariant analysis and the adaptive control analysis. Furthermore, in this method we can easily change various perturbation matrices "numerically" if we receive additional information about the dynamics of system-parameter variations. Therefore this method can be considered as an intermittent-adaptation policy when we study changes in system parameters and we use this information to robustify the system performance with *minimum hardware*.

REMARK 6.7-3 (On the numerical results): To summarize our results we note that if a control system satisfies the basic requirements for the applications of a quadratic-optimization method, then we can apply the results from this section to minimize the large-sensitivity measure of that system. As the numerical examples show, the optimal cost as well as the gain of each controller increase in amplitude for more sophisticated design than that of the plain regulator. Table 6.7-1 confirms our conclusions about these matters. In particular from Algorithm 6.7-1, *per se*, we note that increasing the amplitude of our controller gain by about 25% enables us to tolerate almost 100% variations in A, F, and H, and 50% changes in B. Although one example alone is not sufficient to make a major claim, we believe that this design methodology has considerable engineering applications and must be looked at and compared with a plain optimal regulator design or an advanced adaptive controller design before a final decision on how to control a system with a set of large-varying parameters is being made. Furthermore, this approach can be easily extended to a class of nonlinear regulator systems (cf., [2D. 8]), and for a similar quadratic-optimization design.

Table 6.7-1.

Optimal Regulator Problem	Algorithm 6.7-1	Algorithm 6.7-2
$R_{nominal} = -0.618$ $J_{nominal} = 0.118$	$R_{(6.7-33)} = -0.773$ $J_{(6.7-15)} = 1.551$	$R_{(6.7-82)} = -0.765$ $R_{(6.7-83)} = -0.0016$ $J_{(6.7-62)} = 1.552$
No additional constraint on A. No sensitivity is minimized. Minimum computational efforts.	The initial A must be stable. Sensitivity is minimized. Moderate computational efforts.	No additional system constraint. Sensitivity is minimized. The most computational efforts with complex design structure.

▼

6.8. COMMONALTIES IN OPTIMAL INPUT FOR IDENTIFICATION AND OPTIMAL AUXILIARY INPUT FOR ROBUSTIFICATION

6.8.1. Introduction

Sensitivity optimization plays an important role in a number of seemingly different applied control problems. However, when the underlying sensitivity optimization scheme is carefully analyzed, it becomes quite apparent that this aspect of these control problems can be unified in a manner that the corresponding sensitivity optimization scheme and/or algorithm is concluded from one such algorithm. The purpose of this section is to explore this commonalty and to provide a potential list of such control problems, whose corresponding sensitivity optimization algorithm can be concluded from one general algorithm. But first, we review as succinctly as possible several different points which must be clarified or recalled before we present our analysis.

Generally speaking: One of the most appealing features of any control design is its measure of robustness which means the ability of this design to stand against unwanted changes in system dynamics especially when these changes are large. By and large the prominent source of variations in system dynamics is the inevitable changes which are taking place for various reasons in system parameters, resulting in a discrepancy between the nominal and actual trajectories of a dynamical system. To study this discrepancy, we often select an appropriate system-sensitivity measure, depending on the measuring devices and the system model. Then proper procedures are developed to minimize the unwanted effects of parameter variations on system performance. Indeed, this concept has been the main thrust of a vast literature on state-sensitivity minimization, partially reported in § 6.12.D. On the other hand, there are equally important problems in parameters identification which require sensitivity maximization for accurate identification, as these issues are reported in references of § 6.12.G. Another problem which may also cause unnecessary changes in system operation is the possible lack of experience on the part of its user that can be classified somewhere between the preceding two concepts. Here, we study the commonalties among these concepts and present a unified method to treat their corresponding sensitivity matters. The following study refers to the system of *Case 2* in Table 6.1-1.

On sensitivity matters: Two major issues deserve special attentions.

(i) What is the system-sensitivity measure? It is clear by now that there is no definitive answer to this question, since it depends heavily on the particular application. The sensitivity measures resulting from parameter variations, however, are classified into two groups. The first group concerns with *small*-parameter variations which corresponds to a case in which the parameter change, say Δp, is "small", i.e., $|\Delta p| \ll 1$ or $|\Delta p|^2 \approx 0$. In this case the sensitivity measure is the first derivative of a particular signal with respect to this varying parameter. On the other hand, if $|\Delta p|^2 \neq 0$, then we have the second group of sensitivity measures which are concerned with *large*-parameter variations. Here, of course, the first-order sensitivity function is not useful, as we have extensively elaborated on these issues in this book. For these cases we directly use the difference between the new- and the old-value of a particular signal and we call that *a large*-sensitivity measure of the system. Most of the above comments on *small*- and *large*-sensitivity measures are valid whether the system is described in time domain or in frequency domain. There are, however, many interesting results, stemming from the so-called sensitivity-comparison matrix (cf., Section 6.2), which are more suitable for systems that are described in the frequency domain than those in the time domain. Thus the first step in this entire analysis is to choose an appropriate measure of system sensitivity. In this section, however, we use a (time-domain) response variation as our sensitivity measure in order to be consistent with the trends in this book so far.

(ii) What is the proper type of system-sensitivity optimization? The answer to this question also depends on the encountered application. Here we choose the quadratic optimization, because the results fit well with our response-variation sensitivity measure which in effect is a modified state-sensitivity measure. We note, however, that the sensitivity optimizations using the H_∞-norms are extensively developed for systems whose sensitivity measures are expressed in terms of sensitivity-comparison matrices. For these measures the H_∞-norm optimization is more suitable than the quadratic-norm optimization. However, the H_∞-norm optimization is not carried out for system whose large-sensitivity measure is given with a time-domain response variation. Certainly, this study would be an interesting research problem to investigate.

On identification matters: It is well known that the accuracy of system identification improves with an appropriate choice of input function, called the *optimal* input. For efficient identification the magnitude of this input function must be constrained, such that it does not become excessively large, i.e., the optimal input must be derived under the assumption of either energy or power being constrained, although such assumption always complicates the analysis. In general, the performance criterion for selecting input signals (for minimum variance parameter estimation), is that concluded from the so-called Fisher information matrix (a matrix whose entries are appropriate weighted norms of *small*-trajectory sensitivity functions, cf., Remark 6.5-2). The maximization of this sensitivity measure increases the accuracy of parameter estimation. In the following references the system input synthesis has been formulated, in general, as maximization of the trace of Fisher information matrix, subject to some constraints on system input. Implementation of this input function as an open-loop or a closed-loop design is the next measure of distinction among the following methods of identification. Lopez-Toledo and Athans [G. 16] consider the closed-loop implementation of system input for identification, while most of the earlier researchers in this area such as Aoki and Staley [G. 1], Bonivento [G. 3], Gagliardi [G. 7], Georganas [G. 8], Kalaba and Spingarn [G. 10 & 11], Schweppe [G. 24], Levadi [G. 12], Mehra [G. 17 to 19], Nahi and Napjus [G. 21], Nahi and Wallis [G. 22], and many other papers in Mehra and Lainiotis [G. 20] treat the open-loop system input for enhancing the performance of certain parameter specification procedures. The optimality measure for input selection in most of these papers is the trace of Fisher

6.8. Commonalties in Identification and Robustification

information matrix. The maximization of this measure gives an insight and a natural connection to the main thrust of the above papers which demonstrates the conflicting nature of identification and control in a system. On the other hand, the same algorithm, with a slight change, can be used to minimize system-sensitivity measure by a method that we call the plant-operator concept and is described subsequently.

On optimum supervisory input: Often we face the question of how to account for supervisory "input" of a manager or an operator in a closed-loop system. In either case, it is preferred to have minimum involvement of these people. Because, a problem which may also cause unnecessary changes in system operation is the possible lack of experience on the part of its user. Therefore in spite of advancement in automation and progress in design of an autonomous system, it is imperative that we incorporate the experience and know how (or lack of it) of the plant manager or plant operator in the overall design of an automated manufacturing plant. Certainly, we prefer to have minimum involvement of these external people, because, on the one hand, the plant manager cannot afford to look after each problem, and, on the other hand, the plant operator may not have the experience and knowledge about details and all the intricacies of plant operations, in order to respond effectively when sudden changes are taking place in system operation. Although it seems natural when studying this situation, to shift the burden of maintaining system performance to the automatic part of the system such that involvement of supervisory input(s) becomes minimum, this does not mean that the role of supervisory input will be diminished.[1] Therefore the automatic controllers must be designed with sufficient flexibilities in order to accommodate for changes as new operating condition will develop. Here we model the supervisory input (of either plant manager or plant operator), by an additive auxiliary input which is used for the purpose of improving the robustness of system dynamics when this system is subject to large-parameter variations. Since the modeling of this supervisory input that is provided by someone who is using the system is very much similar to that of the auxiliary input that is used for system identification and/or sensitivity reduction, we should expect that much of the analyses in these two optimization problems to be similar.

On the unification matters: The algorithm which is presented subsequently considers a system with two input vectors. The first input vector serves for automatic regulation and is implemented in a closed-loop manner. This input which is based on our knowledge about the system is designed such that the minimum energy is required in order to steer the system from a real-operating condition to the nominal trajectory. The design philosophy of the second (or auxiliary) input vector is based on adopting an additional operator which can be a human being

---

[1] Indeed, as a result of this phenomenon we are convinced that relying on the automatic feedback controller alone for robustification may not be the most advisable policy. We need another *layer* of control and/or supervision that must be incorporated outside the main operating loop in the plant in order to robustify its performance.

(man-in-the-loop), whose energy (or effort) (or experience) is used for identification and/or correction of the real-operating trajectory. This input is derived with the consideration for a possible final closed-loop implementation (outside the main loop) and it serves either for identification purposes (which in this case the corresponding system-sensitivity measure is maximized); or as an additional input called "plant-operator", that serves for robustification purposes (resulting in minimum system sensitivity measure). This additional (or auxiliary) input can also be viewed as an additional component of system input function which primarily serves for an identification purpose or a sensitivity reduction procedure, because it is well known that the optimal closed-loop control, in general, spends some "energy" or some "effort" for identification purposes. Indeed this effort for learning about the unknown system state and/or system parameter is the main thrust of our study and the point of convergence of the two algorithms: system identification versus system sensitivity reduction.

Historically speaking: At this juncture we may point out that according to Jury and Tsypkin [*N.* 19] this concept of utilizing an auxiliary input to enhance the performance of a closed-loop system has been apparently introduced by Rutman [in Russian] in 1964, and in the following manner. Namely, when synthesizing an optimal [discrete] control system, if we combine a discrete correction policy that changes the shape of control pulses with some external control action, then we get the "best" result (this external control may be resulted from the sensory information *per se*). This simple observation has been also extensively studied in one form or another by a number of other researchers in several different disciplines under the auxiliary inputs or signals. For example, in a very interesting paper on nonlinear discrete-time systems, Geromel and da Cruz [*G.* 9] show that the "additive perturbation (the same as our external input) is much 'richer' than the gain perturbation" (that is designed using automatic feedback policies). They also show that "a system with high tolerance to gain perturbation may present a low tolerance to additive perturbation and *vice versa.*" Therefore a compromise must be made between weighting these two inputs in a given application and this is, in our opinion, a very interesting result. In our earlier work we have used the terms auxiliary input or plant-operator to emphasize similar situations and the outcomes were noticeably improved [*G.* 5 & 6 and 13 to 15].

In summary: Having established these facts, we take a path on the periphery of parameters identification and robust controller design in order to investigate the commonalties among these issues. Using a combined-quadratic cost as the optimality measure, we present an algorithm for derivation of an optimal controller gain for the automatic regulation in a linear system (a closed-loop policy), and an additive optimal auxiliary input (or plant-operator) for the purpose of: either system identification, resulting in maximum sensitivity measure thus increased accuracy; or system sensitivity reduction, resulting in minimum sensitivity measure thus reliable operation. The variation of the developed algorithm from one application to another is done by changing the sign of one scalar parameter. Also the derivation of the optimal external input is carried out under the assumption of a possible closed-loop implementation, although that is not always desirable, unless it is

6.8. Commonalties in Identification and Robustification

outside the main loop. As our numerical examples demonstrate, the controllers which are designed stemming from this algorithm exhibit a very high degree of robustness. The above results are generated for the case in which we impose no constraint on the input vector. Similar but much more involved problem of finding the constrained optimal input vector is proposed for a future research. Our analysis closely follows [G. 6].

6.8.2. Problem Statement

Consider a real-operating system state and response described, respectively, by

$$\dot{\xi}(t) = (A + \Delta A)\xi(t) + (B + \Delta B)\bar{u}(t) + Cv(t), \quad \xi(0) = \xi_0, \quad (6.8\text{-}1)$$

$$\zeta(t) = (H + \Delta H)\xi(t) + (F + \Delta F)\bar{u}(t), \quad (6.8\text{-}2)$$

where $\xi \in R^n$ is the system state vector; $\bar{u} \in R^m$ is the system perturbed (or real-) input vector (that is used for the purpose of automatic regulation), and another auxiliary input $v \in R^\mu$ is employed by a human being operator (or otherwise), for identification or sensitivity reduction. The nominal trajectory and response in a one to one correspondence with (6.8-1) and (6.8-2) are described, respectively, by

$$\dot{x}(t) = Ax(t) + Bu(t), \quad x(0) = x_0, \quad (6.8\text{-}3)$$

$$z(t) = Hx(t) + Fu(t). \quad (6.8\text{-}4)$$

Correspondingly, $A' \triangleq A + \Delta A$, $B' \triangleq B + \Delta B$, $H' \triangleq H + \Delta H$, and $F' \triangleq F + \Delta F$ are disturbed system parameter coefficients associated with the nominal system parameter coefficients, namely A, B, H, and F, respectively.

The design philosophy is that, by an appropriate choice of the feedback automatic controller gain K (such that $u(t) = Kx(t)$ and $\bar{u}(t) = K\xi(t)$), we minimize the operator's cumulative "effort" ($\int_0^\infty \|v(t)\|_{W_{po}}^2 dt$), in the sense of the following combined-quadratic cost.

$$J_I = \int_0^\infty (\|x(t)\|_{W_0}^2 + \|u(t)\|_W^2 + \alpha\|\zeta(t) - z(t)\|_{W_s}^2 + \beta\|v(t)\|_{W_{po}}^2)dt, \quad (6.8\text{-}5)$$

where $W_0 = W_0^T \geq 0$, $W = W^T > 0$, $W_s = W_s^T \geq 0$, and $W_{po} = W_{po}^T > 0$ are the weighting matrices; α and β are scalar quantities introduced for computational purposes provided that, of course, certain fundamental requirements for such optimization have been already met by (6.8-1) to (6.8-4) (cf., [2C. 35]). For the minimization of J_I (robustification) α is chosen a positive number and for the maximization of J_I (identification) α is chosen a negative number. The upper limit for the above integral is chosen so that it simplifies the following analysis and thus enables us to describe the methodology. In this case the feedback automatic controller gain is chosen such that it facilitates all future efforts of system-auxiliary inputs. The fact that an external "effort", or "energy", described by Cv(t), must be used for system-error compensation or the cost for on-line system identification is well known in engineering practice.

REMARK 6.8-1: At this juncture we note that whether we are studying an identification or a control problem, it is imperative to select an appropriate sensitivity measure. Since we want to present as general as possible all aspects of our methodology, we have selected a particular sensitivity measure which is perhaps more suitable for control problems than identification problems. Nevertheless, this choice is general enough to demonstrate the scope and the applicability of the results. Here, the sensitivity measure which has been used in (6.8-5), i.e., $\{\alpha \| \zeta(t) - z(t) \|_{W_s}^2\}$, is an appropriate weighted norm of our system-response variation or *large*-sensitivity measure of the system. The following procedure for deriving the optimal controller gain and the optimal auxiliary input for this sensitivity measure is exactly the same as that for the trace of the Fisher information matrix for a multi-parameter system used in identification problems (cf., Remark 6.5-2). However, using this *large*-sensitivity measure yields a low-dimensional augmented set of differential equations (for generating system state and system-perturbed trajectory) regardless of the number of varying parameters. This is in contrast with the Fisher information matrix which requires a large-dimensional augmented set of differential equations be generated, where we recall that using these sensitivity functions we must also investigate the degree of redundancy in the corresponding system model (cf., Sections 3.5 and 4.3) before making any attempt to use them in control design. Thus for the sake of presentation of analytical procedures involved in the construction of the new results, we use this *large*-sensitivity measure instead of the sensitivity cost resulted from the Fisher information matrix.

REMARK 6.8-2: For those interested to study the *small*-sensitivity measure, as we have extensively reviewed these issues in Sections 6.5 and 6.6, we should emphasize that from the controller design point of view, however, there are a number of shortcomings in minimizing *small*-sensitivity measures of a multi-parameter system. In particular, if the difference between the nominal (model) and actual (real-operating) trajectory becomes *large,* then the conventional *small*-sensitivity analysis and synthesis are no longer useful (cf., § 6.7.1). Thus we are convinced that $\{\alpha \| \zeta(t) - z(t) \|_{W_s}^2\}$ is a sound measure of system sensitivity which has application in control problem and this measure allows us to demonstrate the forthcoming methodology. An interested reader can duplicate easily this method for the case in which we have trajectory-sensitivity functions instead of response-variation. Generation of these trajectory-sensitivity functions are needed to develop the high-dimensional Fisher information matrix.

REMARK 6.8-3: The combined-quadratic cost (6.8-5) can be written as follows.

$$J_I = J_p + \alpha J_s + \beta J_{po}; \qquad (6.8\text{-}6)$$

where J_p, J_s, and J_{po} are the costs due to the plant (or standard regulator cost), system-sensitivity reduction (a weighted norm of system-response variation), and on-line system identification (or plant-operator), respectively. The above cost is

6.8. Commonalties in Identification and Robustification

also subject to the augmented system-state differential equations; and in general the optimization problem is considered as one of the following two problems.

PROBLEM 6.8-1: Consider the extremization of J_I given in (6.8-6), subject to the following constraints

$$\dot{\eta}_I(t) = A_I \eta_I(t) + C_o v(t), \tag{6.8-7}$$

$$\zeta(t) = (H' + F'K)\xi(t), \tag{6.8-8}$$

$$z(t) = (H + FK)x(t), \tag{6.8-9}$$

with

$$\eta_I(t) \triangleq [\xi^T(t), x^T(t)]^T, \tag{6.8-10}$$

$$C_o^T \triangleq (C^T, 0^T), \tag{6.8-11}$$

$$A_I \triangleq \begin{bmatrix} A' + B'K & 0 \\ 0 & A + BK \end{bmatrix}, \tag{6.8-12}$$

with $\bar{u}(t) = K\xi(t)$ and $u(t) = Kx(t)$. Also from (6.8-2) and (6.8-4) we have

$$\zeta(t) - z(t) = (H' + F'K)\xi(t) - (H + FK)x(t)$$

$$= (H' + F'K, -H - FK)\eta_I(t). \tag{6.8-13}$$

On using $u(t) = Kx(t)$ and $\bar{u}(t) = K\xi(t)$, and (6.8-13), J_I of (6.8-5) becomes

$$J_I = \int_0^\infty [\|x(t)\|^2_{(W_0 + K^T W K)} + \alpha \|\eta_I(t)\|^2_{\tilde{W}_s} + \beta \|v(t)\|^2_{W_{po}}] dt; \tag{6.8-14}$$

where

$$\tilde{W}_s \triangleq (H' + F'K, -H - FK)^T W_s (H' + F'K, -H - FK)$$

$$\triangleq \begin{bmatrix} (H' + F'K)^T W_s (H' + F'K) & -(H' + F'K)^T W_s (H + FK) \\ -(H + FK)^T W_s (H' + F'K) & (H + FK)^T W_s (H + FK) \end{bmatrix}. \tag{6.8-15}$$

Imbedding the first two terms of J_I in (6.8-14), yields

$$J_I = \int_0^\infty [\|\eta_I(t)\|^2_{\hat{W}_s} + \beta \|v(t)\|^2_{W_{po}}] dt, \tag{6.8-16}$$

where

$$\hat{W}_s \triangleq \begin{bmatrix} \alpha(H' + F'K)^T W_s (H' + F'K) & -\alpha(H' + F'K)^T W_s (H + FK) \\ -\alpha(H + FK)^T W_s (H' + F'K) & W_0 + K^T W K + \alpha(H + FK)^T W_s (H + FK) \end{bmatrix}.$$

$$\tag{6.8-17}$$

Thus, Problem 6.8-1 is equivalent to extremizing (6.8-16), subject to (6.8-7) to (6.8-12). This extremization is performed by selecting the optimal-auxiliary input v(t) (or the plant-operator strategy) and the optimal-controller gain K such that $\bar{u}(t) = K\xi(t)$ and $u(t) = Kx(t)$.

PROBLEM 6.8-2: Consider the extremization in J_I of Problem 6.8-1, subject to the following additional constraints

$$\int_0^\infty \|v(t)\|_{W_{po}}^2 dt \le E, \quad \text{and/or} \quad \int_0^\infty \|u(t)\|_W^2 dt \le E'. \tag{6.8-18}$$

From the engineering point of view, Problem 6.8-2 is perhaps more desirable than Problem 6.8-1. For instance, it is well known in robotics that each actuator (input function) can provide only a finite or bounded torque, but to account formally for this boundedness in the design of controllers tremendously complicates the design analysis. This is why, for example, we have considered in [G. 13 to 15], this boundedness by simulating the problem at several levels of input torques, i.e., we have "manually" changed the upper limit of the actuator torques to avoid analytical complications. Therefore, in general, and from an analytical point of view, Problem 6.8-2 is more involved than Problem 6.8-1. However, most of the procedures in solving these two problems are the same. Thus in the following we only consider the first problem, because the solution to Problem 6.8-1 shows the main approach for solving this kind of optimization problems. Depending on how we look at this solution; either this solution explains the similar procedure for solving Problem 6.8-2, or it shows the difficulty of solving this problem with its corresponding computational algorithm.

6.8.3. Solution of Problem 6.8-1

6.8.3.1. Derivation of the Optimal Auxiliary Input v(t) by the Maximum Principle

The complete solution of Problem 6.8-1 consists of two parts. The first part involves the derivation of the optimal-auxiliary input v(t) (functional optimization), and the second part entails derivation of the optimal-automatic controller gain K^o (algebraic optimization). In the following we start the analysis by establishing the system Hamiltonian associated with the necessary condition for an optimal v(t) first. From (6.8-16) and (6.8-7) we have

$$H = \lambda^T[A_I\eta_I(t) + C_o v(t)] + \text{Tr}[\|\eta_I(t)\|_{W_s}^2 + \beta\|v(t)\|_{W_{po}}^2]. \tag{6.8-19}$$

It is well known that the optimal v(t) must satisfy [2C. 35].

$$\left(\frac{\partial H}{\partial \eta_I(t)}\right)^T = -\dot{\lambda}(t), \quad \text{with} \quad \lambda(\infty) = 0, \tag{6.8-20}$$

6.8. Commonalties in Identification and Robustification

$$\left(\frac{\partial H}{\partial v(t)}\right)^T = 0, \tag{6.8-21}$$

$$\left(\frac{\partial H}{\partial \lambda(t)}\right)^T = \dot{\eta}_I(t). \tag{6.8-22}$$

Of course, as we describe in the next section, the optimality of K requires that

$$\frac{\partial J_I}{\partial K} = 0. \tag{6.8-23}$$

Perhaps, it is now clear that, in spirit, we are looking for the minimization of (6.8-16). From (6.8-19) and (6.8-20), we conclude that

$$-\dot{\lambda}(t) = A_I^T \lambda(t) + 2\hat{W}_s \eta_I(t). \tag{6.8-24}$$

Similarly, from (6.8-19) and (6.8-21) we have

$$C_o^T \lambda(t) + 2\beta W_{po} v(t) = 0. \tag{6.8-25}$$

Recalling that $W_{po} = W_{po}^T > 0$, the optimal strategy for plant-operator is

$$v(t) = -\frac{1}{2\beta} W_{po}^{-1} C_o^T \lambda(t). \tag{6.8-26}$$

Substituting v(t) into (6.8-7) yields

$$\dot{\eta}_I(t) = A_I \eta_I(t) - \frac{1}{2\beta} C_o W_{po}^{-1} C_o^T \lambda(t), \tag{6.8-27}$$

with (6.8-24) and (6.8-27) forming a set of homogeneous two-point boundary value problems (TPBVP's). The arrangement of Problem 6.8-1 has been such that this set of homogeneous TPBVP's can be solved via a Riccati-type algebraic-matrix equation. Interested reader may consult [2A. 21] for further information on this type of TPBVP's.

In order to make the derivations of the next part manageable we to simplify the optimal auxiliary input by providing a formal solution for (6.8-24) and (6.8-27). First, we let

$$\lambda(t) = R\eta_I(t). \tag{6.8-28}$$

Here, $\dot{R}(t) = 0$, because of the choice on the upper limit of the integral given in (6.8-5). Then, (6.8-27) on using (6.8-28) becomes

$$\dot{\eta}_I(t) = (A_I - \frac{1}{2\beta} C_o W_{po}^{-1} C_o^T R)\eta_I(t) \triangleq A_h \eta_I(t). \tag{6.8-29}$$

Similarly, it is seen that

$$\dot{\lambda}(t) = R\dot{\eta}_I(t) = RA_h \eta_I(t). \tag{6.8-30}$$

On the other hand, from (6.8-24) and (6.8-28), we have

$$\dot{\lambda}(t) = -(A_I^T R + 2\hat{W}_s)\eta_I(t). \tag{6.8-31}$$

Equating (6.8-30) and (6.8-31) with $\eta_I(t)$ from (6.8-29) results in

$$[R(A_I - \frac{1}{2\beta}C_o W_{po}^{-1} C_o^T R) + A_I^T R + 2\hat{W}_s]\eta_I(t) = 0. \qquad (6.8\text{-}32)$$

Equation (6.8-32) yields the following Riccati-type algebraic matrix equation for R

$$A_I^T R + RA_I + 2\hat{W}_s - \frac{1}{2\beta} RC_o W_{po}^{-1} C_o^T R = 0. \qquad (6.8\text{-}33)$$

This equation on substituting for A_I, from (6.8-12), and C_o, from (6.8-11), becomes

$$\begin{bmatrix} (A' + B'K)^T & 0 \\ 0 & (A + BK)^T \end{bmatrix} \begin{bmatrix} R_{11} & R_{12} \\ R_{12}^T & R_{22} \end{bmatrix} + \begin{bmatrix} R_{11} & R_{12} \\ R_{12}^T & R_{22} \end{bmatrix} \begin{bmatrix} A' + B'K & 0 \\ 0 & A + BK \end{bmatrix}$$

$$+ 2\hat{W}_s - \frac{1}{2\beta} \begin{bmatrix} R_{11} & R_{12} \\ R_{12}^T & R_{22} \end{bmatrix} \begin{bmatrix} CW_{po}^{-1} C^T & 0 \\ 0 & 0 \end{bmatrix} \begin{bmatrix} R_{11} & R_{12} \\ R_{12}^T & R_{22} \end{bmatrix} = 0. \qquad (6.8\text{-}34)$$

The above $2n \times 2n$ nonlinear-matrix equation, on substituting for $\hat{W}_s$ from (6.8-17), and for

$$D \triangleq \frac{1}{2\beta} CW_{po}^{-1} C^T, \qquad (6.8\text{-}35)$$

leads us to the following three sets of $n \times n$ nonlinear-matrix equations.

$$(A' + B'K)^T R_{11} + R_{11}(A' + B'K) + 2\alpha(H' + F'K)^T W_s(H' + F'K) - R_{11} D R_{11} = 0,$$

$$(6.8\text{-}36)$$

$$(A' + B'K)^T R_{12} + R_{12}(A + BK) - 2\alpha(H' + F'K)^T W_s(H + FK) - R_{11} D R_{12} = 0,$$

$$(6.8\text{-}37)$$

or equivalently

$$(A' + B'K - DR_{11})^T R_{12} + R_{12}(A + BK) - 2\alpha(H' + F'K)^T W_s(H + FK) = 0, \quad (6.8\text{-}38)$$

and finally

$$(A + BK)^T R_{22} + R_{22}(A + BK) + 2(W_0 + K^T WK)$$

$$+ 2\alpha(H + FK)^T W_s(H + FK) - R_{12}^T D R_{12} = 0. \qquad (6.8\text{-}39)$$

Thus, the optimal-auxiliary input becomes

$$v(t) = -\frac{1}{2\beta} W_{po}^{-1} C_o^T R \eta_I(t). \qquad (6.8\text{-}40)$$

Correspondingly, substituting the above v(t) in (6.8-16) yields the following optimal cost:

$$J_I = \int_0^\infty [\|\eta_I(t)\|_{\hat{W}_s}^2 + \beta \| - \frac{1}{2\beta} W_{po}^{-1} C_o^T R \eta_I(t) \|_{W_{po}}^2] dt \qquad (6.8\text{-}41)$$

6.8. Commonalties in Identification and Robustification

$$\triangleq \int_0^\infty \|\eta_I(t)\|_{W_I}^2 dt. \tag{6.8-42}$$

Here,

$$W_I \triangleq \hat{W}_s + \frac{1}{4\beta} RC_o W_{po}^{-1} C_o^T R = \begin{bmatrix} W_{I11} & W_{I12} \\ W_{I12}^T & W_{I22} \end{bmatrix}. \tag{6.8-43}$$

Substituting for $\hat{W}_s$ from (6.8-17) into (6.8-43) results in

$$W_{I11} \triangleq \alpha(H' + F'K)^T W_s(H' + F'K) + \tfrac{1}{2} R_{11} D R_{11}, \tag{6.8-44}$$

$$W_{I12} \triangleq -\alpha(H' + F'K)^T W_s(H + FK) + \tfrac{1}{2} R_{11} D R_{12}, \tag{6.8-45}$$

$$W_{I22} \triangleq W_0 + K^T W K + \alpha(H + FK)^T W_s(H + FK) + \tfrac{1}{2} R_{12}^T D R_{12}. \tag{6.8-46}$$

On the other hand, using (6.8-11), (6.8-12) and (6.8-35), we note that (6.8-29) can be equivalently rewritten as follows.

$$\dot{\eta}_I(t) \triangleq A_h \eta_I(t) = \begin{bmatrix} A' + B'K - DR_{11} & -DR_{12} \\ 0 & A + BK \end{bmatrix} \eta_I(t), \tag{6.8-47}$$

with

$$\eta_I(0) \triangleq \begin{Bmatrix} \xi(0) \\ x(0) \end{Bmatrix}. \tag{6.8-48}$$

Thus, the cost J_I of (6.8-42) is now equivalent to

$$J_I = \eta_I^T(0) \int_0^\infty \exp(A_h^T t) W_I \exp(A_h t) dt \, \eta_I(0) \triangleq \text{Tr}(P_I H_I), \tag{6.8-49}$$

where $P_I = P_I^T > 0$ satisfies the following Lyapunov equation

$$A_h^T P_I + P_I A_h + W_I = 0. \tag{6.8-50}$$

Also, in order that the performance index does not depend directly on the initial condition which is not generally available, we have used a statistical average of this initial condition, namely

$$H_I = \text{cov}(\eta_I(0), \eta_I^T(0)), \tag{6.8-51}$$

instead of $(\eta_I(0)\eta_I^T(0))$ in (6.8-49).

Before proceeding further, we must note that if we substitute for A_h from (6.8-47) and

$$P_I = \tfrac{1}{2} R, \tag{6.8-52}$$

into (6.8-50), then (6.8-50) becomes the same as (6.8-34). This relationship, which was not noticed in the original work [G. 5] substantially simplifies our analysis. However, to avoid mistakes we keep both P_I and R as they appear in various equations.

6.8.3.2. Derivation of the Optimal Controller Gain

The main objective of the remaining analysis is to derive the optimal-controller gain K such that $\partial J_I/\partial K = 0$. Now that the performance index is in the compact form of (6.8-49), we proceed to compute $\partial J_I/\partial K$ as follows. Suppose that $K \to K + \delta K$, with δK being "small", then $A_h \to A_h + \delta A_h$, where $A_h + \delta A_h$ remains stable and $\delta A_h \to 0$ as $\delta K \to 0$. Similarly $P_I \to P_I + \delta P_I$, where (cf., Lemma 6.4-1)

$$\delta P_I = \int_0^\infty \exp(A_h^T t)(\delta A_h^T P_I + P_I \delta A_h + \delta W_I)\exp(A_h t)dt. \tag{6.8-53}$$

Correspondingly, as $K \to K + \delta K$, with δK being "small", then $J_I \to J_I + \delta J_I$, where

$$\delta J_I = \mathrm{Tr}(\delta P_I H_I). \tag{6.8-54}$$

Substituting (6.8-53) into (6.8-54) and using the trace operator, (6.8-54) becomes

$$\delta J_I = \mathrm{Tr}(\Pi_I E_I). \tag{6.8-55}$$

Here,

$$\Pi_I \triangleq \delta A_h^T P_I + P_I \delta A_h + \delta W_I, \tag{6.8-56}$$

and

$$E_I \triangleq \int_0^\infty \exp(A_h t) H_I \exp(A_h^T t)dt. \tag{6.8-57}$$

When $K \to K + \delta K$ with δK being "small", then $R \to R + \delta R$ and $A_h \to A_h + \delta A_h$. In particular, from (6.8-36), if $K \to K + \delta K$ with δK being "small", then $R_{11} \to R_{11} + \delta R_{11}$ such that

$$\delta R_{11} \triangleq \int_0^\infty \exp(A_{cp}^T t)[\delta K^T \sigma_1 + \sigma_1^T \delta K] \exp(A_{cp} t)dt, \tag{6.8-58}$$

$$A_{cp} \triangleq A' + B'K - DR_{11}, \tag{6.8-59}$$

$$\sigma_1 \triangleq B'^T R_{11} + 2\alpha F'^T W_s(H' + F'K). \tag{6.8-60}$$

Similarly, from (6.8-38), if $K \to K + \delta K$ with δK being "small", then $R_{12} \to R_{12} + \delta R_{12}$ such that

$$\delta R_{12} \triangleq \int_0^\infty \exp(A_{cp}^T t)[\delta K^T \sigma_2 + \sigma_3 \delta K - \delta R_{11} DR_{12}] \exp(A_c t)dt, \tag{6.8-61}$$

$$A_c \triangleq A + BK, \tag{6.8-62}$$

$$\sigma_2 \triangleq B'^T R_{12} - 2\alpha F'^T W_s(H + FK), \tag{6.8-63}$$

6.8. Commonalties in Identification and Robustification

$$\sigma_3 \triangleq R_{12}B - 2\alpha(H' + F'K)^T W_s F. \tag{6.8-64}$$

Also from (6.8-47) when $K \to K + \delta K$, then we have

$$\delta A_h \triangleq \begin{bmatrix} B'\delta K - D\delta R_{11} & -D\delta R_{12} \\ 0 & B\delta K \end{bmatrix}. \tag{6.8-65}$$

We now search for the partitions of Π_I in (6.8-56). Substituting (6.8-65) into (6.8-56) results in

$$\Pi_I = \Pi_I^T \triangleq \begin{bmatrix} B'\delta K - D\delta R_{11} & -D\delta R_{12} \\ 0 & B\delta K \end{bmatrix}^T \begin{bmatrix} P_{I11} & P_{I12} \\ P_{I12}^T & P_{I22} \end{bmatrix}$$

$$+ \begin{bmatrix} P_{I11} & P_{I12} \\ P_{I12}^T & P_{I22} \end{bmatrix} \begin{bmatrix} B'\delta K - D\delta R_{11} & -D\delta R_{12} \\ 0 & B\delta K \end{bmatrix} + \delta W_I. \tag{6.8-66}$$

The partitions of Π_I are as follows.

$$\pi_{I11} \triangleq \delta K^T B'^T P_{I11} - \delta R_{11} D P_{I11} + P_{I11} B' \delta K - P_{I11} D \delta R_{11} + \delta W_{I11}, \tag{6.8-67}$$

$$\pi_{I12} \triangleq \delta K^T B'^T P_{I12} - \delta R_{11} D P_{I12} + P_{I12} B \delta K - P_{I11} D \delta R_{12} + \delta W_{I12}, \tag{6.8-68}$$

$$\pi_{I22} \triangleq \delta K^T B^T P_{I22} - \delta R_{12}^T D P_{I12} + P_{I22} B \delta K - P_{I12}^T D \delta R_{12} + \delta W_{I22}. \tag{6.8-69}$$

Here on using (6.8-44) to (6.8-46), the partitions of δW_I as $K \to K + \delta K$, with δK being "small", are:

$$\delta W_{I11} \triangleq \alpha \delta K^T F'^T W_s(H' + F'K) + \alpha(H' + F'K)^T W_s F' \delta K$$

$$+ (\tfrac{1}{2})[\delta R_{11} D R_{11} + R_{11} D \delta R_{11}], \tag{6.8-70}$$

$$\delta W_{I12} \triangleq -\alpha \delta K^T F'^T W_s(H + FK) - \alpha(H' + F'K)^T W_s F \delta K$$

$$+ (\tfrac{1}{2})[\delta R_{11} D R_{12} + R_{11} D \delta R_{12}], \tag{6.8-71}$$

$$\delta W_{I22} \triangleq \delta K^T[WK + \alpha F^T W_s(H + FK)] + [K^T W + \alpha(H + FK)^T W_s F] \delta K$$

$$+ (\tfrac{1}{2})[\delta R_{12}^T D R_{12} + R_{12}^T D \delta R_{12}]. \tag{6.8-72}$$

Substituting for δW_{Iij} (i, j = 1, 2), from (6.8-70) to (6.8-72), into (6.8-67) to (6.8-69) and considering the fact that from (6.8-52) we have $P_I = \tfrac{1}{2}R$, the partitions of $\Pi_I = \Pi_I^T$ become

$$\pi_{I11} \triangleq \delta K^T[B'^T P_{I11} + \alpha F'^T W_s H' + \alpha F'^T W_s F'K]$$
$$+ [P_{I11}B' + \alpha H'^T W_s F' + \alpha K^T F'^T W_s F']\delta K, \tag{6.8-73}$$

$$\pi_{I12} \triangleq \delta K^T[B'^T P_{I12} - \alpha F'^T W_s H - \alpha F'^T W_s FK]$$
$$+ [P_{I12}B - \alpha H'^T W_s F - \alpha K^T F'^T W_s F]\delta K, \tag{6.8-74}$$

$$\pi_{I22} \triangleq \delta K^T[B^T P_{I22} + \alpha F^T W_s H + (W + \alpha F^T W_s F)K]$$
$$+ [P_{I22}B + \alpha H^T W_s F + K^T(W + \alpha F^T W_s F)]\delta K. \tag{6.8-75}$$

On the other hand, δJ_I of (6.8-55) yields

$$\delta J_I = \mathrm{Tr}(\pi_{I11}E_{I11} + \pi_{I12}E_{I12}^T + \pi_{I12}^T E_{I12} + \pi_{I22}E_{I22}). \tag{6.8-76}$$

Thus substituting for π_{Iij} (i, j = 1, 2), from (6.8-73) to (6.8-75), into δJ_I and using certain trace operations yields

$$\delta J_I = \mathrm{Tr}[\delta K^T \sigma + \sigma^T \delta K]; \tag{6.8-77}$$

where

$$\sigma \triangleq [B'^T P_{I11} + \alpha F'^T W_s H' + \alpha F'^T W_s F'K]E_{I11}$$
$$+ [B'^T P_{I12} - \alpha F'^T W_s H - \alpha F'^T W_s FK]E_{I12}^T$$
$$+ [P_{I12}B - \alpha H'^T W_s F - \alpha K^T F'^T W_s F]^T E_{I12}$$
$$+ [B^T P_{I22} + \alpha F^T W_s H + (W + \alpha F^T W_s F)K]E_{I22}. \tag{6.8-78}$$

Because $\dfrac{\partial \mathrm{Tr}(A^T B)}{\partial A} = B$ and $\dfrac{\partial \mathrm{Tr}(AB)}{\partial A} = B^T$, (6.8-77) results in

$$\frac{\partial J_I}{\partial K} = \lim_{\delta K \to 0} \frac{\delta J_I}{\delta K} = 2\sigma. \tag{6.8-79}$$

The optimality of K requires that $\partial J_I / \partial K = 0$; i.e., the optimal controller gain must necessarily satisfy the following matrix equation

$$[B'^T P_{I11} + \alpha F'^T W_s H' + \alpha F'^T W_s F'K]E_{I11}$$
$$+ [B'^T P_{I12} - \alpha F'^T W_s H - \alpha F'^T W_s FK]E_{I12}^T$$
$$+ [P_{I12}B - \alpha H'^T W_s F - \alpha K^T F'^T W_s F]^T E_{I12}$$

6.8. Commonalties in Identification and Robustification

$$+ [B^TP_{I22} + \alpha F^TW_sH + (W + \alpha F^TW_sF)K]E_{I22} = 0. \qquad (6.8\text{-}80)$$

The scalar α can always be chosen such that $W + \alpha H^TW_sH$ remains a nonsingular matrix. Recall that $W > 0$, $W_s \geq 0$, and α can be either positive (sensitivity reduction *via* plant-operator), or negative (sensitivity maximization, *optimal* input for identification). Denoting $E_{I22}^{\dagger}$ as a generalized inverse matrix of E_{I22}, then the optimal controller gain must necessarily satisfy the following linear matrix equation in K.

$$K^o = -(W + \alpha F^TW_sF)^{-1}\{B^TP_{I22} + \alpha F^TW_sH$$

$$+ \{[B'^TP_{I11} + \alpha F'^TW_sH' + \alpha F'^TW_sF'K]E_{I11}$$

$$+ [B'^TP_{I12} - \alpha F'^TW_sH - \alpha F'^TW_sFK]E_{I12}^T$$

$$+ [P_{I12}B - \alpha H'^TW_sF - \alpha K^TF'^TW_sF]^TE_{I12}\}E_{I22}^{\dagger}\}. \qquad (6.8\text{-}81)$$

Note that here a generalized inverse of E_{I22} is used only for symbolic reasons and does not play any other role.

▼

The main interpretation of the above result is as follows. Suppose that $\alpha > 0$. By choosing K^o as an automatic controller gain, then the plant-operator's cumulative "effort" ($\int_0^\infty \beta \|v(t)\|_{W_{po}}^2 dt$), for bringing the real-operating trajectory close to the nominal value, is minimum in the sense of the overall quadratic cost given by (6.8-5). On the other hand, if we set $\alpha < 0$, then the above K results in an optimal $v(t)$ for system identification, provided that β remains a positive constant. Certainly, if we are using this algorithm for identification purposes, then we need to redefine appropriately our sensitivity measures. Also all the relevant equations corresponding to A_I in (6.8-12) must be redeveloped for the new sensitivity measure. The new A_I becomes an $n(r+1) \times n(r+1)$ matrix, where r is the number of parameters and n is the system dimension. This change will certainly affect the rest of our analysis.

Although it may seem that the present procedure, from both identification and sensitivity reduction points of view, is derived under the assumption for a possible closed-loop implementation of both regulator part (i.e., u(t)) and the auxiliary-input v(t), we do not recommend to use a closed-loop policy for v(t) (in the sense of incorporating that with u(t)), as we have indicated in the introduction. Also the pair (u(t), v(t)) may be considered as one system input, where the portion u(t) ∈ R^m is for regulating and stabilizing the system and the portion v(t) ∈ R^μ may serve to desensitize or to overregulate the system operation.

6.8.4. Numerical Algorithm

For the sake of discussion, we assume that our problem is formulated so as to design controllers using the previous *large*-sensitivity measure. Thus we describe the forthcoming algorithm from the control point of view, although we call that for both identification and robustification. If this sensitivity measure is being replaced by another one, or if we want to use this algorithm for identification, then the following steps must be modified accordingly.

ALGORITHM 6.8-1 (Identification versus Robustification with Auxiliary Input): To derive numerically the solution of Problem 6.8-1 and for a controller design we propose the following steps, provided that the initial perturbed system satisfies the basic requirements for the quadratic optimization such as the establishment of the positive (semi-) definiteness of various weighting matrices with perturbed variables. Indeed, maintaining the positive definiteness of weighting matrices (as in the previous case) provides us certain measure for the largest perturbations of the system matrices. For example, in (6.8-17), when we substitute for K (perhaps, at the initial regulator value), then we must maintain the positivity of $\hat{W}_s$ by choosing a set of appropriate initial perturbation matrices.

STEP 0 (To initialize): Depending on the proposed application, all the relevant parameters must be specified first. Let $J = 1, 2, ..., N$ be the iteration number.

STEP 1 (Sign of α): First we choose the appropriate sign of α. If $\alpha > 0$, then we are minimizing sensitivity using the plant-operator concept. On the other hand, if $\alpha < 0$, then we are maximizing sensitivity which is useful for system identification. In that case the sensitivity measures must be replaced with a set of appropriate *small-sensitivity measures*. Correspondingly, the following steps must be changed accordingly.

Group I (Steps 2 to 5): Equations that directly depend on K

STEP 2 (Computation of the R): From (6.8-36) to (6.8-39), and with the *current* value of K and all the relevant system parameters we compute the *current* R.

STEP 3 (Generating P_I): From (6.8-50), and with A_h from (6.8-47), W_I from (6.8-44) to (6.8-46), D from (6.8-35), A_{cp} from (6.8-59), and A_c from (6.8-62), we can show that

$$0 = A_{cp}^T P_{I11} + P_{I11} A_{cp} + W_{I11}, \tag{6.8-82}$$

$$0 = A_{cp}^T P_{I12} + P_{I12} A_c - P_{I11} D R_{12} + W_{I12}, \tag{6.8-83}$$

$$0 = A_c^T P_{I22} + P_{I22} A_c - R_{12}^T D P_{I12} - P_{I12}^T D R_{12} + W_{I22}. \tag{6.8-84}$$

However, from the computational point of view, we use the fact that $P_I = \frac{1}{2}R$.

STEP 4 (Computation of the optimal cost): Using the the current values of P_I we compute the optimal cost $J_I = \text{Tr}(P_I H_I)$.

STEP 5 (Generating E_I): From (6.8-57), we have

$$A_h E_I + E_I A_h^T + H_I = 0, \tag{6.8-85}$$

6.8. Commonalties in Identification and Robustification

and without any loss of generality and unless stated otherwise, we assume that H_I in (6.8-51) is

$$H_I \triangleq \begin{bmatrix} X & 0 \\ 0 & X \end{bmatrix}; \quad (6.8\text{-}86)$$

where $X = \text{cov}(\xi(0), \xi^T(0))$. Therefore, from (6.8-85), after substituting for A_h, D, and H_I, from (6.8-47), (6.8-35) and (6.8-86), respectively, we can show that

$$0 = A_c E_{I22} + E_{I22} A_c^T + X, \quad (6.8\text{-}87)$$

$$0 = A_{cp} E_{I12} + E_{I12} A_c^T - DR_{12} E_{I22}, \quad (6.8\text{-}88)$$

$$0 = A_{cp} E_{I11} + E_{I11} A_{cp}^T - DR_{12} E_{I12}^T - E_{I12} R_{12}^T D + X. \quad (6.8\text{-}89)$$

Group II (Steps 6 to 9): Equations which yield the initial K
(Only for the first iteration)

For symbolic reasons, let us show the optimal controller gain K^o in (6.8-81) as $K^o = K^o (J_I \triangleq J_p + \alpha J_s + \beta J_{po})$, and let us call $K_p = K_p(J_p)$ and $K_s = K_s(J_p + \alpha J_s)$. We now suggest that first we search for K_p, which corresponds to the standard regulator problem and it can be easily computed. Then, using this K_p as a starting value, we search for K_s, which incorporates the sensitivity measure. Subsequently, this new K_s is used as a starting value to search for the optimal controller gain K^o in steps of Groups I and III.

STEP 6 (Computation of the initial regulator gain): In (6.8-82) to (6.8-84), if we let $\alpha = 0$ and $C = 0$ (i.e., $D = 0$), then $P_{I11} = P_{I12} = 0$ and

$$(A + BK)^T P_{I22} + P_{I22}(A + BK) + W_0 + K^T WK = 0. \quad (6.8\text{-}90)$$

Correspondingly, the optimal controller gain (6.8-81) in this case becomes

$$K_p = -W^{-1} B^T P_{I22}. \quad (6.8\text{-}91)$$

Substituting (6.8-91) into (6.8-90) results in

$$A^T P_{I22} + P_{I22} A + W_0 - P_{I22} BW^{-1} B^T P_{I22} = 0. \quad (6.8\text{-}92)$$

We now search for an optimal controller gain incorporating sensitivity measure and without plant-operator

STEP 7 (Computation of the initial $R = 2P_I$): Now in Step 2, we propose to let $C = 0$ (i.e., $D = 0$), and $K \equiv K_p$ as computed in Step 6. This substitution yields the following sets of matrix equations for $R (= 2P_I)$.

$$(A' + B'K)^T R_{11} + R_{11}(A' + B'K) + 2\alpha(H' + F'K)^T W_s (H' + F'K) = 0, \quad (6.8\text{-}93)$$

$$(A' + B'K)^T R_{12} + R_{12}(A + BK) - 2\alpha(H' + F'K)^T W_s (H + FK) = 0, \quad (6.8\text{-}94)$$

$$(A + BK)^T R_{22} + R_{22}(A + BK) + 2(W_0 + K^T WK) +$$

$$2\alpha(H + FK)^T W_s(H + FK) = 0. \qquad (6.8\text{-}95)$$

STEP 8 (Computation of the initial E_I): From (6.8-87) to (6.8-89), and with the above R and $K \equiv K_p$, we compute E_I.

STEP 9 (Computation of the initial truncated optimal controller gain): Substituting the P_I and E_I generated in Steps 7 and 8 into (6.8-81) results in an algebraic-matrix equation for K_s, which is a truncated-optimal-controller gain, in the sense that we have only considered $(J_p + \alpha J_s)$ in our analysis. We may repeat locally Steps 7 to 9, until we get the "best" K_s. Generally speaking, by the best K_s we mean a controller gain whose corresponding cost is close to its last value. We now let $K \equiv K_s$.

Group III (Step 10): Equations that yield the optimal K^o

STEP 10 (Computation of a new value for controller gain K): We use the best truncated optimal controller gain K generated in Step 9 in order to invoke Steps 2 to 5 in Group I. Then we have a set of *current* values for all the quantities on the right-hand side of (6.8-81). From this linear-algebraic-matrix equation we can generate a new value for controller gain K which satisfies *both sides* of (6.8-81). Using this new K, we go back to the beginning of the algorithm and update the iteration number. In a given iteration, if the difference between the current optimal cost and its last value (Step 4) is large, then we continue to invoke the above steps until this cost converges to a fixed value within some reasonable tolerance and computational effort. Otherwise we stop. ∎

6.8.5. Numerical Examples

In this section we present two tutorial examples to show the applicability of the results developed in this section.

EXAMPLE 6.8-1: Consider a real-operating system and its corresponding nominal system with the following parameters.

$A = -1, \quad \Delta A = -1, \quad A' \triangleq A + \Delta A = -2,$

$B = 2, \quad \Delta B = 1, \quad B' \triangleq B + \Delta B = 3,$

$H = 1, \quad \Delta H = 1, \quad H' \triangleq H + \Delta H = 2,$

$F = 1, \quad \Delta F = 1, \quad F' \triangleq F + \Delta F = 2,$

$W_s = 1, \quad W_0 = 1, \quad W = 1,$

6.8. Commonalties in Identification and Robustification

$$C = 1, \quad \beta = 1, \quad W_{po} = 1, \quad X = 1. \quad (6.8\text{-}96)$$

Apply Algorithm 6.8-1 to this system in order to generate the optimal controller gain K^o for robustification purposes.

SOLUTION: From the above parameters we have $D = \frac{1}{2}$. Clearly, our intention is to use this algorithm to minimize the *large-sensitivity measure* in order to design a robust controller gain.

STEP 1: We let $\alpha = 1$ in order to study the sensitivity minimization.

Iteration 1:

STEP 6: From (6.8-92) we have $P_{I22} = 0.309$, and from (6.8-91) we have $K_p = -0.618$.

STEP 7: With $K \equiv K_p = -0.618$, $C = 0$ (i.e., $D = 0$) and from (6.8-93) to (6.8-95), we have

$$R = \begin{bmatrix} 0.157 & -0.098 \\ -0.098 & 0.683 \end{bmatrix} = 2P_I. \quad (6.8\text{-}97)$$

STEP 8:

$$E_I = \begin{bmatrix} 0.135 & 0.000 \\ 0.000 & 0.224 \end{bmatrix}. \quad (6.8\text{-}98)$$

STEP 9: $K \equiv K_s = -0.960$.

Here, we may repeat locally the above steps in order to improve K by using its last value in Steps 7 and 8, then generating a new value for the controller gain in Step 9. However, we branch out to steps in Group I.

STEP 2: With $K = -0.960$, we have

$$R = \begin{bmatrix} 0.0011 & -0.0007 \\ -0.0007 & 0.6586 \end{bmatrix} = 2P_I. \quad (6.8\text{-}99)$$

STEP 4: $J_I = \text{Tr}(P_I H_I) = 0.32985$.

STEP 5:

$$E_I = \begin{bmatrix} 0.087 & 0.000 \\ 0.000 & 0.171 \end{bmatrix}. \quad (6.8\text{-}100)$$

STEP 10: From (6.8-81) and in this example we have

$$K = -\frac{1}{2}\{2P_{I22} + 1 + \{[3P_{I11} + 4 + 4K]E_{I11}\}E_{I22}^\dagger\}. \quad (6.8\text{-}101)$$

Substituting for P_I and E_I from (6.8-99) and (6.8-100) into the above algebraic matrix equation, yields $K = -0.9162$.

Iteration 2:

STEP 2: Let K = − 0.9162. Then

$$R = \begin{bmatrix} 0.0059 & -0.0037 \\ -0.0037 & 0.6518 \end{bmatrix} = 2P_I. \quad (6.8\text{-}102)$$

STEP 4: $J_I = \text{Tr}(P_I H_I) = 0.32885$.

STEP 5:

$$E_I = \begin{bmatrix} 0.105 & 0.000 \\ 0.000 & 0.176 \end{bmatrix}. \quad (6.8\text{-}103)$$

STEP 10: K = − 0.922.

We stop at this iteration because neither J_I nor K is substantially changing when compared to its last value. Thus the optimal controller gain is $K^o = -0.922$ corresponding to $J_I = 0.329$.

The optimal strategy for the auxiliary input is

$$v(t) = -\tfrac{1}{2}(1, 0) \begin{bmatrix} 0.0059 & -0.0037 \\ -0.0037 & 0.6518 \end{bmatrix} \eta_I(t) = (-0.003, 0.0019)\eta_I(t). \quad (6.8\text{-}104)$$

Clearly, and as we have intended, the above results suggest that the plant-operator's gain is indeed small.

▼

EXAMPLE 6.8-2: Consider a real-operating system and its corresponding nominal system with the following parameters.

$$A = -2, \quad \Delta A = +1, \quad A' \triangleq A + \Delta A = -1,$$

$$B = 2, \quad \Delta B = -1, \quad B' \triangleq B + \Delta B = 1,$$

$$H = 2, \quad \Delta H = -1, \quad H' \triangleq H + \Delta H = 1,$$

$$F = 2, \quad \Delta F = -1, \quad F' \triangleq F + \Delta F = 1,$$

$$W_s = 1, \quad W_0 = 1, \quad W = 1,$$

$$C = 1, \quad \beta = 1, \quad W_{po} = 1, \quad X = 1. \quad (6.8\text{-}105)$$

Our intention is again to compute controller gain K^o such that the plant-operator's cumulative "effort" for maintaining the system operation becomes minimum.

6.8. Commonalties in Identification and Robustification

SOLUTION: From the above parameters we have $D = \frac{1}{2}$. Using the algorithm that is suggested earlier, we can compute the optimal controller gain K^o as follows.

Within our computational accuracy and at the second iteration the optimal controller gain becomes $K^o = -0.927$ corresponding to $J = 0.24535$.

The optimal strategy for the auxiliary input is

$$v(t) = -\tfrac{1}{2}(1, 0) \begin{bmatrix} 0.0025 & -0.0034 \\ -0.0034 & 0.4882 \end{bmatrix} \eta(t) = (-0.00125, 0.0017)\eta(t). \quad (6.8\text{-}106)$$

The plant-operator who knows about the system and in fact has "memorized" by experience the ideal trajectory of the system and is observing its real-operating trajectory, will make decisions on how to implement the necessary changes in order to robustify system performance. Note that here we have not been feeding back the model x(t) into the real-operating system. This nominal model enters into the system only through the plant-operator strategy with a small gain.

▼

REMARK 6.8-4 (On the procedure): In this section, we have studied certain commonalties between control and identification problems. In particular, we have investigated the possibility of having an optimal-auxiliary input which can be computed for either sensitivity minimization or parameter identification. This particular input can be also used for modeling the supervisory effect of a plant manager or a plant operator in an automated-manufacturing plant. The algorithm can be switched from one application to another by changing one scalar parameter. The main feature of this algorithm, when used for sensitivity minimization, is that the plant-operator gain becomes only a fraction of the automatic-controller gain. In other words, most of the task for system robustification will be accomplished by the feedback-automatic controller resulting in the minimum involvement of outside operator. This particular approach in designing system controller has many potentials in practical problems, although in many such problems the actual output instead of state is available for measurement. However, we note that this basic research is the first step toward such an application or extension. There are a number of issues, regarding the inverse optimal problems when parameter changes, and comparisons of this method with other sensitivity-optimization methods, which need to be investigated. The most appealing feature of this formulation is that this method of analysis can be extended easily to a class of nonlinear regulator systems which are of interest in engineering applications such as robotics control. Study of these issues and other applications of these results, including consideration of cases in which we place constraints on the input function(s), merit further efforts.

REMARK 6.8-5 (The final thought on Examples 6.8-1 & 2): The fundamental assumption in sensitivity analysis is that the parameters are constant or they are changing "slowly". Otherwise to associate "dynamics" to parameter variations will put the analysis in the realm of adaptive control theory. Therefore most of the designs in this area are completed based on some possible *projected worst* scenario. In other words, the system cannot be "optimal" for all possible and perhaps unbounded-parameter variations. On the other hand, when dealing with multiple-parameter variations we must consider a *set* of possible perturbation matrices (which may have the same norm). These perturbation matrices must also maintain the positivity of weighting matrices and that constraint leads us to some acceptable-perturbation range. Having said these facts we may state these perturbation matrices in terms of their norms but that representation does not constitute any mechanism for parameter variations. Furthermore, to design controllers while formally bounding the norm of perturbation matrices complicates the analysis and we gain very little from that approach. However, our results which are developed with the assumptions that parameters are changing "slowly" is equivalent to specifying a particular upper bound for the norms of the perturbation matrices. If needs arise, then we can change the previous set in its entirety based upon the new information on system parameter variations. But in any events we cannot associate any dynamics to these changes when we work in the linear time-invariant setting. If we use adaptive controller in order to specify "dynamics" to the parameter variations, then our implementational cost increases substantially. In this section we have tried to get the best result out of what is called "static" or "passive" parameter variations with minimum *hardware*. Here the controller gains are computed from a set of possible worst-perturbation matrices. The plant-operator will make decisions on how to implement necessary changes in order to robustify system performance. Actually these results are more suitable (and simpler to build) in practice than the real adaptive controllers. Certainly, as the preceding numerical examples show, this method of designing controller is extremely robust with respect to parameter variations which some of them are as high as 100%. In each of these two examples K^o is almost the same (up to the third decimal). A careful look at these two numerical examples also reveals that, generally speaking, Example 6.8-2 starts *almost* where Example 6.8-1 ends. In fact we could have selected *exactly* various extreme corners of Example 6.8-1 as the starting point for the next example, in order ultimately to construct a hyperbox in the *PVS*, surrounding the initial operating point. The implication of this constructional approach is that when we find each one of the K^o's which brings us to these extreme points, then we can judiciously make decision on how to select the final optimal-controller gain K^o which is valid for the entire hyperbox surrounding the optimal system, allowing each and every parameter changes in that hyperbox. We invite the reader to continue this construction.

6.9. SENSITIVITY MINIMIZATION IN HARDY SPACES

6.9.1. Introduction

A concise account of sensitivity minimization in Hardy spaces – H^2 and H^∞ – is presented. Originally, the H^∞-sensitivity minimization was introduced by Zames in his 1981-pioneering work [H2. 66].[2] Before we initiate any discussion on this topic, we note that in order to facilitate the future utilization of our references in § 6.12.H, we categorize them according to the following groups: *[H1], [H2 to H5]* and *[H6]*. The mathematical references in group *[H1]* provide the necessary background for research in this area. In particular, Nevanlinna-Pick interpolation theory

---

[2] Although the theoretical aspects of this methodology are well established and we briefly elaborate on that, the following excerpts from the original source in this area seems appropriate for historical reasons. On the motivation behind development of this methodology, Zames wrote that "One way of attenuating disturbances is to introduce a filter of the WHK (Wiener – Hopf – Kalman) type in the feedback path. Despite the unquestioned success of the WHK and the state-space approaches, the classical methods, which rely on lead-lag 'compensators' to reduce sensitivity, have continued to dominate many areas of design. On and off, there have been attempts to develop analogous methods for multivariable systems. However, the classical techniques have been difficult to pin down in a mathematical theory, partly because the purpose of compensation has not been clearly stated. One of our of objectives is to formulate the compensation problem as the solution to a well defined optimization problem. Another motivating factor is the gradual realization that classical theory is not just an old-fashioned way of doing WHK, but is concerned with a different category of mathematical problems. In a typical WHK problem, the quadratic norm of response to a disturbance d is minimized by a projection method; in a deterministic version, the power spectrum $|\hat{d}(j\omega)|$ is a *single, known* vector in, e.g., the space $L_2(-\infty, \infty)$; in stochastic version, d belongs to a *single*, random process of *known* covariance properties. However, there are many practical problems in which $|\hat{d}(j\omega)|$ is unknown but belongs to a prescribed set, or d belongs to a class of random processes whose covariances are uncertain but belong to a prescribed set. For example, in audio design, d is often one of a set of narrow-band signals in the 20-20K Hz interval, as opposed to a single, wide-band signal in the same interval. Problems involving such more general disturbance sets are not tractable by WHK or projection techniques. In a feedback context, they are now usually handled by empirical methods resembling those of classical sensitivity. One objective here is to find a systematic approach to problems involving such sets of disturbances. Another observation is that many problems of plant uncertainty cannot be stated easily in classical theory, e.g., in terms of a tolerance-band on a frequency response, but are difficult to express in a linear-quadratic-state-space framework. One reason for this is that frequency-response descriptions and, more generally, input-output descriptions preserve the operations of system addition and multiplication, whereas state-space description do not. Another reason is that the quadratic norm is hard to estimate for system products, whereas the induced norm (or 'gain') that is implicit in the classical theory is easier to estimate. We would like to exploit these advantages in the study of plant uncertainty. Finally, sensitivity theory is one of the few tools available for the study of organization structure: feedback versus open-loop, aggregated versus disaggregated, etc. For example, feedback reduces complexity of identification roughly for the same reason that it reduces sensitivity. However, it is hard to draw definitive conclusions about the effects of organization without some notion of optimality, and such a notion is missing in the old theory [H2. 66]."

is discussed in [*H1*. 7], the operator theory is discussed in [*H1*. 1 & 2 and 20], and the geometric theory is discussed in [*H1*. 3]. The Duren classical book on H^p -spaces has a rich bibliography and is indispensable [*H1*. 10]. Harmonic analysis is also required for a thorough understanding of this topic [*H1*. 22]. The pertinent work on feedback system design in this group of references is responsible in one form or another for the development of this methodology.

Further regrouping of *[H2 to H5]*, as indicated in § 6.12.*H*, however, should be viewed very cautiously, in the sense that these are quite interrelated as far as this methodology is concerned. We refer to references in group *[H2]* as complex-domain approach, because these references, while covering a wide spectrum of problems ranging from linear time-invariant to discrete-time and delay systems, use *mainly* input-output relationships [operators] in the form of Laplace or z-transform of various system quantities. In this group [*H2*. 66 to 68] establish the core of this methodology. Reference [*H2*. 66] uses analytic functions to formulate the sensitivity-reduction problem with feedback as an optimization problem. Most of the early approaches in this group use Nevanlinna-Pick interpolation techniques (cf., [*H1*. 7]), or the operator theory of [*H1*. 1 & 2 and 20]. The earlier research in H_∞ -norm optimization is reported in [*H2*. 12 & 13], in particular, the Nehari problem. To present these complex-domain approaches requires a different setting which is outside the scope of this book.

On the other hand, references in group *[H3]* discuss *primarily* systems which are described by state-space representations. This kind of representations has brought a new dimension to the H_∞ -norm optimization. The earlier exposition of this approach is given in [*H2*. 12 & 13]. The computational complexity of this early approach, however, was overwhelming, resulting in several Riccati equations of increasing dimension and involved control algorithms. To alleviate some of these difficulties in the H_∞ -norm optimization problem a new look provided by Doyle, Glover, Khargonekar, and Francis [*H3*. 6], which is based on the familiar concepts from the H^2 -control problems such as a separation argument, state-feedback, and observer-based methods. In [*H3*. 6], it is shown that a controller exists if and only if the unique stabilizing solutions for two Riccati algebraic-matrix equations are positive-definite and their product has a spectral radius less than a given positive number. Although references in group *[H3]* are mostly recent (since 1987), and they represent the new trend in this area, the distinction between references in *[H2]* and *[H3]* is also made in accordance with the overall theme of this book regarding system classifications, in order to imbed the H^∞ -literature on sensitivity minimization in the overall sensitivity literature pertaining to dynamical systems.

References in group *[H4]* stand mainly as an advancement of that reported in groups *[H3]* and *[H2]*. We note that this group is heavily based on references [*H1*. 1 & 2 and 20] which provide the relation between operator theory and complex-domain approach. References in group *[H5]* provide certain computational aspects of this methodology and certain pertinent optimization algorithms. The references in group *[H6]* emphasize the forthcoming applications and extensions of this methodology and this list is by no means complete.

6.9. Sensitivity Minimization in Hardy Spaces

At this juncture, it is imperative to explain our *sensitivity measure* and the rationale behind its choice. We recall from Algorithm 6.2-1 (for two nominally equivalent open-loop and closed-loop systems) that the closed-loop system performs better than the open-loop system if we choose the feedback controller gain K such that $S_c^T(-j\omega)QS_c(j\omega) - Q < 0$ holds for a large band of frequencies, where $S_c(s) \triangleq [I + (sI - A)^{-1}BK]^{-1}$ is our sensitivity measure. We can develop a similar sensitivity measure in terms of system output resulting in $S(s) \triangleq [I + G(s)F(s)]^{-1}$, where G(s) is system transfer matrix and F(s) is the feedback or compensator transfer matrix. Here we want to select F(s) such that the overall system is stabilized and a sensitivity measure of the type $S(s) \triangleq [I + G(s)F(s)]^{-1}$ satisfies $S^T(-j\omega)QS(j\omega) - Q < 0$ (with $Q = Q^T > 0$) for as large a frequency-band as possible. Clearly, this measure of sensitivity is generated when *small*-parameter changes are taking place in G(s) and is valid only *under* this condition and thus for large-parameter variations we need to develop a new sensitivity measure reflecting such changes in G(s). Meanwhile, when the above closed-loop system is subject to an exogenous signal which appears before its output (cf., *Case 3* of Table 6.1-1), the transfer matrix between the underlying input and output becomes $[I + G(s)F(s)]^{-1}$, which is the same as $S(s)$. Thus in this sense we may consider $S(s)$ as our sensitivity measure whose reduction may reduce the undesirable effects of both external disturbance and *small*-parametric uncertainty. Based on this thought, the main premise of a sensitivity minimization in the H_2- or H_∞-norm optimization is to minimize the corresponding norm of a certain transfer matrix (that is our sensitivity measure), resulting in the "best" compensator which reduces *primarily* the worst effect of exogenous signal in system output.

In the classical sensitivity-comparison scheme (cf., Algorithm 6.2-1, *per se*), no optimization using feedback is performed, however, it is expected naturally that the outcome of this algorithm can be improved if somehow this sensitivity measure is optimized and that optimization is the point of departure from the classical sensitivity-comparison approach. The implication of this sensitivity optimization is that the system-frequency response will be shaped to meet new design criteria and that frequency shaping is at the *heart* of this type sensitivity optimization process (cf., [H5. 25 to 29]). The main thrust of this section is to study how this kind of sensitivity measures can be optimized and how to incorporate the corresponding optimization procedure in the overall design process in order to select an optimal compensator. In other words, the purpose of this study is to elaborate on the meaning of the "best" compensator resulted from the feedback optimization and the related issues in *one* typical application using an H_2- or H_∞-norm optimization methodology.

It is thus clear that the fundamental issue in this study is the development of a systematic procedure to optimize H_2- or H_∞-norm of a *certain* transfer matrix called system-sensitivity measure. In the current literature, however, an equivalent or perhaps *generic* form of this transfer matrix is being considered which corresponds to the system described in *Case 1* of Table 2.6-4 (cf., Fig. 6.9-1). We note that the preceding F(s) is replaced by K(s) to emphasize that we no longer are in the realm of classical sensitivity-comparison configuration. This figure is also intended to convey the basic aspects of this methodology and if needs arise this configuration can be modified to accommodate a new situation, or we must improvise. Now, the main problem that is being studied in the literature is as follows. Referring to Fig. 6.9-1 and for a given *finite-dimensional,* linear, time-invariant system, with a stable plant transfer matrix G(s), find the class of all controllers K(s) such that the $\| G_{zw}(s) \|_\infty < \gamma$, where $G_{zw}(s) = z(s)/w(s)$ and $\gamma > 0$ is a prespecified number [*H3*. 6].

In summary, in this section an exploratory presentation of this topic is reviewed by studying a linear time-invariant system which is known and is considered to be fixed. This approach is mainly developed for the treatment of exogenous signals (*Case 3* of Table 6.1-1). Since the introduction of this topic efforts are being made to recast much of the theory, originally developed for a system described in the frequency or complex-domain, to a system described in the state-space or time-domain representation. In particular, the computational issues associated with applications of this methodology are studied extensively using time-domain analysis. Like most newly developed methodology this topic is continuing to evolve, however, we maintain our presentation at the level which remains essential for future pursuit of this topic.

A researcher who is interested to study this topic should consult Section 2.5 on the preliminary mathematical techniques; and Section 2.6 on certain aspects of system theory as pertain to this methodology; followed by a review of Section 2.7 on standard [analysis and/or synthesis] problems using H_∞-norm optimization techniques. In the remaining parts of this section we review the fundamentals of sensitivity analysis stemming from the H_∞-norm optimization techniques. This presentation closely follows [*H3*. 6 and 37].

6.9.2. Riccati Algebraic – Matrix Equation

In § 6.8.3.1 we utilize a Riccati algebraic-matrix equation, which plays the central role in quadratic optimal control theory. Now, we study this equation in conjunction with its corresponding Hamiltonian matrix (cf., Definition 2.6-2). A standard n × n Riccati algebraic-matrix equation is as follows.

$$A^T P + PA - PW_1 P + W = 0. \qquad (6.9\text{-}1)$$

Here we choose both weighting matrices $W_1 = W_1^T > 0$ and $W = W^T > 0$, in order to have a unique solution $P = P^T > 0$ for (6.9-1), which this solution also yields an asymptotically stable $A - W_1 P$, i.e., all eigenvalues of this matrix are on the left-hand side of the jω–axis.

Meanwhile, (6.9-1) corresponds to the following 2n × 2n Hamiltonian matrix

6.9. Sensitivity Minimization in Hardy Spaces

$$H \triangleq \begin{bmatrix} A & -W_1 \\ -W & -A^T \end{bmatrix}. \qquad (6.9\text{-}2)$$

Clearly (6.9-2) satisfies the structure condition of a typical Hamiltonian matrix, however, to show that the 2n eigenvalues of H are of the form $\pm\lambda_i$, $i = 1, \cdots, n$, we propose to use the following similarity transformation.

$$T = \begin{bmatrix} 0 & -I \\ I & 0 \end{bmatrix}, \quad T^{-1} = \begin{bmatrix} 0 & I \\ -I & 0 \end{bmatrix}, \qquad (6.9\text{-}3)$$

resulting in $T^{-1}HT = -H^T$, which means both H and $-H$ have the same set of eigenvalues and that can only happen when the 2n eigenvalues of H are of the form $\pm\lambda_i$, $i = 1, \cdots, n$. This property can also be shown by using the following similarity transformation.

$$T = \begin{bmatrix} I & 0 \\ P & I \end{bmatrix}, \quad T^{-1} = \begin{bmatrix} I & 0 \\ -P & I \end{bmatrix}, \qquad (6.9\text{-}4)$$

resulting in an upper triangular matrix

$$T^{-1}HT = \begin{bmatrix} A - W_1 P & -W_1 \\ 0 & -(A - W_1 P)^T \end{bmatrix}. \qquad (6.9\text{-}5)$$

Clearly (6.9-5) confirms that the eigenvalues of H are $\pm\lambda_i(A - W_1 P)$, $i = 1, \cdots, n$. Because of our earlier choices on W_1 and W the unique solution for P becomes $P = P^T > 0$, and the eigenvalues of H are distributed symmetrically with respect to both real and imaginary axes, but none on the $j\omega$–axis.

There are several widely-known methods and computer programs available to solve (6.9-1) for the unique $P = P^T > 0$. For instance, in [2C. 34] the Potter's method, which is widely used to solve a Riccati algebraic-matrix equation, is described. We only consider a few properties of (6.9-2), which are also used in the Potter's method, in order to explore them for our subsequent applications. Using (6.9-1) we have

$$H \begin{bmatrix} I \\ P \end{bmatrix} = \begin{bmatrix} A - W_1 P \\ -W - A^T P \end{bmatrix} = \begin{bmatrix} I \\ P \end{bmatrix} (A - W_1 P). \qquad (6.9\text{-}6)$$

Equation (6.9-6) resembles the standard equation for computing matrix eigenvalues and eigenvectors. This fact together with (6.9-5) suggests that, when H has no eigenvalues on the $j\omega$–axis, we can partition H into two n-dimensional spectral subspaces $S_-(H)$, corresponding to its eigenvalues on the left-hand side of the s-plane, and $S_+(H)$, corresponding to its eigenvalues on the right-hand side of the s-plane. When a basis for $S_-(H)$ is found (which corresponds to the eigenvalues of $A - W_1 P$ that are stable), this set of eigenvectors can be partitioned into $\binom{X}{Y}$, where both X and Y are $n \times n$ matrices. Under these conditions we can always arrange these eigenvectors such that X^{-1} exists, then $P = YX^{-1}$ [2C. 34].

From these well-known results it is inferred that the mapping $H \to P$ can be generalized to a wider class of Riccati algebraic-matrix equations than that of (6.9-1). For notational convenience it is customary to show the mapping $H \to P$ with "Ric" in (6.9-2), and its domain with dom(Ric). In particular, we let dom(Ric) to consist of all Hamiltonian matrices H which have no eigenvalues on the $j\omega$-axis, and for such Hamiltonian matrices we again define the two spectral subspaces $S_-(H)$ and $S_+(H)$, while the two subspaces $S_-(H)$ and $\text{Im}\binom{0}{1}$ are *complementary* (they form a basis). This property is called complementarity. In other words, if $H \in \text{dom(Ric)}$, and if there exists a unique $P^T = P \triangleq \text{Ric}(H) > 0$, then that solution stabilizes *accordingly* the corresponding "A-matrix" of H. For instance, one such Hamiltonian matrix can be defined as

$$H \triangleq \begin{bmatrix} A & W_1 \\ W & -A^T \end{bmatrix} \quad \Rightarrow \quad P \triangleq \text{Ric}(H), \qquad (6.9\text{-}7a)$$

where depending on $W_1 = W_1^T \geq 0$ and $W = W^T \geq 0$, the $P = P^T \geq 0$ (not strictly positive) stabilizes $A + W_1 P$ and is the unique solution of

$$A^T P + PA + PW_1 P - W = 0. \qquad (6.9\text{-}7b)$$

There are other alternatives to the above representation for a Hamiltonian matrix and each results in a new P. For instance the following matrix

$$H \triangleq \begin{bmatrix} A & -BB^T \\ -C^T C & -A^T \end{bmatrix}, \qquad (6.9\text{-}8a)$$

is a Hamiltonian matrix and if the pair (A, B) is stabilizable and the pair (C, A) is observable, then $P^T = P \triangleq \text{Ric}(H) > 0$.

The (6.9-8a) has been modified for other applications subsequently. For instance, if we define

$$H \triangleq \begin{bmatrix} A & \gamma^{-2} BB^T \\ -C^T C & -A^T \end{bmatrix}, \qquad (6.9\text{-}8b)$$

then it is shown in [H3. 6] that for a $G \triangleq (A, B, C, 0)$, the following four conditions are equivalent. (a) $\| G(s) \|_\infty < \gamma$; (b) H has no eigenvalues on the $j\omega$-axis; (c) $H \in \text{dom(Ric)}$; and (d) $H \in \text{dom(Ric)}$ and $\text{Ric}(H) \geq 0$ (or $\text{Ric}(H) > 0$ if the pair (C, A) is observable). This result is used to compute H_∞-norm of $G(s)$. To prove this result γ is set identically equal to one since this is always possible by scaling the transfer matrix. Then $[I - G^T(-s)G(s)]^{-1}$ is generated which shows the above H at $\gamma = 1$ is its "A-matrix". The rest of the proof is straightforward. ∎

6.9.3. The H^2 – and H^∞ – Optimization Schemes

6.9.3.1. Introduction

The system under consideration is *Case 3* of Table 6.1-1 which has an equivalent block diagram depicted in *Case 1* of Table 2.6-4 and is shown in Fig. 6.9-1 with a minor change of notation to reflect the current literature. Here G(s) represents the transfer matrix corresponding to a *finite-dimensional,* linear, time-invariant plant where the input and output vectors are referred to this plant; and K(s) is the associated finite-dimensional [admissible] controller or compensator that is designed subsequently. Both matrices are real-rational and proper functions of s. The signal w(t) is that portion of input vector which contains all external (and perhaps open-loop) inputs, disturbances, sensor noise, and reference inputs or commands [we call them mostly *troublesome inputs*]; z(t) is that portion of output vector which contains all the information we are seeking from the system, and we call that the *desirable output;* y(t) is that portion of output vector which is measured and fed back to the plant *(sensors);* and finally u(s) = K(s)y(s) is the feedback-controlled input *(actuators).* The objective of this study, as formulated in the next two *generic* problems, is to find [an admissible] K(s) such that either the H_2– or H_∞–norm of the transfer matrix between the troublesome input and the desirable output is minimized, while the *internal stability* for the underlying system has been guaranteed. Here the transfer matrix between w and z is shown by $G_{zw}(s)$, similar symbols are introduced for transfer matrices between other variables, subsequently. Also internal stability means that the states of G(s) and K(s) go to zero for all initial conditions in the absent of external inputs. Our presentation closely follows the original contributions of Doyle, Glover, Khargonekar, and Francis [H3. 6]; and Safonov, Limebeer, and Chiang [H3. 37].

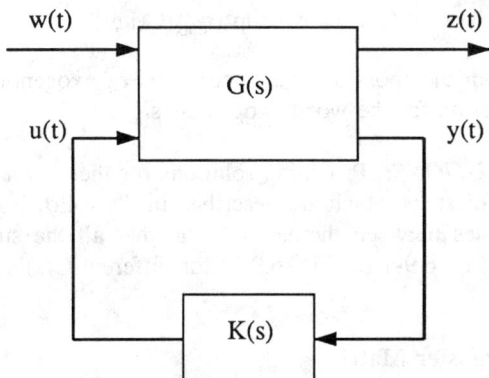

Fig. 6.9-1. Linear fractional transformation.

PROBLEM 6.9-1 (The H_2–Norm): Consider the system of Fig. 6.9-1 (corresponding to *Case 3* of Table 6.1-1), and find a linear-feedback control law K(s) which stabilizes the transfer matrix between the troublesome input w and the desirable output z given by

$$G_{zw}(s) \triangleq G_{11}(s) + G_{12}(s)K(s)[I - G_{22}(s)K(s)]^{-1}G_{21}(s), \qquad (6.9\text{-}9)$$

and minimizes the H_2–norm of this transfer matrix which is shown by $\| G_{zw}(s) \|_2$.

■

Recall from § 2.6.4.1 that the H_2–norm of a stable transfer matrix G(s) for the purpose of this study is given by

$$\langle G, G \rangle = \| G \|_2^2 \triangleq \frac{1}{2\pi} \int_{-\infty}^{+\infty} \text{Tr}[G^*(j\omega)G(j\omega)]\, d\omega, \qquad (6.9\text{-}10)$$

which arises when the exogenous signals are either fixed or have a fixed power spectrum. Although this type of problems and in its time-domain setting either as a regulator or as an estimator has been extensively reviewed in the literature, this frequency-domain setting elucidates many intricate relationships and dualities which exist among these various optimization schemes.

PROBLEM 6.9-2 (The H_∞–Norm): Consider the system of Fig. 6.9-1, find a linear-feedback control law K(s) such that it stabilizes the transfer matrix between the troublesome input and the desirable output given in (6.9-9) and minimizes the H_∞–norm of this transfer matrix which is shown by $\| G_{zw}(s) \|_\infty$. (This and the H_2–norm may be weighted by frequency-dependent weighting matrices).

■

Recall from § 2.6.4.2 that the H_∞–norm of a stable transfer matrix G(s) is given by

$$\| G \|_\infty = \underset{\omega}{\text{ess sup}}\, \sigma_{\max}[G(j\omega)]. \qquad (6.9\text{-}11)$$

This norm arises from the (possibly weighted) balls of exogenous signals and this reflects a design criterion for the worst exogenous signal.

CONCEPTUAL SOLUTIONS: Providing solutions for these problems requires utilization of a number of steps which are described in the following after we develop various common issues between them. We note that all the subsequent analyses pertain to system of Fig. 6.9-1 (or Fig. 6.9-2) for different G(s)'s.

6.9.3.2. General Transfer Matrix

Consider Fig. 6.9-1 with the following transfer matrix.

6.9. Sensitivity Minimization in Hardy Spaces

$$G(s) \triangleq \left[\begin{array}{c:cc} A & B_1 & B_2 \\ \hdashline C_1 & 0 & D_{12} \\ C_2 & D_{21} & 0 \end{array} \right]. \tag{6.9-12}$$

The subsequent analyses depend on this specific partitioning of G(s) which is as general as possible modulo $D_{11} \equiv 0$ (no direct transmission between w and z) and $D_{22} \equiv 0$, otherwise the analyses will become very involved. Various transformational procedures are reported in the literature to structure a given G(s) into the above form *if possible* (cf., § 6.9.3.8). Additionally we assume that (6.9-12) meets the following conditions:

(i) The pair (A, B_1) is *stabilizable* (i.e., its unstable subspace is contained in its controllable subspace) and the pair (C_1, A) is *detectable* (i.e., its unreconstructible, here the same as its unobservable, subspace is contained in its stable subspace).

(ii) The pair (A, B_2) is stabilizable and the pair (C_2, A) is detectable.
Comments: This and the previous condition can be relaxed, although that may complicate the analysis. These two conditions are imbedded in the assumptions that there exist positive-definite solutions for the two subsequent Riccati algebraic-matrix equations.

(iii) $D_{12}^T (C_1, D_{12}) = (0, I)$.
Comments: This means C_1 and D_{12} (or $C_1 x$ and $D_{12} u$) are orthogonal (i.e., z is projected into two orthogonal vectors), and this condition simplifies the analysis by removing the crossing terms, and requires that D_{12} to have more rows than columns, i.e., the dim(z) $\geq$ dim(u). If D_{12} is of maximum rank, then rank $(D_{12}) = $ dim(u).

(iv) $\begin{bmatrix} B_1 \\ D_{21} \end{bmatrix} D_{21}^T = \begin{bmatrix} 0 \\ I \end{bmatrix}$.
Comments: This means B_1 and D_{21} are orthogonal which is the dual of *(iii)*. Furthermore, this condition indicates how the undesirable input enters G(s), and it requires that D_{21} to have more columns than rows, i.e., dim(w) $\geq$ dim(y). If D_{21} is of maximum rank, then rank $(D_{21}) = $ dim(y). ∎

For the purpose of demonstration, (6.9-12) (also known as output-feedback configuration), and in conjunction with Fig. 6.9-1, conveys all principal aspects of this methodology. Furthermore, if the above conditions *(i), (iii)*, and *(iv)* hold, then an *admissible* controller K(s) can be found if and only if the corresponding $G_{zw}(s)$ (cf., (6.9-9)) is a real-rational transfer matrix belonging to the H^∞-space (cf., Lemma 6.9-5 for establishing a proof). Here the internal stability is essentially equivalent to the input-output stability, while condition *(ii)* above is a necessary and sufficient condition for G(s) to be internally stabilizable.

6.9.3.3. Special Transfer Matrices

Studying the next four special structures facilitate the overall presentation and understanding of this topic. Each case plays the role of a special G(s) with its corresponding set of Problems 6.9-1 & 2, which we subsequently call them Sub-Problems 6.9-1a & 2a to 6.9-1d & 2d. We revisit these special structures in the remaining parts of this section. Certainly each one of these special cases should be viewed as a new case, though similar to that of (6.9-12), is not its subset.

Full – Information Structure: Consider a situation where

$$G(s) \triangleq \begin{bmatrix} A & : & B_1 & B_2 \\ \cdots & : & \cdots & \cdots \\ C_1 & : & 0 & D_{12} \\ \begin{bmatrix} I \\ 0 \end{bmatrix} & : & \begin{bmatrix} 0 \\ I \end{bmatrix} & \begin{bmatrix} 0 \\ 0 \end{bmatrix} \end{bmatrix}. \qquad (6.9\text{-}13)$$

This choice of G(s) gives full information on both state x and the external input w, although this information may not be fully utilized. This setting is more natural than the state feedback alone, particularly, for parameterization of all suboptimal controllers, and when $D_{11} \neq 0$, and when we seek the optimal H^∞–controllers instead of suboptimal ones. Here we let $D_{11} \equiv 0$ and $D_{22} \equiv 0$, additionally we assume that similar conditions as in (6.9-12) are met by (6.9-13), namely:

(i) The pair (A, B_1) is stabilizable and the pair (C_1, A) is detectable.
(ii) The pair (A, B_2) is stabilizable.
(iii) $D_{12}^T (C_1, D_{12}) = (0, I)$.
(iv) This condition is considered to be stronger than its counterpart in (6.9-12), in the sense that particular structures for C_2 and D_{21} are specified.

∎

Here, the internal stability is not equivalent to $G_{zw}(s)$ to be a real-rational transfer matrix belonging to the H^∞–space, rather the K(s) being admissible means internally stabilizing.

∎

Full – Control Structure: Consider a transfer matrix

$$G(s) \triangleq \begin{bmatrix} A & : & B_1 & (I, 0) \\ \cdots & : & \cdots & \cdots \\ C_1 & : & 0 & (0, I) \\ C_2 & : & D_{21} & (0, 0) \end{bmatrix}, \qquad (6.9\text{-}14)$$

which is the dual of (6.9-13) and the controller has full access to both state through the output injection and output z. Here also we let $D_{11} \equiv 0$ and $D_{22} \equiv 0$, and

6.9. Sensitivity Minimization in Hardy Spaces

assume that similar conditions as in (6.9-12) are met by (6.9-14), which are the dual of those given for (6.9-13), namely:

(i) The pair (A, B_1) is stabilizable and the pair (C_1, A) is detectable.
(ii) The pair (C_2, A) is detectable.
(iii) Specific structures for B_2 and D_{12} are given.
(iv) $\begin{bmatrix} B_1 \\ D_{21} \end{bmatrix} D_{21}^T = \begin{bmatrix} 0 \\ I \end{bmatrix}.$

■

Here, the same remarks regarding the internal stability can be made as in the preceding case of full-information structure.

■

Disturbance – Feedforward Structure: The corresponding transfer matrix becomes

$$G(s) \triangleq \begin{bmatrix} A & : & B_1 & B_2 \\ \cdots & : & \cdots & \cdots \\ C_1 & : & 0 & D_{12} \\ C_2 & : & I & 0 \end{bmatrix}. \qquad (6.9\text{-}15)$$

For this structure we also let $D_{11} \equiv 0$ and $D_{22} \equiv 0$, again we assume that similar conditions as in (6.9-12) are met by (6.9-15), namely:

(i) Now, $A - B_1 C_2$ is *stable* and the pair (C_1, A) is detectable.
(ii) The pair (A, B_2) is stabilizable.
(iii) $D_{12}^T (C_1, D_{12}) = (0, I)$.
(iv) Specific structure for D_{21} is given.

■

The change from (A, B_1) being stabilizable to $A - B_1 C_2$ being stable is to guarantee the internal stability, which is equivalent to $G_{zw}(s)$ being a real-transfer matrix belonging to the H^∞-space. Here we note that when there is no feedback ($C_2 = 0$), the measurement is w and when there is a feedback $C_2 \neq 0$, because of $A - B_1 C_2$ being stable, the transfer matrix between w and y is stable. In fact, this case is very similar to the full-information structure and therefore we should expect that they behave analogously.

■

Output – Estimation Structure: This situation is the dual of the preceding case with the following transfer matrix.

$$G(s) \triangleq \left[\begin{array}{c:cc} A & B_1 & B_2 \\ \hdashline C_1 & 0 & I \\ C_2 & D_{21} & 0 \end{array} \right]. \tag{6.9-16}$$

Here again we let $D_{11} \equiv 0$ and $D_{22} \equiv 0$, and assume the similar conditions as in (6.9-12) are met by (6.9-16), namely:

(i) The pair (A, B_1) is stabilizable and *now* $A - B_2 C_1$ is *stable*.

(ii) The pair (C_2, A) is detectable.

(iii) Specific structure for D_{12} is given.

(iv) $\begin{bmatrix} B_1 \\ D_{21} \end{bmatrix} D_{21}^T = \begin{bmatrix} 0 \\ I \end{bmatrix}.$

■

Conditions *(i)* and *(iv)* mean that internal stability is equivalent to that corresponding to (6.9-12). Here we can prove that under these conditions K(s) is admissible if and only if $G_{zw}(s)$ is a real-rational transfer matrix belonging to the H^∞-space (cf., Lemma 6.9-5 for establishing a proof). This is a very special estimation problem that has direct application in solving Problems 6.9-1 & 2.

6.9.3.4. The Preliminary Analysis

First we recall the notation introduced in § 2.6.3.2 and § 2.6.3.3. All time-domain signals are considered to be Lebesgue-measurable functions on R^n and form appropriate complete mathematical spaces. In particular, a square-integrable signal is shown by $L^2(-\infty, \infty)$ (or L^2) which forms a Hilbert space. This space can be partitioned into two subspaces *corresponding* to $L_+^2[0, \infty)$ that represents all signals in L^2 which are zero for $t < 0$ *(causal functions)*, and its orthogonal complement which is shown by $L_-^2(-\infty, 0]$ that represents all signals in L^2 which are zero for $t > 0$ *(anticausal functions)*. The frequency-domain signals are considered to be members of Hardy spaces, in particular, H^2- and H^∞-spaces. The square-integrable time- and frequency-domain signals are equivalent due to the Hilbert space isomorphism. Here L_+^2 maps *onto* H_+^2 and L_-^2 maps *onto* H_-^2 (cf., § 2.6.3.3), and these mappings are also *(1-1)*. In the following we use H^2 and $H_\perp^2$ which consist of square-integrable functions, respectively, on the imaginary axis with analytic continuation into the right- and the left-half plane. For the H^∞-space, we use the same notation as in § 2.6.3.3. The prefix R denotes the real-rational and the prefix B denotes the open unit-ball (cf., Definition 2.5-3). For instance, $w \in BH^2$ means that we assume w is a square-integrable signal and its H_2-norm is bounded by 1. (This assumption is reasonable and widely-used to model a disturbance acting on a system, however, in Section 6.10, the same disturbance is modeled by its *maximum amplitude*, and that modeling results in a whole

6.9. Sensitivity Minimization in Hardy Spaces

different way of looking at the disturbance-reduction problems.) Similarly, a transfer matrix $G(s) \in RH^\infty$ means that $G(s)$ is a real-rational matrix belonging to the H^∞-space.

∎

The formal solutions of Problems 6.9-1 & 2 are presented in § 6.9.3.6, and § 6.9.3.7, respectively. To establish various steps toward completion of these solutions, the following results and those in § 6.9.3.5 are studied. These results which have many other applications in control theory play the central role in establishing solutions for Problems 6.9-1 & 2.

∎

Consider a linear system described by

$$\dot{x}(t) = Ax(t) + Bw(t), \quad x(0) = x_o, \quad (6.9\text{-}17a)$$

$$z(t) = Cx(t), \quad (6.9\text{-}17b)$$

where A is assumed stable and $w(t)$ is considered as disturbance or an external input. The transfer matrix for this system is $G(s) \triangleq (A, B, C, 0) = z(s)/w(s)$. We assume that for the first reading the pair (A, B) is controllable and the pair (C, A) is observable, although not necessarily completely. The controllability and observability Gramian matrices for system (6.9-17) are shown by L_c and L_o which satisfy the following two equations.

$$AL_c + L_c A^T + BB^T = 0, \quad (6.9\text{-}18a)$$

$$A^T L_o + L_o A + C^T C = 0. \quad (6.9\text{-}18b)$$

In passing, we recall that in (6.9-17) if $w \in L^2$ (with A stable) then $z \in L^2$, and $\|G\|_\infty$ is the induced norm of the multiplication operator shown by M_G, as well as the Toeplitz operator $P_+ M_G : H^2 \to H^2$. It is well-known that

$$\|G\|_\infty^2 = \sup_{w \in BL_+^2} \|P_+ z\|_2^2 = \sup_{w \in BH^2} \|P_+ M_G w\|_2^2 = \rho(L_o L_c), \quad (6.9\text{-}19)$$

where $\rho(\cdot)$ is the spectral radius of $(\cdot)$ [H3. 6].

∎

If $\|G\|_\infty < 1$, then the Hamiltonian matrix H given in (6.9-8b) (with $\gamma = 1$) is in dom(Ric) and $P \triangleq \text{Ric}(H) \geq 0$, with $A + BB^T P$ stable and

$$A^T P + PA + PBB^T P + C^T C = 0. \quad (6.9\text{-}20)$$

This information leads us to the next important lemma.

LEMMA 6.9-1 [H3. 6]: Consider system (6.9-17) and suppose that $\|G\|_\infty < 1$, then

$$\sup_{w \in L_+^2} (\|z\|_2^2 - \|w\|_2^2) = x_o^T P x_o. \quad (6.9\text{-}21)$$

PROOF: Consider the following derivative along (6.9-17) and (6.9-20).

$$\frac{d}{dt}[x^T(t)Px(t)] = \dot{x}^T(t)Px(t) + x^T(t)P\dot{x}(t)$$

$$= x^T(t)(A^TP + PA)x(t) + w^T(t)B^TPx(t) + x^T(t)PBw(t)$$

$$= x^T(t)(-C^TC - PBB^TP)x(t) + w^T(t)B^TPx(t) + x^T(t)PBw(t)$$

$$= -z^T(t)z(t) + w^T(t)w(t) - [w(t) - B^TPx(t)]^T[w(t) - B^TPx(t)].$$

$$(6.9\text{-}22)$$

Since $w \in L_+^2$ and A is stable, $x \in L_+^2$ (or $x(\infty) = 0$), and integrating both sides of (6.9-22) from $t = 0$ to $t = \infty$ yields

$$-x_o^T P x_o = -\|z\|_2^2 + \|w\|_2^2 - \|w - B^TPx\|_2^2, \qquad (6.9\text{-}23)$$

or

$$\|z\|_2^2 - \|w\|_2^2 \le x_o^T P x_o. \qquad (6.9\text{-}24)$$

By letting $w(t) = B^TPx(t)$ (which is also known from the linear optimal control theory), and because $A + BB^TP$ is stable, i.e., $x(t) \in L_+^2$, we conclude that $w \in L_+^2$ and (6.9-24) becomes an equality. One final note that $w(t)$ is assumed to be a linear function of $x(t)$ throughout this analysis. ∎

Consider a partitioned system (6.9-17) with $B \triangleq [B_1, B_2]$, and conformally $G(s) \triangleq [G_1(s), G_2(s)]$. Then $\|G_2\|_\infty < 1$ if and only if the Hamiltonian matrix

$$H_W \triangleq \begin{bmatrix} A & B_2B_2^T \\ -C^TC & -A^T \end{bmatrix} \in \text{dom}(\text{Ric}) \Rightarrow W \triangleq \text{Ric}(H_W) \ge 0. \qquad (6.9\text{-}25)$$

Suppose that

$$w \in W \triangleq \left\{ \begin{bmatrix} w_1 \\ w_2 \end{bmatrix} \,\middle|\, w_1 \in H_\perp^2,\, w_2 \in L_+^2 \right\}, \qquad (6.9\text{-}26)$$

and let $\Gamma \triangleq P_+[M_{G_1}, M_{G_2}] : W \to H^2$ be a mixed Hankel-Toeplitz operator,

$$\Gamma \begin{bmatrix} w_1 \\ w_2 \end{bmatrix} = P_+[G_1, G_2] \begin{bmatrix} w_1 \\ w_2 \end{bmatrix}, \qquad w_1 \in H_\perp^2,\, w_2 \in L_+^2, \qquad (6.9\text{-}27a)$$

whose adjoint is defined as $\Gamma^* : H^2 \to W$ and is given by

6.9. Sensitivity Minimization in Hardy Spaces

$$\Gamma^* z = \begin{Bmatrix} P_-(G_1^T(-s)z) \\ G_2^T(-s)z \end{Bmatrix} = \begin{Bmatrix} P_- G_1^T(-s) \\ G_2^T(-s) \end{Bmatrix} z, \quad (6.9\text{-}27b)$$

where $P_- Gz = P_-(Gz) = (P_- M_G)z$. Then does $\sup_{w \in BW} \| P_+ z \|_2 < 1$, or

$$\sup_{w \in BW} \| \Gamma w \|_2 < 1? \quad (6.9\text{-}28)$$

LEMMA 6.9-2 [H3. 6]: Inequality (6.9-28) holds if and only if the following two conditions are met:

(i) $H_W \in \text{dom (Ric)}$.

(ii) $\rho(WL_c) < 1$, where $\rho(\cdot)$ is the spectral radius of $(\cdot)$.

PROOF: Condition (i) is necessary for (6.9-28) and this inequality holds if and only if condition (ii) holds. If condition (ii) does not hold, then $\rho(WL_c) \geq 1$ and if we show that this situation is equivalent to $\sup_{w \in BW} \| \Gamma w \|_2 \geq 1$, then that completes the proof. Using the previous lemma and the fact L_c^{-1} exists (system is assumed to be completely controllable) and

$$\inf_{w \in L_-^2} \{ \| w \|_2^2 \mid x(0) = x_o \} = x_o^T L_c^{-1} x_o, \quad (6.9\text{-}29)$$

(this can be shown by computing $\frac{d}{dt}[x^T(t) L_c^{-1} x(t)]$ and following the same steps which lead to (6.9-22)), we can show that

$$\sup_{w \in W} \{ \| P_+ z \|_2^2 - \| w \|_2^2 \mid x(0) = x_o \} = x_o^T (W - L_c^{-1}) x_o. \quad (6.9\text{-}30)$$

If (and only if) there exists an $x_o \neq 0$ such that the right-hand side of (6.9-30) becomes positive, then $\rho(WL_c) \geq 1$, which implies that there exists a $w \in W$, $w \neq 0$, such that $\| P_+ z \|_2^2 \geq \| w \|_2^2$, and that implies $\sup_{w \in BW} \| \Gamma w \|_2 \geq 1$.
■

DEFINITION 6.9-1 (Inner): Given $G(s) \in RH^\infty$, if $G^T(-s)G(s) = I$, then $G(s)$ is called an inner matrix. On the $j\omega$-axis this means $G^*(j\omega)G(j\omega) = I$.

LEMMA 6.9-3 [H4. 5]: Given $G(s) \triangleq (A, B, C, D)$, if the pair (C, A) is detectable and L_o is the corresponding solution of (6.9-18b), then

(i) $L_o \geq 0$ if and only if A is stable.

(ii) $D^T C + B^T L_o = 0 \Rightarrow G^T(-s)G(s) = D^T D$.

(iii) $L_o \geq 0$, the pair (A, B) is controllable and $G^T(-s)G(s) = D^T D$
$\Rightarrow D^T C + B^T L_o = 0$.

LEMMA 6.9-4 [H3. 6]: Consider system of Fig. 6.9-1, where $G(s) \in RH^\infty$ and is an inner matrix and in particular $G_{21}^{-1}(s) \in RH^\infty$, and $K(s)$ is a proper rational transfer matrix of s. Then the following two conditions are equivalent:

(i) This system is internally stable and well-posed with $\| G_{zw}(s) \|_\infty < 1$.

(ii) $K(s) \in RH^\infty$ with $\| K(s) \|_\infty < 1$.

PROOF [H3. 6]: To sketch a proof first we show (ii) $\Rightarrow$ (i). Since $G(s) \in RH^\infty$ and is an inner matrix, the $\| G_{22}(s) \|_\infty < 1$ and together with (ii), the internal stability and well-posedness are true. To show that also $\| G_{zw}(s) \|_\infty < 1$, we also note that since $G(s)$ is an inner matrix, $\| z \|^2 + \| y \|^2 = \| w \|^2 + \| u \|^2$, where $u = Ky$. Therefore

$$\| z \|^2 \leq \| w \|^2 + (\| K \|^2 - 1) \| y \|^2 = [1 - (1 - \| K \|^2)\frac{\| y \|^2}{\| w \|^2}] \| w \|^2$$

$$\leq \| w \|^2 \quad \Rightarrow \quad \| G_{zw}(s) \|_\infty < 1. \tag{6.9-31}$$

Note that in (6.9-31) and by (ii), $\| K \| < 1$, and it can be shown that $\| y \|^2 / \| w \|^2$ is bounded by studying $G_{wy}(s)$.

Next, we show that (i) $\Rightarrow$ (ii). Suppose that at a given frequency ω, $\| Ky \| > \| y \|$, then by reversing the preceding argument we can show that this situation cannot occur and therefore $\| K \|_\infty < 1$ for all ω and K is a rational matrix. To show that $K \in RH^\infty$, we consider a right-coprime factorization $K = NM^{-1}$ where both N and $M \in RH^\infty$. Then it can be shown that $M^{-1} \in RH^\infty$. Using $G_{wy}(s)$ and $u = Ky$, we have $G_{uw}G_{21}^{-1} = K(I - G_{22}K)^{-1} = NM^{-1}(I - G_{22}NM^{-1})^{-1} = N(M - G_{22}N)^{-1}$. Since $G_{uw}G_{21}^{-1} \in RH^\infty$ and $N \in RH^\infty$, we have $(M - G_{22}N)^{-1} \in RH^\infty$. The fact that this inverse exists (i.e., $\det(M - G_{22}N) \neq 0$) and both M and N are in RH^∞ and $\| G_{22} \|_\infty < 1$ and $\| K \|_\infty < 1$ implies that $M^{-1} \in RH^\infty$ and that completes the proof. ∎

LEMMA 6.9-5 [H3. 6 and 24]: Consider (6.9-12) and Fig. 6.9-1, with $A \in R^{n \times n}$, $C_1 \in R^{\omega_1 \times n}$, $C_2 \in R^{\omega_2 \times n}$, $B_1 \in R^{n \times m_1}$, $B_2 \in R^{n \times m_2}$, and define similarly D_{12} and D_{21}. Let

$$n_{12}(\lambda) \triangleq \text{rank} \begin{bmatrix} A - \lambda I & B_2 \\ C_1 & D_{12} \end{bmatrix} \quad \text{and} \quad n_{21}(\lambda) \triangleq \text{rank} \begin{bmatrix} A - \lambda I & B_1 \\ C_2 & D_{21} \end{bmatrix}. \tag{6.9-32}$$

Suppose that we apply a controller K and $n_{12}(\lambda) = n + m_2$ and $n_{21}(\lambda) = n + \omega_2$ for all $\text{Re}(\lambda) \geq 0$, then K is admissible if and only if $G_{zw}(s) \in RH^\infty$. ∎

To establish solutions of the special structure transfer matrices (6.9-13) to (6.9-16) and Problems 6.9-1 & 2, we review the following two cases which

6.9. Sensitivity Minimization in Hardy Spaces

correspond to the H_2- and $H_\infty-$norms. In this regard, we group all the necessary algebraic steps to be utilized for these solutions first, subsequently, we elaborate on these relationships in specific case under study.

CASE I (The H_2- Norm): Consider the following Hamiltonian matrices with their corresponding Riccati algebraic-matrix equation.

$$H_{X_2} \triangleq \begin{bmatrix} A & -B_2B_2^T \\ -C_1^TC_1 & -A^T \end{bmatrix} \Rightarrow X_2 \triangleq \text{Ric}(H_{X_2}) \geq 0, \qquad (6.9\text{-}33a)$$

$$A^TX_2 + X_2A - X_2B_2B_2^TX_2 + C_1^TC_1 = 0. \qquad (6.9\text{-}33b)$$

$$H_{Y_2} \triangleq \begin{bmatrix} A^T & -C_2^TC_2 \\ -B_1B_1^T & -A \end{bmatrix} \Rightarrow Y_2 \triangleq \text{Ric}(H_{Y_2}) \geq 0, \qquad (6.9\text{-}34a)$$

$$AY_2 + Y_2A^T - Y_2C_2^TC_2Y_2 + B_1B_1^T = 0. \qquad (6.9\text{-}34b)$$

Based on some well-known results, we define the following matrices.

$$F_2 \triangleq -B_2^TX_2, \quad A_{F_2} \triangleq A + B_2F_2, \quad C_{1F_2} \triangleq C_1 + D_{12}F_2, \qquad (6.9\text{-}35a)$$

$$L_2 \triangleq -Y_2C_2^T, \quad A_{L_2} \triangleq A + L_2C_2, \quad B_{1L_2} \triangleq B_1 + L_2D_{21}, \qquad (6.9\text{-}35b)$$

$$\hat{A}_2 \triangleq A + B_2F_2 + L_2C_2, \qquad (6.9\text{-}35c)$$

$$G_c(s) \triangleq \begin{bmatrix} A_{F_2} & \vdots & I \\ \cdots & \vdots & \cdots \\ C_{1F_2} & \vdots & 0 \end{bmatrix}, \quad G_f(s) \triangleq \begin{bmatrix} A_{L_2} & \vdots & B_{1L_2} \\ \cdots & \vdots & \cdots \\ I & \vdots & 0 \end{bmatrix}. \qquad (6.9\text{-}35d)$$

Now, consider the following transfer matrix and its equivalent.

$$z(s) = \begin{bmatrix} A_{F_2} & \vdots & B_1 & B_2 \\ \cdots & \vdots & \cdots & \cdots \\ C_{1F_2} & \vdots & 0 & D_{12} \end{bmatrix} \begin{Bmatrix} w(s) \\ v(s) \end{Bmatrix} \triangleq G_c(s)B_1w(t) + U(s)v(t), \qquad (6.9\text{-}36)$$

where $G_c(s)$ is given in (6.9-35d) and $U(s)$ is defined as follows.

$$U(s) \triangleq \begin{bmatrix} A_{F_2} & \vdots & B_2 \\ \cdots & \vdots & \cdots \\ C_{1F_2} & \vdots & D_{12} \end{bmatrix}. \qquad (6.9\text{-}37a)$$

Suppose that $D\perp$ is any matrix such that $(D_{12}, D\perp)$ becomes an orthogonal matrix. Then using the above elements we define

$$U_\perp(s) \triangleq \begin{bmatrix} A_{F_2} & \vdots & -X_2^\dagger C_1^T D_\perp \\ \cdots & \vdots & \cdots \cdots \\ C_{1F_2} & \vdots & D_\perp \end{bmatrix}, \qquad (6.9\text{-}37b)$$

where the superscript "†" denotes generalized inverse.

LEMMA 6.9-6 [H3. 6]: The matrix $(U, U_\perp)$ is a square and inner and a realization for

$$G_c^T(-s)\bigl(U(s), U_\perp(s)\bigr) \triangleq \begin{bmatrix} A_{F_2} & \vdots & -B_2 & X_2^\dagger C_1^T D_\perp \\ \cdots & \vdots & \cdots & \cdots \cdots \\ X_2 & \vdots & 0 & 0 \end{bmatrix} \in RH^2. \qquad (6.9\text{-}38)$$

We can also show that $U(s)$ and $U_\perp(s)$ are both inner matrices, and $U_\perp^T(-s)G_c(s) \in RH_\perp^2$ also $U^T(-s)G_c(s) \in RH_\perp^2$.

PROOF: To sketch a proof, we show that (6.9-37a) is an inner matrix, by using Lemma 6.9-3. Here the corresponding Lyapunov equation of (6.9-18b) and the associated Riccati algebraic-matrix equation of (6.9-33b) become

$$A_{F_2}^T L_o + L_o A_{F_2} + C_{1F_2}^T C_{1F_2} = 0, \qquad (6.9\text{-}39a)$$

$$A_{F_2}^T X_2 + X_2 A_{F_2} + X_2 B_2 B_2^T X_2 + C_1^T C_1 = 0. \qquad (6.9\text{-}39b)$$

Subtracting these two equations and considering that $D_{12}^T C_1 \equiv 0$, we conclude that $X_2 \equiv L_o$, and $D^T C + B^T L_o \equiv D_{12}^T C_{1F_2} + B_2^T L_o = D_{12}^T(C_1 - D_{12}B_2^T X_2) + B_2^T L_o \equiv 0$, and thus from Lemma 6.9-3, $U(s)$ is an inner matrix. Similarly $U_\perp(s)$ of (6.9-37b) is an inner matrix, so is $(U, U_\perp)$, and the rest of the proof is straightforward when applying the following similarity transformation matrix to $G_c^T(-s)\bigl(U(s), U_\perp(s)\bigr)$.

$$T = \begin{bmatrix} I & X_2 \\ 0 & I \end{bmatrix} \qquad T^{-1} = \begin{bmatrix} I & -X_2 \\ 0 & I \end{bmatrix}. \qquad (6.9\text{-}40)$$

CASE II (The H_∞-Norm): Consider the following Hamiltonian matrices with their corresponding Riccati algebraic-matrix equations. Although the $(1,2)$-block partition of these matrices are not sign definite, we *assume* that the corresponding Riccati equations have positive- (semi-) definite solutions.

$$H_{X_\infty} \triangleq \begin{bmatrix} A & \gamma^{-2}B_1 B_1^T - B_2 B_2^T \\ -C_1^T C_1 & -A^T \end{bmatrix} \Rightarrow X_\infty \triangleq \mathrm{Ric}(H_{X_\infty}) \geq 0, \qquad (6.9\text{-}41a)$$

$$A^T X_\infty + X_\infty A + X_\infty(\gamma^{-2}B_1 B_1^T - B_2 B_2^T)X_\infty + C_1^T C_1 = 0. \qquad (6.9\text{-}41b)$$

6.9. Sensitivity Minimization in Hardy Spaces

$$H_{Y_\infty} \triangleq \begin{bmatrix} A^T & \gamma^{-2}C_1^T C_1 - C_2^T C_2 \\ -B_1 B_1^T & -A \end{bmatrix} \Rightarrow Y_\infty \triangleq \text{Ric}(H_{Y_\infty}) \geq 0, \quad (6.9\text{-}42a)$$

$$AY_\infty + Y_\infty A^T + Y_\infty(\gamma^{-2}C_1^T C_1 - C_2^T C_2)Y_\infty + B_1 B_1^T = 0. \quad (6.9\text{-}42b)$$

Here we define the following matrices.

$$F_\infty \triangleq -B_2^T X_\infty, \quad A_{F_\infty} \triangleq A + B_2 F_\infty, \quad C_{1F_\infty} \triangleq C_1 + D_{12} F_\infty, \quad (6.9\text{-}43a)$$

$$L_\infty \triangleq -Y_\infty C_2^T, \quad Z_\infty \triangleq (I - \gamma^{-2} Y_\infty X_\infty)^{-1}. \quad (6.9\text{-}43b)$$

Consider the following transfer matrix

$$G_{rep}(s) \triangleq \begin{bmatrix} A_{F_\infty} & : & B_1 & B_2 \\ \cdots & : & \cdots & \cdots \\ C_{1F_\infty} & : & 0 & D_{12} \\ -B_1^T X_\infty & : & I & 0 \end{bmatrix}, \quad (6.9\text{-}44)$$

where the subscript "rep" stands for replaced (relative to the original transfer matrix) and this notation is different from the original source [H3. 6]. Based on the existence assumption on X_∞, we have the following.

LEMMA 6.9-7 [H3. 6]: If X_∞ exists and $X_\infty \geq 0$, then the above $G_{rep}(s) \in RH^\infty$ and is also an inner matrix. Furthermore the (2,1)–block partition of $G_{rep}(s)$ is invertible and is in RH^∞.

PROOF: If for the time being we assume that A_{F_∞} is stable (this property is proven in Sub-Problems 6.9-1a & 2a, Item FI-4), then we can show that $G_{rep}(s) \in RH^\infty$. To show that $G_{rep}(s)$ is also an inner we note that X_∞ and the observability Gramian matrix corresponding to (6.9-44) are the same (cf., (6.9-39)). Indeed for this system $D^T C + B^T L_o \equiv 0$. Thus according to the Lemma 6.9-3, the $G_{rep}(s)$ is an inner matrix. Finally, we can show that the (2,1)–block partition of $G_{rep}(s)$ is invertible and belongs to the RH^∞ because $A_{F_\infty} + B_1 B_1^T X_\infty$ is stable (cf., Sub-Problems 6.9-1a & 2a, Item FI-4). ∎

We note that $G_{rep}(s)$ in Lemma 6.9-7 satisfies the corresponding condition for transfer matrix in Lemma 6.9-4. Thus if we augment $G_{rep}(s)$ with another transfer matrix $K(s)$ according to Lemma 6.9-4, then additional conclusions can be made (cf., Comment 6.9-5).

6.9.3.5. Special Structures Revisited

For each of the four special structures introduced in § 6.9.3.3 we associate two sub-problems which correspond to Problems 6.9-1 & 2 and Fig. 6.9-1. Subsequently, these sub-problems are solved in response to the following five items.

ITEMS 6.9-1: The main objective in each of the four subsequent cases is to find an admissible K, while the following five items are being evaluated.

1: Find the minimum of $\| G_{zw}(s) \|_2 = \gamma_2^{min}$.

2: Find the unique controller which minimizes $\| G_{zw}(s) \|_2$.

3: Find the family of all controllers such that $\| G_{zw}(s) \|_2 < \gamma_2$, where $\gamma_2 > \gamma_2^{min}$. *Comment:* This is called parameterization of *all* stabilizing controllers and is also defined in the same manner for the remaining cases.

4: Develop the necessary and sufficient conditions for the existence of a controller such that $\| G_{zw}(s) \|_\infty < \gamma_\infty$. Here $\gamma_\infty > \gamma_\infty^{min}$ (the minimum H_∞-norm).

5: Find the family of all controllers such that $\| G_{zw}(s) \|_\infty < \gamma_\infty$. ∎

The rationale for studying these items relative to the following four cases is as follows. First of all, as in the classical optimal control problems, here also the separation arguments hold, although not very straightforward in the case of the H^∞-control, secondly, we should concentrate on the full-information and the output-estimation cases, because the remaining two sub-problems are the dual of these two cases, and that fact reinforces the need for solving these cases prior to solving Problems 6.9-1 & 2. Finally, these solutions can be used to develop solutions for the two main Problems 6.9-1 & 2.

SUB-PROBLEMS 6.9-1a & 2a (Full – Information (FI)): This case corresponds to G(s) given in (6.9-13) with all its additional conditions. The answer to the preceding five items in this case are as follows.

FI-1: $\min \| G_{zw}(s) \|_2 = \| G_c(s)B_1 \|_2 = [\text{Tr}(B_1^T X_2 B_1)]^{1/2}$.

FI-2: $K(s) \triangleq (F_2, 0)$.

FI-3: $K(s) \triangleq (F_2, Q(s))$, where $Q(s) \in RH^2$, and
$\| Q(s) \|_2^2 < \gamma^2 - \| G_c(s)B_1 \|_2^2$.

FI-4: $H_{X_\infty} \in \text{dom}(\text{Ric})$, $X_\infty \triangleq \text{Ric}(H_{X_\infty}) \geq 0$.

FI-5: $K(s) \triangleq (F_\infty - Q(s)\gamma^{-1} B_1^T X_\infty, Q(s))$, where $Q(s) \in RH^\infty$, $\| Q(s) \|_\infty \leq \gamma$.

6.9. Sensitivity Minimization in Hardy Spaces

Comment: It is to be noted that the above γ and Q are different and each becomes clear from the context regarding H_2- or H_∞-norm.

PROOF: The *kernel* of proof is a change of variables and realization of the inner property for certain transfer matrix, which have been successfully developed in [H3. 6]. Suppose that we define a new control variable $v = u - F_2 x$, then the transfer matrix between z and the inputs becomes

$$z(s) = \begin{bmatrix} A_{F_2} & \vdots & B_1 & B_2 \\ \cdots & \vdots & \cdots & \cdots \\ C_{1F_2} & \vdots & 0 & D_{12} \end{bmatrix} \begin{Bmatrix} w(s) \\ v(s) \end{Bmatrix} \triangleq G_c(s) B_1 w(s) + U(s) v(s), \quad (6.9\text{-}45)$$

which is the same as (6.9-36). All elements of this equation are given in (6.9-33) to (6.9-37). If we let $v \equiv 0$, then $u = F_2 x$, the state feedback, and (6.9-45) yields *FI-1*. Clearly, this fact also implies that the optimal controller is a constant gain, i.e., *FI-2* is established.

To prove Item *FI-3*, we let K(s) be an admissible controller such that $\| G_{zw}(s) \|_2 < \gamma$ and $v(s) = Q(s) w(s)$. Then from $v = u - F_2 x$ we have $u(s) = F_2 x(s) + Q(s) w(s) = (F_2, Q(s)) \{ {}^x_w \}$. This is an admissible and suboptimal controller. From (6.9-45), with $v = Qw$ and the fact that U(s) is an inner matrix, we have $\| G_{zw}(s) \|_2^2 = \| G_c(s) B_1 \|_2^2 + \| Q(s) \|_2^2$ resulting in Item *FI-3*. We note that multiplication by an inner matrix is norm preserving.

The remaining two items are closely related and are studied simultaneously. The additional elements used in these two items are defined in (6.9-41) to (6.9-43). The main difference between H^2- and H^∞-controllers is that in H^∞-controller the disturbance comes through B_1. This situation has manifested itself in the core of this methodology, namely, the requirement for the existence of the solution for $X_\infty \geq 0$. When this solution exists, then A_{F_∞} given in (6.9-43a) is stable. The proof of this statement is straightforward and can be shown by (6.9-41b) in terms of A_{F_∞} and C_{1F_∞} as follows.

$$A_{F_\infty}^T X_\infty + X_\infty A_{F_\infty} + C_{1F_\infty}^T C_{1F_\infty} + \gamma^{-2} X_\infty B_1 B_1^T X_\infty = 0. \quad (6.9\text{-}46)$$

By the standard results from Riccati equation, $A_{F_\infty} + \gamma^{-2} B_1 B_1^T X_\infty$ is stable and $(B_1^T X_\infty, A_{F_\infty})$ is detectable, and by the standard results from Lyapunov equation A_{F_∞} is stable if and only if $X_\infty \geq 0$.

To continue the proof for the results of Items *FI-4 & 5*, we assume that an admissible controller exists and $\|G_{zw}(s)\|_\infty < 1$, then $H_{X_\infty} \in \text{dom(Ric)}$, and $X_\infty \triangleq \text{Ric}(H_{X_\infty}) \geq 0$. If this is true, then *only one set* of all admissible controllers such that $\|G_{zw}(s)\| < 1$ is $K(s) \triangleq (F_\infty - Q(s)B_1^T X_\infty, Q(s))$, where we assume $Q(s) \in RH^\infty$ and $\|Q\|_\infty < 1$. Note that this parameterization is *not* the complete set of all possible parameterizations for this structure as pointed out in [*H3*. 27 and 56]. However, for the purpose of ultimately solving Problems 6.9-1 & 2, this parameterization suffices. To show this part of the proof, let $v = u - F_\infty x$ and $r = w - B_1^T X_\infty x$, with the corresponding controller $K_{tmp}(s) \triangleq K(s) - (F_\infty, 0) = Q(s)(-B_1^T X_\infty, I)$. This fact implies that $v = Qr$ and we have an equivalent diagram for Fig. 6.9-1, whose $G(s)$ is replaced by $G_{rep}(s)$ given in (6.9-44) and whose $K(s)$ is replaced by $Q(s) \in RH^\infty$, with $\|Q(s)\|_\infty < 1$. Then by Lemma 6.9-7, when X_∞ exists, $G_{rep}(s) \in RH^\infty$ and is also an inner. Thus the above $K(s)$ is *an* admissible class of controllers. For additional discussions on this equivalence we refer to proof of Theorem 6.9-3.

SUB-PROBLEMS 6.9-1b & 2b (Full − Control (FC)): This case corresponds to the transfer matrix of (6.9-14) and all its pertinent assumptions. The answers to Items 6.9-1 associated with this case are as follows.

FC-1: $\min \|G_{zw}(s)\|_2 = \|C_1 G_f(s)\|_2 = [\text{Tr}(C_1 Y_2 C_1^T)]^{1/2}$.

FC-2: $K(s) \triangleq \begin{bmatrix} L_2 \\ 0 \end{bmatrix}$.

FC-3: $K(s) \triangleq \begin{bmatrix} L_2 \\ Q(s) \end{bmatrix}$, $Q(s) \in RH^2$, and $\|Q(s)\|_2^2 < \gamma^2 - \|C_1 G_f(s)\|_2^2$.

FC-4: $H_{Y_\infty} \in \text{dom(Ric)}$, $Y_\infty \triangleq \text{Ric}(H_{Y_\infty}) \geq 0$.

FC-5: $K(s) \triangleq \begin{bmatrix} L_\infty - \gamma^{-1} Y_\infty C_1^T Q(s) \\ Q(s) \end{bmatrix}$, where $Q(s) \in RH^\infty$ and $\|Q(s)\|_\infty < \gamma$.

PROOF: The above elements are given in (6.9-33) to (6.9-35) and (6.9-41) to (6.9-43). Here the argument to support these answers is dual of that in the preceding sub-problem. ∎

6.9. Sensitivity Minimization in Hardy Spaces

Before presenting the remaining cases, we note that when it comes to establishing Parts 3 & 5 of Items 6.9-1, we use the block diagram in the corresponding Fig. 6.9-2 instead of Fig. 6.9-1. This change basically reflects the strategy for parameterizing the underlying system and is a common procedure in this methodology. Here we propose to use the terms *auxiliary transfer matrix,* because that thought is in accordance with similar notion described in other methodologies of this book. The actual controller now refers to the dashed box in Fig. 6.9-2. This parameterization is not unique (cf., Comment 6.9-1). In each subsequent case we replace $G_{aux}(s)$ with its corresponding transfer matrix.

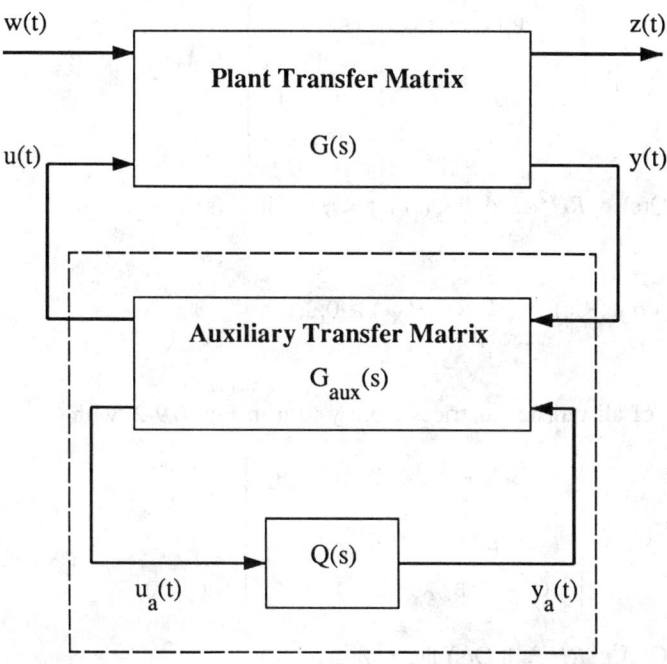

Fig. 6.9-2. Linear fractional transformation for plant and auxiliary transfer matrices.

SUB-PROBLEMS 6.9-1c & 2c (Disturbance – Feedforward (DF)): This case corresponds to the transfer matrix of (6.9-15) and its associated assumptions. The responses to Items 6.9-1 for this case are as follows.

DF-1:
$$\min \| G_{zw}(s) \|_2 = \| G_c(s) B_1 \|_2.$$

DF-2:
$$K(s) \triangleq \left[\begin{array}{c:c} A + B_2 F_2 - B_1 C_2 & B_1 \\ \hdashline F_2 & 0 \end{array} \right].$$

DF-3:
The set of all transfer matrices from y to u in Fig. 6.9-2, with
$$G_{aux}^{DF-H^2}(s) \triangleq \left[\begin{array}{c:cc} A + B_2 F_2 - B_1 C_2 & B_1 & B_2 \\ \hdashline F_2 & 0 & I \\ -C_2 & I & 0 \end{array} \right],$$
where $Q(s) \in RH^2$, and $\| Q(s) \|_2^2 < \gamma^2 - \| G_c(s) B_1 \|_2^2$.

DF-4:
$$H_{X_\infty} \in \text{dom}(\text{Ric}), \quad X_\infty \triangleq \text{Ric}(H_{X_\infty}) \geq 0.$$

DF-5:
The set of all transfer matrices from y to u in Fig. 6.9-2, with
$$G_{aux}^{DF-H^\infty}(s) \triangleq \left[\begin{array}{c:cc} A + B_2 F_\infty - B_1 C_2 & B_1 & B_2 \\ \hdashline F_\infty & 0 & I \\ -C_2 - \gamma^{-1} B_1^T X_\infty & I & 0 \end{array} \right],$$
where $Q(s) \in RH^\infty$, $\| Q(s) \|_\infty < \gamma$.

PROOF [H3. 6]: The following should be studied in conjunction with the proofs for Theorems 6.9-3 & 4. Recall the transfer matrices for full-information and disturbance-feedforward structures from (6.9-13) and (6.9-15), each denoted accordingly as follows.

6.9. Sensitivity Minimization in Hardy Spaces

$$G^{FI}(s) \triangleq \begin{bmatrix} A & : & B_1 & B_2 \\ \cdots & : & \cdots & \cdots \\ C_1 & : & 0 & D_{12} \\ \begin{bmatrix} I \\ 0 \end{bmatrix} & : & \begin{bmatrix} 0 \\ I \end{bmatrix} & \begin{bmatrix} 0 \\ 0 \end{bmatrix} \end{bmatrix}, \quad G^{DF}(s) \triangleq \begin{bmatrix} A & : & B_1 & B_2 \\ \cdots & : & \cdots & \cdots \\ C_1 & : & 0 & D_{12} \\ C_2 & : & I & 0 \end{bmatrix}. \quad (6.9\text{-}47)$$

Each case comes with a set of conditions *(i)* to *(iv)*, where condition *(i)* of disturbance-feedforward requires that $A - B_1 C_2$ being stable. If we show the corresponding admissible controllers in these two cases, with K_{FI} and K_{DF}, respectively, then it is clear that K_{DF} internally stabilizes G^{DF} if and only if $K_{FI} = K_{DF}(C_2, I)$ internally stabilizes G^{FI}. Under this condition $G_{zw}^{FI} = G_{zw}^{DF}$, where subscript "zw" refers to input-output transfer matrix.

Suppose that the controller for full-information case is shown by $\hat{K}_{FI}$; and $\hat{K}_{DF}$ is the transfer matrix of Fig. 6.9-3 with the corresponding $G_{aux}^{DF}(s)$.

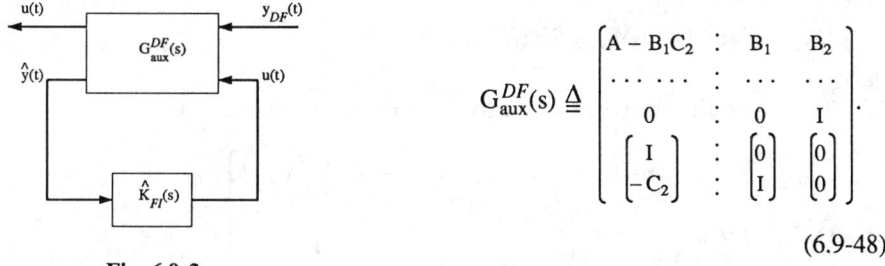

$$G_{aux}^{DF}(s) \triangleq \begin{bmatrix} A - B_1 C_2 & : & B_1 & B_2 \\ \cdots\cdots & : & \cdots & \cdots \\ 0 & : & 0 & I \\ \begin{bmatrix} I \\ -C_2 \end{bmatrix} & : & \begin{bmatrix} 0 \\ I \end{bmatrix} & \begin{bmatrix} 0 \\ 0 \end{bmatrix} \end{bmatrix}.$$

(6.9-48)

Fig. 6.9-3.

Here we can show that $\hat{K}_{FI}$ internally stabilizes $G^{FI}(s)$ if and only if $\hat{K}_{DF}$ internally stabilizes $G^{DF}(s)$, and in this case $G_{zw}^{FI} = G_{zw}^{DF}$ for all s. To show this claim let x denotes the state of G^{DF} and $\hat{x}$ denotes the state of G_{aux}^{DF}, respectively. Then applying $\hat{K}_{DF}$ to G^{DF} yields

$$\dot{e} \triangleq (\dot{x} - \dot{\hat{x}}) = (A - B_1 C_2) e, \quad (6.9\text{-}49a)$$

$$\dot{\hat{x}} = A\hat{x} + B_1 w + B_2 u + B_1 C_2 e, \quad u = K_{FI} \hat{y} = K_{FI} \{\begin{smallmatrix} \hat{x} \\ \hat{w} \end{smallmatrix}\}, \quad (6.9\text{-}49b)$$

$$\hat{w} = y - C_2 \hat{x} = w + C_2 e. \quad (6.9\text{-}49c)$$

This is the same set of equations as that generated when $\hat{K}_{FI}$ is applied to G^{FI}.

SUB-PROBLEMS 6.9-1d & 2d (Output − Estimation (OE)): This case corresponds to transfer matrix of (6.9-16) and the answers to Items 6.9-1 related to this situation are as follows.

OE-1:
$$\min \| G_{zw}(s) \|_2 = \| C_1 G_f(s) \|_2.$$

OE-2:

$$K(s) \triangleq \left[\begin{array}{c:c} A + L_2C_2 - B_2C_1 & L_2 \\ \hdashline C_1 & 0 \end{array}\right].$$

OE-3:

The set of all transfer matrices from y to u in Fig. 6.9-2, with

$$G_{aux}^{OE-H^2}(s) \triangleq \left[\begin{array}{c:cc} A + L_2C_2 - B_2C_1 & L_2 & -B_2 \\ \hdashline C_1 & 0 & I \\ C_2 & I & 0 \end{array}\right],$$

where $Q(s) \in RH^2$, and $\| Q(s) \|_2^2 < \gamma^2 - \| C_1 G_f(s) \|_2^2$.

OE-4:

$H_{Y_\infty} \in \text{dom}(\text{Ric})$, $Y_\infty \triangleq \text{Ric}(H_{Y_\infty}) \geq 0$.

OE-5:

The set of all transfer matrices from y to u in Fig. 6.9-2, with

$$G_{aux}^{OE-H^\infty}(s) \triangleq \left[\begin{array}{c:cc} A + L_\infty C_2 - B_2C_1 & L_\infty & -B_2 - \gamma^{-1} Y_\infty C_1^T \\ \hdashline C_1 & 0 & I \\ C_2 & I & 0 \end{array}\right],$$

where $Q(s) \in RH^\infty$ and $\| Q(s) \|_\infty < \gamma$.

PROOF: All the above elements are defined in (6.9-33) to (6.9-35) and (6.9-41) to (6.9-43). The argument to support these claims is dual of that for the disturbance-feedforward and both optimal estimators are observers with the observer gains L_2 and L_∞. In the H^2–OE, the output estimate is actually the state estimate (multiplied by C_1), while in the H^∞–OE, the output depends explicitly on the output being estimated and this has some implications in the separation argument for the H^∞–output feedback controller.

6.9.3.6. Solution of Problem 6.9-1

The step by step solution of this problem is summarized in the next two theorems. Recall the prior discussions in *Case I* of § 6.9.3.4, in particular (6.9-33) to (6.9-36). The following analyses are strongly based on the separation arguments (similar to the separation theorem in linear optimal control) which reduce the problem

6.9. Sensitivity Minimization in Hardy Spaces

into a combination of full-information and output-estimation cases described in § 6.9.3.5.

THEOREM 6.9-1 [H3. 6]: Recall (6.9-12) and (6.9-33) to (6.9-36). The unique optimal controller that minimizes $\| G_{zw}(s) \|_2$ in Problem 6.9-1 is as follows.

$$K_{optimal}(s) \triangleq \begin{bmatrix} \hat{A}_2 & \vdots & -L_2 \\ \cdots & \vdots & \cdots \\ F_2 & \vdots & 0 \end{bmatrix}, \qquad (6.9\text{-}50)$$

with

$$\min \| G_{zw}(s) \|_2^2 = \| G_c(s) B_1 \|_2^2 + \| F_2 G_f(s) \|_2^2$$

$$= \| G_c(s) L_2 \|_2^2 + \| C_1 G_f(s) \|_2^2. \qquad (6.9\text{-}51)$$

PROOF [H3. 6]: The proof depends on a successful change of control variables such that $v = u - F_2 x$. The transfer matrix between z and inputs becomes that of (6.9-45) with the corresponding separation and inner properties. Given any admissible controller K(s) the new control variable is generated from the system of Fig. 6.9-4 with the transfer matrix given in (6.9-52) whose structure is of the form (6.9-16). This transfer matrix has the same "A-matrix" as G(s) in (6.9-12), and K(s) stabilizes this G(s) if and only if K(s) stabilizes $G_v(s)$.

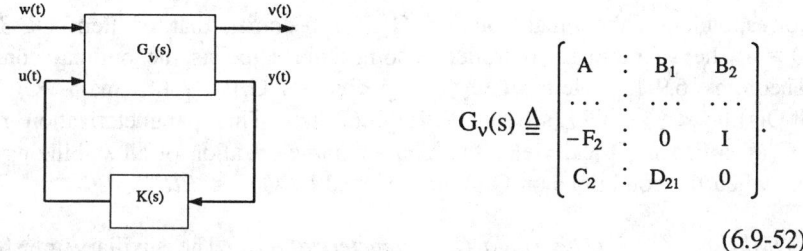

$$G_v(s) \triangleq \begin{bmatrix} A & \vdots & B_1 & B_2 \\ \cdots & \vdots & \cdots & \cdots \\ -F_2 & \vdots & 0 & I \\ C_2 & \vdots & D_{21} & 0 \end{bmatrix}.$$

(6.9-52)

Fig. 6.9-4. Change of variables in Theorem 6.9-1.

Then using $G_{vw}(s) = v(s)/w(s)$, we can show that

$$\min \| G_{zw}(s) \|_2^2 = \| G_c(s) B_1 \|_2^2 + \min \| G_{vw}(s) \|_2^2. \qquad (6.9\text{-}53)$$

By the previous special case, in particular, Item *OE-1*, the $\| G_{vw} \|_2 = \| F_2 G_f(s) \|_2$ and this transfer matrix is uniquely minimized by the controller in *OE-2* when applied to $G_v(s)$ and that completes the proof. ∎

Going back to (6.9-50), F_2 is the optimal state feedback gain in the full-information case and L_2 is the optimal output injection gain in the full-control case.

This optimal controller displays a perfect case of the well-known separation theorem in the quadratic optimal control theory. The controller equations with $\hat{x}$ as an optimal estimate of x can be written in an observer-based form

$$\dot{\hat{x}} = A\hat{x} + B_2 u + L_2(C_2\hat{x} - y), \tag{6.9-54a}$$

$$u = F_2\hat{x}. \tag{6.9-54b}$$

∎

The next theorem extends Problem 6.9-1 to its corresponding case associated with Fig. 6.9-2, when we parameterize the controller and seek the family of all such controllers.

THEOREM 6.9-2 [H3. 6]: The family of all admissible controllers in Theorem 6.9-1 such that $\| G_{zw}(s) \|_2^2 < \gamma$ equals the set of all transfer matrices from y to u in Fig. 6.9-2, with

$$G_{aux}^{H^2}(s) \triangleq \left[\begin{array}{c:cc} \hat{A}_2 & -L_2 & B_2 \\ \hdashline F_2 & 0 & I \\ -C_2 & I & 0 \end{array}\right], \tag{6.9-55}$$

where $Q(s) \in RH^2$, and $\| Q(s) \|_2^2 < \gamma^2 - (\| G_c(s)B_1 \|_2^2 + \| F_2 G_f(s) \|_2^2)$.

PROOF: Here we parameterize the system as in Fig. 6.9-2 by a fixed (independent from γ) linear fractional transformation with a free parameter $Q(s)$. The corresponding suboptimal solution $G_{aux}^{H^2}(s)$ becomes that of Item *OE-3*. When $Q \equiv 0$, the suboptimal controller becomes the same as the optimal controller in Theorem 6.9-1. Here $\| G_{vw}(s) \|_2^2 < \gamma^2 - \| G_c(s)B_1 \|_2^2$, with $Q(s) \in RH^2$, $\| Q(s) \|_2^2 < \gamma^2 - \| G_c(s)B_1 \|_2^2 - \| F_2 G_f(s) \|_2^2$. This parameterization results in $G_{zw}(s)$ affine in Q and yields the Youla parameterization of all stabilizing controllers when the conditions on $Q(s)$ are replaced by $Q(s) \in RH^\infty$.

COMMENT 6.9-1 (The Youla Q–parameterization): The auxiliary transfer matrix in Fig. 6.9-2, itself is subject to a linear fractional transformation with $Q(s)$ as its free parameter. If $Q(s)$ is any stable transfer matrix, then the overall system becomes stable, otherwise it becomes unstable. This free parameter $Q(s)$ must be chosen such that $G_{uy}(s)$ acts as a stabilizing controller for $G(s)$ and this parameterization is not unique. In the case of Theorem 6.9-2, however, this parameterization is special in the sense that when $Q = 0$, it brings us back to the results of Theorem 6.9-1. If we combine accordingly the state-space representation of $G(s)$ and the $G_{aux}^{H^2}(s)$ in (6.9-55) and call that $G_{total}(s)$, and if we keep $Q(s)$ as its linear fractional controller, then we can show that the (2,2)–block partition of $G_{total}(s)$ is identically zero. When that has been accounted for in the expression for $G_{zw}(s)$ (cf., the corresponding (6.9-9) with the (2,2)–block partition set identically zero), as

6.9. Sensitivity Minimization in Hardy Spaces

mentioned in proof of Theorem 6.9-2, the overall $G_{zw}(s)$ becomes affine in $Q(s)$ and the overall system remains stable only when $Q(s)$ is stable (for a similar analysis we refer to Example 6.9-3).

6.9.3.7. Solution of Problem 6.9-2

The following two theorems complement the preceding theorems and give a solution of Problem 6.9-2. The main difference between this and the previous case is that in H^∞-control problem we seek admissible controller $K(s)$ such that $\|G_{zw}(s)\|_\infty < \gamma$, where the prespecified parameter γ is greater than the actual optimal H_∞-norm and the final design is only a *suboptimal* controller. Thus in this sense we are only providing a *suboptimal* solution of Problem 6.9-2. The optimal solution is yet difficult to characterize.

THEOREM 6.9-3 [H3. 6]: Recall (6.9-12), (6.9-41) to (6.9-43) and Fig. 6.9-1. There exists an admissible controller $K(s)$ such that $\|G_{zw}(s)\|_\infty < \gamma$ if and only if the following three conditions are met simultaneously.

(i) $H_{X_\infty} \in \text{dom}(\text{Ric})$ and $X_\infty \triangleq \text{Ric}(H_{X_\infty}) \geq 0$,

(ii) $H_{Y_\infty} \in \text{dom}(\text{Ric})$ and $Y_\infty \triangleq \text{Ric}(H_{Y_\infty}) \geq 0$,

(iii) $\rho(X_\infty Y_\infty) < \gamma^2$, where $\rho(\cdot)$ is the spectral radius of $(\cdot)$.

Then a *suboptimal* controller gain becomes

$$K_{\text{suboptimal}}(s) \triangleq \left[\begin{array}{c:c} \hat{A}_\infty & -Z_\infty L_\infty \\ \cdots & \cdots \\ F_\infty & 0 \end{array}\right], \qquad (6.9\text{-}56a)$$

where

$$\hat{A}_\infty = A + \gamma^{-2} B_1 B_1^T X_\infty + B_2 F_\infty + Z_\infty L_\infty C_2. \qquad (6.9\text{-}56b)$$

PROOF: The proof follows subsequently.

COMMENT 6.9-2 [H3. 6]: The controller in (6.9-56a) is called the *central controller* and is *similar* to that of (6.9-50) for the H^2-controller problem and it seems that the H^∞-controller problem also enjoys some sort of separation property, but that is not so obvious in this case as is in (6.9-50). Conditions (i) and (ii) correspond to Item *FI-4* of Sub-Problems 6.9-1a & 2a and Item *FC-4* of Sub-Problems 6.9-1b & 2b. These two conditions are also necessary for the transfer matrix (6.9-12). In other words, if there exists a controller $K(s)$ such that $\|G_{zw}(s)\|_\infty < \gamma$, then conditions (i) and (ii) must hold as in our special cases. Condition (iii), however, combines all these special cases and is the major contribution of our original source [H3. 6].

COMMENT 6.9-3 [H3. 6]: To differentiate between H^2- and H^∞-controllers, the (6.9-56) is written as follows.

$$\dot{\hat{x}} \triangleq \hat{A}_\infty \hat{x} - Z_\infty L_\infty y$$

$$= (A + \gamma^{-2} B_1 B_1^T X_\infty + B_2 F_\infty + Z_\infty L_\infty C_2)\hat{x} - Z_\infty L_\infty y$$

$$\triangleq A\hat{x} + B_1 \hat{w}_{worst} + B_2 u + Z_\infty L_\infty (C_2 \hat{x} - y), \qquad (6.9\text{-}57a)$$

$$u = F_\infty \hat{x}, \quad \hat{w}_{worst} = \gamma^{-2} B_1^T X_\infty \hat{x}. \qquad (6.9\text{-}57b)$$

Here $\hat{x}$ is the estimated state, and (6.9-57) has the structure of an observer-based compensator. Comparing this set of equations with its H^2-counterpart we note the following.

(i) Where does the $B_1 \hat{w}_{worst}$ come from and what is its interpretation?
(ii) Why $Z_\infty L_\infty$ instead of L_∞?
(iii) What kind of separation, if any, these equations exhibit?

Upon completing the proof of Theorem 6.9-3, subsequently, we note that indeed a very well-defined separation interpretation exists and the $\hat{w}_{worst}$ represents the "worst-case" input for the full-information case. Here the $Z_\infty L_\infty$ is the optimal filter gain for estimating $F_\infty x$, the optimal full-information control input, in the presence of $\hat{w}_{worst}$. This worst disturbance input maximizes $\| z \|_2^2 - \gamma^2 \| w \|_2^2$. Thus here we have $Z_\infty L_\infty$ instead of L_∞. Finally, $\hat{w}_{worst}$ is considered as an "estimate" of the w_{worst} [H3. 6].

COMMENT 6.9-4 [H3. 6]: The final proof of Theorem 6.9-3 depends on a successful change of variables as follows. Let the control variable be $v = u + B_2^T X_\infty x$ and the new disturbance be $r = w - \gamma^{-2} B_1^T X_\infty x$, then

$$\begin{Bmatrix} v \\ y \end{Bmatrix} = \begin{bmatrix} A_{tmp} & \vdots & B_1 & B_2 \\ \cdots & \vdots & \cdots & \cdots \\ -F_\infty & \vdots & 0 & I \\ C_2 & \vdots & D_{21} & 0 \end{bmatrix} \begin{Bmatrix} r \\ u \end{Bmatrix} \triangleq G_{tmp}(s) \begin{Bmatrix} r \\ u \end{Bmatrix}, \qquad (6.9\text{-}58a)$$

$$A_{tmp} \triangleq A + \gamma^{-2} B_1 B_1^T X_\infty. \qquad (6.9\text{-}58b)$$

This transfer matrix is depicted in Fig. 6.9-5 and is similar to the output-estimation structure of (6.9-16). However, to use the results of Sub-Problems 6.9-1d & 2d, we need to verify that indeed $G_{tmp}(s)$ meets all the requirements of this structure, namely:

6.9. Sensitivity Minimization in Hardy Spaces

(i) The pair (A_{tmp}, B_1) is stabilizable and $A_{tmp} + B_2 F_\infty$ is *stable*.

(ii) The pair (C_2, A_{tmp}) is detectable.

(iii) Specific structure for D_{12} is given.

(iv) $\begin{bmatrix} B_1 \\ D_{21} \end{bmatrix} D_{21}^T = \begin{bmatrix} 0 \\ I \end{bmatrix}.$

We must only show that $A_{tmp} + B_2 F_\infty$ is stable and *(ii)* holds, since the remaining conditions are met by the original system. A corresponding set of Hamiltonian matrices for $G_{tmp}(s)$ is as follows.

$$H_{X_\infty^{tmp}} \triangleq \begin{bmatrix} A_{tmp} & \gamma^{-2} B_1 B_1^T - B_2 B_2^T \\ -F_\infty^T F_\infty & -A_{tmp}^T \end{bmatrix}. \tag{6.9-59a}$$

$$H_{Y_\infty^{tmp}} \triangleq \begin{bmatrix} A_{tmp}^T & \gamma^{-2} F_\infty^T F_\infty - C_2^T C_2 \\ -B_1 B_1^T & -A_{tmp} \end{bmatrix}. \tag{6.9-59b}$$

Clearly, if $H_{X_\infty^{tmp}} \in \text{dom}(\text{Ric})$, and $X_\infty^{tmp} \geq 0$, then $A_{tmp} + B_2 F_\infty$ is stable. Also if $H_{Y_\infty^{tmp}} \in \text{dom}(\text{Ric})$, and $Y_\infty^{tmp} \triangleq \text{Ric}(H_{Y_\infty^{tmp}}) \geq 0$, then (C_2, A_{tmp}) is detectable. However, these conclusions are based on the fundamental assumptions regarding the existence of solutions for Riccati algebraic-matrix equation.

Finally, we note that when $K(s)$ is admissible in either general structure of (6.9-12) or the above output-estimation case, then the internal and input-output stability are equivalent. ∎

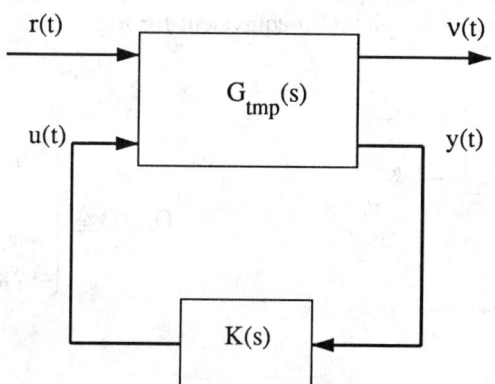

Fig. 6.9-5. Change of variables in Theorem 6.9-3.

The key approach to complete the proof of Theorem 6.9-3 is to deduce all needed information from the changed transfer matrix. For instance, the next lemma compares several properties of Fig. 6.9-5 with those of Fig. 6.9-1.

LEMMA 6.9-8 [H3. 6]: Consider the system corresponding to Figs. 6.9-1 & 5. If we assume that X_∞ exists and $X_\infty \geq 0$, then $K(s)$ is admissible for $G(s)$ in Fig. 6.9-1 and $\| G_{zw}(s) \|_\infty < \gamma$ if and only if $K(s)$ is admissible for $G_{tmp}(s)$ in Fig. 6.9-5 and $\| G_{vr}(s) \|_\infty < \gamma$.

PROOF: First, we assume $X_\infty \triangleq \mathrm{Ric}(H_{X_\infty}) \geq 0$ exists and consider $\frac{d}{dt}[x^T(t) X_\infty x(t)]$ along $\dot{x}(t) = Ax + B_1 w + B_2 u$, and (6.9-41b). We also assume that $x(0) = x(\infty) = 0$ and $w \in L_+^2$. Following the same steps which lead to (6.9-22), we have

$$\| z \|_2^2 - \gamma^2 \| w \|_2^2 \leq \| v \|_2^2 - \gamma^2 \| r \|_2^2. \qquad (6.9\text{-}60)$$

An interpretation of (6.9-60) is that $\| G_{zw}(s) \|_\infty \leq \gamma$ if and only if $\| G_{vr}(s) \|_\infty \leq \gamma$ which reinforces the very appropriate choices for new variables made in [H3. 6].

Second, we recall Comment 6.9-4 which confirms that $G_{tmp}(s)$ satisfies all the relevant assumptions for an output-estimation case. The internal stability of $G_{zw}(s)$ is equivalent to $G_{zw}(s) \in RH^\infty$ and in the output-estimation case also the internal stability requires that $G_{vr}(s) \in RH^\infty$. Together with the preceding discussions we conclude that internal stability is equivalent to input-output stability for both $G(s)$ and $G_{tmp}(s)$ and that completes the proof.

COMMENT 6.9-5 [H3. 6]: Recall $G_{rep}(s)$ of (6.9-44) given below. This transfer matrix is generated using the preceding set of new variables and corresponds to Fig. 6.9-6. Also recall $G_{tmp}(s)$ of (6.9-58) with Fig. 6.9-5. Substituting $G_{vr}(s)$ from Fig. 6.9-5 in Fig. 6.9-6 yields an equivalent for Fig. 6.9-1.

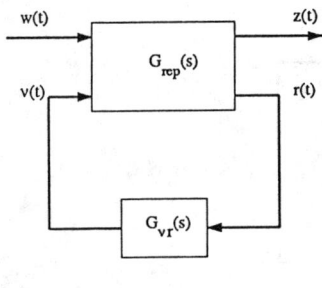

Fig. 6.9-6.

$$G_{rep}(s) \triangleq \left[\begin{array}{c:ccc} A_{F_\infty} & : & B_1 & B_2 \\ \cdots & : & \cdots & \cdots \\ C_{1F_\infty} & : & 0 & D_{12} \\ -B_1^T X_\infty & : & I & 0 \end{array} \right],$$

$$(6.9\text{-}61)$$

PROOF OF THEOREM 6.9-3 CONTINUED (Sufficiency) [H3. 6]: To begin the proof we assume that conditions *(i)* to *(iii)* are met. Then using

6.9. Sensitivity Minimization in Hardy Spaces

$$T = \begin{bmatrix} I & -\gamma^{-2}X_\infty \\ 0 & I \end{bmatrix}, \quad T^{-1} = \begin{bmatrix} I & \gamma^{-2}X_\infty \\ 0 & I \end{bmatrix}, \quad (6.9\text{-}62)$$

together with (6.9-41b) and (6.9-58b), provides a similarity transformation between $H_{Y_\infty^{tmp}}$ and H_{Y_∞} as follows.

$$T^{-1}(H_{Y_\infty^{tmp}})T = \begin{bmatrix} A^T & \gamma^{-2}F_\infty^T F_\infty - C_2^T C_2 - \gamma^{-2}X_\infty(A + \gamma^{-2}B_1 B_1^T X_\infty) \\ -B_1 B_1^T & -A_{tmp} \end{bmatrix} T =$$

$$\begin{bmatrix} A^T & -\gamma^{-2}(A^T X_\infty + X_\infty A) - \gamma^{-4}X_\infty B_1 B_1^T X_\infty + \gamma^{-2}F_\infty^T F_\infty - C_2^T C_2 \\ -B_1 B_1^T & -A \end{bmatrix} = H_{Y_\infty}.$$

(6.9-63)

Thus the spectral subspaces of $H_{Y_\infty^{tmp}}$ and H_{Y_∞} are related as follows.

$$S_-(H_{Y_\infty^{tmp}}) = TS_-(H_{Y_\infty}) = T\text{Im}\begin{pmatrix} I \\ Y_\infty \end{pmatrix} = \text{Im}\begin{bmatrix} I - \gamma^{-2}X_\infty Y_\infty \\ Y_\infty \end{bmatrix}. \quad (6.9\text{-}64)$$

That means

$$Y_\infty^{tmp} \triangleq \text{Ric}(H_{Y_\infty^{tmp}}) = Y_\infty(I - \gamma^{-2}X_\infty Y_\infty)^{-1} = Z_\infty Y_\infty, \quad (6.9\text{-}65)$$

and $\rho(X_\infty Y_\infty) < \gamma^2$ guarantees that $Y_\infty^{tmp} \geq 0$. This fact in turn guarantees that $G_{tmp}(s)$ becomes a candidate for Sub-Problems 6.9-1d & 2d and in turn by Item OE-4 the output-estimation case is solvable. One such solution with $Q = 0$ is as follows.

$$\begin{bmatrix} A + \gamma^{-2}B_1 B_1^T X_\infty - Y_\infty^{tmp} C_2^T C_2 + B_2 F_\infty & \vdots & Y_\infty^{tmp} C_2^T \\ \cdots\cdots\cdots\cdots\cdots\cdots\cdots\cdots\cdots\cdots\cdots\cdots & \vdots & \cdots\cdots \\ F_\infty & \vdots & 0 \end{bmatrix}. \quad (6.9\text{-}66)$$

The (6.9-66) is exactly the same as the $K_{suboptimal}(s)$ in (6.9-56a). Since this controller stabilizes $G_{tmp}(s)$ with $\|G_{vr}(s)\|_\infty < \gamma$, by Lemma 6.9-8 it must stabilizes $G(s)$ and $\|G_{zw}(s)\|_\infty < \gamma$.

PROOF OF THEOREM 6.9-3 CONTINUED (Necessity) [H3. 6]: If $K(s)$ is an admissible controller with $\|G_{zw}(s)\|_\infty < \gamma$, then $H_{X_\infty} \in \text{dom}(\text{Ric})$, $X_\infty \triangleq \text{Ric}(H_{X_\infty}) \geq 0$, also $H_{Y_\infty} \in \text{dom}(\text{Ric})$, $Y_\infty \triangleq \text{Ric}(H_{Y_\infty}) \geq 0$. Then $K(s)$ is admissible for $G_{tmp}(s)$ and $\|G_{vr}(s)\|_\infty < \gamma$. From Comment 6.9-4, $G_{tmp}(s)$ is a

candidate for the output-estimation case, and applying Sub-Problems 6.9-1d & 2d to $G_{tmp}(s)$ yields $H_{Y_\infty^{tmp}} \in \text{dom}(\text{Ric})$ and $Y_\infty^{tmp} \triangleq \text{Ric}(H_{Y_\infty^{tmp}}) \geq 0$. Since also from above $Y_\infty^{tmp} = (I - \gamma^{-2} Y_\infty X_\infty)^{-1} Y_\infty \geq 0$, we have $\rho(X_\infty Y_\infty) < \gamma^2$.
∎

COMMENT 6.9-6 [H3. 6]: It is now apparent that the term $\hat{w}_{worst}$ and $Z_\infty L_\infty$ are the contributions of output-estimation part of estimating the optimal full-information control gain. In the H^2–output feedback case the situation is simpler, because primarily there is no "worst-case" disturbance and the problem of estimating any output is equivalent to the state estimation. While the H^∞–output feedback controller is the output estimator of the full-information control law in the presence of the "worst-case" disturbance w_{worst}.

THEOREM 6.9-4 [H3. 6]: The set of all admissible controllers in Theorem 6.9-3 such that $\|G_{zw}(s)\|_\infty < \gamma$ is the set of all transfer matrices in Fig. 6.9-2, with

$$G_{aux}^{H^\infty}(s) \triangleq \begin{bmatrix} \hat{A}_\infty & \vdots & -Z_\infty L_\infty & Z_\infty B_2 \\ \cdots & \vdots & \cdots & \cdots \\ F_\infty & \vdots & 0 & I \\ -C_2 & \vdots & I & 0 \end{bmatrix}, \quad (6.9\text{-}67)$$

where the free parameter $Q(s) \in RH^\infty$, $\|Q(s)\|_\infty < \gamma$.

PROOF: Here also the suboptimal controllers are parameterized by a fixed linear fractional transformation with a free parameter $Q(s)$. When $Q(s) = 0$, the controller becomes the so-called *central controller* or $K_{suboptimal}(s)$ given in (6.9-56). From Lemma 6.9-8 the set of all admissible controllers for $G(s)$ such that $\|G_{zw}(s)\|_\infty < \gamma$ is the same as the set of all admissible controllers for $G_{tmp}(s)$ such that $\|G_{tmp}(s)\|_\infty < \gamma$. Then applying Item *OE-5* to this case completes the proof.
∎

REMARK 6.9-1 (On γ): Prior to any applications of Theorems 6.9-1 to 4, we must test the sign of the solution for each corresponding Riccati equation, specially in Theorems 6.9-3 & 4 where we specify γ. In the case of Theorems 6.9-3 & 4 we must additionally test $\rho(X_\infty Y_\infty) < \gamma^2$. If any one of these tests fails, then we must adjust γ accordingly. We also note that the H^∞–solution, when exists, guarantees that the corresponding H_∞–norm remains *below* a specified upper bound and this is in contrast with the global optimal solution attained in the H^2–controller. By iterating on γ, we *may* reach the optimal H^∞–solution.

REMARK 6.9-2 (On the B_1 and C_1): The main difference between (6.9-41a) [or (6.9-42a)] and (6.9-33a) [or (6.9-34b)] is in the (1,2)–block partition and in the form of $\gamma^{-2} B_1 B_1^T$ (or $\gamma^{-2} C_1^T C_1$), which manifests itself throughout the

6.9. Sensitivity Minimization in Hardy Spaces

corresponding controller (or estimator) design. In the H^2-controller, the state feedback does not depend on B_1 and the external input does not affect this solution, while in the H^∞-controller we can account for this external input through B_1. Also in the H^2-state estimator we *customarily* apply the statistical-intensity of disturbance equally to each component of the state (cf., Example 6.9-2). While in the H^∞-state estimator, we look at the output feedback through C_1, and therefore we apply different weights to different state components, and this is the most important feature of H^∞-design.

∎

EXAMPLE 6.9-1 *(Regulator):* Given $W_0 = W_0^T > 0$, and $W = W^T > 0$, find an optimal controller gain F such that $u(t) = Fx(t)$ minimizes

$$J = \int_0^\infty [\zeta^T(t)W_0\zeta(t) + u^T(t)Wu(t)]dt, \tag{6.9-68}$$

subject to the following constraints.

$$\dot{x}(t) = Ax(t) + Bu(t), \quad x(0) = x_o, \tag{6.9-69a}$$

$$\zeta(t) = Cx(t). \tag{6.9-69b}$$

SOLUTION: In this and the subsequent example we present a number of steps that must be taken to answer this type of optimization problems.

STEP 0 (The choice of methodology): Clearly, this example corresponds to an H^2-control problem, therefore we need to follow the methods of Problem 6.9-1, in particular, Theorem 6.9-1.

STEP 1 (Frequency-domain variables): It is essential to develop the frequency-domain equivalent of this problem with its corresponding Fig. 6.9-1. The four variables (w, z, y, u) must be selected such that the corresponding $\| G_{zw}(s) \|_2$ becomes *equivalent* to the underlying performance index which herein is (6.9-68). For instance, if we let $w(t) = x_o \delta(t)$, where $\delta(t)$ is a unit impulse function, and consider the following as output vector $z(t)$,

$$z(t) = \begin{Bmatrix} W_0^{\frac{1}{2}} Cx(t) \\ W^{\frac{1}{2}} u(t) \end{Bmatrix}, \tag{6.9-70}$$

where $W_0^{\frac{1}{2}} > 0$, $W_0^{\frac{1}{2}} W_0^{\frac{1}{2}} = W_0$, and $W^{\frac{1}{2}} > 0$, $W^{\frac{1}{2}} W^{\frac{1}{2}} = W$, respectively, then

$$\| z \|_2^2 = \int_0^\infty z^T(t)z(t)dt$$

$$= \int_0^\infty [x^T C^T W_0^{\frac{1}{2}}, u^T W^{\frac{1}{2}}] \begin{Bmatrix} W_0^{\frac{1}{2}} Cx(t) \\ W^{\frac{1}{2}} u(t) \end{Bmatrix} dt$$

$$= \int_0^\infty [\zeta^T(t)W_0\zeta(t) + u^T(t)Wu]dt, \tag{6.9-71}$$

where (6.9-71) is exactly the same as (6.9-68). Thus the original optimization becomes equivalent to minimizing $\| G_{zw}(s) \|_2^2 = \| z \|_2^2 / x_o^T x_o$ for an arbitrary initial condition $x_o \neq 0$. To choose y and u, we keep $u = Fx$ and let $y = x$.

STEP 2 *(Generation of transfer matrix):* Considering all the preceding variables, the system (6.9-69) together with (6.9-68) become

$$\dot{x}(t) = Ax(t) + w(t) + Bu(t), \qquad (6.9\text{-}72a)$$

$$z_1(t) = W_0^{1/2} Cx(t), \qquad (6.9\text{-}72b)$$

$$z_2(t) = W^{1/2} u(t), \qquad (6.9\text{-}72c)$$

$$y(t) = x(t). \qquad (6.9\text{-}72d)$$

Resulting in the following transfer matrix representation for the linear quadratic regulator problem.

$$G(s) \triangleq \begin{bmatrix} A & : & I & B \\ \cdots & : & \cdots & \cdots \\ W_0^{1/2} C & : & 0 & 0 \\ 0 & : & 0 & W^{1/2} \\ I & : & 0 & 0 \end{bmatrix}. \qquad (6.9\text{-}73)$$

STEP 3 *(Evaluation of transfer matrix):* In developing the frequency-domain methodology, we assume that the underlying transfer matrix satisfies certain structural properties. Although not explicitly mentioned at the outset of this example, to have a solution for (6.9-68), we also need to specify certain conditions on (A, B, C). Thus these conditions must be checked before proceeding further. At the moment, however, we note that if in (6.9-72a) we replace $u(t)$ by $u(t) = W^{-1/2} z_2(t)$ from (6.9-72c), then we have an alternative transfer matrix which represents a transformed plant $\tilde{G}(s)$.

$$\tilde{G}(s) \triangleq \begin{bmatrix} A & : & I & BW^{-1/2} \\ \cdots & : & \cdots & \cdots \\ W_0^{1/2} C & : & 0 & 0 \\ I & : & 0 & 0 \end{bmatrix}, \qquad (6.9\text{-}74)$$

which seems more straightforward than the previous one. Now, from the algebraic point of view, $z_2(t)$ acts as our input. This change of variables is perhaps a quick way of eliminating a number of required tests for this evaluation.

STEP 4 *(Final answer):* Now, we apply the method that is chosen in Step 0 to this transfer matrix. The optimal controller gain in terms of $z_2(t)$ is as follows.

6.9. Sensitivity Minimization in Hardy Spaces

$$F_2 = -(BW^{-\frac{1}{2}})^T X_2 = -W^{-\frac{1}{2}} B^T X_2, \quad (6.9\text{-}75)$$

where

$$X_2 \triangleq \text{Ric} \begin{bmatrix} A & -BW^{-1}B^T \\ -C^T W_0 C & -A^T \end{bmatrix}. \quad (6.9\text{-}76)$$

Since $u(t) = W^{-\frac{1}{2}} z_2(t) = W^{-\frac{1}{2}} F_2 x(t) = -W^{-1} B^T X_2 x(t)$, the well-known optimal controller gain $u(t) = F_{\text{optimal}} x(t)$ becomes

$$F_{\text{optimal}} = -W^{-1} B^T X_2. \quad (6.9\text{-}77)$$

▼

EXAMPLE 6.9-2 (Estimator): Consider a standard state estimator described by

$$\dot{x}(t) = Ax(t) + \omega_1(t), \quad (6.9\text{-}78a)$$

$$\zeta(t) = Cx(t) + \omega_2(t), \quad (6.9\text{-}78b)$$

$$\dot{\hat{x}}(t) = A\hat{x}(t) - L(\zeta - C\hat{x}), \quad (6.9\text{-}78c)$$

where ω_1 is a zero-mean plant disturbance with intensity $\text{cov}(\omega_1, \omega_1^T) = \Omega_1 > 0$, and ω_2 is a zero-mean measurement noise with intensity $\text{cov}(\omega_2, \omega_2^T) = \Omega_2 > 0$. We assume that ω_1 is zero correlated with ω_2. We also assume that the initial condition is a zero-mean white noise and is zero correlated with ω_1 and ω_2. The estimate $\hat{x}$ is generated *via* the observer-based optimal estimator such that

$$J = E[(x - \hat{x})^T (x - \hat{x})] \quad (6.9\text{-}79)$$

is minimized. The objective is to find the optimal L.

SOLUTION: To recast this optimization problem in its frequency-domain equivalent we consider the following steps.

STEP 0 (The choice of methodology): This example corresponds to an H^2-type control problem and we should apply the methods of Problem 6.9-1, in particular, Sub-Problem 6.9-1b.

STEP 1 (Frequency-domain variables): The frequency-domain equivalent of this problem with a corresponding Fig. 6.9-1 can be set up using the following variables. Let $w(t) \triangleq [\omega_1^T(t), \omega_2^T(t)]^T$, $z = e \triangleq x - \hat{x}$, $u = Ly$, and $y = \zeta - C\hat{x}$. With these choices, the performance index becomes equivalent to $\min \{ \| G_{zw}(s) \|_2^2 = \| z \|_2^2 / \| w \|_2^2 \}$ of the preceding configuration, and because the exogenous input signals have fixed power spectrum, this is equivalent to (6.9-79).

STEP 2 (Generation of transfer matrix): We rearrange the time-domain equations according to the new set of variables.

$$\dot{e}(t) \triangleq \dot{x}(t) - \dot{\hat{x}}(t) = Ae(t) + \omega_1(t) + u(t), \quad (6.9\text{-}80a)$$

$$z(t) = e(t), \qquad (6.9\text{-}80b)$$

$$y(t) = Ce(t) + \omega_2(t). \qquad (6.9\text{-}80c)$$

These equations are carrying their statistical information in $\omega_1(t)$ and $\omega_2(t)$. It is customary to replace these variables with two new variables such that $\omega_1(t) = \Omega_1^{1/2} w_1(t)$ and $\omega_2(t) = \Omega_2^{1/2} w_2(t)$, where the intensity of $w_1(t)$ and $w_2(t)$ are identity matrices. Then we replace (6.9-80) with the following set.

$$\dot{e}(t) = Ae(t) + \Omega_1^{1/2} w_1(t) + u(t), \qquad (6.9\text{-}81a)$$

$$z(t) = e(t), \qquad (6.9\text{-}81b)$$

$$y(t) = Ce(t) + \Omega_2^{1/2} w_2(t). \qquad (6.9\text{-}81c)$$

The transfer matrix becomes

$$G(s) \triangleq \begin{bmatrix} A & \vdots & \Omega_1^{1/2} & 0 & I \\ \cdots & \vdots & \cdots & \cdots & \cdots \\ I & \vdots & 0 & 0 & 0 \\ C & \vdots & 0 & \Omega_2^{1/2} & 0 \end{bmatrix}. \qquad (6.9\text{-}82)$$

STEP 3 (Evaluation of transfer matrix): Since we have made no assumption regarding $\Omega_2^{1/2}$, to meet condition *(iii)* of a general transfer matrix (6.9-12), we consider a transformed $y(t)$ in (6.9-81c) as follows. Let

$$y_1(t) = \Omega_2^{-1/2} Ce(t) + w_2(t). \qquad (6.9\text{-}83)$$

The corresponding new transfer matrix becomes

$$\tilde{G}(s) \triangleq \begin{bmatrix} A & \vdots & \Omega_1^{1/2} & 0 & I \\ \cdots & \vdots & \cdots & \cdots & \cdots \\ I & \vdots & 0 & 0 & 0 \\ \Omega_2^{-1/2} C & \vdots & 0 & I & 0 \end{bmatrix}. \qquad (6.9\text{-}84)$$

As in the preceding example we assume that all the other relevant conditions are met, however, in applications these tests must be completed.

Before closing this step, we note the duality between (6.9-82) and (6.9-73) which links these two classical problems.

STEP 4 (Final answer): The optimal estimator gain is determined from

$$L_2 = -Y_2 C_2^T = -Y_2 C^T \Omega_2^{-1/2}, \qquad (6.9\text{-}85)$$

6.9. Sensitivity Minimization in Hardy Spaces

where

$$Y_2 \triangleq \text{Ric}\begin{bmatrix} A^T & -C^T\Omega_2^{-1}C \\ -\Omega_1 & -A \end{bmatrix}. \tag{6.9-86}$$

Since $u(t) = Ly(t)$ and the above result is in terms of $y_1(t) = \Omega_2^{-\frac{1}{2}} y(t)$, i.e., $u(t) = L_2 y_1(t) = -Y_2 C^T \Omega_2^{-\frac{1}{2}} y_1(t)$. Thus the optimal estimator gain becomes $L_{\text{optimal}} = -Y_2 C^T \Omega_2^{-1}$ which is a well-known result.

▼

6.9.3.8. Lifting Constraints *via* Scaling and Loop Shifting

For a general transfer matrix (6.9-12), a particular structure associated with D matrix is assumed. The results presented for (6.9-12) are only applicable when the following "not-so-routine" assumptions are met.

$$D_{22} = 0, \quad D_{12}^T D_{12} = I, \quad D_{21} D_{21}^T = I, \tag{6.9-87a}$$

$$D_{11} = 0, \quad D_{12}^T C_1 = 0, \quad B_1 D_{21}^T = 0. \tag{6.9-87b}$$

Certainly, not every system meets these constraints. For instance, in Examples 6.9-1 & 2, we need to make modifications in order to make the initial transfer matrices compatible with (6.9-12) (or perhaps (6.9-13) to (6.9-16)) which satisfies (6.9-87). If the original problem is well-posed and/or solvable in the sense of this optimization methodology, then we should be able to make these changes, otherwise no modifications will alleviate this difficulty. In most applications we may not know *a priori* whether the problem is well-posed or not, therefore we must test these conditions. In the following we review the algorithm proposed by Safonov, Limebeer, and Chiang [H3. 37] which may rectify this compatibility issue.

■

The following algorithm is originally developed for the H^∞-controller with $\gamma \equiv 1$. Here we put back the γ and use that as a varying parameter which changes at each iteration and finally approaches its "optimal" value γ_∞. Clearly, this notion carries a numerical sensitivity analysis of its own.

ALGORITHM 6.9-1 (Lifting The Constraints [H3. 37]): Considering Fig. 6.9-2, the following steps can be taken in order to remove constraints (6.9-87), when possible. All the subsequent steps which modify the system of Fig. 6.9-2 are depicted in Fig. 6.9-7. Each *stage* of the transformation is according to the subsequent step, therefore we use a very simplified set of notation.

STEP 1 (To remove constraint $D_{22} = 0$): This constraint may be removed by the change of variables $y_1 = y - D_{22}u$, which creates a loop in the controller and makes the plant with an effective $D_{22} = 0$.

STEP 2 *(To remove constraints on D_{12} and D_{21}):* Using the singular-value decomposition (cf., § 2.5.8), D_{12} and D_{21} are factored as follows.

$$D_{12} = U_{12}\begin{pmatrix}0\\ \Sigma_{12}\end{pmatrix}V_{12}^T, \qquad (6.9\text{-}88a)$$

$$D_{21} = U_{21}(0, \Sigma_{21})V_{21}^T. \qquad (6.9\text{-}88b)$$

In conjunction with these transformations two operational layers, one above and the other one below the plant transfer matrix, are introduced which simply scale the inputs and outputs of the plant. This trend continues as we progress to introduce more transformations. Ultimately at the end of this stage the scaled D_{12} and D_{21} become $\binom{0}{I}$, and $(0, I)$, respectively. Also at the end of this stage the D_{22} remains zero, but the new D_{11} becomes $U_{12}^T D_{11} V_{21}$.

STEP 3 *(To remove constraint $D_{11} = 0$):* The last transformation generates an effective D_{11} which is partitioned as follows.

$$U_{12}^T D_{11} V_{21} \triangleq \begin{bmatrix} D_{1111} & D_{1112} \\ D_{1121} & D_{1122} \end{bmatrix}, \qquad (6.9\text{-}89)$$

such that the size $(D_{1122}) = \dim(u) \times \dim(y)$. Let the new change of variables be $u_1 = u_2 + K_\infty y_2$, where K_∞ is a constant matrix which is chosen to minimize the maximum singular value of the following matrix.

$$\Delta_{11} \triangleq \begin{bmatrix} D_{1111} & D_{1112} \\ D_{1121} & D_{1122} + K_\infty \end{bmatrix}. \qquad (6.9\text{-}90)$$

This minimization problem is equivalent to solving the H^∞-control problem at $\omega = \infty$ and is a prerequisite for finding a solution for the original problem. The optimal K_∞ is

$$K_\infty = -(D_{1122} + D_{1121}(\gamma^2 I - D_{1111}^T D_{1111})^{-1} D_{1111}^T D_{1112}). \qquad (6.9\text{-}91)$$

This gain is shown in Fig. 6.9-7. If the problem is well-posed, then after this transformation $\sigma_{\max}(\Delta_{11}) < \gamma_\infty$.

STEP 4 *(Step 3 Continued):* To zero out D_{11}, we also need another change of variables as follows.

$$\begin{Bmatrix} z_2 \\ w_2 \end{Bmatrix} = \begin{bmatrix} \theta_{11} & \theta_{12} \\ \theta_{21} & \theta_{22} \end{bmatrix} \begin{Bmatrix} w_1 \\ z_1 \end{Bmatrix} \triangleq \Theta \begin{Bmatrix} w_1 \\ z_1 \end{Bmatrix}, \qquad (6.9\text{-}92)$$

where Θ is a constant unitary matrix with $\sigma_{\max}(\theta_{22}) < \gamma_\infty$, and θ_{12} and θ_{21} are square matrices. The optimal Θ is

$$\Theta \triangleq \begin{bmatrix} -\Delta_{11} & (I - \gamma^{-2}\Delta_{11}\Delta_{11}^T)^{1/2} \\ (I - \gamma^{-2}\Delta_{11}^T\Delta_{11})^{1/2} & \gamma^{-2}\Delta_{11}^T \end{bmatrix}. \qquad (6.9\text{-}93)$$

Carrying out these transformation results in $D_{11} \equiv 0$, however, the remaining terms in D matrix are no longer zeros, these are

6.9. Sensitivity Minimization in Hardy Spaces

$$\tilde{D}_{12} = (I - \gamma^{-2}\Delta_{11}\Delta_{11}^T)^{1/2} \begin{pmatrix} 0 \\ I \end{pmatrix}, \tag{6.9-94a}$$

$$\tilde{D}_{21} = (0, I)(I - \gamma^{-2}\Delta_{11}^T\Delta_{11})^{1/2}, \tag{6.9-94b}$$

$$\tilde{D}_{22} = (0, I)\gamma^{-2}\Delta_{11}^T(I - \gamma^{-2}\Delta_{11}\Delta_{11}^T)^{1/2} \begin{pmatrix} 0 \\ I \end{pmatrix}. \tag{6.9-94c}$$

Clearly, the new $\tilde{D}_{22}$ is no longer zero, but it can be zero out as in Step 1 which is shown in Fig. 6.9-7.

STEP 5 *(To remove constraints on $\tilde{D}_{12}$ and $\tilde{D}_{21}$):* We need to make additional change of variables by factoring $\tilde{D}_{12}$ and $\tilde{D}_{21}$ as follows.

$$\tilde{D}_{12} = \tilde{U}_{12} \begin{pmatrix} 0 \\ \tilde{\Sigma}_{12} \end{pmatrix} \tilde{V}_{12}^T, \tag{6.9-95a}$$

$$\tilde{D}_{21} = \tilde{U}_{21}(0, \tilde{\Sigma}_{21})\tilde{V}_{21}^T. \tag{6.9-95b}$$

The effects of these changes are depicted in Fig. 6.9-7. Note that when, for instance, $\tilde{D}_{12}^T\tilde{D}_{12} \neq I$, it means that we must scale $\tilde{D}_{12}$ and that is always possible when $\tilde{D}_{12}$ is of maximum rank (or $\tilde{\Sigma}_{12}^{-1}$ exists). This idea is used repeatedly to construct Fig. 6.9-7.

STEP 6 *(To remove constraints $D_{12}^TC_1 = 0$ and $B_1D_{21}^T = 0$):* The following generalization of Theorem 6.9-4 enables us to use this theorem for situations in which $D_{12}^TC_1 \neq 0$ and/or $B_1D_{21}^T \neq 0$. This generalization also applies to our *transformed* plant. The corresponding new Hamiltonian matrices are

$$H_{\tilde{X}_\infty} \triangleq \begin{bmatrix} A - B_2D_{12}^TC_1 & \gamma^{-2}B_1B_1^T - B_2B_2^T \\ -\tilde{C}_1^T\tilde{C}_1 & -(A - B_2D_{12}^TC_1)^T \end{bmatrix}, \tag{6.9-96a}$$

$$\tilde{C}_1 \triangleq (I - D_{12}D_{12}^T)C_1. \tag{6.9-96b}$$

$$H_{\tilde{Y}_\infty} \triangleq \begin{bmatrix} (A - B_1D_{21}^TC_2)^T & \gamma^{-2}C_1^TC_1 - C_2^TC_2 \\ -\tilde{B}_1\tilde{B}_1^T & -(A - B_1D_{21}^TC_2) \end{bmatrix}, \tag{6.9-97a}$$

$$\tilde{B}_1 \triangleq B_1(I - D_{21}^TD_{21}). \tag{6.9-97b}$$

The new sub-optimal gains and the associated $\tilde{A}_\infty$ (cf., (6.9-56b)), and $\tilde{Z}_\infty$ become

$$\tilde{F}_\infty \triangleq -(B_2^T\tilde{X}_\infty + D_{12}^TC_1), \tag{6.9-98a}$$

$$\tilde{L}_\infty \triangleq -(\tilde{Y}_\infty C_2^T + B_1D_{21}^T), \tag{6.9-98b}$$

$$\tilde{A}_\infty \triangleq \hat{A}_\infty + \tilde{Z}_\infty \tilde{L}_\infty \gamma^{-2} D_{21} B_1^T \tilde{X}_\infty, \qquad (6.9\text{-}98c)$$

$$\tilde{Z}_\infty \triangleq (I - \gamma^{-2} \tilde{X}_\infty \tilde{Y}_\infty)^{-1}. \qquad (6.9\text{-}98d)$$

Carrying out this analysis for the last value of D matrix yields the corresponding (6.9-67) as follows.

$$\tilde{G}_{aux}^{H^\infty}(s) \triangleq \left[\begin{array}{c|cc} \tilde{A}_\infty & -\tilde{Z}_\infty \tilde{L}_\infty & \tilde{Z}_\infty (B_2 + \gamma^{-2} \tilde{Y}_\infty C_1^T D_{12}) \\ \hline \tilde{F}_\infty & 0 & I \\ -(C_2 + \gamma^{-2} D_{21} B_1^T \tilde{X}_\infty) & I & 0 \end{array} \right]. \qquad (6.9\text{-}99)$$

COMMENT 6.9-7 (On Algorithm 6.9-1): The preceding transformations are generally norm preserving, however, transformation Θ in Step 4 changes the H_∞-norm of the original transfer matrix, but keeps that below its upper bound γ. Referring to Fig. 6.9-7, we group all the transformations between $(\hat{w}, z)$ and $(\hat{z}, w)$ and call that T_1; as well as all those between (y, u_{aux}) and (u, y_{aux}) and call that T_2. Starting with (6.9-92), and referring to Fig. 6.9-7, we have

$$\left\{ \begin{array}{c} (\tilde{U}_{12}^T)^{-1} \hat{z} \\ V_{21}^{-1} w \end{array} \right\} = \left[\begin{array}{cc} \theta_{11} & \theta_{12} \\ \theta_{21} & \theta_{22} \end{array} \right] \left\{ \begin{array}{c} \tilde{V}_{21} \hat{w} \\ U_{12}^T z \end{array} \right\}, \qquad (6.9\text{-}100a)$$

or

$$\left\{ \begin{array}{c} \hat{z} \\ w \end{array} \right\} = \left[\begin{array}{cc} \tilde{U}_{12}^T \theta_{11} \tilde{V}_{21} & \tilde{U}_{12}^T \theta_{12} U_{12}^T \\ V_{21} \theta_{21} \tilde{V}_{21} & V_{21} \theta_{22} U_{12}^T \end{array} \right] \left\{ \begin{array}{c} \hat{w} \\ z \end{array} \right\} \triangleq T_1 \left\{ \begin{array}{c} \hat{w} \\ z \end{array} \right\}. \qquad (6.9\text{-}100b)$$

Substituting for Θ in (6.9-100b) yields.

$$T_1 \triangleq \left[\begin{array}{cc} -\tilde{U}_{12}^T \Delta_{11} \tilde{V}_{21} & \tilde{U}_{12}^T (I - \gamma^{-2} \Delta_{11} \Delta_{11}^T)^{1/2} U_{12}^T \\ V_{21} (I - \gamma^{-2} \Delta_{11}^T \Delta_{11})^{1/2} \tilde{V}_{21} & \gamma^{-2} V_{21} \Delta_{11}^T U_{12}^T \end{array} \right]. \qquad (6.9\text{-}101)$$

6.9. Sensitivity Minimization in Hardy Spaces

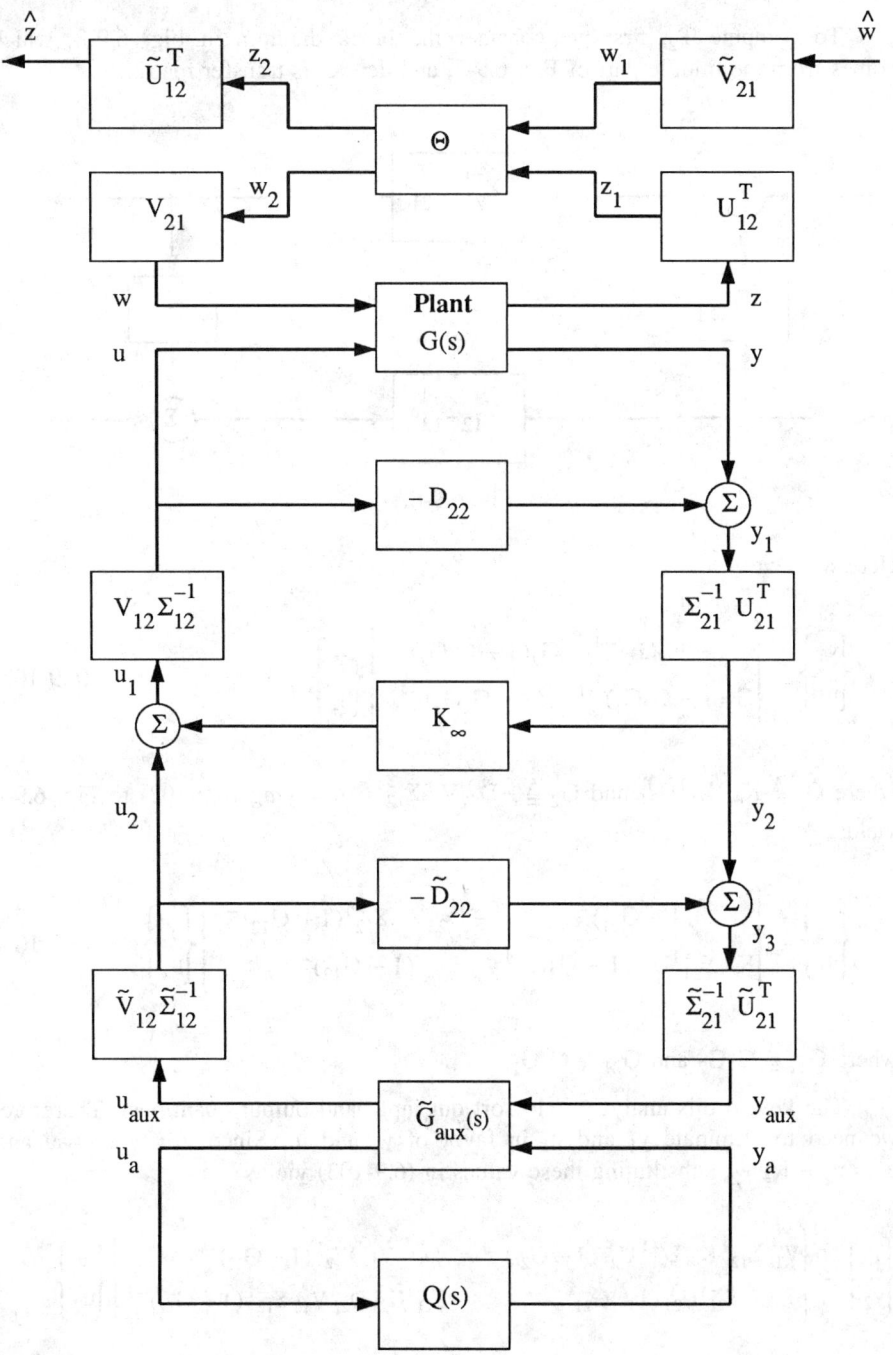

Fig. 6.9-7. Lifting constraints in H^∞–control.

To compute T_2, first we consider the block diagram in Fig. 6.9-8, which comes from the middle part of Fig. 6.9-7, and derive its transfer matrix.

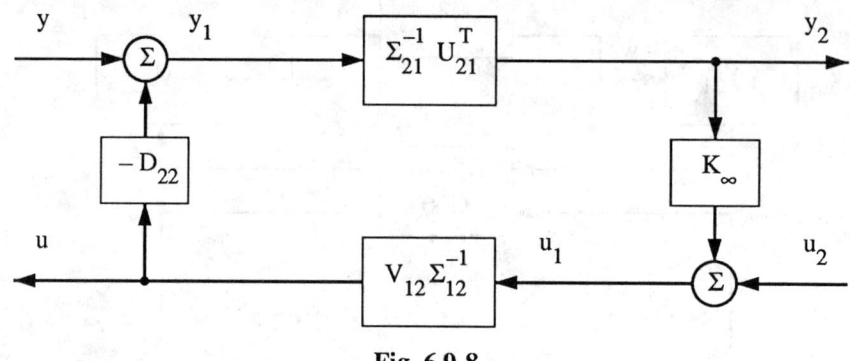

Fig. 6.9-8.

Here we have

$$\begin{Bmatrix} y_1 \\ u_1 \end{Bmatrix} = \begin{bmatrix} (I - G_2 G_1)^{-1} & G_2(I - G_1 G_2)^{-1} \\ G_1(I - G_2 G_1)^{-1} & (I - G_1 G_2)^{-1} \end{bmatrix} \begin{Bmatrix} y \\ u_2 \end{Bmatrix}, \quad (6.9\text{-}102)$$

where $G_1 \triangleq K_\infty \Sigma_{21}^{-1} U_{21}^T$ and $G_2 \triangleq -D_{22} V_{12} \Sigma_{12}^{-1}$. Applying (6.9-102) to Fig. 6.9-8 yields

$$\begin{Bmatrix} y_1 \\ u_1 \end{Bmatrix} = \begin{bmatrix} (I - G_{21})^{-1} & -D_{22} V_{12} \Sigma_{12}^{-1}(I - G_{12})^{-1} \\ K_\infty \Sigma_{21}^{-1} U_{21}^T (I - G_{21})^{-1} & (I - G_{12})^{-1} \end{bmatrix} \begin{Bmatrix} y \\ u_2 \end{Bmatrix}, \quad (6.9\text{-}103)$$

where $G_{12} \triangleq G_1 G_2$ and $G_{21} \triangleq G_2 G_1$.

The key to this analysis is to sort out input and output quantities. Therefore, we need to eliminate y_1 and u_1 in favor of y_2 and u. Since $y_1 = y - D_{22} u$ and $u_1 = u_2 + K_\infty y_2$, substituting these values in (6.9-103) yields

$$\begin{Bmatrix} u \\ y_2 \end{Bmatrix} = \begin{bmatrix} V_{12} \Sigma_{12}^{-1} K_\infty \Sigma_{21}^{-1} U_{21}^T (I - G_{21})^{-1} & V_{12} \Sigma_{12}^{-1}(I - G_{12})^{-1} \\ \Sigma_{21}^{-1} U_{21}^T (I - G_{21})^{-1} & -\Sigma_{21}^{-1} U_{21}^T D_{22} V_{12} \Sigma_{12}^{-1}(I - G_{12})^{-1} \end{bmatrix} \begin{Bmatrix} y \\ u_2 \end{Bmatrix}.$$

(6.9-104)

Considering that $u_2 = \tilde{V}_{12} \tilde{\Sigma}_{12}^{-1} u_{\text{aux}}$ and

6.9. Sensitivity Minimization in Hardy Spaces

$$y_{aux} = \tilde{\Sigma}_{21}^{-1}\tilde{U}_{21}^T y_3 = \tilde{\Sigma}_{21}^{-1}\tilde{U}_{21}^T(y_2 - \tilde{D}_{22}u_2), \tag{6.9-105}$$

we conclude that

$$\begin{Bmatrix} u \\ y_{aux} \end{Bmatrix} \triangleq T_2 \begin{Bmatrix} y \\ u_{aux} \end{Bmatrix} = \begin{bmatrix} T_{211} & T_{212} \\ T_{221} & T_{222} \end{bmatrix} \begin{Bmatrix} y \\ u_{aux} \end{Bmatrix}, \tag{6.9-106}$$

where

$$T_{211} \triangleq V_{12}\Sigma_{12}^{-1} K_\infty \Sigma_{21}^{-1} U_{21}^T (I - G_{21})^{-1}, \tag{6.9-107a}$$

$$T_{212} \triangleq V_{12}\Sigma_{12}^{-1}(I - G_{12})^{-1} \tilde{V}_{12}\tilde{\Sigma}_{12}^{-1}, \tag{6.9-107b}$$

$$T_{221} \triangleq \tilde{\Sigma}_{21}^{-1}\tilde{U}_{21}^T \Sigma_{21}^{-1} U_{21}^T (I - G_{21})^{-1}, \tag{6.9-107c}$$

$$T_{222} \triangleq -\tilde{\Sigma}_{21}^{-1}\tilde{U}_{21}^T (\Sigma_{21}^{-1} U_{21}^T D_{22} V_{12} \Sigma_{12}^{-1}(I - G_{12})^{-1} + \tilde{D}_{22}) \tilde{V}_{12}\tilde{\Sigma}_{12}^{-1}, \tag{6.9-107d}$$

$$G_{12} \triangleq -K_\infty \Sigma_{21}^{-1} U_{21}^T D_{22} V_{12} \Sigma_{12}^{-1}, \tag{6.9-101e}$$

$$G_{21} \triangleq -D_{22} V_{12} \Sigma_{12}^{-1} K_\infty \Sigma_{21}^{-1} U_{21}^T. \tag{6.9-101f}$$

The above computation is essential for the overall study of system under loop shifting and scaling. To put the matter in prospective, Fig. 6.9-7 is redrawn in Fig. 6.9-9 in terms of T_1 and T_2. When a controller $\tilde{K} = \tilde{K}(\tilde{G}_{aux}(s), Q(s))$ is developed for the above transformed plant and according to Algorithm 6.9-1, we need to recast its actual value for the actual plant by a reverse transformation using a linear fractional transformation between T_2 and $\tilde{K}$. ∎

COMMENT 6.9-8 (The H^2–controller): Algorithm 6.9-1 can be specialized to cases corresponding to the H^2–controller. The necessary steps are as follows. These steps are depicted in Fig. 6.9-10.

STEP 1: The constraint $D_{22} = 0$ is removed by the change of variables $y_1 = y - D_{22}u$.

STEP 2: Using singular-value decomposition, we factor D_{12} and D_{21} as follows.

$$D_{12} = U_{12}\begin{pmatrix} 0 \\ \Sigma_{12} \end{pmatrix} V_{12}^T, \tag{6.9-108a}$$

$$D_{21} = U_{21}(0, \Sigma_{21}) V_{21}^T. \tag{6.9-108b}$$

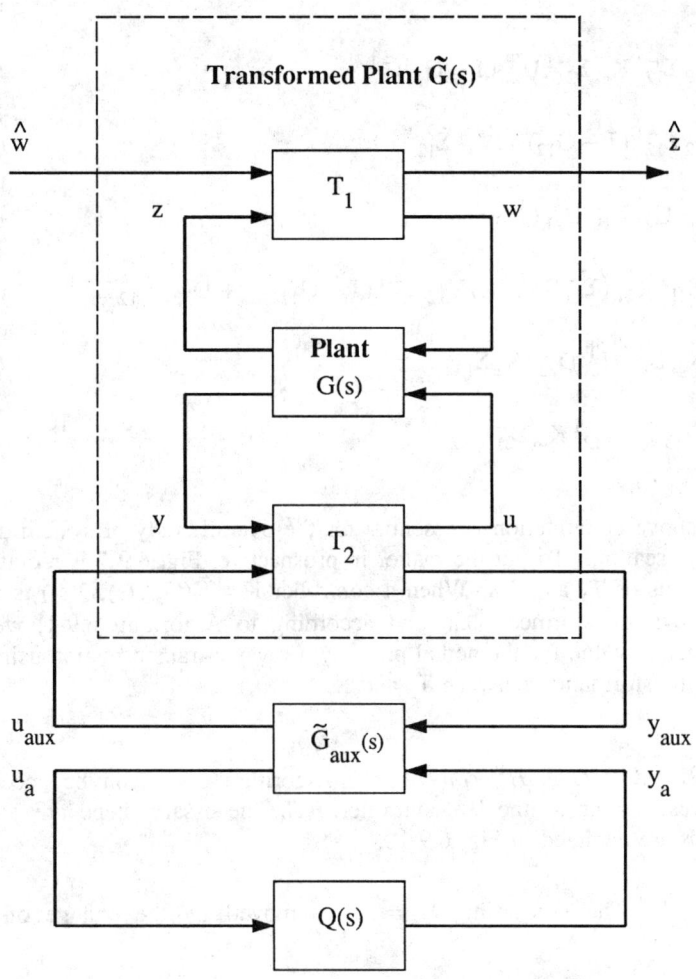

Fig. 6.9-9. Alternative representation of Fig. 6.9-7.

6.9. Sensitivity Minimization in Hardy Spaces

STEP 3: Let

$$\Delta_{11} \triangleq U_{12}^T D_{11} V_{21} + \begin{bmatrix} 0 & 0 \\ 0 & K_2 \end{bmatrix} \triangleq \begin{bmatrix} D_{1111} & D_{1112} \\ D_{1121} & D_{1122} + K_2 \end{bmatrix} \quad (6.9\text{-}109)$$

such that the size $(\Delta_{11})_{22} = \dim(u) \times \dim(y)$, and $K_2 = -D_{1122}$.

STEP 4: The generalized Theorem 6.9-2 becomes

$$H_{\tilde{X}_2} \triangleq \begin{bmatrix} A - B_2 D_{12}^T C_1 & -B_2 B_2^T \\ -\tilde{C}_1^T \tilde{C}_1 & -(A - B_2 D_{12}^T C_1)^T \end{bmatrix}, \quad (6.9\text{-}110a)$$

$$\tilde{C}_1 \triangleq (I - D_{12} D_{12}^T) C_1. \quad (6.9\text{-}110b)$$

$$H_{\tilde{Y}_2} \triangleq \begin{bmatrix} (A - B_1 D_{21}^T C_2)^T & -C_2^T C_2 \\ -\tilde{B}_1 \tilde{B}_1^T & -(A - B_1 D_{21}^T C_2) \end{bmatrix}, \quad (6.9\text{-}111a)$$

$$\tilde{B}_1 \triangleq B_1(I - D_{21}^T D_{21}). \quad (6.9\text{-}111b)$$

Incorporating the above elements and the following optimal gains, we can construct the corresponding $\tilde{G}_{\text{aux}}^{H^2}(s)$ which is similar to (6.9-55).

$$\tilde{F}_2 \triangleq -(B_2^T \tilde{X}_2 + D_{12}^T C_1), \quad (6.9\text{-}112a)$$

$$\tilde{L}_2 \triangleq -(\tilde{Y}_2 C_2^T + B_1 D_{21}^T), \quad (6.9\text{-}112b)$$

$$\tilde{A}_2 \triangleq A + B_2 \tilde{F}_2 + \tilde{L}_2 C_2, \quad (6.9\text{-}112c)$$

$$\tilde{G}_{\text{aux}}^{H^2}(s) \triangleq \begin{bmatrix} \tilde{A}_2 & \vdots & -\tilde{L}_2 & B_2 \\ \cdots & \vdots & \cdots & \cdots \\ \tilde{F}_2 & \vdots & 0 & I \\ -C_2 & \vdots & I & 0 \end{bmatrix}. \quad (6.9\text{-}112d)$$

This transfer matrix is computed for the last value of D matrix.

The preceding steps are depicted in Fig. 6.9-10. Based on this figure, the transformation between $(\hat{w}, z)$ and $(\hat{z}, w)$, and (y, u_{aux}) and (u, y_{aux}) are

$$\left\{\begin{matrix}\hat{z}\\w\end{matrix}\right\} \triangleq T_1 \left\{\begin{matrix}\hat{w}\\z\end{matrix}\right\} \triangleq \begin{bmatrix}0 & U_{12}^T\\V_{21} & 0\end{bmatrix}\left\{\begin{matrix}\hat{w}\\z\end{matrix}\right\}. \qquad (6.9\text{-}113)$$

$$\left\{\begin{matrix}u\\y_{aux}\end{matrix}\right\} \triangleq T_2 \left\{\begin{matrix}y\\u_{aux}\end{matrix}\right\} = \begin{bmatrix}T_{211} & T_{212}\\T_{221} & T_{222}\end{bmatrix}\left\{\begin{matrix}y\\u_{aux}\end{matrix}\right\}. \qquad (6.9\text{-}114)$$

where

$$T_{211} \triangleq V_{12}\Sigma_{12}^{-1}K_2\Sigma_{21}^{-1}U_{21}^T(I - G_{21})^{-1}, \qquad (6.9\text{-}115a)$$

$$T_{212} \triangleq V_{12}\Sigma_{12}^{-1}(I - G_{12})^{-1}, \qquad (6.9\text{-}115b)$$

$$T_{221} \triangleq \Sigma_{21}^{-1}U_{21}^T(I - G_{12})^{-1}, \qquad (6.9\text{-}115c)$$

$$T_{222} \triangleq -\Sigma_{21}^{-1}U_{21}^T D_{22} V_{12}\Sigma_{12}^{-1}(I - G_{12})^{-1}, \qquad (6.9\text{-}115d)$$

$$G_{12} \triangleq -K_2\Sigma_{21}^{-1}U_{21}^T D_{22} V_{12}\Sigma_{12}^{-1}, \qquad (6.9\text{-}115e)$$

$$G_{21} \triangleq -D_{22} V_{12}\Sigma_{12}^{-1}K_2\Sigma_{21}^{-1}U_{21}^T. \qquad (6.9\text{-}115f)$$

Here the transformed D has the following partitions.

$$D_{11} = \begin{bmatrix}D_{1111} & D_{1112}\\D_{1121} & 0\end{bmatrix}, \qquad (6.9\text{-}116a)$$

$$D_{12} = \left\{\begin{matrix}0\\I\end{matrix}\right\}, \quad D_{21} = (0, I), \quad D_{22} = 0. \qquad (6.9\text{-}116b)$$

In closing, we should point out that unless the final $D_{11} \equiv 0$, the H_2–norm is not bounded. Since in the proof of Theorem 6.9-1 & 2, D_{11} does not explicitly appear, we can always solve the H^2–optimization problem, however, that solution is only valid for the original transfer matrix without the D_{11}. As in the preceding algorithm, we need to transform backward from the computed solution for the revised plant to the actual controller values using the above T_2. This is done by a linear fractional transformation of T_2 and the above computed controller. We close this comment by suggesting that in both Examples 6.9-1 & 2, we can apply the preceding steps and reach the same conclusions.

6.9. Sensitivity Minimization in Hardy Spaces

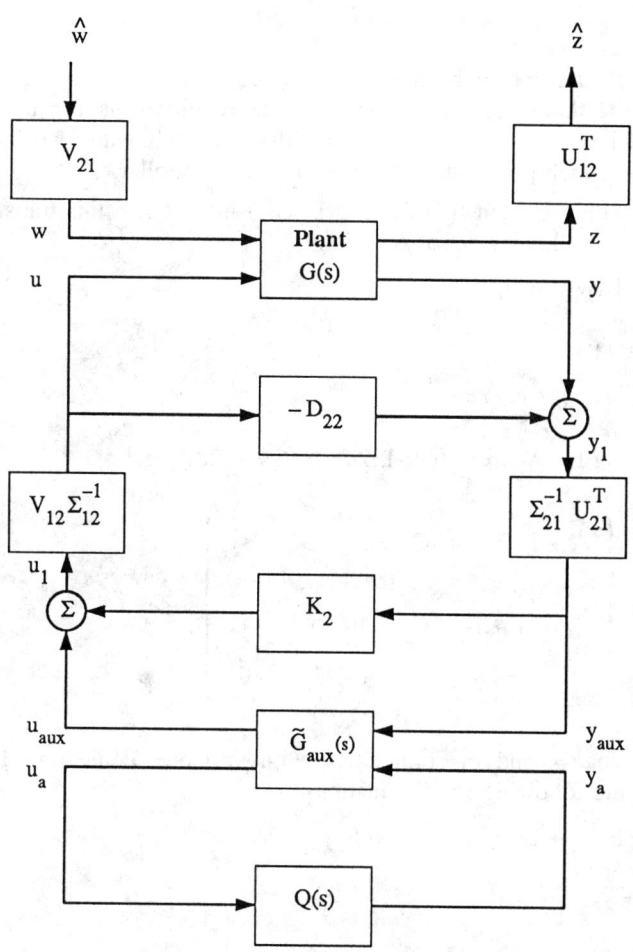

Fig. 6.9-10. Lifting constraints in H^2-control.

EXAMPLE 6.9-3 (The H^2- versus H^∞-controller): Consider the transfer matrix (6.9-12) and the corresponding $\tilde{G}_{aux}^{H^2}(s)$ and $\tilde{G}_{aux}^{H^\infty}(s)$ given in (6.9-112d) and (6.9-99), respectively. These transfer matrices are the most involved cases which we have studied. Draw block-diagram representation of these controllers.

SOLUTION: If and only if all the necessary requirements regarding the existence of solutions for these controllers, namely, the requirements for the generalized Theorems 6.9-1 to 4, are met, then we can draw the following two block-diagram representations which portray the structure of each controller.

Starting with the plant transfer matrix G(s) in (6.9-12), this transfer matrix is equivalent to the following set of equations.

$$\dot{x} = Ax + B_1 w + B_2 u, \quad (6.9\text{-}117a)$$

$$z = C_1 x + D_{12} u, \quad (6.9\text{-}117b)$$

$$y = C_2 x + D_{21} w. \quad (6.9\text{-}117c)$$

Substituting for $\tilde{A}_2$ from (6.9-112c) in (6.9-112d) yields

$$\tilde{G}_{aux}^{H^2}(s) \triangleq \left[\begin{array}{c:cc} A + B_2 \tilde{F}_2 + \tilde{L}_2 C_2 & -\tilde{L}_2 & B_2 \\ \hdashline \tilde{F}_2 & 0 & I \\ -C_2 & I & 0 \end{array} \right]. \quad (6.9\text{-}118)$$

This transfer matrix and in conjunction with variables defined in Fig. 6.9-2 is equivalent to the following set of equations.

$$\dot{\tilde{x}} = (A + B_2 \tilde{F}_2 + \tilde{L}_2 C_2)\tilde{x} - \tilde{L}_2 y + B_2 y_a, \quad (6.9\text{-}119a)$$

$$u = \tilde{F}_2 \tilde{x} + y_a, \quad (6.9\text{-}119b)$$

$$u_a = -C_2 \tilde{x} + y, \quad (6.9\text{-}119c)$$

where $\tilde{x}$ is the state of $\tilde{G}_{aux}^{H^2}(s)$, and $\tilde{F}_2$, $\tilde{L}_2$ are defined in (6.9-112a) and (6.9-112b), respectively.

Equations (6.9-117) and (6.9-119) are shown in Fig. 6.9-11. Next we show that indeed any stable Q(s) is acceptable in this design which yields an overall stable system. But first, let us study this figure and evaluate its structure. Clearly, the dynamics of our auxiliary transfer matrix is similar to an observer-based estimator that estimates the state of plant $\tilde{x}$ by integrating its output together with an estimate of this output which is $C_2 \tilde{x}$. This fact has resulted from using output-

6.9. Sensitivity Minimization in Hardy Spaces

estimation structure to prove Theorem 6.9-1. The estimated state $\tilde{x}$ is then passed through the full-state feedback gain $\tilde{F}_2$ (full-state feedback is the special case of full-information structure which is used to prove Theorem 6.9-1) and together with y_a, the output of Q(s), is fed back to the plant. The input of the free parameter Q(s) is $u_a = y - C_2\tilde{x}$ which is the output-estimation error and this free parameter controls the flow of u_a into the original plant. This feature is common with the corresponding H^∞-controller presented in Fig. 6.9-12. ∎

Looking at Fig. 6.9-11, we may also combine the plant transfer matrix and the auxiliary transfer matrix $\tilde{G}_{aux}^{H^2}(s)$ to get the $\tilde{G}_{total}^{H^2}(s)$. This is easily done by eliminating u and y in (6.9-117) and (6.9-119) in favor of y_a and u_a. The new set of equations becomes

$$\dot{x} = (A + B_2\tilde{F}_2)x - B_2\tilde{F}_2 e + B_1 w + B_2 y_a, \qquad (6.9\text{-}120a)$$

$$\dot{e} \triangleq x - \tilde{x} = (A + \tilde{L}_2 C_2)e + (B_1 + \tilde{L}_2 D_{21})w, \qquad (6.9\text{-}120b)$$

$$z = (C_1 + D_{12}\tilde{F}_2)x - D_{12}\tilde{F}_2 e + D_{12} y_a, \qquad (6.9\text{-}120c)$$

$$u_a = C_2 e + D_{21} w. \qquad (6.9\text{-}120d)$$

The set of equations (6.9-120) can be represented by its equivalent transfer matrix as follows.

$$\tilde{G}_{total}^{H^2}(s) \triangleq \left[\begin{array}{cc:cc} A + B_2\tilde{F}_2 & -B_2\tilde{F}_2 & B_1 & B_2 \\ 0 & A + \tilde{L}_2 C_2 & B_1 + \tilde{L}_2 D_{21} & 0 \\ \hdashline C_1 + D_{12}\tilde{F}_2 & -D_{12}\tilde{F}_2 & 0 & D_{12} \\ 0 & C_2 & D_{21} & 0 \end{array} \right].$$

$$(6.9\text{-}121)$$

From the analyses in § 2.6.2, we can conclude and/or confirm several interesting properties of (6.9-121). First of all, the corresponding "A-matrix" is triangular and its stability depends on the stability of the regulator part $A + B_2\tilde{F}_2$ as well as the state estimator part $A + \tilde{L}_2 C_2$. This fact is precisely the statement of the well-known separation theorem in the linear optimal control theory. Secondly, straightforward computation of $\tilde{G}_{total}^{H^2}(s)$ = "$C(sI - A)^{-1}B + D$" reveals that the (2,2)–block partition of this transfer matrix is identically zero. This gain is the effective gain from input to output of Q(s) and that is why ant stable Q(s) is satisfactory in this design. ∎

Substituting (6.9-98c) with (6.9-56b) into (6.9-99) yields

$\tilde{G}_{aux}^{H^\infty}(s) \triangleq$

$$\begin{bmatrix} A + B_2\tilde{F}_\infty + \gamma^{-2}B_1B_1^T\tilde{X}_\infty + \tilde{Z}_\infty\tilde{L}_\infty(C_2 + \gamma^{-2}D_{21}B_1^T\tilde{X}_\infty) & : & -\tilde{Z}_\infty\tilde{L}_\infty & \tilde{Z}_\infty(B_2 + \gamma^{-2}\tilde{Y}_\infty C_1^T D_{12}) \\ \cdots\cdots\cdots\cdots\cdots\cdots\cdots\cdots\cdots\cdots\cdots\cdots\cdots\cdots\cdots & : & \cdots & \cdots\cdots\cdots\cdots\cdots\cdots \\ \tilde{F}_\infty & : & 0 & I \\ -(C_2 + \gamma^{-2}D_{21}B_1^T\tilde{X}_\infty) & : & I & 0 \end{bmatrix}.$$

(6.9-122)

This transfer matrix is equivalent to the following set of equations

$$\dot{\tilde{x}} = (A + B_2\tilde{F}_\infty + \gamma^{-2}B_1B_1^T\tilde{X}_\infty + \tilde{Z}_\infty\tilde{L}_\infty(C_2 + \gamma^{-2}D_{21}B_1^T\tilde{X}_\infty))\tilde{x}$$
$$- \tilde{Z}_\infty\tilde{L}_\infty y + \tilde{Z}_\infty(B_2 + \gamma^{-2}\tilde{Y}_\infty C_1^T D_{12})y_a, \qquad (6.9\text{-}123a)$$

$$u = \tilde{F}_\infty \tilde{x} + y_a, \qquad (6.9\text{-}123b)$$

$$u_a = -(C_2 + \gamma^{-2}D_{21}B_1^T\tilde{X}_\infty)\tilde{x} + y, \qquad (6.9\text{-}123c)$$

where $\tilde{x}$ is the state of $\tilde{G}_{aux}^{H^\infty}(s)$, and the sub-optimal gains $\tilde{F}_\infty$ and $\tilde{L}_\infty$ given in (6.9-98a) and (6.9-98b) with $\tilde{Z}_\infty$ given in (6.9-98d), respectively.

Equations (6.9-117) and (6.9-123) are shown in Fig. 6.9-12. Here also, any choice of stable Q(s) is acceptable for this controller design, because we can show, in exactly the same manner as in the preceding case, that the $(2,2)$–block partition of the corresponding $\tilde{G}_{total}^{H^\infty}(s)$ (augmenting G(s) and $\tilde{G}_{aux}^{H^\infty}(s)$) is identically zero. However, the overall structure of this $\tilde{G}_{total}^{H^\infty}(s)$ is more complicated than (6.9-121).

Looking at Fig. 6.9-12, *similar* discussions as in Comments 6.9-3 can be made regarding the separation property of this controller. Finally, it is very instructive to compare Fig. 6.9-11 with Fig. 6.9-12 to see what changes must be made in the design of an H^2-controller in order to make it like an H^∞-controller.
▼

6.9.3.9. General Remarks

The main focus of the methodology presented in this section is on the H^∞–controller design for system of *Case 3* in Table 6.1-1. The solution provided is a parameter-dependent sub-optimal solution which is imbedded in a family of problems with a varying parameter γ. This outcome instigates a sensitivity analysis of its own, but that analysis is not trivial and indeed it may be even more difficult than the original problem itself. Recall that Theorem 6.9-3 does not give an explicit solution in terms of γ. As in the case of computing the actual H_∞–norm, it is proposed in the literature to iterate on this parameter until it reaches its "optimal" value. It is also known that when γ approaches its "optimal" value the solution of the corresponding Riccati equations and a host of other variables become numerically very sensitive. There are alternative formulations, for instance, the descriptor forms of [H3. 37], to lessen the effect of this parameter dependency on the sub-optimal solution, but in the final analysis we must realize that the preceding results are most effective when used in the sub-optimal mode. Since most of these results are centered around the assumption on the existence of positive-definite solutions for two Riccati equations, we need to pay special attention to eigenvalues of the corresponding Hamiltonian matrices or the complementarity properties. When this property fails, i.e., the eigenvalues move toward the $j\omega$–axis, the solutions of Riccati equations no longer stabilizes the system and all the numerical results lose their significance. Therefore a good starting point to look at the numerical sensitivity analysis with respect to γ is the eigenvalues of Hamiltonian matrices.

It is also well-known that as $\gamma \to \infty$, the H^∞–controller approaches the H^2–controller because primarily $X_\infty \to X_2$ and $Y_\infty \to Y_2$ (the matrices X_∞ and Y_∞ are decreasing functions of γ). The $K_{\text{suboptimal}}(s)$ in (6.9-56) is the natural generalization of the $K_{\text{optimal}}(s)$ in (6.9-50), and this solution have many other interesting properties, for instance, this is the maximum entropy solution and the minimax control for $\| z \|_2^2 - \gamma^2 \| w \|_2^2$. However, the most important feature of this controller is its significant impact in the utilization of H^∞–methodology and the fact that it has opened many interesting opportunities for future research. Clearly, the trend is to utilize these methods to their fullest potentials. Finally, we recall that Problems 6.9-1 & 2 have been solved when G(s) is subject to exogenous signals *alone*. It is interesting to study these problems in the context of *Case 4* in Table 6.1-1, when G(s) is subject to large-parameter variations as described in Sections 6.7 and 6.8.

In closing, while this methodology has been flourishing in the West, since 1981, a very active group of researchers in the East have been studying the problem of compensating for exogenous signals differently under the title of theory of invariance, since 1939. In the following section we take you back to the beginning of that theory and give you an overview of its current status.

■

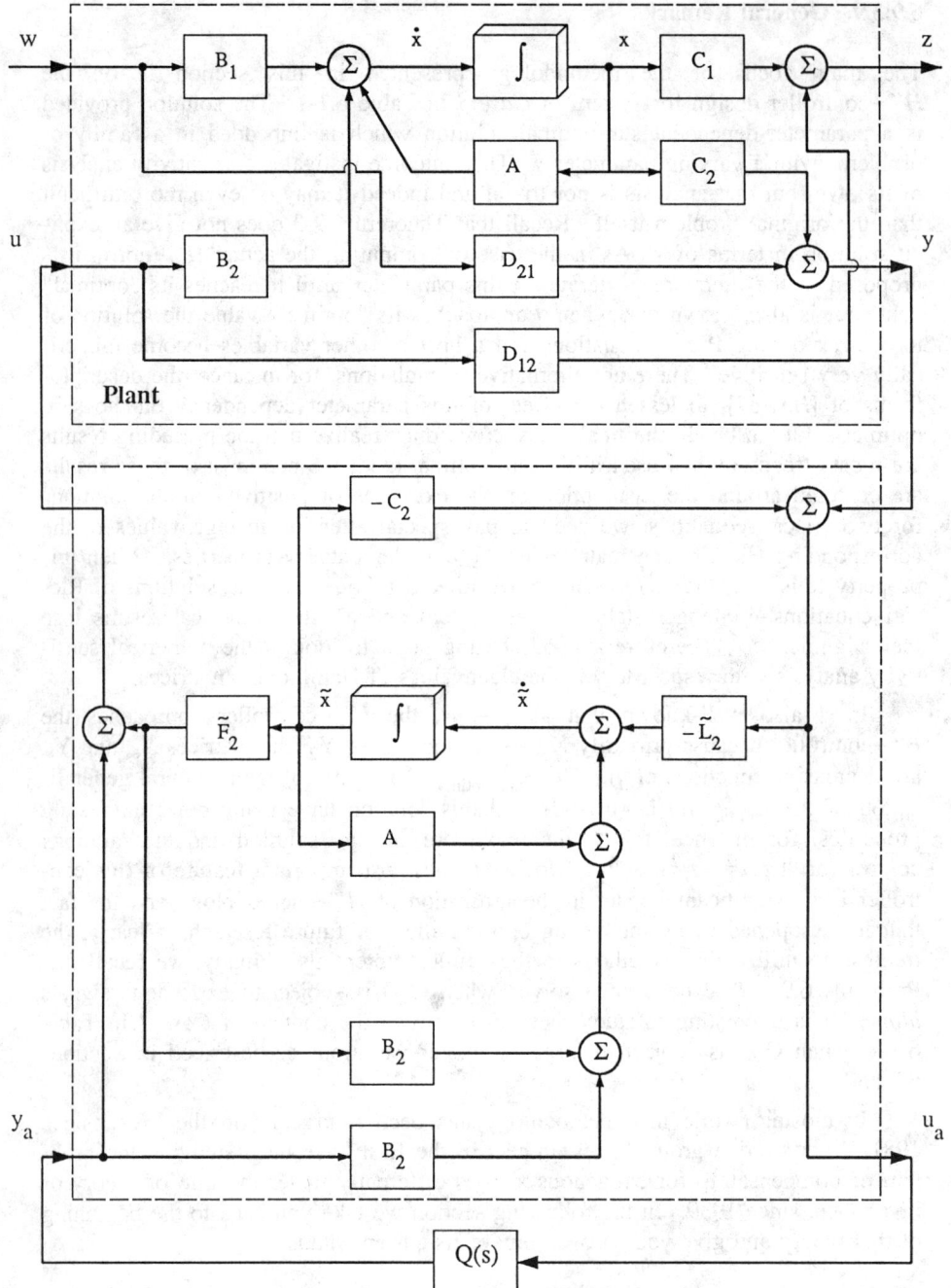

Fig. 6.9-11. Block-diagram representation of the H^2–controller.

6.9. Sensitivity Minimization in Hardy Spaces

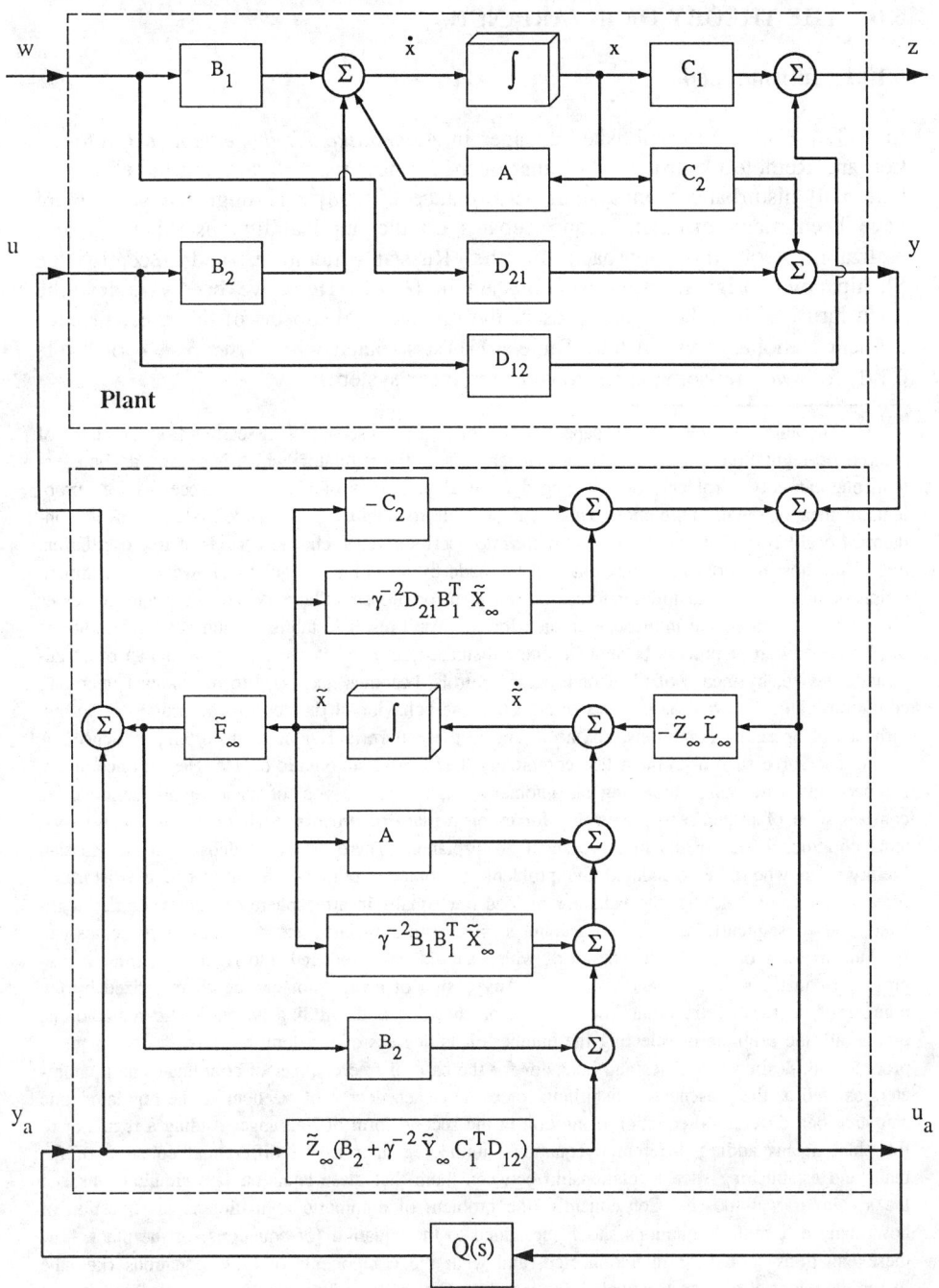

Fig. 6.9-12. Block-diagram representation of the H^∞-controller.

6.10. THE THEORY OF INVARIANCE

6.10.1. Introduction

In 1939, Shchipanov published a paper in *Avtomatika i Telemekhanika* (Automation and Remote Control − *ARC*) that opened a new way of studying the effects of [external] disturbances on system performance [*J*. 58][3]. Through the years there have been many discussions and debates on the applicability, usefulness and/or realizability of this approach in the Russian literature, and recently the Shchipanov's original paper is published in [*I*. 8]. Here, we briefly review the main thrust of this theory and present the fundamental aspects of this area; first for a linear stationary system (cf., Table 3.1-1) associated with *Cases 3 & 4* of Table 6.1-1, followed by some extensions to nonlinear systems.

---

[3] The author of this historic paper [*J*. 58] (or [*I*. 8]) presented his assessment on the status of "automatic regulation", in the following manner: "Using the regulator, which is a new attached system, one can solve problems of changing dynamical properties of plants and processes, i.e., problems of their automatic regulation. Thus, it is possible to stabilize an unstable system, convert non-damped oscillating changes of any parameter into aperiodic ones, change periods of free oscillation, *etc.*. The most important advance that can be made by attaching a regulator, however, is compensation of influences of disturbing forces on the plant or process. Plants do not work and processes do not occur without the influences of such forces. As a result, established and desired conditions of plant operation or process behavior become disturbed, and an intervention for a human or an automatic device, by means of plant or process controls, becomes necessary to re-establish operating conditions. This intervention consists of adjusting the behavior of the controls according to the disturbance of operating conditions, i.e., according to the pattern and value of disturbing influence. A human can solve such a problem less completely than can an automatic device. Hence, once again, the necessity emerges of attaching an automatic regulator. The goal of an automatic regulator is compensation of influence of disturbing forces on a plant or process, such as load in energy systems or atmospheric squalls in the case of an airplane. [There are some debates in the Russian literature on who first considered the problem of compensation for the influence of disturbing forces.] Of course, neither the behavior of load nor squalls in atmosphere can be arbitrarily regulated, and consequently, taking into account such a disturbing influence on plants or processes can be done in form of general function f(t), which should be introduced into right-hand sides of appropriate equations of the plant or process. Any design of a regulator can be characterized by the number of degrees of freedom. The problem of choosing and building an automatic regulator is, first of all, the problem of selecting the number of its degrees of freedom that provides the plant or process with desired properties and behavior for the case of free changes of coordinates and parameters as well as the presence of disturbing forces. For each degree of freedom of the regulator, one can state one differential equation of motion in the second form of Lagrange. Adding a regulator to the plant means adding differential equations of the regulator to the differential equation of the plant and establishing such a relationship between them that these equations are simultaneous and the system is well posed. Consequently, the problem of automatic regulation is the question of how many differential equations should be added to the equation (or equations) of the plant, how these equations should be interconnected, and what the components of these equations (i.e., the forces they consist of) constructively correspond to. For any force, there is a certain device for generating it. Hence, by determining which forces should be present in the regulator and in its connections to the plant, we learn what the design of the regulator should be."

6.10. The Theory of Invariance

One of the early survey papers in this area is written by two veteran contributors Boychuk and Voronin [*J*. 9], wherein some research guidelines and different channels for investigating the theory of invariance are stated. In this reference, it is emphasized that this study originated from the work of Shchipanov and complemented by Luzin's work [*J*. 35]. Since 1958, several annual conferences are being held in the former Soviet Union (cf., § 6.12.*I*), and many scores of papers are being published (cf., § 6.12.*J*) in this area. In particular, the publications of [*I*. 8] and the three recent survey papers which are written by Kukhtenko [*J*. 30 to 32] who is another major contributor in this area, have made it easier to follow the essentials of this topic that has many interesting engineering applications. The complete investigation of invariance theory is not possible in this exploratory study. Instead, this section is prepared in order to bring to the attentions of our readers those results which have potential applications in meeting uncertainties and in conjunction with the whole body of the current work reported herein. The vast research documented in this area makes it clear that a brief familiarization with this topic is inevitable for a comprehensive research in sensitivity theory. The author is indebted to Professor T.J. Higgins who initially suggested this inclusion. Since the purpose of invariance theory is also to provide techniques that reduce the undesirable effects of [external] disturbances, we study this topic in this chapter together with other sensitivity reduction methodologies.

Invariance theory differs from other branches of automatic control theory in compensating for the effects of external disturbances, shown by f(t), on the forced components of system variable (or on functions or functionals of these forced components), by introducing into the control law constraints with respect to disturbances which are measured directly or indirectly [*J*. 9].

The most important aspect and perhaps the most appealing feature of invariance theory is that, in general, we do not know the functional relationship of f(t) characterizing the action of external disturbance on the controlled object nor do we know the statistical characteristics of that disturbance or for that matter we do not model this external disturbance statistically. We only assume that the disturbance is bounded in magnitude and we want to know the worst-possible effect of such external disturbance on the underlying system, in other words, the time variation of f(t) is not so important as is its absolute value or magnitude. Therefore this worst-case analysis (in spirit very similar to that in other sections of this chapter) is only a limiting condition which we expect the system should tolerate. The philosophy behind this concept is that since every control system has only a certain finite (or bounded) energy resource, the problem of compensating for the disturbances makes sense only when $|f(t)| \leq$ constant (i.e., its L_∞-norm is bounded). According to [*J*. 9], this situation represents the point of convergence between invariance theory and optimum system theory. For instance, to achieve the invariance of a quantity is the ideal situation from the invariance study point of view, but under the joint action of disturbances the system elements change and to maintain them within certain bounds requires a typical optimum control analysis. Thus we see that to have a high-performance control system requires a judicious blending of these two methodologies.

The ultimate goal of this theory is to provide synthesis procedures in order that the encountered system will have zero-trajectory error under the effects of external disturbances. The list of problems which can benefit from this research area has a long established history: starting perhaps with the old windmill, and ending with the landing approach of a space vehicle with a prescribed landing point in a disturbed surrounding or atmosphere. All these cases involve compensations for unknown external disturbances or loads. Here the basic requirement is to achieve zero-trajectory error, or in the second example to obtain zero error in final condition (or a specific point) of trajectory.

In scanning the past we notice the following remarks which show how strongly some of our colleagues feel about these matters: "Invariance is the property of automatic control systems that consists of the independence (invariance) of forced motions of one or several controllable coordinates from the variation of the external perturbations that act on the system. The theory of invariance of automatic control systems was proposed in 1938 by Professor G.V. Shchipanov, but, like genetics, it was undeservedly subjected to sharp censure with charges of nonrealizability and idealism. However, rehabilitated by the works of Academicians V.S. Kulebakin, B.N. Petrov, N.N. Luzin, B.A. Ryabov, A.I. Kukhtenko, A.G. Ivakhnenko and many other scientists, it acquired a firm foundation and has served as the topic of a discovery recognized only after the death of its author [*I*. 12]."

6.10.2. The Preliminary Analysis

We first review the Shchipanov's work [*I*. 8], that is now referred to as *strong invariance,* in order to become familiar with the basic ideas in this area. We should point out that the symbols which are used in references of § 6.12.*J* are not often consistent, and certainly not with ours. Thus we choose those symbols which do not contradict our earlier definitions. For instance, in many of these references "p" is used as a time-domain differentiator, while we use "p" as a vector of physical parameters. Therefore here we choose instead $D \equiv d/dt$. On the other hand, in [*I*. 8] and many other subsequent papers in this area, ϕ is used as a "plant coordinate" or a "process parameter," which in effect refers to our output, or perhaps state variable x, but we temporarily use this ϕ as is. [In this regard, our motivation has been to minimize the transitional efforts for those readers who wish to consult some of the articles in § 6.12.*J*.] Next, we proceed to present a set of equations that describe both system and regulator. By introducing generalized symbols for differentiation we can extract a fundamental and general part of automatic regulation problem. Here we must determine a symbolic equation for plant or process and one for the regulator, then we must find out how to interconnect these equations with all the pertinent parameters. When this task is completed, we can determine the necessary relationships among these variables for invariance, and ultimately we must be able to synthesize the regulators (controllers). To that effect,

6.10. The Theory of Invariance

we need to select numerical values for all the coefficients and proper devices to meet the design objectives.

At this juncture, and because we are only concerned with the mathematical development of this theory, we do not elaborate on the realization issues. Certainly an interested reader may consult § 6.12.*I* and § 6.12.*J*, for instance [*I*. 8] as a starter, for a few examples of elements for structural synthesis of regulators. To appreciate, however, the frame of reference in which [*I*. 8] was developed, we look into the following two types of servomotors:

(*i*) Servomotors whose coordinates are subject to change as a function of regulated parameter.

(*ii*) Servomotors whose coordinates do not change as a function of other system coordinates.

In symbolic form, as shown by [·], the equation of the regulated parameter for these two servomotors is:

$$[JD^2]\phi + [MD^2 + L_m D]\sigma \triangleq \Phi\phi + S_\phi \sigma = 0; \qquad (6.10\text{-}1)$$

where, $\Phi\phi$ is called *"free action,"* and $S_\phi \sigma$ is the *"action"* of the servomotor. (In general, most of these symbolic operators shown by [·] are of second-order polynomials in D, corresponding to typical mass (or inertial), friction (or dissipative), and spring (or quasi-elastic) properties of the control object). In other words, these operators correspond to forces that are associated with acceleration-, velocity-, and position-type force.) For a servomotor of type (*i*), $S\sigma$ + (actions of other elements of regulator) = 0, while for a servomotor of type (*ii*): $S\sigma = F(t)$, where $S = S_\phi$, and F(t) is a function of time that determines independently from the other regulator coordinates the ratio of change in σ, in other words, a mechanism will be activated to counteract the disturbance.

∎

PROBLEM 6.10-1 [I. 8] (Regulation of A Single Parameter): How do we compensate for the influence of disturbing forces on the regulated parameter, or in other words, how do we maintain the desired operating conditions of the plant, under the influence of disturbing forces?

SOLUTION [I. 8]: The solution to this problem is the main thrust of this analysis and is the most difficult question to answer. Depending on the desired degrees of complexity for the solution to this problem, we look very briefly into three proposed sub-problems in the following.

But first we study some general ideas. The generalized symbolic method for presenting systems yields the following equation for the regulated parameter:

$$\Phi\phi + [\text{action of the regulator (servomotor)}] = f(t), \qquad (6.10\text{-}2)$$

where f(t) is the disturbing force. It is proposed in [*I*. 8] that compensation problem can be obtained from the following equation:

$$\Phi'\phi = 0, \tag{6.10-3}$$

where Φ' (cf., (6.10-6)), is a new symbolic coefficient which must satisfy the stability condition. In order to convert (6.10-2) into (6.10-3) it is suggested that the regulator must exert a force that is opposite to f(t). That is

$$\Phi\phi = f(t) - (\text{action of the servomotor}) = f(t) - F(t) \approx 0. \tag{6.10-4}$$

Therefore the proposed method of solution is to find out how F(t) must be generated, in order to counter the effects of f(t) which is often unknown. To that effect, and depending on the desired degrees of complexity for the regulator, various solutions to this problem are proposed in [*I. 8*] which correspond to the following three sub-problems. We again emphasize that in these analyses, the constraints on the control signal or phase coordinates are not taken into account, nor is the system noise.

SUB-PROBLEM 6.10-1a (One-Degree-of-Freedom Regulator) (One-DOFR): The symbolic equations for a regulated parameter with *One-DOFR*, and in a matrix notation, that we have adopted herein for the sake of brevity and space, are:

$$\begin{bmatrix} \Phi & T_\phi \\ \Phi_\theta & T \end{bmatrix} \begin{Bmatrix} \phi \\ \theta \end{Bmatrix} = \begin{Bmatrix} f(t) \\ 0 \end{Bmatrix}. \tag{6.10-5a}$$

Here, $T_\phi \theta$ is the "*action*" of regulator on the regulated parameter, and $\Phi_\theta \phi$ is the "*action*" of the regulated parameter on regulator, and f(t) is the external disturbance. As mentioned above, these symbolic operators shown by [·] are at most second-order polynomials in D. From the above equation we have

$$\begin{Bmatrix} \phi \\ \theta \end{Bmatrix} = \frac{1}{\det \begin{bmatrix} \Phi & T_\phi \\ \Phi_\theta & T \end{bmatrix}} \begin{bmatrix} T & -T_\phi \\ -\Phi_\theta & \Phi \end{bmatrix} \begin{Bmatrix} f(t) \\ 0 \end{Bmatrix} = \frac{1}{\det \begin{bmatrix} \Phi & T_\phi \\ \Phi_\theta & T \end{bmatrix}} \begin{Bmatrix} T \\ \Phi_\theta \end{Bmatrix} f(t). \tag{6.10-5b}$$

An obvious condition which results in full compensation for the influence of disturbing force on the regulated parameter is when $T [\triangleq JD^2 + L_m D + L] \equiv 0$, for all D. Here J is the moment of inertia [of mechanical regulator such as a bell-crank] (or as appropriate mass M), L_m is the coefficient of a damping-dissipative force, and L is the coefficient of a position-type force [corresponding to a linear spring]. Thus for fully compensating the influence of disturbing force when using *One-DOFR*, it is necessary to have $T \equiv 0$, for all D, i.e., $L = 0$, $L_m = 0$, and $J = 0$. Resulting in:

$$\Phi_\theta T_\phi \phi \triangleq \Phi'\phi = 0,$$

$$\Phi_\theta T_\phi \theta = \Phi_\theta f(t). \tag{6.10-6}$$

Here, the above coefficients are $\Phi_\theta \triangleq AD^2 + BD + C$ and $T_\phi =$ a constant.

6.10. The Theory of Invariance

Substituting these values in (6.10-6) yields two second-order differential equations for ϕ and θ: ϕ changes as free oscillation depending on our choices for A, B, C; and θ follows f(t) accordingly. However, meeting the above conditions is an unrealistic task (for instance, $J \neq 0$), and thus an "ideal and universal" solution, as was desired by Shchipanov, is not realizable with *One-DOFR*. In other words, an ideal oscillation for ϕ and a pattern for θ that follows changes in f(t) will not occur because the assumptions which we have made are not realizable. Further analysis reveals that in this case there is a conflict between stability and complete compensation for external disturbances.

Equation (6.10-5) is clearly a set of coupled [differential] equations and eliminating either one of the two variables (ϕ, θ) yields a corresponding fourth-order differential equation for each one of these two variables. There are quite a few clever suggestions in [*I*. 8] on how to solve this equation for several special cases. For instance, how to choose elements of a regulator in order that a particular pattern for oscillation frequencies in $D^2\phi$ and/or $D^2\theta$ will occur. Here by [differential] we mean, of course, upon applying the symbolic operators in D to time-varying functions whose initial conditions are assumed to be zero.

SUB-PROBLEM 6.10-1b (Two-Degrees-of-Freedom Regulator) (Two-DOFR): A design with *Two-DOFR* can be constructively implemented in different ways. Here we let one degree of freedom in the regulator correspond to a servomotor (the coordinate σ) and the other degree to a mechanical regulator such as a bell-crank, or to a pendulum (the coordinate ρ). Then the set of augmented [differential] equations of regulator and regulated parameter becomes

$$\begin{bmatrix} \Phi & R_\phi & S_\phi \\ \Phi_\rho & R & S_\rho \\ \Phi_\sigma & R_\sigma & S \end{bmatrix} \begin{Bmatrix} \phi \\ \rho \\ \sigma \end{Bmatrix} = \begin{Bmatrix} f(t) \\ 0 \\ 0 \end{Bmatrix}. \qquad (6.10\text{-}7)$$

Here, $R_\phi\rho$ and $S_\phi\sigma$ are "*actions*" of mechanical regulator (bell-crank) and the servomotor on the regulated parameter; $S_\rho\sigma$ and $\Phi_\rho\phi$ are "*actions*" of servomotor and the regulated parameter on the bell-crank; and $R_\sigma\rho$ and $\Phi_\sigma\phi$ are "*actions*" of bell-crank and the regulated parameter on the servomotor. Each one of these "*actions*", in general, consists of three components, or forces corresponding to mass, friction, and spring. From (6.10-7) we have:

$$\begin{Bmatrix} \phi \\ \rho \\ \sigma \end{Bmatrix} = \frac{1}{\det \begin{bmatrix} \Phi & R_\phi & S_\phi \\ \Phi_\rho & R & S_\rho \\ \Phi_\sigma & R_\sigma & S \end{bmatrix}} \text{Adj} \begin{bmatrix} \Phi & R_\phi & S_\phi \\ \Phi_\rho & R & S_\rho \\ \Phi_\sigma & R_\sigma & S \end{bmatrix} \begin{Bmatrix} f(t) \\ 0 \\ 0 \end{Bmatrix}. \qquad (6.10\text{-}8)$$

Clearly, the solution to (6.10-8) requires only determination of the first column in the Adj (adjoint) matrix. To compensate for the influence of disturbing force on the regulated parameter, it is necessary, to let the (1, 1) entry of the Adj matrix be equal to zero, i.e., $\Delta_{11}(D) \equiv 0$, by letting all the coefficients accompanying symbols for differentiation "D" to be equal to zero. Hence, for a *Two-DOFR*, the equation of compensation, i.e., $\Delta_{11}(D) \equiv 0$, becomes

$$\det \begin{bmatrix} R & S_\rho \\ R_\sigma & S \end{bmatrix} = 0. \qquad (6.10\text{-}9)$$

We note that these operators are polynomials in D and we need not to transpose (6.10-9) for computing this determinant.

Equation (6.10-9) is the fundamental design and calculational equation. Since all the elements in this determinant are related to characteristic equation of the regulator, we have freedom in selecting them as we please, in order to improve stability and/or other aspects of our control design, and to accommodate perhaps other design criteria.

As in the previous case, here we can consider (6.10-9) and eliminate various terms in order to simplify the design, as we have done so in generating (6.10-6), resulting in equations that can be further simplified by proper choices for design elements or variables, but we leave the further pursuit of this issue to an interested reader.

Upon an elaborate analysis in [*I*. 8], it has been concluded that an "ideal and universal" regulator can be developed herein if the friction influences are ignored. Since such an assumption cannot be met in practice, the final conclusion is that a *Two-DOFR* also needs further improvement, and it cannot be accepted as an "ideal and universal" regulator.

SUB-PROBLEM 6.10-1c (Three-Degrees-of-Freedom Regulator) (Three-DOFR): A two-level bell-crank, or a double pendulum, or an elliptic pendulum, together with a servomotor establish a base for a *Three-DOFR*. A *Three-DOFR* reduces the stagnancy, independent from friction in the servomotor and control devices. Denoting the first coordinate of our regulator by ρ, the second one by θ, and the coordinate of our servomotor by σ, we obtain the following set of equations:

$$\begin{bmatrix} \Phi & 0 & 0 & S_\phi \\ \Phi_\rho & R & T_\rho & S_\rho \\ \Phi_\theta & R_\theta & T & S_\theta \\ 0 & R_\sigma & T_\sigma & S \end{bmatrix} \begin{Bmatrix} \phi \\ \rho \\ \theta \\ \sigma \end{Bmatrix} = \begin{Bmatrix} f(t) \\ 0 \\ 0 \\ 0 \end{Bmatrix}. \qquad (6.10\text{-}10)$$

Proceeding in the same manner as before, we conclude that the condition for invariance is that the $\Delta_{11}(D)$ of the adjoint matrix (corresponding to the coefficient matrix in (6.10-10)) must be identically equal to zero. The remaining analyses are similar to the previous two cases, namely, by letting $\Delta_{11}(D) \equiv 0$, and eliminating

6.10. The Theory of Invariance

variables we have equations which are similar to (6.10-6) and these equations can be used to find various regulators' parameters. In this regard we must simultaneously consider system stability and compensation for the influence of external disturbances. We can show that in this case of *Three-DOFR*, we are very close to having an "ideal and universal" regulator.

The preceding analyses are developed for system with a "single parameter" and it has been suggested that this approach can be easily generalized to systems with multiple parameters and complicated structure [*I*. 8]. This trend of analyses has been pursued and advanced by many other scientists. For instance, in [*I*. 12], it is shown that in a linear treatment of invariance problem, an automatic control system can be described by a system of augmented [differential] equations $A(D)x(t) = f(t)$, where $x(t)$ and $f(t)$ are column vectors of system parameters and the perturbations, respectively. Note that herein we assume $f(t) \triangleq [f_1(t), 0, ..., 0]^T$. A necessary and sufficient condition, for example, for $x_1(t)$ to be independent from the external perturbation $f_1(t)$ is that the corresponding $\Delta_{11}(D) \equiv 0$. This is called the invariance condition.

The question of the physical realizability in a system which satisfies the invariance condition is at the heart of this theory. In many cases the preceding [absolute] invariance condition (zero initial conditions) cannot be realized exactly, but only up to some quantity ε. The main drawback is that the invariance condition may be in conflict with the system stability requirements.

Certainly these issues, which some of them seem so true now, and some of them are still very puzzling, constitute the basis for a research discipline in control theory that is maturing very rapidly.

6.10.3. The Basic Steps in the Development of Invariance Theory

6.10.3.1. Introduction

Since the publication of Shchipanov's paper, many papers have appeared, mainly in the Russian literature, which primarily discuss the invariance theory in one form or another. In this section we are initially concentrating on the three survey papers which are written by Kukhtenko [*J*. 30] who is a prominent contributor in this area, and occasionally we incorporate other information from different sources which we find useful.

Consider the system of Sub-Problem 6.10-1b which has three variables $x \triangleq [\phi, \rho, \sigma]$ as follows:

$$\Phi\phi + R_\phi\rho + S_\phi\sigma = f(t), \text{ the equation for control object;}$$

$$\left.\begin{array}{l}\Phi_\rho\phi + R\rho + S_\rho\sigma = 0 \\ \Phi_\sigma\phi + R_\sigma\rho + S\sigma = 0\end{array}\right\} \quad \begin{array}{l}\text{equations for a regulator whose motion} \\ \text{has two degrees of freedom,}\end{array} \quad (6.10\text{-}11)$$

where, Φ [$\triangleq$ MD^2 + LD + k], R [$\triangleq$ mD^2 + lD + k], *etc.*, are all second-order operators in D ≡ d/dt; ϕ is the variable that is being regulated, and ρ and σ are coordinates characterizing a two-degrees-of-freedom regulator *(Two-DOFR)*. Here again our task is to design a regulator which compensates for external disturbances and guarantees system stability.

As discussed in [*I*. 8], to improve dynamical properties of an automatic control system, we need to increase its dimension by adding various controllers. In other words, to attain a better performance than before we must increase the complexity of a dynamical system. For instance, we start with *One-DOFR*, then we advance to *Two-DOFR,* and subsequently to *Three-DOFR,* and in each case we evaluate the overall system performance, in order to make sure that the final choice meets our objectives. We know from Sub-Problem 6.10-1b that a *Two-DOFR* is not sufficient to meet all of our demands. On the other hand, the Sub-Problem 6.10-1c shows that the use of a *Three-DOFR* is sufficient to ensure stability and to compensate for external disturbance f(t) acting on the system. In general, the main question is how many extra equations (or links) and of what structures must be added to the original system in order to meet a particular dynamical behavior.

Since the problem of compensating for disturbances is studied under the assumption that $|f(t)| \leq$ constant, it was in this sense that Shchipanov spoke of constructing an *"ideal and universal"* regulator capable of controlling an object, for instance the one which is described by (6.10-11) with a *Two-DOFR,* and in the case of arbitrarily changing disturbances which are by assumptions bounded.

As we show in § 6.10.2, if we choose the coefficients such that (6.10-9) is satisfied, then the regulated parameter ϕ does not depend on f(t). According to [*J*. 30] condition (6.10-9) is known as Shchipanov-Luzin invariance criterion. Of course, this condition makes the force response to vanish, but the transient response vanishes if the initial conditions are zero and when both of these responses vanish we have *absolute invariance*. We should also point out that in certain applications we may wish *selectively* to eliminate the effects of disturbances on certain regulated variable, but we should not expect to have the complete independence of all such variables from the effects of all disturbances.

6.10.3.2. Partial Compensation for Disturbances and Conditions for Invariance up to ε, and Realizability Condition

The invariance condition of (6.10-9) yields $RS - S_\rho R_\sigma \equiv 0$. Substituting for each one of these operators the corresponding polynomial in D yields.

6.10. The Theory of Invariance

$$\varepsilon_0 D^4 + \varepsilon_1 D^3 + \varepsilon_2 D^2 + \varepsilon_3 D + \varepsilon_4 \equiv 0. \qquad (6.10\text{-}12)$$

Here, each ε_i can be expressed in terms of system parameters as we have used to define the associated operators. The condition for "*complete*" compensation for ϕ with respect to f(t) is when each and every ε_i's vanishes. The condition for "*partial*" compensation is defined accordingly by letting, for instance $\varepsilon_4 \equiv 0$, or ε_4, and $\varepsilon_3 \equiv 0$. In these situations the higher-order derivatives remain uncompensated. Clearly if ε_i's are not identically zero, which is a typical situation in practice, then the invariance conditions must be reexamined.

To generalize the preceding discussions in Sub-Problem 6.10-1c to an augmented system with m [differential] equations and m variables, results in a corresponding equation of the form $A(D)x(t) = f(t)$, where $A(D)$ is an $m \times m$ matrix whose entries are operators that are at most second-order polynomials in D, and $x(t)$ is an m-vector of system variables with $f(t) \triangleq [f_1(t), 0, ..., 0]^T$ is the external disturbance. To establish invariance for $x_1(t)$ with respect to $f_1(t)$, the same steps as before lead us to $\Delta_{11}(D) \equiv 0$. The invariance to some small quantity ε is referred to a case that simultaneously for each and every *stable* solution $[x_1(t), ..., x_m(t)]^T$, there correspond ξ_i's, $i = 1, ..., m$, such that $|\xi_i(t) - x_i(t)| < \varepsilon$, for $t > T$ (a prespecified real number) and $[\xi_1(t), ..., \xi_m(t)]^T$ is the only extremal bounded solution when $f_1(t)$ and its first $2(m-1)$ derivatives on $(-\infty, \infty)$ are continuous and bounded by $\pm M$, and M is a finite constant [J. 30]. For additional information on how to maintain the ε-invariant and/or how to estimate the actual ε when the mathematical modeling of the system does not seem accurate refer to [J. 55].

In this section we group *partial* compensation versus *complete* compensation; and invariance up to ε in one place. The latter can be also called *partial* invariance in analogy with the former. We note that when we have a complete compensation we mean in terms of the forced component. When we have invariance in terms of both forced and transient components we have *absolute invariance*, which is possible only for zero initial conditions. Thus it is imperative that we become specific on the type of invariance that we are seeking.

In the earlier research attentions were given to the measurement of deviations in the single-regulated parameter (the Polzunov-Watt principle). Subsequently, attentions were given to both the deviations in the regulated parameter and the external disturbances acting on the system (combination of the Polzunov-Watt principle and Poncelet principle). Clearly, for stability we must look into both types of deviations since design which are based on the Polzunov-Watt principle *alone* may not be stable. [As these ideas are now well-known facts in the adaptive control theory literature.] Clearly, when conditions for complete invariance are met, then small-parameter variations can be tolerated. This line of research is referred to as invariance in *combined control systems* [J. 30], also known as *control by deviation and by perturbation* [J. 64].

According to [J. 9], B.N. Petrov (cf., [I. 1]) has introduced the so-called "*two-channel principle*," which means that an invariant system (linear or nonlinear) must contain at least two channels for propagation of disturbance from the point of entry to the point where controlled variable is measured. This principle is used to find the structure of both linear and nonlinear invariant systems, and is used to state certain realizability conditions for these schemes. No difficulties will arise when solving invariance problem for a system whose invariance conditions include the object parameters, since there is no contradiction between requirements which are posed by invariance and stability conditions; in other words, the system is physically realizable. This is, of course, an obvious advantage of using "*combined control systems,*" which uses more information about system uncertainties than otherwise. On the other hand, in this case the realization complexity of invariance conditions which requires an on-line information about system perturbations prohibits its wide applicabilities. In order to create a second channel, we use an indirect measurement of perturbations with a "fork" that is formed by a negative and a positive feedback connected towards each other and encompassing our control object. When invariance conditions do not include the object parameters, then their realization requires special tuning of loops which do not include the object parameters, but as a rule is accompanied by the necessity of using additional means to ensure stability. Here, the synthesis of an invariant system is possible in the absence of information about changes in both structure and object parameters [*I.* 12].

It is to be noted that Shchipanov's original system is also referred to as a single-channel invariant system. The realization of this system is further impaired by parasitic pure delays which always exist in so many different kinds of hardware which are used for its implementation. In [*J.* 60 to 62], the robustness issues associated with the realization of absolute invariance are extensively studied. These references study various questions which will naturally arise when we will not meet the invariance conditions and their consequences. A major aspect of this study is to guarantee system stability despite many practical considerations in design of a system.

We now proceed to give a summary of procedures which are suggested in [*J.* 30] in order to construct an invariant control system.

STEP 1: The choice of control scheme depending on our knowledge about the physics of the underlying system must be made.

STEP 2: The mathematical conditions for invariance as defined appropriately (such as complete invariance, or up to ε, *etc.*) must be established.

STEP 3: The realizability condition of results developed in Step 2 must be studied. In particular, we must make sure that these conditions do not violate other system criteria.

STEP 4: Implementation of the final results with all the other pertinent justifications as deemed necessary, for instance cost, must be sought.

6.10.3.3. Extensions to Multidimensional Systems

For a system with several object parameters and a total of m variables, the set of augmented [differential] equations becomes $A(D)x(t) = f(t)$, where $x(t)$ is an m-vector of system variables and $f(t)$ is an m-vector of external disturbances with perhaps a fewer nonzero entries than $x(t)$. The structure of a typical $A(D)$ is as follows [J. 31].

$$\begin{bmatrix} \text{Control object} & \vdots & \text{Action of the regulator on the object} \\ \cdots\cdots\cdots\cdots\cdots\cdots & & \cdots\cdots\cdots\cdots\cdots\cdots \\ \text{Action of the object on the regulator} & \vdots & \text{Regulator (measuring part + servoderiver)} \end{bmatrix}. \qquad (6.10\text{-}13)$$

In this analysis we have assumed that all initial conditions are zero, therefore $A(D)x(t) = f(t)$ can be also written as $A(D)x(D) = f(D)$, where now D becomes the Laplace variable. (When this equation is written in operator form, then this will be called the K(D)–transform of disturbance [J. 16].) Depending on the number of regulated parameters, we have possibly that many [or less] external disturbances. Therefore the nonzero entries of $f(t)$ are usually those which are associated with the number of regulated parameters. As stated in § 6.10.3.1 we should not expect to eliminate completely the effects of every disturbance on every regulated parameter. It is very instructive to investigate the maximum number of independent disturbances that a system can tolerate versus the dimension of that system. We assume that this investigation requires an analysis which is similar to that of § 4.3.2, or § 4.4.2.

Note that the number of variables is chosen as m, and the degree of the corresponding differential equation for each variable is chosen as n_m, and this degree may be different for each variable. From the original set of augmented [differential] equations we have $x(t) = \{A(D)\}^{-1}f(t)$. The procedure to find the inverse of $A(D)$ is the same as before, and depending on the number of external disturbances we may have: $\{\det[A(D)]\}x_1(t) = \Delta_{11}(D)f_1(t) + \Delta_{12}(D)f_2(t) + \ldots$ In order that an invariance of $x_1(t)$ with respect to the external disturbances takes place we need to let the right-hand side of the previous equation be identically equal to zero, but that equation must be carefully analyzed before a final conclusion is made. For instance, since there is no unique way of expanding a determinant, we should expect to have different ways of generating the intermediate steps for reaching the final invariance conditions. However, these conditions must not contradict the overall stability condition of the system. Here, we recall that system stability is related to the $\det[A(D)]$ and this determinant is related to the above $\Delta_{ij}(D)$, but this relationship should not be taken literally. In [J. 31] a system with two regulated parameters and a *Two-DOFR* with two servomotors is studied.

As a result of this specially structured A(D), a slightly simple set of invariance conditions is generated that is more of an art than a science. But the key problem is to meet simultaneously both conditions for invariance and stability, in order that the invariance conditions be realizable. Here the "*two-channel principle*" of Petrov simplifies the realization for the invariance conditions.

Extensions of these topics to: tracking systems, resulting in an invariant with respect to input control signal of the tracking system error; and invariance in discrete-time systems are also discussed in [*J*. 31].

REMARKS 6.10-1: *(i)* Shortly before the appearance of [*I*. 8] and in a very interesting paper, Boychuk [*J*. 6] reexamines some of the issues that are discussed in the Shchipanov's original work [*J*. 58]. In the following we give a brief account of these matters.

(ii) The argument in [*J*. 6] starts with a simple example of a natural system entailing three variables and three different inputs which are acting at different points of the system and are being activated, one at a time, as a unit step. It is shown that in such a dynamical system, the influence of a disturbance does not disappear, but rather shifts from one sub-system to another and based on this observation the absolute invariance in a linear system is being reexamined. The proposed set of augmented equations for the (closed) system is similar to (6.10-13) (in the Laplace transform settings for functions whose initial conditions are zero). In this regard, and to give a point of view for one typical (6.10-13), consider

$$\begin{bmatrix} a_{11}(D) & a_u^T(D) \\ a_x(D) & A_{11}(D) \end{bmatrix} x(D) = \begin{Bmatrix} f_1(D) \\ 0 \end{Bmatrix}, \quad (6.10\text{-}14)$$

where $a_u^T(D) \triangleq [a_{12}(D), ..., a_{1m}(D)]$ is called the control operator, and $a_x^T(D) \triangleq [a_{21}(D), ..., a_{m1}(D)]$ is called the feedback operator. The matrix $A_{11}(D)$ is called the interaction matrix (also called the buffer block of the regulator). Here, the first equation in (6.10-14) can be written as $a_{11}(D)x_1(D) - u_1(D) \triangleq a_{11}(D)x_1(D) + a_u^T(D)\bar{x}(D) = f(D)$, where $\bar{x}(D) \triangleq [x_2(D), ..., x_m(D)]^T$.

From our earlier analysis we know that $x_1(D) = \{\Delta_{11}(D)/\Delta(D)\}f_1(D)$, where $\Delta_{11}(D) = \det[A_{11}(D)]$, and $\Delta(D) = a_{11}(D)\Delta_{11}(D) + \Delta_0(D)$, with $\Delta_0(D)$ indicates the remaining terms. For invariance we need to have $\Delta_{11}(D) \equiv 0$, for all D. This confirms that we must have at least two equations (links). Therefore the necessary and sufficient conditions for invariance are: $\Delta_{11}(D) \equiv 0$, resulting in $\deg[\Delta(D)] = \deg[\Delta_0(D)]$, which to meet the stability requirement we need to have $\text{Re}\{\Delta(D)\} < 0$, and $\text{Re}\{\Delta_0(D)\} < 0$. In other words, both $\Delta(D)$ and $\Delta_0(D)$ must be Hurwitz polynomials.

Because of these conditions the change in control becomes $u_1(D) \triangleq -a_u^T(D)\bar{x}(D) = -\sum_{i=2}^m a_{1i}(D)x_i(D) = -\sum_{i=2}^m \{a_{1i}(D)\{\Delta_{1i}(D)\}/\Delta(D)\}f_1(D)$. Using the fact that $\Delta(D) = a_{11}(D)\Delta_{11}(D) + \Delta_0(D)$, we have

6.10. The Theory of Invariance

$u_1(D) = -\{\Delta_0(D)/[a_{11}(D)\Delta_{11}(D) + \Delta_0(D)]\}f_1(D) = -f_1(D)$, when the invariance condition is met, i.e., $\Delta_{11}(D) \equiv 0$. This relation means that the input which is generated by the regulator produces a strategy that opposes the disturbance. On the other hand, from $A_{11}(D)\bar{x}(D) = 0$, where $\bar{x}(D) = [x_2(D), ..., x_m(D)]^T$, we conclude that $\bar{x}(D) \neq 0$, because $\Delta_{11}(D) \triangleq \det[A_{11}(D)] \equiv 0$, and that is a sufficient condition for a nontrivial solution for $\bar{x}(D)$. "Consequently, the disturbance has not disappeared at all; its influence has been shifted to other links of the system. With such an interpretation of invariance conditions, one cannot detect any trace of idealism, or the 'perpetual motor,' *etc.*, as were suggested at the first discussion on invariance theory [*J.* 6]."

(iii) There are several other important features of an invariant system that are discussed in [*J.* 6]. For instance, the effect of absolute invariance, on the measurement of the highest derivative. This is an important issue which is used to indicate the instant when disturbance is applied to the system, because in most cases the signal with the highest derivative and the disturbance are related with each other through a finite (nondifferential) relationship. When the invariance is ensured for the phase coordinates $x^{(i)}(t) \equiv 0$, for $i \in [0, n-1]$, then we can show that in most cases (i.e., when $x^{(n)}(t) = F(\dot{x}, ..., x^{(n-1)}(t)) + f(t)$, and $F(0) = 0$) $x^{(n)}(t) = f_1(t)$, i.e., the highest derivative is the earliest system response to a noise. Therefore a sufficient condition for invariance is to have $x^{(n)}(t) \equiv 0$.

(iv) Two general approaches to control synthesis are proposed: (a) the compensation approach, in which, to improve the dynamics, we must immediately compensate for disturbances by measuring them or using a mechanism to interact with the object; and (b) the structural approach, which refers to improving the system dynamics by constructing a rigid structure that ensures the required system properties for wide variations in object parameters. The method of compensation for external disturbances in multidimensional systems, or the two-channel compensation, also uses the Shchipanov's basic condition: however, the corresponding invariance condition contains the object parameters, and that fact makes such a solution become a realization of this compensation approach [*J.* 6].

(v) The applicability of the Shchipanov's scheme depends on two factors: (a) measurement of the highest derivative in the quantity being regulated or the output variable introduced into special links of control scheme; and (b) multiloop control circuit which are ensured by inclusion of supplementary links, i.e., to increase system dimension. The degree to which the condition of absolute invariance is satisfied depends on the accuracy of measuring these derivatives, and on the exactness, or on the modeling accuracy, for links which are introduced. These issues which were the impediment for using Shchipanov's scheme are now manageable to some extent. For instance, the contemporary technology enables us to build differentiators for up to the first two derivatives (up to D^2), however, this fact limits the use of invariance conditions for a very high-order system. Finally we should bear in mind that the amplitude of external disturbances must not be excessively large, in order to maintain the invariance [*J.* 6]. For additional information we refer to [*J.* 15 to 17].

6.10.3.4. Extensions to Nonlinear Systems

The preceding analyses are developed for compensating the external disturbances in linear systems. To construct an invariant nonlinear system is considerably more involved than that of the linear system. In this section we outline some basic aspects of research in this area which are gathered from [J. 32]. These procedures are mostly developed in the Russian literature and thus they may not be easily accessible.

As is the case in most nonlinear problems, to develop a unified theory that encompasses all aspects of invariance theory as applied to this class of systems is a formidable task. The formulation of invariance condition for [any] system is the main issue in this theory, and is more so for a general nonlinear system. However, the available procedures which are developed for nonlinear systems treat only one particular class of such systems at a time. The following verbatim classification of invariance problems in nonlinear systems is only a sample of problems under study [J. 32].

(i) Allowance for Dry [Coulomb] Friction: Consider a system with sign-variable dry friction force nonlinearity and a *Two-DOFR*.

$$A(D)\begin{Bmatrix}\phi \\ \rho \\ \sigma\end{Bmatrix} \triangleq \begin{bmatrix}\Phi & 0 & S_\phi \\ \Phi_\rho & R & S_\rho \\ 0 & R_\sigma & S\end{bmatrix}\begin{Bmatrix}\phi \\ \rho \\ \sigma\end{Bmatrix} = \begin{Bmatrix}f(t) \\ \xi F_\rho \\ \eta F_\sigma\end{Bmatrix}, \qquad (6.10\text{-}15)$$

Here the symbolic operators are the same as in (6.10-7) (also note that here $R_\phi \equiv 0$ and $\Phi_\sigma \equiv 0$); F_ρ is a dry friction force toward the coordinate ρ, and F_σ is that toward σ; and ξ and η indicate the sign-variability of friction forces.

The proposed methods for solving these equations are slightly different from that given for (6.10-7). If we use that approach, however, one of the conditions for invariance of ϕ with respect to $f(t)$ becomes the same as (6.10-9) resulting in $\{\det[A(D)]\}\phi = S_\phi(\xi F_\rho R_\sigma - \eta F_\sigma R)$. But this condition is not acceptable because the sign changes in ξ and η are not accounted for. The case for *Three-DOFR* also suffers the same way. One proposal has been to introduce a new variable μ and to let the dry friction be modeled by $\{-\mu\xi\,\text{sign}\dot\rho\}$ or $\{-\mu\eta\,\text{sign}\dot\sigma\}$ (here 'sign' refers to sign function and "dot" means d/dt). For $\mu = 0$, the system becomes linear and the solution is easy to find. Then it is proposed to use this solution with an averaging method for a differential equation that has a small parameter (cf., [5A. 64]), in order to find the solution for the original nonlinear system. The final conclusion, however, is that no satisfactory solution for invariance of this class of nonlinear systems is available.

(ii) Nonlinear Invariant Systems with Combined Control Principle: This topic is discussed based upon specific references which are in Russian, and their discussions are apparently for a special class of nonlinear systems where the

corresponding equations can be linearized section by section over the entire operating conditions and subject to additional constraints on the type of nonlinearity. When these constraints which are too numerous to count here are met, the complete removal of regulation error is possible. However, the problem of compensating for an external disturbance in this setting of nonlinear systems has not been satisfactorily solved.

(iii) Quasi-invariant Systems: This concept refers to real situations where conditions for complete invariance cannot be met, partly due to many approximations which we usually make in analysis and system modeling, and partly because of the constructional defects in various regulators and measuring devices. In this regard, quasi-invariance is a broader concept than invariance up to ε. All systems that are ε invariant are quasi-invariant, but the converse is not true.

(iv) Methods of Compensating for Nonlinearity: This is yet another set of specialized nonlinear problems that are grouped together under this title and can be studied by using a set of ideas on how to counteract for the effects of nonlinearities; and how to measure the deviations of invariance conditions in order to respond to these changes accordingly. A major part of this trend of studies is to reduce the analysis of a nonlinear system to an equivalent set of linear systems in the sense of attaining invariance with respect to external disturbances. However, this linearization approach has its own inherent limitations.

(v) Use of the Method of Harmonic Linearizations: In this case the behavior of a nonlinear system subject to a sinusoidal disturbance is studied.

(vi) Criteria for Invariance for Nonlinear System: Under this group of procedures, it is suggested that in certain class of nonlinear differential equations the so-called disturbance "u(t)" can be replaced by "$\alpha u(t)$", $0 \leq \alpha \leq 1$. Assuming that the corresponding solution depends continuously on α, then for a regulated parameter $x_1(t)$ to be invariant with respect to "u(t)", it is necessary and sufficient that $\partial x_1(t)/\partial \alpha \equiv 0$ for all t and α. This idea has led some researchers to seek decomposition for the corresponding set of differential equations, into two or more subsets, such that the external disturbance appears only in certain subset of these equations. Therefore the principal question is how to determine conditions such that this decomposition for a set of coupled differential equations is feasible and that is the subject of future research.

For another approach to study invariance in nonlinear systems we refer to the following papers that are based on the variational method of the next section [*J.* 52], and [*J.* 49, 59, 63, 68, and 72].

6.10.4. Variational Approach to Invariance Problem

6.10.4.1. Introduction

In § 6.10.3 we study the classical approaches of invariance theory, based upon the three pertinent survey papers which are written by Kukhtenko [*J.* 30 to 32]. In the following we review the Rozonoer's variational approach [*J*. 51 & 52] and several subsequent papers which are stemming from this methodology.

The invariance problem that has been studied in [*J*. 51 & 52] has the same premise as Problem 6.10-1, namely, we must establish a set of conditions in order that the external disturbances do not affect the behavior of a *stable* system (stability is, of course, the underlying issue). The methodology and scope of this approach is, however, different from the preceding one. In developing this variational approach, Rozonoer has also extensively used his three classical papers [*2D*. 13 to 15] which appeared shortly before [*2D*. 4]. We now present the Rozonoer's variational approach for invariance problems.

PROBLEM 6.10-2 (Rozonoer [J. 51]): Consider the following set of continuous nonlinear differential equations:

$$\dot{x}(t) = f(x(t), u(t), t; x_o), \quad x(0) = x_o. \qquad (6.10\text{-}16)$$

Here, $x(t) \in R^n(t)$ is system state vector, $u(t) \in R^m(t)$ is external action; $f(\cdot)$ is a continuous function of its arguments and it has continuous partial derivatives with respect to its arguments, and t is the real variable or time. The initial condition x_o is assumed to be fixed throughout the analyses. Given a functional of $x(t)$, such as

$$J(t) \equiv \phi(x(t), t), \qquad (6.10\text{-}17)$$

then it is required finding conditions on f such that J does not depend on the external action u(t). [Here the trajectory remains a function of u(t) and changes accordingly when u(t) changes and ϕ is temporarily chosen as a scalar.]

SOLUTION: The above problem as stated is very general, therefore we present solutions only for some specific cases of (6.10-16) and (6.10-17) in the subsequent sub-problems.

But first we review the following definition preceded by some general comments.

DEFINITION 6.10-1 (Rozonoer [J. 51]): If the above J(t) = 0 for several t's in [0, T], then this is called "*weak invariance*," and if $J(t) \equiv 0$ for all t's in [0, T], then this is called "*strong invariance*." [The latter is also called invariance in the sense of Shchipanov or "*complete invariance*".]

COMMENT 6.10-1: Given $x(t) \in R^n$ as system state vector, a list of functionals which are mostly used in automatic control theory is

$$J(t) = x_k(t), \qquad (6.10\text{-}18a)$$

$$J(t) = \int_0^t g(x(\tau))\,d\tau, \qquad (6.10\text{-}18b)$$

$$J(t) = h(x(t)), \qquad (6.10\text{-}18c)$$

6.10. The Theory of Invariance

$$J(t) = \sum_{i=1}^{n} c_i x_i(t) = c^T x(t), \qquad c_i\text{'s are constant.} \tag{6.10-18d}$$

The Shchipanov's work is related to the *complete invariance* of functionals which are of the type (6.10-18a).

COMMENT 6.10-2: Suppose that u(t) in (6.10-16) is scalar and is changed to u(t) + Δu(t), then x(t) $\to$ x(t) + Δx(t) and as a result of that J(t) becomes J(t) + ΔJ(t). [Here it is assumed that u(t) + Δu(t) remain piecewise continuous functions of time. Otherwise, we must consider generalized functions when we have discontinuity or jump in system input or trajectory.] The objective of invariance theory is to make ΔJ(t) (as a function of Δu(t)) vanishes in the sense of Definition 6.10-1. That means for the weak invariance we must have ΔJ(t) = 0, for some t = t_i, and for the strong invariance we must have

$$\Delta J(t) \equiv 0, \text{ for } t \in [0, T], \text{ and for all } \Delta u(t). \tag{6.10-19}$$

This condition is analogous to having $\Delta J \geq 0$ (or $\Delta J \leq 0$), which is a criterion for the minimum (or the maximum) of J(t). Therefore the study of invariance in an automatic control system may be accomplished by the same methods which are used for extremization of a functional; i.e., the techniques from Calculus of Variations. In fact, the invariance theory that has been developed within the framework of control theory attempts to ensure that the investigated performance indices is indeed independent from disturbances during the entire system operation. Because system performance such as (6.10-18d) is related to the highest derivative of a functional with respect to the phase-coordinates of the plant which must not depend on disturbances, i.e., this derivative must be either identically zero or independent from disturbances. We also note that (6.10-18a) to (6.10-18c) can be cast in the form of (6.10-18d).

COMMENT 6.10-3: The first step toward establishing an invariance condition for a system quantity is generating its large-sensitivity measure with respect to extreme changes in external disturbance and this is a nontrivial task. In this regard, the role of sensitivity functions generation presented in Chapter Three becomes as follows. Very simply stated, when we face a situation that an external disturbance changes by a large amount resulting in certain "output" to change by a large amount, the theory of invariance seeks ways of controlling the undesirable effects of this large-output change. But this large-output change can also be modeled by (or resulted from) certain parameter variations [internal disturbances] multiplied by system state. If so, then these sensitivity functions can be utilized only for cases wherein the corresponding Taylor-series expansion of original signal provides sufficient information about the incremental changes that are appearing in system output. However, for those cases where the system parameter variations are not "small", and/or the external stimuli disturb the system by a large amount, then we may have to use some other techniques in order to generate the large-sensitivity measure of our system. This determination is the first step toward accomplishing the main objective of invariance theory which is the compensation for incremental changes in system output caused by the external disturbances. As stated in § 6.10.3.2, when conditions for complete invariance are met, the small-parameter variations can be tolerated, otherwise the system ceases to be invariant. Therefore the parametric insensitivity that is responsible for elimination of the undesirable effects resulted from the [internal] disturbance, and the fulfillment of the invariance conditions which eliminates the undesirable effects resulted from the external disturbance in a system that is subject to either or both [internal] and [external] disturbances, are complementing each other. Since we model the effects of these two disturbances by an additive input to the set of differential equations which characterizes our system, we may use similar analysis to design controllers for our system that is subject to both types of disturbances. This explanation should clarify the relation between sensitivity and invariance analyses.

COMMENT 6.10-4: As far as ΔJ's corresponding to (6.10-18) are concerned, the Rozonoer's main contribution has been to develop a method to compute this large variation in terms of other system quantities [J. 51 & 52]. This analysis also depends on [2D. 13 to 15] and is as follows.

Consider (6.10-18d) and let $x(t) \to x(t) + \Delta x(t)$, then $J(t) \to J(t) + \Delta J(t)$, where

$$\Delta J(T) = \sum_{i=1}^{n} c_i \Delta x_i(T) = - \int_0^T [H(x, \lambda, u+\Delta u, t) - H(x, \lambda, u, t)]dt - \eta. \quad (6.10\text{-}20)$$

Here $H(x, \lambda, u, t) = \lambda^T(t) f(x(t), u(t), t) \triangleq <\lambda, f>$ (an inner product), and $\lambda(t)$ satisfies the following differential equation:

$$\dot{\lambda}(t) = -(\frac{\partial H}{\partial x})^T, \quad \lambda(T) = -[c_1, \cdots, c_n]^T, \quad (6.10\text{-}21)$$

and η is a residual term which is bounded as follows.

$$|\eta| \leq \alpha (\int_0^T |\Delta u(\tau)| d\tau)^{\frac{1}{2}}, \quad \alpha > 0 \text{ is a constant.} \quad (6.10\text{-}22)$$

6.10.4.2. Rozonoer's Analysis for Linear System

In the following specific sub-problem we specialize the preceding general relationships to the corresponding linear system, in order to demonstrate a typical approach for solving an invariance problem.

SUB-PROBLEM 6.10-2a (Invariance in DCLTI – GS): Consider a DCLTI – GS as follows.

$$\dot{x}(t) = Ax(t) + bu(t), \quad x(0) = x_o, \quad (6.10\text{-}23a)$$

$$J(t) = c^T x(t). \quad (6.10\text{-}23b)$$

Here A is a stable matrix and u(t) represents the external disturbance which is

6.10. The Theory of Invariance

assumed as a scalar quantity. Determine the necessary and sufficient conditions for establishing the corresponding invariance requirements in (6.10-23).

SOLUTION: The first step toward establishing the invariance condition is to generate the corresponding large-sensitivity measure $\Delta J(t)$ (cf., (6.10-20)). In this linear case the residual term η vanishes identically and

$$\Delta J(T) = -\int_0^T [\lambda^T(Ax + bu + b\Delta u) - \lambda^T(Ax + bu)]dt$$

$$= -\int_0^T \lambda^T(t)b\,\Delta u(t)dt \triangleq -\int_0^T \langle\lambda, b\rangle \Delta u(t)dt, \quad (6.10\text{-}24)$$

with

$$\dot\lambda(t) = -A^T\lambda(t), \quad \lambda(T) = -c. \quad (6.10\text{-}25)$$

The condition for *weak invariance* is satisfied if and only if

$$\langle\lambda(t), b\rangle = 0, \quad \text{for } t \in [0, T]. \quad (6.10\text{-}26)$$

Certainly, when (6.10-26) is met, from (6.10-24) we have $\Delta J(T) = 0$. On the other hand, when $\Delta J = 0$, then (6.10-26) must be satisfied, otherwise by continuity of $\langle\lambda(t), b\rangle$, we can show that if $\langle\lambda, b\rangle$ is not zero, then either $(\langle\lambda, b\rangle) > 0$, or $(\langle\lambda, b\rangle) < 0$, and in both cases there will be a contradiction. Therefore (6.10-26) is a necessary and sufficient condition for *weak invariance*.

Rozonoer has shown that (6.10-26) can be cast in terms of a set of conditions involving the coefficient matrices A, b and c as follows. Let $\zeta_a(t) \triangleq \langle\lambda(t), b\rangle$, then we can show that

$$\dot\zeta_a(t) = \langle\dot\lambda(t), b\rangle = \langle-A^T\lambda(t), b\rangle = -\langle\lambda(t), Ab\rangle. \quad (6.10\text{-}27)$$

Similarly, we can show that for the first $(n-1)$ derivatives we have

$$\zeta_a^{(k)}(t) = (-1)^k\langle\lambda(t), A^kb\rangle, \quad k = 0, 1, \dots, n-1. \quad (6.10\text{-}28)$$

Because of $\zeta_a(t) = 0$, for $t \in [0, T]$, then that is the same as letting $\zeta_a^k(t) = 0$ for all k in that interval. In particular, when $t = T$, $\lambda(T) = -c$ and we have

$$\zeta_a^{(k)}(T) = (-1)^k\langle\lambda(T), A^kb\rangle = (-1)^{k+1}\langle c, A^kb\rangle \equiv 0, \quad (6.10\text{-}29)$$

resulting in the following necessary conditions for *weak invariance*. (Recall that $c \neq 0$.) We can also show that these are sufficient conditions as well.

$$\langle c, A^kb\rangle \equiv 0, \quad k = 0, 1, \dots, n-1. \quad (6.10\text{-}30)$$

Equation (6.10-30) can be written as follows.

$$\langle A^{T^k}c, b\rangle \equiv 0, \quad k = 0, 1, \dots, n-1. \quad (6.10\text{-}31)$$

▼

We can easily prove that the above necessary and sufficient conditions for *weak invariance*, are also the necessary and sufficient conditions for *strong invariance*. Because when strong invariance holds, the weak invariance is guaranteed necessarily; and on the other hand, when (6.10-30) or (6.10-26) is identically equal to zero, by the fundamental lemma from the Calculus of Variations that is a sufficient condition for $\Delta J \equiv 0$ for all t.

REMARKS 6.10-2: *(i)* It is straightforward to show that for the system of (6.10-23) and when $J(t) = c^T x(t) \triangleq <c, x(t)>$ is invariant (i.e., it is independent from disturbance u(t) and in fact its behavior is determined by x(0)), then J(t) satisfies an sth-order differential equation, where $s \leq n-1$. The same can be stated for the above $\zeta_a(t)$, with the boundary conditions (6.10-29).

(ii) In a similar analysis and for the case where the external disturbance is modeled by the output of a linear system with known parameters and unknown initial conditions some results are reported in [*J. 67*], however, these results are heavily depended on the measurability of the external disturbance and/or information about its structure.

EXAMPLE 6.10-1 (A comparative study of invariance techniques [J. 51]): Consider the system of Sub-Problem 6.10-2a, where

$$\frac{d}{dt}\begin{Bmatrix}x_1(t)\\x_2(t)\end{Bmatrix} = \begin{bmatrix}a_{11} & a_{12}\\a_{21} & a_{21}\end{bmatrix}\begin{Bmatrix}x_1(t)\\x_2(t)\end{Bmatrix} + \begin{Bmatrix}b_1\\b_2\end{Bmatrix}u(t), \quad \begin{Bmatrix}x_1(0)\\x_2(0)\end{Bmatrix} \equiv 0, \quad (6.10\text{-}32a)$$

$$J(t) \triangleq c^T x(t) = [c_1, c_2]\begin{Bmatrix}x_1(t)\\x_2(t)\end{Bmatrix}. \quad (6.10\text{-}32b)$$

Determine the invariance conditions for J(t), using both the method of Sub-Problem 6.10-2a and that of § 6.10.3.3.

SOLUTION: From (6.10-30) and for n = 2, we have the invariance conditions as follows.

$$<c, b> = 0, \quad <c, Ab> = 0. \quad (6.10\text{-}33)$$

Substituting for A, b, c, from (6.10-32) into (6.10-33) results in

$$c_1 b_1 + c_2 b_2 = 0, \quad (6.10\text{-}34)$$

$$c_1(a_{11}b_1 + a_{12}b_2) + c_2(a_{21}b_1 + a_{22}b_2) = 0. \quad (6.10\text{-}35)$$

On the other hand, using the method of § 6.10.3.3 results in

$$\begin{bmatrix}D - a_{11} & -a_{12}\\-a_{21} & D - a_{22}\end{bmatrix}x(D) = \begin{Bmatrix}b_1\\b_2\end{Bmatrix}u(D). \quad (6.10\text{-}36)$$

From (6.10-36) we have

6.10. The Theory of Invariance

$$\begin{Bmatrix} x_1(D) \\ x_2(D) \end{Bmatrix} = \frac{1}{\Delta(D)} \begin{Bmatrix} (D - a_{22})b_1 + a_{12}b_2 \\ a_{21}b_1 + (D - a_{11})b_2 \end{Bmatrix} u(D). \quad (6.10\text{-}37)$$

Substituting these values in the performance index results in:

$$J(D) = \frac{(c_1 b_1 + c_2 b_2)D^1 + [c_1(a_{12}b_2 - a_{22}b_1) + c_2(a_{21}b_1 - a_{11}b_2)]D^0}{\Delta(D)} u(D). \quad (6.10\text{-}38)$$

For invariance we let the coefficients of D^1 and D^0 be identically equal to zero, resulting in the same condition as (6.10-34) corresponding to D^1 and the following condition corresponding to D^0:

$$c_1(a_{12}b_2 - a_{22}b_1) + c_2(a_{21}b_1 - a_{11}b_2) \equiv 0. \quad (6.10\text{-}39)$$

Using the facts that $a_{22}c_1 b_1 + a_{22}c_2 b_2 = 0$, and $a_{12}c_1 b_1 + a_{12}c_2 b_2 = 0$, (6.10-39) becomes the same as (6.10-35). Thus as expected under the same set of constraints both approaches result in the same invariance conditions when applied to a given problem.

▼

REMARK 6.10-3: For the case wherein $u(t)$ is not scalar, then (6.10-23a) becomes

$$\dot{x}(t) = Ax(t) + Bu(t), \quad x(0) = x_o, \quad (6.10\text{-}40a)$$

The corresponding invariance conditions for this system with (6.10-23b) are

$$<c, A^k B> \equiv 0, \quad k = 0, 1, \ldots, n-1. \quad (6.10\text{-}40b)$$

6.10.4.3. Extensions of Rozonoer's Analysis for Linear System

Shortly after the introduction of Rozonoer's variational approach several papers have appeared which relate the controllability and observability properties of the underlying system to that of invariance properties. In this regard, Wang [J. 72] is the first who has suggested this relationship for a general dynamical system; followed by Cruz and Perkins [J. 11 & 12] whose work depends on the above results of Rozonoer, and Wang, and generalizes the work by Rutman and Epelman [J. 54]. Here the output becomes a vector, thus the corresponding ϕ is not scalar.

Consider a linear time-invariant system described by

$$\dot{x}(t) = A_1 x(t) + Bu(t) + Ny(t), \quad (6.10\text{-}41)$$

$$z(t) = Cx(t), \quad (6.10\text{-}42)$$

where $x(t) \in R^n(t)$ is system state-vector, $u(t) \in R^m(t)$ is input vector, $z(t) \in R^\omega(t)$ is observation vector, and $y(t) \in R^v(t)$ is external disturbance. A_1, B, C, and N are constant matrices of appropriate dimensions. Suppose that we use a feedback control law such as

$$u(t) = F_1 x(t) + F_2 z(t), \tag{6.10-43}$$

then

$$\dot{x}(t) = (A_1 + BF_1 + BF_2 C)x(t) + Ny(t)$$

$$\triangleq Ax(t) + Ny(t). \tag{6.10-44}$$

DEFINITION 6.10-2 (Cruz and Perkins [J. 12]): An output component $z_k(t)$ is said to be *signal invariant* with respect to a disturbance signal $y_l(t)$ if the zero-state response z_k is identically zero (or does not depend on y_l) for all $y_l \in \{y_l \mid \int_0^\infty |y_l(t)|dt < \infty\}$ and for all $t \geq 0$.

DEFINITION 6.10-3 (Cruz and Perkins [J. 12]): If none of the output components depend on any of the disturbance signals, then it is said that $z(t)$ is *totally signal invariant* with respect to $y(t)$.

■

Here Definition 6.10-2, is equivalent to the uncontrollability defined by Wang [*J.* 72], and Definition 6.10-3 is equivalent to output uncontrollability defined by Wang and strong invariance defined by Rozonoer [*J.* 51].

In Sub-Problem 6.10-2a, our functional, which is the performance index $J(t)$, is chosen as a scalar quantity. In (6.10-42), our functional is the output which is a vector. Therefore here the previous results are extended to cover this situation.

THEOREM 6.10-1 (Cruz and Perkins [J. 12]): A necessary and sufficient condition for system described by (6.10-42) and (6.10-44) to be totally signal invariant with respect to $y(t)$ is that

$$CA^k N \equiv 0, \quad \text{for } k = 0, 1, \ldots, n-1. \tag{6.10-45}$$

■

Equation (6.10-45) implies that for $N \neq 0$ the system must be unobservable, in order to meet the invariance condition.

Consider the case wherein A in (6.10-44), becomes $A + \Delta A$, due to parameter variations. Then the following definition is in order.

DEFINITION 6.10-4 (Cruz and Perkins [J. 12]): The system (6.10-42) and (6.10-44) is *totally parameter invariant* with respect to a parameter change ΔA and for all initial states $x(0) \in \{x(0) \mid \sum_{i=1}^n x_i^2(0) < \infty\}$, if the zero-input responses of the unperturbed and the perturbed systems are equal for all $t \geq 0$.

THEOREM 6.10-2 (Cruz and Perkins [J. 12]): A necessary and sufficient condition that the zero-input response of system (6.10-44) be totally parameter invariant with respect to a perturbation ΔA is that

6.10. The Theory of Invariance

$$CA^k \Delta A \equiv 0, \quad \text{for} \quad k = 0, 1, \ldots, n-1. \tag{6.10-46}$$

∎

Equation (6.10-46) implies that for $\Delta A \neq 0$ the system is unobservable. It is to be noted that this ΔA may take any one of the following forms: ΔA_1, ΔB, ΔF_1, ΔF_2, and ΔC, or a combination of these perturbations.

REMARK 6.10-4: We recall that all the preceding parameter variations are considered to be *static* variations in this section. Therefore it has been in this spirit that we have grouped some of these references from § 6.12.J, in § 3.7.G.

∎

Considering that any linear system such as $\dot{x}(t) = Ax(t) + Bu(t)$, $x(0) = x_o$ can be written as $\dot{x}(t) = Ax(t) + (B, \ x(0))\begin{Bmatrix} u(t) \\ \delta(t) \end{Bmatrix}$, with zero initial condition, where the dimension of input vector is increased by one to augment the initial conditions to the system input, through the Dirac function $\delta(t)$. Therefore we may replace $CA^k N \equiv 0$ by $CA^k \Delta x(0) \equiv 0$, as the condition for invariant under perturbations in the initial conditions.

DEFINITION 6.10-5 (Cruz and Perkins [J. 12]): The system (6.10-42) and (6.8-44) is *simultaneously totally invariant* with respect to y(t) and perturbation ΔA if for all x(0) and all y(t) the complete response of the perturbed system with disturbance input is equal to the response of the unperturbed system with no disturbance input and for all $t \geq 0$.

THEOREM 6.10-3 (Cruz and Perkins [J. 12]): System (6.10-42) and (6.10-44) is simultaneously invariant to y(t) and ΔA if and only if it is individually invariant with respect to y(t) and ΔA as in Theorems 6.10-1 & 2.

∎

If the system exhibit simultaneous invariance, then it is necessarily required be unobservable, and columns of both ΔA and N must be in the unobservable subspace of the original system. This clarification pertinent to these results of Cruz and Perkins is given by Fam and Meditch [J. 14], wherein it is mentioned that the input-output invariance in linear systems under parametric perturbations can be made related to the canonical decomposition of that system into: a controllable and observable part; a controllable and unobservable part; an uncontrollable and observable part; and an uncontrollable and unobservable part. Neither can these requirements be generally achieved in practice, nor can the simultaneous occurrence of signal and parameter-invariance.

6.10.4.4. Extensions to Linear Time-Varying Systems

In this section we specialize Problem 6.10-2 to the following sub-problem.

SUB-PROBLEM 6.10-2b (Invariance in Time-Varying System): Consider the following system which is the time-varying counterpart of system in Sub-Problem 6.10-2a.

$$\dot{x}(t) = A(t)x(t) + b(t)u(t), \quad x(0) = x_o, \quad (6.10\text{-}47a)$$

$$J(t) = c^T(t)x(t). \quad (6.10\text{-}47b)$$

Determine the necessary and sufficient conditions, in order to establish the corresponding invariance conditions for (6.10-47).

SOLUTION: The residual term η in the corresponding (6.10-20) is zero, and the associated ΔJ becomes as follows.

$$\Delta J(T) = -\int_0^T \lambda^T(\tau)b(\tau)\Delta u(\tau)d\tau \triangleq -\int_0^T <\lambda(\tau), b(\tau)>\Delta u(\tau)d\tau, \quad (6.10\text{-}48)$$

$$\dot{\lambda}(t) = -A^T(t)\lambda(t), \quad \lambda(T) = -c. \quad (6.10\text{-}49)$$

Here, also the condition for *weak invariance* is satisfied if and only if

$$\zeta_b(t) \triangleq <\lambda(t), b(t)> \equiv 0, \quad \text{for } t \in [0, T]. \quad (6.10\text{-}50)$$

Differentiating $\zeta_b(t)$ and using (6.10-49) yields

$$\dot{\zeta}_b(t) = <\dot{\lambda}(t), b(t)> + <\lambda(t), \dot{b}(t)>$$

$$= <\lambda(t), \dot{b}(t) - A(t)b(t)> \triangleq <\lambda(t), Q(t)b(t)>, \quad (6.10\text{-}51)$$

where the new operator $Q(t) \triangleq ID - A(t)$ (I is an identity matrix and $D \equiv d/dt$). Keeping in mind the noncommutativeness of ID and $A(t)$, we can show that

$$\zeta_b^{(k)}(t) = <\lambda(t), Q^k(t)b(t)>, \quad k = 0, 1, \ldots, n-1. \quad (6.10\text{-}52)$$

From (6.10-50) and (6.10-52) and for $t \in [0, T]$, we have

$$<\lambda(t), Q^k(t)b(t)> \equiv 0 \quad k = 0, 1, \ldots, n-1. \quad (6.10\text{-}53)$$

At $t = T$, $\lambda(T) = -c(T)$ and (6.10-53) yields

$$<c(T), [Q^k(t)b(t)]|_{t=T}> \equiv 0, \quad k = 0, 1, \ldots, n-1. \quad (6.10\text{-}54)$$

Obviously, (6.10-53) shows that vectors $b, Qb, \ldots, Q^{n-1}b$ are linearly dependent, i.e., there are a set of parameters $\{\alpha_k(t)\}$ which do not vanish simultaneously for all t such that $\sum_{k=0}^{n-1}\alpha_k(t)Q^k(t)b(t) = 0$. Therefore (6.10-53) with (6.10-54) are the necessary and conditions for *weak invariance*. It is easy to show that for the same set of $\{\alpha_k(t)\}$, $\sum_{k=0}^{n-1}\alpha_k(t)\zeta_b^{(k)}(t) = 0$, which together with (6.10-54), yields $\zeta_b(t) \equiv 0$, and thus (6.10-50) is satisfied.

We can show that the above necessary and sufficient conditions for *weak invariance,* are also the necessary and sufficient conditions for *strong invariance*. In other words, the following is the necessary and sufficient conditions for *strong invariance,* of Sub-Problem 6.10-2b.

6.10. The Theory of Invariance

$$\langle c(t), Q^k(t)b(t)\rangle \equiv 0, \qquad k = 0, 1, ..., n-1. \qquad (6.10\text{-}55)$$

▼

REMARKS 6.10-5: *(i)* The system of Sub-Problem 6.10-2b is similar to a standard linear time-varying system with $z(t) = c^T(t)x(t)$, instead of (6.10-47b), as its output. Using this type system with u(t) as its input (instead of external disturbance) Wang [*J.* 72] has indicated that the notions of controllability and observability can be interpreted in the variational framework. He has extended (6.10-55) to the corresponding condition for multi-input multi-output systems, where vectors c(t) and b(t) are replaced by appropriate matrices.

(ii) Extensions of the results presented in § 6.10.4.3. to discrete-time systems and that in § 6.10.4.4. to systems with additional constraints on the phase-coordinates in order to belong to a specified set at certain fixed points in time are discussed in [*J.* 56 and 57].

6.11. BOUNDED UNCERTAINTY AND DISTURBANCE REJECTION

6.11.1. Introduction

In the preceding sections we study methodologies to meet *Cases 2 & 3* of Table 6.1-1. In particular, in Sections 6.9 and 6.10 we consider systems subject to exogenous signals whose boundedness are implied although within different metrics. In this section we study issues which are along the same line, but we allow plant parameters to change and that inclusion brings us to *Case 4* of Table 6.1-1. However, due to generality of this case we review only systems which are subject to a special structured disturbance and a special class of exogenous signals as described subsequently.

For a class of nonlinear ordinary differential equations whose uncertainties are modeled by additive inputs, some powerful techniques stemming from the Lyapunov theory can be provided which enables us to design robust controllers that maintain system operation in spite of such uncertainties. In this approach we confront a set of differential equations with discontinuous right-hand sides, thus the solution of state equation may not exist in the classical sense. The underlying assumption is to model uncertainty as a deterministic quantity whose value belongs to a known and compact set. There are a host of techniques from various interdisciplinary areas of sciences to study these kinds of problems. We present only a very brief review of certain activities in control community and limited to pertinent references in § 6.12.*K1*. This section ends with additional remarks on disturbance rejection techniques corresponding to the references in § 6.12.*K2*.

6.11.2. Deterministic Approaches to Meet Uncertainty

Extensions of Lyapunov's second method to dynamical systems that are subject to uncertainties and modeled in a deterministic manner are presented. In this regard, the early work of Gutman [K1. 33] is our leading source. The organization and presentation of materials in this section closely follows this source and its references. Here we use a set of hybrid notation from these sources which is compatible with our earlier definitions.

Consider a dynamical system represented by

$$\dot{x}(t) = f(x,t) + B(x,t)u(t), \quad x(t_0) = x_o, \quad (6.11\text{-}1)$$

where $x(t) \in R^n(t)$ is system state, $u(t) \in R^m(t)$ is system control input, and both $f(\cdot)$ and $B(\cdot)$ are, in general, nonlinear but continuous functions of their arguments with appropriate dimensions. There are many possibilities regarding the anatomy of uncertainties associated with (6.11-1). As our knowledge about these uncertainties improves, we can provide a more accurate model for their propagations into the system than before. In this school of thought, however, we assume that system uncertainty belongs to a known and compact set, therefore we do not model the uncertainty statistically. Yet again, the underlying system must remain asymptotically stable as the uncertainty sweeps through that set. In this regard, "one is led naturally to the notion of 'worst case' design. It is important, perhaps, to stress that: 1) we do *not* suppose that 'nature' is a true adversary and hence will do its worst, and 2) 'worst case' design is not unduly pessimistic, for if the worst circumstances do not arise, the consequent controller will do 'better' under more favorable circumstances [K1. 33]." In the following a particular class of uncertainties (corresponding to system (6.11-1)) and a method of analysis for compensating their undesirable effects is presented. The underlying system corresponds to *Case 4* of Table 6.1-1.

Suppose that uncertainties in (6.11-1) appear as follows.

$$\dot{x} = f(x,t) + \Delta f(x,t,v) + B(x,t)u + \Delta B(x,t,\nu)u + C(x,t)w, \quad x(t_0)=x_o, \quad (6.11\text{-}2a)$$

where (v, ν, w) is a bounded-uncertainty vector (that may depend on t) and belongs to a compact set Ω; $\Delta f(\cdot)$, $\Delta B(\cdot)$ and $C(\cdot)$ are nonlinear but continuous functions of their arguments (here the dependence on t is dropped for notational convenience). We assume that the perturbed nonlinear dynamical system (6.11-2a) corresponds to the nominal or unperturbed system (6.11-1), and may be exactly rewritten as

$$\dot{x} = f(x,t) + B(x,t)u + \Delta f(x,t,v) + \Delta B(x,t,\nu)u + C(x,t)w. \quad (6.11\text{-}2b)$$

It is, of course, clear that $x(t)$ in (6.11-1) and (6.11-2) are not exactly the same. Since from practical point of view and in most dynamical systems we prefer to keep system structure unchanged, we expect that uncertainty will be compensated by changes which can be made in control action $u(t)$. Here we let $u \to u + \eta$, where η becomes responsible for cancelling out the undesirable effects of disturbances. On the other hand, since in most dynamical systems the control action is bounded (finite energy), i.e., $u \in U$, where U is a known and compact set, η must

6.11. Bounded Uncertainty and Disturbance Rejection

also be bounded, which means in this approach we can compensate for a finite disturbance that is bounded in the same manner as is the control action u(t). This thought has led to the following assumptions that if our uncertainties are members of the following class

$$\Delta f(x, t, v) = B(x, t)h(x, t, v), \quad h \in R^m, \tag{6.11-3a}$$

$$\Delta B(x, t, v) = B(x, t)E(x, t, v), \quad E \in R^{m \times m}, \tag{6.11-3b}$$

$$C(x, t) = B(x, t)D(x, t), \quad D \in R^{m \times r}, \tag{6.11-3c}$$

then (6.11-2b) becomes

$$\dot{x} = f(x, t) + B(x, t)(u + \eta), \tag{6.11-4a}$$

where

$$\eta = h(x, t, v) + E(x, t, v)u + D(x, t)w. \tag{6.11-4b}$$

Therefore the perturbed dynamical system (6.11-4) is of the type which can be managed by variations in control action u(t). Here, the assumptions (6.11-3) are called the *matching conditions* [K1. 33]; and η in (6.11-4b) is referred to as "lumped"-uncertain element [K1. 25].

PROBLEM 6.11-1 (Asymptotic Stability in Uncertain Dynamical Systems [K1. 33]): Let x(t) be the solution of (6.11-4) at t which is generated by {u,η}. Consider system (6.11-4) with a norm-bounded uncertainty η, such that $\eta \in V = \{\eta \in R^m : \|\eta\| \le \rho(x,t)\}$, where ρ(·) is continuous on $R^n \times R^1_+$. (The R^1_+ is the same as our R^+). Find a feedback-control policy u(t) = u(x(t), t), $u(\cdot) : R^n \times R^1_+ \to R^m$ such that the origin becomes uniformly asymptotically stable in the large for all e(x, t) = η(t) ∈ V.

CONCEPTUAL SOLUTION: Before presenting a formal solution of Problem 6.11-1, we note certain characteristics of this problem, and anticipate (and soon verify) certain properties of its solution. The uncertainty is modeled by the maximum value of its norm and the type of dynamical systems which we face is such that its input changes as a set-valued function. Indeed, it is well known that a candidate for solution to this problem is a discontinuous function. Thus the evolution of this system is not uniquely determined by its initial condition (x_o, t_0) alone, but rather by (x_o, t_0) and the control action u(t) ∈ U. The implication of uniform asymptotic stability in this problem is that, given any initial condition (x_o, t_0) a corresponding solution always exists which brings the solution to the origin in a finite time and each solution is different from the previous one, that is generated by another input. Hence we seek a *uniform ultimate boundedness* for each solution as these issues are discussed subsequently. It is clear that the treatment of this situation when input changes as a set-valued function requires an especial theoretical foundation. Using the concept of generalized dynamical systems *(GDS's)* in the mathematical literature, Gutman [K1. 33] has established the required mathematical justification for a solution (often discontinuous) of this problem that has been resulted from a specialized application of Lyapunov method [*K1*. 2, 27 & 28, and 1]. In the following we first review a number of issues which are developed in conjunction with generalization of solution for a system of ordinary differential equations to that containing discontinuous inputs, *vis a vis, GDS's*. Subsequently, we return to this solution in § 6.11.2.3.

6.11.2.1. Generalized Dynamical Systems

The notion of ordinary dynamical systems is synonymous with a generalization of solutions for a set of ordinary differential equations *(ODE's)*. A systematic generalization of this notion to dynamical systems with no unique solutions was developed by Barbashin [*K1*. 5] and Roxin [*K1*. 56]. The theoretical foundation for this generalization is, however, based on the early work of Marchaud [*K1*. 48 & 49] and Zaremba [*K1*. 70], during 1934 to 1936. In this period, the theory of *ODE's* was generalized by Marchaud and Zaremba to include systems whose inputs can be changed as set-valued functions. These early results on *ODE's* show how a set of possible tangent directions defining a family of solutions [trajectories] can be used to generalize the notion of a set of *ODE's* to that "with many-valued right-hand side [*K1*. 27]". For instance, in $\dot{x}(t) = f(x, t, u)$, where $u \in U$ is the input function (U is a known and compact set), each solution is determined by (x_0, t_0) *and* u. The set of all possible tangent directions for x(t) is called "*contingent*" and the corresponding differential equation is called "*contingent equation*" [*K1*. 57]. A set of *ODE's* which has no unique solution is called a set of "*generalized dynamical systems,*" *(GDS's)*. This is the same as saying that a contingent equation defines a generalized dynamical system. Marchaud and Zaremba proved that a contingent equation defines a family of solutions [trajectories] and in that sense defines a *GDS,* however, the problem on how a contingent equation defines a set of *GDS's* and issues regarding the existence and uniqueness theorems for the corresponding solution are developed in [*K1*. 57]. In the following we briefly review the underlying concerns.

DEFINITIONS 6.11-1 *(Definition 2.5-3):* (i) The *distance* between two points $x, y \in R^n$ is shown by $d(x, y) = \| x - y \|$.

(ii) The *distance* between a *point* $x \in R^n$ and a *set* $A \subset R^n$ is defined by $d(x, A) \triangleq \inf \{ \| x - a \| : a \in A \}$.

(iii) The *separation* of $A \subset R^n$ from $B \subset R^n$ is defined by $d^*(A, B) \triangleq \sup \{ d(a, B) : a \in A \}$.

(iv) The *distance between two sets* A and B is defined by $d(A, B) = d(B, A) \triangleq \max \{ d^*(A, B), d^*(B, A) \}$. For compact sets this distance is, of course, finite and it defines the Hausdorff metric.

6.11. Bounded Uncertainty and Disturbance Rejection

(v) The *ε-neighborhood* of a set $A \subset R^n$ is $N_\varepsilon(A) \triangleq \{x \in R^n : d(x, A) < \varepsilon\}$.

(vi) The variable set $E(\alpha) \subset R^n$, where $\alpha \in R^m$, is *continuous* at α_0 if, given any $\varepsilon > 0$, there is a $\delta > 0$ such that $\alpha \in N_\delta(\alpha_0) \Rightarrow d(E(\alpha), E(\alpha_0)) < \varepsilon$.

(vii) The variable set $E(\alpha) \subset R^n$, where $\alpha \in R^m$, is *upper semicontinuous* at α_0 if, given any $\varepsilon > 0$, there is a $\delta > 0$ such that $\alpha \in N_\delta(\alpha_0) \Rightarrow d^*(E(\alpha), E(\alpha_0)) < \varepsilon$.

COMMENT 6.11-1 (Generalized Dynamical Systems [K1. 57 and 33]): An ordinary dynamical system is described by a function $F(x_0, t_0, t) : R^n \times R_+^1 \times R_+^1 \to R^n$, which gives the motion $x(t)$ based on (x_0, t_0). A generalized dynamical system may also be given by a function $F(x_0, t_0, t)$, which describes the movement of a given point $x(t)$ from its initial condition $x_0 = x(t_0)$. This function is called the *attainability function*. As stated before, however, the evolution of a generalized dynamical system is determined by its initial state (x_0, t_0), the time $t > t_0$, and some control action $u \in U$ (U is a known and compact set), and as the control action u *changes* in U, a family of end points $x(t_1)$ can be determined that each corresponds to one control action u.

ASSUMPTIONS 6.11-1 [K1. 57 and 33]: It is assumed that $F(x_0, t_0, t)$ satisfies the following axioms. *(i)* $F(x_0, t_0, t)$ is a closed nonempty subset of R^n, that is defined for every $x \in R^n$ and $t_0 \leq t$.

(ii) $F(x_0, t_0, t_0) = \{x_0\}$ for every x_0, and t_0.

(iii) Semigroup property: for $t_0 \leq t_1 \leq t_2$, $F(x_0, t_0, t_2) = \bigcup F(x_1, t_1, t_2)$, where $x_1 \in F(x_0, t_0, t_1)$.

(iv) Given $x_1 \in R^n$; t_0, t_1, $t_0 \leq t_1$, there exists an $x_0 \in R^n$ such that $x_1 \in F(x_0, t_0, t_1)$.

(v) $F(x_0, t_0, t)$ is continuous in t.

(vi) $F(x_0, t_0, t)$ is upper semicontinuous in (x_0, t_0), uniformly in any finite interval $[t_1, t_2]$.

■

From these axioms, several other properties may be generated which are intuitively apparent and are not repeated herein.

DEFINITION 6.11-2 [K1. 57 and 33]: A mapping $\phi : [t_0, t_1] \to R^n$ such that $t_0 \leq \tau_0 \leq \tau_1 \leq t_1$ implies $\phi(\tau_1) \in F(\phi(\tau_0), \tau_0, \tau_1)$, is called a *motion*. The corresponding curve in the R^n–space is called a *trajectory*.

COMMENTS 6.11-2 [K1. 57]: *(i)* Based on the preceding definition, we can prove that for *GDS's*, if $x_2 \in F(x_1, t_1, t_2)$, then there is a trajectory $\phi(t)$ such that $\phi(t_1) = x_1$ and $\phi(t_2) = x_2$, which means there is a trajectory going from x_1 to x_2 corresponding to $[t_1, t_2]$. Also we can show that $F(x_0, t_0, t)$ can be extended backward in time (for $t < t_0$).

(ii) We can define *GDS's* with *contingent equations* and in turn these equations are defined by *contingent derivatives* as follows. Here we can interpret the contingent in more or less the same way as a set of ordinary tangent lines.

DEFINITIONS 6.11-3 (Contingent [K1. 57 and 33]): *(i)* Let $c = \{x(t): x(\cdot): [t', t''] \to R^n\}$ be a curve defined on $[t', t'']$, and let

$$y(t_i) = \frac{x(t_i) - x(t_0)}{t_i - t_0}, \quad t_0, t_i \in [t', t'']. \quad (6.11\text{-}5)$$

The set of all $y_0 \in R^n$, such that there exists a sequence $\{t_i\}$, $i = 1, 2, 3, \ldots$, $t_i \to t_0$, $t_i \neq t_0$, and $\lim_{i \to \infty} y(t_i) = y_0$, is called the *contingent derivative* of c at $x(t_0)$. At $x(t)$ this derivative is designated by $D^*x(t)$.

(ii) An expression $D^*x \subset E(x, t)$, where the set $E(x, t)$ depends on (x, t) and is defined on $R^n \times R_+^1$, is called a *contingent equation*.

(iii) A *solution* $x(\cdot)$ of $D^*x \subset E(x, t)$ is any curve such that: $x(\cdot)$ is absolutely continuous; and $\dot{x}(t) \in E(x(t), t)$ for almost all $t \in [t', t'']$.

THEOREM 6.11-1 [K1. 70, 57 and 33]: Let $E(x, t) \subset R^n$ be defined on some closed neighborhood $\overline{N}$ of (x_0, t_0) and be compact, convex, and upper semicontinuous in (x, t). Then there exists at least one solution $x = \phi(t)$ to $D^*x \subset E(x, t)$ passing through a given point (x_0, t_0) (i.e., $\phi(t_0) = x_0$), and this solution can be continued until reaching the boundary of $\overline{N}$. ∎

If for every $x \in R^n$ and $t \in R_+^1$, the set $E(x, t)$ is contained in a fixed compact set E_0, i.e., $E(x, t) \subset E_0$, then for every trajectory $x = \phi(t)$ and $t_1, t_2 \in R_+^1$, $\| \phi(t_2) - \phi(t_1) \| \leq k | t_2 - t_1 |$, where $k = \sup \{ \| x \| : x \in E_0 \}$. Here, a finite-escape time is not possible, since for $t \to t_f$ (a finite time) $\phi(t)$ remains finite. Also here the condition $E(x, t) \subset E_0$ can be relaxed to $\sup \{ \| y \| : y \in E(x, t) \} \leq k_1 + k_2 \| x \|$, which becomes a "condition for *finite-escape time*" [K1. 57].

THEOREM 6.11-2 [K1. 57 and 33]: If, in addition, finite-escape time is impossible for $x(t)$, then the function $F(x_0, t_0, t) \subset R^n$ given by

$$\begin{cases} x_1 \in F(x_0, t_0, t_1) \Longleftrightarrow \text{there exists a trajectory } x = \phi(t) \\ \text{of } D^*x \subset E(x, t) \text{ such that } x_0 = \phi(t_0), \ x_1 = \phi(t_1) \end{cases} \quad (6.11\text{-}6)$$

defines a generalized dynamical system. ∎

Extensive discussions on the existence and uniqueness theorems and alternative definitions for contingent in *GDS's* are given in [*K1*. 57].

6.11.2.2. Asymptotic Stability in Generalized Dynamical Systems

The following definitions and theorems provide the necessary mathematical tools to complete a solution of Problem 6.11-1.

DEFINITION 6.11-4 [K1. 58 and 33]: A function $V(\cdot): R^n \times R_+^1 \to R^1$ is *positive definite* if, *(i)* $V(0, t) = 0$ for all $t \geq 0$, and *(ii)* there exists continuous, increasing scalar functions $\gamma(\cdot)$, $\beta(\cdot)$ with $\gamma(0) = 0$, $\beta(0) = 0$, such that for all $t \in R_+^1$ and all $x \in R^n$, $\gamma(|x|) \leq V(x, t) \leq \beta(|x|)$. Similarly $V(\cdot)$ is called *negative definite* if $-V(\cdot)$ is positive definite. ∎

Let us define

$$D^+V(x,t) = \overline{\lim_{\tau \to t^+}} \sup \left\{ \frac{V(y,\tau) - V(x,t)}{\tau - t}; \ y \in F(x,t,\tau) \right\} \quad (6.11\text{-}7)$$

$$D^-V(x,t) = \overline{\lim_{\tau \to t^-}} \sup \left\{ \frac{V(y,\tau) - V(x,t)}{\tau - t}; \ y \in F(x,t,\tau) \right\} \quad (6.11\text{-}8)$$

$$D^0V(x,t) = \max(D^+V(x,t), D^-V(x,t)). \quad (6.11\text{-}9)$$

DEFINITIONS 6.11-5 [K1. 58 and 33]: *(i)* The origin is *strongly (Lyapunov) stable* if, for every $\varepsilon > 0$, $t_0 \geq 0$, there is $\delta(\varepsilon, t_0) > 0$ such that for all $|x_0| < \delta$, and $t > t_0$, $F(x_0, t_0, t) \subset \{x \in R^n : |x| < \varepsilon\}$.

(ii) The origin is *uniformly strongly stable* if, in the above definition, $\delta(\varepsilon)$ is independent from t.

(iii) The origin is *asymptotically strongly stable* if it is strongly stable, and for every $t_0 \geq 0$, there is some $\eta(t_0) > 0$ such that if $|x_0| \leq \eta$ for every motion $\phi(t)$ starting at $\phi(t_0) = x_0$, then $\lim_{t \to \infty} \phi(t) \equiv 0$.

(iv) The origin is *uniformly asymptotically strongly stable* if it is uniformly strongly stable and there is some $\eta > 0$ and a function $\tau(\varepsilon)$ defined for sufficiently small $\varepsilon > 0$ such that if $|x_0| \leq \eta$, and $t_0 \geq 0$, then for every motion $\phi(t)$ starting at $\phi(t_0) = x_0$, $|\phi(t)| < \varepsilon$ holds for all $t \geq t_0 + \tau(\varepsilon)$. For sufficiently large η, the property is *in the large*.

THEOREM 6.11-3 [K1. 58 and 33]: If $V(x,t)$ is a real function defined on $R^n \times R_+^1$, if $V(x,t)$ and $-D^0V(x,t)$ are positive definite, and if $\gamma(|x|) \to \infty$ as $|x| \to \infty$, then the origin is uniformly asymptotically strongly stable in the large.

6.11.2.3. Solution of Problem 6.11-1

We now proceed to give a formal solution of Problem 6.11-1. First we assume that there exists a scalar function $V(\cdot): R^n \times R_+^1 \to R^1$ such that $V(x,t)$ is positive definite according to Definition 6.11-4 with $\gamma(\|x\|) \to \infty$ as $\|x\| \to \infty$, and $W_0(t) \equiv W_0(x,t) = (\partial V/\partial t) + \nabla_x V \cdot f$ is negative definite. Here "·" refers to the inner product. In other words, the existence of this Lyapunov function implies that the free motion $\dot{x} = f(x,t)$ is uniformly asymptotically stable in the large. It is assumed that the free motion of initial system already has this property, and if not, we should be able to stabilize that motion first.

Secondly, we choose the same $V(\cdot)$ of the free motion $\dot{x} = f(x,t)$ as a Lyapunov function for the perturbed system (6.11-4) and *impose* conditions on $\{u, \eta\}$ such that the system remains stable along any solution $x(\cdot)$ as follows.

$$W(t) = \frac{\partial V}{\partial t} + \nabla_x V \cdot \dot{x}$$

$$= \frac{\partial V}{\partial t} + \nabla_x V \cdot f + \nabla_x V \cdot B(u + \eta)$$

$$= W_0(t) + \nabla_x V \cdot B(u + \eta). \tag{6.11-10}$$

The remaining procedure is now standard. If we choose $\{u, \eta\}$ such that

$$\min_{u \in U} \max_{\eta \in V} \nabla_x V \cdot B(u + \eta) \leq 0, \tag{6.11-11}$$

then $V(\cdot)$ decreases along a solution $x(\cdot)$ of (6.11-4). In other words, to make $V(\cdot)$ as negative as possible guarantees system stability. This trend of analyses has laid out a foundation for applications of Lyapunov technique in design of model-reference adaptive control. (Incidentally, on a related matter the reader may find [2E. 9] a very informative source on min-max optimization.) The $u(\cdot)$ that is generated from (6.11-11), however, may be discontinuous, and that means (6.11-4) may not have a solution in the classical sense. This situation can be rectified by letting (6.11-4) to be a contingent equation where the control action $u(\cdot)$ selects values from a set-valued function. This approach was first used by Alimov [K1. 2] to extend Lyapunov stability to a single-relay control problem, and subsequently was extended by Gutman to include multivariable systems with uncertainty.

To proceed further, we let $U = V$, i.e., we assume that the uncertainty can be managed within our resources in control action, thus the above min-max optimization is satisfied if $u(t) = u(x(t), t)$, where

$$u(x,t) = -\rho(x,t) \frac{\alpha(x,t)}{\|\alpha(x,t)\|}, \tag{6.11-12}$$

and

$$\alpha(x,t) \triangleq B^T(x,t) \nabla_x^T V(x,t). \tag{6.11-13}$$

Depending on any given dynamical system (6.11-1), we determine the corresponding $\rho(x,t)$. We must now clarify an issue which has been raised in the literature since the introduction of this procedure, namely, what if $\alpha(x,t) \to 0$? Responding to this question we let

6.11. Bounded Uncertainty and Disturbance Rejection

$$N \triangleq \{(x,t) : \alpha(x,t) = 0\}, \tag{6.11-14}$$

and consider the set-valued function $u^*(\cdot) : R^n \times R_+^1 \to U'$, where U' is a nonempty subset of U, such that

$$u^*(x,t) = \begin{cases} -\rho(x,t) \dfrac{\alpha(x,t)}{\|\alpha(x,t)\|} & \text{for all } (x,t) \notin N \\ \{u \in R^m : \|u\| \le \rho(x,t)\} & \text{for all } (x,t) \in N. \end{cases} \tag{6.11-15}$$

Using (6.11-15) we write (6.11-4) as follows.

$$\dot{x} \in K(x,t) \tag{6.11-16}$$

$$K(x,t) = \{z \in R^n : z = f(x,t) + B(x,t)(u+\eta), \quad u \in u^*(x,t)\}. \tag{6.11-17}$$

To complete the above solution of Problem 6.11-1, the following assumptions are formally made.

ASSUMPTIONS 6.11-2: (i) $\rho(\cdot)$, $v(\cdot)$, $\nu(\cdot)$, $w(\cdot)$ are continuous on $R^n \times R_+^1$.

(ii) $f(0,t) = 0$, for all $t \in R_+^1$.

THEOREM 6.11-4 [K1. 33]: Consider system (6.11-4), and suppose that the Assumptions 6.11-1 & 2 are met. Then the feedback controller $u \in u^*(x,t)$ given by (6.11-13) to (6.11-15) assures uniform asymptotic stability in the large of $x = \{0\}$ for all admissible uncertainties.

PROOF [K1. 33]: By construction $K(x,t)$ is compact and convex for $(x,t) \in R^n \times R_+^1$. Furthermore $K(\cdot)$ is upper semicontinuous on $R^n \times R_+^1$, because outside N, $K(\cdot)$ is continuous. To sketch a proof that $K(\cdot)$ is upper semicontinuous at $(\overline{x}, \overline{t}) \in N$, we let $\phi(\cdot) : R^n \times R_+^1 \to R^n$ be given by

$$\phi(x,t) = f(x,t) + B(x,t)e(x,t), \tag{6.11-18}$$

where $\eta = e(x,t)$. For notational convenience, we also let each $(\cdot)_1$ and $(\overline{\cdot})$ correspond to (x_1, t_1) and $(\overline{x}, \overline{t})$, respectively.

Given $(x_1, t_1) \notin N$, then for a point K_1 in R^n,

$$d^*(K_1, \overline{K}) = d(K_1, \overline{K}) = \inf_{\overline{u} \in U} \|\overline{\phi} - \phi_1 + \overline{B}\,\overline{u} - B_1 u_1^*\|$$

$$\le \|\overline{\phi} - \phi_1\| + \|\overline{\rho}\overline{B} - \rho_1 B_1\|. \tag{6.11-19}$$

Since $\phi(\cdot)$, $e(\cdot)$, and $B(\cdot)$ are continuous, given any $\varepsilon_1, \tilde{\varepsilon}_1 > 0$, there exist δ_1 and $\tilde{\delta}_1$ such that $\|(\overline{x}, \overline{t}) - (x_1, t_1)\| < \delta_1 \Rightarrow \|\overline{\phi} - \phi\| < \varepsilon_1$ and $\|(\overline{x}, \overline{t}) - (x_1, t_1)\| < \tilde{\delta}_1 \Rightarrow \|\overline{\rho}\overline{B} - \rho_1 B_1\| < \tilde{\varepsilon}_1$. Thus for $\varepsilon = \varepsilon_1 + \tilde{\varepsilon}_1$ and $\delta = \min(\delta_1, \tilde{\delta}_1)$ we have $\|(\overline{x}, \overline{t}) - (x_1, t_1)\| < \delta \Rightarrow d^*(K_1, \overline{K}) \le \varepsilon$. For a point $(x_2, t_2) \in N$ we can also show that $d^*(K_2, \overline{K}) < \varepsilon$.

Now, given (x_0, t_0), since there exists a solution for (6.11-17), and because of (6.11-11), $V(\cdot)$ cannot increase along any solution of (6.11-17). Thus the finite-escape time is not possible for $x(\cdot)$. By Theorem 6.11-2, (6.11-17) defines a generalized dynamical solution, and applying Theorem 6.11-3 and using Assumptions 6.11-1 & 2, yields

$$W(t) = \frac{\partial V}{\partial t} + \nabla_x V \cdot \dot{x} = W_0(x,t) + \alpha^T(x,t)(u+\eta). \tag{6.11-20}$$

Upon simplification we have

$$W(t) \begin{cases} = W_0(t) & \text{for all } (x,t) \in N \\ \leq W_0(t) & \text{for all } (x,t) \notin N. \end{cases} \tag{6.11-21}$$

Since all requirements of Theorem 6.11-3 are satisfied, this concludes the proof of Theorem 6.11-4 [K1. 33].

REMARKS 6.11-1 [K1. 33]: *(i)* If (6.11-4b) is replaced by

$$V = \{\eta : \eta^T R^{-1} \eta \leq \rho^2(x,t)\}, \quad R = M^T M > 0, \tag{6.11-22}$$

then (6.11-15) becomes

$$u^*(x,t) = \begin{cases} -\rho(x,t) \dfrac{M\alpha(x,t)}{\|M\alpha(x,t)\|} & \text{for all } (x,t) \notin N \\ \{u \in R^m : u^T R^{-1} u \leq \rho^2(x,t)\} & \text{for all } (x,t) \in N. \end{cases} \tag{6.11-23}$$

(ii) If (6.11-4b) is replaced by

$$V \triangleq \{\eta \in R^m : |\eta_i| \leq \rho_i(x,t)\}, \tag{6.11-24}$$

then, (6.11-15) becomes

$$u_i(x,t) = \begin{cases} -\rho_i(x,t)\, \text{sgn}[\alpha_i(x,t)] & \text{for all } (x,t) \notin N_i \\ \{u_i \in R^1 : |u_i| \leq \rho_i(x,t)\} & \text{for all } (x,t) \in N_i. \end{cases} \tag{6.11-25}$$

The proof of upper semicontinuity is similar to that of norm-bounded uncertainty where

$$\alpha_i(x,t) = b_i^T \nabla_x^T V(x,t), \tag{6.11-26}$$

and b_i's are the columns of B, with

$$N_i = \{(x,t) : \alpha_i(x,t) = 0\}. \tag{6.11-27}$$

The feedback controller is then $u_i \in u_i(x,t)$.

6.11. Bounded Uncertainty and Disturbance Rejection

(iii) If $\rho(x,t)$ in (6.11-15) is replaced by any continuous function $\bar{\rho}(x,t) > \rho(x,t)$, then the previous stability results hold with a faster convergence rate to the origin outside N than before.

COMMENT 6.11-3 (Implementation of Control Policy, [K1. 33]): To close the preceding analysis, which is reported here almost verbatim from [*K1*. 33 and 56 to 58] and their references, we need to elaborate on the actual implementation of a set-valued function such as $u^*(\cdot)$ in (6.11-15). First of all, $u^*(\cdot)$ is unique and smooth for all $(x,t) \notin N$, and in N where discontinuity exists, $u^*(\cdot)$ selects its value from a compact set $\{u: \| u \| \leq \rho\}$. In other words, in N, u is well defined (in the usual sense) in a "ball" whose radius, however, depends on the size of uncertainty η. This uncertain radius creates a new implementation problem for us. Here, since on N, $W(t) < 0$ for any admissible pair $\{u, \eta\}$, in the (x,t)–space the vectogram of $\dot{x}$ at $(\bar{x}, \bar{t}) \in N$ is *always* directed into the $B = \{(x,t): V(x,t) < V(\bar{x}, \bar{t})\}$. So it is suggested by Gutman that if u becomes any *admissible* value, say zero, for *a very "short-time interval"*, then the corresponding Lyapunov function decreases. At the end of that period, if state is outside N, then it continues according to $u^*(\cdot)$ given in (6.11-15); and, on the other hand, if state remains in N at the end of this "short-time interval", then we maintain $u \equiv 0$ for another "short-time interval" until we are out of N. To rectify this situation and in certain circumstances we may utilize an open-loop control called an "injection control" for a "short-time interval", in order to get out of N. Based on this observation Gutman has suggested that $u^*(\cdot)$ in (6.11-15) be replaced by $\hat{u}(\cdot)$ as follows.

$$\hat{u}(x,t) = \begin{cases} -\rho(x,t)\dfrac{\alpha(x,t)}{\| \alpha(x,t) \|} & \text{for all } (x,t) \notin N \\ \text{zero (or any other admissible value) for a} \\ \text{short-time interval once } N \text{ is reached.} \end{cases} \qquad (6.11\text{-}28)$$

This modification is in accordance with the inherent delay which exists in every physical system when performing a discontinuity.

∎

In summary, it is shown how a min-max optimization, in a Lyapunov setting, can be used to develop control policies which stand against large uncertainties in a dynamical system resulted possibly from either exogenous signals and/or internal plant uncertainty. In general, this procedure results in system with discontinuous control action that corresponds to a set of ordinary differential equations whose "right-hand sides are discontinuous". The solution of this set is usually defined by introducing the so-called generalized dynamical systems. To rectify the effects of discontinuity which depend on the unknown uncertainty, it is proposed to use an injection control for a "short-time interval" which coincides with the inherent delay time that exists in any given physical system responding to a discontinuity. Thus the overall system remains asymptotically stable in the usual sense, provided that the above "short-time interval" goes to zero. "Only at that limit can we state that in the worst case, the convergence of the uncertain system to the origin is as fast as the nominal one" [*K1*. 33].

6.11.2.4. Specializations, Applications and Extensions

We briefly describe a glossary of applications for the preceding analytical methodology. But *first* we note that a min-max optimization is at the *heart* of any sensitivity consideration. To many researchers, and quite rightfully, sensitivity consideration means the ability to tackle uncertainties. On the other hand, if we model system uncertainties any other way than what we call *static variations*, then our analysis moves outside the realm of linear, time-invariant systems that we often use to work with. So there is a very fine and delicate boundary which must be observed as when we can or cannot use this analysis.

Secondly, we recall that the preceding analysis is developed based on the existence of *matching conditions* (6.11-3) and in that sense is very specialized.

Thirdly, we should also recall that in the preceding analysis we have not specified the nature of η nor have we elaborated on $f(x,t)$, although we have suggested that this situation corresponds to *Case 4* of Table 6.1-1. The results of Theorem 6.11-4 apply to as general $f(\cdot)$ and η as that described in (6.11-4), but these functions can be further specialized to a few cases of interest which are described in [*K1*. 33] and in the literature § 6.12.*K1*. A few such special cases are reviewed next, and in each case we assume that all pertinent assumptions regarding continuity and boundedness of uncertainty matrices are met. In each situation our main task becomes to determine the corresponding $\rho(x,t)$ (or control policy) based upon system variables.

(i) Systems with parameter and input uncertainties: This case refers to the following situation, where

$$\dot{x} = f(x,t) + B(x,t)[E(x,t,v) + I]u + B(x,t)\eta_1, \qquad (6.11\text{-}29a)$$

$$\eta_1 = h(x,t,v) + D(x,t)w. \qquad (6.11\text{-}29b)$$

Here I is an identity matrix.

(ii) Applications to quasi-linear systems: This situation refers to

$$\begin{cases} f(x,t) = A(t)x, \quad \text{where } \dot{x} = A(t)x \\ \text{is uniformly asymptotically stable.} \\ \Delta f(x,t,v) = \Delta A(x,t,v)x = B(x,t)F(x,t,v)x \\ \quad \Delta B = 0, \quad w = 0. \end{cases} \qquad (6.11\text{-}30)$$

In particular, we may choose $A(t)$ to be a constant matrix.

6.11. Bounded Uncertainty and Disturbance Rejection

(iii) Linear systems with time-varying uncertainties [K1. 41 to 45]: Consider a dynamical system

$$\dot{x}(t) = [A + \Delta A(r(t))]x(t) + [B + \Delta B(s(t))]u(t) + Cv(t), \quad x(0) = x_o. \quad (6.11\text{-}31a)$$

Here, $x(t) \in R^n(t)$ is system state, $u(t) \in R^m(t)$ is system input; $v(t) \in R^m(t)$, $r(t) \in R^\rho$ (t), and $s(t) \in R^\sigma(t)$ are uncertainties vectors affecting system dynamics as indicated in above. Furthermore these uncertainties are belong to known and compact sets; A, B and C are constant parameter matrices, with $\Delta A(r)$ and $\Delta B(s)$ are referring to uncertainties in A and B which are continuous functions of their arguments. We suppose that the measured state is

$$y(t) = x(t) + w(t), \quad (6.11\text{-}31b)$$

where w(t) is the measurement error. Clearly, this system is a special case of (6.11-1), therefore the corresponding analysis of § 6.11.2.3 is applicable to this situation very straightforwardly. However, in [*K1*. 41 to 45, 25 & 26, and 30] additional insights and simplifications are provided which complement our preceding analysis. In the following we briefly outline some of the contributions made in these references.

If we let u = Ky + u(x(t), t), where K is such that A + BK is (or becomes) a stable matrix, then an analysis similar to the preceding section results in a control policy u(x(t), t) in exactly the same manner as before. In this analysis an estimate for the radius of uncertainty is provided and is shown how a measured feedback control which guarantees ultimate boundedness of every system response (within a known neighborhood of zero state), can be found that also assures when every solution enters this set in a finite time remains within this set. The goal in this analysis is to guarantee that all solutions are *ultimately bounded* [*K1*. 43], no matter what the uncertainties are. It is proposed in [*K1*. 43] that the use of saturation control may simplify the theoretical treatment which is suggested in [*K1*. 1] for similar cases.

Subsequently in [*K1*. 44], it is shown that for system (6.11-31) a nonlinear feedback control (which is a nonlinear function of the measured or estimated state) may provide a better performance than a linear controller. (Here we should be cautious not to exceed our fundamental assumptions regarding the nature of *variations* in the underlying system. In order that, we simply avoid the need for establishing a rigorous stability analysis if in effect an adaptive and on-line feedback of measured states are used. Recall that the system is considered to be under large variations.) This control input is generated from a Lyapunov function for the uncertain system. Here the proposed nonlinear feedback approximates arbitrarily closely a minimax control in the absence of state measurement error by guaranteeing that the most negative derivative of the associated Lyapunov function is established. The guaranteed behavior of the system under this nonlinear control is found to be better than that of the corresponding linear control when the state uncertainty is sufficiently small. The standard procedure to make this comparison is to compute an estimate for the radius of uncertainty variations, then it is to be shown that using a nonlinear feedback control will result in a smaller radius.

Finally, we close this group of problems with the following two definitions.

DEFINITION 6.11-5 (Leitmann [K1. 44]): Given a solution $x(\cdot):[t_0,t_1] \to R^n$, $x(t_0) = x_0$, of (6.11-31a), this solution is *uniformly bounded* if there is a positive constant $d(x_0) < \infty$, possibly dependent on x_0 but not on t_0, such that $\|x\| < d(x_0)$, for all $t \in [t_0,t_1]$.

DEFINITION 6.11-6 (Leitmann [K1. 44]): Given a solution $x(\cdot):[t_0,\infty) \to R^n$, $x(t_0) = x_0$, of (6.11-31a), this solution is *uniformly ultimately bounded* with respect to set S if there is a non-negative constant $T(x_0,S) < \infty$, possibly dependent on x_0 and S but not on t_0, such that $x(t) \in S$ for all $t \geq t_0 + T(x_0,S)$.

These definitions can be stated for system (6.11-1) as well.

(iv) Applications in model-reference control [K1. 33]: As a special case of (ii) above, with A matrix constant, we may look at a model-reference control scheme where a real-operating plant and its corresponding model are, respectively, as follows.

$$\dot{\xi} = [A + \Delta A(v)]\xi + B(u+r) \qquad (6.11\text{-}32a)$$

$$\dot{x} = Ax + Br. \qquad (6.11\text{-}32b)$$

Note that v is a vector of uncertainty. If we let $e \triangleq \xi - x$, then

$$\dot{e} = Ae + \Delta A(v)\xi + Bu. \qquad (6.11\text{-}33)$$

Considering the matching conditions $\Delta A(v) = BF(v)$, (6.11-33) becomes

$$\dot{e} = Ae + B[u + F(v)\xi]. \qquad (6.11\text{-}34)$$

A stabilizing control policy for this system is

$$u^*(\xi,t) = \begin{cases} -\max_v \|F(v)\xi\| \dfrac{\alpha(e)}{\|\alpha(e)\|} & \text{for all } (e) \notin N \\ \{u \in R^m: \|u\| \leq \max_v \|F(v)\xi\|\} & \text{for all } (e) \in N, \end{cases} \qquad (6.11\text{-}35)$$

where $\alpha = B^T Pe$, $A^T P + PA + Q = 0$, and $N = \{e: \alpha(e) = 0\}$.

(v) Applications in robotics [G. 13 to 15]: As another application of this methodology, we may consider the model-following adaptive controller design with particular application in robotics. We have extensively applied this approach to design robust adaptive controllers for robot manipulators (cf., [G. 13 to 15] for numerical examples). This method to robustify a system, that is nonlinear and subject to large uncertainties, has been found to be very promising. Here, we propose to replace a state-space representation of the nonlinear system by a *set* of linear, time-invariant systems (provided that the nonlinear system is sampled sufficiently

6.11. Bounded Uncertainty and Disturbance Rejection

fast), then we use the preceding approach to design controllers for each member of this set of linear systems using Lyapunov technique.

Finally, we should point out that one practical way to bypass the set-valued function in (6.11-35), *per se*, is to let $u^*(\xi,t) = -\max_v \| F(v)\xi \| \dfrac{\alpha(e)}{\| \alpha(e) \| + \delta}$, where $\delta > 0$ is specified [K1. 34], and [G. 13 to 15].

(vi) Extensions to systems without matching conditions: One of the underlying assumptions in our preceding analysis has been the occurrence of matching conditions (6.11-3) (or its equivalent for linear systems). In many applications, however, these conditions are not met and the question then becomes what if (6.11-3) does not hold? Here a new measure called the measure of *mismatch* (shown by M) is introduced in [K1. 10], which is used to provide certain tolerance for meeting (6.11-3). In other words, if the norm of M does not exceed some upper bound M^*, then it is still possible to meet the uniform ultimate boundedness of *(iii)* above. The analysis that has led to this determination requires a particular decomposition of uncertainty which if it is met, then it may be possible to have the above M^*, and even to maximize this M^*. However, this analysis requires more information about the nature of system uncertainty than what we can mostly offer. There are several other interesting papers in this area such as [K1. 35, and 51] which should be consulted for different points of views on this topic.

Finally, there is a wealth of results in this area, but due to the scope of this book we suggest that an interested reader consults § 6.12.K1.

6.11.3. Disturbance Rejection

The main thrust of the following discussion is to emphasize the fact that there are several other classes of disturbances studied in the literature. We open this presentation with some excerpts from [K2. 34] in the following footnote, which describes the very essence of this topic[4]. To consider different spaces to describe

---

[4] "One of the main stimuli in the development of control theory has always been the disturbance decoupling (attenuation, isolation, quenching) problem in its many facets. One finds this theme in classical control theory, in feedback stabilization algorithms (keeping the effect of disturbances bounded), in the *LQG* problem (minimizing the L_2-norm of the closed-loop impulse response), and in the H_∞-problem (minimizing the H_∞-norm of the closed-loop transfer matrix. The question of when the disturbance can be completely decoupled by feedback control from the to-be-controlled outputs led to the development of geometric control theory. With its important concepts as controlled invariant subspaces, controllability subspaces, (C, A, B)–pairs, *etc.*, this theory led in the late 1970's to a deeper understanding of the intricate fine structure of linear systems. Many of these ideas, suitably generalized, are instrumental for the present developments of nonlinear system theory and the control of discrete-event systems. An interesting generalization of these geometric notions went into the direction of what could be achieved by high gain feedback. This led to the introduction of almost invariant subspaces and a number of interesting disturbance decoupling problems were solved in this connection. However, one important question defied solution up to now: the almost disturbance decoupling with internal stability. Precisely, when will there exist, for a given finite-dimensional linear time-invariant system with control and disturbance

disturbances has already attracted a very active group of researchers who introduce new ways of looking at disturbances and new ways of designing controllers to meet these disturbances. We sample very briefly some outcomes of their work subsequently. To highlight the essence of issues confronting us in any disturbance-rejection problem, we note that upon defining the disturbance and the corresponding sensitivity measure (or performance index), we should minimize [or reject, if possible] the undesirable effects of this disturbance on system "output" [measured with its own metric], *while* the system remains internally stable.

To present an alternative norm and optimization problem, we refer to papers by Vidyasagar which present some interesting results concerning these issues and serve as the starting point for future research [*K2*. 32 & 33].

Consider the system of *Case 3* in Table 6.1-1 with its equivalent *LFT* representation of Fig. 6.9-1. Suppose that the disturbance w is the output of a system W, which in turn is driven by an input v that is bounded in *time* by 1. The objective is to design a feedback controller that stabilizes this plant (possibly unstable to begin) and at the same time optimally rejects the disturbance (or results in the smallest possible maximum output amplitude in response to this disturbance). Conceptually, this problem represents an advancement with respect to the classical optimization problems. Here, we are attempting to minimize the worst possible adverse impact of a *class* of disturbances, rather than just a single fixed disturbance. Aspects of this optimizational problem are presented in Section 6.9. Since we have only discussed the corresponding time-domain results in that section, the following hybrid discussion seems appropriate.

The idea of minimax optimization, i.e., minimizing the worst-possible impact of a class of disturbances, represents an important conceptual advancement relative to the classical approaches. A mathematical framework for studying such problems is given in [*H2*. 66], in terms of multiplicative seminorms; and explicit solutions are given for the case where the disturbance input is a square-integrable function and the plant is scalar and has a single simple zero on the open right-half plane. From these results of [*H2*. 68 and 3] on minimax interpolation, we can obtain explicit solutions for the case of L^2-disturbances and scalar plants with multiple right-half plane zeros. The type of cost function that can be minimized using the methods of [*H2*. 66, 68, 13 and 14], [*K2*. 3] and [*H1*. 23] is

$$J = \max_{w \in BL^2} \| z \|_2. \tag{6.11-36}$$

---

inputs and to-be-controlled and measured outputs, a feedback controller such that the closed-loop system is internally stable and such that it attenuates the disturbance to any degree of accuracy which may be desired, measured for example, in the sense of the H_∞, the L_2-induced, norm. [Referring to Fig. 6.9-1], A typical well-motivated design objective is to minimize the influence of disturbance [w] on output [z] in the closed-loop system while requiring the closed-loop system to be internal stable. The notion of internal stability can be formalized in several ways. Essentially the internal stability requires that the closed-loop spectrum [system poles] belongs to a prespecified nonempty subset of the complex plane [*K2*. 34]."

6.11. Bounded Uncertainty and Disturbance Rejection

From the analysis in Section 6.9, if G_{zw} denotes the transfer matrix from w to z, then

$$\max_{\|w\|_2 \leq 1} \|z\|_2 = \|G_{zw}\|_\infty. \tag{6.11-37}$$

Having stated these facts, Vidyasagar proposes to minimize the L_∞-norm of z as follows.

$$J = \max_w \|z\|_\infty. \tag{6.11-38}$$

The implication of this choice is that we are interested in a uniformly disturbance rejection at all instants of time. If $w \in BL^2$, then

$$J = \max_{\|w\|_2 \leq 1} \|z\|_\infty = \|G_{zw}\|_2. \tag{6.11-39}$$

Thus, the problem of minimizing the maximum L_∞-norm of z in response to a set of L_2-norm bounded disturbances can be solved using Wiener-Hopf methods [H1. 25 & 26].

Now, suppose that the disturbances themselves are entering persistently as L_∞-norm bounded functions, and the associated cost to be minimized is

$$J = \max_{\|w\|_\infty \leq 1} \|z\|_\infty. \tag{6.11-40}$$

While this problem in spirit belongs to the general class of problems proposed in [H2. 66], its actual solution requires a new approach. In [K2 32] the minimization of cost functions of the type (6.11-40) is developed. Complete solutions for problems of optimal controller synthesis in several important situations, including those where the plant has minimum phase or a single unstable zero are developed as well. It is also shown that in the general case, methods which provide estimates for the optimal performance can be given. The L^∞-optimal controller design is shown to be different from that resulted from the methods of L^2-optimal controller design presented in [H2. 66, 68, 13 and 14], [K2. 3] and [H1. 23]. "Even in cases where the optimal achievable performances are the same, the methods used to arrive at the end results are quite different. The result is a theory that complements the H^∞-optimization theory and draws on it in many ways, but is fundamentally different [K2. 32]." Clearly, this analysis merits further study, however, due to the scope of this book is not pursued here. This study is extended in [K2. 33] for systems with zeros or poles on the stability boundary, and it also clarifies some of the issues discussed in [K2. 8].

Several applications and extensions of the Vidyasagar's original formulation are discussed by Dahleh and Pearson in [K2. 7 & 8]. The problem of minimizing the maximum amplitude of the system error when system inputs are bounded in amplitude and the induced norm is the l^1-norm (discrete-time norm) of system error is solved. The transfer function of the system error becomes $\hat{\Phi} = \hat{H} - \hat{U}\hat{Q}\hat{V}$, where $\hat{H}, \hat{U}, \hat{V}$ are given stable rational matrices, and $\hat{Q}$ is an arbitrary stable rational matrix. In [K2. 7] this problem is solved under the assumption that $\hat{U}$ has full row rank, and $\hat{V}$ has full column rank. In [K2. 8] this problem is solved when

$\hat{U}$ has column rank and/or $\hat{V}$ has row rank. Here, it is no longer possible to compute the optimal rational solutions for these problems as in [K2. 7], rather it is only possible to compute rational *suboptimal* solutions which are arbitrarily close to the optimal solutions. This computation is done by solving a sequence of linear programming problems involving an increasing number of constraints. Under these conditions, the underlying norm of $\hat{\Phi}$ is a nonincreasing function of the number of constraints. When an acceptable value of this norm is reached, the rational compensator that achieves this norm can be found by solving a set of linear equations. However, how to truncate the problem in order to bring it within a prespecified bound of the optimal solution is stated as an open question in [K2. 8]. Here, the application of the l^1-optimization problem in designing closed-loop systems that are robustly stable for all possible additive or multiplicative stable perturbation to the nominal plant are considered. These perturbations are required to be l^∞-stable, and whose l^1-norm provides a measure for such robustness. Based on this result, a mixed sensitivity optimization problem to reject the l^∞-disturbances and simultaneously guarantee robust stability is developed in [K2. 8].

REMARK 6.11-2 (On Table 6.1-1): Concerning all cases described in Table 6.1-1, we offer the following steps which help to formulate a sensitivity-reduction and/or disturbance-rejection problem.

STEP 1 (Mathematical modeling of "disturbance" and the subsequent optimization): We choose an appropriate metric or mathematical space for defining the corresponding "disturbance", followed by the choice of a physically meaningful sensitivity measure (or performance index) described in its own metric which may not necessarily be the same as that for disturbance. Here, we must justify these choices from the physical point of view.

STEP 2 (The optimization scheme): We may either select or devise an optimizational procedure for the corresponding sensitivity measure (or performance index), in terms of the chosen mathematical space with consideration on the overall system stability. Certainly, some of these optimizational problems may not be easy to solve, indeed a very few methodologies are known for optimization in spaces other than H^2 and H^∞. Even in these spaces, we cannot solve every optimizational problem (i.e., different sensitivity measure). Thus the opportunities exist, although efforts are needed to detect them.

STEP 3 (Realizability): No algorithm can be of any use if the final controller cannot be realized. For instance, methods which rely on brute force rejection of disturbances while yielding unrealizable controllers must be avoided. Yet again when realized this controller must maintain the overall system stability.

STEP 4 (Cost): Every design must be cost effective. This aspect of the design can put the very fabric of an elaborate methodology in danger. It is always instructive to compare the outcomes of several controller designs versus their costs for implementation and maintenance before finalizing a design choice. This thought does not imply that we should not explore uncharted areas, rather it emphasizes that we should study the cost as well.

6.12. REFERENCES

[A] Trajectory – Sensitivity Comparison
Certain Feedback Properties of Linear Systems

[1] B.D.O. Anderson, "Sensitivity improvement using optimal design," *Proc. IEE*, vol. 113, pp. 1084-1086, 1966.

[2] D. Bensoussan, "Sensitivity reduction in single-input single-output systems," *IJC*, vol. 39, no. 2, pp. 321-335, 1984.

[3] _____, "Decentralized control and sensitivity reduction of weakly coupled plants," *IJC*, vol. 40, no. 6, pp. 1099-1118, 1984.

[4] R.W. Brockett and M.D. Mesarovic, "The reproducibility of multivariable systems," *JMAA*, vol. 11, pp. 548-563, 1965.

[5] B.M. Chen, A. Saberi, and P. Sannuti, "A new stable compensator design for exact and approximate loop transfer recovery," *Automatica*, vol. 27, no. 2, pp. 257-280, 1991.

[6] _____, "Necessary and sufficient conditions for a nonminimum phase plant to have a recoverable target loop – A stable compensator design for LTR," *Automatica*, vol. 28, no. 3, pp. 493-507, 1992.

[7] J.B. Cruz, Jr., and W.R. Perkins, "A new approach to the sensitivity problem in multivariable feedback system design," *IEEE-TAC*, vol. AC-9, pp. 216-223, July 1964.

[8] _____, "Criteria for system sensitivity to parameter variations," *Proc. IFAC*, vol. 1 (of the 3rd Congress), pp. 18C.1-18C.7, 1966.

[9] J.B. Cruz, Jr., and N. Sundararajan, "Sensitivity improvement of semi-closed loop systems," in *Proc. 2nd IFAC Symp. Multivariable Technical Control Systems*, pp. 1-15, 1971.

[10] R.M. DeSantis and S. Lefebvre, "Comparative sensitivity and absolute invariant compensators," in *Proc. Allerton Conf.*, pp. 964-973, 1976.

[11] R.M. DeSantis and W.A. Porter, "Circle type conditions for sensitivity reduction," *ACTA*, vol. 2, no. 2, pp. 26-36, May 1974.

[12] C.A. Desoer and W.S. Chan, "The feedback interconnection of lumped linear time-invariant systems," *JFI*, vol. 300, nos. 5 & 6, pp. 335-351, 1975.

[13] M. Eslami, "On sensitivity comparison of linear systems," in *Proc. MECO' 77* (M.H. Hamza, Ed.), pp. 119-123, Zurich, June 20-21, 1977.

[14] R.L. Gonzales, "Synthesis for minimum sensitivity under worst case conditions," in *Proc. Allerton Conf.*, pp. 527-537, 1966.

[15] O.R. Gonzalez and P.J. Antsaklis, "Sensitivity considerations in the control of generalized plants," *IEEE-TAC*, vol. AC-34, no. 8, pp. 885-889, Aug. 1989.

[16] G. Grubel and G. Kreisselmeier, "A generalized comparison sensitivity concept for sensitivity reduction in control system design," in *Proc. Joint Auto. Contr. Conf.*, pp. 328-332, 1974.

[17] Y. Hontoir and J.B. Cruz, Jr., "Sensitivity reduction in linear systems," *Automatica*, vol. 8, pp. 445-449, 1972.

[18] _____, "A probabilistic likelihood approach to trajectory sensitivity," *Automatica*, vol. 10, pp. 49-60, 1974.

[19] W.F. Horton and C.T. Leondes, "Sensitivity in multivariable control systems," *ASME-JBE*, pp. 246-250, June 1969.

[20] O.L.R. Jacobs, "Cost of uncertainty about controlled objects," *Proc. IEE*, vol. 136, Pt. D, no. 4, pp. 177-187, July 1989.

[21] A.V. Knyazev, "Frozen-parameter method for integral equations of feedback system dynamics," *ARC*, vol. 41, no. 10, pp. 1347-1348, Oct. 1980.

[22] E. Kreindler, "On the definition and application of the sensitivity function," *JFI*, vol. 285, pp. 26-36, Jan. 1968.

[23] _____, "Closed-loop sensitivity reduction of linear optimal control systems," *IEEE-TAC*, vol. AC-13, no. 3, pp. 254-262, June 1968.

[24] _____, "On trajectory sensitivity in optimal control," *Proc. IEEE*, vol. 57, no. 4, pp. 695-696, April 1969.

[25] G. Kreisselmeier and G. Grubel, "The design of optimally parameter insensitive control systems," in *Proc. IFAC*, vol. 3 (of the 5th Congress), pp. 31.1.1-31.1.6, 1972.

[26] B. Krogh and J.B. Cruz, Jr., "Design of sensitivity-reducing compensators using observers," *IEEE-TAC*, vol. AC-23, no. 6, pp. 1058-1062, Dec. 1978. Addendum, *ibid.*, vol. AC-24, no. 2, p. 353, April 1979.

[27] D.P. Looze and J.S. Freudenberg, "Limitations of feedback properties imposed by open-loop right half plane poles," *IEEE-TAC*, vol. AC-36, no. 6, pp. 736-739, 1991.

[28] A.G.J. MacFarlane, "Return-difference and return-ratio matrices and their use in analysis and design of multivariable feedback control systems," *Proc. IEE*, vol. 117, no. 10, pp. 2037-2049, Oct. 1970.

[29] _____, "Return-difference and return-ratio matrices and their use in analysis and design of multivariable feedback control systems," *Proc. IEE*, vol. 118, no. 7, pp. 946-947, July 1971.

[30] P.J. Marino, "On the synthesis of insensitive linear feedback control systems," *IJC*, vol. 6, no. 1, pp. 33-50, 1967.

[31] W.R. Perkins and J.B. Cruz, Jr., "The parameter variation problem in state feedback control systems," *ASME-JBE*, vol. 87, pp. 120-124, 1965.

[32] _____, "Feedback properties of linear regulators," *IEEE-TAC*, vol. AC-16, no. 6, pp. 659-663, Dec. 1971.

[33] W.R. Perkins, J.B. Cruz, Jr., and R.L. Gonzales, "Design of minimum sensitivity systems," *IEEE-TAC*, vol. AC-13, no. 2, pp. 159-167, April 1968.

[34] H.J. Perlis, "On the residue of a sensitivity function," *IEEE-TAC*, vol. AC-10, pp. 496-497, Oct. 1965.

[35] W.A. Porter, "A design technique for improving system sensitivity," in *Proc. Allerton Conf.*, pp. 517-526, 1966.

[36] _____, "On the reduction of sensitivity in multivariate systems," *IJC*, vol. 5, no. 1, pp. 1-9, 1967.

[37] V.R. Sule and V.V. Athani, "Directional sensitivity tradeoffs in multivariable feedback systems," *Automatica*, vol. 27, no. 5, pp. 869-872, 1991.

[38] N. Sundararajan and J.B. Cruz, Jr., "Trajectory insensitivity of optimal feedback systems," *IEEE-TAC*, vol. AC-15, pp. 663-665, Dec. 1970.

[39] C. Verde and P.M. Frank, "Sensitivity reduction of linear quadratic regulator by matrix modification," *IJC*, vol. 48, no. 1, pp. 211-223, 1988.

[B] Performance – Index Sensitivity Comparison

[1] P. Courtin and J. Rootenberg, "Performance index sensitivity of optimal control systems," *IEEE-TAC*, vol. AC-16, no. 3, pp. 275-277, June 1971.

[2] P. Dorato, "On sensitivity in optimal control systems," *IEEE-TAC*, vol. AC-8, pp. 256-257, July 1963.

[3] J.C. Dunn, "Further results on the sensitivity of optimally controlled systems," *IEEE-TAC*, vol. AC-12, pp. 324-326, June 1967.

[4] H.S. Kang and A.K. Mahalanabis, "Sensitivity of the performance of optimal stochastic systems," *Proc. IEEE*, vol. 61, no. 3, pp. 389-390, 1973.

[5] P.V. Kokotovic, J. Heller, and P. Sannuti, "Sensitivity comparison of optimal controls," *IJC*, vol. 9, no. 1, pp. 111-115, Jan. 1969.

[6] E. Kreindler, "On performance sensitivity of optimal control systems," *IJC*, vol. 15, no. 3, pp. 481-486, 1972.

[7] A. Orbach and R. Fischl, "Performance index sensitivity of optimal control of first order in time- and space-distributed parameter systems," *IEEE-TAC*, vol. AC-25, no. 2, pp. 314-317, April 1980.

[8] B. Pagurek, "Sensitivity of the performance of optimal linear control systems to parameter variations," *IJC*, vol. 1, pp. 33-45, 1965.

[9] _____, "Sensitivity of the performance of optimal control systems to plant parameter variations," *IEEE-TAC*, AC-10, pp. 178-180, April 1965.

[10] W.A. Porter, "Sensitivity problems in linear systems," *IEEE-TAC*, vol. AC-10, July 1965.

[11] N.K. Sinha and S.R. Atluri, "Sensitivity of optimal control systems," in *Proc. Allerton Conf.*, pp. 508-516, 1966.

[12] N.K. Sinha, S.R. Atluri, and H.S. Witsenhausen, "On the sensitivity of optimal control systems with quadratic performance criteria," *IEEE-TAC*, vol. AC-12, pp. 208-209, April 1967.

[13] F.E. Thau, "A Comparison of closed-loop and open-loop optimum systems," *IEEE-TAC*, vol. AC-11, pp. 619-620, July 1966.

[14] H.S. Witsenhausen, "On the sensitivity of optimal control systems," *IEEE-TAC*, vol. AC-10, pp. 495-496, Oct. 1965.

[15] D.C. Youla and P. Dorato, "On the comparison of the sensitivities of open-loop and closed-loop optimal control systems," *IEEE-TAC*, vol. AC-13, pp. 186-188, April 1968.

[C] Optimality – Index Sensitivity: Analysis and Synthesis

[1] Y.V. Aleksandrov, "Sensitivity of the quality criterion of linear optimal systems," *ECy*, vol. 9, no. 5, pp. 948-953, Sept./Oct. 1971.

[2] _____, "Guaranteed sensitivity of linear optimal systems," *ECy*, vol. 13, no. 2, pp. 140-145, March/April 1975.

[3] M. Aoki, "On performance losses in some adaptive control systems: I," *ASME-JBE*, pp. 90-94, March 1965.

[4] N. Becker, "A note on performance index sensitivity of time optimal control systems," *IEEE-TAC*, vol. AC-25, no. 4, pp. 819-821, Aug. 1980.

[5] D.S. Bernstein and W.M. Haddad, "Robust stability and performance analysis for linear dynamic systems," *IEEE-TAC*, vol. AC-34, no. 7, pp. 751-758, July 1989.

[6] B.Z. Bobrovsky and D. Graupe, "Analysis of optimal-cost sensitivity to parameter changes," *IEEE-TAC*, vol. AC-16, pp. 487-488, Oct. 1971.

[7] W.A. Brown and W.J. Vetter, "Sub-optimal design of the linear regulator with incomplete state feedback *via* second-order sensitivity," *IJC*, vol. 16, no. 1, pp. 1-7, 1972.

[8] R.J. Burns and K.S.P. Kumar, "Sensitivity considerations in specific optimum controls," *IJC*, vol. 5, no. 3, pp. 289-296, 1967.

[9] B.N. Chatterji, "Sensitivity of performance of control systems to unintentional coupling signals," *IJC*, vol. 12, no. 2, pp. 265-272, 1970.

[10] M. Eslami, "Sensitivity analysis and synthesis in automatic control systems," Ph.D. Dissertation, The University of Wisconsin – Madison, May 1978.

[11] M. Goujon and M. Lecrique, "Influence de la precision des regulateurs sur les performances des systemes de reglage," *Automatisme*, vol. XVI, no. 2, pp. 100-110, Feb. 1971.

[12] D. Graupe, "Optimal linear control subject to sensitivity constraints," *IEEE-TAC*, vol. AC-19, no. 5, pp. 593-594, Oct. 1974.

[13] A.J. Koivo, "Performance Sensitivity of dynamic systems," in *Proc. Joint Auto. Contr. Conf.*, pp. 444-453, 1968.

[14] _____, "Performance sensitivity of dynamical systems," *Proc. IEE*, vol. 117, no. 4, pp. 825-830, April 1970.

[15] B. Kurtaran and M. Sidar, "Analysis of cost sensitivity for linear-quadratic stochastic problems with instantaneous output feedback," *IEEE-TAC*, vol. AC-19, no. 5, pp. 589-590, Oct. 1974.

[16] W.S. Levine and M. Athans, "On the determination of the optimal constant output feedback gain for linear multivariable systems," *IEEE-TAC*, vol. AC-15, pp. 44-48, Feb. 1970.

[17] E.P. Maslov and A.M. Petrovsky, "Sensitivity of linear systems with respect to random disturbances," *Automatica*, vol. 5, pp. 275-278, 1969.

[18] N.H. McClamroch, L.G. Clark, and J.K. Aggarwal, "Sensitivity of linear control systems to large parameter variations," *Automatica*, vol. 5, pp. 257-263, 1969.

[19] J. Medanic, "Three segment method in the sensitivity design of control systems," in *Proc. Allerton Conf.*, pp. 439-450, 1967.

[20] M. Midy, "Influence d'une non-linearite sur les parametres de reglage d'une boucle de regulation analogique (application a une regulation de debit)," *Automatisme*, vol. XVI, no. 2, pp. 116-121, Feb. 1971.

[21] L. Platzman and M. Athans, "Explicit cost sensitivity analysis for linear systems with quadratic criteria," *IEEE-TAC*, vol. AC-20, no. 2, pp. 252-254, April 1975.

[22] J.J. Rissanen, "Performance deterioration of optimum systems," *IEEE-TAC*, vol. AC-11, pp. 530-532, July 1966.

[23] R.A. Rohrer and M. Sobral, Jr., "Sensitivity considerations in optimal system design," *IEEE-TAC*, vol. AC-10, pp. 43-48, Jan. 1965.

[24] Y.E. Sagalov, "Analysis of sensitivity of an optimal system with parameter variation," *ARC*, no. 2, pp. 180-184, Feb. 1974.

[25] M.P. Sakharov, "Parametric sensitivity in the problem of control with an incomplete plant model," *ARC*, no. 6, pp. 877-883, June 1973.

[26] D.M. Salmon, "Minimax controller design," *IEEE-TAC*, vol. AC-13, no. 4, pp. 369-376, Aug. 1968.

[27] N.K. Sinha and S.H. Dai, "Reduction of the sensitivity of an optimal control system to plant parameter variations," *IEEE-TAC*, vol. AC-15, pp. 589-590, Oct. 1970.

[28] J.S. Tyler, Jr., and F.B. Tuteur, "The use of a quadratic performance index to design multivariable control systems," *IEEE-TAC*, vol. AC-11, no. 1, pp. 84-92, Jan. 1966.

[29] A.A. Vengerov, V.L. Rozhanskii, and G.M. Ulanov, "Estimation of sensitivity of integral performance criteria of systems of variable structure," *ARC*, no. 2, pp. 245-250, Feb. 1971.

[30] R.A. Werner, "Feedback control with magnitude constraints for systems with unknown parameters," in *Proc. Allerton Conf.*, pp. 418-427, 1967.

[31] R.A. Werner and J.B. Cruz, Jr., "Feedback control which preserves optimality for systems with unknown parameters," *IEEE-TAC*, vol. AC-13, no. 6, pp. 621-629, Dec. 1968.

[32] M.A. Zohdy and J.D. Aplevich, "Output feedback controllers optimal for time-multiplied performance indices," *Elect. Lett.*, vol. 11, no. 16, pp. 360-361, Aug. 1975.

[D] Trajectory – Sensitivity Optimization

[1] M.A. Abouelwafa and M.H. Hamza, "Design of an adaptive controller using multi-level and sensitivity concepts," *IJSS*, vol. 10, no. 3, pp. 243-250, 1979.

[2] A.J. Bradt, "Sensitivity functions in the design of optimal controllers," *IEEE-TAC*, vol. AC-13, pp. 110-111, Feb. 1968.

[3] P.C. Byrne and M. Burke, "Optimization with trajectory sensitivity considerations," *IEEE-TAC*, vol. AC-21, no. 2, pp. 282-283, April 1976. Comments, by P.J. Fleming and M.M. Newmann, *ibid.*, vol. AC-22, no. 1, p. 151, Feb. 1977.

[4] J.F. Cassidy, Jr., and I. Lee, "On the optimal feedback control of a large launch vehicle to reduce trajectory sensitivity," in *Proc. Joint Auto. Contr. Conf.*, pp. 587-595, June 1967.

[5] V.V. Ciric and J.V. Leeds, "Further results on sensitivity consideration of multiple-input controller design for dynamic optimization," *IJC*, vol. 15, no. 5, pp. 849-863, 1972.

[6] R.N. Crane and A.R. Stubberud, "Minimum sensitive linear feedback compensators," in *Proc. Asilomar Conf.*, pp. 405-409, 1971.

[7] _____, "Closed-loop formulations of optimal control problems for minimum sensitivity," in *Control and Dynamic Systems advances in theory and applications*, vol. 9 (C.T. Leondes, Ed.). New York: Academic Press, pp. 375-505, 1973.

[8] H. D'Angelo, M.L. Moe, and T.C. Hendricks, "Trajectory sensitivity of an optimal control system," in *Proc. Allerton Conf.*, pp. 489-498, Oct. 1966.

[9] H.J. Dougherty, I. Lee, and P.M. DeRusso, "Synthesis of optimal feedback control systems subject to parameter variations," preprints of *Joint Auto. Contr. Conf.*, pp. 125-133, 1967.

[10] M. Eslami, "Sensitivity analysis and synthesis in automatic control systems," Ph.D. Dissertation, The University of Wisconsin – Madison, May 1978.

[11] _____, "On sensitivity measure optimization with large parameter variation," The University of Wisconsin-Madison, Dept. of ECE, Report ECE-79-3, 32 pp., Jan. 1979.

[12] _____, "On sensitivity minimization algorithms with quadratic performances," in *Proc. 21st IEEE Conf. on Decision and Contr.*, pp. 637-641, Dec. 1982.

[13] _____, "Robust quadratic optimization of systems with large parameter variations," *JOCAM*, vol. 12, no. 1, pp. 33-48, 1991. Corrections, *ibid.*, vol. 15, 1994.

[14] M. Eslami and R.S. Marleau, "On sensitivity minimization with an adaptive controller," in *Proc. Joint Auto. Contr. Conf.*, pp. 219-228, Oct. 1978.

[15] _____, "On trajectory sensitivity minimization with an adaptive controller," in *Advances in Control* (D.G. Lainiotis and T.S. Tzannes, Eds.). Boston: D. Reidel Pub. Company, pp. 342-350, 1980.

[16] P.J. Fleming and M.M. Newmann, "Trajectory sensitivity reduction in the optimal linear regulator," in *Recent Mathematical Development in Control* (D.J. Bell, Ed.). New York: Academic Press, pp. 137-151, 1973.

[17] _____, "Design algorithms for a sensitivity constrained suboptimal regulator," *IJC*, vol 25, no. 6, pp. 965-978, 1977.

[18] I.-K. Fong, T.-S. Kuo, K.-C. Kuo, C.-F. Hsu, and M.-Y. Wu, "Sensitivity analysis of linear uncertain systems and its application in the synthesis of an insensitive linear regulator," *IJSS*, vol. 18, no. 1, pp. 43-55, 1987.

[19] G. Fronza and A. Locatelli, "Insensitivity by linear feedback," *JFI*, vol. 296, no. 4, pp. 237-247, Oct. 1973.

[20] S. Fu, M.E. Sawan, Y. Fu, and M.T. Tran, "Trajectory sensitivity analysis: A new criterion," *IJSS*, vol. 16, no. 6, pp. 769-775, 1985.

[21] M. Gopal and P. Pratapachandran Nair, "Sensitivity reduced optimal linear regulator with prescribed closed-loop eigenvalues," *IEEE-TAC*, vol. AC-29, no. 7, pp. 661-664, July 1984.

[22] _____, "On the design of a sensitivity-reducing optimal dead-beat controller," *IJC*, vol. 42, no. 4, pp. 877-886, 1985.

[23] G. Grubel and G. Kreisselmeier, "Effective parameter sensitivity reduction through minimization of sensitivity measure," in *Proc. Joint Auto. Contr.*, pp. 79-86, 1971.

[24] R.P. Hamalainen and T. Eirola, "Trajectory sensitivity reduction in non-zero-sum differential games," *IJSS*, vol. 11, no. 2, pp. 207-222, 1980.

[25] A.R. Hanafy, "Two-level optimization technique to minimize trajectory sensitivity," in *Proc. Joint Auto. Contr. Conf.*, pp. 568-575, 1976.

[26] E. Higginbotham, "Optimal sensitivity and state control of regulators containing plant and measurement noise," in *Proc. Allerton Conf.*, pp. 677-680, 1969.

[27] Y. Hontoir and J.B. Cruz, Jr., "Minimum trajectory sensitivity design of systems with random parameters," *IJC*, vol. 20, no. 3, pp. 353-362, 1974.

[28] S.J. Kahne, "Low sensitivity design of optimal linear control systems," *IEEE-TAES*, vol. AES-4, pp. 374-379, May 1968.

[29] Y. Kamiya, "Construction of a low parameter and disturbance sensitivity system by a model-following method," *IJC*, vol. 23, no. 4, pp. 515-524, 1976.

[30] I.H. Khalifa and A.A.R. Hanafy, "A note on trajectory sensitivity reduction using a three-term controller," *IEEE-TAC*, vol. AC-29, no. 8, pp. 739-740, Aug. 1984. Errata, *ibid.*, vol. AC-31, no. 1, pp. 93-94, Jan. 1986.

[31] D.L. Kleinman and P. Krishna Rao, "An information matrix approach for aircraft parameter-insensitive control," in *Proc. IEEE Conf. on Decision and Contr.*, 1977.

[32] E. Kreindler, "On minimization of trajectory sensitivity," *IJC*, vol. 8, no. 1, pp. 89-96, 1968.

[33] _____, "Formulation of the minimum trajectory sensitivity problem," *IEEE-TAC*, vol. AC-14, pp. 206-207, April 1969.

[34] C.T. Leondes and P. Pezet, "Sensitivity requirements for suboptimal controllers," *IEEE-TAC*, vol. AC-20, no. 3, pp. 426-428, June 1975.

[35] C.T. Leondes and T.K. Sui, "Payoff sensitivity of linear quadratic differential games to parameter change," *ASME-JDSMC*, vol. 103, pp. 36-38, March 1981.

[36] J.F. Lowinger and J.A. Gibson, "Desensitizing algorithms for state-restrained optimal control assessments," *IJC*, vol. 21, no. 3, pp. 353-373, 1975.

[37] J.Y.S. Luh and E.R. Cross, "Optimal controller design for minimum trajectory sensitivity," *IJC*, vol. 7, no. 6, pp. 557-568, 1968.

[38] M.M. Missaghie and F.W. Fairman, "Sensitivity reducing observers for optimal feedback control," *IEEE-TAC*, vol. AC-22, no. 6, pp. 952-957, Dec. 1977.

[39] _____, "Desensitizing observers for LQG feedback control," *IJSS*, vol. 12, no. 2, pp. 161-175, 1979.

[40] R.B. Newell and D.B. Fisher, "Experimental evaluation of optimal, multivariable regulatory controllers with model-following capabilities," *Automatica*, vol. 8, pp. 247-262, 1972.

[41] M.M. Newmann, "On attempts to reduce the sensitivity of the optimal linear regulator to a parameter change," *IJC*, vol. 11, no. 6, pp. 1079-1084, 1970.

[42] K. Okada and R.E. Skelton, "Sensitivity controller for uncertain systems," *JGCD*, vol. 13, no. 2, pp. 321-329, 1990.

[43] J. O'Reilly, "Low-sensitivity feedback controllers for linear systems with incomplete state information," *IJC*, vol. 29, no. 6, pp. 1047-1058, 1979.

6.12.D. Trajectory – Sensitivity Optimization

[44] A.I. Petrov, A.G. Zubov, and V.V. Minin, "Analysis of trajectory sensitivity of adaptive stochastic control systems," *SJAIS*, vol. 18, no. 2, pp. 50-59, 1985.

[45] S.G. Rao and A.C. Soudack, "Synthesis of optimal control systems with near sensitivity feedback," *IEEE-TAC*, vol. AC-16, pp. 194-196, April 1971.

[46] J.H. Rillings and R.J. Roy, "Analog sensitivity design of Saturn V launch vehicle," *IEEE-TAC*, vol. AC-15, no. 4, pp. 437-442, Aug. 1970.

[47] M. Saif, "Stability constrained robust linear regulator," *CAC*, vol. 16, no. 3, pp. 66-69, 1988.

[48] _____, "Design of a trajectory insensitive regulator with prescribed degree of stability," *ASME-JDSMC*, vol. 112, pp. 513-516, Sept. 1990.

[49] P. Sannuti and J.B. Cruz, Jr., "A note on trajectory sensitivity of optimal control systems," (with reply by A.J. Bradt) *IEEE-TAC*, vol. AC-13, pp. 111-113, Feb. 1968.

[50] V.V.S. Sarma and B.L. Deekshatulu, "Sensitivity design of optimal linear systems," *IJC*, vol. 8, no. 6, pp. 653-658, 1968.

[51] M.E. Sezer and D.D. Siljak, "Sensitivity of large-scale control systems," *JFI*, vol. 312, nos. 3/4, pp. 179-197, Sept./Oct. 1981.

[52] P. Stavroulakis and P.E. Sarachik, "Low sensitivity feedback gains deterministic and stochastic control systems," *IJC*, vol. 19, no. 1, pp. 15-31, 1974.

[53] R. Subbayyan, V.V.S. Sarma, and M.C. Vaithilingam, "Trajectory sensitivity modification in optimal linear systems," *IEEE-TAC*, vol. AC-22, no. 4, pp. 657-659, Aug. 1977.

[54] _____, "An approach for sensitivity-reduced design of linear regulators," *IJSS*, vol. 9, no. 1, pp. 65-74, Jan. 1978.

[55] R. Subbayyan and M.C. Vaithilingam, "Sensitivity-reduced design of linear regulators," *IJC*, vol. 29, no. 3, pp. 435-440, 1979.

[56] S.D. Weinrich and L. Lapidus, "Optimally sensitive and adaptive control systems," *AIChe J.*, vol. 17, no. 6, pp. 1471-1480, Nov. 1971.

[57] C.A. Winsor and R.J. Roy, "The application of specific optimal control to the design of desensitized model following control systems," *IEEE-TAC*, vol. AC-15, no. 3, pp. 326-333, June 1970.

[58] S.C. Yang and W.L. Garrard, "A low sensitivity, modern approach to the longitudinal control of automated transit vehicles," *ASME-JDSMC*, vol. 96, pp. 218-228, June 1974.

[E] Sensitivity with Respect to System – Auxiliary Parameters

[*] Also consult Section *[C]* of the references in this chapter.

[1] Y. Bar-Ness, "Pole sensitivity of the quadratic optimal regulator," *Elect. Lett.*, vol. 12, no. 13, pp. 341-343, June 1976.

[2] Y.V. Kosyuk, "Analysis of nonstationary control systems by the method of 'frozen' coefficients," *SAC*, vol. 3, no. 6, pp. 19-27, 1970.

[3] K. Sugimoto and Y. Yamamoto, "Generalized robustness of optimality of linear quadratic regulators," *IJC*, vol. 51, no. 3, pp. 521-533, 1990. Corrections, *ibid.*, vol. 52, no. 5, pp. 1277-1278, 1990.

[4] D.F. Wilkie and H.M. Van Schieveen, "On the sensitivity of the linear state regulator," *IJC*, vol. 12, no. 4, pp. 709-719, 1970.

[F] Sensitivity Reduction and/or Optimization: Other Methods and Applications

[1] P.R. Belanger, "Some aspects of control tolerances and first-order sensitivity in optimal control systems," *IEEE-TAC*, vol. AC-11, no. 1, p. 77-83, 1966.

[2] _____, "A paper machine color control system design using modern techniques," *IEEE-TAC*, vol. AC-14, no. 6, pp. 610-616, Dec. 1969.

[3] C.S. Berger, "Robust controller design by minimisation of the variation of the coefficients of the closed-loop characteristic equation," *Proc. IEE*, vol. 131, Pt. D, no. 3, pp. 103-107, May 1984.

[4] _____, "Robust control of discrete systems," *Proc. IEE*, vol. 136, Pt. D, pp. 165-170, July 1989.

[5] S.P. Bingulac, "On the role of orthonormality of sensitivity functions in parameter optimization problems," *Automatica*, vol. 5, pp. 513-517, 1969.

[6] V.A. Bodner, V.I. Vasil'ev, and F.A. Shaimardanov, "An algorithmic method of synthesis of a low-sensitive automatic control system," *ARC*, no. 4, pp. 529-533, April 1974.

[7] A. Burzio and D.D. Siljak, "Minimization of sensitivity with stability constraints in linear control systems," *IEEE-TAC*, vol. AC-11, no. 3, pp. 567-569, July 1967.

[8] E.M. Butler and R.A. Rohrer, "On relative sensitivity for certain linear optimal control problems," in *Proc. Asilomar Conf.*, vol. 279-286, 1968.

[9] G.W. Carlock and A.P. Sage, "Sensitivity and error analysis algorithms for combined estimation and control systems," *IJC*, vol. 21, no. 3, pp. 417-441, 1975.

[10] S.S.L. Chang and P.E. Barry, "Optimal control of systems with uncertain parameters," in *Proc. IFAC*, vol. 3 (of the 5th Congress), pp. 31.6.1-31.6.5, 1972.

[11] D.J. Cloud and B. Kouvaritakis, "Weighting sequences, optimal truncation and optimal frequency-response uncertainty bounds," *Proc. IEE*, vol. 134, Pt. D, no. 3, pp. 153-170, May 1987.

[12] J.B. Cruz, Jr., "Probabilistic sensitivity properties of neighboring optimal feedback systems," *Iran. J. Sci & Tech.*, vol. 3, no. 4, pp. 271-282, 1975.

[13] A.R. Daniels, Y.B. Lee, and M.K. Pal, "Nonlinear power-system optimisation using dynamic sensitivity analysis," *Proc. IEE*, vol. 123, no. 4, pp. 365-370, 1976.

[14] _____, "Combined suboptimal excitation control and governing of a.c. turbogenerators using dynamic sensitivity analysis," *Proc. IEE*, vol. 124, no. 5, pp. 473-478, May 1977.

[15] M. Darwish, J.D. Delacour, and J. Fantin, "Sensitivity analysis of optimal regulators with application to large scale power systems," *JASE*, vol. 2, no. 4, pp. 259-268, Dec. 1977.

[16] R.M. DeSantis and J. Conan, "Practical sensitivity reduction tests with application to power systems," *IJSS*, vol. 8, no. 9, pp. 1067-1080, 1977.

[17] T.S. Dillon, K. Morsztyn, and T. Tun, "Sensitivity analysis of the problem of economic dispatch of thermal power systems," *IJC*, vol. 22, no. 2, pp. 229-248, 1975.

[18] T.S. Dillon and T. Tun, "Application of sensitivity methods to the problem of optimal control of hydro-thermal power systems," *JOCAM*, vol. 2, pp. 117-143, 1981.

[19] P. Dorato and A. Kestenbaum, "Application of game theory to the sensitivity design of systems with optimal controller structures," in *Proc. Allerton Conf.*, pp. 35-45, 1965.

[20] _____, "Application of game theory to the sensitivity design of optimal systems," *IEEE-TAC*, vol. AC-12, pp. 85-87, 1967.

[21] M. El-Hodiri, "Sensitivity analysis for an optimal control problem," *IEEE-TAC*, vol. AC-20, no. 2, pp. 251-252, April 1975.

6.12.F. Sensitivity Reduction: Other Methods and Applications

[22] M.M. Elmetwally and N.D. Rao, "Sensitivity analysis in the optimal design of sychronous machine regulators," *IEEE-PAS,* vol-PAS 93, no. 5, pp. 1310-1317, Sept./Oct. 1974.

[23] _____, "Low-sensitivity control of nuclear reactors," *Elect. Lett.,* vol. 11, no. 13, pp. 269-270, June 1975.

[24] A.N. Ermachenko and R.M. Yusupov, "The use of sensitivity functions in synthesizing linear multicoupled control systems," *ECy,* vol. 14, pp. 146-154, March/April 1976.

[25] T. Fuji and N. Mizushima, "Robustness of the optimality property of an optimal regulator: Multi-input case," *IJC,* vol. 39, no. 3, pp. 441-453, 1984.

[26] A.T. Fuller and A.S.I. Zinober, "On the existence of constant-ratio trajectories in nominally time-optimal control systems subject to parameter variation," *JFI,* vol. 303, pp. 359-369, April 1977.

[27] M. Gavrilovic, R. Petrovic, and D.D. Siljak, "Adjoint method in sensitivity analysis of optimal systems," *JFI,* vol. 276, no. 1, pp. 26-38, July 1963.

[28] V.I. Gorodetskiy, F.M. Zakharin, V.M. Ponomarev, and R.M. Yusupov, "Direct and inverse problems of sensitivity theory," *ECy,* vol. 9, no. 5, pp. 935-942, Sept./Oct. 1971.

[29] P.L. Graf and R. Shoureshi, "Gain-sensitivity augmentation for near-optimal control of linear parameter-dependent plants," *JGCD,* vol. 13, no. 2, pp. 310-320, 1990.

[30] A.W.J. Griffin, R.J. Paul, and C.G. Legge, "Direct-sensitivity method of solving boundary-value problems in optimal-control studies," *Proc. IEE,* vol. 116, no. 9, pp. 1611-1612, Sept. 1969.

[31] R.E. Griffin and A.P. Sage, "Sensitivity analysis of fixed point linear smoothing algorithms," *IJC,* vol. 8, no. 4, pp. 321-337, 1968.

[32] _____, "Sensitivity analysis of discrete filtering and smoothing algorithms," *AIAA J.,* vol. 7, no. 10, pp. 1890-1897, 1969.

[33] G. Guardabassi, A. Locatelli, C. Maffezzoni, and N. Schiavoni, "Computer-aided design of structurally constrained multivariable regulators Part 1: Problem statement, analysis and solution," *Proc. IEE,* vol. 130, Pt. D, no. 4, pp. 155-164, July 1983.

[34] _____, "Computer-aided design of structurally constrained multivariable regulators *via* parameter optimisation Part 2: Applications," *Proc. IEE,* vol. 130, Pt. D, no. 4, pp. 165-172, July 1983.

[35] N.J. Guinzy and A.P. Sage, "Identification and modelling of large-scale systems using sensitivity analysis," *IJC,* vol. 17, no. 5, pp. 1073-1087, 1973.

[36] A.H. Haddad, J.B. Cruz, Jr., and P.V. Kokotovic, "Design of control systems with random parameters," *IJC,* vol. 13, no. 5, pp. 981-992, 1971.

[37] M. Hassan and M.G. Singh, "A hierarchical model-following controller for certain non-linear systems," *IJSS,* vol. 7, no. 7, p. 727-730, 1976.

[38] _____, "Synchronous machine control using a two level model follower," *Automatica,* vol. 13, pp. 173-176, March 1977.

[39] D.R. Howard and Z.V. Rekasius, "Error analysis with the maximum principle," *IEEE-TAC,* vol. Ac-9, pp. 223-229, July 1964.

[40] C.L. Irwin and V. Komkov, "Sensitivity analysis and model optimization for reaction-diffusion systems," *JOTA,* vol. 44, no. 4, pp. 569-584, 1984.

[41] H. Ishitani and S. Yamamura, "Sensitivity analysis of optimal control systems," *EEJ,* vol. 87, no. 12, pp. 63-74, 1967.

[42] Y.B. Kadimov, E.Y. Kuliyev, and S.I. Myachin, "Some questions concerning the application of the sensitivity theory to an analysis of transient processes while simulating systems with distributed parameters on analog computers," *Proc. IFAC,* vol. 3 (of the 5th Congress), pp. 31.4.1-31.4.9, 1972.

[43] H.S. Kang, "Minimum sensitivity design of M.V. systems," *IJC*, vol. 11, no. 5, pp. 791-801, 1970.

[44] B.Y. Katkovnik, "Sensitivity of gradient networks," *ARC*, no. 12, pp. 1983-1989, Dec. 1970.

[45] J.M. Kelly, G. Leitmann, and A.G. Soldatos, "Robust control of base-isolated structures under earthquake excitation," *JOTA*, vol. 53, no. 2, pp. 159-180, 1980.

[46] D.L. Kleinman, "Solving the optimal attention allocation problem in manual control," *IEEE-TAC*, vol. AC-21, no. 6, pp. 813-822, Dec. 1976.

[47] P.V. Kokotovic, J.B. Cruz, Jr., J.E. Heller, and P. Sannuti, "Synthesis of optimally sensitive systems," *Proc. IEEE*, vol. 56, pp. 1318-1324, 1968.

[48] V. Komkov and N. Coleman, "Optimality of design and sensitivity analysis of beam theory," *IJC*, vol. 18, no. 4, pp. 731-740, 1973.

[49] D.V. Lebedev, "Sensitivity of parameters of plant motion to errors of inertial navigation and control system," *SAC*, vol. 9, pp. 41-47, Nov./Dec. 1976.

[50] C.-K. Lee and C.-T. Chen, "Sensitivity comparisons of various analogue computer simulations," *IJC*, vol. 10, no. 2, pp. 227-233, 1969.

[51] N.G. Malek, O.T. Tan, P.M. Julich, and E.C. Tacker, "Trajectory-sensitivity design of load-frequency control systems," *Proc. IEE*, vol. 120, no. 10, pp. 1273-1277, 1973.

[52] J.E. Marshall and S.V. Salehi, "Improvement of system performance by the use of time-delay elements," *Proc. IEE*, vol. 129, Pt. D, no. 5, pp. 177-181, Sept. 1982. Comments, with reply, by P.H. Landers, *ibid.*, vol. 130, Pt. D, no. 2, p. 92, March 1983.

[53] K. Nordstrom, "Trade-off between noise sensitivity and robustness for LQG regulators," *IJC*, vol. 46, no. 5, pp. 1689-1714, 1987.

[54] D.E. Olson and I.M. Horowitz, "Design of dominant-type control systems with large parameter variations," *IJC*, vol. 12, no. 4, pp. 545-554, 1970.

[55] C.S. Padilla and J.B. Cruz, Jr., "A linear dynamic feedback controller for stochastic systems with unknown parameters," *IEEE-TAC*, vol. AC-22, no. 1, pp. 50-55, 1977.

[56] _____, "Sensitivity adaptive feedback with estimation redistribution," *IEEE-TAC*, vol. AC-23, no. 3, pp. 445-451, 1978.

[57] C.S. Padilla, J.B. Cruz, Jr., and R.A. Padilla, "A simple algorithm for SAFER control," *IJC*, vol. 32, no. 6, pp. 1111-1118, 1980.

[58] P.N. Paraskevopoulos, "Sensitivity reduction in exact model – matching of linear multivariable systems," *JFI*, vol. 305, no. 2, pp. 99-118, Feb. 1978.

[59] _____, "Decoupling controller design *via* exact model – matching techniques," *Proc. IEE*, vol. 125, no. 11, pp. 1285-1289, Nov. 1978.

[60] H.J. Payne, E. Polak, D.C. Collins, and W.S. Meisel, "An algorithm for bicriteria optimization based on the sensitivity function," *IEEE-TAC*, vol. AC-20, no. 4, pp. 546-548, Aug. 1975.

[61] D.C. Reddy, "General sensitivity-reduction criterion for single-input multiple-output systems," *Elect. Lett.*, vol. 6, pp. 86-87, Feb. 1970.

[62] J. Rissanen, "Drift compensation of linear systems by parameter adjustments," *ASME-JBE*, pp. 415-418, June 1966.

[63] P. Ronge, "Performance index sensitivity of optimal control systems with uncertain parameters," *JOCAM*, vol. 6, pp. 359-384, 1985.

[64] J. Rootenberg, "The sensitivity of optimally designed control systems, with minimum fuel performance index," *IJC*, vol. 20, no. 1, pp. 101-112, 1974.

[65] J. Rootenberg and P. Courtin, "System sensitivity for optimal problems with singular control," *IJC*, vol. 20, no. 5, pp. 787-800, 1974.

6.12.F. Sensitivity Reduction: Other Methods and Applications

[66] M.V. Rybashov, "Insensitivity of gradient systems in the solution of linear problems on analog computers," *ARC*, no. 10, pp. 1679-1687, Oct. 1969.

[67] M.P. Sakharov, "Simplification of sensitivity models in the design of self-adjusting and adaptive systems," *ARC*, vol. 29, no. 5, pp. 743-747, May 1968.

[68] _____, "Application of simplification conditions of sensitivity models in the design of nonscanning adaptive systems," *ARC*, vol. 32, pp. 1080-1086, July 1971.

[69] _____, "On parameter adaptation algorithms for simplified sensitivity models," *ARC*, vol. 32, no. 10, pp. 1664, 1669, Oct. 1971.

[70] A.A. Sbaiti and A.P. Sage, "System optimization using quasilinearization and sensitivity analysis," *IJC*, vol. 16, no. 2, pp. 343-352, 1972.

[71] D.B. Schaechter, "Closed-loop control performance sensitivity to parameter variations," *JGCD*, vol. 6, no. 5, pp. 399-402, 1983.

[72] L.A. Shirokov, "Parametric optimization with an ideal reference model," *SAC*, vol. 15, no. 5, pp. 26-30, 1970.

[73] D.D. Siljak and R.C. Dorf, "On the minimization of sensitivity in optimal control systems," in *Proc. Allerton Conf.*, pp. 225-229, 1965.

[74] C.S. Sims and J.L. Melsa, "Sensitivity reduction in specific optimal control by the use of a dynamical controller," *IJC*, vol. 8, pp. 491-502, 1968.

[75] R.T. Stefani, "Reducing the sensitivity to parameter variations of a minimum-order reduced-order observer," *IJC*, vol. 35, no. 6, pp. 983-995, 1982.

[76] M.G. Strintzis and B. Liu, "Sensitivity minimization in the design of linear multivariable feedback systems," in *Proc. Allerton Conf.*, pp. 163-172, 1970.

[77] A. Swierniak, "State-inequalities approach to control systems with uncertainty," *Proc. IEE*, vol. 129, Pt. D, no. 6, pp. 271-275, Nov. 1982.

[78] _____, "Control laws for systems with inequality models of uncertainty," *Proc. IEE*, vol. 133, Pt. D, no. 4, pp. 153-158, July 1986.

[79] D.R. Towill and Z. Mehdi, "A new approach to system transient response sensitivity," *IJC*, vol. 15, no. 2, pp. 319-331, 1972.

[80] T.-P. Tsai and T.-S. Wang, "Optimal design of non-minimum-phase control systems with large plant uncertainty," *IJC*, vol. 45, no. 6, pp. 2147-2159, 1987.

[81] A.K. Tugcu, O. Coskunoglu, and R.E. Reid, "Performance criterion sensitivity analysis of ship-steering models with respect to shaping filter design parameters," *JOCAM*, vol. 6, pp. 77-90, 1985.

[82] S.G. Tzafestas, "Model-matching multicontroller design of reduced sensitivity," *ACTA*, vol. 3, no. 3, pp. 67-69, Sept. 1975.

[83] S.G. Tzafestas and P.N. Paraskevopoulos, "A sensitivity approach to the decoupling of linear systems with parameter disturbances," in *Proc. Joint Auto. Contr.*, pp. 92-100, 1973.

[84] S.S. Venkata, W.J. Eccles, and J.H. Noland, "Multi-parameter sensitivity analysis of power-system stability by Popov's method," *IJC*, vol. 17, no. 2, pp. 291-304, 1973.

[85] M.J. Vilenius, "The application of sensitivity analysis to electrohydraulic position control servos," *ASME-JDSMC*, vol. 105, pp. 77-82, June 1983.

[86] J.C. Wauer, J.M.H. Bruckner, and C.H. Humphrey, "Airplane performance sensitivities to lateral and vertical profiles," *JGCD*, vol. 4, no. 6, pp. 606-613, 1981. Errata, *ibid.*, vol. 5, no. 2, p. 224, 1982.

[87] A.I. Yermachenko, "Synthesis of linear control systems with limited sensitivity," *ECy*, vol. 9, no. 5, pp. 943-948, Sept./Oct. 1971.

[88] K.-K. D. Young, "Near insensitivity of linear feedback systems," *JFI*, vol. 314, no. 2, pp. 129-142, 1982.

[89] A.S.I. Zinober and A.T. Fuller, "The sensitivity of nominally time-optimal control systems to parameter variation," *IJC,* vol. 17, no. 4, pp. 673-703, 1973.

[G] Optimal Inputs for System Identification/ Auxiliary Inputs for System Control (The Effects of the Human Operator)

[1] M. Aoki and R. Staley, "On input signal synthesis in parameter identification," *Automatica,* vol. 6, pp. 431-440, 1970.

[2] O. Berman, E. Modiano, and J.A. Schnabel, "Sensitivity analysis and robust regression in investment performance evaluation," *IJSS,* vol. 15, no. 5, pp. 481-489, 1984.

[3] C. Bonivento, "Structural insensitivity versus identifiability," *IEEE-TAC,* vol. AC-18, no. 2, pp. 190-192, April 1973.

[4] V.N. Bukov, "Optimal algorithms in problems with bounded controllable coordinates," *ECy,* vol. 20, no. 2, pp. 133-140, 1982.

[5] M. Eslami, "Optimal input (policy) for system identification versus system sensitivity reduction with plant-operator," The Univ. of Wisconsin-Madison, Dept. of Elec. and Computer Eng'g, Tech. Report ECE-79-2, 27 pp., Jan. 1979.

[6] _____, "Optimal input for identification versus sensitivity reduction of systems with large parameter variations," *JOTA,* vol. 77, no. 3, pp. 591-612, June 1993.

[7] R.M. Gagliardi, "Input selection for parameter identification in discrete systems," *IEEE-TAC,* vol. AC-12, pp. 597-599, Oct. 1967.

[8] N.D. Georganas, "Optimal inputs and sensitivities for nonlinear process parameter estimation using imbedding techniques," *IEEE-TAC,* vol. AC-21, no. 3, pp. 415-417, June 1976.

[9] J.C. Geromel and J.J. da Cruz, "On the robustness of optimal regulators for nonlinear discrete-time systems," *IEEE-TAC,* vol. AC-32, no. 8, pp. 703-712, Aug. 1987.

[10] R.E. Kalaba and K. Spingarn, "Optimal inputs and sensitivities for parameter estimation," *JOTA,* vol. 11, no. 1, pp. 56-67, 1973.

[11] _____, "Optimal input system identification for nonlinear dynamic systems," *JOTA,* vol. 21, no. 1, pp. 91-102, Jan. 1977.

[12] V.S. Levadi, "Design of input signals for parameter estimation," *IEEE-TAC,* vol. AC-11, pp. 205-211, April 1966.

[13] K.Y. Lim and M. Eslami, "Adaptive controller designs for robot manipulator systems using Lyapunov direct method," *IEEE-TAC,* vol. AC-30, pp. 1229-1233, Dec. 1985. Comments, with reply, by Y.H. Chen, *ibid.,* vol. AC-32, no. 2, pp. 190-192, Feb. 1987, and, with reply, by R.H. Middelton, *ibid.,* vol. AC-32, no. 8, p. 749, Aug. 1987.

[14] _____, "Adaptive controller design for robot manipulator systems yielding reduced Cartesian error," *IEEE-TAC,* vol. AC-32, no. 2, pp. 184-187, Feb. 1987. Correction, *ibid.,* vol. AC-38, no. 2, p. 384, 1993.

[15] _____, "Robust adaptive controller designs for robot manipulator systems," *IEEE-JRA,* vol. RA-3, no. 1, pp. 54-66, 1987. Correction, *ibid.,* vol. RA-9, no. 1, p. 119, 1993.

[16] A.A. Lopez-Toledo and M. Athans, "Optimal policies for identification of stochastic linear systems," *IEEE-TAC,* vol. AC-20, pp. 754-765, Dec. 1975.

[17] R.K. Mehra, "Synthesis of optimal inputs for multiinput/multioutput systems with process noise, Parts I and II," Division of Engineering and Applied Physics, Harvard University, Cambridge, MA, Tech. Rep. TR 649, Feb. 1974.

[18] _____, "Optimal inputs for linear system identification," *IEEE-TAC*, vol. AC-19, no. 3, pp. 192-200, June 1974. Comments, by M.B. Zarrop and G.C. Goodwin, *ibid.*, vol. AC-20, no. 2, pp. 299-300, April 1975.

[19] _____, "Optimal input signals for parameter estimation in dynamic systems-survey and new results," *IEEE-TAC*, vol. AC-19, no. 6, pp. 753-768, Dec. 1974.

[20] R.K. Mehra and D.G. Lainiotis, Eds., *System Identification: Advances and Case Studies*. New York: Academic Press, 1976. Reviewed by P. Eykhoff, *IEEE-TAC*, vol. AC-23, no. 4, p. 766, 1978.

[21] N.E. Nahi and G. Napjus, "Design of optimal probing signals for vector parameter estimation," in *Proc. IEEE Conf. on Decision Contr.*, pp. 162-168, 1971.

[22] N.E. Nahi and D.E. Wallis, Jr., "Optimal inputs for parameter estimation in dynamic systems with white observation noise," in *Proc. Joint Auto. Contr. Conf.*, pp. 506-512, 1969.

[23] Y. Sawaragi and K. Ogino, "Sensitivity approach to optimal input synthesis for parameter identification of bilinear system," in *Proc. IFAC*, vol. 3 (of the 5th Congress), pp. 31.2.1-31.2.6, 1972.

[24] F.C. Schweppe, "On the accuracy and resolution of radar signals," *IEEE-TAES*, vol. AES-1, pp. 235-245, Dec. 1965.

[25] Y. Stepanenko and J. Yuan, "Robust adaptive control of a class of nonlinear mechanical systems with unbounded and fast-varying uncertainties," *Automatica*, vol. 28, no. 2, pp. 265-276, 1992.

[26] K. Watanabe and D.M. Himmelblau, "Instrument fault detection in systems with uncertainties," *IJSS*, vol. 13, no. 2, pp. 137-158, 1982.

[27] C. Yaling, "Sensitivity operator and approximate algorithm for parameter estimation," *IJC*, vol. 46, no. 2, pp. 537-546, 1987.

[H] Sensitivity Considerations in Hardy Spaces

[*] Also consult references on disturbance rejection in § 6.12.*K2;* and on sensitivity issues in selected multidisciplinary optimization methods, and numerical methods in § 6.12.*P;* as well as entries on stability robustness analysis in § 5.12.*E*, and § 5.12.*G*.

[H1] Sensitivity Considerations in Hardy Spaces
Pertinent References on Feedback System Design and Mathematics

[*] Also consult references in § 6.12.*A*, § 2.9.*A*, and § 2.9.*E*.

[1] V.M. Adamjan, D.Z. Arov, and M.G. Krein, "Analytic properties of Schmidt pairs for a Hankel operator and the generalized Schur-Takagi problem," *Math. USSR Sbornik*, vol. 15, pp. 31-73, 1971.

[2] _____, "Infinite Hankel block matrices and related extension problems," *Amer. Math. Soc. Translations*, series 2, vol. 111, pp. 133-156, 1978.

[3] J.A. Ball and J.W. Helton, "A Beurling-Lax theorem for the Lie group U(m, n) which contains most classical interpolation theory," *JOT*, vol. 9, pp. 107-142, 1983.

[4] J.J. Bongiorno, Jr., "Minimum sensitivity design of linear multivariable feedback control systems by matrix spectral factorization," *IEEE-TAC*, vol. AC-14, pp. 665-573, 1969.

[5] J.B. Conway, *A Course in Functional Analysis*. New York: Springer – Verlag, 1985.

[6] P. Delsarte, Y. Genin, and Y. Kamp, "Schur parameterization of positive definite block-Toeplitz systems," *SIAM-JAM*, vol. 36, no. 1, pp. 34-46, 1979.

[7] _____, "The Nevanlinna-Pick problem for matrix-valued functions," *SIAM-JAM*, vol. 36, no. 1, pp. 47-61, 1979.

[8] C.A. Desoer and W.S. Chan, "The feedback interconnection of lumped linear time-invariant systems," *JFI*, vol. 300, nos. 5 & 6, pp. 335-351, 1975.

[9] C.A. Desoer, R.-W. Liu, J. Murray, and R. Saeks, "Feedback system design: The fractional representation approach to analysis and synthesis," *IEEE-TAC*, vol. AC-25, no. 3, pp. 399-412, 1980.

[10] P.L. Duren, *Theory of H^p Spaces*. New York: Academic Press, 1970.

[11] P. Koosis, *Introduction to H_p Spaces*. Cambridge, England: Cambridge Univ. Press, 1980.

[12] S.-Y. Kung and D.W. Lin, "Optimal Hankel-norm model reductions: Multivariable systems," *IEEE-TAC*, vol. AC-26, no. 4, pp. 832-852, 1981.

[13] Z. Nehari, "On bounded bilinear forms," *Annals of Mathematics*, vol. 65, pp. 153-162, 1957.

[14] R. Nevanlinna, *Analytic Funstions*. [Translated from German by P. Emig.] New York: Springer – Verlag, 1970.

[15] R. Redheffer, "Inequalities for a matrix Riccati equation," *JMM*, vol. 8, pp. 349-367, 1959.

[16] _____, "Supplementary notes on matrix Riccati equations," *JMM*, vol. 9, pp. 745-748, 1960.

[17] R.M. Redheffer, "On a certain linear fractional transformation," *JMP*, vol. 39, pp. 269-286, 1960. [This is the same author as the preceding one.]

[18] W.T. Reid, "Solutions of a Riccati matrix differential equation as functions of initial values," *JMM*, vol. 8, pp. 221-230, 1959.

[19] _____, "Properties of solutions of a Riccati matrix differential equation," *JMM*, vol. 9, pp. 749-769, 1960.

[20] D. Sarason, "Generalized interpolation in H^∞," *Trans. of American Mathematical Society*, vol. 127, no. 2, pp. 179-203, 1967.

[21] G. Stein and M. Athans, "The LQG/LTR procedure for multivariable feedback control design," *IEEE-TAC*, vol. AC-32, no. 2, pp. 105-114, 1987.

[22] B. Sz.-Nagy and C. Foias, *Harmonic Analysis of Operators on Hilbert Space*. Amsterdam: North-Holland Publ., 1970.

[23] M. Vidyasagar, *Control System Synthesis: A Factorization Approach*. Cambridge, MA: MIT Press, 1985. Reviewed by B.F. Wyman, *IEEE-TAC*, vol. AC-31, no. 11, p. 1085, 1986, and by F.M. Callier, *Automatica*, vol. 22, no. 4, pp. 500-501, 1986.

[24] J.C. Willems, "Least squares stationary optimal control and the algebraic Riccati equation," *IEEE-TAC*, vol. AC-16, no. 6, pp. 621-634, 1971.

[25] D.C. Youla, J.J. Bongiorno, Jr., and H.A. Jabr, "Modern Wiener-Hopf design of optimal controllers Part I: The single-input-output case," *IEEE-TAC*, vol. AC-21, pp. 3-13, 1976.

[26] D.C. Youla, H.A. Jabr, and J.J. Bongiorno, Jr., "Modern Wiener-Hopf design of optimal controllers–Part II: The multivariable case," *IEEE-TAC*, vol. AC-21, pp. 319-338, 1976.

[H2] Sensitivity Considerations in Hardy Spaces
Complex – Domain Approach

[1] B. Bamieh, J.B. Pearson, B.A. Francis, and A. Tannenbaum, "A lifting technique for linear periodic systems with applications to sampled-data control," *SCL*, vol. 17, pp. 79-88, 1991.

[2] P. Boekhoudt, "Solution of polynomial equations in H_∞ optimal control," *CTAT*, vol. 6, no. 3, pp. 321-337, 1990.

6.12.H. Sensitivity Considerations in Hardy Spaces

[3] M.J. Englehart and M.C. Smith, "A four-block problem for H_∞ design: Properties and applications," *Automatica*, vol. 27, no. 5, pp. 811-818, 1991.

[4] N.A. Fairbairn and M.J. Grimble, "H_∞ robust controller for self-tuning applications Part 3. Self-tuning controller implementation," *IJC*, vol. 52, no. 1, pp. 15-36, 1990.

[5] D.S. Flamm and S.K. Mitter, "H^∞ sensitivity minimization for delay systems," *SCL*, vol. 9, pp. 17-24, 1987.

[6] C. Foias and A. Tannenbaum, "Weighted optimization theory for nonlinear systems," *SIAM-JCO*, vol. 27, no. 4, pp. 842-860, July 1989.

[7] C. Foias, A. Tannenbaum, and G. Zames, "On decoupling the H^∞-optimal sensitivity problem for products of plants," *SCL*, vol. 7, pp. 239-245, 1986.

[8] _____, "Weighted sensitivity minimization for delay systems," *IEEE-TAC*, vol. AC-31, no. 8, pp. 763-766, Aug. 1986.

[9] _____, "On the H^∞-optimal sensitivity problem for systems with delays," *SIAM-JCO*, vol. 25, no. 3, pp. 686-705, May 1987.

[10] Y.K. Foo and I. Postlethwaite, "An H^∞-minimax approach to the design of robust control systems," *SCL*, vol. 5, pp. 81-88, 1984.

[11] _____, "An H^∞-minimax approach to the design of robust control systems, Part II: All solutions, all-pass form solutions and the 'best' solution," *SCL*, vol. 7, pp. 261-268, 1986.

[12] B.A. Francis, *A Course in H_∞ Control Theory*. New York: Springer – Verlag, 1987. Reviewed by G. Conte, *IEEE-TAC*, vol. AC-32, no. 12, pp. 1144-1145, 1987, and by P.P. Khargonekar, *SIAM R.*, vol. 30, no. 2, pp. 335-336, 1988.

[13] B.A. Francis and J.C. Doyle, "Linear control theory with an H_∞ optimality criterion," *SIAM-JCO*, vol. 25, no. 4, pp. 815-844, July 1987.

[14] B.A. Francis, J.W. Helton, and G. Zames, "H^∞-optimal feedback controllers for linear multivariable systems," *IEEE-TAC*, vol. AC-29, no. 10, pp. 888-900, Oct. 1984.

[15] B.A. Francis and G. Zames, "On H^∞-optimal sensitivity theory for SISO feedback systems," *IEEE-TAC*, vol. AC-29, no. 1, pp. 9-16, Jan. 1984.

[16] D. Fragopoulos, M.J. Grimble, and U. Shaked, "H_∞ controller design for the SISO case using a Wiener approach," *IEEE-TAC*, vol. AC-36, no. 10, pp. 1204-1208, 1991.

[17] K. Glover, D.J.N. Limbeer, J.C. Doyle, E.M. Kasenally, and M.G. Safonov, "A characterization of all solutions to the four block general distance problem," *SIAM-JCO*, vol. 29, no. 2, pp. 283-324, 1991.

[18] M. Green, "H_∞ controller synthesis by J–lossless coprime factorization," *SIAM-JCO*, vol. 30, no. 3, pp. 522-547, 1992.

[19] M.J. Grimble, "H_∞ robust controller for self-tuning control applications Part 1. Controller design," *IJC*, vol. 46, no. 4, pp. 1429-1444, 1987.

[20] _____, "H_∞ robust controller for self-tuning control applications Part 2. Self-tuning and robustness," *IJC*, vol. 46, no. 5, pp. 1819-1840, 1987.

[21] _____, "Optimal H_∞ multivariable robust controllers and the relationship to LQG design problems," *IJC*, vol. 48, no. 1, pp. 33-58, 1988.

[22] _____, "Extensions to H_∞ multivariable robust controllers and the relationship to LQG design problems," *IJC*, vol. 50, no. 1, pp. 309-338, 1989.

[23] _____, "Predictive H_∞ model reference optimal control law for SISO systems," *Proc. IEE*, vol. 136, Pt. D, no. 6, pp. 273-284, Nov. 1989.

[24] _____, "Generalised H_∞ multivariable controllers," *Proc. IEE*, vol. 136, Pt. D, no. 6, pp. 285-297, Nov. 1989.

[25] _____, "Comments on Stein's postulated high-frequency behavior of H^∞ optimal controllers," *IEEE-TAC*, vol. AC-35, no. 6, pp. 762-765, June 1990. Comments by J.M. Krause, *ibid.*, vol. AC-37, no. 5, p. 702, 1992.

[26] _____, "H_∞ controllers with a PID structure," *ASME-JDSMC*, vol. 112, pp. 325-336, Sept. 1990.

[27] _____, "H_∞ observations weighted control law," *ASME-JDSMC*, vol. 112, pp. 337-348, Sept. 1990.

[28] S. Hara and T. Sugie, "Inner-outer factorization for strictly proper functions with $j\omega$-axis zeros," *SCL*, vol. 16, pp. 179-185, 1991.

[29] S. Hara, T. Sugie, and R. Kondo, "H_∞ control problem with $j\omega$-axis zeros," *Automatica*, vol. 28, no. 1, pp. 55-70, 1992.

[30] J.W. Helton, "Worst case analysis in the frequency domain: The H^∞ approach to control," *IEEE-TAC*, vol. AC-30, no. 12, pp. 1154-1170, Dec. 1985.

[31] H. Ito, H. Ohmori, and A. Sano, "Design of stable controllers attaining low H^∞ weighted sensitivity," *IEEE-TAC*, vol. AC-38, no. 3, pp. 485-488, 1993.

[32] P.P. Khargonekar and A. Tannenbaum, "Non-Euclidian matrices and robust stabilization of systems with parameter uncertainty," *IEEE-TAC*, vol. AC-30, pp. 1005-1013, Oct. 1985.

[33] R. Kondo and S. Hara, "On cancellation in H_∞ optimal controllers," *SCL*, vol. 13, pp. 205-210, 1989.

[34] H. Kwakernaak, "A polynomial approach to minimax frequency domain optimization of multivariable feedback systems," *IJC*, vol. 44, no. 1, pp. 117-156, 1986.

[35] D.K. Le and A.E. Frazho, "A numerical procedure for a non-rational H^∞-optimization problem in control design," *SCL*, vol. 16, pp. 9-15, 1991.

[36] K.E. Lenz, "Simple mixed sensitivity optimal controllers," *SCL*, vol. 17, pp. 363-373, 1991.

[37] K.E. Lenz, P.P. Khargonekar, and J.C. Doyle, "When is a controller H^∞-optimal?" *MCSS*, vol. 1, pp. 107-122, 1988.

[38] G.M.H. Leung, T.P. Perry, and B.A. Francis, "Performance analysis of sampled-data control systems," *Automatica*, vol. 27, no. 4, pp. 699-704, 1991.

[39] D.W. Luse and J.A. Ball, "Frequency-scale decomposition of H^∞-disk problems," *SIAM-JCO*, vol. 27, no. 4, pp. 814-835, July 1989.

[40] R.J. Ober and J.A. Sefton, "Stability of control systems and graphs of linear systems," *SCL*, vol. 17, pp. 265-280, 1991.

[41] Y. Ohta, H. Maeda, and S. Kodama, "Unit interpolation in H_∞: Bounds of norm and degree of interpolants," *SCL*, vol. 17, pp. 251-256, 1991.

[42] Y. Ohta, G. Tadmor, and S.K. Mitter, "Sensitivity reduction over a frequency band," *IJC*, vol. 48, no. 5, pp. 2129-2138, 1988.

[43] S.D. O'Young and B.A. Francis, "Sensitivity tradeoffs for multivariable plants," *IEEE-TAC*, vol. AC-30, no. 7, pp. 625-632, July 1985.

[44] S.D. O'Young, I. Postlethwaite, and D.-W. Gu, "A treatment of $j\omega$-axis model-matching transformation zeros in the optimal H^2 and H^∞ control designs," *IEEE-TAC*, vol. AC-34, no. 5, pp. 551-553, May 1989.

[45] H. Ozbay and A. Tannenbaum, "On the structure of suboptimal H^∞ controllers in the sensitivity minimization problem for distributed stable plants," *Automatica*, vol. 27, no. 2, pp. 293-305, 1991.

[46] J.R. Partington and K. Glover, "Robust stabilization of delay systems by approximation of coprime factors," *SCL*, vol. 14, pp. 325-331, 1990.

[47] L. Qiu and E.J. Davison, "Feedback stability under simultaneous gap metric uncertainties in plant and controller," *SCL*, vol. 18, pp. 9-22, 1992.

6.12.H. Sensitivity Considerations in Hardy Spaces

[48] S. Raman and E.-W. Bai, "A linear, robust and convergent interpolatory algorithm for quantifying model uncertainties," *SCL*, vol. 18, pp. 173-177, 1992.

[49] R. Ravi, A.M. Pascoal, and P.P. Khargonekar, "Normalized coprime factorizations for linear time-varying systems," *SCL*, vol. 18, pp. 455-465, 1992.

[50] M. Saeki, "Methods of solving a polynomial equation for an H^∞ optimal control problem for a single-input single-output discrete-time system," *IEEE-TAC*, vol. AC-34, no. 2, pp. 166-168, Feb. 1989.

[51] _____, "H^∞/LTR procedure with specified degree of recovery," *Automatica*, vol. 28, no. 3, pp. 509-517, 1992.

[52] M.G. Safonov and V.X. Le, "An alternative solution to the H_∞-optimal control problem," *SCL*, vol. 10, pp. 155-158, 1988.

[53] J. Sefton and K. Glover, "Pole/zero cancellations in the general H_∞ problem with reference to a two block design," *SCL*, vol. 14, pp. 295-306, 1990.

[54] U. Shaked, "The explicit structure of inner matrices and its applications in H^∞-optimization," *IEEE-TAC*, vol. AC-34, no. 7, pp. 734-738, July 1989.

[55] M.C. Smith, "Well-posedness of H^∞ optimal control problems," *SIAM-JCO*, vol. 28, no. 2, pp. 342-358, March 1990.

[56] T. Sugie and S. Hara, "H_∞-suboptimal control problem with boundary constraints," *SCL*, vol. 13, pp. 93-99, 1989.

[57] G. Tadmor, "An interpolation problem associated with H^∞-optimal design in systems with distributed input lags," *SCL*, vol. 8, pp. 313-319, 1987.

[58] M.C. Tsai, E.J.M. Geddes, and I. Postlethwaite, "Pole–zero cancellations and closed-loop properties of an H^∞ mixed sensitivity design problem," *Automatica*, vol. 28, no. 3, pp. 519-530, 1992.

[59] L.Y. Wang and G. Zames, "Lipschitz continuity of H^∞ interpolation," *SCL*, vol. 14, pp. 381-387, 1990.

[60] N.E. Wu and G. Gu, "Discrete Fourier transform and H^∞-approximation," *IEEE-TAC*, vol. AC-35, no. 9, pp. 1044-1046, Sept. 1990.

[61] Q.-H. Wu and M. Mansour, "Robust output regulation for a class of linear multivariable systems," *SCL*, vol. 13, pp. 227-232, 1989.

[62] _____, "Decentralized robust output regulation using H^∞-optimization techniques," *CTAT*, vol. 6, no. 2, pp. 187-214, 1990.

[63] _____, "H^∞-optimal solutions of robust regulator problem for linear MIMO systems," *IJC*, vol. 52, no. 5, pp. 1241-1262, 1990.

[64] I. Yaesh and U. Shaked, "Nondefinite least squares and its relation to H_∞-minimum error state estimation," *IEEE-TAC*, vol. AC-36, no. 12, pp. 1469-1472, 1991.

[65] J.-S. Young, C.E. Lin, and F.-B. Yeh, "Characterization of the sub-layers for 2-block H^∞-optimal control problem," *SCL*, vol. 15, pp. 193-198, 1990.

[66] G. Zames, "Feedback and optimal sensitivity: Model reference transformation, multiplicative seminorms, and approximate inverses," *IEEE-TAC*, vol. AC-26, no. 2, pp. 301-320, April 1981.

[67] G. Zames and D. Bensoussan, "Multivariable feedback, sensitivity, and decentralized control," *IEEE-TAC*, vol. AC-28, no. 11, pp. 1030-1035, Nov. 1983.

[68] G. Zames and B.A. Francis, "Feedback, minimax sensitivity, and optimal robustness," *IEEE-TAC*, vol. AC-28, no. 5, pp. 585-601, May 1983.

[H3] Sensitivity Considerations in Hardy Spaces
State – Space Approach

[1] D.S. Bernstein and W.M. Haddad, "LQG control with an H^∞ performance bound: A Riccati approach," *IEEE-TAC*, vol. AC-34, no. 3, pp. 293-305, March 1989.

[2] P.M.M. Bongers and O.H. Bosgra, "Normalized coprime factorizations for systems in generalized state-space form," *IEEE-TAC*, vol. AC-38, no. 2, pp. 348-350, 1993.

[3] B.-C. Chang, "A stable state-space realization in formulation of H^∞ norm computation," *IEEE-TAC*, vol. AC-32, no. 9, pp. 811-815, Sept. 1987.

[4] C.E. de Souza, M. Fu, and L. Xie, "H_∞ analysis and synthesis of discrete-time systems with time-varying uncertainty," *IEEE-TAC*, vol. AC-38, no. 3, pp. 459-462, 1993.

[5] C.E. de Souza and L. Xie, "On the discrete-time bounded real lemma with application in the characterization of static state feedback H_∞ controllers," *SCL*, vol. 18, pp. 61-71, 1992.

[6] J.C. Doyle, K. Glover, P.P. Khargonekar, and B.A. Francis, "State-space solutions to standard H^2 and H^∞ control problems," *IEEE-TAC*, vol. AC-34, pp. 831-847, Aug. 1989. [cf., [27] and 56].]

[7] M.K.H. Fan and A.L. Tits, "A measure of worst-case H_∞ performance and of largest acceptable uncertainty," *SCL*, vol. 18, pp. 409-421, 1992.

[8] M.K.H. Fan, A.L. Tits, and J.C. Doyle, "Robustness in the presence of mixed parametric uncertainty and unmodeled dynamics," *IEEE-TAC*, vol. AC-36, no. 1, pp. 25-38, 1991.

[9] B.A. Francis and J.C. Doyle, "Linear control theory with an H_∞ optimality criterion," *SIAM-JCO*, vol. 25, no. 4, pp. 815-844, July 1987.

[10] K. Glover and J.C. Doyle, "State-space formulae for all stabilizing controllers that satisfy an H_∞–norm bound and relations to risk sensitivity," *SCL*, vol. 11, pp. 167-172, 1988.

[11] M. Green, K. Glover, D.J.N. Limebeer, and J.C. Doyle, "A J–spectral factorization approach to H^∞ control," *SIAM-JCO*, vol. 28, pp. 1350-1371, 1990.

[12] W.M. Haddad and D.S. Bernstein, "Generalized Riccati equations for the full- and reduced-order mixed-norm H_2/H_∞ standard problem," *SCL*, vol. 14, pp. 185-197, 1990.

[13] _____, "Robust stabilization with positive real uncertainty: Beyond the small gain theorem," *SCL*, vol. 17, pp. 191-208, 1991.

[14] H.S. Hvostov, "Simplifying H^∞ controller synthesis via classical feedback system structure," *IEEE-TAC*, vol. AC-35, no. 4, pp. 485-488, April 1990.

[15] Y.S. Hung and B. Pokrud, "An H^∞ approach to feedback design with two objective functions," *IEEE-TAC*, vol. AC-37, no. 6, pp. 820-824, 1992.

[16] P.A. Iglesias, D. Mustafa, and K. Glover, "Discrete time H_∞ controllers satisfying a minimum entropy criterion," *SCL*, vol. 14, pp. 275-286, 1990.

[17] P.P. Khargonekar, K.M. Nagpal, and K.R. Poolla, "H_∞ control with transients," *SIAM-JCO*, vol. 29, no. 6, pp. 1373-1393, 1991.

[18] P.P. Khargonekar, I.R. Petersen, and M.A. Rotea, "H_∞–optimal control with state-feedback," *IEEE-TAC*, vol. AC-33, no. 8, pp. 786-788, 1988.

[19] P.P. Khargonekar, I.R. Petersen, and K. Zhou, "Robust stabilization of uncertain linear systems: Quadratic stabilizability and H^∞ control theory," *IEEE-TAC*, vol. AC-35, no. 3, pp. 356-361, 1990.

[20] P.P. Khargonekar and M.A. Rotea, "Multiple objective optimal control of linear systems: The quadratic norm case," *IEEE-TAC*, vol. AC-36, no. 1, pp. 14-24, 1991.

[21] _____, "Mixed H_2/H_∞ control: A convex optimization approach," *IEEE-TAC*, vol. AC-36, no. 7, pp. 824-837, 1991.

[22] H. Kimura, Y. Lu, and R. Kawatani, "On the structure of H^∞ control systems and related extensions," *IEEE-TAC*, vol. AC-36, no. 6, pp. 653-667, 1991.

[23] V.X. Le and M.G. Safonov, "Rational matrix GCD's and the design of squaring-down compensators – A state-space theory," *IEEE-TAC*, vol. AC-37, no. 3, pp. 384-392, 1992.

[24] D.J.N. Limebeer and G.D. Halikias, "A controller degree bound for H^∞-optimal control problems of the second kind," *SIAM-JCO*, vol. 26, no. 3, pp. 646-677, 1988.

[25] D.J.N. Limebeer and Y.S. Hung, "An analysis of the pole-zero cancellations in H^∞-optimal problems of the first kind," *SIAM-JCO*, vol. 25, no. 6, pp. 1457-1493, 1987.

[26] A.N. Madiwale, W.M. Haddad, and D.S. Bernstein, "Robust H^∞ control design for systems with structured parameter uncertainty," *SCL*, vol. 12, pp. 393-407, 1989.

[27] T. Mita, K.Z. Liu, and S. Ohuchi, "Correction of the FI results in H_∞ control and parameterization of H_∞ state feedback controllers," *IEEE-TAC*, vol. AC-38, no. 2, pp. 343-347, 1993.

[28] J.B. Moore and T.T. Tay, "Loop recovery *via* H^∞/H^2 sensitivity recovery," *IJC*, vol. 49, no. 4, pp. 1249-1271, 1989.

[29] D. Mustafa, "Relations between maximum-entropy /H_∞ control and combined H_∞/LQG control," *SCL*, vol. 12, pp. 193-203, 1989.

[30] D. Mustafa and K. Glover, "Controller reduction by H_∞-balanced truncation," *IEEE-TAC*, vol. AC-36, no. 6, pp. 668-682, 1991.

[31] _____, *Minimum Entropy H_∞ Control*. Berlin: Springer – Verlag, 1990. Reviewed by D.S. Bernstein, *IEEE-TAC*, vol. AC-37, no. 8, pp. 1276-1277, 1992, and by J.A. Ball, *SIAM R.*, vol. 35, no. 4, pp. 652-655, Dec. 1993.

[32] D. Mustafa, K. Glover, and D.J.N. Limebeer, "Solutions to the H_∞ general distance problem which minimize an entropy integral," *Automatica*, vol. 27, no. 1, pp. 193-199, 1991.

[33] I.R. Petersen, "Disturbance attenuation and H^∞ optimization: A design method based on the algebraic Riccati equation," *IEEE-TAC*, vol. AC-32, no. 5, pp. 427-429, May 1987.

[34] I.R. Petersen and C. Hollot, "High gain observers applied to problems in the stabilization of uncertain linear systems, disturbance attenuation and H^∞ optimization," *IJACSP*, vol. 2, pp. 347-369, 1988.

[35] R. Ravi, K.M. Nagpal, and P.P. Khargonekar, "H^∞ control of linear time-varying systems: A state-space approach," *SIAM-JCO*, vol. 29, no. 6, pp. 1394-1413, 1991.

[36] M.A. Rotea and P.P. Khargonekar, "H^2-optimal control with an H^∞-constraint: The state feedback case," *Automatica*, vol. 27, no. 2, pp. 307-316, 1991.

[37] M.G. Safonov, D.J.N. Limebeer, and R.Y. Chiang, "Simplifying the H^∞ theory *via* loop-shifting, matrix-pencil and descriptor concepts," *IJC*, vol. 50, no. 6, pp. 2467-2488, 1989.

[38] C. Scherer, "H^∞-control by state-feedback and fast algorithms for the computation of optimal H^∞-norms," *IEEE-TAC*, vol. AC-35, no. 10, pp. 1090-1099, Oct. 1990.

[39] G. Shi, Y. Zou, and C. Yang, "An algebraic approach to robust H^∞ control *via* state feedback," *SCL*, vol. 18, pp. 365-370, 1992.

[40] A.A. Stoorvogel, "The singular H_2 control problem," *Automatica*, vol. 28, no. 3, pp. 627-631, 1992.

[41] G. Tadmor, "Uncertain feedback loops and robustness in general linear systems," *Automatica*, vol. 27, no. 6, pp. 1039-1042, 1991.

[42] H.T. Toivonen, "Sampled-data control of continuous-time systems with an H_∞ optimality criterion," *Automatica*, vol. 28, no. 1, pp. 45-54, 1992.

[43] M.-C. Tsai, D.-W. Gu, and I. Postlethwaite, "A state-space approach to super-optimal H^∞ control problems," *IEEE-TAC*, vol. AC-33, no. 9, pp. 833-843, Sept. 1988.

[44] K. Uchida and M. Fujita, "Finite horizon H^∞ control problems with terminal penalties," *IEEE-TAC*, vol. AC-37, no. 11, pp. 1762-1767, 1992.

[45] A.J. van der Schaft, "On a state space approach to nonlinear H_∞ control," *SCL*, vol. 16, pp. 1-8, 1991.

[46] _____, "L_2-gain analysis of nonlinear systems and nonlinear state feedback H_∞ control," *IEEE-TAC*, vol. AC-37, no. 6, pp. 770-784, 1992.

[47] D.J. Walker, "Relationship between three discrete-time H^∞ algebraic Riccati equation solutions," *IJC*, vol. 52, no. 4, pp. 801-809, 1990.

[48] L. Xie and C.E. de Souza, "State feedback H_∞ optimal control problems for non-detectable systems," *SCL*, vol. 13, pp. 315-319, 1989.

[49] _____, "Robust H_∞ control for linear time-invariant systems with norm-bounded uncertainty in the input matrix," *SCL*, vol. 14, pp. 389-396, 1990.

[50] _____, "Robust H_∞ control for linear systems with norm-bounded time-varying uncertainty," *IEEE-TAC*, vol. AC-37, no. 8, pp. 1188-1191, 1992.

[51] I. Yaesh and U. Shaked, "Minimum H^∞-norm regulation of linear discrete-time systems and its relation to linear quadratic discrete games," *IEEE-TAC*, vol. AC-35, no. 9, pp. 1061-1064, Sept. 1990.

[52] _____, "H_∞-optimal one-step-ahead output feedback control of discrete-time systems," *IEEE-TAC*, vol. AC-37, no. 8, pp. 1245-1250, 1992.

[53] H.-H. Yeh, S.S. Banda, and B.-C. Chang, "Necessary and sufficient conditions for mixed H_2 and H_∞ optimal control," *IEEE-TAC*, vol. AC-37, no. 3, pp. 355-358, 1992.

[54] H.-H. Yeh, S.S. Banda, S.A. Heise, and A.C. Bartlett, "Robust control design with real-parameter uncertainty and unmodeled dynamics," *JGCD*, vol. 13, no. 6, pp. 1117-1125, 1990.

[55] K. Zhou, "Comparison between H_2 and H_∞ controllers," *IEEE-TAC*, vol. AC-37, no. 8, pp. 1261-1265, 1992.

[56] _____, "On the parameterization of H_∞ controllers," *IEEE-TAC*, vol. AC-37, no. 9, pp. 1442-1446, 1992.

[57] K. Zhou and P.P. Khargonekar, "On the weighted sensitivity minimization problem for delay systems," *SCL*, vol. 8, pp. 307-312, 1987.

[58] _____, "An algebraic Riccati equation approach to H^∞ optimization," *SCL*, vol. 11, pp. 85-91, 1988.

[H4] Sensitivity Considerations in Hardy Spaces Operator Approach

[1] J.A. Ball and N. Cohen, "Sensitivity minimization in an H^∞ norm: Parametrization of all suboptimal solutions," *IJC*, vol. 45, no. 3, pp. 785-816, 1987.

[2] J.A. Ball and A.C.M. Ran, "Optimal Hankel norm model reductions and Wiener-Hopf factorization I: The canonical case," *SIAM-JCO*, vol. 25, no. 2, pp. 362-382, 1987.

[3] _____, "Optimal Hankel norm model reductions and Wiener-Hopf factorization II: The noncanonical case," *IEOT*, vol. 10, pp. 416-436, 1987.

[4] F. Fagnani, "An operator-theoretic approach to the mixed-sensitivity minimization problem," *SCL*, vol. 17, pp. 227-235, 1991.

[5] K. Glover, "All optimal Hankel-norm approximations of linear multivariable systems and their L^∞-error bounds," *IJC*, vol. 39, no. 6, pp. 1115-1193, 1984.

[6] K. Glover and D. McFarlane, "Robust stabilization of normalized coprime factor plant descriptions with H^∞-bounded uncertainty," *IEEE-TAC*, vol. AC-34, pp. 821-830, 1989.

[7] B. Hanzon, "The area enclosed by the (oriented) Nyquist diagram and the Hilbert-Schmidt-Hankel norm of a linear system," *IEEE-TAC*, vol. AC-37, no. 6, pp. 835-839, 1992.

[8] E.A. Jonckheere and J.-C. Juang, "Fast computation of achievable feedback performance in mixed sensitivity H^∞ design," *IEEE-TAC*, vol. AC-32, no. 10, pp. 896-906, 1987.

[9] D. McFarlane and K. Glover, "A loop shaping design procedure using H_∞ synthesis," *IEEE-TAC*, vol. AC-37, no. 6, pp. 759-769, 1992.

[10] G. Michaletzky, "Hankel-norm approximation of a rational function using stochastic realizations," *SCL*, vol. 13, pp. 211-216, 1989.

[11] H. Ozbay and A. Tannenbaum, "A skew Toeplitz approach to the H^∞ optimal control of multivariable distributed systems," *SIAM-JCO*, vol. 28, no. 3, pp. 653-670, May 1990.

[12] M.G. Safonov and M.S. Verma, "L_∞ optimization and Hankel approximation," *IEEE-TAC*, vol. AC-30, no. 3, pp. 279-280, 1985.

[13] C.-D. Yang and F.-B. Yeh, "An efficient algorithm on optimal Hankel-norm approximation for multivariable systems," *IEEE-TAC*, vol. AC-37, no. 6, pp. 815-820, 1992.

[14] J.-S. Young and C.E. Lin, "Construction of the maximal Schmidt pair for the 4-block H^∞-optimal control problem," *IEEE-TAC*, vol. AC-37, no. 8, pp. 1250-1252, 1992.

[H5] *Sensitivity Considerations in Hardy Spaces Computational Issues and Optimization Approaches*

[1] T. Auba and Y. Funahashi, "Upper and lower bounds of Gramian for a class of perturbed linear systems," *IEEE-TAC*, vol. AC-37, no. 10, pp. 1659-1661, 1992.

[2] A.C. Antoulas, "On minimal realization: A polynomial approach," *SCL*, vol. 14, pp. 319-324, 1990.

[3] S. Boyd and V. Balakrishnan, "A regularity result for the singular values of a transfer matrix and a quadratically convergent algorithm for computing its L_∞-norm," *SCL*, vol. 15, pp. 1-7, 1990.

[4] S. Boyd, V. Balakrishnan, and P. Kabamba, "On computing the H_∞ norm of a transfer matrix," in *Proc. Amer. Contr. Conf.*, pp. 396-397, 1988.

[5] _____, "A bisection method for computing the H_∞ norm of a transfer matrix and related problems," *MCSS*, vol. 2, pp. 207-219, 1989. [cf., [P. 11 & 12].]

[6] N.A. Bruinsma and M. Steinbuch, "A fast algorithm to compute the H_∞-norm of a transfer function matrix," *SCL*, vol. 14, pp. 287-293, 1990.

[7] B.-C. Chang and S.S. Banda, "Optimal H^∞ norm computation for multivariable systems with multiple zeros," *IEEE-TAC*, vol. AC-34, no. 5, pp. 553-557, May 1989.

[8] B.-C. Chang, S.S. Banda, and T.E. McQuade, "Fast iterative computation of optimal two-block H^∞-norm," *IEEE-TAC*, vol. AC-34, no. 7, pp. 738-743, July 1989.

[9] B.M. Chen, A. Saberi, and U.-L. Ly, "Exact computation of the infimum in H_∞-optimization via output feedback," *IEEE-TAC*, vol. AC-37, no. 1, pp. 70-78, 1992.

[10] B.M. Chen, A. Saberi, P. Sannuti, and Y. Shamash, "Construction and parameterization of all static and dynamic H_2-optimal state feedback solutions, optimal fixed modes, and fixed decoupling zeros," *IEEE-TAC*, vol. AC-38, no. 2, pp. 248-261, 1993.

[11] G. Chen and R.J.P. de Figueiredo, "Construction of the left coprime fractional representation for a class of nonlinear control systems," *SCL*, vol. 14, pp. 353-361, 1991.

[12] C.-C. Chu, J.C. Doyle, and E.B. Lee, "The general distance problem in H_∞ optimal control theory," *IJC*, vol. 44, no. 2, pp. 565-596, 1986.

[13] D.F. Enns, "Model reduction with balanced realizations: An error bound and a frequency weighted generalization," in *Proc. 23rd IEEE Conf. on Decision and Contr.*, pp. 127-132, Dec. 1984.

[14] C.H. Fang and F.R. Chang, "A connection between state-space and doubly coprime matrix-fraction descriptions of multivariable systems," *SCL*, vol. 14, pp. 261-265, 1990.

[15] Y.K. Foo, "Convergent inequalities for four-block γ-iteration in H^∞ optimization problems," *CTAT*, vol. 6, no. 3, pp. 463-472, 1990.

[16] T.T. Georgiou and P.P. Khargonekar, "A constructive algorithm for sensitivity optimization of periodic systems," *SIAM-JCO*, vol. 25, no. 2, pp. 334-340, March 1987.

[17] D.-W. Gu and D.Q. Mayne, "A new problem in control system design," *IEEE-TAC*, vol. AC-35, no. 1, pp. 114-117, Jan. 1990.

[18] D.-W. Gu, M.C. Tsai, and I. Postlethwaite, "Improved formulae for the 2-block H^∞ superoptimal solution," *Automatica*, vol. 26, no. 2, pp. 437-440, 1990.

[19] _____, "A frame approach to the H^∞ superoptimal solution," *IEEE-TAC*, vol. AC-35, no. 7, pp. 829-835, 1990.

[20] L. Guo, L. Xia, and Y. Liu, "Recursive algorithm for the computation of the H^∞-norm of polynomials," *IEEE-TAC*, vol. AC-33, no. 12, pp. 1154-1158, Dec. 1988.

[21] L. He and E. Polak, "Effective diagonalization strategies for solution of a class of optimal design problems," *IEEE-TAC*, vol. AC-35, no. 3, pp. 258-267, March 1990.

[22] J. Lam and B.D.O. Anderson, "L_1 impulse response error bound for balanced truncation," *SCL*, vol. 18, pp. 129-137, 1992.

[23] Y. Liu and K.L. Teo, "Convergence rate for an approximation approach to H_∞-norm optimization problems with an application to controller order reduction," *Automatica*, vol. 28, no. 3, pp. 617-621, 1992.

[24] P. Pandey, C. Kenney, A. Packard, and A.J. Laub, "A gradient method for computing the optimal H_∞ norm," *IEEE-TAC*, vol. AC-36, no. 7, pp. 887-890, 1991.

[25] E. Polak and D.Q. Mayne, "Algorithm models for nondifferentiable optimization," *SIAM-JCO*, vol. 23, no. 3, pp. 477-491, May 1985.

[26] E. Polak and S.E. Salcudean, "On the design of linear multivariable feedback systems *via* constrained nondifferentiable optimization in H^∞ spaces," *IEEE-TAC*, vol. AC-34, no. 3, pp. 268-276, 1989.

[27] E. Polak, S.E. Salcudean, and D.Q. Mayne, "Adaptive control of ARMA plants using worst-case design by semi-infinite optimization," *IEEE-TAC*, vol. AC-32, no. 5, pp. 388-396, May 1987.

[28] E. Polak and E.J. Wiest, "Variable-metric technique for the solution of affinely parameterized nondifferentiable optimal design problems," *JOTA*, vol. 66, no. 3, pp. 391-414, 1990.

[29] E. Polak and T.L. Wuu, "On the design of stablizing compensators *via* semiinfinite optimization," *IEEE-TAC*, vol. AC-34, no. 2, pp. 196-200, Feb. 1989.

[30] F. Reza, "The application of Caratheodory–Schur optimization," *Computers Elect. Engng*, vol. 17, no. 1, pp. 49-53, 1991.

[31] G. Robel, "On computing the infinity norm," *IEEE-TAC*, vol. AC-34, no. 8, pp. 882-884, 1989.

[32] U. Shaked and I. Yaesh, "A simple method for deriving J–spectral factors," *IEEE-TAC*, vol. AC-37, no. 6, pp. 891-895, 1992.

[33] K. Sugimoto and Y. Yamamoto, "A polynomial matrix method for computing stable rational doubly coprime factorization," *SCL*, vol. 14, pp. 267-273, 1990.

[34] C.P. Therapos, "Balance realization of stable transfer function matrices," *IEEE-TAC*, vol. AC-37, no. 2, pp. 281-285, 1992.

[35] R.J. Veillette and J.V. Medanic, "H_∞-norm bounds for ARE-based designs," *SCL*, vol. 13, pp. 193-204, 1989.

[36] C.-D. Yang and F.-B. Yeh, "A simple algorithm on minimal balanced realization for transfer function matrices," *IEEE-TAC*, vol. AC-34, no. 8, pp. 879-882, 1989.

[37] F.-B. Yeh and L.-F. Wei, "Inner-outer factorizations of right-invertible real-rational matrices," *SCL*, vol. 14, pp. 31-36, 1990.

[H6] Sensitivity Considerations in Hardy Spaces
Applications and Extensions

[*] Publications in this area are pouring in and these are just representative.

[1] T. Basar and P. Bernhard, H_∞-*Optimal Control and Related Minimax Design Problems, a dynamic game approach*. Boston: Birkhauser, 1991. Reviewed by D. Ghose, *J. Indian Sciences*, vol. 72, March/April 1992, and by A. Halanay and V. Ionescu, *J. Rev. Romanian Sciences, Techn.-Electrotech et Energie*, vol. 37, no. 2, p. 245, Bucharest 1992.

[2] A.R. Guesalaga and H.W. Kropholler, "Improved temperature and humidity control using H_∞ synthesis," *Proc. IEE*, vol. 137, Pt. D, no. 6, pp. 374-380, 1990.

[3] H. Kazerooni, T.-I. Tsay, and K. Hollerbach, "A controller design framework for telerobotic systems," *IEEE-TCST*, vol. 1, no. 1, pp. 50-62, 1993.

[4] M. Morari and E. Zafiriou, *Robust Process Control*. Englewood Cliffs, NJ: Prentice-Hall, 1989. Reviewed by J.C. Kantor, *AIChE J.*, vol. 37, no. 12, pp. 1905-1906, 1991.

[5] D.E. Rivera and M. Morari, "Low-order SISO controller tuning methods for the H_2, H_∞ and μ objective functions," *Automatica*, vol. 26, no. 2, pp. 361-369, 1990.

[6] _____, "Plant and controller reduction problems for closed-loop performance," *IEEE-TAC*, vol. AC-37, no. 3, pp. 398-404, 1992.

[7] A. Yue and I. Postlethwaite, "Improvement of helicopter handling qualities using H^∞-optimization," *Proc. IEE*, vol. 137, Pt. D, no. 3, pp. 115-129, 1990.

[I] Conference Proceedings / Reports
and Book Reviews on Invariance Theory

[1] *Invariance Theory and Its Use in Control Devices*. Proc. First All-Union Conf. on Invariance Theory and its use in Control Systems. Moscow: Izd-vo AN SSSR, 1959. [A commentary on this meeting and its resolutions can be found in *ARC*, vol. 20, no. 8, pp. 1109-1115, 1959.]

[2] *Invariance Theory in Control Systems*. Proc. Second All-Union Conf. on Invariance Theory and its use in Control Systems. Moscow: Nauka Press, 1964.

[3] *Theory of Invariance of Control Systems*. Proc. Third All-Union Conf. on Invariance Theory and Its Use in Control Systems. Moscow: Nauka Press, 1970.

[4] *Invariance Theory and the Theory of Sensitivity of Control Systems*. Materials of Fourth All-Union Conf., Parts 1-3, AN SSSR and AN Ukr SSR, 1971.

[5] *News Items, Fifth All-Union Conference on Invariance Theory, Sensitivity Theory, and Their Applications. Republican School-Seminar on Invariance, Stability*, and Sensitivity. *SAC*, vol. 9, pp. 73-74, Jan./Feb. 1976.

[6] Invariance Theory, Sensitivity Theory, and Applications, Six All-Union Conference (Reports of Papers) [in Russian], Inst. of Control Problems, Moscow (1982).

[7] Seventh All-Union Conference on the Theory of Invariance and Sensitivity of Automatic Systems (Reports of Papers) (Academician V.A. Trapeznikov, General Chairman), Baku, June 1987. [The author could not locate any other information on this and the previous entry.]

[8] The fiftieth anniversary of publishing Professor G.V. Shchipanov's work: "Theory and methods of design of automatic regulators," *SJAIS*, vol. 22, no. 2, pp. 82-95, 1989.

[9] N.M. Chumakov, "Fourth all-union conference on the theory of invariance and the theory of sensitivity of automatic systems," *ARC*, no. 3, pp. 511-514, March 1972.

[10] N.A. Kachanova and P.I. Chinaev, "Conference on invariance theory and its applications to automatic devices," *ARC*, vol. 20, no. 8, pp. 1109-1114, Resolutions of this conference, *ibid*, pp. 1114-1115, Aug. 1959.

[11] V.S. Kulebakin, *Teoriia Invariantnosti i Ee Primenenie v Avtomati-Cheskikh Ustroistvakh*. Moskva: 1959.

[12] V.M. Kuntsevich, "Conclusion of discussion on the invariance of automatic control systems," *SJAIS*, vol. 21, no. 2, pp. 98-99, 1988.

[13] L.N. Mikhailov, "Some remarks on the theory of complete compensation of disturbances," *Avtomatika i Telemekhanika*, no. 5, [pp. 145-154 in Russian], 1940.

[14] T. Soveshch, *Theory of Invariance and Its Application to Automatic Control [in Russian]*. Izd-vo AN Uk. SSR, 1959.

[15] M. Ulanov, "(Book Review) 'The Problem of Invariance Theory In Automatic Control'," By A.I. Kukhtenko, *ARC*, vol. 27, no. 4, pp. 729-731, April 1966.

[16] S.D. Zemlyakov, "Third all-union meeting on invariance theory and its applications in automatic control systems," *ARC*, no. 4, pp. 681-684, April 1967.

[J] Selected Techniques from Invariance Theory

[1] J.K. Aggarwal, H.T. Banks, and N.H. McClamroch, "Invariance in linear systems," *JMAA*, vol. 29, pp. 498-506, 1970.

[2] S. Barnett, C. Storey, J.B. Cruz, and W.R. Perkins, "Comment on 'Invariance and sensitivity'," *IEEE-TAC*, vol. AC-12, pp. 210-211, April 1967.

[3] C. Bonivento, R. Gudorzi, and G. Marro, "Parametric insensitivity and controlled invariance," in *Proc. 3rd IFAC Symp. on Sensitivity, Adaptivity and Optimality*, pp. 177-182, Ischia, Italy, 1973.

[4] V.G. Borisov, S.N. Diligenskii, and A.Y. Efremov, "Synthesis of invariant control systems using eigenstructures," *ARC*, vol. 51, no. 7, pp. 855-866, 1990.

[5] L.M. Boychuk, "Necessary and sufficient conditions for absolute invariance of control systems with indirect measurement of disturbances," *SAC*, vol. 2, no. 4, pp. 13-18, 1969.

[6] _____, "Was there a mistake in G.V. Shchipanov's work?" *SJAIS*, vol. 19, no. 3, pp. 82-94, 1986. Comments by A.G. Ivakhnenko, *ibid.*, pp. 95-97, and in vol. 20, no. 3, p. 87, 1987.

[7] _____, "Invariant filtering of discrete processes," *SJAIS*, vol. 20, no. 2, pp. 9-15, 1987.

[8] L.M. Boychuk and G.S. Finin, "A method of controlling objects with variable parameters based on using coordinating systems," *SJAIS*, vol. 22, no. 6, pp. 37-44, 1989.

[9] L.M. Boychuk and A.M. Voronin, "Invariance theory in control systems (a survey)," *SAC*, vol. 2, no. 4, pp. 1-13, 1969.

[10] L.M. Boychuk and Y.P. Yurachkovskiy, "Static stability and homeostatism of coordinating automatic control systems," *SJAIS*, vol. 23, no. 6, pp. 33-43, 1990.

[11] J.B. Cruz, Jr., and W.R. Perkins, "On invariance and sensitivity," in *IEEE Int. Conv. Record*, vol. 14, Pt. 7, pp. 159-162, 1966.

[12] _____, "Conditions for signal and parameter invariance in dynamical systems," *IEEE-TAC*, vol. AC-11, pp. 614-615, July 1966.

[13] V.I. Elkin, "Realization, invariance, and autonomy of nonlinear controllable dynamic systems," *ARC*, vol. 42, no. 7, pp. 878-885, 1981.

[14] A.T. Fam and J.S. Meditch, "On input-output parametric invariance in linear systems," *IEEE-TAC*, vol. AC-21, No. 6, pp. 870-871, Dec. 1976.

[15] A.R. Gaiduk, "Analytic design of invariant control systems for one-dimensional plants," *ARC*, vol. 42, no. 5, pp. 557-565, 1981.

[16] _____, "Design of control systems with a given form of inputs," *ARC*, vol. 45, no. 6, pp. 692-699, 1984.

[17] _____, "Selecting the feedback in a control system of minimum complexity," *ARC*, vol. 51, no. 5, pp. 593-600, 1990.

[18] C. Gori-Giogi and O.M. Grasselli, "A new approach to the study of parameter insensitivity," *Automatica*, vol. 11, pp. 181-188, 1975.

[19] J.M. Ham, "Parameter invariance through the use of nonlinear comparators," in *Proc. Allerton Conf.*, pp. 578-588, 1965.

[20] J.M. Ham and M.A. Hassan, "Appropriate parameter invariance with nonlinear feedback," *IEEE-TAC*, vol. AC-10, no. 1, pp. 87-89, 1965.

[21] A.G. Ivakhnenko and L.M. Boychuk, "Ensuring a given motion of an invariant servo system (as applied to the construction of differentiators)," *SJAIS*, vol. 20, no. 5, pp. 62-64, 1987.

[22] V.I. Ivanenko, "Some questions regarding indeterminacy and dynamics in control problems," *SJAIS*, vol. 20, no. 5, pp. 31-41, 1987.

[23] C.D. Johnson, "Invariant hyperplanes for linear dynamical systems," *IEEE-TAC*, vol. AC-11, pp. 113-116, Jan. 1966.

[24] Y.S. Kan and A.I. Kibzun, "Stabilization of a dynamic system which is under the action of undetermined and random disturbances," *ARC*, vol. 51, no. 12, pp. 1665-1673, 1990.

[25] Y.S. Kharin, "Training of invariant recognition systems," *SAC*, vol. 11, no. 1, pp. 17-26, 1978.

[26] M.M. Khrustalev and M.V. Azanov, "A parametric family of generating functions and sufficient conditions for the optimality of discontinuous systems," *SJAIS*, vol. 22, no. 6, pp. 78-84, 1989.

[27] Y.R. Kotta, "Designing nonlinear discrete-time systems which are invariant under disturbances," *ARC*, vol. 51, no. 3, pp. 294-300, 1990.

[28] A.I. Kukhtenko, "Criteria for absolute invariance for control systems with variable parameters," *Izv. AN SSSR, Otd. tekhn., Energetika i Avtomatika*, no. 2, 1961.

[29] _____, "What can the 'abstract theory of systems' offer control science?" *SAC*, vol. 12, no. 4, pp. 1-8, 1979.

[30] _____, "The basic steps in the development of invariance theory," *SAC*, vol. 17, no. 2, pp. 1-10, 1984.

[31] _____, "Basic steps in the development of invariance theory Part 2. Extension of the subject area of investigations," *SJAIS*, vol. 18, no. 2, pp. 1-11, 1985.

[32] _____, "The basic steps in the development of invariance theory Part III. Nonlinear invariant systems," *SJAIS*, vol. 18, no. 6, pp. 1-11, 1985.

[33] R.M. Kukuliyev, "Application of invariance theory to the synthesis of inertial damped navigational systems for complex motion of an object," *SJAIS*, vol. 22, no. 1, pp. 31-39, 1989.

[34] M.B. Leitman, "Use of the invariance principle for improving the accuracy and dynamic characteristics of compensation data-measurement transducers," *ARC*, vol. 39, no. 7, pp. 1077-1087, 1978.

[35] N.N. Luzin, "Study of matrix theory of differential equations," *Avtomatika i Telemekhanika*, no. 5, 1940.

[36] N.H. McClamroch, J.K. Aggarwal, and L.G. Clark, "On parameter invariance in linear control systems," *IJC*, vol. 5, no. 4, pp. 361-367, 1967.

[37] _____, "Sensitivity of linear control systems to large parameter variations," *Automatica*, vol. 5, pp. 257-263, 1969.

[38] V.S. Mechetnyy, "Double invariance of control systems," *SAC*, vol. 2, no. 4, pp. 21-27, 1969.

[39] _____, "Generalized conditions for absolute invariance of horizontal flight coordinates under atmospheric disturbances," *SAC*, vol. 4, no. 1, pp. 5-9, 1971.

[40] B.G. Mel'nikov, "Sensitivity of the optimum probability filters," *ECy*, vol. 23, no. 5, pp. 138-144, 1985.

[41] A.N. Michel and V. Vittal, "On the mechanism of transient instability of power systems," *CSSP*, vol. 4, no. 3, pp. 413-434, 1985.

[42] T. Mita, "Design of a zero-sensitive systems," *IJC*, vol. 24, no. 1, pp. 75-81, 1976. Comment on this and a sequence of other related papers of this author by B. Porter, *ibid.*, vol. 27, no. 2, pp. 325-326, 1978. Author's reply, *ibid.*, vol. 29, no. 3, p. 535, 1979.

[43] T. Mita and K. Hasegawa, "Zeroing and the design of a zero-sensitivity system," *EEJ*, vol. 95, no. 5, pp. 118-125, Sept./Oct. 1975.

[44] V.A. Orlov, "Technique for the analysis of invariant and optimal systems with periodic parameters," *ARC*, vol. 27, no. 9, pp. 1522-1534, Sept. 1966.

[45] T.K.C. Peng, "Invariance and stability for bounded uncertain systems," *SIAM J. Contr.*, vol. 10, no. 4, pp. 679-690, Nov. 1972.

[46] B.N. Petrov, V.V. Dement'yeva, and M.M. Khrustalev, "Synthesis of an invariant dynamic system with program control," *ECy*, vol. 19, no. 4, pp. 148-151, 1981.

[47] B.N. Petrov and P.D. Krut'ko, "Sensitivity theory in automatic control," *ECy*, vol. 8, no. 2, pp. 380-389, 1970.

[48] B.N. Petrov, V.Y. Rutkovski, and I.N. Krutova, "The problems of invariance, sensitivity and optimization in the theory of one class of self-adapting systems," *Proc. IFAC*, vol. 1 (of the 3rd Congress), pp. 18D.1-18D.10, 1966.

[49] V.P. Pichkurenko and O.V. Fokin, "Invariance conditions in stationary and nonstationary systems," *SAC*, vol. 14, no. 4, pp. 27-38, 1969.

[50] J. Preminger and J. Rootenberg, "Some considerations relating to control systems employing the invariance principle," *IEEE-TAC*, vol. AC-9, pp. 209-215, July 1964.

[51] L.I. Rozonoer, "A variational approach to the problem of invariance of automatic control systems. I," *ARC*, vol. 24, no. 6, pp. 680-691, Nov., 1963.

[52] _____, "A variational approach to the problem of invariance II," *ARC*, vol. 24, no. 7, pp. 793-800, Dec. 1963. [cf., [63] and [49] for comments.]

[53] R.S. Rutman, "Conditions of zero sensitivity in linear control systems," *ECy*, vol. 7, no. 2, pp. 137-145, March/April 1969.

[54] R.S. Rutman and M.S. Epelman, "Parametric invariance of linear dynamic systems," *Dokl. Adad. Nauk. USSR*, vol. 159, no. 4, pp. 764-766, 1964.

[55] B.A. Ryabov and G.P. Sachkov, "Estimating the value of ε–invariant systems," *SJAIS*, vol. 23, no. 4, pp. 23-30, 1990.

[56] P.H. Shac, "Invariance and controllability in certain linear processes," *ARC*, vol. 37, no. 7, pp. 994-1003, 1976. [The author of this paper seems to be the same as the next one.]

[57] F.H. Shak, "Invariance in linear discrete processes," *ARC*, vol. 34, no. 6, pp. 991-976, June 1973.

[58] G.V. Shchipanov, "Theory and methods of design of control systems," *Avtomatika i Telemekhanika*, no. 1, 1939. [This entry which has been cited (with a minor variation) in a number of papers in this area is included here only for historical reasons. The actual paper

has been recently reappeared, with perhaps a more accurate title than the earlier translation, in [*I.* 8].]

[59] Y.B. Shtessel and A.Y. Evnin, "Invariant control of output of nonlinear systems," *ARC*, vol. 51, no. 3, pp. 315-323, 1990.

[60] E.M. Solnechnyi, "Making the dynamic properties of a single-channel control system approach absolute invariance," *ARC*, vol. 46, no. 4, pp. 449-456, 1985.

[61] E.M. Solnechnyy, "Investigation of the stability of a Shchipanov system of intentionally introduced small stabilizing inertialities," *SJAIS*, vol. 20, no. 6, pp. 79-83, 1987.

[62] E.M. Solnechnyi, "Conditions of stability and robustness of a single-channel invariant system of any order," *ARC*, vol. 49, no. 7, pp. 864-873, 1988.

[63] R.O. Spas'kyy, "Absolute invariance conditions for nonlinear systems of differential equations," *SAC*, vol. 14, no. 4, pp. 39-44, 1969.

[64] A.F. Starikov, "Invariance in linear systems," *ECy*, vol. 23, no. 2, pp. 131-133, 1985.

[65] A.V. Ushakov, "Conditions of zero parametric sensitivity in the tracking problem," *ARC*, vol. 42, no. 9, pp. 1157-1163, Sept. 1981.

[66] A.V. Ushakov and A.A. Dzhamanbaev, "A geometrical approach in the problem of finding a class of regulators that ensure zero parametric sensitivity," *SJAIS*, vol. 20, no. 5, pp. 84--, 1987. [The author regretfully could not locate the rest of this article, it has been apparently deleted in the course of production.]

[67] V.A. Utkin and V.I. Utkin, "Design of invariant systems by the method of separation of motions," *ARC*, vol. 44, no. 12, pp. 1559-1566, Dec. 1983.

[68] V.V. Velichenko, "The variational method in invariance theory," *SAC*, vol. 5, no. 3, pp. 1-16, May/June 1972.

[69] _____, "Invariance of discontinuous systems," *ARC*, no. 8, pp. 1336-1341, Aug. 1973.

[70] A.M. Voronin, "Structural analysis of a class of invariant control systems," *SAC*, vol. 3, no. 4, pp. 8-10, 1970.

[71] M. Vukobratovic, "Invariance and sensitivity of multivariate dynamic systems," *SAC*, vol. 13, no. 4, pp. 1-7, 1968.

[72] P.K.C. Wang, "Invariance, uncontrollability, and unobservability in dynamical systems," *IEEE-TAC*, vol. AC-10, pp. 366-367, July 1965.

[73] S.V. Yemel'yanov, S.K. Korovin, and B.V. Ulanov, "Control of nonstationary dynamic systems with coordinate-parametric feedback," *ECy*, vol. 20, no. 6, pp. 120-130, 1982.

[K] Bounded Uncertainty and Disturbance Rejection
[K1] Bounded Uncertainty

[1] M.A. Aizerman and Y.S. Pyatnitskiy, "Theory of dynamic systems which incorporate elements with incomplete information and its relation to the theory of discontinuous systems," *JFI*, vol. 306, no. 6, pp. 379-408, 1978.

[2] Y.I. Alimov, "On the application of Lyapunov's direct method to differential equations with ambiguous right sides," *ARC*, vol. 22, no. 7, pp. 713-725, July 1961.

[3] G.M. Bakan and V.T. Strashko, "A minimax approach to static plant control with incomplete information," *SAC*, vol. 11, no. 5, pp. 1-6, 1978.

[4] _____, "Minimax control of multivalued controlled processes," *SAC*, vol. 12, no. 3, pp. 46-55, 1979.

[5] E.A. Barbashin, "On the theory of generalized dynamical systems," *Uch. Zap. Moskov. Gosudarst. Univ.* no. 135, 1949, pp. 110-133.

[6] B.R. Barmish, "Robust solution of perturbed dynamical equations from within a convex restraint set," *IEEE-TAC*, vol. AC-24, no. 6, pp. 921-926, 1979.

[7] _____, "Stabilization of uncertain systems via linear control," *IEEE-TAC*, vol. AC-28, no. 8, pp. 848-850, Aug. 1983.

[8] _____, "Necessary and sufficient conditions for quadratic stablizability of an uncertain system," *JOTA*, vol. 46, no. 4, pp. 399-408, 1985.

[9] B.R. Barmish, L.E. Blume, and S.D. Chikte, "Robustness of systems with uncertainties in the input," *JMAA*, vol. 84, pp. 208-234, 1981.

[10] B.R. Barmish, M. Corless, and G. Leitmann, "A new class of stabilizing controllers for uncertain dynamical systems," *SIAM-JCO*, vol. 21, no. 2, pp. 246-255, March 1983.

[11] B.R. Barmish and G. Leitmann, "On ultimate boundedness control of uncertain systems in the absence of matching assumptions," *IEEE-TAC*, vol. AC-27, no. 1, pp. 153-158, Feb. 1982.

[12] B.R. Barmish, W.E. Schmitendorf, and G. Leitmann, "A note on avoidance control," *ASME-JDSMC*, vol. 103, pp. 69-70, March 1981.

[13] G. Basile and G. Marro, "On the robust controlled invariant," *SCL*, vol. 9, pp. 191-195, 1987.

[14] A.S. Belen'kii, "Min-max problems with monotone functions on polyhedral sets," *ARC*, vol. 43, no. 10, pp. 1304-1314, Oct. 1982.

[15] _____, "Search for min-max of two monotone functions in polyhedral set," *ARC*, vol. 43, no. 11, pp. 1389-1393, Nov. 1982.

[16] J. Bernussou, P.L.D. Peres, and J.C. Geromel, "A linear programming oriented procedure for quadratic stabilization of uncertain systems," *SCL*, vol. 13, pp. 65-72, 1989.

[17] V.N. Bukov, "Optimal algorithms in problems with bounded controllable coordinates," *ECy*, vol. 20, no. 2, pp. 133-140, 1982.

[18] C.H. Chao and H.L. Stalford, "On the robustness of linear stabilizing feedback control for linear uncertain systems: Multi-input case," *JOTA*, vol. 64, no. 2, pp. 229-244, 1990.

[19] _____, "Necessary and sufficient condition in Lyapunov robust control: Multi-input case," *JOTA*, vol. 66, no. 1, pp. 1-21, 1990.

[20] Y.H. Chen, "Robust output feedback controller: Direct design," *IJC*, vol. 46, no. 3, pp. 1083-1091, 1987.

[21] _____, "Robust output feedback controller: Indirect design," *IJC*, vol. 46, no. 3, pp. 1093-1103, 1987.

[22] _____, "On the robustness of mismatched uncertain dynamical systems," *ASME-JDSMC*, vol. 109, pp. 29-35, 1987.

[23] _____, "Design of robust controllers for uncertain dynamical systems," *IEEE-TAC*, vol. AC-33, no. 5, pp. 487-491, May 1988.

[24] Y.H. Chen and G. Leitmann, "Robustness of uncertain systems in the absence of matching assumptions," *IJC*, vol. 45, no. 5, pp. 1527-1542, 1987.

[25] M.J. Corless and G. Leitmann, "Continuous state feedback guaranteeing uniform ultimate boundedness for uncertain dynamic systems," *IEEE-TAC*, vol. AC-26, no. 5, pp. 1139-1144, Oct. 1981. Erratum, *ibid.*, vol. AC-28, no. 2, p. 249, Feb. 1983.

[26] _____, "Controller design for uncertain systems via Lyapunov function," in *Proc. American Contr. Conf.*, pp. 2019-2025, 1988.

[27] A.F. Filippov, "Differential equations with many-valued discontinuous right-hand side," *Soviet Mathematics*, vol. 4, no. 4, pp. 941-945, 1963.

[28] _____, *Differential Equations with Discontinuous Righthand Sides*. Dordrecht, the Netherlands: Kluwer, 1988. Reviewed by S.R. Bernfeld, *SIAM R.*, vol. 32, no. 2, pp. 312-315, 1990.

[29] A.R. Galimidi and B.R. Barmish, "The constrained Lyapunov problem and its application to robust output feedback stabilization," *IEEE-TAC*, vol. AC-31, pp. 410-419, May 1986.

[30] F. Garofalo and G. Leitmann, "Guaranteeing ultimate boundedness and exponential rate of convergence for a class of nominally linear uncertain systems," *ASME-JDSMC*, vol. 111, pp. 584-588, Dec. 1989.

[31] D.P. Goodall and E.P. Ryan, "Feedback controlled differential inclusions and stabilization of uncertain dynamical systems," *SIAM-JCO*, vol. 26, no. 6, pp. 1431-1441, 1988.

[32] G. Gu, "Stabilizability conditions of multivariable uncertain systems *via* output feedback control," *IEEE-TAC*, vol. AC-35, no. 8, pp. 925-927, Aug. 1990.

[33] S. Gutman, "Uncertain dynamical systems – A Lyapunov min-max approach," *IEEE-TAC*, vol. AC-24, no. 3, pp. 437-443, June 1979. Correction, *ibid.*, vol. AC-25, no. 3, p. 613, June 1980.

[34] _____, "Synthesis of min-max strategies," *JOTA*, vol. 46, no. 4, pp. 515-523, 1985.

[35] I.-J. Ha, "New matching conditions for output regulation of a class of uncertain nonlinear systems," *IEEE-TAC*, vol. AC-34, no. 1, pp. 116-119, Jan. 1989.

[36] T.H. Hopp and W.E. Schmitendorf, "Design of a linear controller for robust tracking and model following," *ASME-JDSMC*, vol. 112, pp. 552-558, Dec. 1990.

[37] F. Jabbari and W.E. Schmitendorf, "Effects of using observers on stabilization of uncertain linear systems," *IEEE-TAC*, vol. AC-38, no. 2, pp. 266-271, 1993.

[38] K. Khorasani, "Robust stabilization of non-linear systems with unmodelled dynamics," *IJC*, vol. 50, no. 3, pp. 827-844, 1989.

[39] N.F. Kirichenko, "Minimax control and estimation in dynamic systems," *SAC*, vol. 15, no. 1, pp. 31-39, 1982.

[40] N.A. Lehtomaki, D.A. Castanon, B.C. Levy, G. Stein, N.R. Sandell, Jr., and M. Athans, "Robustness and modeling error characterization," *IEEE-TAC*, vol. AC-29, no. 3, pp. 211-220, March 1984.

[41] G. Leitmann, "Guaranteed ultimate boundedness for a class of uncertain linear dynamical systems," *IEEE-TAC*, vol. AC-23, no. 6, pp. 1109-1110, Dec. 1978.

[42] _____, "Guaranteed asymptotic stability for a class of uncertain linear dynamical systems," *JOTA*, vol. 27, no. 1, pp. 99-106, 1979.

[43] _____, "Guaranteed asymptotic stability for some linear systems with bounded uncertainties," *ASME-JDSMC*, vol. 101, pp. 212-216, Sept. 1979.

[44] _____, "On the efficacy of nonlinear control in uncertain linear systems," *ASME-JDSMC*, vol. 102, pp. 95-102, 1981.

[45] G. Leitmann, E.P. Ryan, and A. Steinberg, "Feedback control of uncertain systems: Robustness with respect to neglected actuator and sensor dynamics," *IJC*, vol. 43, no. 4, pp. 1243-1256, 1986.

[46] T.-L. Liao, L.-C. Fu, and C.-F. Hsu, "Output tracking control of nonlinear systems with mismatched uncertainties," *SCL*, vol. 18, pp. 39-47, 1992.

[47] A. Linnemann, I. Postlethwaite, and B.D.O. Anderson, "Almost disturbance decoupling with stabilization by measurement feedback," *SCL*, vol. 12, pp. 225-234, 1989.

[48] A. Marchaud, "Sur les champs de demi-droites et les equations differentielles du premier ordre," *Bulletin de la Societe Mathematique de France*, vol. 62, pp. 1-38, 1934.

[49] _____, "Sur les champs continus de demi-cones convexes et leurs integrales," *Compositio Mathematica*, vol. 3, pp. 89-127, 1936.

[50] P. Myszkorowski, "Practical stabilization of a class of uncertain nonlinear systems," *SCL*, vol. 18, pp. 233-236, 1992.

[51] I.R. Petersen, "Structural stabilization of uncertain systems: Necessity of the matching condition," *SIAM-JCO*, vol. 23, no. 2, pp. 286-296, March 1985.

[52] S. Phoojaruenchanachai and K. Furuta, "Memoryless stabilization of uncertain linear systems including time-varying state delays," *IEEE-TAC*, vol. AC-37, no. 7, pp. 1022-1026, 1992.

[53] Z. Qu, "Robust control of a class of nonlinear uncertain systems," *IEEE-TAC*, vol. AC-37, no. 9, pp. 1437-1442, 1992.

[54] _____, "Global stabilization of nonlinear systems with a class of unmatched uncertainties," *SCL*, vol. 18, pp. 301-307, 1992.

[55] M.A. Rotea and P.P. Khargonekar, "Stabilization of uncertain systems with norm bounded uncertainty – A control Lyapunov function approach," *SIAM-JCO*, vol. 27, no. 6, pp. 1462-1476, 1989.

[56] E. Roxin, "Stability in general control systems," *JDE*, vol. 1, no. 2, pp. 115-150, 1965.

[57] _____, "On generalized dynamical systems defined by contingent equations," *JDE*, vol. 1, no. 2, pp. 188-205, 1965.

[58] _____, "On asymptotic stability in control systems," *Rendiconti del Circolo Matematico, di Palermo*, series II, vol. XV, pp. 193-208, 1966.

[59] W.E. Schmitendorf, "Stability controllers for uncertain linear systems with additive disturbances," *IJC*, vol. 47, no. 1, pp. 85-95, 1988.

[60] W.E. Schmitendorf and B.R. Barmish, "Null controllability of linear systems with constrained controls," *SIAM-JCO*, vol. 18, no. 4, pp. 327-345, 1980.

[61] J.-C. Shen, B.-S. Chen, and F.-C. Kung, "Memoryless stabilization of uncertain dynamic delay systems: Riccati equation approach," *IEEE-TAC*, vol. AC-36, no. 5, pp. 638-640, 1991.

[62] S.N. Singh, "Ultimate boundedness control of uncertain robotic systems," *IJSS*, vol. 17, no. 6, pp. 859-863, 1986.

[63] H.L. Stalford and C.H. Chao, "A necessary and sufficient condition in Lyapunov robust control," *JOTA*, vol. 63, no. 2, pp. 191-203, 1989.

[64] _____, "On the robustness of linear stabilizing feedback control for linear uncertain systems," *JOTA*, vol. 63, no. 2, pp. 205-212, 1989.

[65] J.S. Thorp and B.R. Barmish, "On guaranteed stability of uncertain linear systems *via* linear control," *JOTA*, vol. 35, no. 4, pp. 559-579, 1981.

[66] A. Trofino Neto, J.M. Dion, and L. Dugard, "Robustness bounds for LQ regulators," *IEEE-TAC*, vol. AC-37, no. 9, pp. 1373-1377, 1992.

[67] _____, "On the robustness of LQ regulators for discrete-time systems," *IEEE-TAC*, vol. AC-37, no. 10, pp. 1564-1568, 1992.

[68] S.-C. Tsay, "Robust control for linear uncertain systems *via* linear quadratic state feedback," *SCL*, vol. 15, pp. 199-205, 1990.

[69] D.J. Wilson and G. Leitmann, "Minimax control of systems with uncertain state measurements," *AMO*, vol. 2, no. 4, pp. 315-336, 1976.

[70] S.C. Zaremba, "Sur les equations au paratingent," *Bulletin des Sciences Mathematiques*, (2nd series), vol. 60, pp. 139-160, 1936.

[K] Bounded Uncertainty and Disturbance Rejection
[K2] Disturbance Rejection

[1] T. Basar, "A dynamic games approach to controller design: Disturbance rejection in discrete-time," *IEEE-TAC*, vol. AC-36, no. 8, pp. 936-952, 1991.

[2] S.P. Bhattacharyya, A.C. del Nero Gomes, and J.W. Howze, "The structure of robust disturbance rejection control," *IEEE-TAC*, vol. AC-28, no. 9, pp. 874-881, Sept. 1983.

[3] B.C. Chang and J.B. Pearson, Jr., "Optimal disturbance reduction in linear multivariable systems," *IEEE-TAC*, vol. AC-29, pp. 880-887, Oct. 1984.

[4] H. Chapellat and M. Dahleh, "Analysis of time-varying control strategies for optimal disturbance rejection and robustness," *IEEE-TAC*, vol. AC-37, no. 11, pp. 1734-1745, 1992.

[5] C.-M. Chien and B.-C. Wang, "An SISO uncertain system designed by an equivalent disturbance attenuation method," *CTAT*, vol. 6, no. 2, pp. 257-271, 1990.

[6] C. Commault, J.-M. Dion, and A. Perez, "Disturbance rejection for structured systems," *IEEE-TAC*, vol. AC-36, no. 7, pp. 884-887, 1991.

[7] M.A. Dahleh and J.B. Pearson, Jr., "l^1-Optimal feedback controllers for MIMO discrete-time systems," *IEEE-TAC*, vol. AC-32, pp. 314-322, April 1987.

[8] _____, "Optimal rejection of persistent disturbances, robust stability, and mixed sensitivity minimization," *IEEE-TAC*, vol. AC-33, no. 8, pp. 722-731, Aug. 1988. Comments, with reply, by J. Wu, H.-J. Fang, J.-Y. Li, and J.-J. Chen, *ibid.*, vol. AC-38, no. 5, p. 831, 1993.

[9] M.A. Dahleh and J.S. Shamma, "Rejection of persistent bounded disturbances: Nonlinear controllers," *SCL*, vol. 18, pp. 245-252, 1992.

[10] M.A. Dahleh, P.G. Voulgaris, and L.S. Valavani, "Optimal and robust controllers for periodic and multirate systems," *IEEE-TAC*, vol. AC-37, no. 1, pp. 90-99, 1992.

[11] R.J. de Figueiredo and G. Chen, "Optimal disturbance rejection for nonlinear control systems," *IEEE-TAC*, vol. AC-34, no. 12, pp. 1242-1248, Dec. 1989.

[12] G. Deodhare and M. Vidyasagar, "Every stabilizing controller is l_1-and H_∞-optimal," *IEEE-TAC*, vol. AC-36, no. 9, pp. 1070-1073, 1991.

[13] M. Fujita, K. Uchida, and F. Matsumura, "Asymptotic H^∞ disturbance attenuation based on perfect observation," *IEEE-TAC*, vol. AC-36, no. 7, pp. 875-880, 1991.

[14] O.M. Grasselli and S. Longhi, "Robust output regulation under uncertainties of physical parameters," *SCL*, vol. 16, pp. 33-40, 1991.

[15] A. Isidori and A. Astolfi, "Disturbance attenuation and H_∞-control *via* measurement feedback in nonlinear systems," *IEEE-TAC*, vol. AC-37, no. 9, pp. 1283-1293, 1992.

[16] M.H. Khammash, "Necessary and sufficient conditions for the robustness of time-varying systems with applications to sampled-data systems," *IEEE-TAC*, vol. AC-38, no. 1, pp. 49-57, 1993.

[17] E. Noldus, "Disturbance rejection using dynamic output feedback," *Proc. IEE*, vol. 129, Pt. D, no. 3, pp. 76-80, May. 1982.

[18] H. Ozbay, "On L^1 optimal control," *IEEE-TAC*, vol. AC-34, pp. 884-885, Aug. 1989.

[19] P.N. Paraskevopoulos, F.N. Koumboulis, and K.G. Tzierakis, "Disturbance rejection of left-invertible systems," *Automatica*, vol. 28, no. 2, pp. 427-430, 1992.

[20] A.G. Parlos, A.F. Henry, F.C. Schweppe, L.A. Gould, and D.D. Lanning, "Nonlinear multivariable control of nuclear power plants based on the unknown-but-bounded disturbance model," *IEEE-TAC*, vol. AC-33, no. 2, pp. 130-137, 1988.

[21] J.B. Pearson and B. Bamieh, "On minimizing maximum errors," *IEEE-TAC*, vol. AC-35, no. 5, pp. 598-601, May 1990.

[22] W.A. Porter, "Concerning disturbance measures in linear systems," *IEEE-TAC*, vol. AC-11, pp. 532-534, July 1966.

[23] I. Rhee and J.L. Speyer, "A game theoretic approach to a finite-time disturbance attenuation problem," *IEEE-TAC*, vol. AC-36, no. 9, pp. 1021-1032, 1991.

[24] J.S. Shamma, "Performance limitations in sensitivity reduction for nonlinear plants," *SCL*, vol. 17, pp. 43-47, 1991.

[25] J.S. Shamma and M. Athans, "Guaranteed properties of gain scheduled control for linear parameter-varying plants," *Automatica*, vol. 27, no. 3, pp. 559-564, 1991.

[26] J.S. Shamma and M.A. Dahleh, "Time-varying versus time-invariant compensation for rejection of persistent bounded disturbances and robust stabilization," *IEEE-TAC*, vol. AC-36, no. 7, pp. 838-847, 1991.

[27] N. Sivashankar and P.P. Khargonekar, "Robust stability and performance analysis of sampled-data systems," *IEEE-TAC*, vol. AC-38, no. 1, pp. 58-69, 1993.

[28] R.E. Skelton and G. Zhu, "Optimal L_∞ bounds for disturbance robustness," *IEEE-TAC*, vol. AC-37, no. 10, pp. 1568-1572, 1992.

[29] A.A. Stoorvogel and J.W. van der Woude, "The disturbance decoupling problem with measurement feedback and stability for systems with direct feedthrough matrices," *SCL*, vol. 17, pp. 217-226, 1991.

[30] K.C.Q. Tsai, D.M. Auslander, "A statistical methodology of designing controllers for minimum sensitivity of parameter variations," *ASME-JDSMC*, vol. 110, pp. 126-133, June 1988.

[31] P.B. Usoro, F.C. Schweppe, D.N. Wormley, and L.A. Gould, "Ellipsoidal set-theoretic control synthesis," *ASME-JDSMC*, vol. 104, pp. 331-336, Dec. 1982.

[32] M. Vidyasagar, "Optimal rejection of persistent bounded disturbances," *IEEE-TAC*, vol. AC-31, no. 6, pp. 527-534, June 1986.

[33] _____, "Further results on the optimal rejection of persistent bounded disturbances," *IEEE-TAC*, vol. AC-36, no. 6, pp. 642-652, 1991.

[34] S. Weiland and J.C. Willems, "Almost disturbance decoupling with internal stability," *IEEE-TAC*, vol. AC-34, no. 3, pp. 277-286, March 1989.

[35] G. Zhu and R.E. Skelton, "Robust discrete controllers guaranteeing l_2 and l_∞ performances," *IEEE-TAC*, vol. AC-37, no. 10, pp. 1620-1625, 1992.

[L] Sensitivity Analyses in Nonlinear, Time-Varying Systems, System Sensitivity Theory – Statistically Oriented, Certain Statistical Problems, and Certain Social and Economical Matters

[1] B.D.O. Anderson and J.B. Moore, "Tolerance of nonlinearities in time-varying optimal systems," *Elect. Lett.*, vol. 3, no. 6, pp. 250-251, June 1967.

[2] D.P. Atherton, H.T. Dorrah, and S.T. Nichols, "The algebra of λ matrices: An effective tool for sensitivity analysis," *IEEE-TAC*, vol. AC-21, no. 6, pp. 881-882, 1976.

[3] O. Berman, E. Modiano, and J.A. Schnabel, "Sensitivity analysis and robust regression in investment performance evaluation," *IJSS*, vol. 15, no. 5, pp. 481-489, 1984.

[4] X.-R. Cao and Y.-C. Ho, "Sensitivity analysis and optimization of throughput in a production line with blocking," *IEEE-TAC*, vol. AC-32, no. 11, pp. 959-967, Nov. 1987.

[5] _____, "Estimating the sojourn time sensitivity in queueing networks using perturbation analysis," *JOTA*, vol. 53, no. 3, pp. 353-375, 1987.

6.12.L. Sensitivity Analyses in Nonlinear, Time-Varying Systems

[6] C.G. Cassandras and S.G. Strickland, "On-line sensitivity analysis of Markov chains," *IEEE-TAC*, vol. AC-34, no. 1, pp. 76-86, Jan. 1989.

[7] _____, "Observable augmented systems for sensitivity analysis of Markov and semi-Markov processes," *IEEE-TAC*, vol. AC-34, no. 10, pp. 1026-1037, Oct. 1989.

[8] P.I. Dekhtyarenko and A.Z. Zakharyan, "Effect of parameter deviations of nonlinear loops on equivalent transfer function," *SAC*, vol. 6, pp. 5-12, Nov./Dec. 1973.

[9] R.M. DeSantis and W.A. Porter, "A generalized Nyquist plot and its use in sensitivity analysis," *IJSS*, vol. 5, no. 12, pp. 1143-1153, 1974.

[10] D. Efthymiatos and S. Tzafestas, "A generalized sensitivity approach to feedback systems in Hilbert space," *IJSS*, vol. 4, no. 1, pp. 87-95, 1973.

[11] D.L. Erickson and F.E. Norton, "Application of sensitivity constrained optimal control to national economic policy formulation," in *Control and Dynamic Systems advances in theory and applications*, vol. 9 (C.T. Leondes, Ed.). New York: Academic Press, pp. 131-237, 1973.

[12] M.A. Eyler, "Sensitivity of sample values to parameter changes," *JOTA*, vol. 45, no. 1, pp. 159-163, 1985.

[13] A. Feintuch, P.P. Khargonekar, and A. Tannenbaum, "On the sensitivity minimization problem for linear time-varying periodic systems," *SIAM-JCO*, vol. 24, no. 5, pp. 1076-1085, 1986.

[14] I.V. Filatov and S.N. Sharov, "An investigation of parametric sensitivity of nonlinear dynamic correcting devices," *ECy*, vol. 15, pp. 166-169, 1977.

[15] A. Ford and P. Gardiner, "A new measure of sensitivity for social system simulation models," *IEEE-TSMC*, vol. SMC-9, no. 3, pp. 105-114, 1979.

[16] J.S. Gibson and L.G. Clark, "Sensitivity analysis for a class of evolution equations," *JMAA*, vol. 58, pp. 22-31, 1977.

[17] W.-B. Gong, C.G. Cassandras, and J. Pan, "Perturbation analysis of a multiclass queueing system with admission control," *IEEE-TAC*, vol. AC-36, no. 6, pp. 707-723, 1991.

[18] D.A. Hanson, W.R. Perkins, and J.B. Cruz, Jr., "Public investment strategies for regional development: An analysis based on optimization and sensitivity results," *IEEE-TSMC*, vol. SMC-6, no. 3, pp. 165-176, March 1976.

[19] R.J. Herbert, "Structural sensitivity of feedback controls for second-order nonlinear systems," *JOTA*, vol. 30, no. 3, pp. 395-421, 1980.

[20] Y.-C. Ho, "Parametric sensitivity of a statistical experiment," *IEEE-TAC*, vol. AC-24, no. 6, pp. 982-983, Dec. 1979.

[21] J.M. Holtzman, "On using perturbation analysis to do sensitivity analysis: Derivatives versus differences," *IEEE-TAC*, vol. AC-37, no. 2, pp. 243-247, 1992.

[22] Y.B. Kadimov, E.Y. Kuliyev, and S.I. Myachin, "Some questions concerning the application of the sensitivity theory to an analysis of transient processes while simulating systems with distributed parameters on analog computers," *Proc. IFAC*, vol. 3 (of the 5th Congress), pp. 31.4.1-31.4.9, 1972.

[23] V. Kaitala and G. Leitmann, "Stabilizing employment in a fluctuating resource economy," *JOTA*, vol. 67, no. 1, pp. 1-16, 1990.

[24] K.-I. Kanatani, "Generalized global sensitivity and correlation analysis," *IC*, vol. 47, pp. 37-58, 1980.

[25] Y.S. Kharin, "Robustness of decision rules in the presence of errors of classification of the training sample," *ARC*, vol. 44, no. 11, pp. 1470-1479, Nov. 1983.

[26] D.G. Lainiotis and F.L. Sims, "Sensitivity analysis of discrete Kalman filters," *IJC*, vol. 12, no. 4, pp. 657-669, 1970.

[27] D.H. Martin, "Prediction sensitivity to functional perturbations in modelling with ordinary differential equations," *AMO*, vol. 6, pp. 123-137, 1980.

[28] N.H. McClamroch, "Evaluation of suboptimality and sensitivity in control and filtering processes," *IEEE-TAC*, vol. AC-14, pp. 282-285, June 1969.

[29] L.H. Meyer, F.Q. Raines, T.J. Tarn, and S.K. Gupta, "Optimal coordination of aggregate stabilization policy and price controls: A sensitivity analysis," *Proc. IFAC*, vol. 3 (of the 6rd Congress), pp. 62.4.1-62.4.8, 1975.

[30] K.S. Miller and F.J. Murray, "A mathematical basis for an error analysis of differential analyzers," *JMP*, vol. 32, nos. 2 and 3, pp. 136-163, July/Oct. 1953.

[31] R.W. Newcomb and B.D.O. Anderson, "A distributional approach to time-varying sensitivity," *SIAM-JAM*, vol. 15, no. 4, pp. 1001-1010, 1967.

[32] W.R. Perkins and J.B. Cruz, Jr., "Sensitivity operator for linear time-varying systems," in *Sensitivity Methods in Control Theory* (L. Radanovic, Ed.). New York: Pergamon Press 1966. (Proc. Int. Symp. Dubrovink, Aug. 1964.)

[33] W.A. Porter, "Some theoretical limitations of system sensitivity reduction," in *Proc. Allerton Conf.*, pp. 241-251, 1965.

[34] W.A. Porter and R.M. DeSantis, "Sensitivity analysis in multilinear systems," *IJSS*, vol. 7, no. 2, pp. 191-205, 1976.

[35] J. Rootenberg and P. Courtin, "Sensitivity of optimal control systems with bang-bang control," *IJC*, vol. 18, no. 3, pp. 537-543, 1973.

[36] A.P. Sage, "Sensitivity analysis in systems for planning and decision support," *JFI*, vol. 312, nos. 3/4, pp. 265-291, Sept./Oct. 1981.

[37] F.R. Shupp, "Financing fiscal policy: A sensitivity analysis," *JFI*, vol. 312, nos. 3/4, pp. 293-306, Sept./Oct. 1981.

[38] A.A. Siapkara, "Human operator system representation *via* cross-sensitivity minimization," *JFI*, vol. 312, nos. 3/4, pp. 307-326, 1981.

[39] L.H. Sibul, "Sensitivity analysis of linear control systems with random plant parameters," *IEEE-TAC*, vol. AC-15, pp. 459-462, Aug. 1970.

[40] P. Stavroulakis and P.E. Sarachik, "Low sensitivity feedback gains for deterministic and stochastic control systems," *IJC*, vol. 19, no. 1, pp. 15-31, 1974.

[41] N. Sundararajan and J.B. Cruz, Jr., "Policy modification for macro-economic systems resulting in reduced sensitivity to parameter perturbations," *IJSS*, vol. 5, no. 12, pp. 1193-1205, 1974.

[42] S.G. Tzafestas, "Sensitivity in the decoupling of non-linear control systems," *IJC*, vol. 18, no. 6, pp. 1249-1266, 1973.

[43] M. Vukobratovic, "Note on the sensitivity of non-linear dynamic systems," *IEEE-TAC*, vol. AC-13, pp. 453-454, Aug. 1968.

[44] M. Vukobratovic and D. Juricic, "Sensitivity of nonlinear systems," *ARC*, no. 9, pp. 1381-1388, Sept. 1970.

[45] J. Yang and H.J. Kushner, "A Monte Carlo method for sensitivity analysis and parametric optimization of nonlinear stochastic systems," *SIAM-JCO*, vol. 29, no. 5, pp. 1216-1249, 1991.

[46] G.E. Young and D.M. Auslander, "A design methodology for nonlinear systems containing parameter uncertainty," *ASME-JDSMC*, vol. 106, pp. 15-20, March 1984.

[47] K.-K. D. Young, "Near insensitivity of linear feedback systems," *JFI*, vol. 314, no. 2, pp. 129-142, Aug. 1982.

[48] A.V. Yudayev, A.M. Lisitsyn, and N.P. Baranov, "Optimization of controllers with respect to minimum sensitivity to disturbances," *ECy*, vol. 13, no. 1, pp. 141-145, 1975.

[M] Sensitivity Considerations in Nonlinear Systems

[1] J.L. Aravena and W.A. Porter, "Finite memory partial inverses," *IEEE-TCAS*, vol. CAS-28, pp. 287-294, April 1981.

[2] _____, "System partial inverses for sensitivity, adaptivity and estimation," *JFI*, vol. 312, nos. 3/4, pp. 141-165, Sept./Oct. 1981.

[3] J.B. Cruz, D.P. Looze, and W.R. Perkins, "Sensitivity analysis of nonlinear feedback systems," *JFI*, vol. 312, nos. 3/4, pp. 199-215, Sept./Oct. 1981.

[4] S.A. Doganovskii, "Compensation of perturbations in nonlinear systems," *ARC*, vol. 23, no. 6, pp. 676-690, Dec. 1962.

[5] O.N. Fomenko, "Accuracy of nonlinear control systems with random parameters," *ECy*, vol. 5, no. 1, pp. 131-138, Jan./Feb., 1967.

[6] W. Hejmo, "Sensitivity to switching-function variations in a time-optimal positional system," *IJC*, vol. 39, no. 1, pp. 19-30, 1984.

[7] E. Kreindler, "On the closed-loop sensitivity reduction of non-linear systems," *IJC*, vol. 6, no. 2, pp. 171-178, 1967.

[8] _____, "On sensitivity of closed-loop nonlinear optimal control systems," *SIAM J. Contr.* vol. 7, no. 3, pp. 512-520, Aug. 1969.

[9] _____, "Sensitivity of time-varying linear optimal control systems," *JOTA*, vol. 3, no. 2, pp. 98-106, 1969.

[10] V.A. Mayorov, Y.G. Borisenko, O.B. Kerber, V.V. Pavlov, and O.S. Yakovlev, "The problem of synthesizing a nonlinear control law," *SJAIS*, vol. 21, no. 2, pp. 41-49, 1988.

[11] K. Pedersen and L. Nardizzi, "Synthesis of optimally sensitive control for systems with time-varying parameters," *IJC*, vol 15, no. 1, pp. 1-20, 1972.

[12] W.A. Porter, "Minimizing system sensitivity through feedback," *JFI*, vol. 286, pp. 225-240, 1968.

[13] _____, " "On sensitivity in multivariable non-stationary systems," *IJC*, vol. 7, pp. 481-491, 1968.

[14] _____, "The Interrelationship between observers and system sensitivity," *IEEE-TAC*, vol. AC-22, no. 2, pp. 144-146, Feb. 1977.

[15] _____, "Partial inverses for parameter sensitivity reduction," *IJC*, vol. 29, no. 6, pp. 949-961, 1979.

[16] S.S. Rao, T.-S. Pan, and V.B. Venkayya, "Robustness improvement of actively controlled structures through structural modifications," *AIAA J.*, vol. 28. no. 2, pp. 353-361, Feb. 1990.

[17] N. Sundararajan and J.B. Cruz, Jr., "Sensitivity reduction in time-varying linear and nonlinear systems," *IJC*, vol. 15, no. 5, pp. 937-943, 1972.

[18] V.D. Tourassis and C.P. Neuman, "Robust nonlinear feedback control for robotic manipulators," *Proc. IEE*, vol. 132, Pt. D, no. 4, pp. 134-143, July 1985.

[N] Sensitivity Considerations in Discrete-Time (Digital) Systems
Sensitivity Reduction of Systems with Time Lag,
and Multi-Dimensional Digital Systems

[1] S. Barnett, "Insensitivity of optimal linear discrete-time regulators," *IJC*, vol. 21, no. 5, pp. 843-848, 1975.

[2] A.W. Bennett and A.P. Sage, "Discrete system sensitivity and variable increment sampling," in *Proc. Joint Auto. Contr. Conf.*, pp. 603-612, 1967.

[3] G.A. Bekey and R. Tomovic, "Sensitivity of discrete systems to variation of sampling interval," *IEEE-TAC*, vol. AC-11, no. 2, pp. 284-287, April 1966.

[4] K.C. Cheok, N.K. Loh, and M.A. Zohdy, "Cost sensitivity analysis for discrete-time optimal feedback controllers with time-multiplied performance indexes," *IEEE-TAC*, vol. AC-31, no. 3, pp. 262-263, March 1986.

[5] S.-M. Chiang and Y.-P. Shih, "Low sensitivity feedback control of non-linear discrete systems," *IJC*, vol 14, no. 4, pp. 659-668, 1971.

[6] _____, "Low-sensitivity design of optimal linear control systems with transportation lag," *Proc. IEE*, vol. 120, no. 7, pp. 810-813, July 1973.

[7] F.H. Clarke and P.R. Wolenski, "The sensitivity of optimal control problems to time delay," *SIAM-JCO*, vol. 29, no. 5, pp. 1176-1215, 1991.

[8] M.A. Connor, "Minimization of performance sensitivity for time-lag systems," *IEEE-TAC*, vol. AC-16, no. 5, pp. 496-497, Oct. 1971.

[9] J.B. Cruz, Jr., "Sensitivity considerations for time-varying sampled-data feedback systems," *IRE-TAC*, vol. AC-6, no. 2, pp. 228-236, 1961.

[10] J.B. Cruz, Jr., and M.E. Sawan, "Low-sensitivity optimal feedback control for linear discrete-time systems," *IEEE-TAC*, vol. AC-24, no. 1, pp. 119-122, Feb. 1979.

[11] M.A. Dahleh and J.B. Pearson, Jr., "l^1–Optimal feedback controllers for MIMO discrete-time systems," *IEEE-TAC*, vol. AC-32, no. 4, pp. 314-322, April 1987.

[12] C. Foias, A. Tannenbaum, and G. Zames, "Weighted sensitivity minimization for delay systems," *IEEE-TAC*, vol. AC-31, no. 8, pp. 763-766, Aug. 1986.

[13] J.S. Freudenberg and D.P. Looze, "A sensitivity tradeoff for plants with time delay," *IEEE-TAC*, vol. AC-32, no. 2, pp. 99-104, Feb. 1987.

[14] S. Fukata and M. Takata, "On sampling period sensitivities of the optimal stationary sampled-data linear regulator," *IJC*, vol. 29, no. 1, pp. 145-158, 1979.

[15] I. Gumowski, "Sensitivity of certain dynamic systems with respect to a small delay," *Automatica*, vol. 10, pp. 659-674, 1974.

[16] S. Hara and H.-K. Sung, "Sensitivity improvement by a stable controller in SISO digital control systems," *SCL*, vol. 12, pp. 123-128, 1989.

[17] K. Inoue, H. Akashi, K. Ogino, and Y. Sawaragi, "Sensitivity approaches to optimization of linear systems with time delay," *Automatica*, vol. 7, pp. 671-679, 1971.

[18] T. Ishihara, "Sensitivity properties of a class of discrete-time LQG controllers with computation delays," *SCL*, vol. 11, pp. 299-307, 1988.

[19] E.I. Jury and Y. Z. Tsypkin, "On the theory of discrete systems," *Automatica*, vol. 7, pp. 89-107, 1971.

[20] M. Koda, "Sensitivity analysis of time-delay systems," *IJSS*, vol. 12, no. 11, pp. 1389-1397, 1981.

[21] A.J. Koivo, "Performance sensitivity of sampling systems – A unified approach," *JFI*, vol. 287, no. 3, pp. 209-221, March 1969.

[22] L.P. Kukhtenkov and A.I. Ruban, "The sensitivity of discontinuous systems described by differential-difference equations with lagging argument," *ECy*, vol. 17, no. 6, pp. 136-141, 1979.

[23] D.P. Lindorff, "Sensitivity in sampled-data systems," *IEEE-TAC*, vol. AC-8, pp. 120-125, 1963.

[24] A. Locatelli and S. Rinaldi, "Controllability versus sensitivity in linear discrete systems," *IEEE-TAC*, vol AC-15, pp. 254-255, April 1970.

[25] S.M. Melzer and B.C. Kuo, "Sampling period sensitivity of the optimal sampled data linear regulator," *Automatica*, vol. 7, pp. 367-370, 1971.

[26] C.P. Neuman and D.I. Schonbach, "Discrete weighted residual methods: A sensitive non-linear boundary-value problem," *JOTA*, vol. 22, no. 2, pp. 239-249, June 1977.

[27] L. Pandolfi and A.W. Olbrot, "On the minimization of sensitivity to additive disturbances for linear-distributed parameter MIMO feedback systems," *IJC*, vol. 43, no. 2, pp. 389-399, 1986.

[28] E.N. Rozenvasser, "General sensitivity equations of discontinuous systems," *ARC*, no. 3, pp. 400-404, March 1967.

[29] E.N. Rozenvasser and R.M. Yusupov, "Sensitivity equations of pulse control systems," *ARC*, no. 4, pp. 526-536, April 1969.

[30] E.P. Ryan, "On the sensitivity of a time-optimal switching function," *IEEE-TAC*, vol. AC-25, no. 2, pp. 275-277, April 1980.

[31] M. Saeki, "Methods of solving a polynomial equation for an H^∞ optimal control problem for a single-input single-output discrete-time system," *IEEE-TAC*, vol. AC-34, no. 2, pp. 166-168, Feb. 1989.

[32] A.H. Shevelyev, "Analysis of sampled-data systems with variable parameters," *SAC*, vol. 13, no. 3, pp. 14-22, 1968.

[33] P. Stavroulakis, "Low sensitivity feedback law implementation for 2-D digital systems," *JFI*, vol. 312, nos. 3/4, pp. 217-229, Sept./Oct. 1981.

[34] P. Stavroulakis and P.N. Paraskevopoulos, "Low-sensitivity observer-compensator design for two-dimensional digital systems," *Proc. IEE*, vol. 129, Pt. D, no. 5, pp. 193-200, Sept. 1982.

[35] P. Stavroulakis and S.G. Tzafestas, "State reconstruction in low-sensitivity design of 3-dimensional systems," *Proc. IEE*, vol. 130, Pt. D, no. 6, pp. 333-340, Nov. 1983.

[36] K.L. Suryanarayanan and A.C. Soudack, "Method for the generation of sensitivity functions for nonlinear sampled-data systems," *Elect. Lett.*, vol. 6, no. 19, pp. 611-613, Sept. 1970.

[37] K.E. Tait, "Sensitivity considerations and comparisons of sampling interval criteria for discrete-continuous feedback control systems," *IJC*, vol. 6, no. 2, pp. 101-145, 1967.

[38] V.I. Teverovskii, "On one particular case of a sampled-data system with variable parameters which change by jumps," *ARC*, vol. 21, no. 1, pp. 42-46, Aug. 1960.

[39] R. Tomovic and G.A. Bekey, "Adaptive sampling based on amplitude sensitivity," *IEEE-TAC*, vol. AC-11, no. 2, pp. 282-284, April 1966.

[40] R.K. Varshney and W.R. Perkins, "Sensitivity of linear time invariant sampled data systems to sampling period," *Automatica*, vol. 10, pp. 317-319, 1974.

[41] W.-Y. Yan and J.B. Moore, "On L^2-sensitivity minimization of linear state-space systems," *IEEE-TCAS*, Part I, vol. CAS-39, no. 8, pp. 641-648, 1992.

[O] Sensitivity Considerations in Distributed [Parameter] Systems

[1] S. Abu El Ata-Doss, "Cost function sensitivity to small parameter variations for a class of distributed control systems," *JMAA*, vol. 72, pp. 106-113, 1979.

[2] ———, "New technique of sensitivity reduction for distributed control systems," *AMO*, vol. 5, pp. 217-229, 1979.

[3] A. Bamberger, G. Chavent, and P. Lailly, "About the stability of the inverse problem in 1-D wave equations – Application to the interpretation of seismic profiles," *AMO*, vol. 5, pp. 1-47, 1979.

[4] C.-K. Chu and Y.-P. Shih, "Low sensitivity optimal control of a class of linear distributed systems," *IJC*, vol. 16, no. 2, pp. 325-336, 1972.

[5] J.M. Davis and W.R. Perkins, "Comparison sensitivity of distributed parameter systems," *IEEE-TAC*, vol. AC-17, no. 1, pp. 100-105, Feb. 1972.

[6] N.F. Degtyareva, "Use of sensitivity functions for optimal control synthesis in systems with distributed parameters," *SA*, vol. 19, no. 4, pp. 24-28, 1976.

[7] C. Foias, A. Tannenbaum, and G. Zames, "Sensitivity minimization for arbitrary SISO distributed plants," *SCL*, vol. 8, pp. 189-195, 1987.

[8] M. Kelemen, Y. Kannai, and I. Horowitz, "Arbitrarily low sensitivity (ALS) in linear distributed systems using pointwise linear feedback," *IEEE-TAC*, vol. AC-35, no. 9, pp. 1071-1075, Sept. 1990.

[9] T. Kobayashi, "Controllability and stabilizability of sensitivity combined systems for distributed parameter systems," *IJC*, vol. 35, no. 2, pp. 309-321, 1982.

[10] V. Komkov, "Sensitivity techniques for systems with distributed parameters," *JMAA*, vol. 128, pp. 443-455, 1987.

[11] M.C.Y. Kuo and M.N.B. Ayiku, "Synthesis of low-sensitivity optimal control in distributed parameter systems," *Proc. IEE*, vol. 125, no. 6, pp. 550-554, June 1978.

[12] I. Lasiecka and J. Sokolowski, "Sensitivity analysis of optimal control problems for wave equations," *SIAM-JCO*, vol. 29, no. 5, pp. 1128-1149, 1991.

[13] K. Malanowski and J. Sokolowski, "Sensitivity of solutions to convex, control constrained optimal control problems for distributed parameter systems," *JMAA*, vol. 120, pp. 240-263, 1986.

[14] A. Orbach and R. Fischl, "Performance index sensitivity of optimal control of first order in time- and space-distributed parameter systems," *IEEE-TAC*, vol. AC-25, no. 2, pp. 314-317, April 1980.

[15] K.C. Pedersen and L.R. Nardizzi, "Optimally sensitive control for distributed parameter systems," *IJC*, vol. 16, no. 4, pp. 723-735, 1972.

[16] H.J. Perlis, "Open-loop performance index sensitivity in a class of distributed optimal water management systems," in *Proc. Joint Auto. Contr.*, pp. 118-123, 1973.

[17] S. Pohjolainen, "Robust controller for systems with exponentially stable strongly continuous semigroups," *JMAA*, vol. 111, pp. 622-636, 1985.

[18] W.A. Porter, "Sensitivity problems in distributive systems," *IJC*, vol. 5, no. 5, pp. 393-412, 1967.

[19] _____, "Parameter sensitivity in distributive feedback systems," *IJC*, vol. 5, no. 5, pp. 413-423, 1967.

[20] A.I. Ruban, "Identification of distributed dynamic objects on the basis of a sensitivity algorithm," *ECy*, vol. 9, no. 6, pp. 1137-1142, Nov./Dec. 1971.

[21] J. Sokolowski, "Sensitivity analysis of control constrained optimal control problems for distributed parameter systems," *SIAM-JCO*, vol. 25, no. 6, pp. 1542-1556, 1987.

[22] _____, "Shape sensitivity analysis of boundary optimal control problems for parabolic systems," *SIAM-JCO*, vol. 26, no. 4, pp. 763-787, 1988.

[23] J. Sokolowski and J.P. Zolesio, "Shape design sensitivity analysis of plates and plane elastic solids under unilateral constraints," *JOTA*, vol. 54, no. 2, pp. 361-382, 1987.

[24] S.G. Tzafestas, "Eigenvalue and eigenfunction sensitivity of distributed-parameter systems (DPS)," *IEEE-TAC*, vol, AC-20, no. 1, pp. 172-174, 1975.

[25] _____, "Sensitivity functions of distributed parameter eigenvalue control systems," *Elect. Lett.*, vol. 12, no. 1, pp. 4-6, Jan. 1976.

[26] _____, "Eigenvalue control in distributed-parameter systems with parameter variations," *Proc. IEEE*, vol. 64, no. 9, pp. 1444-1446, Sept. 1976.

[27] S.G. Tzafestas and P.N. Paraskevopoulos, "Sensitive decoupling control of linear distributed-parameter systems," *IEEE-TAC*, vol. AC-20, no. 1, pp. 151-153, Feb. 1975.

*[P] Sensitivity Issues in Selected Multidisciplinary
Optimization Methods, and Numerical Methods*

[1] D. Aze and M. Volle, "A stability result in quasi-convex programming," *JOTA*, vol. 67, no. 1, pp. 175-184, 1990.

[2] J.-F. M. Barthelemy and J. Sobieszczanski-Sobieski, "Optimum sensitivity derivatives of objective functions in nonlinear programming," *AIAA J.*, vol. 21, no. 6, pp. 913-915, June 1983.

[3] R.G. Batson, "Extensions of Radstrom's lemma with application to stability theory of mathematical programming," *JMAA*, vol. 117, pp. 441-448, 1986.

[4] C.S. Berger, "Numerical method for the design of insensitive control systems," *Proc. IEE*, vol. 120, no. 10, pp. 1283-1292, Oct. 1973.

[5] _____, "Numerical comparison of two methods of designing insensitive controllers," *Elect. Lett.*, vol. 11, no. 20, pp. 489-490, Oct. 1975.

[6] M.J. Best and N. Chakravarti, "Stability of linearly constrained convex quadratic programs," *JOTA*, vol. 64, no. 1, pp. 43-53, 1990.

[7] J.H. Bigelow and N.Z. Shapiro, "Optimization problems with large parameters," *SIAM-JAM*, vol. 24, no. 2, pp. 152, 163, March 1973.

[8] R.N. Buie and J. Abraham, "Post-optimality sensitivity analysis in abstract spaces with applications to continuous-time programming problems," *JOTA*, vol. 45, no. 3, pp. 347-373, 1985.

[9] J.D. Buys and R. Gonin, "The use of augmented Lagrangian functions for sensitivity analysis in nonlinear programming," *MP*, vol. 12, no. 2, p. 281-284, 1977.

[10] R. Byers, "A Hamiltonian QR algorithm," *SIAM-JSSC*, vol. 7, no. 1, pp. 212-229, 1986.

[11] _____, "A bisection method for measuring the distance of a stable matrix to the unstable matrices," *SIAM-JSSC*, vol. 9, no. 5, pp. 875-881, 1988.

[12] D. Chenais, "Optimal design of midsurface of shells: Differentiability proof and sensitivity computation," *AMO*, vol. 16, pp. 93-133, 1987.

[13] F.H. Clarke and P.D. Loewen, "The value function in optimal control: sensitivity, controllability, and time-optimality," *SIAM-JCO*, vol. 24, no. 2, pp. 243-263, 1986.

[14] W. Cook, A.M.H. Gerards, A. Schrijver, and E. Tardos, "Sensitivity theorems in integer linear programming," *MP*, vol. 34, pp. 251-264, 1986.

[15] S. Dafermos and A. Nagurney, "Sensitivity analysis for the asymmetric network equilibrium problem," *MP*, vol. 28, pp. 174-184, 1984.

[16] A. Dax, "The smallest point of a polytope," *JOTA*, vol. 64, no. 2, pp. 429-432, 1990.

[17] A. Deif, *Sensitivity Analysis in Linear Systems*. New York: Springer – Verlag, 1986.

[18] R.S. Dembo, "Sensitivity analysis in geometric programming," *JOTA*, vol. 37, no. 1, pp. 1-21, 1982.

[19] J.W. Demmel, "A counterexample for two conjectures about stability," *IEEE-TAC*, vol. AC-32, pp. 340-342, April 1987. [cf., [74].]

[20] A.H. Evers, "Sensitivity analysis in dynamic optimization," *JOTA*, vol. 32, no. 1, pp. 17-37, 1980.

[21] S.D. Fassois, K.F. Eman, and S.M. Wu, "Sensitivity analysis of the discrete-to-continuous dynamic system transformation," *ASME-JDSMC*, vol. 112, pp. 1-9, March 1990.

[22] A.V. Fiacco, "Sensitivity analysis for nonlinear programming using penalty methods," *MP*, vol. 10, no. 3, pp. 287-311, 1976.

[23] _____, *Introduction to Sensitivity and Stability Analysis in Nonlinear Programming*. New York: Academic Press, 1983. Reviewed by C.E. Lemke, *SIAM R.*, vol. 27, no. 1, pp. 114-115, 1985.

[24] A.V. Fiacco, Ed., *Sensitivity, Stability, and Parametric Analysis.* (A publication of the Mathematical Programming Society.) Amsterdam: North-Holland, 1984.

[25] P.M. Gahinet, A.J. Laub, C.S. Kenney, and G.A. Hewer, "Sensitivity of the stable discrete-time Lyapunov equation," *IEEE-TAC*, vol. AC-35, no. 11, pp. 1209-1217, Nov. 1990.

[26] F.W. Gembicki and Y.Y. Haimes, "Approach to performance and sensitivity multiobjective optimization: The goal attainment method," *IEEE-TAC*, vol. AC-20, no. 6, pp. 769-771, 1975.

[27] H. Gerencser, "Stability theorems for 2×2 hypermatrices," *JOTA*, vol. 35, no. 1, pp. 1-7, 1981.

[28] J.M. Goethals, "Le controle des erreurs dans les transmissions numeriques," *Automatisme*, vol. XIX, no. 1, pp. 17-23, 1974.

[29] P.H. Hammond and M.J. Duckenfield, "Automatic optimization by continuous perturbation of parameters," *Automatica*, vol. 1, pp. 147-175, 1963.

[30] G. Hewer and C. Kenney, "The sensitivity of the stable Lyapunov equation," *SIAM-JCO*, vol. 26, no. 2, pp. 321-344, 1988.

[31] W.J. Hopp, "Sensitivity analysis in discrete dynamic programming," *JOTA*, vol. 56, no. 2, pp. 257-269, 1988.

[32] R.H.F. Jackson and G.P. McCormick, "Second-order sensitivity analysis in factorable programming: Theory and applications," *MP*, vol. 41, pp. 1-27, 1988.

[33] R. Janin, "On sensitivity in an optimal control problem," *JMAA*, vol. 60, no. 3, pp. 631-657, Oct. 1977.

[34] V.M. Karas, "Stability of approximate solutions of the graph partitioning problem," *SJAIS*, vol. 20, no. 3, pp. 55-59, 1987.

[35] V.A. Kas'yanov and Y.P. Udartsev, "Increasing the stability of estimates by the least squares methods," *SJAIS*, vol. 20, no. 3, pp. 44-48, 1987.

[36] C. Kenney and G. Hewer, "The sensitivity of the algebraic and differential Riccati equations," *SIAM-JCO*, vol. 28, no. 1, pp. 50-69, Jan. 1990.

[37] S.W. Kim, P.G. Park, and W.H. Kwon, "Lower bounds for the trace of the solution of the discrete algebraic Riccati equation," *IEEE-TAC*, vol. AC-38, no. 2, pp. 312-314, 1993.

[38] V.G. Kolobov, "A method for increasing the stability of the numerical solution of differential equation systems," *SJAIS*, vol. 22, no. 3, pp. 74-81, 1989.

[39] N. Komaroff, "Upper bounds for the solution of the discrete Riccati equation," *IEEE-TAC*, vol. AC-37, no. 9, pp. 1370-1373, 1992.

[40] N. Komaroff and B. Shahian, "Lower summation bounds for the discrete Riccati and Lyapunov equations," *IEEE-TAC*, vol. AC-37, no. 7, pp. 1078-1080, 1992.

[41] M.M. Konstantinov and G.B. Pelova, "Sensitivity of the solutions to differential Riccati equations," *IEEE-TAC*, vol. AC-36, no. 2, pp. 213-215, 1991.

[42] V.S. Kouikoglou and Y.A. Phillis, "Trace bounds on the covariance of continuous-time systems with multiplicative noise," *IEEE-TAC*, vol. AC-38, no. 1, pp. 138-142, 1993.

[43] V.Y. Krivonozhko and A.I. Propoy, "The method of successive improvement of control in dynamic linear programming problems. I," *ECy*, vol. 16, no. 3, pp. 1-12, 1978.

[44] J. Kyparisis, "Sensitivity analysis in posynomial geometric programming," *JOTA*, vol. 57, no. 1, pp. 85-121, 1988.

[45] K. Malanowski, "Differential stability of solutions to convex, control constrained optimal control problems," *AMO*, vol. 12, pp. 1-14, 1984.

[46] _____, "Stability and sensitivity of solutions to optimal control problems for systems with control appearing linearly," *AMO*, vol. 16, pp. 73-91, 1987.

[47] _____, "Differential sensitivity of solutions of convex constrained optimal control problems for discrete systems," *JOTA*, vol. 53, no. 3, pp. 429-449, 1987.

[48] _____, "Sensitivity analysis of optimization problems in Hilbert space with application to optimal control," *AMO*, vol. 21, pp. 1-20, 1990.

[49] V.V. Mal'tsev, "Sufficient conditions for ε–sensitivity in mathematical programming problems," *SAC*, vol. 12, no. 5, pp. 35-39, 1979.

[50] J.M. Martin and G.A. Hewer, "Smallest destablizing perturbations for linear systems," *IJC*, vol. 45, no. 5, pp. 1495-1504, 1987. Comments by C.B. Soh, *ibid.*, vol. 49, no. 5, pp. 1813-1814, 1989.

[51] T. Masuda, "Hierarchical sensitivity analysis of priority used in analytic hierarchy process," *IJSS*, vol. 21, no. 2, pp. 415-427, 1990.

[52] V.S. Mel'nik, "Nonconvex optimization problems for quasilinear distributed-parameter systems," *SJAIS*, vol. 21, no. 2, pp. 80-83, 1988.

[53] M. Mrabti and A. Hmamed, "Bounds for the solution of the Lyapunov matrix equation – A unified approach," *SCL*, vol. 18, pp. 73-81, 1992.

[54] P.H. Naccache, "Stability in multicriteria optimization," *JMAA*, vol. 68, pp. 441-453, 1979.

[55] A. Neumaier, "Rigorous sensitivity analysis for parameter-dependent systems of equations," *JMAA*, vol. 144, pp. 16-25, 1989.

[56] K.C. Park and J.C. Chiou, "Stabilization of computational procedures for constrained dynamical systems," *JGCD*, vol. 11, no. 4, pp. 365-370, 1988.

[57] D.W. Peterson, "On sensitivity in optimal control problems," *JOTA*, vol. 13, no. 1, pp. 56-73, 1974.

[58] B.D. Prudovskiy, "Stability of solutions in certain nonlinear optimization problems," *ECy*, vol. 17, no. 1, pp. 151-153, 1979.

[59] V. Rupnik, "Stability conditions on continuous dynamic linear programming," *JMAA*, vol. 119, pp. 171-181, 1986.

[60] J.K. Sengupta, "Sensitivity analysis for a linearized method of geometric programming," *IJSS*, vol. 8, no. 2, pp. 153-161, 1977.

[61] A. Shapiro, "Second order sensitivity analysis and asymptotic theory of parameterized nonlinear programs," *MP*, vol. 33, pp. 280-299, 1985.

[62] _____, "Sensitivity analysis of nonlinear programs and differentiability properties of metric projections," *SIAM-JCO*, vol. 26, pp. 628-645, 1988.

[63] _____, "Perturbation theory of nonlinear programs when the set of optimal solutions is not a singleton," *AMO*, vol. 18, pp. 215-229, 1988.

[64] _____, "On concepts of directional differentiability," *JOTA*, vol. 66, no. 3, pp. 477-487, 1990.

[65] S. Shiraishi, "First-order and second-order ε–directional derivatives of a marginal function in convex programming with linear inequality constraints," *JOTA*, vol. 66, no. 3, pp. 489-502, 1990.

[66] J. Sokolowski, "Sensitivity analysis of contact problems with prescribed friction," *AMO*, vol. 18, pp. 99-117, 1988.

[67] A.M. Steinberg, "Application of relaxed solutions to minimum sensitivity optimal control," *JOTA*, vol. 10, no. 4, pp. 178-186, 1972.

[68] T. Tanino, "Stability and sensitivity analysis in convex vector optimization," *SIAM-JCO*, vol. 26, no. 3, pp. 521-536, 1988.

[69] _____, "Sensitivity analysis in multiobjective optimization," *JOTA*, vol. 56, no. 3, pp. 479-499, 1988.

[70] R.L. Tobin, "Sensitivity analysis for variational inequalities," *JOTA*, vol. 48, no. 1, pp. 191-204, 1986.

[71] M.D. Troutt, "A stability concept for matrix game optimal strategies and its application to linear programming sensitivity analysis," *MP*, vol. 36, pp. 353-361, 1986.

[72] K.C.Q. Tsai, D.M. Auslander, "A statistical methodology of designing controllers for minimum sensitivity of parameter variations," *ASME-JDSMC*, vol. 110, pp. 126-133, June 1988.

[73] C. Van Loan, "How near is a stable matrix to an unstable matrix," in *Linear Algebra and Its Role in Systems Theory*, in Proc. AMS-IMS-SIAM Conf. held July 29 to Aug. 4, 1984 (R.A. Brualdi, et al., Eds.), *Contemporary Mathematics*, American Mathematical Society, vol. 47, pp. 465-478, 1985. [cf., [20].]

[74] J. Varah, "On the separation of two matrices," *SIAM-JNA*, vol. 16, pp. 216-222, 1979.

[75] R.B. Vinter, "Optimality and sensitivity of discrete time processes," *CCy*, vol. 17, nos. 2/3, pp. 191-211, 1988.

[76] W. Von Dinkelbach, *Sensitivitatsanalysen und parametrische Programmierung*. Berlin: Springer – Verlag, 1969. Reviewed by S.I. Gass, *SIAM R.*, vol. 12, no. 1, pp. 165-166, 1970.

[77] A.N. Voronin, "A multicriteria variational problem for deterministic and statistically specified actions," *SJAIS*, vol. 21, no. 5, pp. 37-42, 1988.

[78] A.N. Voronin and V.V. Pavlov, "Solution of multicriterial variational problems under indeterminacy conditions," *SJAIS*, vol. 21, no. 2, pp. 23-31, 1988.

[79] K. Watanabe and H. Shimizu, "Parameter optimizing software system using Pade's computation method of transient response," *EEJ*, vol. 95, no. 2, pp. 119-125, March/April 1975.

[80] P. Whittle, "Entropy-minimising and risk-sensitive control rules," *SCL*, vol. 13, pp. 1-7, 1989.

[81] _____, "A risk-sensitive maximum principle," *SCL*, vol. 15, pp. 183-192, 1990.

[82] _____, "A risk-sensitive maximum principle: The case of imperfect state observation," *IEEE-TAC*, vol. AC-36, no. 7, pp. 793-801, 1991.

[83] P. Whittle and J. Kuhn, "A Hamiltonian formulation of risk-sensitive linear/quadratic/Gaussian control," *IJC*, vol. 43, no. 1, pp. 1-12, 1986.

[84] L.A. Wolsey, "Integer programming duality: Price functions and sensitivity analysis," *MP*, vol. 20, pp. 173-195, 1981.

[85] J.-H. Xu and R.E. Skelton, "An improved covariance assignment theory for discrete systems," *IEEE-TAC*, vol. AC-37, no. 10, pp. 1588-1591, 1992.

[86] J. Zowe and S. Kurcyusz, "Regularity and stability for the mathematical programming problem in Banach spaces," *AMO*, vol. 5, pp. 49-62, 1979.

CHAPTER SEVEN

THE THEORY OF SENSITIVITY IN NETWORKS

7.1. INTRODUCTION

This chapter primarily presents a concise account of the applications of sensitivity theory in network analysis and synthesis and is an up-dated and revised version of [*IC*. 1 & 2]. Here we have categorized the pertinent literature into 14 different groups of active research in network sensitivity theory. This survey on various aspects of sensitivity theory provides only an entry guide for an interested researcher.

Network theory is the first branch of electrical engineering to have received extensive attentions by researchers in applications and development of sensitivity theory. Generally speaking, in network analysis and synthesis we are mostly concerned with the accuracy of the electrical network elements and the effects which the variations in these elements may have on the overall behavior of the network. As is well known, each electrical circuit element is fabricated with a given nominal value and subject to a specified range of tolerances. Since in a synthesis process the designer cannot assume the nominal value to be as the exact value of the element. Hence this fact is the main stimulus behind "static" sensitivity analysis. This analysis involves determining specifications on the range of different circuit elements such that certain network properties hold.

The principal aspects of circuit sensitivity theory are the same as that for a general dynamical system sensitivity theory. However, a significant difference between these two areas is the method of approach, using mainly frequency-domain techniques for circuits, as compared to both frequency- and time-domain techniques for a dynamical system sensitivity analysis. Circuit sensitivity theory remains apart from the whole body of system sensitivity theory due to its early development and to the fact that the corresponding functionals have a characteristic different from those of system theory. Indeed, we note that many ideas in system sensitivity theory are direct generalizations of those used in circuit theory. However, the trends of, and the concerns for, sensitivity of a network are different enough to be studied separately, as the large number of existing papers pertinent to sensitivity analysis and synthesis in circuit theory suggest. We must emphasize that the topological sensitivity of a network is well handled by signal flow graph procedure, and this topic is not discussed further.

A commonly used sensitivity definition in network theory is Bode's logarithmic sensitivity function, i.e., given any function such as T=T(s, x), where x is an arbitrary parameter and s is the Laplace variable, then Bode's sensitivity function for T in terms of x, is expressed as $S_x^T \triangleq (x/T)(\partial T/\partial x) = (\partial \ln T/\partial \ln x)$ (cf., Remark 2.2-4). Certain extensions of this sensitivity function to multi-parameter, multi-loop systems, are reported in § 6.2.2, under sensitivity comparison matrices. Discussions pertinent to the effectiveness of the feedback theory in multi-parameter, multi-loop networks have been developed by Sandberg in [A. 14], and are advanced by Kuh and Rohrer in [B. 9]. Sensitivity analysis in network theory, concerned with variations of several parameters, is similar to that of a general multi-parameter system, in which certain scattered results are reported, and these are cited subsequently.

The main goal of the present study is to locate and to classify, as feasible, some main channels of investigations in the theory of sensitivity as applied to circuit theory.

References in § 7.16.A, are chosen so as to emphasize the importance of the time-domain analysis of a network. Such analysis gives network theory the benefits of extensive results available from system sensitivity theory. Historically speaking, Kuh and Rohrer [A. 6] wrote an early paper on the time-domain analysis of network. Dervisoglu [A. 4] extends this work to an active network. MacFarlane and Sabouni [A. 8 to 11] contribute matrix theory applications pertinent to network analysis. Chen's survey paper [A. 3] serves as starting point in research in active network analysis. Bordewijk [A. 1] gives a useful paper in network sensitivity analysis: the concept of adjoint network. The readers are also urged to review the Van Valkenburg's lucid paper [A. 15] on certain historical aspects of teaching circuit theory.

In § 7.16.B, some recent and historical textbooks on the theory of sensitivity pertinent to network theory are reported. The research in this area can be further categorized into the 14 following divisions as discussed next.

7.2. GENERAL NETWORK SENSITIVITY: DEFINITIONS AND ANALYSES

Sensitivity analysis of linear circuits has been extensively carried out from the point view of pole-zero sensitivity, i.e., to investigate the variations in the locations for the zeros of a rational polynomial due to changes in its coefficients. Perhaps this study is one of the oldest research area in sensitivity theory. The early work of Papoulis [C. 15] is still inspiring, so is the general approach of root locus given in [C. 23]. However, we must recall that only rational equations of the form A(s) + KB(s) = 0, where A and B are polynomials of the complex variable s, and K is the varying parameter, is in a form amenable to treatment by the root locus. A generalization of root locus to multi-input system is discussed by MacFarlane [A. 7];

and recently to multivariable is given by [5J. 30]. Certain *ad hoc* definitions and/or results for sensitivity measures in a linear network exist, a subset of these are reported in § 7.16.C. The extensive results on the stability robustness analysis in Chapter Five are welcome addition to this research area. In fact due to such an extensive development, regarding stability under multiple-large parameter variations, it seems that many new opportunities exist in order to evaluate the impact of these results in the corresponding circuit theory problems.

7.3. TOLERANCES OR SENSITIVITY IN THE LARGE OF NETWORKS

In system sensitivity parlance "sensitivity in the large," refers to a situation wherein a particular system "quantity" changes by a "large" amount, resulting in a certain system "output" changes, perhaps by a "large" amount. To investigate whether the variation in this "output" from its nominal value, caused by the earlier change in the system "quantity" is acceptable referred to as "sensitivity in the large" or robustness analysis. The actual acceptable variation is a measure of "sensitivity in the large" for this system. This is similar to the study of the tracking error between a nominal and a disturbed system. If this variation (depending on the problem) is qualitatively and/or quantitatively acceptable, then this system is called "robust", in terms of the varying element(s). Of course, the true meaning of robustness is that instead of having one system with *a* particular performance we have a class of systems which performs similarly. Roughly speaking, the larger is this class, the more robust is the system. In other words, when a system is robust with respect to *a* particular property, then there is room for certain degree of maneuverability around its operating condition. Construction of this class entails an analysis that we call "sensitivity in the large" or "robustness" analysis. In this situation the system performance exhibits certain "tolerances" to changes that are not necessarily "small". In many cases "sensitivity in the large" is much more desirable than conventional "small" sensitivity analysis. Because ultimately we prefer to have a system that tolerates large parameter variations.

In general, to perform a "large" sensitivity analysis for a network is simpler than that of a general system, because there are several simple relations for the analysis of a network, which serve to simplify the "large" sensitivity analysis.

Network sensitivity analysis, in general, can be performed in one of the following manners: *(i)* network sensitivity model (which is derived from the application of the compensation theorem [B. 14]) and is discussed in Section 7.4, and *(ii)* the concept of an adjoint network that is discussed in Section 7.13. Network sensitivity analysis in the "large" can be performed in the same manner as that of a general dynamical system. Here we must up-date the network function involved for each level of changes in the varying element, which is not an efficient process [D. 10]. In general, most of the existing approaches require the application of extra auxiliary functions, up to as many as four. The method of Parker, *et al.* [D. 18] and Parker [D. 17] for calculating sensitivity in the "large" of a driving point impedance is based on the bilinear formulation of the network and requires only one auxiliary function. This result of Parker, et al. [D. 18] is then revised by Sorenson [D. 23], for the case of a transfer ratio. Sorenson proved that "any network function executed by a linear network is a bilinear function of any bilateral or unilateral element immittance embedded in the network [D. 23]." This result is then used to derive difference quotient theorems expressing the exact sensitivity in the "large" of the network. In an instructive paper by Fidler [D. 8], several other useful formulae are derived which basically relate the network sensitivity in the "large" to that of differential or "small" sensitivity. Fidler's result [D. 8] uses the result of Goddard, et al. [D. 10] in correlating the sensitivity in the "large" of a function to the "small" sensitivity for that function.

The Fidler's result is as follows. Consider a rational function F=N/D, where F=F(s, e) is a function of Laplace variable s, and e is the element value of interest. This e might be an immittance, amplifier gain, or any other quantity for which it is known that F is a bilinear function of e. If $e \to e + \Delta e$, then $F \to F + \Delta F$, where $F + \Delta F = (N + \Delta e \times N')/(D + \Delta e \times D')$, here prime denotes $\partial/\partial e$. Thus it is easy to show that $\Delta F = \Delta e \times F'/(1 + \Delta e \times \frac{D}{D'})$, or $(\Delta F/\Delta e) = F'D'/(D' + \Delta e \times D)$. Using this equation Fidler developed several other expressions for the relations between the "small"-sensitivity measure and sensitivity in the "large" of a network. This result is general, and as long as any network function F has bilinear relationship in terms of a particular element, holds without requiring any auxiliary functions. El-Turky has used Fidler's result to generate network sensitivity functions for a general network [D. 7]. Haley's paper [D. 12] shows a new method to determine the large-change multiparameter sensitivity of networks using scattering theory. Additional results are also reported in [D. 26].

7.4. SENSITIVITY MODEL OF CIRCUITS

When we model a circuit with a differential equation, it is easy to show that the homogeneous parts of the sensitivity differential equation and the dynamical differential equation of the network are indeed similar. The only difference between these two differential equations is in the forcing function. The forcing function for the sensitivity differential equation includes the solution of the network differential equation (cf., Section 2.4). Parker [D. 17] notes that Tomovic [B. 17] originally introduced this concept for analog computation of linear systems, and the results were extended to discrete-time calculation for linear circuits by Leeds [H. 8] and Leeds and Urgon [H. 9]. However, we note that the above mentioned property of the perturbed differential equation had been discussed by Miller and Murray [E. 1] many years earlier than [B. 17].

As is pointed out by Parker [D. 17], the idea of a sensitivity model is indeed an extension of the compensation theorem recently reviewed in [B. 14]. In short the concept of sensitivity model is an alternative approach for generating, in general on-line, the sensitivity functions of a network which is a special application of the procedures given in Chapter Three. Also Thiele's sensitivity measure which gives the transfer function sensitivity of a linear system described in state-space model, can be used to construct a new sensitivity model for a circuit [E. 4].

7.5. NETWORK ELEMENTS SENSITIVITY AND ENERGY STORED IN THE NETWORK, VRATSANOS' THEOREM

It is desired to study the relationships between various network-sensitivity functions and certain physical and/or topological properties of a network. In this section the relationship between a network-element sensitivity and energy stored in a network with invariant properties is reviewed. The main *tool* for this study is the well-known theorem of Vratsanos.

THEOREM 7.5-1 (Vratsanos' (Cohn's) theorem [F. 10]): If $z^{(n)}$ is an element of a one-port network with $f_n = I^{(n)}/I_o$, where $I^{(n)}$ is the current through $z^{(n)}$ and I_o is the input current to the one-port z_o, then for reciprocal networks Vratsanos' theorem gives $f_n^2 = \partial z_o / \partial z^{(n)}$.

This theorem has been studied by several researchers. Using the above theorem Smith [F. 10] relates the sensitivity of a passive impedance (admittance) to variations in individual elements of the network. He has also shown that for reciprocal realizations, the sensitivity magnitude sums, over resistances and reactances, are separately related to the total power dissipation and the energy storage in the network. To minimize sensitivity, the energy storage must be minimized. The wider class of networks containing gyrators (using an extension of Vratsanos' theorem), is also discussed in [F. 10]. In this reference Smith has additionally derived bounds for the sensitivity magnitude sums over resistances and reactances in terms of total power dissipation and the energy storage in the network. Blostein [F. 2], presents properties of a classical sensitivity function in resistance-terminated LC two ports, for studying the effects of element variations, incidental dissipation, and stray terminations on transmission characteristics. The above results which are extended to resistively terminated LC n-port networks with reactive energy stored in each L or C under harmonic excitation, are given by Kishi and Kida [F. 5]. This paper shows that the mean value of the energy stored in a reactive element can be expressed in terms of the imaginary parts of the sensitivity functions for the scattering parameters with respect to the elements, and provides further remarks about upper bounds of the sensitivity functions.

7.6. NETWORK SENSITIVITY INVARIANTS AND BOUNDS ON SENSITIVITY FUNCTIONS

Network sensitivity invariance refers to the algebraic relations which may exist between various combinations of network sensitivity functions. These relations, not so easy or straightforward to derive, give many algebraic and/or topological constraints for a network which must be considered in any design algorithm incorporating sensitivity considerations. There are many such relations known and perhaps many more to be discovered. A survey paper by Neirynck and Bastelaer [G. 12] has a large number of these relationships, including and in particular the work by Swamy, et al. [G. 19 to 21]]. Swamy, et al. [G. 19], find certain bounds on the sum of the element sensitivity magnitudes for transfer immittance, transfer voltage, and transfer current ratios for networks consisting of resistors, capacitors, inductors, and gyrators with further discussions about sensitivity invariance of RC and LC networks.

The main thrust of these activities is to derive an expression for the sum of the squared sensitivity functions, or certain other network sensitivity measures. It turns out that either this expression is a constant (numerically), or it is bounded (below or above) in terms of other network quantities. These expressions, in general, serve as additional side conditions for sensitivity-measure optimizations. The result stated in Theorem 3.6-1, is a useful result in this field, since many other existing results on linear network sensitivity invariants can be easily deduced from this theorem.

7.7. CONCEPT OF CONTINUOUSLY EQUIVALENT NETWORKS SYNTHESIS PROCEDURES WITH SENSITIVITY INVARIANTS

This topic initiated by Schoeffler [H. 14], who first introduced the concept of continuously equivalent networks. This concept appears to be of considerable practical importance. The supporting theory is essentially based upon the use of a congruence transformation. Cauer [A. 2] showed that by means of a congruence transformation one physically realizable network could be generated from another in such a way that specified driving point and/or transfer functions were held invariant [H. 14]. Schoeffler's method of continuously generating equivalent networks by means of node-admittance matrix transformation, which in fact is an extension of Howitt theory [H. 5], is revised by Cheetham [H. 2]. Cheetham [H. 2] proposes a new approach which is computationally simpler to apply and conceptually easier to grasp than that of Schoeffler's method. Schoeffler's technique is used for optimizing the continuously equivalent network with respect to a sensitivity criterion defined as the sum magnitudes squared of the sensitivity functions. However, Newcomb [H. 12], has pointed out that the continuously equivalent network technique is incomplete, i.e., there may exist other equivalent network realization with lower sensitivity criterion, or in other words not all the equivalent networks can be found. Shirakawa, et al. [H. 15] have discussed other classes of transformations in which the port structures of a network are invariant.

The paper by Schoeffler [H. 14] was the beginning of a series, to be followed up by several researchers such as Leeds and Urgon [H. 9], who postulated three hypotheses concerning continuously equivalent networks [D. 17]. "The first is that the optimum network resulting from minimizing Schoeffler's sensitivity criteria at a

7.7. Concept of Continuously Equivalent Networks

given frequency is *also* optimal in the same sense at all frequencies. The second hypothesis is that the sum of the magnitudes squared of the sensitivity functions decreases as the number of elements increases. The third hypothesis is that the sum of the sensitivity function is invariant [D. 17]." Schmidt and Kasper [H. 13] have demonstrated that the sum of sensitivities to changes in all elements of one kind is a constant for all continuously equivalent networks (though a function of frequency). Kuh and Lau [H. 7] have also shown that sum of the sensitivities to all elements in a continuously-equivalent network is a constant. Kuh and Lau [H. 7] also point out the minor error of the paper by Blostein [H. 1], on generation of minimum sensitivity networks. The second result of Kuh and Lau [H. 7] states that under continuously equivalent transformation the individual sensitivity for capacitances and inductances is invariant if there are no capacitive loops or inductive cut-sets in the network. But, Neirynck and van Bastelaer [H. 11] show that this result of Kuh and Lau [H. 7] is not generally true! Sohal and Singh [H. 16] have attempted to show that the sum of the sensitivity functions over all components of each continuously equivalent network as is true for impedance and admittance functions, is not true for transfer functions. Based on the following theorem, it is easy to show by some numerical examples that the argument in [H. 16] is not complete.

THEOREM 7.7-1 (Theorem 3.6-1): Consider the state equation of an open-loop linear system

$$\dot{x}(t) = M^{-1}N\,x(t) + Q^{-1}Pu(t), \quad x(0) = x_o, \tag{7.7-1}$$

where $x(t) \in R^n$ is the state vector; $u(t) \in R^m$ is the input vector; $M = (m_{kl})$, $N = (n_{rs})$, $Q = (q_{vw})$, and $P = (p_{ef})$ are coefficient matrices of appropriate dimensions. If $\{\alpha_{ij}\} = \{m_{kl}, n_{rs}, q_{vw}, p_{ef}\}$, then

$$\sum_{i,j} \alpha_{ij} \frac{\partial x(t)}{\partial \alpha_{ij}} = 0. \tag{7.7-2}$$

7.8. NETWORK SYNTHESIS WITH MINIMUM OR PRESCRIBED SENSITIVITY MEASURE

It is the ultimate goal in any sensitivity study to include information obtained from sensitivity analysis in the design procedures and to develop algorithms resulting in networks with minimum sensitivity to parameter variations or external disturbances. To accomplish this goal requires that we establish the following general steps: *(i)* the definition of an appropriate sensitivity measure; *(ii)* the calculation of this measure; and *(iii)* the incorporation of this measure in design algorithms in order to reduce its troublesome effects. This concept closely follows that for a general system sensitivity reduction. We note that a network which is minimum sensitive in terms of one sensitivity measure may not, of course, be the best design. A commonly used network sensitivity measure is the sum of the squared magnitude of network sensitivity functions in terms of network elements. This is not the only useful sensitivity measure, as many other researchers have demonstrated this fact (cf., Smith [F. 10]).

The scattered research in network sensitivity minimization enjoys the benefits of classical network synthesis as well as new trends in optimization theory. In particular, various algorithms for sensitivity minimization of a general system that are developed in Chapter Six, provide considerable opportunities for network designers to incorporate these results in order to have a pertinent optimal design. In this regard, network sensitivity minimization still lags that of the general system theory.

7.9. SENSITIVITY OF THE COST FUNCTION IN THE OPTIMAL CIRCUIT DESIGN

The main aspect of effectual automated design and optimization of nonlinear networks is the sensitivity vector [J. 4]. The sensitivity vector is a measure of network performance with respect to a change in the design parameter. The sensitivity vector, in automated circuit design and optimization areas, is playing an important role due to the availability of extensive computing facilities. Generally speaking, and upon specifying the cost functions in optimal circuit design, which usually is a given network-sensitivity measure, the designer has to use various optimization techniques to complete the optimal circuit design process.

El-Turky and Vlach give a method to generate both passive and active networks with a minimum sensitivity measure [J. 2]. Ho's paper [J. 4] has several references pertinent to the present topic and considers derivation of the sensitivity vector for circuits containing transmission lines, by using variational techniques. Kida and Osumi [J. 5] present another optimization method to investigate the parasitic effects on the impulse responses of linear time-invariant networks. Bachler and Guggenbuhl [J. 1] use an optimization criterion (a two-dimensional object function) to deal either with noise or sensitivity of a (second-order active filter) single-amplifier biquad. This result shows that a slight sacrifice of the overall sensitivity properties is needed for the optimum noise design, and the output noise increases significantly if the circuit is designed for minimum sensitivity. We note that the sensitivity analysis of the cost function as applied in general control system theory is more advanced than that which has been considered in network sensitivity theory. A list of additional references on this topic is given in § 6.12.C.

7.10. SENSITIVITY ANALYSIS AND SYNTHESIS IN ACTIVE NETWORKS

The difference between active and passive networks is well known. Correspondingly, sensitivity analysis associated with this trend in circuit design has been the subject of extensive research. The trends are exactly the same as that of a general network or system. In particular, if an active network were represented in the

7.10. Sensitivity Analysis and Synthesis in Active Networks

time-domain, then the existing literature on system sensitivity theory can be directly applied to the encountered problem.

Active network sensitivity involves the same steps as other cases which we discuss herein, namely we must again establish: *(i)* definition of an appropriate sensitivity measure; *(ii)* computations and/or derivation of this measure; and *(iii)* incorporation of the results obtained from the above analysis in the active network design as feasible. In particular, in active RC-networks and second-order filters the two factors Q and ω_o (reciprocal of twice of damping ratio, and natural (or center) frequency, respectively) are very crucial. The sensitivity analyses for these two factors in terms of the active network parameters received much attentions. Because of space limitation we cannot discuss all the papers that are reported in § 7.16.K. The outstanding tutorial paper of Moschytz [K. 78] on the sensitivity problem in active filters is an excellent source for further research and is a paper that must be read. So is the paper by Laker, et al. [K. 60], that reviews active circuits based on first- or second-order building blocks in a feedback configuration. The following research categorizations are reported in active networks.

7.10.1. The Pole-Zero Sensitivity and Certain Other General Sensitivity Measures (Gain and Phase) of Active Network

Hakim [K. 41] develops synthesis procedures for RC-active filters with prescribed pole sensitivity. Newcomb further advances Hakim's theory [K. 85]. Hirano, *et al*. [K. 45] discuss the frequency-domain analysis of the sensitivity function in a two-port active network, using graphical techniques. Moschytz [K. 75] gives an algorithm for selecting the pole-zero pair of an nth-order minimum sensitivity network by selecting that pair for the second-order realization of prescribed nth-order network. Shieu and Chan [K. 104] give topological formulations for symbolic network functions with corresponding sensitivity analysis of active networks. Leuder [K. 63] and Premoli [K. 88] discuss various polynomial decompositions and the corresponding sensitivity problems. Horowitz [K. 49] presents algorithms for achieving various combinations of zero sensitivity integrated filter circuits. Bruton and Hasse [K. 14] discuss sensitivity of a generalized immittance converter-embedded ladder structures. Brice [K. 11] presents sensitivity and feedback in amplifiers. Saha [K. 98] shows the synthesis procedure for an RC-active network with prescribed root sensitivity. Soderstrand [K. 106] uses the pole of an operational amplifier to produce the state variables without employing external L's and C's, and the resulting filters offer low sensitivity. In [K. 55] the frequency dependence of sensitivities of gain and phase is studied and a statistical variability measure for optimal filter design is introduced.

7.10.2. The Q and ω_o As Sensitivity Measures of the Active Networks With Corresponding Comparison and Minimization of These Measures

Bialko and Sienko [K. 7] give a synthesis procedure with zero Q-sensitivity to active parameters. This algorithm provides the realization of high Q-factors for the transfer function poles, and an optimal decomposition of third-degree polynomials is also given. Bruton [K. 13], discusses sensitivity comparison of high-Q second-order active filter synthesis procedures. Mackay and Sedra [K. 65], comprehensively give a systematic procedure for evaluating the sensitivity of the transfer function magnitude, and hence the transmission deviation for high-order active filters. This method is illustrated by carrying out a comparative analysis for a band pass filter realization by: Cascade, leap-frog, and the primary resonator block approaches. Fleischer [K. 28] discusses sensitivity minimization in a single amplifier biquadratic circuit. Gadenz [K. 30] and Laker and Ghausi [K. 56 & 57] treat low sensitivity for *(i)* realizations of band-elimination filters, and *(ii)* multiloop feedback active RC-filters, respectively. Mitra and Aatre [K. 69] work on low sensitivity high-frequency active R-filters. Reddy [K. 92 & 93] presents an analysis of low sensitivity active RC-filters, for high Q and high frequencies. The paper by Schaumann [K. 101] presents a low-sensitivity high-frequency tunable active filter without external capacitors, simply by considering the non-ideal transfer function of an operational amplifier, instead of assuming either infinite gain or constant gain, as do Rao, *et al.* [K. 90 & 91]. Soderstrand and Mitra [K. 111] show a design of an active filter with zero passive Q-sensitivity, and in [K. 112], they obtain a zero gain-sensitivity product. Geiger and Budak [K. 32] give a general characterization of active RC-filters using one to three operational amplifiers with low (zero) sensitivity. Sedra and Espinoza [K. 103] study sensitivity and frequency limitations of biquadratic active filters. Weyten [K. 121] gives a new sensitivity measure for comparison between optimization of second-order filter cells. Also this measure can be related to Q and ω_o of the network. Wilson, *et al.* [K. 124] discuss RC-active networks with reduced sensitivity to amplifier gain-bandwidth. They assume a single-pole representation of the complex gain amplifier, to derive the transfer function, and they use a perturbation technique to derive an approximate expression for frequency sensitivity.

7.10.3. Active Distributed Parameter RC Networks

Bello and Ghausi [K. 5] study the low sensitivity design for a class of active distributed RC-networks including the sensitivity of ω_o, the real and imaginary parts of the dominant poles, and the Q of the dominant poles. Bialko and Guzinski [K. 6] consider the properties of an active filter with a four-layer RC line in a feedback loop. In fact they proceed to show that it is possible to obtain a filter with zero Q-sensitivity to active-element gain variations.

7.10.4. Active Network Sensitivity Analysis – Statistically Oriented

Parallel to the main ideas which are presented in Section 7.12, for a general circuit, this topic has received much attentions. The paper by Rosenblum and Ghausi [K. 95] shows sensitivity minimization in active RC-networks. Laker and Ghausi [K. 57] extend the statistical work of Rosenblum and Ghausi [K. 95] to the "sensitivity in the large" for active networks. Also [K. 55] has discussions on statistical variability measure of an optimal filter design. Additional references for the above mentioned categories are also given in § 7.16.K.

7.11. SENSITIVITY ANALYSIS IN DIGITAL FILTERS

This is yet another area of active research in network sensitivity analysis. An extensive work has been done in this field, samples of them are reported in § 7.16.L. These papers clearly show the need for a separate study of sensitivity analysis in digital filters.

In general, there are two major problems of concern in the sensitivity analysis of a digital filter: *(i)* truncation or rounding the variation of the multiplier coefficients due to parameter changes or precision, and *(ii)* the round off noise due to the finite arithmetic operations.

Agarwal and Burrus [L. 3] study the recursive digital filter structures having both low sensitivity in term of variation in parameters and round off noise, where Bruton [L. 8] discusses a class of low-sensitivity digital ladder filters. Ali uses a linear transformation to design digital filter structures from analog doubly terminated lossless ladder filter [L. 4]. Crochiere applies optimization scheme to minimize the statistical word length of a given filter structure and a given set of maximum error constraints [L. 11]. Fettweis [L. 17] discusses the round off noise and attenuation sensitivity in the digital filter, so does Jackson [L. 19] who discovers a lower bound on the round-off noise output in terms of a coefficient sensitivity for the transfer function of a digital filter. Ku and Ng [L. 20] have set out the floating-point coefficient sensitivity and round-off noise for a recursive digital filter. They have shown that the ladder structures first proposed by Fettweis have a much lower coefficient sensitivity than was thought to be possible. This new Fettweis ladder structure is found to exhibit the lowest of all round off errors. Ledbetter and Yarlagadda [L. 22], using results available from eigenvalue sensitivity of Singer [L. 35] and Mantey [L. 27], develop algorithms to introduce a system matrix to realize a second-order digital filter with minimum eigenvalue sensitivity. In [L. 2] a new class of recursive digital filters with low sensitivity and low roundoff noise is proposed, and in [L. 39] a new sensitivity measure is defined and is optimized to reduce the noise power gain.

Sedlmeyer [L. 34], and Renner and Gupta [L. 32], discuss the sensitivity of wave digital filters with various design considerations. In [L. 12] a class of new low-sensitivity second-order digital-filter structures is presented. The results that are presented in Section 5.10 provide a comprehensive method to determine the largest variations in coefficients of a convergent polynomial (such as the characteristic equation of a digital filter), which does not deteriorate the system stability (or convergent). This section is certainly very relevant to the sensitivity analysis in a digital filter.

7.12. SENSITIVITY THEORY – STATISTICALLY ORIENTED

Parallel to the probabilistic sensitivity analysis in system theory, many researchers in circuit theory describe the parameter variations in network by statistical measures. It is definitely more appropriate to consider probabilistic sensitivity analysis than the usual deterministic approach, in particular when the number of varying parameters are very large and some of them are correlated. However, that statistical approach is more rigorous than the methods which are developed in this manuscript.

Liu and Van Valkenburg [M. 17] introduce some probabilistic measures for parameter deviations in a network and they have developed certain bounds for these measures. Acar, et al. [M. 1 to 3] study the statistical multiparameter sensitivity measures for gain and phase functions with applications to a high Q, RC-active network. Rosenblum and Ghausi [M. 21] consider the statistical multiparameter sensitivity for transfer functions of a linear system. Balaban and Golembeski [M. 6] investigate statistical analysis for practical circuit design *via* the Monte Carlo method. Hilberman [M. 13] presents an approach to sensitivity and statistical variability of biquadratic filters. Laker and Ghausi [M. 16] study the statistical multiparameter sensitivity of notch filters. Styblinski [M. 23] extends the Rosenblum-Ghausi statistical sensitivity measure [M. 21] to complex parameters. Rezai-Fakhr and Temes [M. 20] study *via* Monte Carlo analysis the statistical large-tolerance analysis of nonlinear circuits in the time-domain. Also in [M. 7] a new figure of statistical parametric sensitivity for large uncorrelated parameter changes is presented. The Poliscuk and Rojo give a modified Tellegen's theorem to generate sensitivity functions for switched-capacitor networks using statistical analysis [M. 19].

Overall this is an exciting research area and several of the references in § 7.16.M and [6L. 45] can be consulted, in order to find additional information for launching a very promising research direction.

7.13. TELLEGEN'S THEOREM: THE GENERALIZED ADJOINT NETWORK AND NETWORK SENSITIVITY

This topic, well-developed by Director and Rohrer [N. 2], advances the concept of a sensitivity model in a linear system. The idea is based on Tellegen's theorem, which a novel discussion of that and its application in sensitivity analysis, has been presented by Penfield, et al. [N. 2]. The subsequent discussion, detailed here for the purpose of illustration, closely follows the original work of Director and Rohrer.

7.13.1. The General Adjoint Network

Consider a network N, which may contain arbitrary two-terminal elements, the only requirement is that each element possesses a parametric representation. Let the branch voltages and currents of N be denoted by $v_B(t)$ and $i_B(t)$, respectively. Now let $\overline{N}$ be the adjoint network related to N; i.e., both N and $\overline{N}$ have the same topology, but not necessarily the same element types in corresponding branches. This requirement is essential for application of Tellegen's theorem. Let $\psi_B(\tau)$ and $\phi_B(\tau)$ denote the branch voltages and currents in the adjoint network $\overline{N}$; then by the use of Tellegen's theorem we can obtain

$$\sum_B v_B(t)\phi_B(\tau) \equiv 0, \quad \text{and} \quad \sum_B i_B(t)\psi_B(\tau) \equiv 0, \qquad (7.13\text{-}1)$$

where the summation is taken over all branches of N and $\overline{N}$. Tellegen's theorem is based on the topology of the circuit, and it is independent from the element values. If the element values of N change yet the topology of the network remains the same, Tellegen's theorem still holds, and it is independent of the time origin. If the element values of N change, then logically we expect that the currents through and the voltages across the elements of N to change. Tellegen's theorem still holds; i.e.,

$$\sum_B [v_B(t) + \Delta v_B(t)]\phi_B(\tau) \equiv 0, \quad \text{and} \quad \sum_B [i_B(t) + \Delta i_B(t)]\psi_B(\tau) \equiv 0. \qquad (7.13\text{-}2)$$

Equations (7.13-1) and (7.13-2) can be jointly manipulated to yield

$$\sum_B [\Delta v_B(t) \cdot \phi_B(\tau) - \Delta i_B(t) \cdot \psi_B(\tau)] \equiv 0. \qquad (7.13\text{-}3)$$

Now, consider a resistive-type branch whose $v_R(t)$ and $i_R(t)$ are

$$v_R(t) = f_R[x_R(t), p_R, t], \quad \text{and} \quad i_R(t) = g_R[x_R(t), p_R, t], \qquad (7.13\text{-}4)$$

where $x_R(t)$ is the chosen representation parameters, the scalar p_R is a typical design parameter, and the subscript R denotes a resistive element.

Assuming small changes in the parameters of (7.13-4), a Taylor series expansion of (7.13-4), with only the first two terms shown, is

$$\Delta v_R(t) = \frac{\partial f_R}{\partial x_R} \cdot \Delta x_R(t) + \frac{\partial f_R}{\partial p_R} \cdot \Delta p_R, \qquad (7.13\text{-}5a)$$

and

$$\Delta i_R(t) = \frac{\partial g_R}{\partial x_R} \cdot \Delta x_R(t) + \frac{\partial g_R}{\partial p_R} \cdot \Delta p_R. \qquad (7.13\text{-}5b)$$

Substituting (7.13-5) into (7.13-3) that are associated with the resistive elements yields

$$\left[\frac{\partial f_R}{\partial x_R} \phi_R(t) - \frac{\partial g_R}{\partial x_R} \psi_R(\tau)\right] \Delta x_R(t) + \left[\frac{\partial f_R}{\partial p_R} \phi_R(\tau) - \frac{\partial g_R}{\partial p_R} \psi_R(\tau)\right] \Delta p_R(t). \qquad (7.13\text{-}6)$$

If we assume that the adjoint network is such that

$$\frac{\partial f_R}{\partial x_R} \phi_R(\tau) - \frac{\partial g_R}{\partial x_R} \psi_R(\tau) = 0, \quad \text{or} \quad \phi_R(\tau) = \left[\frac{\partial f_R}{\partial x_R}\right]^{-1} \left[\frac{\partial g_R}{\partial x_R}\right] \psi_R(\tau). \qquad (7.13\text{-}7)$$

Then this relation between the resistive branch of the adjoint network causes that the coefficient of $\Delta x_R(t)$ in (7.13-6) to vanish, and the remaining terms are those which include only the design-parameter variations. In particular, for the special case of a branch with a linear time-invariant resistor, where (7.13-4) becomes

$$f_R[i_R(t), p_R, t] = Ri_R(t), \quad \text{and} \quad g_R[i_R(t), p_R, t] = i_R(t). \qquad (7.13\text{-}8)$$

Then we have

$$\psi_R(\tau) = R\phi_R(\tau). \qquad (7.13\text{-}9)$$

This account should make the generality of this approach and its main idea apparent.

For capacitive elements let us consider the following values.

$$v_C(t) = f_C[x_C(t), p_C, t], \qquad (7.13\text{-}10a)$$

$$q_C(t) = g_C[x_C(t), p_C, t], \qquad (7.13\text{-}10b)$$

$$i_C(t) = \frac{d}{dt} q_C(t), \qquad (7.13\text{-}10c)$$

where q_C is the capacitance charge. Following the same steps as we take for the resistive elements and letting $t = \zeta$ as a dummy variable, we have

$$\frac{\partial f_C}{\partial x_C}(\zeta)\phi_C(\tau)\Delta x_C(\zeta) - \frac{d}{d\zeta}\left[\frac{\partial g_C}{\partial x_C}(\zeta)\Delta x_C(\zeta)\right]\psi_C(\tau)$$

7.13. The Generalized Adjoint Network and Network Sensitivity

$$+ \left\{ \frac{\partial f_C}{\partial p_C}(\zeta)\phi_C(\tau) - \frac{d}{d\zeta}\left[\frac{\partial g_C}{\partial p_C}(\zeta)\right]\psi_C(\tau) \right\} \Delta p_C, \tag{7.13-11}$$

as the corresponding expression for a term in (7.13-3) associated with the capacitive elements. Let $\zeta \in (0, t)$ and assume $\tau = t - \zeta$. Then integrating (7.13-11), results in

$$\int_0^t \left\{ \frac{\partial f_C}{\partial x_C}(\zeta)\phi_C(t - \zeta)\Delta x_C(\zeta) - \frac{d}{d\zeta}\left[\frac{\partial g_C}{\partial x_C}(\zeta)\Delta x_C(\zeta)\right]\psi_C(t - \zeta) \right\} d\zeta$$

$$+ \Delta p_C \int_0^t \left\{ \frac{\partial f_C}{\partial p_C}(\zeta)\phi_C(t - \zeta) - \frac{d}{d\zeta}\left[\frac{\partial g_C}{\partial p_C}(\zeta)\right]\psi_C(t - \zeta) \right\} d\zeta. \tag{7.13-12}$$

If we choose the adjoint branch relation from the first integral expression in (7.13-12) and we integrate by parts the second term of the first integral expression in (7.13-12). Then we have ascertained the adjoint branch relation as follows.

$$0 = \int_0^t \left[\frac{\partial f_C}{\partial x_x}(\zeta)\phi_C(t - \zeta) + \frac{\partial g_C}{\partial x_C}(\zeta)\frac{d}{d\zeta}\psi_C(t - \zeta) \right] \Delta x_C(\zeta) d\zeta \tag{7.13-13}$$

$$- \frac{\partial g_C}{\partial x_C}(\zeta)\Delta x_C(\zeta)\psi_C(t - \zeta) \bigg|_{\zeta=0}^{\zeta=t}. \tag{7.13-13}$$

The transversal conditions in (7.13-13) for unspecified initial and final conditions are considered the sensitivity components for these boundary conditions. If we let $\Delta x_C(0) = 0$, which is generally the case, then the transversal conditions become zero if $\psi_C(0) = 0$: i.e., the initial condition for the adjoint network is also zero. Now, (7.13-13) holds for nonzero values of $\Delta x_C(\zeta)$ (otherwise trivial) if

$$\frac{\partial f_C}{\partial x_C}(\zeta)\phi_C(t - \zeta) + \frac{\partial g_C}{\partial x_C}(\zeta)\frac{d}{d\zeta}\psi_C(t - \zeta) \equiv 0, \tag{7.13-14}$$

or,

$$\frac{\partial f_C}{\partial x_C}(t - \zeta)\phi_C(\tau) = \frac{\partial g_C}{\partial x_C}(t - \zeta)\frac{d}{d\tau}\psi_C(\tau). \tag{7.13-15}$$

This is a relation between the voltage and the current of the element corresponding to a capacitive branch in the adjoint network, which in the case of linear time-invariant elements, the adjoint branch relation simplifies to $\phi_C(\tau) = C\frac{d}{d\tau}\psi_C(\tau)$.

Following this general flow of analysis, and with similar assumptions, we can derive the relations between the current and the voltage for the element corresponding to an inductive branch in the adjoint network, as well as for the class of memoryless nonlinear coupling elements which are described by two representation parameters.

7.13.2. General Network Sensitivity

Before the general sensitivity coefficients are presented, and from (7.13-6) for $t = \zeta$, $\zeta \in (0, t)$ and $\tau = t - \zeta$, we define:

$$-\Delta p_R S_R \triangleq \int_0^t \Delta p_R \left[\frac{\partial f}{\partial p_R}(\zeta) \phi_R(t - \zeta) - \frac{\partial g_R}{\partial p_R}(\zeta) \psi_R(t - \zeta) \right] d\zeta, \qquad (7.13\text{-}16)$$

in order to simplify the corresponding expression. From (7.13-12):

$$-\Delta p_C S_C \triangleq \int_0^t \Delta p_C \left\{ \frac{\partial f_C}{\partial p_C}(\zeta) \phi_C(t - \zeta) - \frac{d}{d\zeta} \left[\frac{\partial g_C}{\partial p_C}(\zeta) \right] \psi_C(t - \zeta) \right\} d\zeta. \qquad (7.13\text{-}17)$$

Similarly $-\Delta p_L S_L$ and $-\Delta p_D S_D$ can be defined for the terms corresponding to the inductive and the coupling elements, respectively. Now, from (7.13-3), we deduce that

$$\sum_B \int_0^t [\Delta v_B(\zeta) \cdot \phi_B(t - \zeta) - \Delta i_B(\zeta) \cdot \psi_B(t - \zeta)] \, d\zeta = 0. \qquad (7.13\text{-}18)$$

If we classify and separate the terms in above according to the sources and different kinds of branch elements contained, then by using (7.13-16), (7.13-17) and similar expression from (7.13-18), for the inductive and the coupling elements, we have

$$\sum_V \Delta v_V(t) - \sum_I \Delta i_I(t) = \sum_R S_R \Delta p_R + \sum_C S_C \Delta p_C$$

$$+ \sum_L S_L \Delta p_L + \sum_{\substack{\text{coupling} \\ \text{elements}}} S_D \Delta p_D. \qquad (7.13\text{-}19)$$

The left-hand side of the above equation are resulted from the special choices for impulse voltage and impulse current as sources in $\overline{N}$. In the original work of

Director and Rohrer [N. 2] they left out the integral in (7.13-16) and (7.13-18), as this omission was pointed out by Seth and Singhal [N. 6].

The above illustration presents a comprehensive picture of the applications stemming from the Tellegen's theorem, the concept of adjoint network, and some description for the sensitivity in nonlinear, as well as, time-varying circuits. Another extension of this approach for use in gradient calculations of non-commensurate networks is given by Bandler, et al. [O. 9].

One other published work, which closely follows the above technique, is presented by Swamy, et al. [G. 20], wherein a class of nonlinear networks is considered and the explicit expressions for sensitivity of a response due to nonlinear elements are derived. Subsequently, using these expressions certain invariance relations are established in [G. 20].

Another interesting research reported in this connection is the work by Therrien [O. 40], in which the adjoint network concept has been employed to compute the exact changes in response due to changed parameters in a class of cascaded two-port networks.

For additional references the reader may consult § 7.16.N and § 7.16.O.

7.14. NETWORK SENSITIVITY FUNCTION GENERATION: COMPUTATIONAL PROCEDURES

We recall that two general approaches for network sensitivity analysis exist: *(i)* the concept of circuit model; and *(ii)* the concept of adjoint network. From a practical point of view, however, every problem has to be specially treated and often it is difficult to determine *a priori* which one of the above two approaches is easier. In particular, several researchers insist that, although the concept of adjoint network is theoretically very appealing, it is not very efficient computationally. The extensive research on the symbolic network functions (cf., a survey paper by Lin [O. 28] and Fidler, et al. [O. 22]) coupled with the fact that digital computers are commonly used in electrical circuit analysis, have shown that roundoff and loss-of-significance value may become important sources for numerical errors. Thus the symbolic network functions can serve as an alternative for sensitivity computations of a network which is more efficient than the procedures based upon the adjoint network concept. However, as mentioned before, it is not easy to predict *a priori* which one of the several computational techniques is more efficient than the other. Perhaps further studies are required to classify the type of each encountered problem, in order to choose the most efficient computational procedure for that problem. As an example, consider the paper by Therrien [O. 40]. In this paper it appears that the adjoint network is an efficient procedure, while Branin [O. 12], insists that this is not so. However, Vallese [O. 41] and Parker [P. 4] extends the adjoint model concept so as to encompass larger variations, and nonlinear networks, respectively. Bandler and El-Kady [O. 5] present an algebraic notation to evaluate sensitivity of power network expressed in a complex mode.

Several other such discussions and applications of the adjoint network model are presented in various papers in § 7.16.*O* and § 7.16.*P*.

The applications of signal flow graph in network sensitivity function determination and the general system sensitivity-function generation are two other approaches for network sensitivity function determinations.

Also several other sample works in network sensitivity computations such as Ceyhun [*O*. 15 & 16], Neill [*O*. 29 & 30], Neumann and Agnew [*O*. 31], *etc.*, can be found for some specific applications.

7.15. SENSITIVITY ANALYSIS OF NONLINEAR NETWORKS

Generally speaking, since there does not exist any unique representation of nonlinear circuits, it is not possible to provide a more specific technique to analyze these circuits than what we have already obtained for the sensitivity analysis of a general system (cf., Section 2.4). Each circuit problem depends on its structure and has to be treated specially. The method of a generalized adjoint network seems conceptually promising for certain classes of nonlinear problems [*P*. 4]. For the usual nonlinear problem there is no general solution, unless we have an explicit knowledge about the nonlinearity. The Wyatt's paper [*P*. 7] represents one approach to tackle the sensitivity of a special nonlinear circuit with RC elements. Jain *et al.*, follow through this work in order to study the sensitivity of delay-time in linear RC network portion of a nonlinear network [*P*. 2]. The following theorem due to Parker, [*P*. 4] and [*D*. 17] is of some value in this regard.

THEOREM 7.15-1 (Parker [P. 4]): Sensitivity functions for a nonlinear circuit may be obtained by calculating the corresponding responses of a dependent circuit, topologically identical to the original, in which each component is replaced by a dependent linear equivalent given, at any instant, by the slopes (small-signal equivalent) of the voltage (current) versus current (voltage) characteristic for resistive (conductive) elements, the charge versus voltage characteristic for capacitive elements, and the flux versus current characteristic for inductive elements. The sensitivity model for the driving function depends upon the sensitivity parameter and is a voltage source in series with the component when the sensitivity parameter is explicit in resistive or inductive elements, and a current source in parallel with the component when the sensitivity parameter is explicit in conductive or capacitive elements. It is always directed to cause a current flow in the sensitivity model element opposite to the direction of current through the element in the original circuit. The value of the source function is determined by the partial derivative of element voltage (current) with respect to the sensitivity parameter for the resistive (conductive) elements; by the time rate of change of the partial derivatives of charge with respect to the sensitivity parameter for capacitive elements, and by flux with respect to the sensitivity parameter for inductive elements.

Another report on this topic, which closely follows the idea of generalized adjoint network, is given by Swamy, *et al.* [*G*. 20].

7.16. REFERENCES

[A] Selected References on Network Theory

[1] J.L. Bordewijk, "Inter-reciprocity applied to electrical networks," *Appl. Sci. Res.*, vol. 6, pp. 1-74, 1956.

[2] W. Cauer, "Vierpole," *Elek. Nach. Tech.*, vol. 6, pp. 272-282, July 1929.

[3] W.K. Chen, "Topological formulations and the order of complexity of active networks: A unified survey," *Networks*, vol. 2, pp. 237-260, 1972.

[4] A. Dervisoglu, "Formulation of state equations with initial values in linear active networks and distributions," in *Proc. 1974 IEEE Int. Symp. on Circuits and Systems*, pp. 495-499, April 1974.

[5] E.S. Kuh, "Stability of linear time-varying networks – the state space approach," *IEEE-TCT*, vol. CT-12, pp. 150-157, June 1965.

[6] E.S. Kuh and R.A. Rohrer, "The state-variable approach to network analysis," *Proc. IEEE*, vol. 53, pp. 672-686, July 1965.

[7] A.G.J. MacFarlane, "Multivariable Nyquist-Bode and multivariable root-locus techniques," in *Proc. IEEE Conf. on Decision and Contr.*, pp. 342-347, 1976.

[8] A.G.J. MacFarlane and R. Sabouni, "Functional matrix theory for the general linear electrical network. Pt. 1-The linear functional matrix," *Proc. IEE*, vol. 112, no. 4, pp. 754-762, 1965.

[9] _____, "Functional matrix theory for the general linear electrical network. Pt. 2-The general functional matrix," *Proc. IEE*, vol. 112, no. 4, pp. 763-770, 1965.

[10] _____, "Functional matrix theory for the general linear electrical network. Pt. 3-Eigenvector method for inversion of the general functional matrix," *Proc. IEE*, vol. 113, no. 7, pp. 1268-1276, 1966.

[11] _____, "Functional matrix theory for the general linear electrical network. Pt. 4-Bounds on eigenvalues and network reduction," *Proc. IEE*, vol. 115, no. 5, pp. 755-757, May 1968.

[12] J.M. Manley and H.E. Rowe, "Some general properties of nonlinear elements – Part I. General energy relations," *Proc. IRE*, vol. 44, no. 7, pp. 904-913, July 1956.

[13] F. Masutomi, "CAD to calculate the numerical solution for the linear electrical circuit with variable parameters," *EEJ*, vol. 98, no. 4, pp. 121-133, 1978.

[14] I.W. Sandberg, "On the theory of linear multiloop feedback systems," *Bell System Tech. J.*, vol. 42, no. 2, pp. 355-382, 1963.

[15] M.E. Van Valkenburg, "Teaching circuit theory : 1934 – 1984," *IEEE-TCAS*, vol. CAS-31, no. 1, pp. 133-138, Jan. 1984.

[B] Representative Textbooks (or Reports) with Network Sensitivity Discussions: Definitions, Analyses, Computational Procedures, and Compensation Theorem

[1] H.W. Bode, *Network Analysis and Feedback Amplifier Design.* New York: Van Nostrand, 1945.

[2] M.L. Bykhovsky, *Fundamentals of Dynamic Accuracy of Electrical and Mechanical Circuits [in Russian].* Moscow: Izd-vo AN SSSR, 1958.

[3] D.A. Calahan, *Computer Aided Network Design.* New York: McGraw-Hill, 1972.

[4] K. Geher, *Theory of Network Tolerances*. Budapest, Hungary: Akademiai Kiado, 1971.

[5] M.S. Ghausi, *Principles and Design of Linear Active Circuits*. New York: McGraw-Hill, 1965.

[6] B. Gold and C.M. Rader, *Digital Processing of Signals*. New York: McGraw-Hill, 1969.

[7] L.P. Huelsman, *Theory and Design of Active RC Circuits*. New York: McGraw-Hill, 1968.

[8] _____, *Active Filters*. New York: McGraw-Hill, 1970.

[9] E.S. Kuh and R.A. Rohrer, *Theory of Linear Active Networks*. San Francisco: Holden-Day, Inc., 1967.

[10] S.K. Mitra, *Analysis and Synthesis of Linear Active Networks*. New York: Wiley, 1969.

[11] J.B. Murdoch, *Network Theory*. New York: McGraw-Hill, 1970.

[12] R.W. Newcomb, *Active Integrated Circuit Synthesis*. Englewood Cliffs, NJ: Prentice-Hall, 1968.

[13] L.R. Rabiner and C.M. Rader, Eds., *Digital Signal Processing*. New York: IEEE Press, 1972.

[14] D.D. Siljak, *Nonlinear Systems, the parameter analysis and design*. New York: Wiley, 1969.

[15] R. Spence, *Linear Active Networks*. New York: Wiley-Interscience, 1970.

[16] G.C. Temes and S.K. Mitra, Eds., *Modern Filter Theory and Design*. New York: Wiley, 1973.

[17] R. Tomovic, *Sensitivity Analysis of Dynamic Systems*. New York: McGraw-Hill, 1964.

[18] T.N. Trick, *Introduction to Circuit Analysis*. New York: Wiley, 1977.

[19] J.G. Truxal, *Automatic Feedback, Control System Synthesis*. New York: McGraw-Hill, 1955.

[20] M.E. Van Valkenburg, *Analog Filter Design*. New York: Holt Rinehart and Winston, 1982.

[21] J.V. Wait, L.P. Huelsman, and G.A. Korn, *Introduction to Operational Amplifier Theory and Application*. New York: McGraw-Hill, 1975.

[C] General Network Sensitivity: Definitions and Analyses

[1] C. Belove, "Sensitivity sums for homogeneous functions," *IEEE-TCT*, vol. CT-11, pp. 171, March 1964.

[2] D.P. Brown, "General sensitivity relations for zeros," in *Proc. 19th Midwest Symp. on Circuits and Systems*, pp. 378-383, Aug. 1976.

[3] W.J. Butler and S.S. Haykin, "Multiparameter sensitivity problems in network theory," *Proc. IEE*, vol. 117, pp. 2228-2235, Dec. 1970.

[4] M. Chandrashekar and H.K. Kesavan, "Network sensitivity simplified," *Proc. IEEE*, vol. 62, no. 8, pp. 1179-1180, Aug. 1974.

[5] R. DeCarlo, "Sensitivity calculations using the component connection model," *IJCTA*, vol. 12, pp. 288-291, 1984.

[6] S.B. Haley, "Pole sensitivity to network component change," *IEEE-TCAS*, vol. CAS-33, no. 11, pp. 1128-1132, Nov. 1986.

[7] V.I. Ivanov, "Rate errors in parametric frequency transducers," *ARC*, vol. 26, no. 5, pp. 912-914, May 1965.

[8] H.K. Kesavan, I.G. Sarma, and U.R. Prasad, "Sensitivity-state models for linear systems," *IJC*, vol. 9, no. 3, pp. 291-310, 1969.

[9] R.L. Koo and M. Sobral, Jr., "Pole sensitivity in coupling networks," *JFI*, vol. 300, no. 3, pp. 203-207, Sept. 1975.

[10] S.C. Lee, "Non-series-parallel realization of symmetrical and bisymmetrical two-element-kind two-ports to minimize multiparameter sensitivity," *IEEE-TCT*, vol. CT-14, pp. 159-166, June 1967.

[11] G. Martinelli, "On the matrix analysis of network sensitivities," *Proc. IEEE*, vol. 54, no. 1, pp. 72-73, Jan. 1966.

[12] F. Masutomi, "CAD to calculate the numerical solution for linear electrical circuit with variable parameters," *EEJ*, vol. 98, no. 4, pp. 121-133, 1978.

[13] V.A. Monaco and P. Tiberio, "Two properties for circuit sensitivity in terms of scattering parameters," *Elect. Lett.*, vol. 8, no. 15, pp. 382-383, July 1972.

[14] S.P. Papazov and I.A. Maslarov, "Analysis of the sensitivity of circuits with photoresistors," *ARC*, vol. 24, No. 3, pp. 363-367, Oct. 1963.

[15] A. Papoulis, "Displacement of the zeros of the impedance Z(s) due to an incremental variation in the network elements," *Proc. IRE*, vol. 43, pp. 79-82, Jan. 1955.

[16] C.P. Phan and H.K. Kim, "New definition of multiple-root sensitivity," *Elect. Lett.*, vol. 8, no. 20, p. 497, Oct. 1972.

[17] M.J. Remec, "On the upper and lower bounds of excursions of critical-frequencies of a perturbed network," in *Proc. Allerton Conf.*, pp. 502-508, 1963.

[18] D. Rabrenovic and V. Jovanovic, "Sensitivity analysis in the transformed plane," *IJE*, vol. 59, no. 2, pp. 217-223, 1985.

[19] F.N.H. Robinson, "Stability and distortion in tuneable low frequency sinusoidal oscillators," *IJE*, vol. 65, no. 5, pp. 971-982, 1988.

[20] G.W. Swift, R.W. Menzies, and A.M. Gole, "Sensitivity analysis for a regulating transformer connected to a high impedance source," *IEEE-TCAS*, vol. CAS-38, no. 2, pp. 227-229, 1991.

[21] A.F. Tereschchenko, "Transistorized high-sensitivity pulse-modulators," *ARC*, vol. 26, no. 9, pp. 1594-1596, Sept. 1965.

[22] W.J. Troop and E. Peskin, "The transfer function and sensitivity of a network with n-variable elements," *IEEE-TCT*, vol. CT-16, no. 2, pp. 242-244, May 1969.

[23] H. Ur, "Root locus properties and sensitivity relations in control systems," *IRE-TAC*, vol. AC-5, pp. 57-65, Jan. 1960.

[24] M.E. Valtonen, "Measurement of network sensitivities using voltage and impedance measurements," *Proc. IEEE*, vol. 65, no. 11, pp. 1601-1602, Nov. 1977.

[25] P. Valtonen, "Multiparameter sensitivity via a regression model," *IEEE-TCAS*, vol. CAS-24, no. 5, pp. 269-270, May 1977.

[D] Tolerances or Sensitivity in the Large of Networks

[1] D.M. Bohling and L.A. O'Neill, "An interactive computer approach to tolerance analysis," *IEEE Trans. Comput.*, vol. C-19, pp. 10-16, Jan. 1970.

[2] E.M. Butler, "Realistic design using large-change sensitivities and performance contours," *IEEE-TCT*, vol. CT-18, no. 1, pp. 58-66, Jan. 1971.

[3] W.J. Butler and S.S. Haykin, "Multiparameter sensitivity problem in network theory," *Proc. IEE*, vol. 117, no. 12, pp. 2228-2236, Dec. 1970.

[4] J. Cajka, "Computer aided network function sensitivity analysis and the tolerance field determination of complicated active 2-ports," in *Proc. Summer School on Circuit Theory*, 1971, Tale, Czechoslovakia, pp. (6-1)-(6-11).

[5] T. Downs, "An approach to the computation of network sensitivities," *TMR*, vol. 35, Third Quart., pp. 204-213, 1972.

[6] _____, "A note on the computation of large-change sensitivities," *IEEE-TCT*, vol. CT-20, no. 6, pp. 741-742, Nov. 1973.

[7] F.M. El-Turky, "Efficient computation of network sensitivities," *IEEE-TCAS*, vol. CAS-33, no. 7, pp. 659-664, July 1986.

[8] J.K. Fidler, "Network sensitivity calculation," *IEEE-TCAS*, vol. 23, no. 9, pp. 567-571, Sept. 1976.

[9] R.N. Gadenz, M.G. Rezai-Fakhr, and G.C. Temes, "A method for computation of large tolerance effects," *IEEE-TCT*, vol. CT-20, pp. 704-708, Nov. 1973.

[10] P.J. Goddard, P.A. Villalaz, and R. Spence, "Method for the efficient computation of the large change sensitivity of linear non-reciprocal networks," *Elect. Lett.*, vol. 7, pp. 112-113, Feb. 1971.

[11] S.L. Hakimi and J.B. Cruz, Jr., "Measure of sensitivity for linear systems with large multiple parameter variations," in *IRE WESCON Conv. Rec.*, Pt. 2, pp. 109-115, Aug. 1960.

[12] S.B. Haley, "Large change response sensitivity of linear networks," *IEEE-TCAS*, vol. CAS-27, no. 4, pp. 305-310, April 1980.

[13] R. Krishman and T. Downs, "A note on the computation of large-change multiparameter sensitivities," *IJCTA*, vol. 4, pp. 307-310, 1976.

[14] K.H. Leung and R. Spence, "Multiparameter large-change sensitivity analysis and systematic exploration," *IEEE-TCAS*, vol. CAS-22, no. 10, pp. 796-804, Oct. 1975.

[15] V.K. Manaktala and G.L. Kelly, "Computer aided worst case sensitivity analysis of electrical networks over a frequency interval," *IEEE-TCT*, vol. CT-19, no. 1, pp. 91-93, Jan. 1972.

[16] M.I. Mathew and C.P. Lewis, "Noise performance comparison for third and fourth order cascaded sigma-delta modulators having parameter tolerance errors," *Elec. Lett.*, vol. 26, pp. 158-159, 1990.

[17] S.R. Parker, "Sensitivity: Old questions, some new answers," *IEEE-TCT*, vol. CT-18, no. 1, pp. 27-35, Jan. 1971.

[18] S.R. Parker, E. Peskin, and P.M. Chirlian, "Application of a bilinear theorem to network sensitivity," *IEEE-TCT*, vol. CT-12, pp. 448-450, Sept. 1965.

[19] A.F. Schwarz, "Large-change and differential network sensitivity," *IEEE-TCAS*, vol. CAS-24, no. 11, pp. 662-663, Nov. 1977.

[20] K. Singhal, J. Vlach and P.R. Bryant, "Efficient computation of large-change multiparameter sensitivity," *IJCTA*, vol. 1, pp. 237-247, 1973.

[21] M. Sobral, Jr., "Sensitivity considerations in the synthesis of doubly-terminated coupling networks," *IEEE-TCT*, vol. CT-12, no. 2, pp. 272-274, June 1965. Erratum, *ibid.*, p. 633, Dec. 1965. Comments, by H. Matthes, *ibid.*, vol. CT-15, pp. 145-147, June 1968.

[22] J. Solymosi and T. Trony, "Generalized interpretation of sensitivity function," *IJCTA*, vol. 4, pp. 75-80, 1976.

[23] E.V. Sorenson, "General relations governing the exact sensitivity of linear networks," *Proc. IEE*, vol. 114, no. 9, pp. 1209-1212, Sept. 1967.

[24] G.C. Temes and H.J. Orchard, "First-order sensitivity and worst case analysis of doubly-terminated reactance two-ports," *IEEE-TCT*, vol. CT-20, no. 5, pp. 567-571, Sept. 1973.

[25] H. Tromp, "The generalized tolerance problem and worst-case search," in *Proc. Conf. Computer-Aided Design of Electronic and Microwave Circuits and Systems*, Hull, England, pp. 72-77, 1977.

[26] L. Weyten, "Bounds on tolerance sensitivity of resistive networks," *IEEE-TCAS*, vol. CAS-31, no. 4, pp. 390-393, April 1984.

[E] Sensitivity Model of Circuits

[1] K.S. Miller and F.J. Murray, "A Mathematical basis for an error analysis of differential analyzers," *JMP*, vol. 32, nos. 2 and 3, pp. 136-163, July/Oct. 1953.

[2] S.R. Parker, "The use of models for sensitivity calculations," in *Proc. Summer School on Circuit Theory*, 1971, Tale, Czechoslovakia, vol. 1, pp. 109-113.

[3] P.S. Satsangi, G.H. Hashemi, and J.B. Ellis, "Sensitivity-state modelling for systems with multi-terminal components," *IJC*, vol. 14, no. 2, pp. 209-232, 1971.

[4] L. Thiele, "On the sensitivity of linear state-space systems," *IEEE-TCAS*, vol. CAS-33, no. 5, pp. 502-510, May 1986.

[F] Network Elements Sensitivity and Energy Stored in the Network, Vratsanos' Theorem

[1] H.G. Ansell, "Vratsanos' theorem," *IRE-TCT*, vol. CT-5, no. 2, p. 143, June 1958.

[2] M.L. Blostein, "Sensitivity analysis of parasitic effects in resistance-terminated LC two-ports," *IEEE-TCT*, vol. CT-14, no. 1, pp. 21-25, March 1967.

[3] S.R. Deards, "Vratsanos' theorem," *IRE-TCT*, vol. CT-5, pp. 143-144, June 1958.

[4] R.G. de Buda, "Vratsanos' theorem and two-port reciprocity," *IEEE-TCT*, vol. CT-9, p. 87, March 1962.

[5] G. Kishi and T. Kida, "Energy theory of sensitivity in LCR networks," *IEEE-TCT*, vol. CT-14, no. 4, pp. 380-387, Dec. 1967.

[6] G. Kishi and K. Nakazawa, "Relation between reactive energy and group delay in lumped constant networks," *IEEE-TCT*, vol. CT-10, pp. 67-71, March 1963.

[7] W.E. Smith, "Electric and magnetic energy storage in passive nonreciprocal networks," *Elect. Lett.*, vol. 3, pp. 389-391, Aug. 1967.

[8] _____, "Average energy storage by a one-port and minimum energy synthesis," *IEEE-TCT*, vol. CT-17, pp. 427-430, Aug. 1970.

[9] _____, "An extension of Vratsanos' theorem to nonreciprocal networks," *IEEE-TCT*, vol. CT-18, pp. 276-277, March 1971.

[10] _____, "Element sensitivity and energy storage of a passive impedance," *IEEE-TCT*, vol. CT-18, no. 3, pp. 337-342, May 1971.

[11] J. Vratsanos, "Zur Berechnung der Stromverteilung in Einem Linearen Netzwerk," *Arch. Elek. Ubertragung*, vol. 11, pp. 76-80, Feb. 1957.

[G] Network Sensitivity Invariants and Bounds on Sensitivity Functions

[1] T. Ae, "On the sensitivity summations of one-port RLS networks," *IJE*, vol. 36, pp. 113-120, 1974.

[2] M.L. Blostein, "Sensitivity analysis of parasitic effects in resistance-terminated LC two-ports," *IEEE-TCT*, vol. CT-14, no. 1, pp. 21-25, March 1967.

[3] M. Bukowski and A. Filipkowski, "A new large – change sensitivity invariant for linear networks and new relations between sensitivities," *IEEE-TCAS*, vol. CAS-27, no. 10, pp. 978-980, Oct. 1980.

[4] Y. Ceyhun, "Sensitivity invariants of certain class of networks," *Elect. Lett.*, vol. 7, no. 3, pp. 85-86, Feb. 1971.

[5] J.C. Giguere, "Simple bounds on the variance of system functions of two-elements-kind networks having randomly chosen elements," *IEEE-TCT*, vol. CT-19, no. 5, pp. 504-505, Sept. 1972.

[6] K.L. Grudev, "Realization of parametric invariance in direct-current amplifier on the basis of sensitivity models," *SAC*, vol. 13, no. 2, pp. 52-58, 1980.

[7] H.W. Hanneman and H.N. Linssen, "A relation between sensitivity and variance of network functions," *IEEE-TCT*, vol. CT-19, no. 5, pp. 499-502, Sept. 1972.

[8] A.G.J. Holt and J.K. Fidler, "Summed sensitivity of network functions," *Elect. Lett.*, vol. 4, no. 5, pp. 85-87, 1968.

[9] A.G.J. Holt and M.R. Lee, "A relationship between sensitivity and noise," *IJE*, vol. 26, no. 6, pp. 591-594, 1969.

[10] T. Kida and K. Kurogochi, "New sensitivity measures for resistively terminated LC filters," *IJCTA*, vol. 11, pp. 219-234, 1983.

[11] M. Lal, J.S. Sohal, and H. Singh, "Sensitivity invariance using state variables," in *Proc. Allerton Conf.*, pp. 967-974, Oct. 1974.

[12] J. Neirynck and P.V. Bastelaer, "Tables on sensitivity invariants and bounds for lossless two-ports," *IJCTA*, vol. 3, pp. 285-291, 1975.

[13] H.J. Orchard, "Inductorless filters," *Elect. Lett.*, vol. 2, pp. 224-225, June 1966.

[14] _____, "Loss sensitivities in singly and doubly terminated filters," *IEEE-TCAS*, vol. CAS-26, no. 5, pp. 293-297, May 1979.

[15] H.J. Orchard, G.C. Temes, and T. Cataltepe, "Sensitivity formulas for terminated lossless two-ports," *IEEE-TCAS*, vol. CAS-32, no. 5, pp. 459-466, May 1985.

[16] M. Piedade and M.M. Silva, "An improved method for the evaluation of filter sensitivity performance," *IEEE-TCAS*, vol. CAS-33, no. 3, pp. 332-335, March 1986.

[17] T. Roska, "Summed-sensitivity invariants and their generation," *Elect. Lett.*, vol. 4, no. 14, pp. 281-282, July 1968.

[18] M. Sablatash and R. Seviora, "Sensitivity invariants for scattering matrices," *IEEE-TCT*, vol. CT-18, pp. 282-284, March 1971.

[19] M.N.S. Swamy, C. Bhushan, and K. Thulasiraman, "Bounds on the sum of element sensitivity magnitudes for network functions," *IEEE-TCT*, vol. CT-19, no. 5, pp. 502-504, Sept. 1972.

[20] _____, "Sensitivity invariants for nonlinear networks," *IEEE-TCT*, vol. CT-19, no. 6, pp. 599-606, Nov. 1972.

[21] _____, "Sensitivity invariants for linear time-invariants networks," *IEEE-TCT*, vol. CT-20, no. 1, pp. 21-24, Jan. 1973.

[22] L. Weyten, "Lower bounds on the summed absolute and squared voltage transfer sensitivities in RLC networks," *IEEE-TCAS*, vol. CAS-25, no. 2, pp. 70-73, Feb. 1978. Comments, by J. Tow, *ibid.*, vol. CAS-26, no. 3, pp. 209-211, March 1979. Author's reply, *ibid.*, vol. CAS-27, no. 7, pp. 654-655, July 1980.

[H] Concept of Continuously Equivalent Networks Synthesis Procedures with Sensitivity Invariants

[1] M.L. Blostein, "Generation of minimum sensitivity networks," *IEEE-TCT*, vol. CT-14, pp. 87-88, March 1967.

[2] B.M.G. Cheetham, "A new theory of continuously equivalent networks," *IEEE-TCAS*, vol. CAS-21, no. 1, pp. 17-20, Jan. 1974.

[3] R.G. de Buda, "About sensitivity invariants of equivalent networks," *IEEE-TCT*, vol. CT-17, no. 2, pp. 248-249, May 1970.

[4] G.R. Haack, "On the noncompleteness of continuously equivalent networks," *IEEE-TCT*, vol. CT-17, no. 4, pp. 619-620, Nov. 1970.

[5] N. Howitt, "Group theory and the electric circuit," *Phys. Rev.*, vol. 37, pp. 1583-1595, June 1931.

[6] O. Huseyin, "Application of equivalent network theory," *IEEE-TCT*, vol. CT-19, pp. 376-378, July 1972.

[7] E.S. Kuh and C.G. Lau, "Sensitivity invariants of continuously equivalent networks," *IEEE-TCT*, vol. CT-15, no. 3, pp. 175-177, Sept. 1968.

[8] J.V. Leeds, Jr., "Transient and steady-state sensitivity analysis," *IEEE-TCT*, vol. CT-13, pp. 288-289, Sept. 1966.

[9] J.V. Leeds, Jr. and G.I. Urgon, "Simplified multiple parameter sensitivity calculation and continuously equivalent network," *IEEE-TCT*, vol. CT-14, pp. 188-191, June 1967.

[10] B.J. Leon and C.F. Yokomoto, "Generation of a class of equivalent networks and its sensitivities," *IEEE-TCT*, vol. CT-19, no. 1, pp. 2-8, Jan. 1972.

[11] J. Neirynck and P. van Bastelaer, "On sensitivity invariants for continuously equivalent networks," *IEEE-TCAS*, vol. CAS-22, no. 6, pp. 563-564, June 1975.

[12] R.W. Newcomb, "The noncompleteness of continuously equivalent networks," *IEEE-TCT*, vol. CT-13, pp. 207-208, June 1966.

[13] G. Schmidt and R. Kasper, "On minimum sensitivity networks," *IEEE-TCT*, vol. CT-14, pp. 438-440, Dec. 1967.

[14] J.D. Schoeffler, "The synthesis of minimum sensitivity networks," *IEEE-TCT*, vol. CT-11, pp. 271-276, June 1964.

[15] I. Shirakawa, T. Temma, and H. Ozaki, "Synthesis of minimum-sensitivity networks through some classes of equivalent transformations," *IEEE-TCT*, vol. CT-17, no. 1, pp. 2-8, Feb. 1970.

[16] J.S. Sohal and H. Singh, "On sensitivity invariance of multi-input multi-output networks," *Proc. IEEE*, vol. 64, no. 4, p. 560, April 1976.

*[I] Network Synthesis with Minimum or
Prescribed Sensitivity Measures*

[1] O.G. Alekseyev and S.M. Gayev, "Choice of optimal tolerances in circuit components," *ECy*, vol. 5, no. 4, pp. 167-173, July/Aug. 1967.

[2] J.W. Bandler and P.C. Liu, "Automated network design with optimal tolerances," *IEEE-TCAS*, vol. CAS-21, no. 2, pp. 219-222, March 1974.

[3] M.G. Govindarajulu Naidu, K. Thulasiraman, and M.N.S. Swamy, "The n-Port resistive network synthesis from prescribed sensitivity coefficients," *IEEE-TCAS*, vol. CAS-22, no. 6, pp. 482-485, June 1975.

[4] W.J. Lutz and S.L. Hakimi, "Design of multi-input multi-output systems with minimum sensitivity," *IEEE-TCAS*, vol. 35, no. 9, pp. 1114-1122, Sept. 1988.

[5] G. Martinelli and M. Salerno, "Iterative synthesis of low-sensitive RC earthed networks," *Proc. IEE*, vol. 121, no. 7, pp. 568-572, July 1974.

[6] E.C. Strycula and N.K. Bose, "Simultaneous realization of transfer and sensitivity functions," *Elect. Lett.*, vol. 7, no. 26, pp. 769-770, Dec. 1971.

[7] _____, "Theory and applications of simultaneous realization of transfer and sensitivity functions," *IEEE-TCAS*, vol. CAS-21, no. 3, pp. 382-387, May 1974.

[J] Sensitivity of the Cost Function in the Optimal Circuit Design

[1] H.J. Bachler and W. Guggenbuhl, "Noise and sensitivity optimization of a single-amplifier biquad," *IEEE-TCAS*, vol. CAS-26, no. 1, pp. 30-36, Jan. 1979.

[2] F.M. El-Turky and J. Vlach, "Generation of equivalent active networks with minimized sensitivities," *IEEE-TCAS*, vol. CAS-28, no. 10, pp. 941-946, Oct. 1981.

[3] A. Fettweis and A. Khalil, "Optimal low-sensitivity anti-aliasing conversion filters," *IEEE-TCAS*, vol. CAS-27, no. 6, pp. 559-566, June 1980.

[4] C.W. Ho, "Time-domain sensitivity computation for networks containing transmission lines," *IEEE-TCT*, vol. CT-18, no. 1, pp. 114-122, Jan. 1971.

[5] T. Kida and N. Osumi, "New time-domain sensitivity formulas and their applications," *IEEE-TCAS*, vol. CAS-27, no. 3, pp. 176-184, March 1980.

[6] D.A. Pierre, "Sensitivity measures for optimally selected parameters," *Proc. IEEE*, vol. 54, no. 2, pp. 321-322, Feb. 1966.

[7] J. Vlach and K. Singhal, "Sensitivity minimization of networks with operational amplifier and parasitics," *IEEE-TCAS*, vol. CAS-27, no. 8, pp. 688-697, Aug. 1980. Corrections, *ibid.*, vol. CAS-28, pp. 358, April 1981.

[K] Sensitivity Analysis and Synthesis in Active Networks

[1] P.V. Anandamohan, "Q-sensitivity of certain RC active filters to gain-bandwidth product," *Elect. Lett.*, vol. 15, no. 3, pp. 95-97, Feb. 1979.

[2] A. Barua, "Novel parasitic insensitive SC large pole selectivity (Q) bandpass filter with less capacitor spread and low pole 'Q' sensitivity," *IJE*, vol. 62, no. 4, pp. 637-639, 1987.

[3] Y. Bedri, "Elimination of restriction on the position of zeros in certain low-sensitivity RC active network," *IJE*, vol. 35, no. 3, pp. 321-334, 1973.

[4] Y. Bedri and T. Deliyannis, "Selectivity improvement in a useful second-order active RC section," *IJE*, vol. 13, pp. 243-248, 1971.

[5] V. Bello and M.S. Ghausi, "An analytic approach to low sensitivity design of active distributed RC networks," *JFI*, vol. 290, no. 4, pp. 365-375, Oct. 1970.

[6] M. Bialko and A. Guzinski, "Active filter with distributed RC line having zero Q-sensitivity," *IEEE-TCAS*, vol. CAS-21, no. 1, pp. 87-90, Jan. 1974.

[7] M. Bialko and W. Sienko, "Zero Q-sensitivity active RC circuit synthesis," *IEEE-TCAS*, vol. CAS-21, no. 2, pp. 239-244, March 1974.

[8] M. Biey, "Design of twin-T single amplifier building block with prescribed values of capacitors and minimum gain-sensitivity product," *Elect. Lett.*, vol. 17, no. 7, pp. 249-250, April 1981.

[9] S.A. Boctor and A.S. Sedra, "Realization of nonminimum phase transfer functions using twin-T RC networks," *IEEE-TCT*, vol. CT-18, no. 4, pp. 471-475, July 1971.

[10] A. Borys, "Slew-induced distortion of single-amplifier active filters using gain-sensitivity product concept," *IEEE-TCAS*, vol. CAS-31, no. 3, pp. 306-308, March 1984.

[11] N.M. Brice, "Sensitivity and feedback in amplifiers," *IEEE-TCT*, vol. CT-12, pp. 274-275, June 1965.

[12] D. Bruckmann and U. Kline, "Novel voltage inverter switches with minimum sensitivity properties," *IEEE-TCAS*, vol. CAS-32, pp. 732-726, July 1985.

[13] L.T. Bruton, "Sensitivity comparison of high-Q second-order active filter synthesis techniques," *IEEE-TCAS*, vol. CAS-22, no. 1, pp. 32-38, Jan. 1975. Comments, with reply, by L. Weyten, *ibid.*, vol. CAS-26, no. 1, pp. 77, Jan. 1979.

[14] L.T. Bruton and A.B. Hasse, "Sensitivity of generalized immittance converter-embedded ladder structures," *IEEE-TCAS*, vol. CAS-21, no. 2, pp. 245-250, March 1974.

[15] A. Budak and D.M. Petrela, "Frequency limitations of active filters using operational amplifiers," *IEEE-TCT*, vol. CT-19, pp. 322-328, July 1972.

[16] A. Budak, G. Wullink, and R.L. Geiger, "Active filters with zero transfer function sensitivity with respect to the time constants of operational amplifiers," *IEEE-TCAS*, vol. CAS-27, no. 10, pp. 849-854, Oct. 1980.

[17] A. Dabrowski and G.S. Moschytz, "Direct by-inspection derivation of signal-flow graphs for multiphase stray-insensitive switched-capacitor filters," *Elect. Lett.*, vol. 25, no. 6, pp. 387-389, 1989.

[18] K.E. Daggett and J. Vlach, "Sensitivity-compensated active networks," *IEEE-TCT*, vol. CT-16, no. 4, pp. 416-422, Nov. 1969.

[19] R.B. Datar and A.S. Sedra, "Exact design of strays-insensitive switched capacitor ladder filters," *IEEE-TCAS*, vol. CAS-30, pp. 888-898, Dec. 1983.

[20] T. Deliyannis, "Sensitivity study of five RC active networks using the method of single inversion," *IJE*, vol. 22, no. 3, pp. 197-213, 1967.

[21] _____, "High-Q Factor Circuit with Reduced Sensitivity," *Elect. Lett.*, vol. 4, p. 577, Dec. 1968.

[22] P.S.R. Diniz and L.P. Caloba, "Zero pole sensitivity active filters," *IJE*, vol. 53, no. 4, pp. 341-348, 1982.

[23] S.C. Dutta Roy and D.K. Bhargava, "Sensitivity of third-and higher-order filters," *IJCTA*, vol. 5, no. 3, pp. 235-238, July 1977.

[24] E.I. El-Masry, "Strays-insensitive state-space switched-capacitor filters," *IEEE-TCAS*, vol. CAS-30, no. 7, pp. 474-488, July 1983.

[25] E.I. El-Masry and H.-L. Lee, "Low-sensitivity realization of switched-capacitor filters," *IEEE-TCAS*, vol. CAS-34, no. 2, pp. 510-523, May 1987.

[26] A. Fabre and M. Alami, "Insensitive current-mode bandpass implementations-based nonideal gyrators," *IEEE-TCAS*, Part I, vol. CAS-39, no. 2, pp. 152-155, 1992.

[27] A. Fabre, F. Martin, and M. Hanafi, "Current mode allpass/notch and bandpass filters with reduced sensitivities," *Elec. Lett.*, vol. 26, no. 18, pp. 1495-1496, 1990.

[28] P.E. Fleischer, "Sensitivity minimization in a single amplifier biquad circuit," *IEEE-TCAS*, vol. CAS-23, no. 1, pp. 45-55, Jan. 1976.

[29] N. Fuji, "On the minimum gain pole-sensitivity product of single amplifier RC active networks," *IEEE-TCAS*, vol. CAS-24, no. 9, pp. 504-510, Sept. 1977.

[30] R.N. Gadenz, "On low-sensitivity realizations of band-elimination active filters," *IEEE-TCAS*, vol. 24, no. 4, pp. 175-183, April 1977.

[31] P.R. Geffe, "Passive Q-sensitivities of the twin-T selective amplifier," *IEEE-TCT*, vol. CT-19, no. 6, pp. 685-686, Nov. 1972.

[32] R.L. Geiger and A. Budak, "Active filters with zero amplifier sensitivity," *IEEE-TCAS*, vol. CAS-26, no. 4, pp. 277-288, April 1979.

[33] P. Gillingham, "Strays-insensitive switched capacitor biquads with reduced number of capacitors," *Elect. Lett.*, vol. 17, pp. 171-171, Feb. 1981.

[34] F.E.J. Girling and E.F. Good, "Active filters Parts 7 and 8: The two-integrator loop," *WW*, vol. 76, pp. 117-119, March 1970, and pp. 134-139, April 1970.

[35] _____, "Active filters Part 12: The Leap frogs or active ladder synthesis," *ibid.*, pp. 341-345, July 1970.

[36] _____, "Active filters Part 13: Applications of active ladder synthesis," *ibid.*, pp. 445-450, Sept. 1970.

[37] _____, "Active filters Part 14: Bandpass types," *ibid.*, pp. 505-510, Oct. 1970.

[38] J.J. Golembeski, M.S. Ghausi, J.H. Mulligan, Jr., and S.S. Shamis, "A class of minimum sensitivity amplifiers," *IEEE-TCT*, vol. CT-14, pp. 69-74, March 1967.

[39] G. Grainger and D.K. Bedford, "Low noise charge-sensitive preamplifier for fast pulse counting," *Elect. Lett.*, vol. 14, no. 5, pp. 126-127, March 1978.

[40] M.S. Gupta, "Stability measure for negative-resistance amplifiers," *IJE*, vol. 45, no. 3, pp. 241-245, 1978.

[41] S.S. Hakim, "Synthesis of RC active filters with prescribed pole sensitivity," *Proc. IEE*, vol. 112, pp. 2235-2242, Dec. 1965.

[42] T.A. Hamilton and A.S. Sedra, "Some new configurations for active filters," *IEEE-TCT*, vol. CT-19, no. 1, pp. 25-33, Jan. 1972.

[43] K. Haug, F. Maloberti, and G.C. Temes, "Switched-capacitor integrators with low finite-gain sensitivity," *Elect. Lett.*, vol. 21, pp. 1156-1157, Nov. 1985.

[44] A. Hiltgen and G.S. Moschytz, "A pole-zero pairing strategy for minimum-gain-sensitivity nth-order biquadratic filter cascades," *IJCTA*, vol. 18, pp. 145-164, 1990.

[45] K. Hirano, I. Unami, and M. Hashimoto, "Sensitivity of active two-ports," *IEEE-TCT*, vol. CT-16, no. 4, pp. 489-495, Nov. 1969.

[46] E. Hokenek and G. Moschytz, "Design of parasitic-insensitive bilinear-transformed admittance-scaled (BITAS) SC ladder filters," *IEEE-TCAS*, vol. CAS-30, pp. 873-888, Dec. 1983.

[47] A.G.J. Holt and P. Bowron, "Sensitivity of active RC distributed network," *IEEE-TCT*, vol. CT-16, no. 1, pp. 99-101, Feb. 1969.

[48] A.G.J. Holt and M.R. Lee, "Summed sensitivity of active RC networks," *Elect. Lett.*, vol. 4, pp. 385-386, June 1968.

[49] I. Horowitz, "Design of zero-sensitivity frequency-selective integrated circuits," *IEEE-TCT*, vol. CT-15, pp. 440-446, Dec. 1968.

[50] T. Inoue and F. Ueno, "Design of very low sensitivity low-pass switched-capacitor ladder filters," *IEEE-TCAS*, vol. CAS-34, pp. 524-532, May 1987.

[51] N.K. Jain, G.S. Visweswaran, and A.B. Bhattacharyya, "Time-domain sensitivity analysis of dynamic sense amplifier of an n-MOS dynamic RAM," *IEEE-TCAS*, vol. CAS-33, pp. 77-82, Jan. 1986.

[52] A. Kaelin, J. Goette, W. Guggenbuhl, and G.S. Moschytz, "A novel capacitance assignment procedure for the design of sensitivity- and noise-optimized SC-filters," *IEEE-TCAS*, vol. CAS-38, pp. 1255-1268, Nov. 1991.

[53] H.K. Kim and C.S. Phan, "Polynomial decomposition for minimization of quality factor sensitivity," *IEEE-TCT*, vol. CT-19, pp. 397-398, July 1972.

[54] A. Knob, "Novel strays-insensitive switched-capacitor integrator realising the bilinear z-transform," *Elect. Lett.*, vol. 16, pp. 173-174, Feb. 1980.

[55] T. Kunieda, Y. Hiramatsu, and A. Fukui, "Frequency dependance of sensitivities in second-order RC active filters," *IEEE-TCAS*, vol. CAS-27, pp. 77-84, Feb. 1980.

[56] K.R. Laker and M.S. Ghausi, "Synthesis of a low-sensitivity multiloop feedback active RC filter," *IEEE-TCAS*, vol. CAS-21, no. 2, pp. 252-259, 1974, Erratum, *ibid.*, p. 811, Nov. 1974.

[57] _____, "Large change sensitivity – a dual pair of approximate statistical sensitivity measures," *JFI*, vol. 298, nos. 5 and 6, pp. 395-413, 1974.

[58] _____, "Design of minimum sensitivity multiple loop feedback bandpass active filters," *JFI*, vol. 310, no. 1, pp. 51-64, July 1980.

[59] K.R. Laker, M.S. Ghausi, and J.J. Kelly, "Minimum sensitivity active (leapfrog) and passive ladder bandpass filters," *IEEE-TCAS*, vol. CAS-22, no. 8, pp. 670-677, 1975.

[60] K.R. Laker, R. Schaumann, and M.S. Ghausi, "Multiple-loop feedback topologies for the design of low-sensitivity active filters," *IEEE-TCAS*, vol. CAS-26, pp. 1-21, Jan. 1979.

[61] E.A. Laksberg, "On the sensitivity analysis of linear active networks," *Elect. Lett.*, vol. 14, pp. 221-223, March 1978.

[62] S.C. Lee, "Synthesis of tapered distributed RCG networks," *IEEE-TCT*, vol. CT-16, no. 1, pp. 57-67, Feb. 1969.

[63] E. Leuder, "Decomposition of a transfer function minimizing sensitivity," *IEEE-TCT*, vol. CT-17, pp. 426-427, Aug. 1970.

[64] P. Lutz, "Sensitivities of impedance scaled SC filters," *Elect. Lett.*, vol. 17, pp. 24-25, Jan. 1981.

[65] R. Mackay and A.S. Sedra, "Sensitivity evaluation and comparison for high-order active filters," *IJCTA*, vol. 4, no. 4, pp. 345-356, 1976.

[66] _____, "Generation of low-sensitivity state-space active filters," *IEEE-TCAS*, vol. CAS-27, no. 10, pp. 863-870, Oct. 1980.

[67] N. Mijat and G.S. Moschytz, "Sensitivity limitations of some FLF-type active filters," *IJCTA*, vol. 14, pp. 153-161, 1986.

[68] V.B. Misic, "Sensitivity evaluation for multiple feedback active filters," *Elect. Lett.*, vol. 17, pp. 240-241, March 1981.

[69] A.K. Mitra and V.K. Aatre, "Low sensitivity high-frequency active R filters," *IEEE-TCAS*, vol. 23, pp. 670-676, Nov. 1976.

[70] S.K. Mitra, P.P. Vaidyanathan, and B.D.O. Anderson, "A general theory and synthesis procedure for low-sensitivity active RC filters," *IEEE-TCAS*, vol. CAS-32, pp. 687-699, July 1985.

[71] M. Miura, Y. Fukui, and S. Yoneda, "On the stability of an active gyrator operating at high frequencies," *IJE*, vol. 51, no. 5, pp. 653-662, 1981.

[72] P.V.A. Mohan, V. Ramachandran, and M.N.S. Swamy, "General stray insensitive first-order active SC network," *Elect. Lett.*, vol. 18, pp. 1-2, Jan. 1982.

[73] C. Moore, B. Eads, and J.V. Leeds, "Efficient parameter modification for state-variable formulations," *IEEE-TCT*, vol. CT-18, pp. 190-192, Jan. 1971.

[74] G.S. Moschytz, "Active RC filter building blocks using frequency emphasizing networks," *IEEE-JSSC*, vol. SC-2, pp. 59-62, June 1967.

[75] _____, "Second-order pole-zero pair selection for nth-order minimum sensitivity networks," *IEEE-TCT*, vol. CT-17, pp. 527-534, Nov. 1970.

[76] _____, "Gain-sensitivity product: A figure of merit for hybrid integrated filters using single Op. Amp.," *IEEE-JSSC*, vol. SC-6, pp. 103-110, June 1971.

[77] _____, "A note on pole, frequency, and Q sensitivity," *IEEE-JSSC*, vol. SC-6, pp. 267-269, Aug. 1971.

[78] _____, "The sensitivity problem in active filters," *Scientia Electrica*, vol. 21, pp. 81-105, 1975.

[79] _____, "Single-amplifier active filters: A review," *Sc. E*, vol. 26, pp. 1-46, 1980.

[80] G.S. Moschytz and P. Horn, "Reducing nonideal Op-Amp effects in active filters by minimizing the gain-sensitivity product (GSP)," *IEEE-TCAS*, vol. CAS-24, pp. 437-445, Aug. 1977.

[81] E. Moustakas and S.P. Chan, "Sensitivity considerations in a multiple-feedback universal filter," *IEEE-TCAS*, vol. CAS-24, pp. 695-703, Dec. 1977.

[82] K. Nagaraj, "Switched-capacitor delay circuit that is insensitive to capacitor mismatch and stray capacitance," *Elect. Lett.*, vol. 20, pp. 663-664, Aug. 1984.

[83] K. Nagaraj, J. Vlach, T.R. Viswanathan, and K. Singhal, "Switched-capacitor integrator with reduced sensitivity to amplifier gain," *Elect. Lett.*, vol. 22, pp. 1103-1105, Oct. 1986.

[84] S. Natarajan, "Active sensitivity minimization in SAB's with active compensation and optimization," *IEEE-TCAS*, vol. CAS-29, pp. 239-245, April 1982.

[85] R.W. Newcomb, "Hakim-theory transfer function sensitivity," *Elect. Lett.*, vol. 6, pp. 98-100, Feb. 1970.

[86] A. Petraglia, J. Szczupak, and S. Mitra, "Active RC filters with very low sensitivity," *IJCTA*, vol. 18, pp. 209-214, 1990.

[87] D.A. Petrela and A. Budak, "Design of single-voltage amplifier active filters for minimum open-loop gain sensitivity," *IEEE-TCT*, vol. CT-18, pp. 631-635, Nov. 1971.

[88] A. Premoli, "The MUCROMAF polynomials: An approach to the maximally flat approximation of RC active filters with low sensitivity," *IEEE-TCT*, vol. CT-20, pp. 77-80, Jan. 1973.

[89] H. Qiuting and W. Sansen, "A low-sensitivity, low-capacitance ratio realization of high-Q biquads," *IEEE-TCAS*, vol. CAS-32, pp. 1039-1042, Oct. 1986.

[90] K. Rao and S. Srinivasan, "Low-sensitivity active filters using the operational amplifier pole," *IEEE-TCAS*, vol. CAS-21, pp. 260-262, March 1974.

[91] _____, "A high-Q temperature insensitive band-pass filter using the operational amplifier pole," *Proc. IEEE*, vol. 62, pp. 1713-1714, Dec. 1974.

[92] M.A. Reddy, "An insensitive active-RC filter for high Q and high frequencies," *IEEE-TCAS*, vol. 23, pp. 429-433, July 1976.

[93] _____, "An active-RC filter for high-Q and high-frequencies with zero-Q and zero-frequency-sensitivity to amplifier gain-bandwidth product," *Proc. IEEE*, vol. 65, pp. 814-815, May 1977.

[94] P.A. Regalia and S.K. Mitra, "Low-sensitivity active filter realization using a complex all-pass filter," *IEEE-TCAS*, vol. CAS-34, pp. 390-399, April 1987.

[95] A.L. Rosenblum and M.S. Ghausi, "Multi-parameter sensitivity in active RC networks," *IEEE-TCT*, vol. CT-18, pp. 592-599, Nov. 1971.

[96] _____, "Sensitivity minimization in active RC networks," *JFI*, vol. 294, pp. 95-111, Aug. 1972.

[97] S.B. Roy, D. Patranabis, and P. Kundu, "An insensitive linear single-element-control pulse generator," *IJE*, vol. 46, no. 3, pp. 229-239, 1979.

[98] S.K. Saha, "On the synthesis of RC active networks with prescribed root sensitivity," *IEEE-TCAS*, vol. CAS-21, pp. 799-803, Nov. 1974.

[99] E. Sanchez-Sinencio, R.L. Geiger, and J. Silva-Martinez, "Tradeoffs between passive sensitivity, optimal voltage swing, and total capacitance in biquad SC filters," *IEEE-TCAS*, vol. CAS-31, pp. 984-987, Nov. 1984.

[100] R. Sarpeshkar, J.L. Wyatt, Jr., N.C. Lu, and P.D. Gerber, "Analysis of mismatch sensitivity in a simultaneously latched CMOS sense amplifier," *IEEE-TCAS*, Part II, vol. CAS-39, no. 5, pp. 277-292, 1992.

[101] R. Schaumann, "Low-sensitivity high-frequency tunable active filter without external capacitors," *IEEE-TCAS*, vol. CAS-22, pp. 39-43, Jan. 1975.

7.16.K. Sensitivity Analysis and Synthesis in Active Networks

[102] R. Schaumann, R.E. Chalstrom, and K.R. Laker, "Optimization of sensitivity and dynamic range of I.F.L.F. active filters," *Elect. Lett.*, vol. 13, pp. 367-368, June 1977.

[103] A.S. Sedra and J.L. Espinoza, "Sensitivity and frequency limitations of biquadratic active filters," *IEEE-TCAS*, vol. CAS-22, pp. 122-130, Feb. 1975.

[104] S.-D. Shieu and S.-P. Chan, "Topological formulation of symbolic network functions and sensitivity analysis of active networks," *IEEE-TCAS*, vol. CAS-21, pp. 39-45, Jan. 1974.

[105] N.P. Singh, V. Ramachandran, and B.B. Bhattacharyya, "optimum realizations of single-amplifier networks with zero pole frequency sensitivity and with minimized pole-Q deviation," *IJCTA*, vol. 7, pp. 9-19, 1979.

[106] M.A. Soderstrand, "Active R ladders: High-frequency high-order low-sensitivity active R filters without external capacitors," *IEEE-TCAS*, vol. CAS-25, pp. 1032-1038, Dec. 1978.

[107] M.A. Soderstrand and D.C. Huey, "Sensitivities of fourth-order obtained by a low-pass to band-pass transformation," in *Proc. Midwest Symp. Circuit Theory*, vol. 16, pp. IV.4.1-IV.4..10, 1973.

[108] M.A. Soderstrand and S.K. Mitra, "Very low sensitivity canonic active RC filter," *Proc. IEEE*, vol. 57, pp. 2175-2176, Dec. 1969.

[109] _____, "Sensitivity analysis of third-order filters," *IJE*, vol. 33, pp. 265-272, March 1971.

[110] _____, "Gain and sensitivity limitations of active RC filters," *IEEE-TCT (Special Issue on Active and Digital Networks)*, vol. CT-18, pp. 600-609, Nov. 1971. Comments, by R. Palomera-Garcia, *ibid.*, vol. CAS-26, pp. 897-898, Oct. 1979.

[111] _____, "Sensitivity analysis of third order filters," *IJE*, vol. 30, pp. 265-272, 1971.

[112] _____, "Design of active filters with zero passive Q-sensitivity," *IEEE-TCT*, vol. CT-20, pp. 289-294, May 1973.

[113] _____, "Design of active RC filters with zero gain-sensitivity product," *IEEE-TCT*, vol. CT-20, pp. 441-445, July 1973.

[114] F.W. Stephenson, "Q-sensitivity in optimized RC-NIC filters," *IJE*, vol. 29, no. 3, pp. 261-268, 1970.

[115] J.A. Svoboda and R.J. Wojcik, "Sensitivity analysis of *RLC* Nullor network," *IJCTA*, vol. 10, pp. 139-150, 1982.

[116] R. Tarmi and M.S. Ghausi, "Very high-Q insensitive active RC networks," *IEEE-TCT*, vol. CT-17, pp. 358-366, Aug. 1970.

[117] J. Taylor and J. Mavor, "Exact design of stray-insensitive switched-capacitor LDI ladder filters from unit element prototypes," *IEEE-TCAS*, vol. CAS-33, pp. 613-622, June 1986.

[118] G.H. Tomlinson, "Optimum design of networks that use operational amplifiers," *IEEE-TCT*, vol. CT-17, pp. 440-442, Aug. 1970.

[119] S. Venkateswaran and K.R. Rao, "Power gain, stability and sensitivity of linear active two-ports," *IJE*, vol. 26, no. 1, pp. 17-28, 1969.

[120] J. Vlach and E. Christen, "Poles, zeros, and their sensitivities in switched-capacitor networks," *IEEE-TCAS*, vol. CAS-32, pp. 279-284, March 1985.

[121] L. Weyten, "A useful sensitivity measure for second-order RC active filter design," *IEEE-TCAS*, vol. 23, pp. 506-508, Aug. 1976.

[122] _____, "Passive elements sensitivity of second-order RC active filter sections," *IEEE-TCAS*, vol. CAS-27, pp. 855-862, Oct. 1980.

[123] G. Wilson, "Insensitive very-low-frequency RC oscillator," *Elect. Lett.*, vol. 10, pp. 447-448, Oct. 1974.

[124] G. Wilson, Y. Bedri, and P. Bowron, "RC-active networks with reduced sensitivity to amplifier gain-bandwidth product," *IEEE-TCAS*, vol. CAS-21, pp. 618-626, Sept. 1974.

[L] Sensitivity Analysis in Digital Filters

[1] J.I. Acha and A. Ayerbe, "Sensitivity of second-order two-pair digital filters," *IJE*, vol. 57, no. 4, pp. 543-556, 1984.

[2] J.W. Adams and A.N. Willson, Jr., "A new approach to FIR digital filters with fewer multipliers and reduced sensitivity," *IEEE-TCAS*, vol. CAS-30, pp. 277-283, May 1983.

[3] R.C. Agarwal and C.S. Burrus, "New recursive digital filter structures having very low sensitivity and round off noise," *IEEE-TCAS*, vol. CAS-22, pp. 921-927, Dec. 1975.

[4] A.M. Ali, "Design of low-sensitivity digital filters by linear transformations," *IEEE-TCAS*, vol. CAS-27, pp. 435-444, June 1980.

[5] R. Ansari, "Digital filter network sensitivity with multiple identical transmittances," *IEEE-TCAS*, vol. CAS-32, pp. 1182-1184, Nov. 1984.

[6] M. Bhattacharya, R.C. Agarwal, and S.C. Dutta Roy, "On realization of low-pass and highpass recursive filters with low sensitivity and low roundoff noise," *IEEE-TCAS*, vol. CAS-33, pp. 425-428, April 1986.

[7] M. Bisiacco, "Stabilization theory for single-input single-output two-dimensional systems," *CSSP*, vol. 6, no. 1, pp. 77-93, 1987.

[8] L.T. Bruton, "Low sensitivity digital ladder filters," *IEEE-TCAS*, vol. CAS-22, pp. 168-176, March 1975.

[9] P. Burrascano, G. Martinelli, and G. Orlandi, "Low-sensitivity digital filters based on zero extraction," *IEEE-TCAS*, vol. CAS-34, pp. 1581-1587, Dec. 1987.

[10] S. Colonna, U. DeJulio and M. Salerno, "Determination of sensitivity functions of digital filters," *IEEE-TCT*, vol. CT-19, no. 5, pp. 538-539, 1972.

[11] R.E. Crochiere, "A new statistical approach to the coefficient word length problem for digital filters," *IEEE-TCAS*, vol. 22, pp. 190-196, March 1975.

[12] P.S.R. Diniz and A. Antoniou, "Low-sensitivity digital-filter structures which are amenable to error-spectrum shaping," *IEEE-TCAS*, vol. CAS-32, pp. 1000-1007, Oct. 1985.

[13] M. Domanski and M.S. Piekarski, "Voltage-current digital filters with zero magnitude sensitivity in attenuation zeros," *IEEE-TCAS*, vol. CAS-33, pp. 547-551, May 1986.

[14] M. Eslami, "On sensitivity of nonlinear digital systems," in *Proc. Allerton Conf.*, pp. 424-432, 1980.

[15] C. Eswaran, K. Manivannan, and A. Antoniou, "An alternative sensitivity measure for designing low-sensitivity digital biquads," *IEEE-TCAS*, vol. CAS-38, no. 2, pp. 218-221, 1991.

[16] A. Fettweis, "Pseudopassivity, sensitivity and stability of wave digital filters," *IEEE-TCT*, vol. CT-19, pp. 668-673, Nov. 1972.

[17] _____, "Roundoff noise and attenuation sensitivity in digital filters with fixed-point arithmetic," *IEEE-TCT*, vol. CT-20, pp. 174-175, March 1973.

[18] W.A. Gardner, "Reduction of sensitivity in sampled-data filters," *IEEE-TCT*, vol. CT-17, pp. 660-663, Nov. 1970.

[19] L.B. Jackson, "Roundoff noise bounds derived from coefficient sensitivities for digital filters," *IEEE-TCAS*, vol. 23, pp. 481-485, Aug. 1976.

[20] W.H. Ku and S.-M. Ng, "Floating-point coefficient sensitivity and roundoff noise of recursive digital filters realized in ladder structures," *IEEE-TCAS*, vol. CAS-22, pp. 927-936, Dec. 1975.

[21] S.S. Lawson, "On complementarity and sensitivity of generalized wave digital filters," *IEEE-TCAS*, vol. CAS-33, pp. 1244-1248, Dec. 1986.

7.16.L. Sensitivity Analysis in Digital Filters

[22] J.D. Ledbetter and R. Yarlagadda, "Digital filter synthesis-a low-sensitivity system matrix," *IEEE-TCT*, vol. CT-20, pp. 322-324, May 1973.

[23] A. Lepschy, G.A. Mian, and U. Viaro, "Stability regions for second-order fixed-point digital filters in coupled form," *CSSP*, vol. 9, no. 4, pp. 409-420, 1990.

[24] G. Li, B.D.O. Anderson, M. Gevers, and J.E. Perkins, "Optimal FWL design of state-space digital systems with weighted sensitivity minimization and sparseness consideration," *IEEE-TCAS*, Part I, vol. CAS-39, no. 5, pp. 365-377, 1992.

[25] A. Liberatore and S. Manetti, "On the sensitivity analysis of digital filters," *IJCTA*, vol. 8, pp. 161-166, 1980.

[26] P.H. Lo and Y. Jenq, "Minimum sensitivity realization of second order recursive digital filter," *IEEE Trans. Acoustics, Speech, and Signal Processing*, vol. ASSP-30, no. 6, pp. 930-937, 1982.

[27] P.E. Mantey, "Eigenvalue sensitivity and state-variable selection," *IEEE-TAC*, vol. AC-13, pp. 263-269, June 1968.

[28] S. Nishimura, K. Hirano, and R.N. Pal, "A new class of very low sensitivity and low roundoff noise recursive digital filter structures," *IEEE-TCAS*, vol. CAS-28, pp. 1152-1158, Dec. 1981.

[29] G. Orlandi and G. Martinelli, "Low-sensitivity recursive digital filters obtained *via* the delay replacement," *IEEE-TCAS*, vol. CAS-31, pp. 654-657, July 1984.

[30] M.M. Pakerabo, "Time-domain sensitivity of digital networks," *CSSP*, vol. 2, no. 3, pp. 277-284, 1983.

[31] K.S. Prasad, C. Eswaran, and V.G.K. Murti, "On the stopband sensitivity of digital filters," *IEEE-TCAS*, vol. CAS-34, pp. 582-584, May 1987.

[32] K. Renner and S.C. Gupta, "On the design of wave digital filters with low sensitivity properties," *IEEE-TCT*, vol. CT-20, no. 5, pp. 555-567, 1973.

[33] T. Saramaki, T.-H. Yu, and S.K. Mitra, "Very low sensitivity realization of IIR digital filters using a cascade of complex all-pass structures," *IEEE-TCAS*, vol. CAS-34, pp. 876-886, Aug. 1987.

[34] A. Sedlmeyer, "On the sensitivity of coefficients in wave digital filters with open-and-short circuit filters as reference filters," *J. Applied Science and Engineering*, vol. 1, pp. 281-300, Oct. 1976.

[35] R.A. Singer, "Selecting state variables to minimize eigenvalue sensitivity of multivariable systems," *Automatica*, vol. 5, pp. 85-93, 1969.

[36] D.A. Spaulding, "Optimizing sensitivity in the design of a class of linear filter," *IEEE-TCT*, vol. CT-17, pp. 459-461, Aug. 1970.

[37] M.R. Stojic and R.M. Stojic, "Pole-zero sensitivities of a digital filter due to parameter quantization," *IJCTA*, vol. 5, pp. 299-304, July 1977.

[38] E.C. Tan and C.J. Price, "On the sensitivity study of low-pass wave digital filters," *IJE*, vol. 53, no. 4, pp. 331-340, 1982.

[39] V. Tavsanoglu and L. Thiele, "Optimal design of state-space digital filters by simultaneous minimization of sensitivity and roundoff noise," *IEEE-TCAS*, vol. CAS-31, pp. 884-888, Oct. 1984.

[40] G.C. Temes and K.M. Cho, "Large-change sensitivities of linear digital networks," *IEEE-TCAS*, vol. CAS-25, pp. 113-114, Feb. 1978. Comments, by S.S. Lawson, *ibid.*, vol. CAS-26, pp. 896-897, Oct. 1979.

[41] L. Thiele, "Design of sensitivity and round-off noise optimal state-space discrete systems," *IJCTA*, vol. 12, pp. 39-46, 1984.

[42] M. Toy and P.M. Chirlian, "Low multiplier coefficient sensitivity block digital filters," *IEEE-TCAS*, vol. CAS-31, pp. 993-1001, Dec. 1984.

[43] L.E. Turner and B.K. Ramesh, "Low sensitivity digital LDI ladder filters with elliptic magnitude response," *IEEE-TCAS*, vol. CAS-33, pp. 697-706, July 1986.

[44] P.P. Vaidyanathan and S.K. Mitra, "Passivity properties of low-sensitivity digital filter structures," *IEEE-TCAS*, vol. CAS-32, pp. 217-224, March 1985.

[45] _____, "Very low sensitivity FIR filter implementation using "structural passivity" concept," *IEEE-TCAS*, vol. CAS-32, pp. 360-364, April 1985.

[46] M. Wang, E.B. Lee, and D. Boley, "A simple method to determine the stability and margin of stability of 2-D recursive filters," *IEEE-TCAS*, Part I, vol. CAS-39, no. 3, pp. 237-239, 1992.

[47] G.-T. Yan, "New digital notch filter structures with low coefficient sensitivity," *IEEE-TCAS*, vol. CAS-31, no. 9, pp. 825-828, Sept. 1984.

[48] S.M. Yang, "On the sensitivity of a class of wave digital filters," *IEEE-TCAS*, vol. CAS-28, pp. 1158-1164, Dec. 1981.

[M] Network Sensitivity Theory – Statistically Oriented

[*] Special Issue on "Statistical circuit design," *Bell Syst. Tech. J.*, vol. 50, April 1971.

[1] C. Acar, "Statistical multiparameter sensitivity measures for transfer gain and phase functions," *IJCTA*, vol. 7, pp. 143-150, 1979.

[2] C. Acar and M.S. Ghausi, "Statistical multiparameter sensitivity measure of gain and phase function," *IJCTA*, vol. 5, pp. 13-22, 1977.

[3] C. Acar, M.S. Ghausi, and K.R. Laker, "Statistical multiparameter sensitivity measures in high Q networks," *JFI*, vol. 300, pp. 281-297, Oct. 1975.

[4] P.R. Adby, "Component tolerance assignment using the pseudo-inverse sensitivity matrix," *IJCTA*, vol. 4, pp. 199-201, April 1976.

[5] P.M. Anderson and K.J. Timko, "A probabilistic model of power system disturbances," *IEEE-TCAS*, vol. CAS-29, pp. 789-796, Nov. 1982.

[6] P. Balaban and J.J. Golembeski, "Statistical analysis for practical circuit design," *IEEE-TCAS*, vol. CAS-22, pp. 100-108, Feb. 1975.

[7] P. Barbini, A. Liberatore, and G. Mazzoni, "Large-change sensitivity figure," *IEEE-TCAS*, vol. CAS-25, pp. 380-381, June 1978.

[8] C. Belove, "The sensitivity function in variability analysis," *IEEE-TR*, vol. R-15, pp. 70-76, Aug. 1966.

[9] R.N. Gadenz, M.G. Rezai-Fakhr, and G.C. Temes, "A method for the computation of large tolerance effects," *IEEE-TCT (Special Issue on Computer-Aided Design)*, vol. CT-20, pp. 704-708, Nov. 1973.

[10] R.N. Gadenz and G.C. Temes, "An efficient procedure for the statistical transient analysis of switching circuits," *Comput. Aided Design*, vol. 3, no. 4, 1971.

[11] P.J. Goddard, P.A. Villalaz, and R. Spence, "Method for the efficient computation of the large-change sensitivity of linear reciprocal networks" *Elect. Lett.*, vol. 7, pp. 112-113, Feb. 1971.

[12] A.N. Gonuleren, "Sensitivity analysis and dynamic range of QUAD filters and some comparative results," *IJCTA*, vol. 10, pp. 175-200, 1982.

[13] D. Hilberman, "An approach to the sensitivity and statistical variability of biquadratic filters," *IEEE-TCT,* vol. CT-20, pp. 382-390, July 1973. Comments, by P. Barbini, A. Liberatore, and G. Mazzoni, *ibid.,* vol. CAS-25, pp. 381-382, June 1978.

[14] G.P. Jessel, "Network statistics for computer-aided network analysis," *IEEE-TCT,* vol. CT-20, pp. 635-641, Nov. 1973.

[15] K.R. Laker and M.S. Ghausi, "Large change sensitivity--a dual pair of approximate statistical sensitivity measures," *JFI,* vol. 298, nos. 5 and 6, pp. 395-413, 1974.

[16] _____, "A statistical multiparameter sensitivity measure for notch filters," *IEEE-TCAS,* vol. 23, pp. 173-176, March 1976.

[17] B. Liu and M.E. Van Valkenburg, "A probabilistic approach to the parameter sensitivity problem," in *Proc. Allerton Conf.,* pp. 701-710, 1967.

[18] S.R. Parker, "The use of sensitivity functions for statistical extrapolation in lieu of Monte Carlo methods," in *Proc. 1971 Int. IEEE Conf. Systems, Networks and Computers,* Oaxtepec, Mor., Mexico, Jan. 19-20, 1971.

[19] W.W. Poliscuk and B.L. Rojo, "A note on statistical sensitivity computation in switched-capacitor networks," *IEEE-TCAS,* vol. CAS-35, pp. 423-425, April 1988.

[20] M.G. Rezai-Fakhr and G.C. Temes, "Statistical large-tolerance analysis of nonlinear circuits in time domain," *IEEE-TCAS,* vol. CAS-22, pp. 15-21, Jan. 1975.

[21] A.L. Rosenblum and M.S. Ghausi, "Multiparameter sensitivity in active RC networks," *IEEE-TCT,* vol. CT-18, pp. 592-599, Nov. 1971. Comments, by P. Barbini, A. Liberatore, and G. Mazzoni, *ibid.,* vol. CAS-25, pp. 382-383, June 1978.

[22] _____, "Sensitivity minimization in active RC networks," *JFI,* vol. 294, pp. 95-111, Aug. 1972.

[23] M.A. Styblinski, "An extension of the Rosenblum-Ghausi sensitivity measure to complex parameters," *IEEE-TCAS,* vol. 23, pp. 343-349, June 1976.

[24] A.R. Thorbjornsen and S.W. Director, "Computer-aided tolerance assignment for linear circuits with correlated elements," *IEEE-TCT,* vol. CT-20, no. 5, pp. 518-524, 1973.

[N] Tellegen's Theorem; Generalized Adjoint Network and Network Sensitivity Computations

[1] D.P. Brown, "Some extensions of tellegen's theorem," in *Proc. 17th-Midwest Symp. on Circuits and Systems,* pp. 211-218, Sept. 1974.

[2] S.W. Director and R.A. Rohrer, "The generalized adjoint network and network sensitivities," *IEEE-TCT,* vol. CT-16, pp. 318-323, Aug. 1969.

[3] P. Penfield, Jr., R. Stence, and S. Duinker, *Tellegen's Theorem and Electrical Networks.* Cambridge, MA: MIT Press, 1970.

[4] G.S. Reszka, "General formulae for the sensitivity coefficients in the TR and AC analysis of linear networks," *Bull. de L'Academie Polonaise des Sciences,* Serie des Science Techniques, vol. XXIV, No. 3, pp. 1-[227], 1976.

[5] A.K. Seth, "Network sensitivity results using an algebraic-differential operator notation," *IEEE-TCT,* vol. CT-20, no. 5, pp. 612-614, 1973.

[6] A.K. Seth, S.W. Director, and R.A. Rohrer, "Comments on time-domain network sensitivity using the adjoint network concept." *IEEE-TCT,* vol. CT-19, pp. 367-370, July 1972.

[7] A.K. Seth and H.K. Kesavan, "On the time domain network sensitivity," *IJE,* vol. 35, pp. 81-96, July 1973.

[8] A.K. Seth and K. Singhal, "Time-domain sensitivity using the adjoint network," *Elect. Lett.,* vol. 7, pp. 563-565, Sept. 1971.

[9] B.D.H. Tellegen, "A general network theorem, with applications," *Philips Res. Rept.*, vol. 7, pp. 259-269, Aug. 1952.

[10] _____, "A general network theorem, with applications," *Proc. Inst. Radio Engrs.*, Australia, vol. 14, pp. 265-270, Nov. 1953.

[11] M.E. Valtonen, "Equivalence in sensitivity calculation between direct differentiation and the method based on Tellegen's theorem," *Proc. IEEE*, vol. 65, pp. 1602-1603, Nov. 1977.

[12] J. Vandewalle, H. De Man, and J. Rabaey, "The adjoint switched capacitor network and its application to frequency, noise and sensitivity analysis," *IJCTA*, vol. 9, pp. 77-88, 1981.

[O] Network Sensitivity Function Generation: Computational Procedures

[1] G.E. Alderson and P.M. Lin, "Computer generation of symbolic network functions – a new theory and implementation," *IEEE-TCT*, vol. CT-20, pp. 48-56, Jan. 1973.

[2] J.W. Bandler, S.H. Chen, and S. Daijavad, "Proof and extension of general sensitivity formulas for lossless two-ports," *Elect. Lett.*, vol. 20, pp. 481-482, May 1984.

[3] _____, "Novel approach to multicoupled-cavity filter sensitivity and group delay computation," *Elect. Lett.*, vol. 20, pp. 580-582, July 1984.

[4] _____, "Exact sensitivity analysis for optimization of multi-coupled cavity filters," *IJCTA*, vol. 14, pp. 63-77, 1986.

[5] J.W. Bandler and M.A. El-Kady, "A complex Lagrangian approach with applications to power network sensitivity analysis," *IEEE-TCAS*, vol. CAS-29, pp. 1-6, Jan. 1982.

[6] _____, "A generalized, complex adjoint approach to power network sensitivity," *IJCTA*, vol. 12, pp. 191-222, 1984.

[7] _____, "The complex adjoint approaches to network sensitivities," *JFI*, vol. 318, no. 2, pp. 91-103, 1984.

[8] J.W. Bandler, M.A. El-Kady, and H.K. Grewal, "Sensitivity evaluation of phase-shifting transformers using the complex Lagrangian method," *IJCTA*, vol. 14, pp. 83-87, 1986.

[9] J.W. Bandler and R.E. Seviora, "Computation of sensitivities for noncommensurate networks," *IEEE-TCT*, vol. CT-18, pp. 174-177, Jan. 1971.

[10] J.W. Bandler and Q.J. Zhang, "Large change sensitivity analysis in linear systems using generalized Householder formulae," *IJCTA*, vol. 14, pp. 89-101, 1986.

[11] J.L. Bordewijk, "Comments on 'Automated network design and interreciprocity'," *IEEE-TCT*, vol. CT-18, no. 1, p. 179, Jan. 1971.

[12] F.H. Branin, Jr., "Network sensitivity and noise analysis simplified," *IEEE-TCT*, vol. CT-20, pp. 285-288, May 1973.

[13] R.K. Brayton and S.W. Director, "Computation of delay time sensitivities for use in time domain optimization," *IEEE-TCAS*, vol. CAS-22, pp. 910-920, Dec. 1975.

[14] P.R. Bryant, "Tracking sensitivity: An alternative algorithm for linear nonreciprocal circuits," *Elect. Lett.*, vol. 11, pp. 114-116, 1975.

[15] Y. Ceyhun, "Sensitivity calculations of ladder networks by using continuants," *Elect. Lett.*, vol. 7, pp. 157-158, April 1971.

[16] _____, "On the properties of continuants and sensitivity computations," *IEEE-TCT*, vol. CT-20, pp. 167-169, March 1973.

[17] M. Chandrashekar, "Time-domain network sensitivity; row sensitivity and column sensitivity," *Proc. IEEE*, vol. 62, pp. 1179-1180, Aug. 1974.

[18] R. DeCarlo, "Sensitivity calculations using the component connection model," *IJCTA*, vol. 12, no. 3, pp. 288-291, 1984.

7.16.O. Network Sensitivity Function – Computational Procedures

[19] P.H. Di Mambro, "Calculating transfer function and its first- and second-order sensitivities using one network analysis," *Elect. Lett.*, vol. 19, pp. 421-423, May 1983. Comments, with reply, by J.K. Fidler, *ibid.*, vol. 19, pp. 914-916, Oct. 1983.

[20] S.W. Director, "LU factorization in network sensitivity calculations," *IEEE-TCT*, vol. CT-18, pp. 184-185, Jan. 1971.

[21] S.W. Director and R.A. Rohrer, "Automated network design-the frequency-domain case," *IEEE-TCT*, vol. CT-16, pp. 330-337, Aug. 1969.

[22] J.K. Fidler and J.I. Sewell, "Symbolic analysis for computer-aided circuit design-the interpolative approach," *IEEE-TCT*, vol. CT-20, pp. 738-741, Nov. 1973.

[23] W.W. Happ and D.E. Moody, "Topological techniques for sensitivity analysis," *IEEE-TANE*, vol. ANE-11, pp. 249-254, Dec. 1964.

[24] W.W. Happ and M.C. Weinberg, "Variance analysis applied to performance error from component failure in microcircuits," in *Proc. Southeastern Symp. System Theory*, May 1969.

[25] K.H. Leung and R. Spence, "Tracking sensitivity: An efficient algorithm for linear non-reciprocal circuits," *Elect. Lett.*, vol. 10, pp. 377-378, 1974.

[26] _____, "Multiparameter large-change sensitivity analysis, and systematic exploration," *IEEE-TCAS, vol. CAS-22, pp. 796-804, Oct. 1975.*

[27] P.M. Lin, "Computer generation of symbolic network functions: An overview," in *Proc. Working Conf. Principles of Computer-Aided Design* (J. Vlietstra and R.F. Wielings, Eds.). Amsterdam, The Netherlands: North-Holland, 1973.

[28] _____, "A Survey of applications of symbolic network functions," *IEEE-TCT*, vol. CT-20, pp. 732-737, Nov. 1973.

[29] T.B.M. Neill, "Sensitivity analysis in computer aided design of linear circuits," *Elect. Lett.*, vol. 4, pp. 316-317, July 1968.

[30] _____, "The inefficiency of the adjoint network approach to the calculation of first-order sensitivity coefficients," *Computer Aided Design*, vol. 6, pp. 32-34, June 1974.

[31] T. Neumann and D. Agnew, "Tracking sensitivity: A practical algorithm," *Elect. Lett.*, vol. 13, pp. 371-372, June 1977.

[32] C. Phrydas and J.I. Sewell, "Symbolic network sensitivities using partition methods," *IJE*, vol. 41, no. 1, pp. 25-32, 1976.

[33] J.F. Pinel, "Computer-aided network tuning," *IEEE-TCT*, vol. CT-18, pp. 192-194, Jan. 1971.

[34] G.A. Richards, "Second-derivative sensitivity using the concept of the adjoint network," *Elect. Lett.*, vol. 5, pp. 398-399, Aug. 1969.

[35] A.F. Schwarz, "Semisymbolic analysis of network sensitivity," *Elect. Lett.*, vol. 14, pp. 291-292, April 1978.

[36] A.K. Seth and P.H. Roe, "Hybrid formulation of explicit formulae for higher-order network sensitivities," *IEEE-TCAS*, vol. CAS-22, pp. 475-478, May 1975.

[37] R.E. Stuffle and P.M. Lin, "New approaches to computer-aided determination of oscillator frequency sensitivities," *IEEE-TCAS*, vol. CAS-27, pp. 882-892, Oct. 1980.

[39] E.C. Tan, "Numerical algorithm for pole-sensitivity study of LC ladder networks," *Elect. Lett.*, vol. 19, pp. 7-9, Jan. 1983.

[40] C.W. Therrien, "Use of the adjoint for computing exact changes in response of cascaded two-port networks," *IEEE-TCAS*, vol. CAS-21, pp. 217-218, March 1974.

[41] L.M. Vallese, "Incremental versus adjoint models for network sensitivity analysis," *IEEE-TCAS*, vol. CAS-21, pp. 46-49, Jan. 1974.

[42] C.F. Yokomoto, "A simple bookkeeping scheme for computing sensitivities of symbolic transfer functions," *IEEE-TCAS*, vol. CAS-21, no. 5, pp. 606-608, 1974.

[P] Sensitivity Analysis of Nonlinear Networks

[1] A. Borys, "A simplified analysis of nonlinear distortion in analog electronic circuits using Volterra – Wiener series," *Sc. E*, vol. 30, pp. 77-107, 1984.

[2] N.K. Jain, V.C. Prasad, and A.B. Bhattacharyya, "Delay-time sensitivity in linear RC tree," *IEEE-TCAS*, vol. CAS-34, pp. 443-445, April 1987.

[3] R.M. Lewis and B.D.O. Anderson, "Insensitivity of a class of nonlinear compartmental systems to the introduction of arbitrary time delays," *IEEE-TCAS*, vol. CAS-27, pp. 604-612, July 1980.

[4] S.R. Parker, "Sensitivity analysis and models of nonlinear circuits," *IEEE-TCT*, vol. CT-16, pp. 443-447, Nov. 1969.

[5] P.F. Ridler, "First-order parameter sensitivities of nonlinear device models," *Elect. Lett.*, vol. 8, pp. 506-507, Oct. 1972.

[6] T.N. Trick, F.R. Colon, and S.P. Fan, "Computation of capacitor voltage and inductor current sensitivities with respect to initial conditions for the steady-state analysis of nonlinear periodic circuits," *IEEE-TCAS*, vol. CAS-22, pp. 391-396, May 1975.

[7] J.L. Wyatt, Jr., "Monotone sensitivity of nonlinear nonuniform RC transmission lines, with application to timing analysis of digital MOS integrated circuits," *IEEE-TCAS*, vol. CAS-32, pp. 28-33, Jan. 1985.

APPENDIX A

SELECTED NUMERICAL APPLICATIONS

A.1. INTRODUCTION

To demonstrate the applications of selected analytical results developed in this book a few numerical examples are studied in depth. The system that governs these examples is extracted with modifications from [2C. 28] to accommodate the pertinent applications.[1] The organization of this appendix is as follows. In Section A.2, the physical system is presented. Section A.3 gives preliminary analysis of this system. The corresponding *[parametric-]* sensitivity analysis and synthesis is given subsequently in Section A.4. The numerical computations are performed using another classical reference on numerical computations in modern control theory [2C. 36]. Finally, a project is proposed in Section A.5.

A.2. POSITION SERVO SYSTEM

Suppose that an object is moving in a plane and at the origin of this plane a rotating antenna, driven by an electric motor, is pointing at the object direction at all the time. The control problem is to command the motor such that

$$\theta(t) \approx \theta_r(t), \quad t \geq t_0, \qquad (A.2\text{-}1)$$

---

[1] In Section 1.5, it is mentioned that there are a few references, though some out of print, which had special effects in the training of this author. The textbook by Kwakernaak and Sivan [2C. 28] is one of them. In the Spring of 1974, this reference was the textbook for my graduate course in optimal control theory taught by Professor T.J. Higgins at the University of Wisconsin-Madison. One day, referring to this textbook, one student asked Professor Higgins, why we are applying all this advanced theory to a few (four) simple examples (repeatedly) – what is the point? Professor Higgins replied that in an orchestra, the violin is perhaps one of the "simplest" musical instruments: however, depending on who is playing it, one can make all sorts of sounds using this instrument alone! So the idea of applying a body of knowledge to a simple example for the purpose of emphasizing the new concept as developed, rather than losing the main thrust of the underlying issue in the process of describing a complicated example, became the trend in my presentations and a message that I have tried to convey to my own students. [While preparing this book, I learned that a *theme example* has been also used in [2B. 14], which appeared before [2C. 28], to illustrate different aspects of the developed theory.]

where $\theta(t)$ denotes the angular position of antenna and $\theta_r(t)$ is the angular position of object. Although not deterministic, it is assumed that $\theta_r(t)$ is made available as a mechanical angle by manually pointing the binocular in the object direction.

Here, The plant consists of antenna and motor and disturbance is the torque exerted by wind on this antenna. The observed variable is the output of a potentiometer or other transducer mounted on the shaft of antenna, given by

$$\eta(t) = \theta(t) + v(t), \qquad (A.2\text{-}2)$$

where $v(t)$ is the measurement noise. In this example, the angle $\theta(t)$ is to be controlled and therefore it is the controlled variable. The reference variable is the object direction $\theta_r(t)$. The plant input is μ, which is the input voltage to motor.

The control mechanism for this system is simply the applications of several transducers, in order to measure $\theta(t)$ and $\theta_r(t)$ as electrical signals. Then $\theta_r(t) - \theta(t)$ is used as the input voltage for motor. Thus when $\theta_r(t) - \theta(t)$ is positive, a positive input voltage is produced that makes the antenna to rotate in a positive direction so that the difference between $\theta_r(t)$ and $\theta(t)$ is reduced.

This scheme is obviously a closed-loop controller. An open-loop controller would generate the driving voltage $\mu(t)$ on the basis of reference angle $\theta_r(t)$ alone, which such controller generally has no way of compensating for external disturbances such as wind torques, or plant parameter variations such as different friction coefficients at different temperatures, *etc..*

The motion of antenna can be described by a differential equation

$$J\ddot{\theta}(t) + f\dot{\theta}(t) = \tau(t) + \tau_d(t), \qquad (A.2\text{-}3)$$

where J is the moment of inertia for all the rotating parts, including antenna; f is the coefficient of viscous friction; $\tau(t)$ is the torque applied by motor, and $\tau_d(t)$ is a disturbing torque caused by wind. The motor torque is assumed to be proportional to $\mu(t)$, the motor input voltage, so that

$$\tau(t) = \kappa \mu(t). \qquad (A.2\text{-}4)$$

Defining the state variables $\xi_1(t) \triangleq \theta(t)$ and $\xi_2(t) \triangleq \dot{\theta}(t)$, the state-space representation for this system becomes

$$\dot{\xi}(t) = \begin{bmatrix} 0 & 1 \\ 0 & -f/J \end{bmatrix} \xi(t) + \begin{Bmatrix} 0 \\ \kappa/J \end{Bmatrix} \mu(t) + \begin{Bmatrix} 0 \\ 1/J \end{Bmatrix} \tau_d(t). \qquad (A.2\text{-}5)$$

The control variable or observation $\zeta(t)$ is the angular position of antenna

$$\zeta(t) = [1, 0]\, \xi(t). \qquad (A.2\text{-}6)$$

The nominal values for these parameters are chosen as follows.

$$f = 46 \ \text{kg m}^2/\text{s}, \ J = 10 \ \text{kg m}^2, \ \kappa = 7.87 \ \text{kg m}^2 \ \text{rad}/(vs^2). \qquad (A.2\text{-}7)$$

Appendix A – Selected Numerical Applications

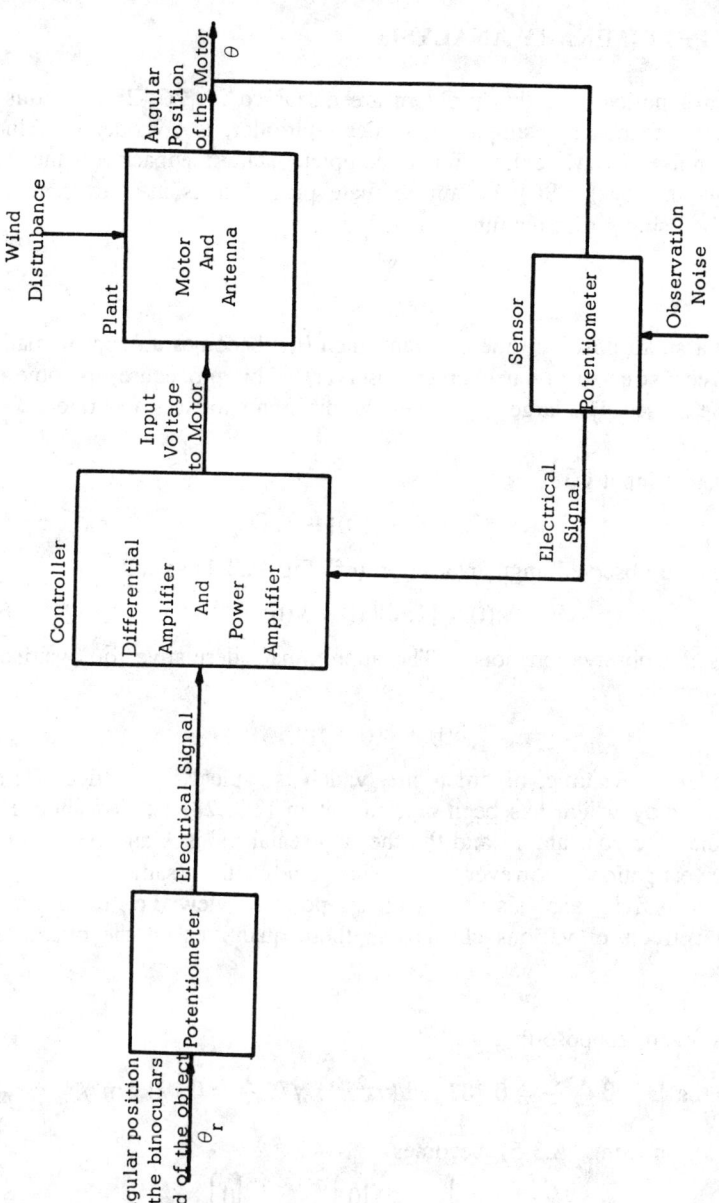

Fig. A.2-1. A position servo system.

A.3. THE PRELIMINARY ANALYSIS

Several control policies for this problem are discussed in [2C. 28]. In this study only a position feedback, using a first-order controller, is considered. Since the observation noise always exists, for a complete state feedback of the system, instead of having $\xi_2(t) = \dot{\theta}(t)$, an approximate procedure is used to generate this $\dot{\theta}(t)$. Namely, using a transfer function such as

$$\frac{s}{T_d s + 1}, \tag{A.3-1}$$

where T_d is a small positive time constant, then $\hat{\dot{\theta}}(t)$ becomes an approximation for $\dot{\theta}(t)$ that is reconstructed (a Luenberger observer). This procedure, to some extent, filters out the noise. The larger T_d makes the differentiator less accurate.

The system input is

$$\mu(t) = \lambda[\theta_r(t) - \eta(t)] - \lambda \rho \delta(t), \tag{A.3-2}$$

where $\eta(t)$ is the observed angular position (cf., Fig. A.3-1) as follows.

$$\eta(t) = [1, 0]\xi(t) + \nu(t). \tag{A.3-3}$$

Here $\nu(t)$ is the observation noise. The approximate derivative, $\delta(t)$, satisfies the following equation

$$T_d \dot{\delta}(t) + \delta(t) = \dot{\eta}(t). \tag{A.3-4}$$

This controller is dynamic, of order one, which is typical in practice. The main thrust of the study which has been carried out in [2C. 28] is: To choose proper values for the time constant T_d and the the two scalar gains λ and ρ. The purpose of this investigation, however, is: To study the same problem from *parametric*-sensitivity analysis and synthesis point of view; i.e., to investigate as feasible, the effects of various changes in these quantities on the overall system performance.

In this regard suppose that

$$\alpha \triangleq \frac{f}{J} = 4.6 \text{ s}^{-1}, \quad \beta \triangleq \frac{\kappa}{J} = 0.787 \text{ rad/(vs}^2), \quad \gamma \triangleq \frac{1}{J} = 0.1 \text{ (kg m}^2)^{-1}. \tag{A.3-5}$$

Then (A.2-5), on using (A.3-5), becomes

$$\dot{\xi}(t) = \begin{bmatrix} 0 & 1 \\ 0 & -\alpha \end{bmatrix} \xi(t) + \begin{Bmatrix} 0 \\ \beta \end{Bmatrix} \mu(t) + \begin{Bmatrix} 0 \\ \gamma \end{Bmatrix} \tau_d(t), \tag{A.3-6}$$

with

$$T_d \dot{\delta}(t) + \delta(t) = \xi_2(t) + \omega(t), \tag{A.3-7}$$

where $\omega(t)$ is the observation noise in reading $\delta(t)$.

Appendix A – Selected Numerical Applications

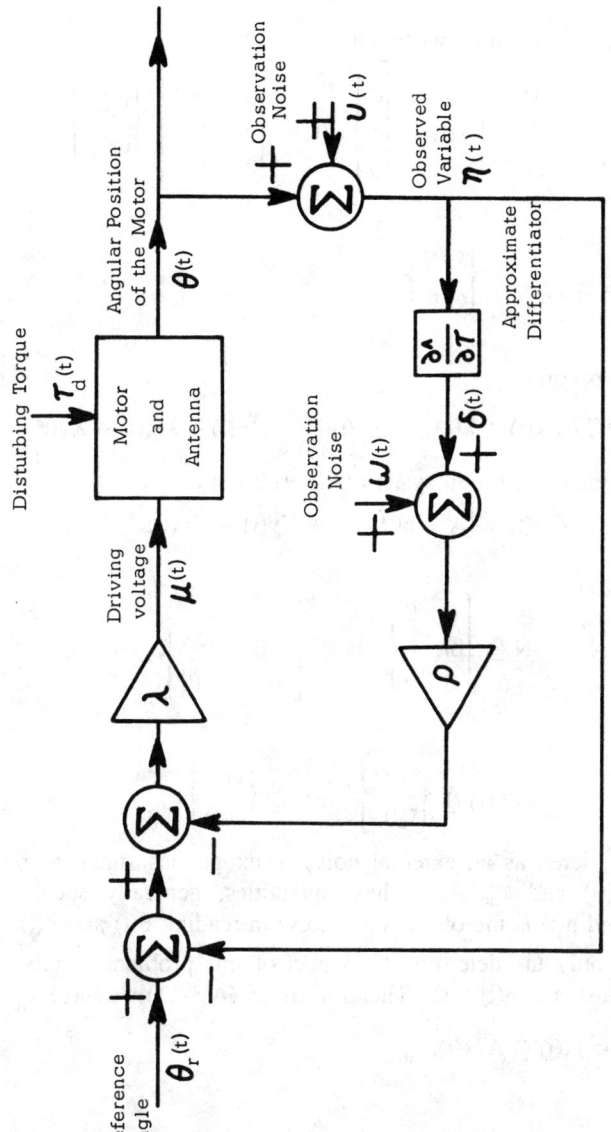

Fig. A.3-1. Simplified block diagram of a position – feedback control system using a first – order controller.

Let
$$x(t) \triangleq [\xi_1(t), \xi_2(t), \delta(t)]^T, \tag{A.3-8}$$

then (A.3-6) and (A.3-7) can be written as

$$\dot{x}(t) = \begin{bmatrix} 0 & 1 & 0 \\ 0 & -\alpha & 0 \\ 0 & 1/T_d & -1/T_d \end{bmatrix} x(t) + \begin{Bmatrix} 0 \\ \beta \\ 0 \end{Bmatrix} \mu(t) + \begin{bmatrix} 0 & 0 \\ \gamma & 0 \\ 0 & 1/T_d \end{bmatrix} \begin{Bmatrix} \tau_d(t) \\ \omega(t) \end{Bmatrix}$$

$$\triangleq Ax(t) + b\mu(t) + C \begin{Bmatrix} \tau_d(t) \\ \omega(t) \end{Bmatrix}. \tag{A.3-9}$$

Equation (A.3-2), becomes

$$\mu(t) = -[\lambda, 0, \lambda\rho] x(t) + \lambda\theta_r(t) - \lambda v(t) \triangleq -k^T x(t) + \lambda\theta_r(t) - \lambda v(t). \tag{A.3-10}$$

Correspondingly, the closed-loop system (A.3-9) becomes

$$\dot{x}(t) = (A - bk^T)x(t) + Ny(t) + Hn(t); \tag{A.3-11}$$

where

$$N \triangleq \begin{bmatrix} 0 & 0 \\ \beta\lambda & \lambda \\ 0 & 0 \end{bmatrix}, \quad H \triangleq \begin{bmatrix} 0 & 0 \\ -\lambda\beta & 0 \\ 0 & 1/T_d \end{bmatrix}, \tag{A.3-12}$$

and

$$y(t) \triangleq \begin{Bmatrix} \theta_r(t) \\ \tau_d(t) \end{Bmatrix}, \quad n(t) \triangleq \begin{Bmatrix} v(t) \\ \omega(t) \end{Bmatrix}. \tag{A.3-13}$$

Here, y(t) is considered as an external noise or exogenous signal to represent the uncertainty in $\theta_r(t)$ and τ_d, since these quantities, generally speaking, are not known exactly; and n(t) is the observation noise in reading $\theta(t)$ and $\delta(t)$.

Considering only the deterministic aspect of this problem in this section, we assume that $y(t) \equiv 0$, and $n(t) \equiv 0$. Therefore from (A.3-11) we have

$$\dot{x}(t) = (A - bk^T)x(t) \triangleq A_1 x(t), \tag{A.3-14}$$

where

$$A_1 \triangleq (A - bk^T) = \begin{bmatrix} 0 & 1 & 0 \\ -\beta\lambda & -\alpha & -\lambda\rho\beta \\ 0 & 1/T_d & -1/T_d \end{bmatrix}. \tag{A.3-15}$$

Here the nominal values of parameters are given by (A.3-5), and

$$T_d = \frac{1}{100} \text{ (second)}. \tag{A.3-16}$$

We let $\lambda = 225$, $\rho = 0.08$ (or $\lambda\rho = 18$), with (A.3-5), then A_1 becomes

$$A_1 = \begin{bmatrix} 0 & 1 & 0 \\ -177.08 & -4.6 & -14.17 \\ 0 & 100 & -100 \end{bmatrix}. \tag{A.3-17}$$

Computing the eigenvalues of A_1, results in

$$\lambda_1 = -82.27, \quad \lambda_{2,3} = -11.17 \pm j9.53. \tag{A.3-18}$$

Thus the system at the neighborhood of these nominal values, and with no external noise present, is asymptotically stable.

Now, based upon this operating condition, in the following a systematic *parametric*-sensitivity analysis and synthesis are performed in order to show the meaning and the applicability of certain selected results presented in this book. But first we recall the actual parametric form of the *ODE's* which represents this system. From (A.3-14), with A_1 given in (A.3-15), and with α and β from (A.3-5) we have

$$\dot{x}(t) \triangleq A_1 x(t) = \begin{bmatrix} 0 & 1 & 0 \\ -\lambda\kappa/J & -f/J & -\lambda\rho\kappa/J \\ 0 & 1/T_d & -1/T_d \end{bmatrix} x(t). \tag{A.3-19}$$

Here the nominal values for parameters f, J, κ, T_d are given in (A.2-5) and (A.3-16), respectively, with $\lambda = 225$ and $\rho = 0.08$.

A.4. SYSTEM SENSITIVITY: ANALYSIS AND SYNTHESIS

For the system of (A.3-19), we let $\{p_j, j = 1, \cdots, 4\} \triangleq \{f, J, \kappa, T_d\}$. Then the following questions regarding these parameters are to be studied from the *parametric*-sensitivity analysis and synthesis point of view.

(a) To generate all the system sensitivity functions with the corresponding system – *TSF*.

(b) To investigate the terminal conditions insensitivity *(TCI)* for the set of differential equations that must be solved in order to study this *TCI* and its complete simultaneity.

(c) To apply the results from Algorithm 5.6-1, in order to determine the "largest" variations in each parameter p_j, *individually*.

(d) To derive the optimal controller gain $k^T = [\lambda, 0, \lambda\rho]$, such that the following combined-quadratic cost is minimized. The weighting matrices are given in the following.

$$J_1 = \min \int_0^\infty \{x^T W_0 x + \mu^T W \mu + \alpha \frac{\partial x^T}{\partial J} W_1 \frac{\partial x}{\partial J}\} dt. \qquad (A.4-1)$$

(e) To compare the regulator cost corresponding to the optimal controller gain; with and without the sensitivity minimization contribution.

A.4.a.

To generate system sensitivity functions, we must solve the following set of differential equations.

$$\begin{Bmatrix} \dot{x} \\ \frac{\partial \dot{x}}{\partial f} \\ \frac{\partial \dot{x}}{\partial J} \\ \frac{\partial \dot{x}}{\partial \kappa} \\ \frac{\partial \dot{x}}{\partial T_d} \end{Bmatrix} = \begin{bmatrix} A_1 & 0 & 0 & 0 & 0 \\ \frac{\partial A_1}{\partial f} & A_1 & 0 & 0 & 0 \\ \frac{\partial A_1}{\partial J} & 0 & A_1 & 0 & 0 \\ \frac{\partial A_1}{\partial \kappa} & 0 & 0 & A_1 & 0 \\ \frac{\partial A_1}{\partial T_d} & 0 & 0 & 0 & A_1 \end{bmatrix} \begin{Bmatrix} x \\ \frac{\partial x}{\partial f} \\ \frac{\partial x}{\partial J} \\ \frac{\partial x}{\partial \kappa} \\ \frac{\partial x}{\partial T_d} \end{Bmatrix}, \qquad (A.4-2)$$

where the numerical values for the above coefficient matrix (i.e., A_1 and each corresponding $\frac{\partial A_1}{\partial p_j}$) with the associated initial conditions are given next. The dimension of this set of *ODE's* is 15, thus to generate these sensitivity functions is a very involved computational task. However, we can show that these equations are redundant.

The system – *TSF*, on the other hand, can be generated from the following set of low-dimensional differential equations.

$$\begin{Bmatrix} \dot{x} \\ \dot{f}_{c1} \end{Bmatrix} = \begin{bmatrix} A_1 & 0 \\ A_1 & A_1 \end{bmatrix} \begin{Bmatrix} x \\ f_{c1} \end{Bmatrix}, \qquad (A.4-3a)$$

where A_1 is given in (A.3-17) and the initial condition for this system is assumed as follows.

$$[0, 1, 1, 0, 0, 0]^T. \qquad (A.4-3b)$$

A careful look at the solution of (A.4-2) (shown in the following graphs) shows that some of these trajectories are almost identical. This observation confirms our expectation that there are redundancies in this set of 15-*ODE's*.

Appendix A — Selected Numerical Applications

SYSTEM MODEL MATRIX A1=A-BK

0.0000E+00	1.0000E+00	0.0000E+00	0.0000E+00	0.0000E+00	0.0000E+00	0.0000E+00	0.0000E+00	0.0000E+00
0.0000E+00	0.0000E+00	0.0000E+00	0.0000E+00	0.0000E+00	0.0000E+00	0.0000E+00	0.0000E+00	0.0000E+00
-1.7708E+02	-4.6000E+00	-1.4170E+01	0.0000E+00	0.0000E+00	0.0000E+00	0.0000E+00	0.0000E+00	0.0000E+00
0.0000E+00	0.0000E+00	-1.0000E+02	0.0000E+00	0.0000E+00	0.0000E+00	0.0000E+00	0.0000E+00	0.0000E+00
0.0000E+00	1.0000E+02	0.0000E+00	0.0000E+00	0.0000E+00	0.0000E+00	0.0000E+00	0.0000E+00	0.0000E+00
0.0000E+00	0.0000E+00	0.0000E+00	0.0000E+00	0.0000E+00	0.0000E+00	0.0000E+00	0.0000E+00	0.0000E+00
0.0000E+00	0.0000E+00	0.0000E+00	1.0000E+00	0.0000E+00	0.0000E+00	0.0000E+00	0.0000E+00	0.0000E+00
0.0000E+00	-1.0000E-01	0.0000E+00	-1.7708E+02	-4.6000E+00	-1.4170E+01	0.0000E+00	0.0000E+00	0.0000E+00
0.0000E+00	0.0000E+00	0.0000E+00	0.0000E+00	0.0000E+00	-1.0000E+02	0.0000E+00	0.0000E+00	0.0000E+00
0.0000E+00	0.0000E+00	0.0000E+00	0.0000E+00	1.0000E+02	0.0000E+00	0.0000E+00	0.0000E+00	0.0000E+00
1.7710E+01	4.6000E-01	1.4200E+00	0.0000E+00	0.0000E+00	0.0000E+00	0.0000E+00	0.0000E+00	0.0000E+00
0.0000E+00	0.0000E+00	0.0000E+00	0.0000E+00	0.0000E+00	0.0000E+00	0.0000E+00	1.0000E+00	0.0000E+00
0.0000E+00	0.0000E+00	0.0000E+00	0.0000E+00	0.0000E+00	0.0000E+00	0.0000E+00	0.0000E+00	0.0000E+00
-1.0000E+00	-1.8000E+00	0.0000E+00	-1.4170E+01	0.0000E+00	0.0000E+00	0.0000E+00	0.0000E+00	-1.7708E+02
-2.2500E+01	0.0000E+00	0.0000E+00	0.0000E+00	0.0000E+00	0.0000E+00	0.0000E+00	0.0000E+00	0.0000E+00
-4.6000E+00	0.0000E+00	0.0000E+00	0.0000E+00	0.0000E+00	0.0000E+00	0.0000E+00	0.0000E+00	0.0000E+00
1.0000E+02	0.0000E+00	0.0000E+00	0.0000E+00	0.0000E+00	1.0000E+00	-4.6000E+00	-1.0000E+00	0.0000E+00
0.0000E+00	0.0000E+00	0.0000E+00	0.0000E+00	0.0000E+00	-4.6000E+00	0.0000E+00	0.0000E+00	0.0000E+00
0.0000E+00	0.0000E+00	-1.7708E+02	-1.4170E+01	0.0000E+00	1.0000E+00	0.0000E+00	0.0000E+00	0.0000E+00
0.0000E+00	-1.0000E+00	1.0000E+04	0.0000E+00	0.0000E+00	0.0000E+00	0.0000E+00	0.0000E+00	0.0000E+00
0.0000E+00	0.0000E+00	0.0000E+00	0.0000E+00	-1.0000E+02	0.0000E+00	0.0000E+00	0.0000E+00	0.0000E+00

INITIAL CONDITION

0.0000E+00	1.0000E+00	0.0000E+00	0.0000E+00	0.0000E+00	0.0000E+00
0.0000E+00	0.0000E+00	0.0000E+00			

Appendix A – Selected Numerical Applications

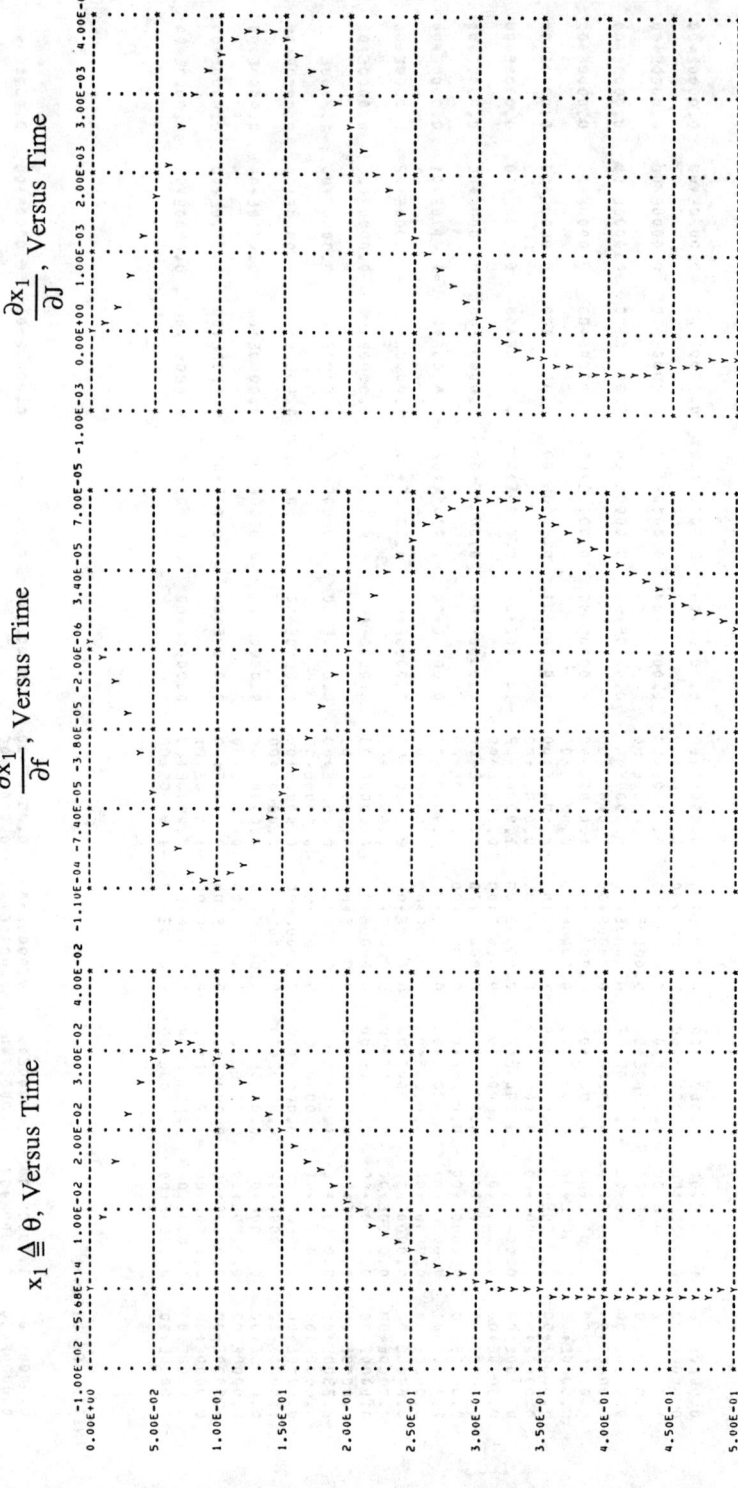

Appendix A – Selected Numerical Applications

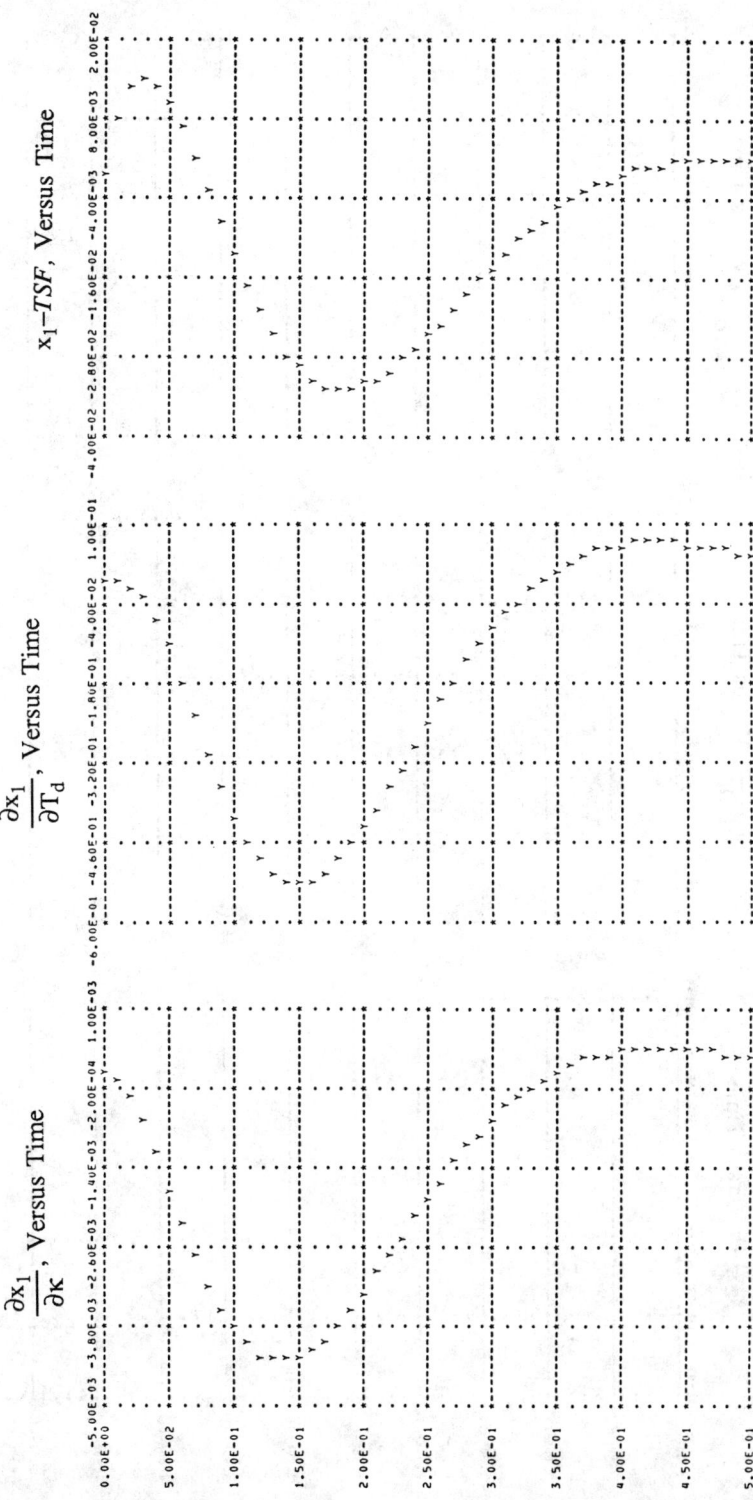

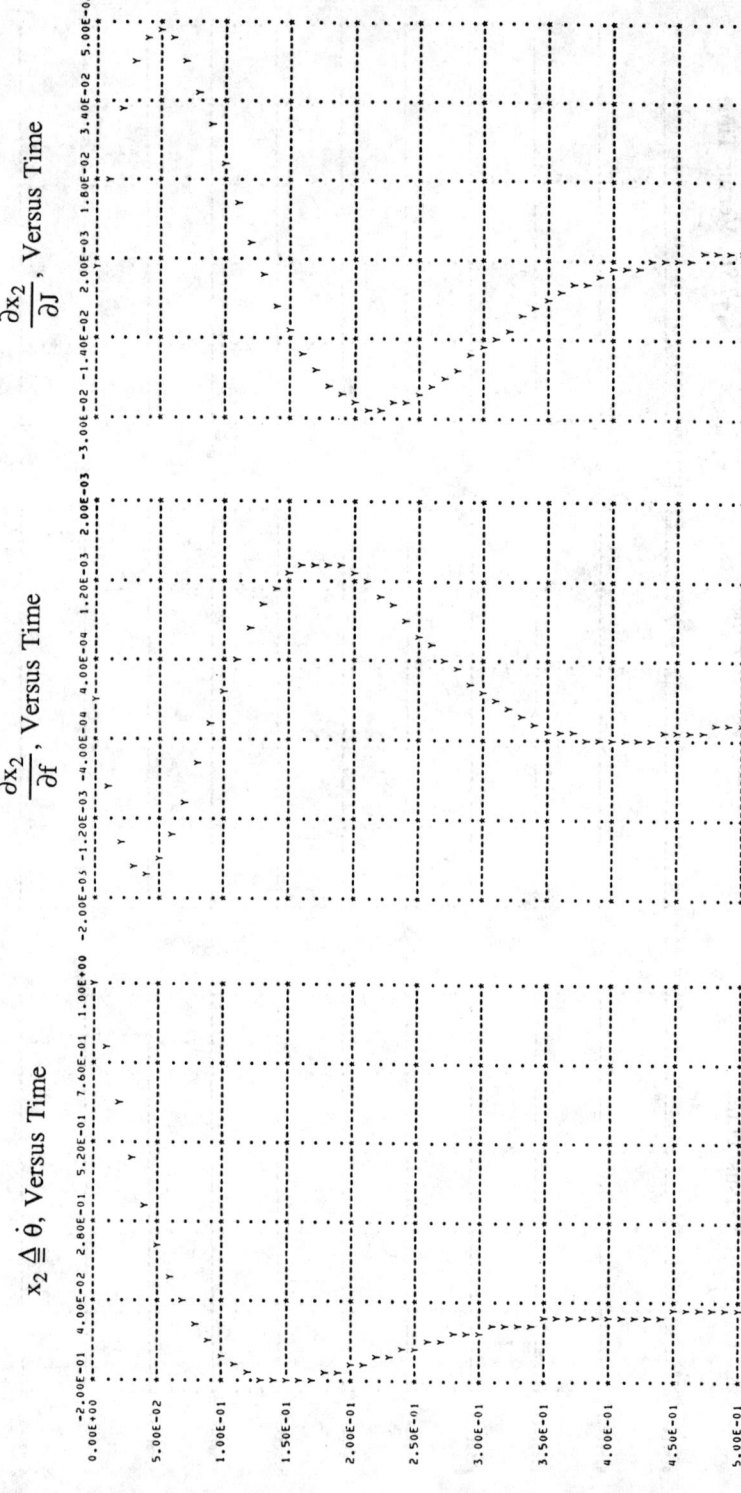

Appendix A – Selected Numerical Applications

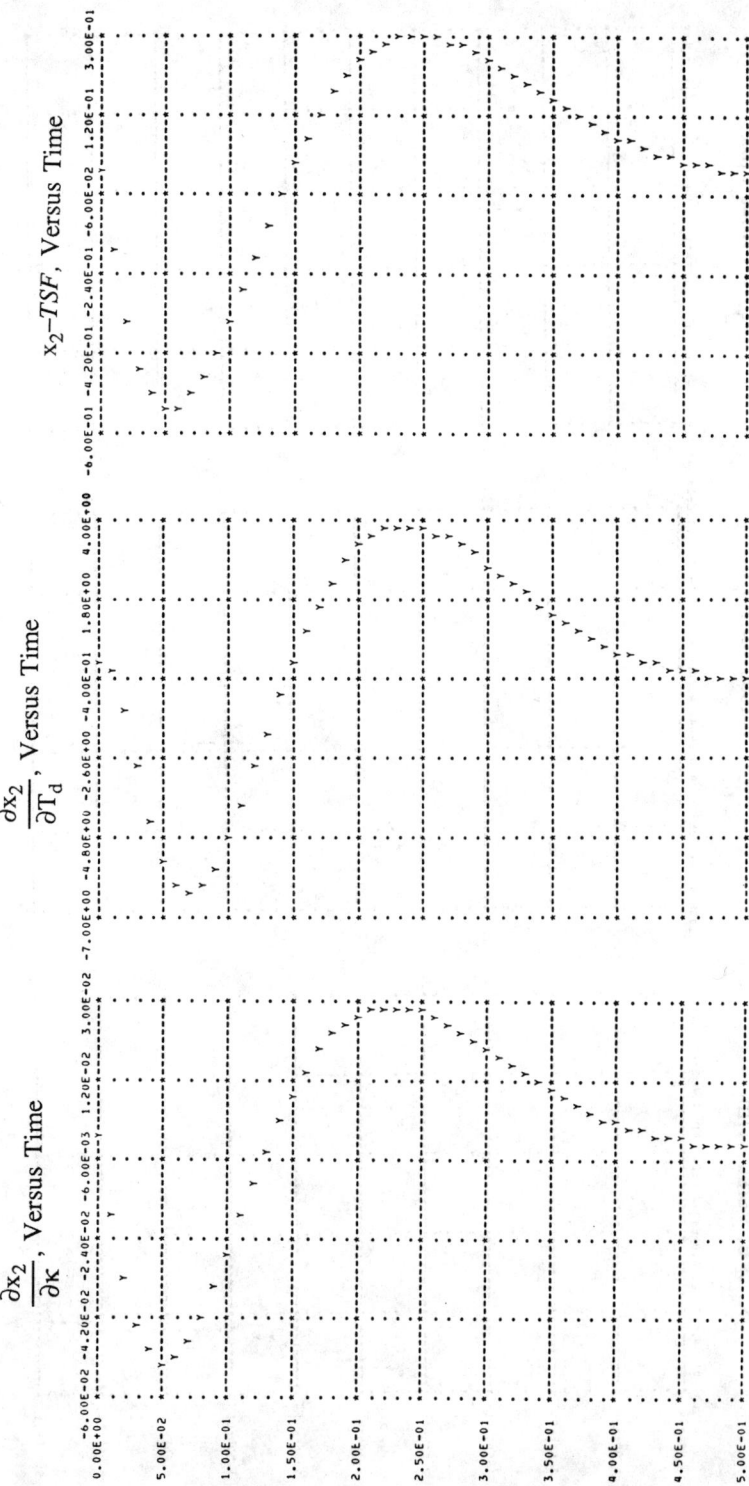

570 Appendix A – Selected Numerical Applications

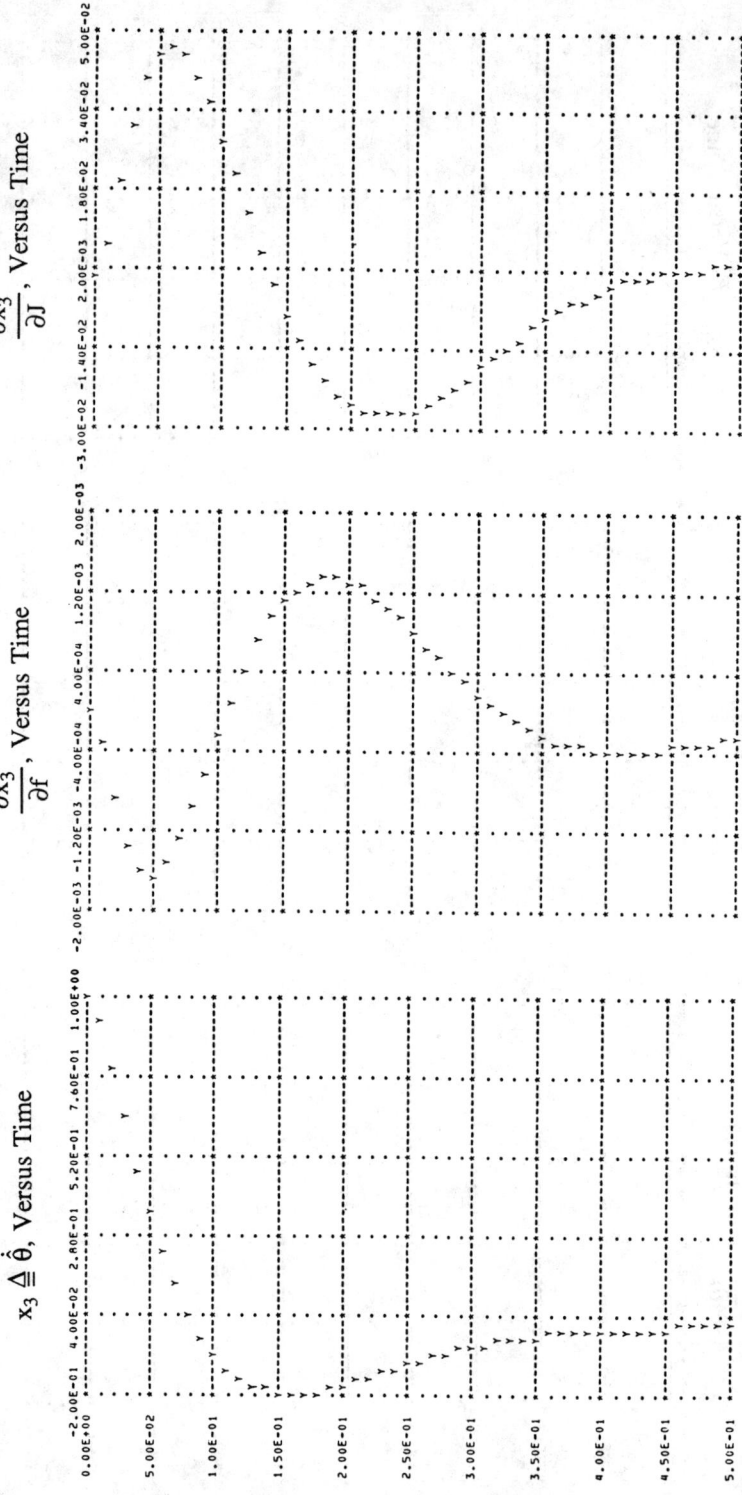

Appendix A – Selected Numerical Applications

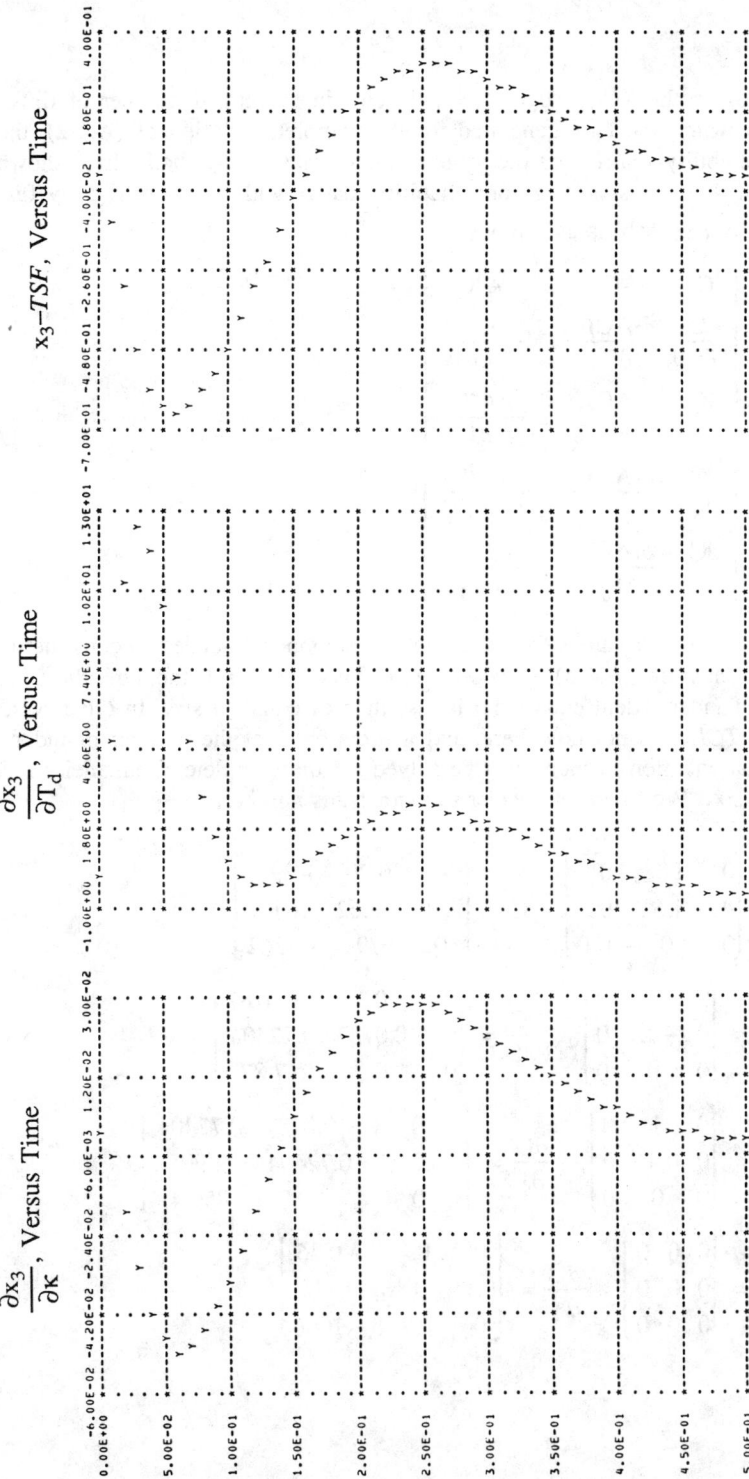

A.4.b.

To investigate the *TCI*, and the redundancies in the actual number of differential equations which must be generated for the complete solution of (A.4-2), the rank of controllability matrix for the system model must be studied. It is shown in § 4.3.2, that the corresponding controllability matrix with all its sensitivity functions and $C \triangleq (b, Ab, A^2b)$ is as follows.

$$\mathcal{C} \triangleq \begin{bmatrix} C & AC & \cdots & A^4C \\ \dfrac{\partial C}{\partial f} & \dfrac{\partial (AC)}{\partial f} & \cdots & \dfrac{\partial (A^4C)}{\partial f} \\ \dfrac{\partial C}{\partial J} & \dfrac{\partial (AC)}{\partial J} & \cdots & \dfrac{\partial (A^4C)}{\partial J} \\ \dfrac{\partial C}{\partial \kappa} & \dfrac{\partial (AC)}{\partial \kappa} & \cdots & \dfrac{\partial (A^4C)}{\partial \kappa} \\ \dfrac{\partial C}{\partial T_d} & \dfrac{\partial (AC)}{\partial T_d} & \cdots & \dfrac{\partial (A^4C)}{\partial T_d} \end{bmatrix}. \qquad (A.4\text{-}4)$$

The numerical values for the essential components needed to construct (A.4-4) are given next in (A.4-5). According to Theorem 4.3-1 the rank of $\mathcal{C}$ (or the number of independent equations) is less than or equal to six. In other words, the complete *TCI* does not hold here, and it turns out that the number of independent differential equations which must be solved for the complete simultaneity is indeed less than six. We invite the reader to extract this number.

$$A = \begin{bmatrix} 0 & 1 & 0 \\ 0 & -4.6 & 0 \\ 0 & 100 & -100 \end{bmatrix}, \quad C = \begin{bmatrix} 0 & 0.79 & -3.62 \\ 0.79 & -3.62 & 16.65 \\ 0 & 79 & -8262 \end{bmatrix}, \qquad (A.4\text{-}5a)$$

$$\frac{\partial A}{\partial f} = \begin{bmatrix} 0 & 1 & 0 \\ 0 & -0.1 & 0 \\ 0 & 0 & 0 \end{bmatrix}, \quad \frac{\partial C}{\partial f} = \begin{bmatrix} 0 & 0 & -0.0787 \\ 0 & -0.0787 & 0.72404 \\ 0 & 0 & -7.87 \end{bmatrix}, \qquad (A.4\text{-}5b)$$

$$\frac{\partial A}{\partial J} = \begin{bmatrix} 0 & 1 & 0 \\ 0 & 0.46 & 0 \\ 0 & 0 & 0 \end{bmatrix}, \quad \frac{\partial C}{\partial J} = \begin{bmatrix} 0 & -0.0787 & 0.72404 \\ -0.0787 & 0.72404 & -4.9958 \\ 0 & -7.87 & 859.4 \end{bmatrix}, \qquad (A.4\text{-}5c)$$

$$\frac{\partial A}{\partial \kappa} = \begin{bmatrix} 0 & 0 & 0 \\ 0 & 0 & 0 \\ 0 & 0 & 0 \end{bmatrix}, \quad \frac{\partial C}{\partial \kappa} = \begin{bmatrix} 0 & 0.1 & -0.46 \\ 0.1 & -0.46 & 2.116 \\ 0 & 10 & -1046 \end{bmatrix}, \qquad (A.4\text{-}5d)$$

$$\frac{\partial A}{\partial T_d} = \begin{bmatrix} 0 & 1 & 0 \\ 0 & 0 & 0 \\ 0 & -10^4 & 10^4 \end{bmatrix}, \quad \frac{\partial C}{\partial T_d} = \begin{bmatrix} 0 & 0 & 0 \\ 0 & 0 & 0 \\ 0 & -7870 & 1.6102 \times 10^6 \end{bmatrix}. \quad \text{(A.4-5e)}$$

A.4.c.

This analysis is performed using Algorithm 5.6-1 (single-parameter variations). Note that we have extensively studied this type of applications in Chapter Five. There are, however, a few additional or special cases that are not yet explored. For instance, to study the stability robustness analysis of (A.3-19) with respect to T_d and J, we note that the corresponding "ΔS" becomes a type E situation (cf., § 5.5.2). Here, we let "epsilon" be chosen as $1/T_d$ or $1/J$ in order to determine the largest-single variations for T_d, and J. The stability of A_1 is found to be *not* sensitive to the variations in T_d so long as T_d remains positive. Similar procedure yields that the convergence process for finding the "largest" variations in J is very slow due to this conservative procedure. We invite the reader to study this issue. Here, A_1 is given in (A.3-17), and with Q = I and the perturbation matrix

$$(\Delta A_1)_\kappa = \varepsilon_\kappa \begin{bmatrix} 0 & 0 & 0 \\ -22.5 & 0 & -1.8 \\ 0 & 0 & 0 \end{bmatrix}, \quad \text{(A.4-6)}$$

the largest ε_κ at iteration seven becomes -9.931. Similarly, at the second iteration the ε_f becomes 95.65.

A.4.d.

The synthesis procedure is performed according to Algorithm 6.5-1. We note that from practical point of view, the actual optimal controller gain is that consisting of $\lambda = 234.34$ and $\lambda \rho = 19.65$, which the latter comes from adding the gains corresponding to $\theta(t)$ and $\dot{\theta}(t)$. This analysis is valid only for *small*-parameter variations.

$$A = \begin{bmatrix} 0 & 1 & 0 \\ 0 & -4.6 & 0 \\ 0 & 100 & -100 \end{bmatrix}, \quad \frac{\partial A}{\partial J} = \begin{bmatrix} 0 & 0 & 0 \\ 0 & 0.46 & 0 \\ 0 & 0 & 0 \end{bmatrix}, \quad \text{(A.4-7a)}$$

$$b = [0, 0.787, 0]^T, \quad \frac{\partial b}{\partial J} = [0, -0.0787, 0]^T, \quad \text{(A.4-7b)}$$

$$W = 2\times 10^{-5}, \quad W_0 = \begin{bmatrix} 1 & 0 & 0 \\ 0 & 0 & 0 \\ 0 & 0 & 0 \end{bmatrix}, \quad W_1 = \begin{bmatrix} 100 & 0 & 0 \\ 0 & 0 & 0 \\ 0 & 0 & 0 \end{bmatrix}, \quad X = \begin{bmatrix} 0 & 0 & 0 \\ 0 & 1 & 0 \\ 0 & 0 & 1 \end{bmatrix}. \quad \text{(A.4-7c)}$$

After six iterations the algorithm stops at the following values:

$$\text{gain matix of regulator} = [226.3, \ 19.382, \ 6.4791\times 10^{-4}]^T, \quad \text{(A.4-8a)}$$

$$\text{gain matix for sensitivity reduction} = [8.0442, \ 0.14665, \ 0.11786]^T, \quad \text{(A.4-8b)}$$

$$\text{The optimal gain} = [234.34, \ 19.529, \ 0.11851]^T. \quad \text{(A.4-8c)}$$

A.4.e.

The regulator cost is $J_p = x_o^T P x_o$; where $P = P^T$ is the unique solution of $A_1^T P + P A_1 + Q = 0$; with $A_1 = A - bk^T$; and $Q = W_0 + kWk^T$.

Without sensitivity reduction (i.e., only the regulator problem with the assumption that $\dot{\theta}(t) \equiv \hat{\dot{\theta}}(t)$), the optimal controller gain becomes

$$k_p^T = [223.61, \ 0, \ 18.699]. \quad \text{(A.4-9)}$$

Correspondingly:

$$A_1 = \begin{bmatrix} 0 & 1 & 0 \\ -176 & -4.6 & -14.72 \\ 0 & 100 & -100 \end{bmatrix}, \quad Q = \begin{bmatrix} 2 & 0 & 0.084 \\ 0 & 0 & 0 \\ 0.084 & 0 & 0.007 \end{bmatrix}, \quad \text{(A.4-10a)}$$

$$P = 10^{-5} \begin{bmatrix} 10998 & 568.18 & 6.2205 \\ 568.18 & 51.175 & -3.3278 \\ 6.2205 & -3.3278 & 3.9898 \end{bmatrix}. \quad \text{(A.4-10b)}$$

The optimal cost is $J_p = x_o^T P x_o = 4.85092 \times 10^{-4}$.

With sensitivity reduction (assuming that $\dot{\theta}(t) \equiv \hat{\dot{\theta}}(t)$), the optimal controller gain becomes

$$k_{ps}^T = [234.34, \ 0, \ 19.65]. \quad \text{(A.4-11)}$$

This solution is found by letting (A.4-9) be an initial optimal controller gain in a process of iteratively solving for the overall optimal solution resulting in

$$A_1 = \begin{bmatrix} 0 & 1 & 0 \\ -184.43 & -4.6 & -15.46 \\ 0 & 100 & -100 \end{bmatrix}, \quad Q = \begin{bmatrix} 2.098 & 0 & 0.092 \\ 0 & 0 & 0 \\ 0.092 & 0 & 0.015 \end{bmatrix}, \quad \text{(A.4-12a)}$$

$$P = 10^{-5} \begin{bmatrix} 14415 & 568.78 & 8.728 \\ 568.78 & 68.704 & -2.5274 \\ 8.728 & -2.5274 & 7.8907 \end{bmatrix}. \quad \text{(A.4-12b)}$$

The optimal cost is $J_{ps} = 7.15399 \times 10^{-4}$. As expected the regulator cost has increased for the price of reducing system sensitivity measure, a typical trade off. Here, we have $k_{ps} \approx 1.05 k_p$, and we may postulate that perhaps 5% variations (that is *small*) in system parameters can be tolerated, before the system loses its "optimality" in the quadratic sense. In other words, we may not be able to find a quadratic *PI* if system parameters change by more than 5%. The resolution of this matter is at the heart of our assumptions in regard to the static-parameter variations. As stated before, in this type of problem formulation (we mean for all *small*-sensitivity formulations, whether they are treated in this book or not), where we have stretched the boundary of the linear time-invariant system to its limit, indeed what we are getting is in response to our limited hardware utilization and to our limited information regarding the nature of our system variations. Thus the system may become overdesigned in some aspects and although certain parameter variations may never take place. A resolution of one issue hinges on the other.

A.5. PROJECT

The main thrust of the preceding study is the *parametric*–sensitivity analysis and synthesis. For the sake of discussion, we discard various non-parametric disturbances in our system. We now propose to consider (A.3-11) with both y(t) and n(t) present. Let n(t) be a zero-mean White noise with a known positive-definite intensity (covariance), and let the exogenous signal y(t) be a "bounded" disturbance. At this juncture, it is clear that there are several ways to interpret the above "bounded" signal. Make your own nontrivial assumptions regarding this signal and in at least two different ways in the context of Table 6.1-1. Design the corresponding controllers and compare their performance and robustness.

Optional[2]: Consider Table 6.1-1 with (A.3-11) as its plant. Design at least one controller for the remaining two cases in that table.

---

[2] Two questions Professor Higgins asked in his tests, beside a few "optionals" (sometimes double and triple optionals)! The first question among several, which he always asked in every final examination, was: write the title of your textbook, its author, its publisher, the place and the year of publication. He used to emphasize how important this information is. Since everyone knew about this question, these parameters were memorized [and often remembered] by most students. However, the second question also among several, which he asked in every written Ph.D. qualifying examination and was by far more difficult and less predictable than the first one, was as follows. "Make your own numerical example in 'a certain area' which shows you have mastered the subject and complete its solution." (No one knew what that area might be: for instance, when I took mine the area was the maximum principle). It is in that spirit that the above question is raised. Although I wonder what would happen these days if someone asked a question like that in a Ph.D. qualifying examination?

APPENDIX B

LIST OF THE REFERENCES: A SENSITIVITY TREE

CHAPTER ONE: INTRODUCTION

[A] *Representative Historical References*, 13
[B] *Representative Monographs, Books, Report and the Conference Proceedings*, 15
[C] *Survey Papers*, 16

CHAPTER TWO: THE PRINCIPAL ASPECTS OF SENSITIVITY THEORY

[A] *General Mathematical References*, 67
[B] *General References on System Theory*, 68
[C] *Representative References on: Advanced System Theory, Optimal Control Systems, Calculus of Variations, and Dynamic Programming*, 69
[D] *Certain Papers on Optimal Deterministic Control Systems*, 71
[E] *Representative References on the Theory of Errors Perturbation, and Approximation Theory*, 72

CHAPTER THREE: SENSITIVITY FUNCTIONS GENERATION

[A] *Direct Computation and/or Simulation of Sensitivity Functions*, 108
[B] *Sensitivity Functions Generation "Low-Order Sensitivity Model"*, 109
[C] *General Sensitivity Approaches with Applications*, 109
[D] *Sensitivity Analysis via Signal Flow Graph*, 110
[E] *Sensitivity Analysis for Oscillatory Systems*, 111
[F] *Sensitivity Functions Generation Linear Retarded Systems*, 111
[G] *Sensitivity in the Large Bounded Variations*, 112
[H] *Selected Sensitivity Analyses with Applications in Mechanical Systems*, 112

CHAPTER FOUR: ALGEBRAIC PROPERTIES OF SYSTEM MODELS

TERMINAL CONDITIONS INSENSITIVITY

[A] *Algebraic Properties of the Matrices of a System Model*, 130
[B] *Terminal Conditions Insensitivity*, 130
[C] *Certain Issues in Guidance Theory As Pertain to Terminal Conditions Insensitivity*, 131

CHAPTER FIVE: STABILITY ROBUSTNESS ANALYSIS

MANEUVERABILITY, AND SENSITIVITY IN THE LARGE

[A] *Theory of Stability: Survey Papers and General References*, 255
[B] *Stability Robustness Analysis, Maneuverability, and Sensitivity in the Large*
   Group I: Matrices and Polynomials, 262
[C] *Group II: Polynomials*, 268
[D] *Group III: Discrete-Time Systems*, 274
[E] *Group IV: The Classical Frequency-Domain / Servomechanism*, 278
[F] *Group V: Time-varying, Nonlinear, and Delayed Systems,*
   The Concept of Absolute Stability, 281
[G] *Group VI: Stabilization Using the H^∞ and Other Optimization Approaches*, 288
[H] *Root Sensitivity and Pole Placement*, 289
[I] *Eigenvalue and Eigenvector Sensitivity*, 291
[J] *Synthesis Considering Certain Sensitivity Equations Resulted From System Stability*, 293

CHAPTER SIX: SENSITIVITY REDUCTION AND ROBUSTNESS

[A] *Trajectory-Sensitivity Comparison/Feedback Properties of Linear Systems*, 477
[B] *Performance-Index Sensitivity Comparison*, 478
[C] *Optimality-Index Sensitivity: Analysis and Synthesis*, 479
[D] *Trajectory-Sensitivity Optimization*, 481
[E] *Sensitivity with Respect to System-Auxiliary Parameters*, 483
[F] *Sensitivity Reduction and/or Optimization: Other Methods and Applications*, 484
[G] *Optimal Inputs for System Identification/*
   Auxiliary Inputs for System Control (The Effects of the Human Operator), 488
[H] *Sensitivity Considerations in Hardy Spaces*, 489
   [H1] *Pertinent References on Feedback System Design and Mathematics*, 489

[H2] *Complex – Domain Approach,* **490**

[H3] *State – Space Approach,* **494**

[H4] *Operator Approach,* **496**

[H5] *Computational Issues and Optimization Approaches,* **497**

[H6] *Applications and Extensions,* **499**

[I] *Conference Proceedings / Reports and Book Reviews on Invariance Theory,* **499**

[J] *Selected Techniques from Invariance Theory,* **500**

[K] *Bounded Uncertainty and Disturbance Rejection,* **503**

[K1] *Bounded Uncertainty,* **503**

[K2] *Disturbance Rejection,* **507**

[L] *Sensitivity Analyses in Nonlinear, Time-Varying Systems, System Sensitivity Theory Statistically Oriented, Certain Statistical Problems, and Certain Social and Economical Matters,* **508**

[M] *Sensitivity Considerations in Nonlinear Systems,* **511**

[N] *Sensitivity Considerations in Discrete-Time (Digital) Systems Sensitivity Reduction of Systems with Time Lag, and Multi-Dimensional Digital Systems,* **511**

[O] *Sensitivity Considerations in Distributed [Parameter] Systems,* **513**

[P] *Sensitivity Issues in Selected Multidisciplinary Optimization Methods, and Numerical Methods,* **515**

CHAPTER SEVEN: THE THEORY OF SENSITIVITY IN NETWORKS

[A] *Selected References on Network Theory,* **537**

[B] *Representative Textbooks (or Reports) with Network Sensitivity Discussions: Definitions, Analyses, Computational Procedures, and Compensation Theorem,* **537**

[C] *General Network Sensitivity: Definitions and Analyses,* **538**

[D] *Tolerances or Sensitivity in the Large of Networks,* **539**

[E] *Sensitivity Model of Circuits,* **541**

[F] *Network Elements Sensitivity and Energy Stored in the Network, Vratsanos' Theorem,* **541**

[G] *Network Sensitivity Invariants and Bounds on Sensitivity Functions,* **541**

[H] *Concept of Continuously Equivalent Networks Synthesis Procedures with Sensitivity Invariants,* **542**

[I] *Network Synthesis with Minimum or Prescribed Sensitivity Measures,* **543**

[J] *Sensitivity of the Cost Function in the Optimal Circuit Design,* **544**

[K] *Sensitivity Analysis and Synthesis in Active Networks,* **544**

[L] *Sensitivity Analysis in Digital Filters,* **550**

[M] *Network Sensitivity Theory – Statistically Oriented,* **552**

[N] *Tellegen's Theorem; Generalized Adjoint Network and Network Sensitivity Computations,* **553**

[O] *Network Sensitivity Function Generation: Computational Procedures,* **554**

[P] *Sensitivity Analysis of Nonlinear Networks,* **556**

APPENDIX C

JOURNAL ABBREVIATIONS AND ACRONYMS

ACCS	Automatic Control and Computer Sciences English Translation of Avtomatika i Vychislitel'naya Tekhnika New York: Allerton Press Inc.
ACTA	Automatic Control Theory and Applications Calgary, Canada: ACTA Press. (Name changed to *CAC* in 1981.)
AIAA J.	Journal of the American Institute of Aeronautics and Astronautics (New York).
AIChE J.	Journal of the American Institute of Chemical Engineers (New York).
AMO	Applied Mathematics and Optimization, An International Journal New York: Springer Verlag.
ARC	Automation and Remote Control English Translation of Avtomatika i Telemekhanika New York: Plenum Publishing Corporation.
ASME-JBE	The American Society of Mechanical Engineers, Journal of Basic Engineering. Name changed to several new transactions in 1972.
ASME-JDSMC	ASME – Journal of Dynamic Systems, Measurement, and Control.
Automatica	The journal of the International Federation of Automatic Control (IFAC) Oxford: Pergamon Press.
Automatisme	Paris, France: Bordas – Dunod. Name changed to *Nouvel Automatisme* in 1978.
BSMM	Boletin de la Sociedad Mathematica Mexicana.
BSTJ	Bell System Technical Journal.
CAC	Control and Computers Calgary, Canada: ACTA Press.
CCy	Control and Cybernetics Polish Academy of Sciences System Research Institute. Warsaw: PWN – Polish Scientific Publishers.
CEEJ	Canadian Electrical Engineering Journal The Canadian Society of Electrical Enginering.
CSSP	Circuits, Systems, and Signal Processing. Boston: Birkhauser Verlag.
CTAT	Control Theory and Advanced Technology Tokyo, Japan: MITA Press.

DEDSTA	Discrete Event Dynamic Systems, Theory and Applications	
	Boston: Kluwer.	
DSS	Dynamics and Stability of Systems, An International Journal	
	Oxford: Oxford University Press.	
ECJ	Electronics and Communications in Japan	
	English Translation of Denshi Tsushin Gakkai Ronbunshi	
	Washington: Scripta Publishing Company.	
ECy	Engineering Cybernetics	
	English Translation of Tekhnicheskaya Kibernetika	
	Washington: Scripta Publishing Company.	
	Name changed to *SJCSS* in 1985.	
EEJ	Electrical Engineering in Japan	
	English Translation of Denki Gakkai Ronbunchi	
	Washington: Scripta Publishing Company.	
Elect. Lett.	Electronics Letters	
	The Institution of Electrical Engineers (London).	
FCE	Foundation of Control Engineering	
	Institute of Control Engineering, Technical University of Poznan, Poland.	
IC	Information and Control	
	New York: Academic Press.	
IEE	The Institution of Electrical Engineers (England).	
IEE, EE Trans.	The Institution of Engineers (Sydney Australia)	
	Electrical Engineering Transactions	
IEEE	Institute of Electrical and Electronics Engineers (New York).	
IEEE-JRA (now TRA)	IEEE – J. (now Trans.) of Robotics and Automation.	
IEEE-JSSC	IEEE – Journal of Solid State Circuits.	
IEEE-TAC	IEEE – Transactions on Automatic Control.	
IEEE-TAES	IEEE – Transactions on Aerospace and Electronic Systems.	
IEEE-TANE	IEEE – Transactions on Aerospace Navigational Electronics.	
IEEE-TCAS	IEEE – Transactions on Circuits and Systems (formerly Circuit Theory).	
IEEE-TCST	IEEE – Transactions on Control Systems Technology.	
IEEE-TCT	IEEE – Transactions on Circuit Theory.	
IEEE-TE	IEEE – Transactions on Education.	
IEEE-TPAS	IEEE – Transactions on Power Apparatus and Systems.	
IEEE-TR	IEEE – Transactions of Reliability.	
IEEE-TRA	IEEE – Transactions on Robotics and Automation.	
IEEE-TSMC	IEEE – Transactions on Systems Man and Cybernetics.	
IEOT	Integral Equations and Operator Theory.	
	Boston: Birkhauser Verlag.	
IJACSP	Int. J. of Adaptive Control and Signal Processing	
	New York: A Wiley – Interscience Publication.	

Appendix C – Journal Abbreviations and Acronyms

IJC	International Journal of Control London: Taylor & Francis Ltd.
IJCTA	Int. J. of CIRCUIT THEORY and APPLICATION London: John Wiley & Sons.
IJE	International Journal of ELECTRONICS Theoretical and Experimental London: Taylor & Francis Ltd.
IJSS	International Journal of Systems Science London: Taylor & Francis Ltd.
IRE	Institute of Radio Engineers (now IEEE).
IRE-TAC	IRE – Transactions on Automatic Control.
IRE-TCT	IRE – Transactions on Circuit Theory.
IRE-TIT	IRE – Transactions on Information Theory.
JAMM	Journal of Applied Mathematics and Mechanics (Translated from Russian) Oxford: Pergamon Press Ltd.
JASE	Journal of Applied Science and Engineering Amsterdam: Elsevier Scientific Pub. Company. (Apparently seized publication.)
JDE	J. of Differential Equations San Diego: Academic Press.
JEM	Journal of Engineering Mathematics Leyden, Netherlands: Noordhoff Int. Pub. Company.
JFI	Journal of the Franklin Institute Oxford: Pergamon Press Ltd.
JGCD	J. of Guidance, Control, and Dynamics. A publication of the *AIAA*.
JMAA	J. of Mathematical Analysis and Applications New York: Academic Press.
JMM	Journal of Mathematics and Mechanics.
JMP	Journal of Mathematics and Physics.
JOCAM	J. of Optimal Control Applications & Methods New York: A Wiley – Interscience Publication.
JOT	Journal of Operator Theory Bucharest, Romania: Romanian Academy of Sciences.
JOTA	Journal of Optimization Theory and Applications New York: Plenum Publishing Company.
MCSS	Mathematics of Control, Signals, and Systems New York: Springer Verlag.
MP	Mathematical Programming Amsterdam: North Holland Publishing Company.
MSSP	Multidimensional Systems and Signal Processing Boston: Kluwer.

Network	An International Journal New York: John Wiley and Sons.
PCIT	Problems of Control and Information Theory Budapest: Hungarian Academy of Sciences.
Proc. (...)	Proceedings of the (...)
RA	Ricerche di Automatica Rome: Oderisi Gubbio. (Apparently seized publication.)
SA	Soviet Aeronautics English Translation of Izvestiya VUZ Aviatsionnaya Tekhnika New York: Allerton Press Inc.
SAC	Soviet Automatic Control English Translation of Avtomatyka Washington: Scripta Publishing Company. Name changed to *SJAIS* in 1985.
SAM	Soviet Applied Mechanics English Translation of Prikladnaya Mekhanika Washington: Scripta Publishing Company.
Sc. E	Scientia Electrica Zurich: Birkhauser Verlag.
Sc. S	Scientia Sinica Bijing (or Peking): Science Press.
SCL	Systems and Control Letters Amsterdam: Elsevier Scientific Pub. Company.
SIAM	Society for Industrial and Applied Mathematics.
SIAM J. Contr.	SIAM – Journal of Control, up to 1975.
SIAM-JCO	SIAM – Journal of Control and Optimization, from 1976.
SIAM-JAM	SIAM – Journal of Applied Mathematics.
SIAM-JNA	SIAM – Journal of Numerical Analysis.
SIAM-JSSC	SIAM – Journal of Scientific and Statistical Computing.
SIAM R.	SIAM Review.
SJAIS	Soviet J. of Automation and Information Sciences English Translation of Avtomatyka Washington: Scripta Technica Publishing Company.
SJCSS	Soviet J. of Computer and System Sciences English Translation of Tekhnicheskaya Kibernetika.
TIMC	Transactions of the Institute of Measurement and Control London: England.
TMR	The Marconi Review Chelmsford, Essex, England. (Absorbed by the GEC J. of Research in 1983.)
WW	Wireless World Electronics, Television, Radio, Audio London, England.

AUTHOR INDEX

Aatre, V.K., 528, 547
Abdul-Wahab, A.A., 262, 274, 296
Abed, E.H., 255, 265, 273
Abouelwafa, M.A., 481
Abraham, J., 515
Abramovici, F., 258
Abu El-Ata-Doss, S., 513
Aburdene, M.F., 109
Acar, C., 530, 552
Acha, J.I., 550
Achieser, N.I., 67, 72
  (cf., Akhiezer, N.I.)
Ackermann, J., 278, 293, 294
Adamjan, V.M., 489
Adams, J.W., 550
Adby, P.R., 109, 552
Ae, T., 541
Agarwal, R.C., 529, 550
Aggarwal, J.K., 112, 255, 480, 500, 502
Agnew, D., 536, 555
Ahmadi, G., 281
Ahmadi, M., 263
Aivazyan, E.Y., 281
Aizerman, M.A., 241, 255, 281, 503
Akashi, H., 512
Akhiezer, N.I.,
  (cf., Achieser, N.I.)
Alami, M., 545
Albrecht, F., 69
Alden, R.H., 293
Alderson, G.E., 554
Alekperov, V.P., 255
Aleksandrov, A.G., 288
Aleksandrov, Y.V., 479
Alekseyev, O.G., 543
Ali, A.M., 529, 550
Alimov, Y.I., 466, 503
Allwright, J.C., 76, 108
Amillo, J., 288
Anagnost, J.J., 268, 272
Anandamohan, P.V., 544

Anderson, B.D.O., 68, 69, 255, 268, 271, 276, 281, 289, 477, 498, 505, 508, 510, 547, 551, 556
Anderson, P.M., 552
Anderson, W.J., 294
Ando, K., 281
Andreev, Y.N., 68, 255
Andres, R.P., 279
Andronov(w), A.A., 5, 13
Andronov, V.G., 293
Andrusevich, V.V., 281
Ansari, R., 550
Ansell, H.G., 541
Antoniou, A., 550
Antoulas, A.C., 497
Antsaklis, P.J., 264, 477
Aoki, M., 69, 358, 479, 488
Apkarian, P.R., 293
Aplevich, J.D., 480
Araki, M., 255
Aravena, J.L., 511
Arbel, A., 293
Arbib, M.(A.), 68, 70
Argoun, M.B., 268, 269, 278
Arora, J.S., 113
Arov, D.Z., 489
Arshad, M., 284
Artzrouni, M., 255
Asai, K., 131
Asher, R.B., 108, 130
Ashworth, M.J., 15, 293
Asner, B.A., 259
Astolfi, A., 507
Athani, V.V., 478
Athans, M., 69, 71, 268, 280, 315, 317, 358, 480, 488, 490, 505, 508
Atherton, D.P., 255, 282, 508
Atluri, S.R., 311, 479
Attarzadeh, F., 255

Auba, T., 497
Aulbach, B., 262
Auslander, D.M., 508, 510, 518
Ayerbe, A., 550
Ayiku, M.N.B., 514
Azanov, M.V., 501
Aze, D., 515
Azema, P., 111
Aziz, A., 274

Baartz, A.P., 257
Babak, S.F., 269
Bacciotti, A., 261, 282
Bachler, H.J., 526, 544
Bachmann, W.K., 72
Badr, R.I., 262
Bahgat, A., 296
Bahill, A.T., 109
Bahnasawi, A.A., 276
Bahri, Z., 284
Bai, E.W., 274, 493
Bainov, D.D., 261
Baizaev, S., 255
Bakan, G.M., 293, 503
Bakri, T.M., 291
Balaban, P., 530, 552
Balakrishnan, A.(V.), 293, 497,
Balas, M.J., 293
Balasubramanian, R., 282
Ball, J.A., 489, 492, 495, 496
Bamberger, A., 513
Bamieh, B., 490, 507
Banda, S.S., 268, 269, 281, 496, 497
Bandler, J.W., 535, 543, 554
Bandoni, J., 272, 294
Banks, H.T., 112, 500
Barabanov, A.T., 255
Barabanov, N.E., 255, 262, 282
Baranov, N.P., 510
Barbashin, E.A., 255, 462, 504
Barbini, P., 552, 553
Bar-Itzhack, I.Y., 112
Bar-Kana, I., 282

Barker, H.A., 109
Barkin, A.I., 282
Barmish, B.R., 131, 164, 203, 205, 262, 263, 264, 269, 270, 272, 273, 274, 281, 282, 286, 504, 505, 506
Bar-Ness, Y., 483
Barnett, S., 68, 112, 116, 130, 134, 142, 219, 255, 256, 262, 267, 272, 277, 282, 500, 511
Barry, P.E., 484
Barthelemy, J.F.M., 515
Bartlett, A.C., 269, 271, 274, 275, 496
Barua, A., 544
Basar, T., 499, 507
Basile, G., 504
Basov, V.P., 258
Bastelaer, P.V.(van), 523, 525, 542, 543
Bateman, H., 5, 13, 14
Batni, R.P., 259
Batson, R.G., 515
Baumeister, J., 255
Baxter, J.R., 109
Bayard, D.S., 69
Bayoumi, M.M., 289
Becker, N., 479
Becker, R.G., 108
Bedford, D.K., 546
Bedri, Y., 544, 549
Beers, Y., 72
Bekey, G.A., 512, 513
Belanger, P.R., 3, 265, 311, 484
Belen'kii, A.S., 504
Belhouari, A., 262
Bell, D.J., 68, 267, 481
Bellman, R., 14, 109
Bello, V., 528, 544
Belousov, Y.G., 293
Belove, C., 538, 552
Benedict, R., 112
Benhabib, R.J., 278
Bennett, A.W., 511
Bensoussan, D., 477, 493
Berger, C.S., 273, 277, 280, 484, 515

Bergman, N.J., 110
Berkovitz, L.D., 70
Berman, O., 488, 508
Bernfeld, S.R., 257, 505
Bernhard, P., 499
Bernstein, D.S., 256, 264, 274, 282, 479, 494, 495
Bernussou, J., 262, 504
Bertele, U., 130
Bertram, J.E., 132, 258
Bertrand, P., 265
Best, M.J., 515
Betz, R.E., 291
Bhargava, D.K., 545
Bhatia, N.P., 256
Bhattacharya, M., 550
Bhattacharyya, A.B., 546, 556
Bhattacharyya, B.B., 549
Bhattacharyya, K.C., 294
Bhattacharyya, S.P., 256, 265, 270, 274, 278, 282, 294, 507
Bhaya, A., 278
Bhushan, C., 542
Bialas, S., 263, 269
Bialko, M., 528, 544
Bien, Z., 264
Biernacki, R.M., 269, 278
Biey, M., 544
Bigelow, J.H., 515
Bilal, A.Y., 262
Bingulac, S.P., 15, 108, 484
Birkhoff, G., 67, 256
Bisiacco, M., 256, 550
Biswas, R.N., 108
Bitmead, R.R., 255, 281, 289
Bjerhammar, A., 72
Blinov, I.N., 109
Bliss, G.A., 69
Bloch, A.M., 282
Blostein, M.L., 523, 525, 541, 542
Blume, L.E., 504
Bobrovsky, B.Z., 479
Bobylev, N.A., 282
Boctor, S.A., 544
Bode, H.W., 4, 14, 537

Bodner, V.A., 484
Boekhoudt, P., 490
Boese, F.G., 289
Bohling, D.M., 539
Boley, D., 552
Boltyanskii, V.(G.), 14, 70
Bongers, P.M.M., 494
Bongiorno, J.J, Jr., 489, 490
Bonivento, C., 358, 488, 500
Bordewijk, J.L., 520, 537, 554
Borisenko, Y.G., 285, 511
Borisov, V.G., 500
Boron, L.F., 67
Borushko, Y.M., 269
Borys, A., 544, 556
Bose, N.K., 269, 271, 274, 543, 544
Bosgra, O.H., 494
Bouguerra, H., 269
Bourles, H., 282
Bowron, P., 546,
Boychuk, L.M., 435, 446, 500
Boyd, S., 497
Bradshaw, E., 14
Bradt, A.J., 481, 483
Brajovic, V., 108, 109
Branin, F.H, Jr., 535, 554
Brasch, F.M, Jr., 289
Brayton, R.K., 554
Breakwell, J.V., 131
Breiner, M., 279
Bremmer, H., 68
Brewer, J.W., 108
Brice, N.M., 527, 544
Brockett, R.W., 68, 70, 114, 115, 256, 258, 281, 477
Brown, D.P., 108, 538, 553
Brown, G.S., 5, 14
Brown, W.A., 479
Brualdi, R.A., 518
Bruckmann, D., 545
Bruckner, J.M.H., 487

Bruevich, N.G., 5, 14
Bruinsma, N.A., 497
Bruton, L.T., 528, 529, 545, 550
Bryant, G.F., 281
Bryant, P.R., 540, 554
Bryson, A.E.(Jr.), 69, 131, 281
Bucy, R.S., 69
Budak, A., 528, 545, 548
Buie, R.N., 515
Bukov, V.N., 488, 504
Bukowski, M., 541
Bure, E.G., 111
Burgat, C., 282
Burke, M., 481
Burns, R.J., 479
Burrascano, P., 550
Burrus, C.S., 529, 550
Burzio, A., 296, 484
Bushaw, D., 70
Buslowicz, M., 267, 282
Butcher, J.C., 276
Butler, E.M., 484, 539
Butler, W.J., 538, 539
Butterworth, S., 14
Buys, J.D., 515
Byers, R., 515
Bykhovskiy, M.L., 4, 14, 537
Byrne, P.C., 481

Cajka, J., 539
Calahan, D.A., 537
Callier, F.M., 71, 490
Caloba, L.P., 545
Campbell, D.P., 5, 14
Cao, X.R., 508
Caratheodory, C., 67
Carlock, G.W., 484
Cassandras, C.G., 509
Cassidy, J.F, Jr., 481
Castanon, D.A., 505
Cataltepe, T., 542
Cauer, W., 524, 537
Cavallo, A., 270
Cavin, R.K., III, 294, 295
Cefola, P.J., 131
Celentano, G., 264, 270

Cesar, M.O., 256
Ceyhun, Y., 536, 541, 554
Chaikin, C.E., 13
(cf., Khaikin, S.E.)
Chakravarti, N., 515
Chakravarti, P.C., 293
Chalstrom, R.E., 549
Chan, C.K., 267
Chan, S.P., 527, 548, 549
Chan, W.S., 477, 490
Chandrashekar, M., 538, 554
Chang, B.C., 269, 494, 496, 497, 507
Chang, C.H., 294
Chang, F.R., 498
Chang, S.S.L., 69, 263, 484
Chao, C.H., 504, 506
Chaparro, L.F., 272, 276
Chapellat, H., 270, 274, 278, 282, 507
Charlet, B., 282
Chatterji, B.N., 479
Chavent, G., 513
Chazan, D., 72
Cheetham, B.M.G., 524, 542
Chegancas, J., 282
Chehab, Y., 276
Chekhovoy, Y.N., 282, 283
Chelpanov, I.B., 271
Chen, B.M., 477, 497
Chen, B.S., 263, 266, 278, 280, 283, 288, 289, 506
Chen, C.F., 256
Chen, C.L., 270
Chen, C.M., 272
Chen, C.T., 68, 256, 291, 486
Chen, D., 295
Chen, G., 283, 497, 507
Chen, J., 263, 276
Chen, J.D., 267
Chen, J.J., 507
Chen, K.H., 290
Chen, M.J., 278
Chen, S.H., 554
Chen, S.S., 291

Author Index

Chen, W.K., 520, 537
Chen, W.L., 263
Chen, W.S., 477, 490
Chen, Y.H., 264, 488, 504
Chenais, D., 112, 515
Cheney, E.W., 72
Cheok, K.C., 512
Cheres, E., 283
Chernousko, F.L., 70
Chernyavs'ka, S.S., 111
Chestnut, H., 5, 14
Chi, H.H., 291
Chiang, C.C., 278, 283
Chiang, H.D., 283
Chiang, R.Y., 281, 385, 417, 495
Chiang, S.M., 112, 512
Chien, C.M., 507
Chikte, S.D., 504
Chimishkyan, S.Y., 283
Chin, P.S.M., 256
Chinaev, P.I., 500
Chiou, J.C., 517
Chiou, K.L., 283
Chirlian, P.M., 540, 552
Cho, K.M.,
Choatai, A., 266, 280, 286
Choi, K.K., 16, 113
Chon, C.T., 112
Chopra, I., 113, 295
Chou, H.H., 288, 289
Chou, J.H., 263
Chouinard, L.G., 263
Chow, P.L., 283
Christen, E., 549
Christie, S.H., 4, 14
Christodoulou, M.A., 259
Chu, C.C., 497
Chu, C.K., 513
Chumakov, N.M., 500
Ciric, V.V., 481
Clark, L.G., 112, 261, 480, 502, 509
Clarke, F.H., 512, 515
Clifford, A.A., 72
Cloud, D.J., 484
Cobb, J.D., 263
Cochran, J.E., Jr., 131, 256

Coddington, E.A., 67, 256
Cohen, N., 496
Cole, J.D., 73
Coleman, N., 486
Collado, J.M., 286
Collins, D.C., 486
Collins, E.G., Jr., 264, 275
Colon, F.R., 556
Colonna, S., 550
Combettes, P.L., 291
Commault, C., 507
Conan, J., 484
Connor, M.A., 512
Conte, G., 491
Conway, J.B., 489
Cook, W., 515
Cooke, K.L., 283
Corless, M.(J.), 276, 283, 296, 504
Coskunoglu, O., 487
Courant, R., 67
Courtin, P., 310, 478, 486, 510
Cowley, D., 276
Crane, R.N., 481
Crochiere, R.E., 529, 550
Cronin, J., 256
Cross, E.R., 482
Crossley, T.R., 139, 291, 294
Crouch, J.G., 294
Cruz, J.B., Jr., 15, 25, 69, 112, 114, 115, 130, 256, 279, 303, 304, 310, 455, 456, 457, 477, 478, 480, 482, 483, 484, 485, 486, 500, 509, 510, 511, 512, 540
Csaki, F.G., 68
Cullmann, G., 72
Cullum, J., 71

Dabke, K.P., 273, 274, 277, 280
Dabrowski, A., 545
da Cruz, J.J., 283, 360, 488

Dafermos, S., 515
Daggett, K.E., 545
Dahleh, M., 270, 274, 282, 507
Dahleh, M.A., 270, 288, 475, 507, 508, 512
Dai, S.H., 480
Daijavad, S., 554
Dailey, R.L., 291
Dandeno, P.L., 294
D'Angelo, H., 481
Daniel, J.W., 71
Daniel, R.W., 278
Daniels, A.R., 484
Dannan, F.M., 256
Daoyi, X., 263
Darkhovskii, B.S., 256
Darwish, M., 484
Dasgupta, S., 274
Datar, R.B., 545
Datko, R., 70
Datko, R.F., 257
Dauer, J.P., 263
Davis, G.W., Jr., 109
Davis, J.M., 513
Davis, P.J., 72
Davison, E.J., 272, 278, 292, 492
Dax, A., 515
De Abreu-Garcia, J.A., 276
Deards, S.R., 541
de Buda, R.G., 541, 543
DeCaro, S.M., 291
DeCarlo, R., 538, 554
Decauline, P., 70
Deekshatulu, B.L., 483
de Figueiredo, R.J.P., 283, 497, 507
de Gaston, R.R.E., 278
Degtyareva, N.F., 514
Deif, A., 15, 515
DeJulio, U., 550
Dekhtyarenko, P.I., 509
Delacour, J.D., 484
Delansky, J.F., 274
Delfour, M.C., 112
Deliyannis, T., 544, 545
del Nero Gomes, A.C., 507

Del Rio Castillo, R.R., 256
Delsarte, P., 489, 490
De Man, H., 554
DeMarco, C.L., 263, 267
De Maria, G., 270
Dembo, R.S., 515
Dement'yeva, V.V., 131, 502
Demmel, J.W., 256, 515
Dem'yanov, V.F., 72
Denery, D.G., 109
Deng, J.L., 268, 277
Deodhare, G., 507
de Oliveira, C., 292
de Oliveira, J.C.F., 257
Dergacheva, E.I., 142, 263, 283
DeRusso, P.M., 481
Dervisoglu, A., 520, 537
Desages, A., 272, 294
DeSantis, R.M., 258, 477, 484, 509, 510
Desoer, C.A., 68, 69, 257, 259, 261, 268, 272, 278, 292, 294, 477, 490
de Souza, C.E., 494, 496
Devanathan, R., 288
Dickman, A., 263
Diligenskii, S.N., 257, 500
Dillon, T.S., 484
Dillow, J.D., 108, 130
Di Mambro, P.H., 555
Diniz, P.S.R., 545, 550
Dion, J.M., 506, 507
Director, S.W., 530, 534, 553, 554, 555
Djaferis, T.E., 270
Doganovskii, S.A., 511
Dolezal, V., 257
Domanski, M., 550
Doraiswami, R., 278, 279
Dorato, P., 15, 303, 304, 310, 478, 479, 484
Dorf, R.C., 108, 291, 487
Dorrah, H.T., 508
Dougherty, H.J., 481
Downing, D.(R.), 294, 296
Downs, T., 539, 540

Doyle, J.(C.), 279, 301, 380, 385, 491, 492, 494, 497
Dreyfus, S.E., 70
Duckenfield, M.J., 516
Dugard, L., 506
Duffield, T.L., 289
Duinker, S., 553
Dunford, N., 67
Dunn, J.C., 479
Durante, C., 111
Duren, P.L., 67, 380, 490
Dutta Roy, S.C., 545, 550
Dzhamanbaev, A.A., 503

Eads, B., 547
Eccles, W.J., 487
Eckhaus, W., 257
Edwards, J.B., 294
Efremov, A.Y., 500
Efthymiatos, D., 509
Eirola, T., 482
Elangovan, S., 289
Elaydi, S., 256
El Ghaoui, L., 294
El-Hodiri, M., 484
El-Kady, M.A., 535, 554
Elkin, V.I., 501
Ellis, J.B., 541
El-Masry, E.I., 545
Elmetwally, M.M., 294, 485
Elsgoc, L.E., 70
El-Turky, F.M., 526, 540, 544
Eman, K.F., 515
Emami-Naeini, A., 295
Emig, P., 490
Engell, S., 16
Englehart, M.J., 491
Enns, D.F., 497
Epelman, M.S., 455, 502
Erickson, D.L., 509
Ermachenko, A.N., 485
Eslami, M., 15, 16, 69, 71, 108, 263, 264, 267, 268, 270, 274, 283, 477, 480, 481, 488, 550
Espinoza, J.L., 528, 549

Eswaran, C., 550, 551
Evans, R.J., 268, 291
Evers, A.H., 515
Evnin, A.Y., 503
Eykhoff, P., 489
Eyler, M.A., 509

Fabian, E., 264
Fabre, A., 545
Fagnani, F., 496
Fahmy, M.M., 258
Faibusovich, L.E., 257
Fairbairn, N.A., 491
Fairman, F.W., 482
Falb, P.(L.), 69, 70
Falk, J.E., 70
Fam, A.T., 457, 501
Fan, M.K.H., 272, 494
Fan, S.P., 556
Fang, C.H., 498
Fang, H.J., 507
Fantin, J., 484
Farison, J.B., 275, 276
Farsi, M., 275
Fassois, S.D., 515
Faxun, L., 270
Fedenia, M.M., 291
Feintuch, A., 509
Feldbaum, A.A., 70
Ferreira, J.M., 283
Ferreira, P.M.G., 279
Fettweis, A., 529, 544, 550
Fiacco, A.V., 15, 515, 516
Fidler, J.K., 110, 522, 535, 540, 542, 555
Filatov, I.V., 509
Filipkowski, A., 541
Filippov, A.F., 257, 504, 505
Finin, G.S., 500
Fischer, J., 14, 15
Fischl, R., 479, 514
Fisher, D.B., 482
Fisher, D.G., 296
Fishwick, W., 13
Flamm, D.S., 491
Flashner, H., 113, 287
Fleischer, P.E., 528, 545

Fleishman, B.A., 258
Fleming, J.A., 131
Fleming, P.J., 323, 481
Fleming, W.H., 70
Foias, C., 490, 491, 512, 514
Fokin, O.V., 502
Fomenko, O.N., 511
Fomin, S.V., 70
Fong, I.K., 271, 481
Foo, N.Y., 257
Foo, Y.K., 262, 264, 270, 272, 279, 290, 491, 498
Foord, T.R., 14
Ford, A., 110, 509
Foss, A.M., 294
Fragopoulos, D., 491
Franca, L.N.F., 257
Francis, B.A., 279, 289, 301, 380, 385, 490, 491, 492, 493, 494
Francis, N.D., 292
Frank, P.M., 15, 110, 478
Franklin, J.N., 67, 257
Franklin, S.N., 294
Frannin, D.R., 287
Frazho, A.E., 492
Freudenberg, J.S., 256, 279, 478, 512
Fronza, G., 130, 481
Fu, L.C., 505
Fu, M., 262, 264, 270, 275, 283, 494
Fu, M.Y., 270
Fu, S., 481
Fu, Y., 481
Fuji, N., 545
Fuji, T., 485
Fujita, M., 279, 495, 507
Fukao, T., 294
Fukata, S., 512
Fukui, A., 546
Fukui, Y., 547
Fukuma, N., 286
Fuller, A.T., 485, 488
Funahashi, Y., 497
Furuta, K., 506

Gadenz, R.N., 528, 540, 545, 552
Gagliardi, R.M., 358, 488
Gahinet, P.M., 516
Gaiduk, A.I., 241, 257
Gaiduk, A.R., 501
Galimidi, A.R., 264, 265, 271, 505
Gamkrelidze, R.(V.), 14, 70
Gandelman, J., 16
Gantmacher, F.R., 67, 281
Gao, W.B., 267
Gao, Z., 264
Garabedian, P.R., 67
Garashchenko, F.G., 264, 283
Gardiner, P., 509
Gardiner, P.C., 110
Gardner, W.A., 550
Garland, B., 112
Garloff, J., 269, 271
Garofalo, F., 264, 505
Garrard, W.L., 113, 483
Gass, S.I., 16, 518
Gavrilovic, M., 110, 485
Gayev, S.M., 543
Geddes, E.J.M., 493
Geffe, P.R., 545
Geher, K., 538
Geiger, R.L., 528, 545, 548
Gelb, A., 70
Gelfand, I.M., 70
Gelig, A.K., 257, 283
Gembicki, F.W., 516
Genesio, R., 257, 264, 283
Genin, Y., 489, 490
Georganas, N.D., 358, 488
Georgiou, T.T., 288, 498
Gerards, A.M.H., 515
Gerasimov, A.N., 111
Gerber, P.D., 548
Gerencser, H., 516
Geromel, J.C., 264, 283, 360, 488, 504
Gevers, M., 551
Ghaemmaghami, P., 292
Ghausi, M.S., 15, 528, 529, 530, 538, 544,

Author Index

546, 547, 548, 549, 552, 553
Ghodekar, J.G., 295
Ghose, D., 499
Gibson, J.A., 292, 482
Gibson, J.S., 263, 266, 509
Giguere, J.C., 542
Gil', M.I., 284
Gilbert, E.G., 68, 290
Gil'bo, E.P., 271
Gille, J.C., 70
Gillingham, P., 545
Girling, F.E.J., 546
Glazman, I.M., 67
Glielmo, L., 264
Glover, K., 288, 301, 380, 385, 491, 492, 493, 494, 495, 496, 497
Godbout, L.F., 110
Goddard, P.J., 522, 540, 552
Goethals, J.M., 516
Goette, J., 546
Goh, B.S., 70
Goknar, I.C., 68
Gold, B., 538
Goldowskaja, N., 13
Gole, A.M., 539
Golembeski, J.J., 530, 546, 552
Gong, L., 259
Gong, W.B., 509
Gonin, R., 515
Gonuleren, A.N., 552
Gonzales, R.L., 477, 478
Gonzalez, O.R., 477
Good, E.F., 546
Goodall, D.P., 505
Goodwin, G.C., 489
Gopal, M., 295, 482
Gordon, S.P., 257
Gori-Giogi, C., 108, 501
Gorodetskii, V.I., 16
Gorodetskiy, V.I., 485
Gostev, V.I., 275
Goujon, M., 480
Gould, L.A., 507, 508
Gourishankar, V., 290, 296

Govil, N.K., 275
Govindarajulu Naidu, M.G., 543
Gradshtein, I.S., 14
Graf, P.L., 485
Grainger, G., 546
Grasselli, O.M., 108, 501, 507
Graupe, D., 70, 479, 480
Gray, J.O., 287
Green, M., 491, 494
Green, W.L., 284
Greenlee, T.L., 294
Greenlee, W.M., 294
Greenwood, D.T., 294
Grenander, U., 72
Greschak, J.P., 263
Grewal, H.K., 554
Griffin, A.W.J., 485
Griffin, R.E., 485
Grigor'ev, V.V., 294
Grimble, M.J., 16, 294, 491, 492
Grimm, W.M., 275
Groom, N.J., 265
Grubel, G., 477, 478, 482
Grudev, K.L., 542
Grujic, L.T., 284
Gu, D.W., 289, 290, 492, 495, 498
Gu, G., 281, 493, 505
Gu, K., 264
Guardabassi, G., 16, 109, 115, 130, 485
Gudorzi, R., 500
Guenther, R.B., 68
Guerin, J.P., 289
Guesalaga, A.R., 499
Guggenbuhl, W., 526, 544, 546
Guilandoust, M., 275
Guinzy, N.J., 485
Guiver, J.P., 271
Gumowski, I., 512
Guo, L., 498
Gupta, M.S., 546
Gupta, N.K., 82, 109, 119, 121, 122, 130, 293
Gupta, S.C., 529, 551
Gupta, S.K., 295, 510

Gurdal, Z., 112
Guruprasada Rau, V., 294
Gutman, S., 271, 283, 460, 466, 505
Guzinski, A., 528, 544

Ha, I.J., 505
Haack, G.R., 543
Hached, M., 284, 285
Haddad, A.H., 485
Haddad, R.A., 279
Haddad, W.M., 479, 494, 495
Hafez, W., 109
Haftka, R.T., 112, 293
Hague, B, 4, 14
Hahn, W., 257
Haimes, Y.Y., 516
Hakim, S.S., 527, 546
Hakimi, S.L., 540, 543
Hakkapakki, S.L., 291
Halanay, A., 499
Hale, J.K., 257
Haley, S.B., 522, 538, 540
Halikias, G.D., 495
Hall, A.C., 14
Halpern, M.E., 295
Ham, J.M., 501
Hamada, N., 271
Hamalainen, R.P., 482
Hamilton, T.A., 546
Hamilton-Jenkins, M.A., 292
Hammond, P.H., 516
Hamza, M.H., 477, 481
Han, K.W., 294
Hanafi, M., 545
Hanafy, A.(A.)R., 258, 482
Hanau, P.Th.L.M., 268
Haneda, H., 259
Hanneman, H.W., 542
Hanson, D.A., 509
Hanzon, B., 496
Happ, W.W., 111, 555
Hara, S., 277, 492, 499, 512
Harada, K., 110, 294
Haraldsdottir, A., 294

Hardy, G.H., 67
Harrison, G.W., 264
Harrison, R.F., 289
Harvey, C.A., 16, 66
Hasegawa, K., 290, 502
Hashemi, G.H., 541
Hashimoto, M., 546
Hashish, M., 109
Hashish, M.A., 110
Hassan, M., 485
Hassan, M.A., 501
Hassan, M.F., 258, 262
Hasse, A.B., 527, 545
Hattis, P.D., 280
Haug, E.J., 16, 112, 113
Haug, K., 546
Hawkins, T., 67
Hayashi, S., 271
Haykin, S.S., 538, 539
Hayland, D.C., 264, 275
Haynes, D.A., 131
He, L., 498
Hedrick, J.K., 113
Heinen, J.A., 264, 274, 275, 293
Heise, S.A., 496
Hejmo, W., 511
Heller, J.(E.), 479, 486
Helton, J.W., 489, 491, 492
Hendricks, T.C., 481
Henry, A.F., 507
Heppenheimer, T.A., 131
Herbert, R.J., 509
Hermes, H., 70, 258, 260
Herrera-Vaillard, A., 294
Hertz, D., 258
Hewer, G.(A.), 264, 265, 266, 516, 517
Higginbotham, E., 482
Higgins, T.J., 4, 14, 435, 557, 575
Hilberman, D., 530, 552
Hilbert, D., 67
Hill, D.J., 279
Hill, R.D., 295
Hiltgen, A., 546
Himmelblau, D.M., 489
Hinrichsen, D., 264, 265
Hiramatsu, Y., 546

Hirano, K., 527, 546, 551
Hirsch, M.W., 283
Hmamed, A., 262, 284, 511
Ho, C.S., 256
Ho, C.W., 526, 544
Ho, T.C., 275
Ho, Y.C., 69, 508, 509
Hochstadt, H., 67
Hoft, R.G., 286
Hokenek, E., 546
Hollerbach, K., 499
Hollot, C.V., 263, 265, 267, 269, 270, 271, 272, 274, 275, 284, 289, 495
Holmes, W.H., 111
Holt, A.G.J., 542, 546
Holtzman, J.M., 130, 131, 258, 509
Hong, Z.C., 265, 290
Hontoir, Y., 477, 482
Hopp, T.H., 505
Hopp, W.J., 516
Horiguchi, K., 271
Horing, S., 13, 131
Horisberger, H.P., 265
Horn, P., 548
Horng, I.R., 263
Horowitz, I.(M.), 5, 15, 279, 486, 514, 527, 546
Horton, W.F., 477
Hosenthien, H.H., 257
Hosoe, S., 110
Hou, J.W., 113
Householder, A.S., 67
Howard, D.R., 311, 485
Howard, R., 70
Howitt, N., 524, 543
Howze, J.W., 256, 265, 295, 507
Hsiao, C.H., 256
Hsiao, F.B., 266
Hsu, C.F., 268, 271, 275, 481, 505
Hsu, J.C., 70
Hsu, L., 284
Hsu, P.T., 69
Hu, H.Z., 278, 293
Huang, C.G., 277
Huang, P.C.K., 268

Huang, Q., 279
Huelsman, L.P., 538
Huey, D.C., 549
Hughes, P.C., 295
Hui, S., 284
Humphrey, C.H., 487
Hung, Y.S., 265, 279, 280, 494, 495
Hurwitz, A., 14
Huseyin, O., 265, 543
Hvostov, H.S., 494
Hwang, H., 278
Hwang, S.Y., 111
Hyman, C.J., 72

Ichikawa, K., 284
Iftar, A., 279
Iglesias, P.A., 494
Ikeda, M., 289, 296
Il'yasov, B.G., 269
Imanishi, H., 259
Imanudin, 131
Immirzi, F., 13
Infante, E.F., 255, 257, 259
Ingoldby, R.N., 131
Inman, D.J., 291
Inoue, K., 131, 512
Inoue, T., 546
Ionescu, V., 499
Irwin, C.L., 485
Ishihara, T., 275, 512
Ishitani, H., 485
Isidori, A., 262, 507
Ito, H., 492
Ito, M., 110
Ivakhnenko, A.G., 501
Ivanenko, V.I., 501
Ivanov, V.I., 538
Iwens, R.P., 278

Jabbari, F., 265, 505
Jabr, H.A., 490
Jackson, L.B., 529, 550
Jackson, R.H.F., 516
Jackson, R.L., 278
Jacobs, O.L.R., 477
Jain, N.K., 536, 546, 556
Jain, R., 110

Jamshidi, M., 258, 263, 274
Janin, R., 516
Jaworska, I., 110
Jayasuriya, S., 272
Jenq, Y., 551
Jessel, G.P., 553
Jiang, C.L., 265, 267, 275
Joannic, Y., 282
Johnson, C.D., 501
Johnson, C.R., Jr., 255
Johnson, T.L., 71
Jonckheere, E.(A.), 281, 497
Jordan, D., 110
Joseph, P.D., 69
Joshi, S.M., 265, 284
Jovanovic, V., 539
Juang, J.C., 497
Juang, J.N., 292
Juang, J.Y., 269
Juang, Y.T., 265, 267, 268, 271, 275, 290, 295
Judd, L.F., 287
Julich, P.M., 486
Junkins, J.L., 292
Juricic, D., 510
Jury, E.I., 258, 268, 269, 271, 272, 273, 274, 275, 276, 360, 512

Kabamba, P., 497
Kabamba, P.T., 131, 294
Kachanova, N.A., 500
Kadimov, Y.B., 485, 509
Kaelin, A., 546
Kaesbauer, D., 278, 293
Kahne, S.J., 69, 482
Kailath, T., 69
Kaitala, V., 509
Kalaba, R.E., 14, 109, 358, 488
Kalita, A.S., 293
Kalman, R.E., 70, 71, 132, 258, 281
Kalouptsidis, N., 258
Kamen, E.W., 275, 284
Kamenetskii, V.A., 284
Kamenetskiy, V.A., 265

Kamiya, Y., 295, 482
Kamp, Y., 489, 490
Kan, Y.S., 284, 501
Kanatani, K.I., 509
Kane, J.H., 113
Kanellakis, A., 275
Kang, H.I., 269
Kang, H.S., 479, 486
Kannai, Y., 514
Kantor, J.C., 16, 279, 499
Kapania, R.K., 112
Kaplan, D., 131
Kaplan, W., 68, 70
Karas, V.M., 516
Karkas, D.A., 108
Karl, W.C., 263, 265, 275
Kasenally, E.M., 289, 491
Kasper, R., 525, 543
Kas'yanov, V.A., 516
Kaszkurewicz, E., 284
Katbab, A., 271, 275
Katkovnik, B.Y., 486
Kautsky, J., 290
Kawatani, R., 495
Kazerooni, H., 499
Keel, L.H., 256, 265
Kekkeris, G.T., 110
Kelemen, M., 514
Kelley, H.J., 131
Kelley, J.L., 67
Kelly, G.L., 540
Kelly, J.J., 547
Kelly, J.M., 486
Kenney, C.(S.), 264, 265, 498, 516
Kerber, O.B., 285, 511
Kesavan, H.K., 538, 553
Kestenbaum, A., 484
Khabarov, V.S., 255
Khaikin, S.E., 13
  (cf., Chaikin, C.E.)
Khalifa, I.H., 482
Khalil, A., 544
Khammash, M.H., 507
Khargonekar, P.P., 268, 269, 275, 279, 284, 285, 288, 301, 380, 385, 491, 492, 493, 494, 495, 496, 498, 506, 508, 509

Author Index

Kharin, Y.S., 501, 509
Kharitonov, V.L., 164, 166, 167, 168, 169, 203, 271, 275
Kheir, N.A., 111
Khorasani, K., 285, 505
Khrustalev, M.M., 131, 501 502
Kibzun, A.I., 284, 501
Kida, T., 523, 526, 541, 542, 544
Kiendl, H., 265
Kihara, M., 296
Kim, H.K., 539, 546
Kim, J.H., 264
Kim, K.D., 269, 271
Kim, S.W., 516
Kimura, H., 280, 281, 290, 495
Kinderlehrer, D., 285
King, R.E., 112
Kirchgraber, U., 258
Kirichenko, N.F., 505
Kirk, D.E., 70
Kishi, G., 523, 541
Klafin, J.F, Jr., 110
Klein, G., 292
Kleinman, D.L., 71, 482, 486
Kline, U., 545
Knob, A., 546
Knyazev, A.V., 258, 478
Kobayashi, T., 514
Koda, M., 512
Kodama, S., 492
Koditschek, D.E., 258
Kogan, J., 271
Kohn, R.V., 16, 113
Koivo, A.J., 480, 512
Kokame, H., 265, 266, 271, 272, 276
Kokotovic, P.V., 5, 15, 16, 108, 255, 290, 479, 485, 486
Kolla, S.R., 275, 276
Kolobov, V.G., 516
Komaroff, N., 516
Komkov, V., 16, 113, 485, 486, 514
Kondo, R., 492

Kondrat'eva, N.V., 295
Konstantinov, M.M., 516
Koo, R.L., 538
Koosis, P., 490
Korchanov, V.M., 296
Korevaar, J., 68
Korn, G.A., 538
Korneyev, V.A., 295
Korovin, S.K., 503
Korshunov, A.I., 285
Kosmidou, O.I., 265
Kosut, R.L., 255, 295
Kosyuk, Y.V., 483
Kotov, E.O., 111
Kotta, Y.R., 501
Kouikoglou, V.S., 516
Koumboulis, F.N., 507
Kouvaritakis, B., 16, 280, 289, 484
Kozorez, V.V., 295
Kozyakin, V.S., 285
Kraiman, A., 70
Krasil'nikov, I.B., 285
Krasnosel'skii, M.A., 258, 285
Kraus, F.J., 269, 271, 276
Krause, J.M., 491
Krein, M.G., 489
Kreindler, E., 69, 131, 307, 323, 478, 479, 482, 511
Kreisselmeier, G., 477, 478, 482
Krishman, R., 540
Krishnan, V., 110
Krishna Rao, P., 482
Krivonozhko, V.Y., 516
Krogh, B., 478
Kropholler, H.W., 499
Krutin, V.I., 258
Krut'ko, P.D., 295, 502
Krutova, I.N., 502
Ku, W.H., 529, 550
Kubiak, E.T., 280
Kudva, P., 290
Kuh, E.S., 106, 108, 520, 525, 537, 538, 543
Kuhn, J., 518
Kukareko, E.P., 285

Kukhtenko, A.I., 435, 436, 441, 450, 500, 501
Kukhtenkov, L.P., 512
Kukuliyev, R.M., 501
Kulebakin, V.S., 436, 500
Kuliyev, E.Y., 485, 509
Kumar, K.S.P., 479
Kundu, P., 548
Kundur, P., 294
Kung, F.C., 506
Kung, S.Y., 490
Kunieda, T., 546
Kuntsevich, A.V., 295
Kuntsevich, V.M., 265, 295, 500
Kuo, B.C., 70, 512
Kuo, K.C., 481
Kuo, M.C.Y., 514
Kuo, T.S., 265, 268, 271, 275, 277, 481
Kurcyusz, S., 518
Kurogochi, K., 542
Kurtaran, B., 480
Kushner, H.J., 510
Kuwahara, M., 286
Kuznetsov, V.P., 285
Kwakernaak, H, 15, 70, 280, 492, 557
Kwon, W.E., 258
Kwon, W.H., 516
Kyparisis, J., 516

Ladikov, Y.P., 295
Lailly, P., 513
Lainiotis, D.G., 358, 481, 489, 509
Laker, K.R., 15, 527, 528, 529, 530, 546, 547, 549, 552, 553
Laksberg, E.A., 547
Lakshmikantham, V., 276, 289
Lal, M., 542
Lam, J., 498
Lancaster, P., 271
Landers, P.H., 486
Lankford, J.G., 258
Lanning, D.D., 507
Lapidus, L., 483
Larin, V.B., 289, 295

LaSalle, J.P., 14, 70, 258
Lasiecka, I., 514
Latchman, H., 280
Latchman, H.A., 289
Latimer, J.R., 109
Lau, C.G., 525, 543
Laub, A.J., 498, 516
Laurent, F., 110
Law, A.M., 70
Lawson, S.S., 550, 551
Lazer, A., 260
Le, D.K., 492
Le, V.X., 493, 495
Leal, M.A., 266
Lebedev, D.V., 486
Lecrique, M., 480
Ledbetter, J.D., 529, 551
Lee, A.Y., 111
Lee, C.K., 486
Lee, E.B., 69, 70, 71, 271, 497, 552
Lee, G.K.F., 291
Lee, H.L., 545
Lee, I., 481
Lee, M.R., 542, 546
Lee, S.C., 539, 547
Lee, Y.B., 484
Lee, Y.S., 290
Leeds, J.V.(Jr.), 481, 522, 524, 543, 547
Lefebvre, S., 477
Lefschetz, S., 13, 258
Legge, C.G., 485
Lehnigh, S.H., 257
Lehtomaki, N.A., 505
Leitman, M.B., 501
Leitmann, G., 263, 283, 296, 472, 486, 504, 505, 506, 509
Lemke, C.E., 15, 515
Lenz, K.E., 492
Leon, B.J., 543
Leondes, C.T., 259, 260, 477, 481, 482, 509
Leonov, G.A., 285, 295
Lepschy, A., 295, 551
Lessman, F., 72
Leuder, E., 527, 547
Leugering, G., 285
Leung, G.M.H., 492

Leung, K.H., 540, 555
Levadi, V.S., 358, 488
Levan, N., 258
Levine, W.S., 70, 71, 480
Levinson, N., 67, 256
Levy, B.C., 505
Levy, H., 72
Lewis, C.P., 540
Lewis, F.L., 285
Lewis, J.B., 70
Lewis, R.M., 556
Lewkowicz, I., 266, 285
Li, G., 551
Li, J.H., 276
Li, J.Y., 507
Li, T., 295
Li, T.H.S., 276
Li, Y., 271
Lian, W., 266
Liao, T.L., 505
Liao, X.X., 266
Liapunov, A.M., 258
Liberatore, A., 551, 552, 553
Liberty, S.R., 69, 259
Liberzon, M.R., 285
Lim, J.W., 113, 295
Lim, K.B., 292
Lim, K.Y., 488
Lim, T.C., 277
Lim, Y.H., 277
Limebeer, D.J.N., 265, 279, 280, 289, 385, 417, 491, 494, 495
Lin, C.E., 493, 497
Lin, C.L., 266
Lin, D.W., 490
Lin, H., 269, 271
Lin, P.M., 535, 554, 555
Lin, S.H., 271
Lin, T.P., 288
Lin, W.G., 293
Lin, Y.H., 268
Lin, Y.P., 288, 289
Lindorff, D.P., 512
Linnemann, A., 505
Linssen, H.N., 542
Linton, F.E.J., 67
Lipatov, A.V., 258, 285
Lisitsyn, A.M., 510

Littlewood, J.E., 67
Liu, B., 487, 530, 553
Liu, C.K., 110
Liu, J., 296
Liu, K., 285
Liu, K.Z., 495
Liu, P.C., 543
Liu, R., 257, 279
Liu, R.W., 490
Liu, X.Y., 276
Liu, Y., 498
Lo, C.H., 283
Lo, P.H., 551
Locatelli, A., 16, 109, 130, 481, 485, 512
Loewen, P.D., 515
Loh, N.K., 264, 296, 512
Longhi, S., 507
Longman, R.W., 295
Looze, D.P., 256, 271, 279, 478, 511, 512
Loparo, K., 109
Lopez-Toledo, A.A., 358, 488
Lorentz, G.G., 73
Louvish, D., 72
Lowinger, J.F., 482
Lu, H.C., 278
Lu, N.C., 548
Lu, Y., 495
Luenberger, D.G., 69, 70, 85
Luh, J.Y.S., 482
Lukes, D.L., 71
Lukes, T., 255
Lukich, M.S., 295
Lunze, J., 16, 280
Luo, Z.H., 265
Lur'e, A.I., 259
Luse, D.W., 492
Luther, W.J., 289
Lutz, P., 547
Lutz, W.J., 543
Luus, R., 131
Luzin, N.N., 435, 436, 442, 501
Ly, U.L., 497
L'yanova, N.I., 283
Lychak, M.M., 259
Lyubushin, A.A., 70

Ma, C.C.H., 285, 289
Mablekos, V.E., 258
MacFarlane, A.G.J., 478, 520, 537
Mackay, R., 528, 547
MacLane, S., 67
Madatov, G.L., 111
Madiwale, A.N., 495
Maeda, H., 492
Maffezzoni, C., 485
Mahalanabis, A.K., 479
Mahmoud, M.S., 276
Makowski, K., 70
Malanowski, K., 514, 516, 517
Malek, N.G., 486
Maloberti, F., 546
Malozemov, V.N., 72
Mal'tsev, V.V., 517
Manaktala, V.K., 540
Manes, E., 68
Manetti, S., 551
Manivannan, K., 550
Manley, J.M., 537
Mansour, M., 265, 266, 268, 271, 272, 273, 275, 276, 493
Mantey, P.E., 292, 295, 529, 551
Marchaud, A., 462, 505
Marden, M., 272
Mareels, I.M.Y., 255
Marino, P.J., 478
Marino, R., 259
Markus, L., 70, 71
Marleau, R.S., 16, 481
Maroulas, J., 272
Marro, G., 500, 504
Marshal, C., 71
Marshall, J.E., 112, 486
Martin, D.H., 510
Martin, F., 545
Martin, J.M., 266, 517
Martinelli, G., 539, 543, 550, 551
Maslarov, I.A., 539
Maslov, E.P., 480
Mason, S.J., 111
Massera, J.L., 259
Massey, J.L., 70

Massey, W.S., 67
Mastascusa, E.J., 110
Masuda, T., 517
Masutomi, F., 537, 539
Mata, F.A., 288
Mathew, M.I., 540
Matsumura, F., 507
Matthes, H., 540
Mavor, J., 549
Maxwell, J.C., 14
Maybeck, P.S., 108, 130
Mayer, R.W., 5, 14
Mayne, D.Q., 498,
Mayorov, V.A., 285, 511
Mazer, W.M., 108
Mazko, A.G., 266, 292
Mazzoni, G., 552, 553
McClamroch, N.H., 112, 131, 282, 480, 500, 502, 510
McCormick, G.P., 516
McDaniel, W.L., 290
McFarlane, D., 496, 497
McMorran, P.D., 110
McQuade, T.E., 497
McRuer, D.T., 291
McTavish, D.J., 295
Mechetnyy, V.S.,
Medanic, J., 480
Medanic, J.V., 498
Medani-Esfahani, S.M., 284, 285
Meditch, J.S., 69, 457, 501
Mee, D.H., 112
Mehdi, Z., 487
Mehra, R.K., 82, 109, 119, 121, 122, 130, 358, 488, 489
Mei, C., 113
Meilakhs, A.M., 266, 285
Meisel, W.S., 486
Meissinger, H.F., 108
Melikyan, A.A., 295
Mel'nik, V.S., 517
Mel'nikov, B.G., 502
Melsa, J.L., 487
Melvin, W.W., 131
Melzer, S.M., 512
Mendel, J.M., 69

Author Index

Menzies, R.W., 539
Mercier, O., 282
Meressi, T., 295
Merriam, C.W., III, 141, 266
Mertzios, B.G., 259
Mesarovic, M.D., 108, 114, 115, 477
Messac, A.,
Messerli, E.J., 72
Meyer, K.R., 256, 257
Meyer, L.H., 295, 510
Mian, G.A., 295, 551
Michael, G.J., 141, 266
Michaletzky, G., 497
Michel, A.N., 70, 259, 285, 502
Middelton, R.H., 488
Midy, M., 480
Miele, A., 131
Mijat, N., 547
Mikhailov, L.N., 500
Miklaszewski, R., 16
Mikulcik, E.C., 113
Milanese, M., 268, 296
Miller, K.S., 5, 15, 74, 108, 510, 522, 541
Miller, R.K., 259, 285
Mills-Curran, W.C., 291
Min, B.J., 286
Mingori, D.L., 295
Minin, V.V., 483
Minnichelli, R.J., 268, 272
Minorsky, N., 14, 259
Miroshnik, I.V., 259
Mishchenko, E.F., 70
Mishulina, O.A., 110
Misic, V.B., 547
Missaghie, M.M., 482
Mita, T., 290, 495, 502
Mitchell, J.R., 290
Mitra, A.K., 528, 547
Mitra, S., 548
Mitra, S.K., 528, 538, 547, 548, 549, 551, 552
Mitter, S.K., 69, 491, 492
Miura, M., 547
Miyagi, H., 259, 286
Mizushima, N., 485

Modiano, E., 488, 508
Moe, M.L., 481
Mohammad, Q.G., 274
Mohan, P.V.A., 547
Molchanov, A.P., 286
Momot, M.E., 287
Monaco, V.A., 539
Montemayer, J.J., 260
Moody, D.E., 555
Moore, B.C., 292
Moore, C., 547
Moore, J.B., 69, 280, 495, 508, 513
Moore, J.C., 259
Moore, R.E., 271
Morari, M., , 16, 280, 499
Morgan, B.S, Jr., 136, 139, 290, 292
Mori, T., 265, 266, 271, 272, 274, 276, 286
Morozov, M.V., 259
Morse, A.S., 68, 69, 256
Morshedi, A.M., 281
Morsztyn, K., 484
Moschytz, G.S., 527, 545, 546, 547, 548
Mossaheb, S., 259
Moustakas, E., 548
Movchan, L.T., 276
Mozgalevskii, A.V., 109
Mrabti, M., 517
Muench, R., 293
Mukhamadiev, E., 255
Mukhopadhyay, V., 280
Muller, P.C., 260
Mulligan, J.H., Jr., 546
Munkres, J.R., 67
Munro, N., 270
Murdoch, J.B., 538
Murotsu, Y., 295
Murray, F.J., 5, 15, 74, 108, 510, 522, 541
Murray, J., 490
Murray, J.J., 113
Murray-Smith, D.J., 109
Murti, V.G.K., 551
Mustafa, D., 494, 495
Mutter, V.M., 286
Myachin, S.I., 485, 509

Myszkorowski, P., 506

Naccache, P.H., 517
Nagaraj, K., 548
Nagpal, K.M., 271, 494, 495
Nagurney, A., 515
Nahi, N.E., 358, 489
Nahvi, M., 263
Nail, J.B., 290
Nakazawa, K., 541
Napjus, G., 358, 489
Nardizzi, L.(R.), 511, 514
Narendra, K.S., 241, 258, 259
Natarajan, S., 548
Nayfeh, A.H., 73
Nehari, Z., 490
Neill, T.B.M., 536, 555
Neirynck, J., 524, 525, 542, 543
Nesbit, R.A., 259
Nesline, F.W., Jr., 296
Nestell, M., 67
Neumaier, A., 517
Neuman, C.P., 81, 109, 113, 511, 513
Neumann, T., 536, 555
Neustadt, L.W., 70
Nevanlinna, R., 490
Newcomb, R.W., 510, 524, 538, 543, 548
Newell, R.B., 482
Newmann, M.M., 323, 481, 482
Ng, S.M., 529, 550
Ngo, N.T., 6, 16
Nichols, M.A., 113
Nichols, N.K., 290
Nichols, S.T., 508
Nicholson, H., 292
Nieuwenhuis, J.W., 259
Nightingale, C., 110
Nishimura, S., 551
Niu, X., 276, 277
Noland, J.H., 487
Noldus, E., 507
Nordstrom, K., 486
Norton, F.E., 509

Nwokah, O.D.I., 280
Nyquist, H., 14

Ober, R.J., 492
Ocali, O., 276
Ogata, K., 259
Ogino, K., 489, 512
Ohba, F., 295
Ohm, D., 295
Ohmori, H., 492
Ohta, Y., 492,
Ohuchi, S., 495
Okada, K., 482
Okada, T., 296
Olbrot, A.W., 275, 283, 286, 513
Olson, D.E., 486
O'Neill, L.A., 539
Opoitsev, V.I., 286
Orbach, A., 479, 514
Orchard, H.J., 540, 542
O'Reilly, J., 296, 482
Orlandi, G., 550, 551
Orlov, V.A., 502
Orlov, V.N., 259
Ortega, R., 284
Osumi, N., 526, 544
Owens, D.H., 16, 266, 280, 286, 287
Owens, T.J., 292, 296
O'Young, S.D., 289, 290, 492
Ozaki, H., 543
Ozbay, H., 492, 497, 507
Ozguner, U., 279

Packard, A., 498
Paden, B., 295
Padilla, C.S., 486
Padilla, R.A., 486
Padmanabhan, P., 272
Paduano, J.(D.), 294, 296
Pagurek, B., 310, 479
Paguta, M.T., 295
Pakerabo, M.M., 551
Pakshin, P.V., 286
Pal, M.K., 484
Pal, R.N., 551

Paley, R.E.A.C., 58, 67
Palmor, Z., 283
Palomera-Garcia, R., 549
Pan, J., 509
Pan, T.S., 113, 511
Pandey, P., 498
Pandolfi, L., 259, 286, 513
Panier, E.R., 272
Pantaliyenko, L.A., 283
Papazov, S.P., 539
Papoulis, A., 290, 520, 539
Paraskevopoulos, P.N., 108, 110, 139, 292, 293, 486, 487, 507, 513, 514
Park, K.C., 517
Park, P.G., 516
Parker, S.R., 521, 522, 535, 536, 540, 541, 553, 556
Parks, P.C., 255, 258
Parlos, A.G., 507
Partington, J.R., 492
Pascoal, A.M., 284, 288, 493
Patel, R.V., 266
Patranabis, D., 548
Paul, R.J., 485
Pavlidis, T., 275
Pavlov, V.V., 285, 511, 518
Payne, H.J., 486
Pearson, A.E., 70, 258
Pearson, C.E., 67
Pearson, J.B. (Jr.,), 71, 289, 475, 490, 507, 512
Pedersen, K.(C.), 511, 514
Pelegrin, M.J., 70
Pelova, G.B., 516
Penchuk, A.N., 280
Penfield, P., Jr., 530, 553
Peng, T.K.C., 259, 263, 502
Peres, P.L.D., 504
Perev, K., 276
Perez, A., 507
Perkins, J.E., 551

Perkins, W.R., 25, 69, 79, 80, 81, 109, 112, 114, 115, 130, 303, 304, 455, 456, 457, 477, 478, 500, 509, 510, 511, 513
Perlis, H.J., 478, 514
Perry, T.P., 492
Perry, W.L., 68
Persek, S.C., 2260
Peskin, E., 539, 540
Petersen, I.R., 266, 267, 286, 289, 291, 494, 495, 506
Peterson, D.W., 517
Petkovski, D.(B.), 267, 284
Petraglia, A., 548
Petrela, D.A., 548
Petrela, D.M., 545
Petrov, A.I., 483
Petrov, B.N., 131, 269, 436, 443, 502
Petrovic, R., 110, 485
Petrovsky, A.M., 480
Pezet, P., 482
Phan, C.P., 539
Phan, C.S., 546
Phillis, Y.A., 516
Phoojaruenchanachai, S., 506
Phrydas, C., 555
Pichkurenko, V.P., 502
Piechottka, U., 275
Piedade, M., 542
Piekarski, M.S., 550
Pierre, C., 290
Pierre, D.A., 544
Pinel, J.F., 555
Pipes, L.A., 108
Platzman, L., 315, 317, 480
Pliss, V.A., 258
Pohjolainen, S., 514
Pokrovskii, A.V., 258, 285
Pokrud, B., 494
Polak, E., 108, 281, 289, 486, 498
Polis, M.P., 275, 283

Poliscuk, W.W., 530, 553
Polivanov, V.I., 275
Polukhin, A.V., 260
Polya, G., 67
Polyak, B.T., 272, 273, 277
Pomet, J.B., 286
Ponomarev, V.M., 485
Pontryagin, L.(S.), 5, 13, 14, 70
Poolla, K.(R.), 285, 494
Pope, R.E., 16
Popov, V.M., 260,
Porter, B., 69, 139, 291, 292, 294, 502
Porter, W.A., 16, 309, 477, 478, 479, 508, 509, 510, 511, 514
Postlethwaite, I., 279, 280, 289, 290, 491, 492, 493, 495, 498, 499, 505
Pradeep, S., 260,
Praly, L., 255, 286
Prasad, K.S., 551
Prasad, U.R., 538
Prasad, V.C., 556
Pratapachandran (Nair), P., 295, 482
Preminger, J., 502
Premoli, A., 527, 548
Price, C.J., 551
Pritchard, A.J., 264, 265
Propoy, A.I., 516
Prudovskiy, B.D., 517
Pujara, L.R., 272, 296
Pyatnitskii, E.S., 286
Pyatnitskiy, Y.S., 265, 503,
Pyatnytsky, E.S., 241, 260, 286

Qian, J.L., 266
Qian, R.X., 267
Qi-Bin, W., 271
Qing-Long, H., 271
Qiu, L., 272, 292, 492
Qiuting, H., 548
Qu, Z., 506

Rabaey, J., 554
Rabiner, L.R., 538
Rabrenovic, D., 539
Rachid, A., 275, 276
Radanovic, L., 16, 510
Rader, C.M., 538
Radzyner, R., 111
Ragazzini, J.R., 14
Raines, F.Q., 295, 510
Ramachandran, V., 547, 549
Raman, S., 493
Ramar, K., 290, 296
Ramesh, B.K., 552
Ran, A.C.M., 496
Rantzer, A., 272
Rao, K., 528, 548
Rao, K.R., 549
Rao, M.R.M., 287
Rao, N.D., 294
Rao, S.G., 483
Rao, S.S., 113, 511
Rapoport, L.B., 287
Rasmussen, M.L., 73
Rasmy, E., 109
Rasmy, M.E.M., 110
Ravi, R., 284, 493, 495
Raya, A., 286
Rebarber, R., 260
Reddy, D.C., 292, 296, 486
Reddy, M.A., 528, 548
Redheffer, R.(M.), 54, 490
Reed, M., 67
Regalia, P.A., 548
Reid, J.G., 91, 93, 108, 117, 130
Reid, R.E., 487
Reid, W.T., 260, 490
Rekasius, Z.V., 276, 311, 485
Rektorys, K., 67
Remec, M.J., 290, 539
Renner, K., 529, 551
Reszka, G.S., 553
Reyer, S.E., 293
Reyhanoglu, M., 282
Reza, F., 136, 142, 260, 263, 267, 272, 290, 498

Author Index

Rezai-Fakhr, M.G., 530, 540, 552, 553
Reztsov, V.P., 260
Rhee, I., 508
Richards, G.A., 555
Ridgely, D.B., 268, 281
Ridler, P.F., 556
Riedle, B.D., 255
Riesz, F., 67
Rillings, J.H., 483
Rinaldi, S., 16, 109, 130, 512
Rishel, R.W., 70
Rissanen, J., 486
Rissanen, J.J., 480
Rivera, D.E., 499
Robel, G., 498
Robinson, F.N.H., 539
Roe, P.H., 555
Roesler, M.D., 287
Rohn, J., 293
Rohrer, R.A., 106, 314, 480, 484, 520, 530, 531, 534, 537, 538, 553, 555
Rojo, B.L., 530, 553
Romagnoli, J., 272, 294
Romanowicz, T.M., 110, 111
Romeo, F., 130
Ronge, P., 486
Rootenberg, J., 310, 478, 486, 502, 510
Rosen, O., 131
Rosenblum, A.L., 529, 530, 548, 553
Rosenbrock, H.H., 69, 293
Roska, T., 542
Rota, G.C., 256
Rotea, M.A., 494, 495, 506
Rotstein, H., 272, 294
Roubellat, F., 111
Rousselet, B., 112, 113
Rowe, H.E., 537
Roxin, E., 260, 462, 506
Roy, R.J., 483
Roy, S.B., 548
Roytman, L.M., 291
Rozenvasser, E.N., 16, 111, 513

Rozhanskii, V.L., 480
Rozonoer, L.I., 8, 72, 450, 452, 453, 455, 456, 502
Ruban, A.I., 512, 514
Rudin, W., 67
Rupnik, V., 517
Russell, D.L., 71, 165, 179, 260, 264, 267
Rutkovski, V.Y., 502
Rutman, R.S., 5, 16, 108, 111, 360, 455, 502,
Ryabov, B.A., 436, 502
Ryabov, B.O., 139, 272, 290
Ryan, E.P., 296, 505, 513
Rybashov, M.V., 487

Sabaev, E.F., 287,
Sabaeva, T.A., 261, 287
Saberi, A., 477, 497
Sablatash, M., 542
Sabouni, R., 520, 537
Sachkov, G.P., 139, 272, 290, 502
Sacker, R.J., 256
Sadykov, F.R., 285
Saeki, M., 280, 281, 493, 513
Saeks, R., 71, 260, 262, 490
Safonov, M.G., 260, 278, 280, 281, 385, 417, 491, 493, 495, 497
Sagalov, Y.E., 480
Sagan, H., 71, 267
Sage, A.P., 6, 16, 71, 484, 485, 487, 510, 511
Saha, S.K., 527, 548
Saif, M., 483
Saigal, S., 113
Sain, M.K., 68, 114, 115, 130
Sakharov, M.P., 480, 487
Salam, F.M.A., 287
Salcudean, S.E., 498
Saleh, S., 262
Salehi, S.V., 486

Salerno, M., 543, 550
Salikov, L.M., 260
Salmon, D.M., 314, 480
Salzwedel, H., 295
Sambandan, A., 291
Samoilenko, Y.I., 296
Samoylenko, Y.I., 111
Sanchez Pena, R.(S.), 272, 280
Sanchez-Sinéncio, E., 548
Sandberg, I.W., 260, 520, 537
Sandell, N.R., Jr., 71, 505
Sandridge, C.A., 293
Sankaran, J., 274
Sannuti, P., 477, 479, 483, 486, 497
Sano, A., 492
Sansen, W., 548
Santo, A.O.E., 264
Sarabudla, N.R., 285
Sarachik, P.E., 483, 510
Saramaki, T., 551
Sarason, D., 490
Sarma, I.G., 538
Sarma, V.V.S., 130, 483
Sarpeshkar, R., 548
Sasagawa, T., 287
Satsangi, P.S., 541
Savkin, A.V., 287
Sawan, M.E., 481, 512
Sawaragi, Y., 131, 489, 512
Saydy, L., 272, 293
Sbaiti, A.A., 487
Scattolini, R., 130
Schaechter, D.B., 487
Schaumann, R., 528, 547, 548, 549
Scherer, C., 495
Schiavoni, N., 485
Schmidt, E.J.P. G., 130
Schmidt, G., 525, 543
Schmitendorf, W.E., 265, 267, 272, 504, 505, 506
Schnabel, J.A., 488, 508
Schneider, H., 268, 291
Schoeffler, J.D., 524, 543

Schonbach, D.I., 513
Schrijver, A., 515
Schultz, D.G., 260
Schwartz, J.T., 67
Schwarz, A.F., 540, 555
Schwendler, L., 15
Schweppe, F.C., 71, 358, 489, 507, 508
Seacat, R.H., 287
Seborg, D.E., 296
Sedlmeyer, A., 529, 551
Sedra, A.S., 528, 528, 544, 545, 546, 547, 549
Sefton, J.(A.), 492, 493
Seibert, P., 260
Seletzky, A.C., 15
Sengupta, J.K., 517
Sesak, J.R., 6, 15, 16, 91
Seth, A.K., 535, 553, 555
Sevely, Y., 111
Seviora, R., 542
Seviora, R.E., 554
Sewell, J.I., 555
Sezer, M.(E.), 265, 267, 276, 483
Shabalin, A.V., 267
Shac, P.H., 502
Shafai, B., 276
Shah, S.L., 296
Shahian, B., 516
Shaimardanov, F.A., 484
Shak, F.H., 502
Shaked, U., 276, 493, 496, 498
Shamash, Y., 497
Shamis, S.S., 546
Shamma, J.S., 507, 508
Shamsa, K., 287
Shanbhag, N., 272
Shane, B.A., 134, 219, 267, 272, 277
Shankar, S., 282
Shao, C.S., 265
Shao, C.X., 269
Shapiro, A., 517
Shapiro, E.Y., 260
Shapiro, N.Z., 515
Sharov, S.N., 509

Shaw, G.L., 113
Shaw, J., 272
Shaw, L., 291
Shchennikov, V.N., 267
Shchipanov, G.V., 434, 435, 436, 442, 444, 446, 447, 500, 502
Shen, C.N., 131
Shen, J.C., 506
Shevelev, A.G., 260
Shevelyev, A.H., 513
Shevlyakov, G.L., 271
Shi, G., 495
Shi, Y.Q., 269, 272, 273, 275
Shi, Z., 282,
Shi, Z.C., 267, 269
Shieu, S.D., 527, 549
Shih, Y.P., 112, 512, 513
Shimemura, E., 279
Shimizu, H., 518
Shimshoni, M., 258
Shiraishi, S., 517
Shirakawa, I., 524, 543
Shirokov, L.A., 108, 487
Shoureshi, R., 287, 485
Shrivastava, S.K., 260
Shtessel, Y.B., 503
Shupp, F.R., 510
Siapkara, A.A., 510
Sibul, L.H., 510
Sidar, M., 480
Sideris, A., 272, 280
Sidi, M., 279
Sienko, W., 528, 544
Siljak, D.D., 16, 110, 111, 256, 260, 265, 267, 276, 289, 290, 296, 483, 484, 485, 487, 538
Silva, M.M., 542
Silva-Madriz, R.I., 286
Silva-Martinez, J., 548
Simeonov, P.S., 261
Simes, J.G., 110
Simon, B., 67
Sims, C.S., 487
Sims, F.L., 509
Singer, R.A., 296, 529, 551

Singh, G., 131
Singh, H., 525, 542, 543
Singh, M.G., 485
Singh, N.P., 549
Singh, S.N., 130, 506
Singh, V., 73
Singhal, K., 535, 540, 544, 548, 553
Sinha, N.K., 311, 479, 480
Siouris, G.M., 71, 255
Sirovich, L., 68
Sivan, R., 70, 266, 285, 557
Sivashankar, N., 508
Sivasundaram, S., 287
Skelton, R.E., 277, 290, 296, 482, 508, 518
Sklavounos, P.G., 108
Skorodinskii, V.I., 286, 287
Slivinsky, C., 286
Smirnov, G.V., 287
Smith, M.C., 491, 493
Smith, R.A., 261
Smith, W.E., 523, 526, 541
Sneddon, I.N., 68
Sobieszczanski-Sobieski, J., 113, 515
Sobral, M., Jr., 5, 16, 314, 480, 538, 540
Soderstrand, M.A., 527, 528, 549
Soh, C.B., 265, 267, 269, 272, 273, 275, 276, 277, 280
Soh, Y.C., 262, 264, 268, 270, 273, 291
Sohal, J.S., 525, 542, 543
Sokolov, V.I., 287
Sokolowski, J., 113, 514, 517
Soldatos, A.G., 486
Soliman, H.M., 296
Solnechnyi, E.M., 273, 287, 296, 503
Solnechnyy, E.M., 261, 503
Soloveichik, G.Y., 285

Solymosi, J., 111, 540
Sondergold, K.P., 261
Sontag, E.D., 261, 268
Sood, A.K., 81, 109
Sorenson, E.V., 522, 540
Soudack, A.C., 483, 513
Soveshch, T., 500
Spanier, E.H., 68
Spas'kyy, R.O., 503
Spaulding, D.A., 551
Spector, V.A., 113
Spence, R., 538, 540, 552, 555
Speyer, J.L., 131, 508,
Spingarn, K., 72, 358, 488
Sridhar, B., 266
Sridhar, R., 109
Srinivasan, G., 289
Srinivasan, S., 548
Stakgold, I., 68
Staley, R., 358, 488
Stalford, H.L., 504, 506
Stapleford, R.L., 291
Starikov, A.F., 503
Starozhilov, Y.F., 255
Stavroulakis, P., 15, 483, 510, 513
Stefani, R.T., 487
Stein, G., 490, 505
Steinberg, A., 505
Steinberg, A.M., 517
Steinbuch, M., 497
Stence, R., 553
Stepanenko, Y., 489
Stephenson, F.W., 549
Stern, R.J., 130, 261
Stojic, M.R., 111, 291, 551
Stojic, R.M., 291, 551
Stoorvogel, A.A., 495, 508
Storey, C., 112, 130, 142, 256, 262, 282, 500
Strashko, V.T., 503
Strickland, S.G., 509
Strintzis, M.G., 487
Strycula, E.C., 543, 544
Stubberud, A.R., 481
Stuffle, R.E., 555

Styblinski, M.A., 530, 553
Su, T.J., 277
Suarez, R., 260
Subbayyan, R., 483
Sugie, J., 288
Sugie, T., 492, 493
Sugimoto, K., 483, 498
Sui, T.K., 482
Sule, V.R., 478
Sun, Y.Y., 277
Sundararajan, N., 477, 478, 510, 511
Sundareshan, M.K., 268
Sung, H.K., 277, 512
Suryanarayanan, K.L., 513
Sussmann, H.J., 281
Suzuki, T., 109
Svoboda, J.A., 549
Swamy, M.N.S., 291, 524, 535, 536, 542, 543, 547
Swierniak, A., 487
Swift, G.W., 539
Szczupak, J., 548
Szego, G., 72
Szego, G.P., 256
Sz.-Nagy, B., 67, 490
Sznaier, M., 277

Tabak, D., 68
Tacker, E.C., 486
Tadmor, G., 492, 493, 495
Tait, K.E., 513
Takahara, Y., 108, 110
Takahashi, T., 113
Takata, M., 512
Tan, E.C., 551, 555
Tan, O.T., 486
Tan, X.L., 289
Taniguchi, T., 277, 287
Tanino, T., 517
Tannenbaum, A., 490, 491, 492, 497, 509, 512, 514
Tardos, E., 515
Tarmi, R., 549
Tarn, T.J., 261, 262, 295, 510
Tartaglia, M., 257, 264, 283
Taussky, O., 268, 291

Author Index

Tavsanoglu, V., 551
Tay, T.T., 495
Taylor, J., 549
Taylor, J.H., 241, 259
Tekawy, J.A., 281
Tellegen, B.D.H., 554
Temes, G.C., 530, 538, 540, 542, 546, 551, 552, 553
Temma, T., 543
Tempo, R., 269
Teo, K.L., 498
Tereschchenko, A.F., 539
Tesi, A., 268, 273, 287, 296
Teverovskii, V.I., 513
Thau, F.E., 479
Theodorou, N., 275
Therapos, C.P., 498
Therrien, C.W., 535, 555
Thiele, L., 523, 541, 551
Thorbjornsen, A.R., 553
Thorp, J.S., 131, 283, 506
Thowsen, A., 273
Thulasiraman, K., 542, 543
Tiberio, P., 539
Tikhonov, A.N., 5, 15
Timko, K.J., 552
Tismenetsky, M., 271
Tissir, E., 262
Titovskiy, I.N., 295
Tits, A.L., 265, 272, 293, 494
Tkachenko, V.A., 273
Tobin, R.L., 518
Toda, M., 266
Toivonen, H.T., 495
Tokarzewski, J., 277
Tomei, P., 259
Tomizuka, M., 280
Tomlinson, G.H., 549
Tomovic, R., 5, 16, 512, 513, 522, 538
Tourassis, V.D., 511
Tow, J., 542
Towill, D.R., 255, 293, 487
Townley, S., 289
Toy, M., 552

Tran, M.T., 481
Trapeznikov, V.A., 499
Trick, T.N., 538, 556
Trigiante, D., 276
Tripathi, A.N., 110
Trirogoff, K.N., 70
Trofino Neto, A., 506
Tromp, H., 540
Tron, T., 111
Trony, T., 540
Troop, W.J., 539
Troost, B.T., 109
Troutt, M.D., 518
Truxal, J.G., 5, 15, 279, 538
Tsai, J.S.H., 291
Tsai, K.C.Q., 508, 518
Tsai, M.C., 493, 495, 498
Tsai, T.P., 487
Tsay, S.C., 506
Tsay, T.I., 499
Tsen, F.S.P., 257
Tsinias, J., 258, 261, 287
Tsonis, C.A., 292
Tsypkin, Y.Z., 69, 272, 273, 277, 360, 512
Tugco, A.K., 487
Tun, T., 484
Tung, S.L., 265, 275
Tunik, A.A., 111
Turgeon, A.B., 291
Turing, A.M., 10
Turner, J.D., 110
Turner, L.E., 552
Turski, K.K., 110
Tuteur, F.B., 480
Tyler, J.S, Jr., 480
Tzafestas, S.G., 275, 292, 293, 487, 509, 510, 513, 514
Tzannes, T.S., 481
Tzierakis, K.G., 507

Uchida, K., 495, 507
Udartsev, Y.P.,
Ueno, F., 546
Ueno, S., 110
Ulanov, B.V., 296, 503
Ulanov, G.M., 480
Ulanov, M., 500

Ulsoy, A.G., 294
Unami, I., 546
Ur, H., 291, 539
Urgon, G.I., 522, 524, 543
Ushakov, A.V., 503
Usher, P.D., 73
Usoro, P.B., 508
Utkin, V.A., 503
Utkin, V.I., 503

Vaidyanathan, P.P., 547, 552
Vaithilingam, M.C., 483
Valavani, L.S., 507
Vallese, L.M., 535, 555
Valsamis, D., 287
Valtonen, M.E., 539, 554
Valtonen, P., 539
Van der Ha, J.C., 131
van der Pol, B, 14, 68 (or Van Der Pol, B.)
van der Schaft, A.J., 496
van der Woude, J.W., 508
Vandewalle, J., 554
Van Loan, C., 518
Vannelli, A., 261, 287
Van Schieveen, H.M., 483
Van Trees, H.L., 71
Van Valkenburg, M.E., 520, 530, 537, 538, 553
van Woerkom, L.M., 268
Vaplyushkin, V.M., 261, 287
Varah, J., 518
Varshney, R.K., 513
Vartanyan, V.M., 269
Vasil'ev, V.I., 484
Vattuone, E.S., 291
Vdovin, S.I., 261, 287
Veillette, R.J., 498
Velichenko, V.V., 503
Vengerov, A.A., 480
Venkata, S.S., 487
Venkatesh, Y.V., 259, 261
Venkateswaran, S., 549
Venkayya, V.B., 113, 511

Verde, C., 478
Veremei, E.I., 296
Verghese, G.C., 263, 275
Verghese, G.V., 265
Verma, M.(S.), 281, 497
Vetter, W.J., 479
Viaro, U., 295, 551
Vicino, A., 257, 264, 268, 273, 283, 287, 296
Vidyasagar, M., 68, 71, 257, 259, 260, 261, 281, 285, 287, 475, 490, 507, 508
Vielsack, P., 113
Vilenius, M.J., 487
Villalaz, P.A., 540, 552
Vinter, R.B., 518
Viswanathan, T.R., 548
Visweswaran, G.S., 546
Vitt, A.A.(A.V.), 13
Vittal, V., 502
Vlach, J., 526, 540, 544, 545, 548, 549
Vlietstra, J., 555
Volgin, L.N., 72
Volle, M., 515
Volosov, V.V., 293
Von Dinkelbach, W., 16, 518
Voronin, A.M., 435, 500, 503
Voronin, A.N., 518
Voronov, A.A., 261, 288
Vorotnikov, V.I., 277
Voulgaris, P.G., 507
Vratsanos, J., 541
Vukcevic, M.B., 260, 261
Vukobratovic, M., 5, 16, 503, 510

Wagie, D.A., 291, 296
Wait, J.V., 538
Wakeland, W.R., 72
Walker, D.J., 289, 496
Wallis, D.E., Jr., 358, 489
Walsh, J.L., 73
Wang, B.C., 507
Wang, L.Y., 493

Wang, M., 552
Wang, P.K.C., 455, 456, 503
Wang, S.D., 268, 271
Wang, S.S., 273, 278, 283, 288, 293
Wang, T., 131
Wang, T.S., 279, 487
Wang, Y.T., 265, 278, 290
Warwick, K., 275
Watanabe, K., 489, 518
Wauer, J.C., 487
Wei, K., 268, 273, 281
Wei, L.F., 499
Weiland, S., 508
Weinberg, M.C., 555
Weinrich, S.D., 483
Weir, A.J., 68
Weiss, L., 68, 257
Wen, J.T., 262
Werner, R.A., 310, 480
Weyten, L., 528, 540, 542, 545, 549
Wheatstone, C., 4, 15
White, C.C., III, 71
Whittle, P., 518
Wie, B., 281
Wielings, R.F., 555
Wiener, N., 58, 67
Wierzbicki, A., 16
Wiest, E.J., 498
Wilkie, D.F., 79, 80, 81, 109, 483
Willbanks, C.E., 110
Willems, J.C., 69, 70, 71, 257, 262, 288, 490, 508
Willems, J.L., 262, 288
Williamson, D., 281
Willson, A.N, Jr., 550
Wilson, D.J., 506
Wilson, G., 528, 549
Winsor, C.A., 483
Witsenhausen, H.S., 310, 479
Wojcik, R.J., 549
Wolenski, P.R., 512
Wolovich, W., 69
Wolsey, L.A., 518

Womack, B.F., 260
Wong, P.K., 268
Wonham, W.M., 69, 70, 71, 264
Wood, J., 68
Wormley, D.N., 508
Wu, C.C., 113
Wu, F.F., 283
Wu, J., 507
Wu, M.Y., 481
Wu, N.E., 493
Wu, Q.H., 493
Wu, S.M., 515
Wullink, G., 545
Wuu, T.L., 108, 289, 498
Wyatt, J.L., Jr., 536, 548, 556
Wyetzner, G., 280
Wyman, B.F., 71, 49

Xia, L., 498
Xie, L., 494, 496
Xin, L.X., 268
Xu, D., 277
Xu, D.Y., 288
Xu, J.H., 277, 518
Xu, S.J., 269
Xue, Y.X., 113

Yadykin, I.B., 111
Yaesh, I., 493, 496, 498
Yakovlev, O.S., 285, 511
Yakubovich, V.A., 288
Yaling, C., 489
Yamamoto, Y., 483, 498
Yamamura, S., 485
Yamashita, K., 259, 286
Yan, G.T., 552
Yan, W., 289
Yan, W.Y., 281, 513
Yang, C., 495
Yang, C.D., 497, 499
Yang, J., 510
Yang, S.C., 113, 483
Yang, S.M., 552
Yaniv, O., 281
Yanowitch, M., 67
Yanushevskiy, R.T., 288
Yarlagadda, R., 529, 551

Yaz, E., 276, 277
Yeh, F.B., 493, 497, 499
Yeh, H.H., 268, 269, 281, 496
Yen, K.K., 269, 272
Yemel'yanov, S.V., 503
Yermachenko, A.I., 487
Yeung, K.S., 273
Yeung, L.F., 281
Yokomoto, C.F., 543, 555
Yoneda, S., 547
Yoneyama, T., 288
Youla, D.C., 69, 479, 490
Young, G.E., 510
Young, J.S., 493, 497
Young, K.K.D., 487, 510
Yu, T.C., 295
Yu, T.H., 551
Yu, Y., 288
Yuan, J., 489
Yudayev, A.V., 510
Yue, A., 499
Yurachkovskiy, Y.P., 500
Yusupov, I.Y., 269
Yusupov, R.M., 16, 485, 513

Zadeh, L.A., 14, 69
Zafiriou, E., 16, 499
Zaitsev, V.V., 288
Zak, S.H., 69, 284, 285
Zakharin, F.M., 16, 485
Zakharyan, A.Z., 509
Zakian, V., 273
Zames, G., 262, 301, 379, 491, 493, 512, 514
Zampieri, G., 256
Zappa, G., 287
Zarchan, P., 296
Zaremba, S.C., 462, 506
Zarrop, M.B., 489
Zaslavskii, B.G., 262
Zavgren, J.R., Jr., 261
Zdor, V.V., 296
Zeheb, E., 258, 269, 273, 274, 291
Zeidan, V., 71
Zein El-Din, H.M., 293
Zelentsovskii, A.L., 282

Zemanian, A.H., 68
Zemlyakov, S.D., 500
Zeng, X., 261
Zhan, W., 262
Zhang, D.N., 281
Zhang, H., 275
Zhang, Q.J., 554
Zhao, K.Y., 275
Zheng, D.Z., 268
Zhong, X.C., 289
Zhou, C.S., 268, 277
Zhou, S.F., 269, 272
Zhou, K., 268, 281, 288, 494, 496
Zhu, G., 277, 508
Zhukov, V.P., 262, 288
Zilovic, M.S., 291
Zimmermann, H.J., 111
Zinober, A.S.I., 485, 488
Zohdy, M.A., 264, 296, 480, 512
Zolesio, J.P., 112, 113, 514
Zou, Y., 495
Zowe, J., 518
Zubov, A.G., 483
Zurcher, L.A., 15

SUBJECT INDEX

Actuator, 364, 385
Adjoint network, 521, 531
Auxiliary
    input, 359, 364, 366, 371, 376, 377
    parameter, 20
    transfer matrix, 401

Bounded
    disturbance, 435
    uncertainty, 459
    variations, 38

Calculus of variations, 84, 451, 454
canonical decomposition, 457
canonical transformation, 81
Cayley-Hamilton theorem, 121
characteristic equation, 121, 136
classical problems, 63, 64, 65
Cohn's theorem, 523
common factor, 136, 137
complementarity, 384
complex domain, 97, 380
Computational Sensitivity, 10
condition number, 47
continuously equivalent network, 524
convergent (discrete) system
    Lyapunov, 134, 135, 218
    matrix, 134
    nonlinear, 253
    polynomial, 224, 225
Controllability
    canonical transformation, 119
    complete, 18
    structural, 125
    system model, 117
Controller
    adaptive, 332, 335
    admissible, 387

central, 407, 412
H^2-, 423, 427, 428, 429, 432
$H^\infty-$, 420, 421, 428, 430, 433
injection, 469
model-following adaptive, 472
sensitivity reduction/optimal, 307, 308, 318, 321, 324, 328, 330, 331, 332, 334, 340, 343, 345, 347, 350, 352, 353, 354, 355, 356, 364, 368, 371, 374, 375, 376, 377

Definitions, 17
Detectable, 387
Digital filter, 529
discrete system (cf., convergent system)
disturbance, 298, 301
    bounded, 435
    feedforward, 389, 402
    external, 301
    internal, 301
    normed-bounded, 461
    rejection, 459, 473
    special structured, 460
    worst-case, 412
Duffing's equation, 242, 249

Eigenvalue, 47
    sensitivity, 139, 140
Eigenvector, 47, 140
Embedding Problems, 11
Energy stored network element, 523
Estimator, 415
    observer-based, 380, 406
Exogenous signal, 17, 298, 381

Feedback amplifier, 4
Fidler's result, 522
finite-escape time, 464
Fisher information matrix, 358
Frequency-domain

sensitivity, 97
Full – Control
    structure, 388
    transfer matrix, 388, 400
Full – Information
    structure, 388
    transfer matrix, 388, 398

Generating system, 17
Generalized dynamical systems, 462, 463, 464
    asymptotic stability, 465
    contingent derivative, 464
    contingent equation, 462, 464
    motion, 463
    trajectory, 463
Gramian matrices
    controllability, 61, 391
    observability, 61, 391
Guidance theory, 129

Hadamard's inequality, 47, 210
Hamiltonian matrix, 62, 382, 383 384, 391, 392, 395, 396, 409, 419, 431
Historical Remarks, 4
hyperbox, 2, 147

Identification, 18, 357, 371
Imbedding Problems, 11
Information matrix, 331
Inner matrix, 393
Input
    auxiliary, 359, 360, 361, 364, 365, 372, 376, 377
    optimal, 358, 371
    supervisory, 359
    troublesome, 385
insensitive, 140
Invariance condition
    absolute, 441, 442, 443
    complete, 443, 450, 451
    partial, 443
    selective, 442
    Shchipanov-Luzin, 442

strong, 436, 450, 453, 456, 458
weak, 450, 451, 453, 458
Invariance Glossary
   action, 437
   combined control system, 443
   control, deviation/perturbation, 443
   dry friction, 448
   fork, 444
   free action, 437
   Harmonic linearization, 449
   interaction matrix, 446
   operator
      buffer block, 446
      control operator, 446
      feedback, 446
      second-order, 437
      symbolic, 448
   parameter invariant, 456
   perpetual motor, 447
   plant coordinate, 436
   process parameter, 436
   signal invariant, 456
   single-channel invariant, 444
   two-channel principle, 443, 446
[*Invariance Glossary Ends*]
Invariance theory, 434
   linear systems, 452, 455, 457
   multidimensional systems, 445
   nonlinear systems, 448, 449
   variational approach, 449, 450, 451

Kharitonov theorem, 167, 203

Ladder RC-network, 105
Large-scale system, 211
Linear fractional transformation, 53, 385, 401
Loop shifting, 417
Luenberger's observer, 85
Lyapunov direct method, 134, 135, 142, 149, 175, 176, 177, 178, 179, 190, 200, 205, 206, 212, 218, 226, 228, 233, 250, 252, 254, 312, 313, 315, 316, 319, 320, 325, 330, 331, 341, 343, 348, 367, 396

Maneuverability, 132, 143
man-in-the-loop, 360
Matching condition, 461, 473
Mathematical Glossary
   Blaschke product, 46
   convergent, 32, 33
   Cauchy sequence, 32
   Determinant, 46
   Field, 30
   Functions
      anticausal, 58, 390
      attainability, 463
      bounded, 39
      causal, 58, 390
      continuous, 37
      contraction, 39
      convexity, 38
      differentiable, 38
      inner, 46
      isometry, 39
      Lebesgue, 40
      Lipschitz, 39
      outer, 46
      set-valued, 461, 462, 466 467, 469, 473
      upper semi-continuous, 37, 463, 467, 468
   Group, 29
   Inequalities, 40
      Hadamard, 47, 210
      Holder, 41
      Jensen, 40
      Minkowski, 41
      matrix, 153
   infimum, 32
   Mappings, 27, 28
      bounded, 39
      continuous, 38
      contraction, 39
      convexity, 38
      differentiable, 38
      domain, 27
      image, 27
      range, 27
   Matrix, 47, 153
   Module, 31
   Ring, 29
   Semigroup, 28
   Set theory, 34
   sequences, 32, 33
   Series
      complete trigonometric, 42
      Fourier, 42
   Spaces, Definitions
      Banach, 33
      complete, 33
      Hardy, 44
      Hausdorff, 36
      Hilbert, 33
      inner product, 32
      linear, 30
      measurable, 36
      metric, 31
      Nevanlinna, 44
      normed, 31
      topological, 36
      vector, 30
   Spaces, Properties
      Banach, 43
      H^p, 44, 45, 46,
      Hilbert, 34, 41, 42, 43
      L^p, 40
   supremum, 32
   Theorems
      contraction mapping, 39
      Hahn–Banach, 43
      implicit function, 39
      mean-value, 39
      open-mapping, 43
      Parseval, 43, 307
      Riesz–Fischer, 42
   Topology, 35
[*Mathematical Glossary Ends*]
matrix
   condition number, 47
   norms, 48
   spectral radius, 47
Maximum Principle, 364
Modeling Inaccuracies, 9
Model-reference control, 472
min max, 466, 474, 475

Subject Index

minimax, 431
mismatch, 473

Nehari problem, 65, 380
Network sensitivity, 6, 519, 535
    active, 526, 527, 528, 529
    adjoint, 521, 531,
    computation, 535
    model, 522
    nonlinear, 536
    statistically, 530
Nevanlinna–Pick
    interpolation, 379
Nominally equivalent, 19
Nominal value, 21
nonlinear system, 75, 241, 253
Norms
    Frobenius, 48
    matrix, 48
    vector, 47, 48
Norms computation
    H_2-, 60
    $H_\infty-$, 61

Observability condition, 18
Observer-based method, 380, 406
Operator
    Hankel, 61
    Hankel-Teoplitz, 392
    Multiplication, 391
    Theory, 380
    Toeplitz, 391
Output
    desirable, 385
    estimation, 389, 403
    uncontrollability, 456

Parameter
    Auxiliary, 20
    constraint, 148, 249
    Coupling, 21
    maneuverability, 132, 143
    Physical, 20
    space, 144
    Structural, 20
    variations space, 144
Parameterization, 398, 400, 406

Performance index
    large-sensitivity analysis, 314, 315, 316, 317, 318
    network, 526
    sensitivity comparison, 310
    small-sensitivity analysis, 312, 313, 314
    small-sensitivity synthesis, 318, 319, 320
Perturbation
    directional, 146, 156, 160, 162, 174, 175, 180, 190, 200, 207, 220, 227, 233, 250
    E-Box, 145, 146, 147, 149, 152, 153, 155, 159, 162, 165, 166, 167, 168, 169, 186, 187, 188, 190, 193, 207, 210, 213, 215, 217, 223, 229
    matrices, 151, 152, 355
    structure, 155, 162
phase-variable canonical, 78
Physical-parameter sensitivity
    function, 90
    alternative, 96
    closed-loop reconstructible, 96
    direct computation, 91, 117
    higher-order, 91
    minimum number, 119, 121
    open-loop reconstructible, 90
Plant-operator, 360, 361, 362, 365, 371
Polynomial
    discrete, 224, 225
    Hurwitz, 164, 165, 167, 169, 173, 174, 203, 205, 206
    large-scale, 171
    minimum, 117
    monic, 170, 171, 172, 174, 200, 224, 225, 240
    set generator, 171, 225
    stability, 133
positive definite, 465
Potter's method, 383

Q sensitivity of network, 528

Reduced-order observer, 85
Regulator, 413
    ideal and universal, 439, 440, 442
    one-degree-of-freedom, 438
    two-degree-of-freedom, 439
    three-degree-of-freedom, 440
reproducibility, 114
Riccati equation, 366, 382, 383, 384, 387, 396, 399, 409
Robotics, 364, 472
robust, 23, 309
Robustification, 361
Robustness, 140

Scaling, 417
Scope, 4
Sensitivity
    active network, 526
    comparison, 25, 101, 302, 307, 308, 310, 381
    adaptive, 308, 332
    matrix, 306, 307
    plant parameter, 307
    performance index, 310
    plant/controller parameters, 308
    trajectory, 302, 322
    complex domain, 97, 380
    Digital filter, 529
    invariance, 100, 523
    measure, 23, 309, 339, 362, 381
    network, 520
    root, 135, 136, 137, 138
    statistically oriented, 530
Sensitivity functions
    Bode, 22, 306
    eigenvalue, 139, 140
    frequency domain, 97
    gain, 527
    generation, 25
    large, 21, 77
    matrix, 22
    nonlinear system, 26

phase, 527
physical parameter, 90
normalized, 22
primitive, 21
root, 138, 520, 527,
simultaneity, 80, 81
small, 21, 77
structural parameter, 76
total *(TSF)*, 82, 83, 84,
   86, 87, 88, 89, 305,
   306, 307, 308
total symmetry, 79
vector, 22
unnormalized, 22
zero, 136, 527
Sensitivity minimization/optimization
   in Hardy spaces, 62, 379
   large-, 337, 341, 348
   mixed, 476
   model matching, 65
   network, 525, 526
   small-, 322, 324, 331
   stabilization, 64
   tracking, 63
sensor, 385
Separation argument (theorem),
   380, 429
servomotor, 437
set generator, 162, 167, 171,
   219, 225
Signals
   time-domain, 57
   frequency-domain, 58
Similarity transformation,
   53, 57, 396, 411
Simultaneous diagonalization,
   153, 154
Spectral
   radius, 47
   subspace, 383
Stability
   absolute, 241
   asymptotic, 134, 461, 465
   complementarity property,
      384
   conditional, 153, 165
   in the large, 465
   large-scale system, 211
   Lyapunov (strong),
      134, 465

matrix, 133
negative definite,
   465
polynomial, 133
positive definite, 465
robustness measure, 144,
   145, 146, 147, 148
nonlinear system, 241
system model, 115
uniform ($\cdots$), 465
Stabilizable, 387
Stabilization, 64
Structural-parameter sensitivity
   function, 76
closed-loop reconstructible,
   87
complete simultaneity,
   80, 81
low-order, 79
open-loop reconstructible,
   76
structural-parameter, 76
total sensitivity, 82
total symmetry, 79
structural controllability, 125
summary of the book, 7
System
   Identification, 18, 357
   Nominally-Equivalent, 19
   Normal, 19
   Realization, 18
   Specification, 18
System classifications (references)
   distributed parameter, 513
   nonlinear, 508
   time lag, 511
   time-varying 508
System model, 115
   algebraic properties, 115
   controllability, 117
   open-loop, 98
   stability, 115

Taylor series, 23, 237
Tellegen's theorem, 530
Terminal conditions
   insensitivity, 125
tolerance, 6, 521
Trace
   differentiation, 313

matrix, 47
Trajectory-sensitivity
   comparison, 302
Transfer matrix, 49, 386
   auxiliary, 401
   disturbance-feedforward,
      389
   full-control, 388
   full-information, 388
   norms, 60
   output-estimation, 389
   realization, 49
Transversal condition, 533

Uncertainty, 64, 297, 298,
   459, 460
   asymptotic stability, 461
   input, 461
   lumped, 461
   normed-bounded, 461
   time-varying, 471
ultimately bounded, 471

Vandermonde matrix, 124
Varatsanos' theorem, 523
Variation
   α, β, γ, 5
   large, 3, 23
   parameter, 21
   nonlinear, 236
   small, 3, 23

Worst-case, 1, 299, 337, 378,
   460
   disturbance, 408, 412
   input, 408

Springer-Verlag and the Environment

We at Springer-Verlag firmly believe that an international science publisher has a special obligation to the environment, and our corporate policies consistently reflect this conviction.

We also expect our business partners – paper mills, printers, packaging manufacturers, etc. – to commit themselves to using environmentally friendly materials and production processes.

The paper in this book is made from low- or no-chlorine pulp and is acid free, in conformance with international standards for paper permanency.